SODIUM–CALCIUM EXCHANGE

PROCEEDINGS OF THE THIRD
INTERNATIONAL CONFERENCE

⌈ANNALS OF THE NEW YORK ACADEMY OF SCIENCES⌉
Volume 779

SODIUM–CALCIUM EXCHANGE

PROCEEDINGS OF THE THIRD INTERNATIONAL CONFERENCE

Edited by Donald W. Hilgemann, Kenneth D. Philipson, and Guy Vassort

The New York Academy of Sciences
New York, New York
1996

Copyright © 1996 by the New York Academy of Sciences. All rights reserved. Under the provisions of the United States Copyright Act of 1976, individual readers of the Annals are permitted to make fair use of the material in them for teaching and research. Permission is granted to quote from the Annals provided that the customary acknowledgment is made of the source. Material in the Annals may be republished only by permission of the Academy. Address inquiries to the Executive Editor at the New York Academy of Sciences.

Copying fees: For each copy of an article made beyond the free copying permitted under Section 107 or 108 of the 1976 Copyright Act, a fee should be paid through the Copyright Clearance Center, 222 Rosewood Drive, Danvers, MA 01923. For articles of more than 3 pages, the copying fee is $1.75.

∞ The paper used in this publication meets the minimum requirements of American National Standard for Information Sciences—Permanence of Paper for Printed Library Materials, ANSI Z39.48-1984.

Cover: A two-dimensional model of the cardiac sodium-calcium exchange protein with several of its proposed functional domains highlighted.

Library of Congress Cataloging-in-Publication Data

Sodium-calcium exchange / edited by Donald W. Hilgemann, Kenneth D. Philipson, and Guy Vassort.
    p. cm. — (Annals of the New York Academy of Sciences ; v. 779)
    ". . . proceedings of a conference entitled Third International Conference on Sodium-Calcium Exchange, sponsored by the New York Academy of Sciences and held at the Marine Biological Laboratory in Woods Hole, Massachusetts, on April 23–26, 1995"—Contents p.
    Includes bibliographical references and index.
    ISBN 1-57331-000-X (cloth : alk. paper). — ISBN 1-57331-001-8 (pbk. : alk. paper)
    1. Calcium channels—Congresses. 2. Sodium channels—Congresses. I. Hilgemann, Donald W. II. Philipson, Kenneth D. III. Vassort, Guy. IV. New York Academy of Sciences. V. International Conference on Sodium-Calcium Exchange (3rd : 1995 : Woods Hole, Mass.) VI. Series.
Q11.N5 vol. 779
[QP535.C2]
500 s—dc20
[574.87'5]
                                                                                        96-5419
                                                                                          CIP

Printed in the United States of America
ISBN 1-57331-000-X (cloth)
ISBN 1-57331-001-8 (paper)
ISSN 0077-8923

ANNALS OF THE NEW YORK ACADEMY OF SCIENCES

Volume 779
April 15, 1996

SODIUM–CALCIUM EXCHANGE

PROCEEDINGS OF THE THIRD INTERNATIONAL CONFERENCE[a]

Editors and Conference Chairs
DONALD W. HILGEMANN, KENNETH D. PHILIPSON, AND GUY VASSORT

CONTENTS

Preface and Dedication. *By* THE EDITORS............................ xiii

Commentaries

Commentary. *By* DENIS NOBLE 1

The Molecular Biology of the Na^+-Ca^{2+} Exchanger and Its Functional Roles in Heart, Smooth Muscle Cells, Neurons, Glia, Lymphocytes, and Nonexcitable Cells. *By* W. J. LEDERER, S. HE, S. LUO, W. DUBELL, P. KOFUJI, R. KIEVAL, C. F. NEUBAUER, A. RUKNUDIN, H. CHENG, M. B. CANNELL, T. B. ROGERS, and D. H. SCHULZE 7

Part I: Molecular Biology

Introduction... 19

Molecular Regulation of the Na^+-Ca^{2+} Exchanger. *By* K. D. PHILIPSON, D. A. NICOLL, S. MATSUOKA, L. V. HRYSHKO, D. O. LEVITSKY, and J. N. WEISS 20

The Structural Basis of Na^+-Ca^{2+} Exchange Activity. *By* HANNAH RAHAMIMOFF, WALTER LOW, ORNA COOK, IAN FURMAN, JUDITH KASIR, and RAFI VATASHSKI.. 29

A New Splicing Variant in the Frog Heart Sarcolemmal Na-Ca Exchanger Creates a Putative ATP-Binding Site. *By* TOMOKO IWATA, ALEXANDER KRAEV, DANILO GUERINI, and ERNESTO CARAFOLI........ 37

[a] This volume represents the proceedings of a conference entitled Third International Conference on Sodium–Calcium Exchange, sponsored by the New York Academy of Sciences and held at the Marine Biological Laboratory in Woods Hole, Massachusetts, on April 23–26, 1995.

Alternative Splicing of the Na^+-Ca^{2+} Exchanger Gene, NCX1. *By* D. H.
SCHULZE, P. KOFUJI, C. VALDIVIA, S. HE, S. LUO, A. RUKNUDIN,
S. WISEL, M. S. KIRBY, W. DUBELL, and W. J. LEDERER.............. 46

The Kidney Sodium-Calcium Exchanger. *By* JONATHAN LYTTON,
SHWU-LUAN LEE, WEN-SEN LEE, J. VAN BAAL, RENÉ J. M. BINDELS,
RACHEL KILAV, TALLY NAVEH-MANY, and JUSTIN SILVER............ 58

Sodium-Calcium Exchange and Calcium Homeostasis in Transfected
Chinese Hamster Ovary Cells. *By* JOHN P. REEVES, GALINA CHERNAYA,
and MADALINA CONDRESCU...................................... 73

Mutagenesis Studies of the Cardiac Na^+-Ca^{2+} Exchanger. *By* D. A.
NICOLL, L. V. HRYSHKO, S. MATSUOKA, J. S. FRANK, and K. D.
PHILIPSON... 86

Antisense Oligodeoxynucleotides Directed against the Na-Ca Exchanger
mRNA: Promising Tools for Studies on the Cellular and Molecular
Level. *By* ERNST NIGGLI, BEAT SCHWALLER, and PETER LIPP......... 93

Molecular Biological Studies of the Cardiac Sodium-Calcium Exchanger.
By ALEXANDER KRAEV, ILYA CHUMAKOV, and ERNESTO CARAFOLI..... 103

Poster Papers

Expression of Na^+-Ca^{2+} Exchanger with Modified C-Terminal
Hydrophobic Domains and Enhanced Activity. *By* N. GABELLINI,
T. IWATA, and E. CARAFOLI..................................... 110

Identification and Antisense Inhibition of Na-Ca Exchange in Renal
Epithelial Cells. *By* K. E. WHITE, F. A. GESEK, and P. A. FRIEDMAN.... 115

Effects of NCX1 Antisense Oligodeoxynucleotides on Cardiac Myocytes
and Primary Neurons in Culture. *By* K. S. BLAND, K. TAKAHASHI,
S. ISLAM, and M. L. MICHAELIS 119

Initial Characterization of the Feline Sodium-Calcium Exchanger Gene.
By KIMBERLY V. BARNES, MYRA M. DAWSON, and DONALD R. MENICK.. 121

Cloning of the Mouse Cardiac Na^+-Ca^{2+} Exchanger and Functional
Expression in *Xenopus* Oocytes. *By* INJUNE KIM and CHIN O. LEE..... 126

Identification of a Novel Alternatively Spliced Isoform of the Na^+-Ca^{2+}
Exchanger (NACA8) in Heart. *By* R. F. REILLY and D. LATTANZI...... 129

Presence of NACA3 and NACA7 Exchanger Isoforms in Insulin-
Producing Cells. *By* F. VAN EYLEN, M. SVOBODA, A. BOLLEN, and
A. HERCHUELZ 132

Part II: Kinetics, Mechanism, and Regulation

Introduction... 135

The Cardiac Na-Ca Exchanger in Giant Membrane Patches. *By* DONALD
W. HILGEMANN.. 136

Multiple Functional States of the Cardiac Na^+-Ca^{2+} Exchanger: Whole-Cell, Native-Excised, and Cloned-Excised Properties. *By* SATOSHI MATSUOKA, KENNETH D. PHILIPSON, and DONALD W. HILGEMANN 159

Mechanism of XIP in Cardiac Sarcolemmal Vesicles. *By* CALVIN C. HALE 171

Cardiac Na-Ca Exchange and pH. *By* ANDREA E. DOERING, DAVID A. EISNER, and W. JON LEDERER 182

In Squid Axons Phosphoarginine Plays a Key Role in Modulating Na-Ca Exchange Fluxes at Micromolar $[Ca^{2+}]_i$. *By* REINALDO DIPOLO and LUIS BEAUGÉ.. 199

A Nerve Cytosolic Factor Is Required for MgATP Stimulation of a Na^+ Gradient-Dependent Ca^{2+} Uptake in Plasma Membrane Vesicles from Squid Optic Nerve. *By* LUIS BEAUGÉ, DANIEL DELGADO, HÉCTOR ROJAS, GRACIELA BERBERIÁN, and REINALDO DIPOLO................ 208

Kinetics and Mechanism: Modulation of Ion Transport in the Cardiac Sarcolemma Sodium-Calcium Exchanger by Protons, Monovalent Ions, and Temperature. *By* DANIEL KHANANSHVILI, EVELYNE WEIL-MASLANSKY, and DAVID BAAZOV............................... 217

Voltage Dependence of Na-Na Exchange in Barnacle Muscle Cells: I. Na-Na Exchange Activated by α-Chymotrypsin. *By* H. RASGADO-FLORES, R. ESPINOSA-TANGUMA, J. TIE, and J. DESANTIAGO 236

Phosphorylation and Modulation of the Na^+-Ca^{2+} Exchanger in Vascular Smooth Muscle Cells. *By* MUNEKAZU SHIGEKAWA, TAKAHIRO IWAMOTO, and SHIGEO WAKABAYASHI 249

Regulation of Expression of Sodium-Calcium Exchanger and Plasma Membrane Calcium ATPase by Protein Kinases, Glucocorticoids, and Growth Factors. *By* JEFFREY BINGHAM SMITH, HYEON-WOO LEE, and LUCINDA SMITH 258

Poster Papers

Enzyme Kinetics: Thermodynamic Constraints on Assignment of Rate Coefficients to Kinetic Models. *By* JONATHAN WAGG and PETER H. SELLERS... 272

Effects of External Monovalent Cations of Na^+-Ca^{2+} Exchange in Cultured Rat Glial Cells. *By* ANDREA HOLGADO and LUIS BEAUGÉ 279

ATP Stimulation of a Na^+ Gradient-Dependent Ca^{2+} Uptake in Cardiac Sarcolemmal Vesicles. *By* GRACIELA BERBERIÁN and LUIS BEAUGÉ 282

Modifications of XIP, the Autoinhibitory Region of the Na-Ca Exchanger, Alter Its Ability to Inhibit the Na-Ca Exchanger in Bovine Sarcolemmal Vesicles. *By* C. GATTO, W-Y. XU, H. A. DENISON, C. C. HALE, and M. A. MILANICK 284

Use of Cysteine Replacements and Chemical Modification to Alter XIP, the Autoinhibitory Region of the Na-Ca Exchanger: Inhibition of the Activated Plasma Membrane Ca Pump. *By* W-Y. XU, C. GATTO, C. J. ALLEN, and M. A. MILANICK 286

Effects of Phe-Met-Arg-Phe-NH$_2$ (FMRFa)-Related Peptides on Na-Ca Exchange and Ionic Fluxes in Rat Pancreatic B Cells. *By* F. VAN EYLEN, P. GOURLET, A. VANDERMEERS, P. LEBRUN, and A. HERCHUELZ 288

Kinetics of Na-Ca Exchange Current after a Ca^{2+} Concentration Jump. *By* M. KAPPL and K. HARTUNG.................................. 290

Calcemic Hormones Regulate the Level of Sodium-Calcium Exchange Protein in Osteoblastic Cells. *By* NANCY S. KRIEGER 293

A Novel Approach for Imaging the Influx of Ca^{2+}, Na^+, and K^+ in the Same Cell at Subcellular Resolution: Ion Microscopy Imaging of Stable Tracer Isotopes. *By* SUBHASH CHANDRA and GEORGE H. MORRISON ... 295

Part III: Sodium-Calcium Exchange in the Neural System

Introduction 299

The Na^+-Ca^{2+} Exchanger in Rat Brain Synaptosomes: Kinetics and Regulation. *By* MORDECAI P. BLAUSTEIN, GIOVANNI FONTANA, and ROBERT S. ROGOWSKI 300

Localization of the Na^+-Ca^{2+} Exchanger in Vascular Smooth Muscle, and in Neurons and Astrocytes. *By* MAGDALENA JUHASZOVA, HIROSHI SHIMIZU, MIKHAIL L. BORIN, RICK K. YIP, ELIGIO M. SANTIAGO, GEORGE E. LINDENMAYER, and MORDECAI P. BLAUSTEIN............. 318

Regulation of the Bovine Retinal Rod Na-Ca+K Exchanger. *By* PAUL P. M. SCHNETKAMP, JOSEPH E. TUCKER, and ROBERT T. SZERENCSEI 336

Turnover Rate and Number of Na^+-Ca^{2+},K^+ Exchange Sites in Retinal Photoreceptors. *By* GIORGIO RISPOLI, ANACLETO NAVANGIONE, and VITTORIO VELLANI 346

Na-Ca Exchange in Ca^{2+} Signaling and Neurohormone Secretion: Secretory Vesicle Contributions in Adrenal Chromaffin Cells. *By* ALLAN S. SCHNEIDER and CHUNG-REN JAN 356

Na^+-Ca^{2+} Exchange in Anoxic/Ischemic Injury of CNS Myelinated Axons. *By* PETER K. STYS and ISABELLA STEFFENSEN 366

The Sodium-Calcium Exchanger and Glutamate-Induced Calcium Loads in Aged Hippocampal Neutrons *In Vitro*. *By* L. R. MILLS 379

Poster Papers

Release of Catecholamines and Enkephalin Peptides Induced by Reversal of the Na^+-Ca^{2+} Exchanger in Chromaffin Cells. *By* E. P. DUARTE, G. BALTAZAR, P. VERÍSSIMO, and A. P. CARVALHO 391

Agents That Promote Protein Phosphorylation Increase Catecholamine Secretion and Inhibit the Activity of the Na^+-Ca^{2+} Exchanger in Bovine Chromaffin Cells. *By* L. F. LIN, L-S. KAO, and E. W. WESTHEAD ... 395

What Mechanisms Are Involved in Ca^{2+} Homeostasis in Hair Cells? *By* CHRISTIAN CHABBERT, A. SANS, and J. LEHOUELLEUR 397

Immunohistochemical Localization of the Cardiac Sodium-Calcium Exchange Protein in the Inner Ear. *By* P. M. MANCINI and P. A. SANTI . 400

Reversed Mode Na^+-Ca^{2+} Exchange Activated by Ciguatoxin (CTX-1b) Enhances Acetylcholine Release from *Torpedo* Cholinergic Synaptosomes. *By* YVETTE MOROT GAUDRY-TALARMAIN, JORDI MOLGO, FREDERIC A. MEUNIER, NATHALIE MOULIAN, and ANNE-MARIE LEGRAND 404

Part IV: Sodium-Calcium Exchange in the Cardiovascular System

Introduction 407

Calcium in the Cardiac Diadic Cleft: Implications for Sodium-Calcium Exchange. *By* G. A. LANGER and A. PESKOFF 408

Action Potential Duration Modulates Calcium Influx, Na^+-Ca^{2+} Exchange, and Intracellular Calcium Release in Rat Ventricular Myocytes. *By* R. B. CLARK, R. A. BOUCHARD, and W. R. GILES 417

Na-Ca Exchange and Ca Fluxes during Contraction and Relaxation in Mammalian Ventricular Muscle. *By* DONALD M. BERS, JOSÉ W. M. BASSANI, and ROSANA A. BASSANI 430

The Roles of the Sodium and Calcium Current in Triggering Calcium Release from the Sarcoplasmic Reticulum. *By* M. B. CANNELL, C. J. GRANTHAM, M. J. MAIN, and A. M. EVANS 443

Evidence That Reverse Na-Ca Exchange Can Trigger SR Calcium Release. *By* SHELDON LITWIN, OSAMI KOHMOTO, ALLEN J. LEVI, KENNETH W. SPITZER, and JOHN H. B. BRIDGE 451

Regulation of Sodium-Calcium Exchange in Intact Myocytes by ATP and Calcium. *By* R. A. HAWORTH AND A. B. GOKNUR 464

Functional Roles of Sodium-Calcium Exchange in Normal and Abnormal Cardiac Rhythm. *By* DENIS NOBLE, JEAN-YVES LEGUENNEC, and RAIMOND WINSLOW 480

The Exchanger and Cardiac Hypertrophy. *By* DONALD R. MENICK, KIMBERLY V. BARNES, USHA F. THACKER, MYRA M. DAWSON, DIANE E. MCDERMOTT, JOHN D. ROZICH, ROBERT L. KENT, and GEORGE COOPER IV 489

Na-Ca Exchange in Circulating Blood Cells. *By* J. P. GARDNER and M. BALASUBRAMANYAM 502

Poster Papers

Effects of External Mg^{2+} on the Na-Ca Exchange Current in Guinea Pig Cardiac Myocytes. *By* JUNKO KIMURA 515

Ca^{2+} Influx via Na-Ca Exchange and I_{Ca} Can Both Trigger Transient Contractions in Cat Ventricular Myocytes. *By* A-M. VITES and J. A. WASSERSTROM... 521

Demonstration of an Inward Na^+-Ca^{2+} Exchange Current in Adult Human Atrial Myocytes. *By* GUI-RONG LI and STANLEY NATTEL 525

Intracellular pH Is Insensitive to Changes in Intracellular Calcium Concentration in Isolated Rat Ventricular Myocytes. *By* K. W. DILLY, A. E. DOERING, W. A. ADAMS, C. AUSTIN, and D. A. EISNER 529

Immunolocalization of the Na^+-Ca^{2+} Exchanger in Cardiac Myocytes. *By* J. S. FRANK, F. CHEN, A. GARFINKEL, E. MOORE, and K. D. PHILIPSON.. 532

Sodium-Calcium Exchange Expression in Ischemic Rabbit Hearts. *By* MALCOLM M. BERSOHN.. 534

Ontogeny and Hormonal Regulation of Cardiac Na^+-Ca^{2+} Exchanger Expression in Rabbits. *By* SCOTT R. BOERTH, WILLIAM A. COETZEE, and MICHAEL ARTMAN 536

Functional Relevance of an Enhanced Expression of the Na^+-Ca^{2+} Exchanger in the Failing Human Heart. *By* M. FLESCH, F. PÜTZ, R. H. G. SCHWINGER, and M. BÖHM............................. 539

Role of the Cardiac Sarcolemmal Na^+-Ca^{2+} Exchanger in End-Stage Human Heart Failure. *By* HANS REINECKE, ROLAND STUDER, ROLAND VETTER, HANJÖRG JUST, JÜRGEN HOLTZ, and HELMUT DREXLER 543

Lanthanum Provides Cardioprotection by Modulating Na^+-Ca^{2+} Exchange. *By* NILANJANA MAULIK, ARPAD TOSAKI, RICHARD M. ENGELMAN, GORACHAND CHATTERJEE, and DIPAK K. DAS 546

Sodium-Calcium Balance in Coronary Angiography: Experimental Experience with Isotonic Iodixanol. *By* PER JYNGE, GEIR FALCK, HANS K. PEDERSEN, JAN O. G. KARLSSON, and HELGE REFSUM 551

Properties of the Na^+-Ca^{2+} Antiport of Heart Mitochondria. *By* DENNIS W. JUNG, KEMAL BAYSAL, and GERALD P. BRIERLEY 553

Na-Ca Exchange Studies in Sarcolemmal Skeletal Muscle. *By* H. GONZALEZ-SERRATOS, D. W. HILGEMANN, M. ROZYCKA, A. GAUTHIER, and H. RASGADO-FLORES............................ 556

Lorin J. Mullins, Professor of Biophysics

Lorin J. Mullins, Professor of Biophysics: A Life Dedicated to the Study of the Interaction of Ions with Excitable Membranes. *By* JAIME REQUENA   562

Subject Index... 583

Index of Contributors 591

Financial assistance was received from:

Supporter
- NATIONAL HEART, LUNG AND BLOOD INSTITUTE—
  NATIONAL INSTITUTES OF HEALTH

Contributors
- AMERICAN HEART ASSOCIATION
- AXON INSTRUMENTS, INC.
- BAYER AG
- BURROUGHS WELLCOME CO.
- CARDIAC MUSCLE SOCIETY
- INSTITUT DE RECHERCHES INTERNATIONALES SERVIER
- GLAXO INC.
- MERCK RESEARCH LABORATORIES
- PACER SCIENTIFIC
- PFIZER CENTRAL RESEARCH
- SUTTER INSTRUMENTS
- A. R. VETTER CO. INC.
- ZENECA PHARMACEUTICALS GROUP

The New York Academy of Sciences believes it has a responsibility to provide an open forum for discussion of scientific questions. The positions taken by the participants in the reported conferences are their own and not necessarily those of the Academy. The Academy has no intent to influence legislation by providing such forums.

FRONTISPIECE. What is sodium-calcium exchange? Sodium-calcium exchange is a powerful plasmalemmal calcium transporter whose activity profoundly influences the function of many cell types. This volume outlines the state-of-the-art of sodium-calcium exchange research as it was presented at the Third International Conference on Sodium-Calcium Exchange. This meeting took place April 23–26, 1995 in Woods Hole, Massachusetts at the Marine Biological Laboratory (MBL), shown in this aerial photograph. The MBL is an independent scientific institution, founded in 1888, which is dedicated to the highest level of creative research and education in biological sciences. The MBL has been the site of a great deal of fundamental research on sodium-calcium exchange, as well as other important ion transport processes. (Photograph generously provided by Richard Howard and the MBL.)

Preface

The second-messenger function of intracellular calcium continues to attract great scientific attention. Most of the established molecular players in calcium signaling have now been cloned and are being characterized with powerful new tools of cell and molecular biology. Among the players are sodium-calcium exchangers, potent calcium transporters in the surface membranes of many cell types.

Why study sodium-calcium exchange? Sodium-calcium exchange is still a remarkably open field of high physiological significance. Physiological functions as diverse as vision, secretion, and cardiac contractility are strongly dependent on sodium-calcium exchange activity. In many cell types, sodium-calcium exchange is the primary mechanism of calcium extrusion, and small changes of sodium-calcium exchange activity have large effects on cell function. In heart and in brain, sodium-calcium exchange activity likely becomes pivotal in pathological settings with possible outcomes of calcium overload, altered electrical activity and ultimately cell death. At the same time, there is still uncertainty as to how sodium-calcium exchangers might be tied to many cell regulatory pathways and might themselves be regulated.

For the third time, scientists with a strong interest in sodium-calcium exchange have met to discuss their latest work and exchange ideas.[a] In the present time of diminished support for meetings such as this, the strong personal commitment of many scientists to this field was the key to a successful conference. It is certain that the free exchange of ideas and data which occurred will contribute to a new round of experimental progress. It is hoped that this volume will find the attention, in particular, of students and young scientists looking for a promising new niche in science.

As documented in these proceedings, our understanding of sodium-calcium exchange is now improving dramatically through a wide range of experimental approaches in laboratories around the world. Successes have been achieved in the arenas of molecular biology, biophysics and cell biology. Structure–function studies of the cloned cardiac exchanger are well advanced. Multiple exchanger types are identified and characterized in different cells types (including brain, blood cells, smooth muscle and kidney). Molecular mechanisms of the actual exchange process and intrinsic exchanger "gating" reactions are being elucidated. Multiple exchanger modulation mechanisms are being characterized in different

---

[a] The Third International Conference on Sodium-Calcium Exchange was held in Woods Hole, Massachusetts in April of 1995. Those who desire a broad background in sodium-calcium exchange will find publications from the previous two meetings to be very valuable references. The First International Conference on Sodium-Calcium Exchange was held in Stowe, England in June of 1987, and the Second International Conference on Sodium-Calcium Exchange was held in Baltimore, Maryland in April of 1991. The two resulting publications were, respectively:

Sodium-Calcium Exchange. 1989. Edited by T. Jeff, A. Allen, Denis Noble and Harald Reuter. Oxford University Press. Oxford, UK.

Sodium-Calcium Exchange: Proceedings of the Second International Conference. 1991. Annals of the New York Academy of Sciences, Volume 639. Edited by Mordecai P. Blaustein, Reinaldo DiPolo and John P. Reeves. The New York Academy of Sciences. New York.

cell types. And the function of sodium-calcium exchange in intact cells is being elucidated with improved electrical methods, confocal microscopy, and anti-sense knockout approaches.

As organizers of the Third International Conference on Sodium-Calcium Exchange, we express out gratitude first to the participants who made personal sacrifices to join this conference. We thank the staff of the New York Academy of Sciences for their fine assistance with all aspects of the organization. Renée Wilkerson was importantly involved at all stages up through the meeting itself. Cook Kimball has been a proficient and helpful editor for this volume, aided by Audrey Herbst who was very helpful in the book's production. We thank LouAnn King and the entire staff of the Marine Biological Laboratory in Woods Hole for their excellent services.

We are indebted to David C. Gadsby for his role as conference advisor to the New York Academy of Sciences, and for his personal assistance with many aspects of the conference organization. Our Advisory Board was a source of much helpful counsel and criticism: Luis Beaugé, Mordecai P. Blaustein, Reinaldo DiPolo, David C. Gadsby, Jaime Requena and John P. Reeves. In addition, we thank Denis Noble and W. Jonathan Lederer for their extra work in preparing Commentaries for this volume.

DEDICATION

This volume is dedicated to the memory of Lorin J. Mullins, whose contributions to the field of membrane biophysics are commemorated by Jaime Requena in the final article of these proceedings. The volume is also dedicated to all young scientists and students whose interest in sodium-calcium exchange may in some way be promoted by reading its contents.

Donald W. Hilgemann
Kenneth D. Philipson
Guy Vassort

Commentary

DENIS NOBLE

University Laboratory of Physiology
Parks Road
Oxford OX1 3PT, United Kingdom

Understanding of the genetic, molecular and subcellular mechanisms of biological systems has reached a critical stage of development. At the molecular level, many of the ionic transporter proteins have been sequenced, cloned and expressed, leading to a much clearer, though still very incomplete, picture of the individual transport processes and to the development of new agents for investigation, such as antisense nucleotides and small specific peptide inhibitors, with which to probe the function of individual proteins in whole cells and organs and in developing possible new forms of therapy. At the genetic level, various types of transgenic cells can now be made, thus opening the way both to genetic manipulation as an experimental tool and to the possibility of gene therapy. At the subcellular level, the measurement of intracellular biophysical and biochemical parameters (such as ion and metabolite levels, contractile state, etc.) has greatly increased in speed and accuracy both at the cellular and whole organ level. Moreover, with the development of techniques for simultaneous measurement of several such parameters, it is possible to characterize the changes involved in important pathological conditions such as ischemia and hypertrophy in the case of the heart. As a result, there is now an avalanche of information at all of these levels, leading to a commensurate and rapidly growing problem. This is that there is a need to integrate this information into an understanding of how the individual transporter and metabolic mechanisms interact to produce both normal and pathological states.

This is a very complex form of interaction: the relationships between individual molecular mechanisms and overall functional or dysfunctional states are rarely one-to-one. They are much more frequently multifactorial. This, together with the lack of clear diagnostic guides in the opposite direction (*i.e.*, from whole organ measurements to inferred molecular action), is the underlying explanation for the fact that we have great trouble predicting therapeutic efficacy from molecular actions. A good example is the fact that some of the important clinical trials of, for example, antiarrhythmic drugs in protecting against heart attack, have given very disappointing results. The solution to this problem must lie in understanding the interactions of the many molecular processes sufficiently well to allow solution of the inverse problem, *i.e.*, inference of molecular defects, and hence appropriate therapy, from overall organ-level measurements.

The benefits to be obtained from work at the integrative level will not only be felt at that level. Successful integrative work can also act as a guide to further work at the molecular level since it identifies what is functionally important in the molecular descriptions. The genotype and its expression in individual proteins succeed or fails, after all, by virtue of what it produces functionally in the whole animal. Many of these issues are explored in much greater detail in a book (*The Logic of Life*[1]) published for the last international congress of Physiological Sciences.

This meeting on sodium-calcium exchange has amply illustrated these themes, as a comparison with the state of knowledge four years ago[2] and, even more so, that of eight years ago[3] will show. A whole family of Na-Ca exchangers and related transporters, such as the rod outer segment Na-Ca,K transporter, have now been cloned.[4,5] Tissue-specific splicing isoforms have been identified,[6] while deletion methods have revealed what is essential to the transporting function[7] and where the calcium regulatory site is located.[4] Yet, we still do not know the functional significance of this site. Measurements of its calcium affinity range all the way from values below the physiological range (meaning that calcium never falls low enough to deregulate the protein) to values near the top of the range. The mystery is deepened by the discovery that an exchanger cloned from Drosophila is actually *inhibited* by this regulatory site![4] A different deletion will remove sodium-dependent inactivation, another phenomenon for which we do not have a functional understanding, though we now know that it is interrelated with the Ca regulatory site. Yet further modifications of the amino acid sequence influence the voltage sensitivity of the exchanger,[8,9] a property that could be important in the involvement of the exchanger in sodium-overload arrhythmias, since it is this voltage dependence that makes the exchanger well suited to its role in restoring calcium balance yet also ensures that such arrhythmias are difficult to suppress.[10] It will take a lot more work, however, at both the molecular and the integrative levels, to determine whether these insights open a path to a new form of therapy. What is certain is that we can now think about these questions with a more refined knowledge of the subtlety of the underlying interactions, rather than simply and crudely asking whether we should make the tissue more or less excitable.

The issue of voltage dependence raises in turn the question where the voltage-dependent steps are located. The introduction of the giant patch technique has opened the way to answering this question by being able to study the partial reactions of the exchanger with control of ion concentrations on both sides. The results are consistent with the view that the great majority of the voltage dependence arises from the sodium translocation and that this occurs in multiple steps.[11] This work also leads to valuable estimates of the exchange rates and densities and to the successful prediction that it should be possible to detect current noise in smaller membrane patches.

The ability to downregulate the exchanger in cells using antisense oligodeoxynucleotides[12-14] is a very powerful tool indeed, since it is important to understand how calcium balance is approached and maintained under a wide variety of stresses on the system. These questions are still poorly understood. How far, for example, is the exchanger from thermodynamic equilibrium? The answer to this question determines whether up- or downregulation will have much influence on steady state ionic concentrations. It was one of Lorin Mullins' insights into the functional role of sodium-calcium exchange to realize that, in a rhythmic tissue like the heart, the resting exchanger reversal potential might lie at an intermediate potential between the extremes and then change rapidly during each cycle. In the retina, the question has an added significance since the energy available to the exchange is greatly increased by having access also to the potassium gradient with the consequence that very much lower steady state levels of intracellular calcium can be achieved. In the heart, it is still not clear what actually determines the low point of free intracellular calcium. Is this the role of the sarcolemmal calcium pump, operating at levels where the exchanger is deactivated (because internal calcium is too low—but this depends, of course, on the answer to the question posed above concerning the affinity of the calcium regulation site)? And, in that case, why do we find so little evidence for a role of this pump? The papers presented to this meeting on the balance between calcium entry via the calcium channel and calcium exit show a close balance even if one assumes that the *only* means by which calcium leaves the cell is via sodium-calcium exchange.[15]

This meeting has identified a number of other regulators of the exchanger, including the mechanism of the exchange inhibitory peptide known as XIP,[16] activation by protein kinase[17] and the regulatory role of ATP and how this can be removed.[18]

Regulation of the exchanger seems also to play an important role in cardiac hypertrophy, with very considerable upregulation occurring in this condition. Stretching the ventricle with increased pressure can upregulate the production of exchanger protein messenger by as much as 2–4 fold.[19,20] Upregulation also occurs during heart failure,[21] while downregulation occurs during ischemia.[22] We are beginning therefore to unravel the role of this protein in more than one form of cardiac pathology. Could upregulation in hypertrophy be a cause of arrhythmias by increasing the current carried by the exchange during late after-depolarizations? The answer to this question depends in turn on the answer to the thermodynamic question. If the system is far from thermodynamic equilibrium, then upregulation of the exchanger would, in itself, be expected to reduce resting free calcium and so reduce the probability of calcium overload. We need also to know which other transporter proteins are up- and downregulated by volume and tension changes. This would enable us to determine whether regulation of the sodium-calcium exchanger is a primary or a secondary factor. Upregulation might, for example, be an adaptation to an increased calcium transient. But the increased calcium transient would then have to arise from other changes since upregulation of the exchanger would, in itself, be expected to shift calcium balance away from the sarcoplasmic reticulum and towards the sarcolemma and so reduce the calcium release during each beat.

Another area explored in some detail at this meeting and which raises important integrative questions is the role of pH regulation of the exchanger. The exciting finding here is that internal and external pH have opposite actions.[23] Intracellular protonation strongly inhibits sodium-calcium exchange, while extracellular protonation moderately increases it. In ischemic conditions both the inside and outside pH change in an acid direction. The question then is which has the greater effect? And are the two effects simply additive? Inhibition of the sodium-calcium exchanger could be an important protective mechanism in ischemia since it would suppress calcium overload arrhythmias. The situation during reperfusion would again be different, since the extracellular acidity would be reduced at a time while the internal pH remains low. It is also important to note that ischemia also reduces the production of exchanger protein messenger. The level of exchanger protein during and following ischemia may therefore also depend on altered transcription.

Many of these functional questions depend greatly on how the exchanger protein is distributed within the cell structure. Both experimental and theoretial studies support the view that a high proportion of the exchanger proteins are located near restricted subsarcolemmal spaces in relation with the junctional SR[24,25] (the so-called diadic cleft space[26]). The changes in sodium and calcium concentrations sensed by the Na-Ca exchange would then be much higher than expected from models that assume a well-mixed uniform intracellular space. The results presented by Fred Fay[27] using a lipophilic calcium indicator are particularly exciting in this context, showing as they do that measuring $[Ca]_i$ simultaneously with normal and lipophilic fura-2 reveals a very much higher calcium transient near the sarcolemmal membrane (up to 2 μM) than in the main cytosol (around 0.5 μM). This kind of organization also favors the possibility that reversed sodium-calcium exchange may play a role in triggering SR calcium release. Indeed, it now seems firmly established that entry of calcium via reversed Na-Ca exchange can be sufficient on its own to trigger substantial release.[28,29] We should, however, be careful not to conclude that this necessarily means that much of the release in normal circumstances is produced in this way.[30] The mechanisms involved are highly nonlinear. The physiological end result is not necessarily therefore a simple sum of the isolated effects of individual mechanisms.

In this summary of the meeting, I have concentrated on the role of sodium-calcium exchange in the heart, and indeed this organ was the object of the great majority of the presentations. The Na-Ca exchanger is, however, very widely distributed and readers of this volume will find its role investigated in synapses,[31,32] retinal rods,[33] kidney,[14,34] smooth muscle,[35] chromaffin cells,[36-38] myelinated axons,[39] inner hair cells,[40] skeletal muscle,[41] mitochondria,[42] osteoblasts[43] and the pancreas.[44,45]

Such a wide distribution naturally raises the question whether there is a general regulator, perhaps like the endogenous circulating ouabain involved in the case of the Na-K pump.[46] A clue to the question comes from the observation that the exchanger is strongly upregulated in perinatal tissue and that it is subsequently downregulated, perhaps by thyroid hormone.[47] Further clues may eventually come from studies on the evolutionary origin of the exchanger.[48]

Finally, what about the possibilities for therapeutic intervention? This is, as yet, almost unexplored. Yet the possibilities for drug development targeting such an important transporter deeply implicated in physiological and pathological states are obviously immense. This was the area that was conspicuous by its complete absence at this meeting (though a contributor from Bayer emphasized its importance in discussion). It will be interesting to see whether this field gets opened up before the 4th international conference on sodium-calcium exchange.

REFERENCES

1. BOYD, C. A. R. & D. NOBLE, Eds. 1993. The Logic of Life: the challenge of integrative physiology. Oxford University Press. Oxford.
2. BLAUSTEIN, M. P., R. DIPOLO & J. D. REEVES, Eds. 1991. Sodium-Calcium Exchange. Annals of the New York Academy of Sciences, Vol. 639. (2nd international conference on Na-Ca exchange. Baltimore.)
3. ALLEN, T. J. A., D. NOBLE & H. REUTER Eds. 1989. Sodium-Calcium Exchange. Oxford University Press. Oxford. (1st international meeting on sodium-calcium exchange, Stowe, UK, 1987.)
4. PHILIPSON, K. D., D. A. NICOLL, S. MATSUOKA, L. V. HRYSHKO, D. O. LEVITSKY & J. N. WEISS. 1996. Molecular regulation of the Na^+-Ca^{2+} exchanger. This volume: 20–28.
5. IWATA, T., A. KRAEV, D. GUERINI & E. CARAFOLI. 1996. A new splicing variant in the frog heart sarcolemmal Na-Ca exchanger creates a putative ATP-binding site. This volume: 37–45.
6. SCHULZE, D. H., P. KOFUJI, C. VALDIVIA, S. HE, S. LUO, A. RUKNUDIN, S. WISEL, M. S. KIRBY, W. DUBELL & W. J. LEDERER. 1996. Alternative splicing of the Na^+-Ca^{2+} exchanger gene, NCX1. This volume: 46–57.
7. RAHAMIMOFF, H., W. LOW, O. COOK, I. FURMAN, J. KASIR & R. VATASHSKI. 1996. The structural basis of Na^+-Ca^{2+} exchange activity. This volume: 29–36.
8. NICOLL, D. A., L. V. HRYSHKO, S. MATSUOKA, J. S. FRANK & K. D. PHILIPSON. 1996. Mutagenesis studies of the cardiac Na^+-Ca^{2+} exchanger. This volume: 86–92.
9. RASGADO-FLORES, H., R. ESPINOSA-TANGUMA, J. TIE & J. DESANTIAGO. 1996. Voltage dependence of Na-Ca exchange in barnacle muscle cells: I. Na-Na exchange activated by α-chymotrypsin. This volume: 236–248.

10. NOBLE, D., J-Y. LEGUENNEC & R. WINSLOW. 1996. Functional roles of sodium-calcium exchange in normal and abnormal cardiac rhythm. This volume: 480–488.
11. HILGEMANN, D. W. 1996. The cardiac Na-Ca exchanger in giant membrane patches. This volume: 136–158.
12. NIGGLI, E., B. SCHWALLER & P. LIPP. 1996. Antisense oligodeoxynucleotides directed against Na-Ca exchanger mRNA: promising tools for studies on the cellular and molecular level. This volume: 93–102.
13. BLAND, K. S., K. TAKAHASHI, S. ISLAM & M. L. MICHAELIS. 1996. Effects of NCX1 antisense oligodeoxynucleotides on cardiac myocytes and primary neurons in culture. This volume: 119–120.
14. WHITE, K. E., F. A. GESEK & P. A. FRIEDMAN. 1996. Identification and antisense inhibition of Na-Ca exchange in renal epithelial cells. This volume: 115–118.
15. CLARK, R. B., R. A. BOUCHARD & W. R. GILES. 1996. Action potential duration modulates calcium influx, Na^+-Ca^{2+} exchange, and intracellular calcium release in rat ventricular myocytes. This volume: 417–429.
16. HALE, C. C. 1996. Mechanism of XIP in cardiac sarcolemmal vesicles. This volume: 171–181.
17. SMITH, J. B., H-W. LEE & L. SMITH. 1996. Regulation of expression of sodium-calcium exchanger and plasma membrane calcium ATPase by protein kinases, glucocorticoids, and growth factors. This volume: 258–271.
18. REEVES, J. P., G. CHERNAYA & M. CONDRESCU. 1996. Sodium-calcium exchange and calcium homeostasis in transfected Chinese hamster ovary cells. This volume: 73–85.
19. MENICK, D. R., K. V. BARNES, U. F. THACKER, M. M. DAWSON, D. E. MCDERMOTT, J. D. ROZICH, R. L. KENT & G. COOPER. IV. 1996. The exchanger and cardiac hypertrophy. This volume: 489–501.
20. FLESCH, M., F. PÜTZ, R. H. G. SCHWINGER & M. BÖHM. 1996. Functional relevance of an enhanced expression of the Na^+-Ca^{2+} exchanger in the failing human heart. This volume: 539–542.
21. REINECKE, H., R. STUDER, R. VETTER, H. JUST, J. HOLTZ & H. DREXLER. 1996. Role of the cardiac sarcolemmal Na^+-Ca^{2+} exchanger in end-stage human heart failure. This volume: 543–545.
22. BERSOHN, M. M. 1996. Sodium-calcium exchange expression in ischemic rabbit hearts. This volume: 534–535.
23. DOERING, A. E., D. A. EISNER & W. J. LEDERER. 1996. Cardiac Na-Ca exchange and pH. This volume: 182–198.
24. JUHASZOVA, M., H. SHIMIZU, M. L. BORIN, R. K. YIP, E. M. SANTIAGO, G. E. LINDENMAYER & M. P. BLAUSTEIN. 1996. Localization of the Na^+-Ca^{2+} exchanger in vascular smooth muscle, and in neurons and astrocytes. This volume: 318–335.
25. FRANK, J. S., F. CHEN, A. GARFINKEL, E. MOORE & K. D. PHILIPSON. 1996. Immunolocalization of the Na^+-Ca^{2+} exchanger in cardiac myocytes. This volume: 532–533.
26. LANGER, G. A. & A. PESKOFF. 1996. Calcium in the cardiac diadic cleft: implications for sodium-calcium exchange. This volume: 408–416.
27. FAY, F. 1995. This conference: oral presentation.
28. LITWIN, S., O. KOHMOTO, A. J. LEVI, K. W. SPITZER & J. H. B. BRIDGE. 1996. Evidence that reverse Na-Ca exchange can trigger SR calcium release. This volume: 451–463.
29. VITES, A-M. & J. A. WASSERSTROM. 1996. Ca^{2+} influx via Na-Ca exchange and I_{Ca} can both trigger transient contractions in cat ventricular myocytes. This volume: 521–524.
30. CANNELL, M. B., C. J. GRANTHAM, M. J. MAIN & A. M. EVANS. 1996. The roles of sodium and calcium current in triggering calcium release from the sarcoplasmic reticulum. This volume: 443–450.
31. REUTER, H. 1996. Regulation of exocytosis in presynaptic boutons by Ca^{2+} channels and Na-Ca exchange. This conference, abstract 8.
32. BLAUSTEIN, M. P., G. FONTANA & R. S. ROGOWSKI. 1996. The Na^+-Ca^{2+} exchanger in rat brain synaptosomes: kinetics and regulation. This volume: 300–317.
33. SCHNETKAMP, P. P. M., J. E. TUCKER & R. T. SZERENCSEI. 1996. Regulation of the bovine retinal rod Na-Ca+K exchanger. This volume: 336–345.

34. LYTTON, J., S-L. LEE, W-S. LEE, J. VAN BAAL, R. J. M. BINDELS, R. KILAV, T. NAVEH-MANY & J. SILVER. 1996. The kidney sodium-calcium exchanger. This volume:
35. SHIGEKAWA, M., T. IWAMOTO & S. WAKABAYASHI. 1996. Phosphorylation and modulation of the Na^+-Ca^{2+} exchanger in vascular smooth muscle cells. This volume: 249–257.
36. SCHNEIDER, A. S. & C-R. JAN. 1996. Na-Ca exchange in Ca^{2+} signaling and neurohormone secretion: secretory vesicle contributions in adrenal chromaffin cells. This volume: 356–365.
37. DUARTE, E. P., G. BALTAZAR, P. VERÍSSIMO & A. P. CARVALHO. 1996. Release of catecholamines and enkephalin peptides induced by reversal of the Na^+-Ca^{2+} exchanger in chromaffin cells. This volume: 391–394.
38. LIN, L. F., L-S. KAO & E. W. WESTHEAD. 1996. Agents that promote protein phosphorylation increase catecholamine secretion and inhibit the activity of the Na^+-Ca^{2+} exchanger in bovine chromaffin cells. This volume: 395–396.
39. STYS, P. K. & I. STEFFENSEN. 1996. Na^+-Ca^{2+} exchange in anoxic/ischemic injury of CNS myelinated axons. This volume: 366–378.
40. CHABBERT, C., A. SANS & J. LEHOUELLEUR. 1996. What mechanisms are involved in Ca^{2+} homeostasis in hair cells? This volume: 397–399.
41. GONZALEZ-SERRATOS, H., D. W. HILGEMANN, M. ROZYCKA, A. GAUTHIER & H. RASGADO-FLORES. 1996. Na-Ca exchange studies in sarcolemmal skeletal muscle. This volume: 556–560.
42. JUNG, D. W., K. BAYSAL & G. P. BRIERLEY. 1996. Properties of the Na^+-Ca^{2+} antiport of heart mitochondria. This volume: 553–555.
43. KRIEGER, N. S. 1996. Calcemic hormones regulate the level of sodium-calcium exchange protein in osteoblastic cells. This volume: 293–294.
44. VAN EYLEN, F., P. GOURLET, A. VANDERMEERS, P. LEBRUN & A. HERCHUELZ. 1996. Effects of Phe-Met-Arg-Phe-NH_2(FMRFa)-related peptides on Na-Ca exchange and ionic fluxes in rat pancreatic B cells. This volume: 288–289.
45. VAN EYLEN, F., M. SVOBODA, A. BOLLEN & A. HERCHUELZ. 1996. Presence of NACA3 and NACA7 exchanger isoforms in insulin-producing cells. This volume: 132–133.
46. HAMLYN, J. M. 1995. Endogenous ouabain: cardiovascular and central nervous system implications. This conference, abstract 22.
47. BOERTH, S. R., W. A. COETZEE & M. ARTMAN. 1996. Ontogeny and hormonal regulation of cardiac Na^+-Ca^{2+} exchanger expression in rabbits. This volume: 536–538.
48. KRAEV, A., I. CHUMAKOV & E. CARAFOLI. 1996. Molecular biological studies of the cardiac sodium-calcium exchanger. This volume: 103–109.

The Molecular Biology of the Na^+-Ca^{2+} Exchanger and Its Functional Roles in Heart, Smooth Muscle Cells, Neurons, Glia, Lymphocytes, and Nonexcitable Cells[a]

W. J. LEDERER,[b,c,d] S. HE,[e] S. LUO,[e] W. DuBELL,[g]
P. KOFUJI,[f] R. KIEVAL,[c] C. F. NEUBAUER,[c]
A. RUKNUDIN,[c,e] H. CHENG,[c] M. B. CANNELL,[c,h]
T. B. ROGERS,[g] AND D. H. SCHULZE[e]

[c]Department of Physiology
[d]Medical Biotechnology Center
[e]Department of Microbiology and Immunology
[f]Department of Pharmacology and Experimental Therapeutics
[g]Department of Biological Chemistry
University of Maryland School of Medicine
660 West Redwood Street
Baltimore, Maryland 21201

[h]Department of Pharmacology
St. George's Hospital Medical School
Cranmer Terrace
London, England

INTRODUCTION

The Na^+-Ca^{2+} exchanger, first identified in 1968 and 1969 in heart muscle,[1,2] has since been identified in virtually every tissue examined[3-5] and in species ranging from human,[3,6] to dog,[7] to squid[1] and to fruit fly.[8,9] This brief commentary is an attempt to review several important findings, identify and provide commentary on critical outstanding questions and suggest lines of work that should prove fruitful in the future.

The questions raised by recent work fall into two categories. (1) Questions about how the known or suspected functions of the Na^+-Ca^{2+} exchanger are brought about by the Na^+-Ca^{2+} exchanger protein. (2) Questions about why different tissues have different primary sequences of the Na^+-Ca^{2+} exchanger. Included in this second class of questions are those that center on why there are several genes encoding for the Na^+-Ca^{2+} exchanger and how or why evolutionary changes may have occurred.

---

[a] Our work has been funded by grants from the NIH and the American and Maryland Heart Associations.
[b] Corresponding author, at his Dept. of Physiology address.

Important Features of the Na^+-Ca^{2+} Exchanger

Local Control of Excitation-Contraction in Heart: What Does It Mean for the Na^+-Ca^{2+} Exchanger?

In heart muscle excitation-contraction coupling occurs as L-type calcium channels are activated by depolarization of the membrane potential.[10] The Ca^{2+} influx via these channels increases local $[Ca^{2+}]_i$ sufficiently to activate the Ca^{2+}-release channels in the sarcoplasmic reticulum (SR) (known as ryanodine receptors, RyR) by the mechanism known as Ca^{2+}-induced Ca^{2+}-release (CICR).[11,12] The elementary events of SR Ca^{2+}-release are known as "Ca^{2+}-sparks."[13-16] These events take place at the transverse-tubule–ryanodine receptor junction[17] (FIG. 1). The Na^+-Ca^{2+} exchanger is responsible for extruding the Ca^{2+} that enters the cell from the extracellular space (including the t-tubules) and is distributed over the entire extracellular surface including the t-tubules[18] (see FIG. 2). Thus just as EC coupling in heart arises under local control of SR Ca^{2+} release, the Na^+-Ca^{2+} exchanger proteins are well placed to extrude the activating Ca^{2+}. It is efficient to have the extrusion mechanisms widely distributed and plentiful (approximately 250–400 Na^+-Ca^{2+} exchanger proteins per square micron) as it appears to be.[19,20]

Having the extrusion mechanisms (sarcolemmal Na^+-Ca^{2+} exchanger and the SR Ca^{2+} ATPase) placed near the co-localized trigger (L-type Ca^{2+} channels[13-15]) and Ca^{2+} release sites (RyR[16,17]) as shown in FIGURES 1 and 2 means that this region of signaling operates at low resting $[Ca^{2+}]_i$ and can thus maintain a good signal to noise ratio. Furthermore this permits the L-type calcium channels to be able to efficiently and rapidly regulate $[Ca^{2+}]_i$ levels by means of local $[Ca^{2+}]_i$ control.[7,21]

The location of the Na^+-Ca^{2+} exchanger (operating in reverse mode) makes it possible for it, in principle, to provide Ca^{2+} to activate RyRs just as the L-type Ca^{2+} channels have been shown to function. This would happen following depolarization and elevated local intracellular sodium that would favor entry of Ca^{2+} via the Na^+-Ca^{2+} exchanger.[22-27] In rat heart cells this does not appear to be the case under our normal experimental conditions[28-30] (see FIG. 3). The best demonstration has been in guinea pig heart cells,[25,27] but recent examination by Cannell and his group raise serious questions (personal communication, unpublished data). Additional work must be completed to settle this issue.

Is There a Local-Control Feature in Neuronal Na^+-Ca^{2+} Exchanger?

Although the Na^+-Ca^{2+} exchanger mRNA has been found in abundance in the CNS[3] and in nerve cells, its specific function remains uncertain. The Na^+-Ca^{2+} exchanger protein certainly acts to extrude Ca^{2+} under normal conditions. Immunofluorescence imaging suggests that there is a particularly rich concentration of the protein at neuronal synapses and on spines (Yip, Blaustein & Lederer, unpublished observations), presynaptically at the neuromuscular junction[31] and in astrocytes.[32] Since local Ca^{2+} signaling appears to be important in neuronal synaptic signaling, it is possible that the Na^+-Ca^{2+} exchanger may also serve the purpose of extruding Ca^{2+} that enters via plasmalemmal Ca^{2+} channels. The issue of "microdomains" of locally high $[Ca^{2+}]_i$ in nerve is suspected but less well established than Ca^{2+}-sparks in heart.[33] The relative roles of the Na^+-Ca^{2+} exchanger and the plasmalemmal Ca^{2+} ATPase in removing Ca^{2+} from the neurons is uncertain as is the relative importance of Ca^{2+}-signal amplification by CICR that may involve

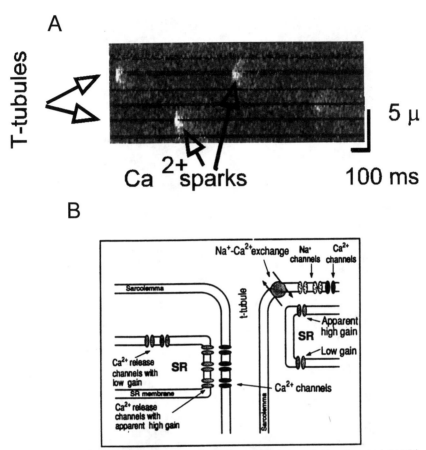

FIGURE 1. Excitation-contraction coupling in heart muscle is under local control. **(A)** Line scan images of transverse tubules and Ca^{2+}-sparks were used to show that Ca^{2+}-sparks occur at the transverse (T) tubules as expected.[10,14,15,58] Sulphorhodamine B was used to fill the transverse tubules and simultaneously Ca^{2+}-sparks were visualized in the flu-3 loaded cells. (Modified from Cheng et al.[17]) See also Shacklock et al. (1995). **(B)** Local control of EC coupling is thought to occur due to the voltage-dependent gating of the L-type calcium channels in the sarcolemmal surface membrane and in the transverse tubules. (Modified from Niggli & Lederer.[10])

RyR of IP3 receptors. There is still murky but mounting evidence of the importance of CICR, its triggering by Ca^{2+} influx and the role played by the Na^+-Ca^{2+} exchanger (see this volume). Nevertheless, the physiological roles of the transport and localization of the Na^+-Ca^{2+} exchanger in the central and peripheral nervous systems are not yet fully defined.

The Kidney and the Na^+-Ca^{2+} Exchanger

The kidney, along with the heart and the brain, is one of the three richest sources of the Na^+-Ca^{2+} exchanger protein in an animal (*e.g.*, human[3]). In the

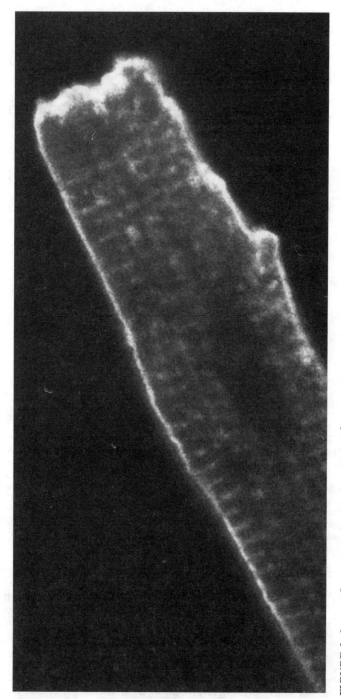

FIGURE 2. Immunofluorescence image of the Na^+-Ca^{2+} exchanger distribution on the surface of guinea pig heart cells (from Kieval et al.;[18] reprinted by permission from the *American Journal of Physiology: Cell*.) Similar results were seen for rat heart cells. See also Frank et al.[59] and Chen et al.[60]

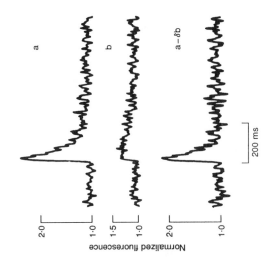
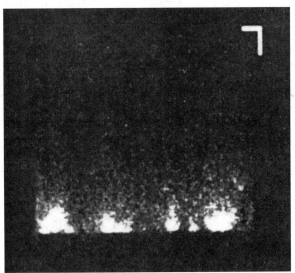

FIGURE 3. The Na^+-Ca^{2+} exchanger does not appear to significantly activate RyR to release Ca^{2+} in rat heart cells. Using normal rat heart cells at room temperature, cells were loaded with fluo-3 (by the AM method) so that we could investigate to what extent the Na^+-Ca^{2+} exchanger contributed to the activation of SR Ca^{2+} release. Cells were stimulated to contract by field stimulations with a normal extracellular solution. Then 10 μM Ca^{2+} was added to the extracellular medium. This is a concentration that largely blocks I_{Ca} but does little to alter Na^+-Ca^{2+} exchanger current of I_{Na}. The $[Ca^{2+}]_i$ transient was reduced from its peak value of around 1–3 μM to about 150 nM (spatially averaged). This confocal line-scan image shows that the normal $[Ca^{2+}]_i$ transient was virtually eliminated by this procedure and all that remained were identifiable Ca^{2+}-sparks that were synchronously activated by the stimulated action potential in this rat heart cell. The *arrow* marked by (a) indicates the region with a local Ca^{2+}-spark while the *arrow* marked by (b) indicates a region that reports a small increase in $[Ca^{2+}]_i$. Fluorescence time courses are shown on the right and also show the difference between the traces at the bottom of the right-hand panel. (From Lederer *et al.*[28] and Cannell *et al.*[29] Reprinted by permission from *Heart Vessels* and the *Journal of Physiology* (London).)

kidney, the Na^+-Ca^{2+} exchanger is thought to be found in richest concentration in the connecting tube of the nephron (see Kofuji et al.[34]).[35-38] Presumably the location of the Na^+-Ca^{2+} exchanger in this region of the tubule is important in reabsorbing Ca^{2+} and in controlling overall renal Ca^{2+} transport. Details regarding this function are not now fully established. The cellular localization of the Na^+-Ca^{2+} exchanger appears to be important in heart and brain, and this importance is established largely by the specific location of this transporting protein in the plasmalemmal membrane. In the kidney—at the preseent time—the important feature is not only where in the plasma membrane the Na^+-Ca^{2+} exchanger is located but also in *which* cells the Na^+-Ca^{2+} exchanger is expressed. Interestingly, different isoforms of the Na^+-Ca^{2+} exchanger are found in the kidney[36,39,40] and they too may provide distinct function for this tissue; but this has not yet been demonstrated.[41]

Smooth Muscle Cells, Lymphocytes, Glia and Many Other Cells

The smooth muscle cell may represent a more typical cell than those of the heart, CNS and kidney. Few if any cells completely lack the Na^+-Ca^{2+} exchanger but there is great variation in its abundance.[3] Smooth muscle has been shown to have Na^+-Ca^{2+} exchanger protein present,[42,43] and there are functional consequences of its presence.[42-45] Glia appear to have enough Na^+-Ca^{2+} exchanger for it to be readily identified[32] but neither local actions nor cell-specific function has been established. It is certainly intriguing to speculate that the glia serve as secondary signaling pathway, monitoring the amount of extracellular $[Ca^{2+}]_i$ and using that to modulate some neuronal function. Smooth muscle in the vasculature reveals that altered extracellular sodium concentrations and ouabain (or other Na,K pump inhibitor) application produces changes in Ca^{2+} metabolism.[46-49] Similar results have been reported for many other cell types.[1,4] While the cellular function of the Na^+-Ca^{2+} exchanger still needs to be further defined in these tissues its contribution to cellular Ca^{2+} metabolism appears to be important. Frequently there is evidence that the Na^+-Ca^{2+} exchanger plays an important role analogous to that in heart muscle. For example, in heart the Na^+-Ca^{2+} exchanger functions to link altered sodium metabolism to altered Ca^{2+} metabolism. Thus the therapeutic agent, digoxin, can be used to increase the strength of contraction of the heart. Since the Na,K ATPase is also a transport protein found in virtually every cell of the body, the actions of digoxin presumably are quite wide.[46,49] That, however, leaves a great deal to be explained. What normal role does the Na^+-Ca^{2+} exchanger subserve in the different tissues? Does the endogenous "digitalis-like" factor, an isomer of ouabain, play an important regulatory role in Ca^{2+} metabolism with the cells in these different tissues? Is there local control of the Na^+-Ca^{2+} exchanger function? These issues are largely unresolved and motivate much of the planned cellular work.

Commentary on Cellular Function of the Na^+-Ca^{2+} Exchanger

Calcium extrusion from the interior of a cell to the extracellular space appears to be handled by two classes of transport proteins: (1) the Na^+-Ca^{2+} exchanger and (2) the Ca^{2+} ATPase. At the present time it is not clear how the two interact in a given cell type and why there are two parallel systems. Since virtually all cells have both systems, the control of the expression and turnover rate of each would seem to be very important. The rule that seems to be followed is that whenever large amounts of Ca^{2+} enter the cell from the extracellular space and must be transported out of the cell, then the Na^+-Ca^{2+} exchanger is a dominant transporter. This clearly applies to cardiac myocytes and nerve terminals and to kidney cells in the connecting tubules. Whenever the calcium flux is lower, there is a larger fraction of Ca^{2+} extrusion that is attributed to the plasmalemmal Ca^{2+} ATPase (*e.g.*, skeletal muscle, smooth muscle, almost all other cells). In none of these cases are we suggesting that the Na^+-Ca^{2+} exchanger is unimportant; rather we are suggesting that the parallel Ca^{2+} transport system is also important.

Molecular Investigations of the Na^+-Ca^{2+} Exchanger

There are three Na^+-Ca^{2+} exchanger genes and the Na^+-Ca^{2+},K transporter gene in mammals[3,6,7,34,39,40,50-55] (see also Philipson at this meeting). From the NCX1 gene, the heart gene, isoforms are found in many tissues including brain, kidney and heart.[39] There are also distinctive features in the 5' untranslated regions of the exchanger mRNA from different tissues.[36] There are five major questions related to the molecular biology of the Na^+-Ca^{2+} exchanger which are tied into the tissue-specific roles of the Na^+-Ca^{2+} exchanger.

1. Why are there differences in the tissue-specific isoforms of NCX1? We do not know what the differences in isoforms mean in terms of the tissue requirements. Do they lead to tissue-specific functional differences? Are there differences in the subcellular targeting? What controls the tissue specific expression?
2. Why are there (at least) three different genes? NCX2 and NCX3 are genes that appear to be expressed at much higher levels in brain than in other tissues but still only complement NCX1 expression in brain. Are these genes expressed in different cells? Do they function differently in different cells?
3. Do the Na^+-Ca^{2+} exchanger gene products function as singular monomers, homomultimers or heteromultimers? Are there other subunits of the Na^+-Ca^{2+} exchanger?
4. How does the final functional protein use the energy from one electrochemical gradient to perform work against a different electrochemical gradient? What are the structural molecular features that permit the Na^+-Ca^{2+} exchanger to work?
5. Are there specific proteins that interact with the Na^+-Ca^{2+} exchanger to provide targeting or to modulate function?

Future of the Na^+-Ca^{2+} Exchanger: Lines of Inquiry

The work over the next five years will certainly make use of the molecular and cellular studies to date. They will probably center on the questions asked above and can be lumped into three areas: (1) molecular, (2) physiological, and (3) biophysical.

Molecular Studies

What regulates the Na^+-Ca^{2+} exchanger genes? How are they activated in a specific tissue? What regulates the isoform expression? How does the expression of the different genes and isoforms change with development? with disease? How many more genes are there? What is the phylogenetic profile of the Na^+-Ca^{2+} exchanger? From drosophila to squid to human? Are there novel genetic motifs that are used for potential subunits of a multimeric Na^+-Ca^{2+} exchanger protein? What genes exist that encode for important regulatory proteins?

Physiological Studies

What purpose(s) does the Na^+-Ca^{2+} exchanger subserve in each tissue? Why do different tissues and cell types have different isoforms? Why do some tissues (*e.g.*, brain) express more than one gene? How is the location of the Na^+-Ca^{2+} exchanger protein determined? What interactions are there among regulatory proteins and the Na^+-Ca^{2+} exchanger? Are there regulatory factors that change intracellular and extracellular ion affinities and turnover rates?

Biophysical Studies

What is the nature of the ion binding sites? What is the character of the ion permeation path(s)? What are the charge translocations reactions of the Na^+-Ca^{2+} exchanger protein? Can the protein function both as a channel and as a transport protein? How are the electrochemical gradients for Ca^{2+} and Na^+ coupled?

ACKNOWLEDGMENTS

We would like to thank Don Hilgemann and Ken Philipson for their consistent work over many years on the Na^+-Ca^{2+} exchanger. Specifically, we would like to acknowledge two very important recent contributions: the giant patch method (Hilgemann[56,57]) and the original cloning of the canine cardiac Na^+-Ca^{2+} exchanger (Philipson[7]). We would like to thank Ernst Niggli, Peter Lipp, Marcel Egger, Mark Nelson, and Mordecai P. Blaustein for many valuable discussions on topics reviewed in this paper.

REFERENCES

1. BAKER, P. F., M. P. BLAUSTEIN, A. L. HODGKIN & R. A. STEINHARDT. 1969. The influence of calcium on sodium efflux in squid axons. J. Physiol. **200:** 431–458.
2. REUTER, H. & N. SEITZ. 1968. The dependence of calcium efflux from cardiac muscle on temperature and external ion composition. J. Physiol. (London) **195:** 451–470.
3. KOFUJI, P., R. W. HADLEY, R. S. KIEVAL, W. J. LEDERER & D. H. SCHULZE. 1992. Expression of the Na-Ca exchanger in diverse tissues: a study using the cloned human cardiac Na-Ca exchanger. Am. J. Physiol. Cell **263:** C1241–C1249.
4. BLAUSTEIN, M. P., R. DIPOLO & J. REEVES. 1991. Sodium-Calcium Exchange. Ann. N. Y. Acad Sci. **639**.
5. ALLEN, T. J. A., D. NOBLE & H. REUTER. 1989. Sodium-Calcium Exchange. Oxford University Press. Oxford.
6. KOMURO, I., K. E. WENNINGER, K. D. PHILIPSON & S. IZUMO. 1992. Molecular cloning and characterization of the human cardiac Na^+-Ca^{2+} exchanger cDNA. Proc. Natl. Acad. Sci. USA **89:** 4769–4773.

7. NICOLL, D. A., S. LONGONI & K. D. PHILIPSON. 1990. Molecular cloning and functional expression of the cardiac sarcolemmal Na^+-Ca^{2+} exchanger. Science **250:** 562–565.
8. VALDIVIA, C., P. KOFUJI, W. J. LEDERER & D. H. SCHULZE. 1995. Characterization of the Na-Ca exchanger cDNA in *Drosophila*. Biophys. J. **68:** A410.
9. HRYSHKO, L. V., D. A. NICOLL, S. MATSUOKA, J. N. WEISS, E. SCHWARZ, S. BENZER & K. D. PHILIPSON. 1995. Anomalous regulation of the Na^+-Ca^{2+} exchanger from *Drosophila*. Biophys. J. **68:** A410.
10. NIGGLI, E. & W. J. LEDERER. 1990. Voltage-independent calcium release in heart muscle. Science **250:** 565–568.
11. BERS, D. M. 1991. Excitation-Contraction Coupling and Cardiac Contractile Force. Kluwer Academic Publishers.
12. FABIATO, A. 1985. Simulated calcium current can both cause calcium loading and trigger calcium release from the sarcoplasmic reticulum of a skinned canine cardiac Purkinje cell. J. Gen. Physiol. **85:** 291–320.
13. CANNELL, M. B., H. CHENG & W. J. LEDERER. 1995. The control of calcium release in heart muscle. Science **268:** 1045–1049.
14. CANNELL, M. B., H. CHENG & W. J. LEDERER. 1994. Spatial non-uniformities in $[Ca^{2+}]_i$ during excitation contraction coupling in cardiac myocytes. Biophys. J. **67:** 1942–1956.
15. CHENG, H., W. J. LEDERER & M. B. CANNELL. 1993. Calcium sparks: elementary events underlying excitation-contraction coupling in heart muscle. Science **262:** 740–744.
16. SANTANA, L. F., H. CHENG, A. M. GÓMEZ, M. B. CANNELL & W. J. LEDERER. 1996. Relation between the sarcolemmal Ca^{2+} current and Ca^{2+} sparks and local control theories for cardiac excitation-contraction coupling. Circ. Res. In press.
17. CHENG, H., M. B. CANNELL & W. J. LEDERER. 1996. Calcium sparks and $[Ca^{2+}]_i$ waves in heart. Am. J. Physiol. Cell. In press.
18. KIEVAL, R. S., R. J. BLOCH, G. E. LINDENMAYER, A. AMBESI & W. J. LEDERER. 1992. Immunofluorescence localization of the Na-Ca exchanger in heart cells. Am. J. Physiol. Cell **263:** C545–C550.
19. HILGEMANN, D. W., D. A. NICOLL & K. D. PHILIPSON. 1991. Charge movement during Na^+ translocation by native and cloned cardiac Na^+-Ca^{2+} exchanger. Nature **352:** 715–718.
20. NIGGLI, E. & W. J. LEDERER. 1991. Molecular operations of the sodium-calcium exchanger revealed by conformation currents. Nature **349:** 621–624.
21. STERN, M. D. 1992. Theory of excitation-contraction coupling in cardiac muscle. Biophys. J. **63:** 497–517.
22. HUME, J. R., P. C. LEVESQUE & N. LEBLANC. 1991. Sodium-calcium exchange (technical comment). Science **251:** 1370–1371.
23. LEDERER, W. J., E. NIGGLI & R. W. HADLEY. 1991. Sodium-calcium exchange (technical comment). Science **251:** 1370–1371.
24. BRIDGE, J. H. B., J. R. SMOLLEY & K. W. SPITZER. 1990. The relationship between charge movements associated with I_{Ca} and $I_{Na\text{-}Ca}$ in cardiac myocytes. Science **248:** 376–378.
25. LEBLANC, N. & J. R. HUME. 1990. Sodium current-induced release of calcium from cardiac sarcoplasmic reticulum. Science **248:** 372–376.
26. LEDERER, W. J., E. NIGGLI & R. W. HADLEY. 1990. Sodium-calcium exchange in excitable cells: fuzzy space. Science **248:** 283.
27. LIPP, P. & E. NIGGLI. 1994. Sodium current-induced calcium signals in isolated guinea-pig ventricular myocytes. J. Physiol. (London) **474:** 439–446.
28. LEDERER, W. J., H. CHENG, S. HE, C. VALDIVIA, P. KOFUJI, D. H. SCHULZE & M. B. CANNELL. 1995. Na-Ca exchanger—role in excitation-contraction coupling in heart muscle and physiological insights from the gene structure. Heart Vessels. Suppl. **9:** 161–162.
29. CANNELL, M. B., H. CHENG & W. J. LEDERER. 1994. Nifedipine decreases the spatial uniformity of the depolarization-evoked Ca^{2+} transient in isolated rat cardiac myocytes. J. Physiol. (London) **477:** 25P.

30. Cheng, H., W. J. Lederer & M. B. Cannell. 1995. Partial inhibition of calcium current by D600 reveals spatial non-uniformities in $[Ca^{2+}]_i$ during excitation-contraction coupling in cardiac myocytes. Circ. Res. **76:** 236–241.
31. Luther, P. W., R. K. Yip, R. J. Bloch, A. Ambesi, G. F. Lindenmayer & M. P. Blaustein. 1992. Presynaptic localization of sodium/calcium exchangers in neuromuscular preparations. J. Neurosci. **12:** 4898–4904.
32. Goldman, W. F., P. J. Yarowsky, M. Juhaszova, B. K. Krueger & M. P. Blaustein. 1994. Sodium/calcium exchange in rat cortical astrocytes. J. Neurosci. **14:** 5834–5843.
33. Bootman, M. D. & M. J. Berridge. 1995. The elemental principles of calcium signalling. Cell **83:** 675–678.
34. Kofuji, P., W. J. Lederer & D. H. Schulze. 1993. Na-Ca exchanger isoforms expressed in kidney. Am. J. Physiol. Renal Fluid Electrolyte Physiol. **265:** F598–F603.
35. Reilly, R. F., C. A. Shugrue, D. Lattanzi & D. Biemesderfer. 1993. Immunolocalization of the Na^+-Ca^{2+} exchanger in rabbit kidney. Am. J. Physiol. Renal Fluid Electrolyte Physiol. **265:** F327–F332.
36. Lee, S-L., A. S. L. Yu & J. Lytton. 1994. Tissue-specific expression of Na^+-Ca^{2+} exchanger isoforms. J. Biol. Chem. **269:** 14849–14852.
37. Yu, A. S. L., S. C. Hebert, B. M. Brenner & J. Lytton. 1992. Molecular characterization and nephron distribution of a family of transcripts encoding the pore-forming subunit of Ca^{2+} channels in the kidney. Proc. Natl. Acad. Sci. USA **89:** 10494–10498.
38. Yu, A. S. L., S. C. Hebert, S-L. Lee, B. M. Brenner & J. Lytton. 1992. Identification and localization of renal Na^+-Ca^{2+} exchanger by polymerase chain reaction. Am. J. Physiol. Renal Fluid Electrolyte Physiol. **263:** F680–F685.
39. Kofuji, P., W. J. Lederer & D. H. Schulze. 1994. Mutually exclusive and cassette exons underlie alternatively spliced isoforms of the Na-Ca exchanger. J. Biol. Chem. **269:** 5145–5149.
40. Reilly, R. F. & C. A. Shugrue. 1992. cDNA cloning of a renal Na^+-Ca^{2+} exchanger. Am. J. Physiol. Renal Fluid Electrolyte Physiol. **262:** F1105–F1109.
41. Matsuoka, S., D. A. Nicoll, R. F. Reilly, D. W. Hilgemann & K. D. Philipson. 1993. Initial localization of regulatory regions of the cardiac sarcolemmal Na^+-Ca^{2+} exchanger. Proc. Natl. Acad. Sci. USA **90:** 3870–3878.
42. Juhaszova, M., A. Ambesi, G. E. Lindenmayer, R. J. Bloch & M. P. Blaustein. 1994. Na^+-Ca^{2+} exchanger in arteries: identification by immunoblotting and immunofluorescence microscopy. Am. J. Physiol. **266:** C234–C242.
43. Blaustein, M. P., A. Ambesi, R. J. Bloch, W. F. Goldman, M. Juhaszova, G. E. Lindenmayer & D. N. Weiss. 1992. Regulation of vascular smooth muscle contractility: roles of the sarcoplasmic reticulum (SR) and the sodium/calcium exchanger. Japan J. Pharmacol. **58**(Suppl. 2): 107P–114P.
44. Slodzinski, M. K., M. Juhaszova & M. P. Blaustein. 1995. Antisense inhibition of Na^+-Ca^{2+} exchange in primary cultured arterial myocytes. Am. J. Physiol. **269:** C1340–C1345.
45. Borin, M. L., W. F. Goldman & M. P. Blaustein. 1993. Intracellular free Na^+ in resting and activated A7r5 vascular smooth muscle cells. Am. J. Physiol. Cell **264:** C1513–C1524.
46. Blaustein, M. P. 1994. Endogenous ouabain: physiological activity and pathophysiological implications. Clin. Invest. **72:** 706–707.
47. Borin, M. L., R. M. Tribe & M. P. Blaustein. 1994. Increased intracellular Na^+ augments mobilization of Ca^{2+} from SR in vascular smooth muscle cells. Am. J. Physiol. **266:** C311–C317.
48. Tribe, R. M., M. L. Borin & M. P. Blaustein. 1994. Functionally and spatially distinct Ca^{2+} stores are revealed in cultured vascular smooth muscle cells. Proc. Natl. Acad. Sci. USA **91:** 5908–5912.
49. Blaustein, M. P. 1993. Physiological effects of endogenous ouabain: control of intracellular Ca^{2+} stores and cell responsiveness. Am. J. Physiol. Cell. **264:** C1367–C1387.
50. Li, Z., S. Matsuoka, L. V. Hryshko, D. A. Nicoll, M. M. Bersohn, E. P. Burke,

R. P. LIFTON & K. D. PHILIPSON. 1994. Cloning of the NCX2 isoform of the plasma membrane Na^+-Ca^{2+} exchanger. J. Biol. Chem. **269:** 17434–17439.
51. FURMAN, I., O. COOK, J. KASIR & H. RAHAMIMOFF. 1993. Cloning of two isoforms of the rat brain Na^+-Ca^{2+} exchanger gene and their functional expression in HeLa cells. FEBS Lett. **319:** 105–109.
52. REILÄNDER, H., A. ACHILLES, U. FRIEDEL, G. MAUL, F. LOTTSPEICH & N. J. COOK. 1992. Primary structure and functional expression of the Na/Ca,K-exchanger from bovine rod photoreceptors. EMBO J. **11:** 1689–1695.
53. LOW, W., J. KASIR & H. RAHAMIMOFF. 1993. Cloning of the rat heart Na^+-Ca^{2+} exchanger and its functional expression in HeLa cells. FEBS Lett. **316:** 63–67.
54. MCDANIEL, L. D., W. J. LEDERER, P. KOFUJI, D. H. SCHULZE, R. KIEVAL & R. A. SCHULTZ. 1993. Mapping of the human cardiac Na^+-Ca^{2+} exchanger gene (NCX1) by fluorescent *in situ* hybridization to chromosome region 2p22→p23. Cytogen. Cell Genet. **63:** 192–193.
55. ACETO, J. F., M. CONDRESCU, C. KROUPIS, H. NELSON, N. NELSON, D. NICOLL, K. D. PHILIPSON & J. P. REEVES. 1992. Cloning and expression of the bovine cardiac sodium-calcium exchanger. Arch. Biochem. Biophys. **298:** 553–560.
56. HILGEMANN, D. W. 1990. Regulation and deregulation of cardiac Na^+-Ca^{2+} exchange in giant excised sarcolemmal membrane patches. Nature **344:** 242–245.
57. HILGEMANN, D. W. 1989. Giant excised cardiac sarcolemmal membrane patches: sodium and sodium-calcium exchange currents. Pflugers Arch. **415:** 247–249.
58. CANNELL, M. B., H. CHENG & W. J. LEDERER. 1995. The control of calcium release in heart muscle. Science **268:** 1045–1050.
59. FRANK, J. S., G. MOTTINO, D. REID, R. S. MOLDAY & K. D. PHILIPSON. 1992. Distribution of the Na^+-Ca^{2+} exchange protein in mammalian cardiac myocytes: an immunofluorescence and immunocolloidal gold-labeling study. J. Cell. Biol. **117:** 337–345.
60. CHEN, F., G. MOTTINO, T. S. KLITZNER, K. D. PHILIPSON & J. S. FRANK. 1995. Distribution of the Na^+-Ca^{2+} exchange protein in developing rabbit myocytes. Am. J. Physiol. Cell **268:** C1126–C1132.

Part I
Molecular Biology
Introduction

Substantial progress has been made in the *Molecular Biology* of the Na^+-Ca^{2+} exchangers. New exchangers have been cloned and extensive alternative splicing has been described. Exons coding for different portions of the large intracellular loop are used in different combinations in different tissues. Initial work on the organization of the exchanger gene has been completed. Mutagenesis has begun to define portions of the exchanger protein involved in both ion translocation and regulation.

New work was presented on the use of cell lines stably transfected with exchanger DNA. This approach should be invaluable in studying the role of the exchanger in calcium signaling pathways. Antisense oligonucleotide technology has now been applied to exchanger research; initial studies indicate this will also be a promising approach.

The findings presented at the conference raise many questions which are just beginning to be addressed. What is the functional significance of the splice variants? What controls the tissue-specific expression of the Na^+-Ca^{2+} exchangers and the different splice variants? Are NCX1 and NCX2 located in spatially distinct domains? Tools are now available to begin to address these and many other questions.

Molecular Regulation of the Na$^+$-Ca^{2+} Exchanger

K. D. PHILIPSON,[a] D. A. NICOLL, S. MATSUOKA,[b]
L. V. HRYSHKO,[c] D. O. LEVITSKY,[d] AND J. N. WEISS

*Departments of Physiology and Medicine
and the
Cardiovascular Research Laboratories
University of California, Los Angeles
School of Medicine
Los Angeles, California 90095-1760*

INTRODUCTION

The use of modern molecular biology to study the Na$^+$-Ca^{2+} exchanger became possible in 1990 with the cloning of the exchanger of canine cardiac sarcolemma (NCX1).[1] Subsequently, NCX1 and several splicing isoforms of NCX1 have been cloned from several species and tissues (see other papers in this volume). NCX1 is by far the most extensively studied and characterized form of the exchanger. In 1992, the Na$^+$-Ca^{2+},K$^+$ exchanger from rod out segments (RetX) was cloned.[2] RetX is different from NCX1 in both functional and molecular characteristics. Functionally, RetX exchanges 4 Na$^+$ ions for 1 Ca^{2+} ion plus 1 K$^+$ ion whereas NCX1 exchanges 3 Na$^+$ ions for 1 Ca^{2+} ion. Molecularly, RetX has little sequence similarity to NCX1.[2] However, topology models based on hydropathy analysis are very similar for the two exchangers.[3] Sequence similarity is present only in regions overlapping putative transmembrane segments (TMSs) 2 and 3 and TMSs 8 and 9 of both exchangers (FIG. 1). In both these regions, amino acid identity is about 50% extending over 30–50 amino acids. In addition to this intermolecular homology, the exchanger also exhibits a limited intramolecular homology. That is, there is homology between the TMSs in the amino terminal half of the protein and the TMSs in the carboxyl terminal half of the protein. Specifically, TMSs 2 and 3 have sequence similarity with TMSs 8 and 9 (FIG. 1). The intramolecular homology was first noted by Schwarz and Benzer,[4] who refer to these regions as the α-repeats. The intramolecular homology indicates that the evolution of exchanger proteins involved a gene duplication event. This paper will first briefly review the proliferation of members of the Exchanger Superfamily and will then focus on recent data from our laboratory on understanding regulation of NCX1 at the molecular level.

---

[a] Address for correspondence: Cardiovascular Research Laboratories, MRL Bldg. 3645, UCLA School of Medicine, Los Angeles, CA 90095-1760.
[b] Current address: Department of Physiology, Faculty of Medicine, Kyoto University, Kyoto, 606-01 Japan.
[c] Current address: Division of Cardiovascular Sciences, St. Boniface Hospital Research Centre, Winnipeg, Manitoba, Canada R2H 2A6.
[d] Current address: URA CNRS 1340, Université de Nantes, 44072 Nantes Cedex 03, France.

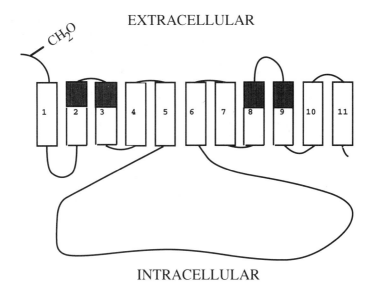

FIGURE 1. Proposed topology of NCX1. *Rectangles* represent proposed TMSs. CH$_2$O represents the known site of glycosylation. The large intracellular loop represents 520 amino acids and is more than half the size of the total protein by itself. The *shaded areas* show similarity to the analogous regions of all members of the proposed Exchanger Superfamily. Also, the shaded area overlapping TMSs 2 and 3 shows similarity to the shaded area overlapping TMSs 8 and 9 for all the related putative exchangers.

Exchanger Superfamily

We divide members of the cation Exchanger Superfamily into two groups: "Na$^+$-Ca^{2+} exchangers" and "Others" (TABLE 1). For the most part, the function of members of the "Others" group is unknown. Members of the Superfamily are being found by traditional cloning techniques but also as a result of genome

TABLE 1. Exchanger Superfamily

	Database Accession Number
Na$^+$-Ca^{2+} Exchangers	
NCX1	gb/M36119
NCX2	gb/U08141
Drosophila	gb/L39835
C. elegans	dbj/D35735
Others	
ROS Na$^+$-Ca^{2+},K$^+$ exchanger	gb/X66481
C. elegans	sp/P34322
Yeast	gb/Z46259
E. coli	gb/U18997

sequencing projects. (The GenBank access numbers for these sequences can be obtained by doing a BLAST search with sequence from other exchanger clones.) The defining characteristics of members of the Superfamily are as follows:

1. Similar hydropathy profiles to that initially seen for NCX1.
2. All members have homology to one another in TMSs 2 and 3 and TMSs 8 and 9; the same regions for which homology was initially seen between NCX1 and RetX.
3. Each member shows intramolecular homology between TMSs 2 and 3 and TMSs 8 and 9. These regions are shown schematically in FIGURE 1 for NCX1.

The Na^+-Ca^{2+} exchanger (NCX) group so far comprises products of two mammalian genes and two genes of invertebrates. The two NCX genes found in mammals are NCX1 (discussed above) and NCX2. NCX1 is expressed at high level in the myocardium but can be detected in several tissues. NCX2 was cloned from rat brain[5] and transcripts can only be detected in brain and skeletal muscle by Northern blot analysis. The significance of this highly specific pattern of tissue expression for NCX2 is unknown. NCX2 has generally similar functional properties to NCX1 and has an amino acid sequence which is 65% identical to NCX1. The *Drosophila* Na^+-Ca^{2+} exchanger (Calx) has been cloned by two groups[4,6] and has some interesting regulatory properties (see below). The sequence of an apparent NCX from *C. elegans* has also appeared in the data banks as the result of a sequencing project. All members of the NCX family have recognizable similarity to NCX1 throughout the protein coding regions.

The members of the "Other" category are, like RetX, distantly related to NCX1 and are all distantly related to each other. Only the properties of RetX, the Na^+-Ca^{2+},K^+ exchanger have been measured functionally. Awareness of the *C. elegans* and *E. coli* clones is due to large-scale sequencing projects. Thus, *C. elegans* has at least two exchanger genes: a member of the NCX group (see above) and a member of the "Other" group. A yeast exchanger homologue was found in a genetic complementation experiment (T. Pozos and M. Cyert, personal communication). We consider the *C. elegans*, yeast, and *E. coli* clones to represent putative cation exchangers. However, sequence divergence is so substantial that it cannot be assumed that any of the gene products are Na^+-Ca^{2+} exchangers. A detailed discussion of the similarities and evolutionary relationships among these proteins can be found in Schwarz and Benzer.[7]

Molecular Regulation: Intracellular

NCX1 and NCX2 transport Ca^{2+} but are also regulated by Ca^{2+} at a high-affinity binding site on the cytoplasmic surface of the exchanger. This regulatory binding site is distinct from the Ca^{2+} transport site. Existence of a Ca^{2+} regulatory site was first observed by DiPolo[8] for the squid axon Na^+-Ca^{2+} exchanger. Ca^{2+} regulation of the exchanger is now studied most easily using the giant excised patch technique of Hilgemann.[9] In this system, Ca^{2+} regulation can be observed most readily when outward exchange currents are measured. With the inside-out giant patches, addition of Na^+ to the bath at the intracellular surface of the exchanger initiates an exchange of bath Na^+ for pipette Ca^{2+} with a resultant outward exchange current (reverse Na^+-Ca^{2+} exchange). Initiation of exchange activity, however, requires the presence of trace, nontransported Ca^{2+} in the bath in addition to Na^+. This is illustrated in FIGURE 2. Ca^{2+} regulation of inward

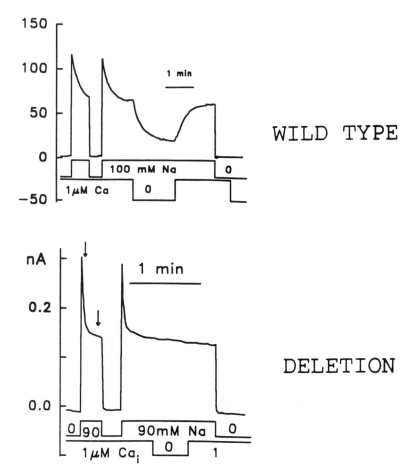

FIGURE 2. Outward Na^+-Ca^{2+} exchange currents from wild type (*top*) and mutant $\Delta 562$–685 (*bottom*) exchangers. Recordings are from giant excised patches from oocytes expressing exchangers. Ca^{2+} is present within the pipette. *Lower tracings* show manipulations of bath Na^+ and Ca^{2+}. The wild type tracing demonstrates two forms of exchanger regulation: Na^+-dependent inactivation and Ca^{2+} regulation. Na^+-dependent inactivation is seen as the partial decline in exchange current from a peak value to a new steady state value after addition of bath Na^+. Ca^{2+} regulation is seen as the decline in current upon removal of bath Ca^{2+}. Note that *trans* Ca^{2+} is present in the pipette; only regulatory Ca^{2+} is being removed. In the deletion mutant, Ca^{2+} regulation is absent. There is no response to the removal of regulatory Ca^{2+} although Na^+-dependent inaction is still present. In the *upper panel* the unit of current is pA and in the *lower panel* the unit is nA. The two exchangers express at similar levels. (Modified from Matsuoka *et al.*[11])

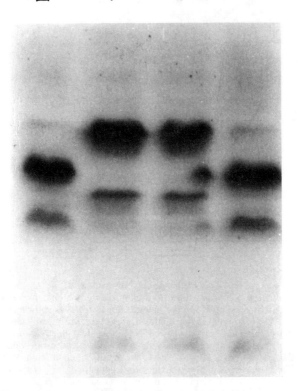

FIGURE 3. $^{45}Ca^{2+}$ overlay of a fusion protein containing amino acids 240–532 of the NCX1 loop. *E. coli* expressing the protein were solubilized, subjected to SDS/PAGE, and transferred onto nitrocellulose. The nitrocellulose was incubated with $^{45}Ca^{2+}$, washed, and exposed to X-ray film. The bands that bind Ca^{2+} on the autoradiogram correspond to the position of the fusion protein as detected by immunoblot. The mobility of the fusion protein shifts depending on the presence or absence of Ca^{2+} in the gel loading buffer (see text). (From Levitsky *et al.*[12] Reprinted by permission from the *Journal of Biological Chemistry*.)

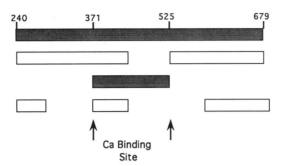

FIGURE 4. Summary of $^{45}Ca^{2+}$ overlay experiments. Each *bar* represents a different fusion protein constructed from NCX1. The *numbers at top* refer to amino acid positions. *Shaded bars* indicate fusion proteins capable of binding $^{45}Ca^{2+}$ in overlay experiments. Thus, for example, the portion of the exchanger loop encompassing amino acids 371–525 was capable of binding Ca^{2+}.

exchange current (forward exchange) has also been observed.[10] The involvement of our laboratory in the study of Ca^{2+} regulation began with the finding that Ca^{2+} regulation could be eliminated when the large intracellular loop of the exchanger was deleted using recombinant DNA techniques.[11] Specifically, a 440-amino acid deletion removed Ca^{2+} regulation but left transport activity intact. A smaller deletion of 124 amino acids from the loop also eliminated Ca^{2+} regulation (FIG. 2). The implication is that the exchanger loop is involved in Ca^{2+} binding and regulation, whereas the proposed TMS portion of the protein carries out ion translocation. In these experiments and in those described below, mutant exchangers were expressed in *Xenopus* oocytes and analyzed in giant excised patches.

To localize the Ca^{2+} binding site on the exchanger loop, we employed the $^{45}Ca^{2+}$ overlay technique and production of fusion proteins containing different portions of the loop.[12] FIGURE 3 shows an example of $^{45}Ca^{2+}$ binding to one such fusion protein on nitrocellulose and a summary of some data is shown in FIGURE 4. Initially, as shown in FIGURE 4, the smallest protein fragment which could bind Ca^{2+} with high affinity encompassed amino acids 371–525. Subsequent work indicated that the minimum fragment to bind Ca^{2+} contained amino acids 371–508. Two domains within amino acids 371–508 are rich in acidic residues. Mutation of specific aspartic acids within these domains decreased the affinity of fusion proteins for $^{45}Ca^{2+}$. We propose that these residues are involved in the high-affinity binding of Ca^{2+} to the exchanger loop.

The autoradiogram from a ^{45}Ca overlay experiment in FIGURE 3 also demonstrates a striking shift in the mobility of fusion proteins during sodium dodecyl sulfate polyacrylamide gel electrophoresis (SDS/PAGE) depending on the presence or absence of Ca^{2+} in the gel loading buffer.[12] Apparently, even in SDS, the fusion proteins retain the ability to bind Ca^{2+}. We interpret the substantial mobility shift as being due to a change in protein conformation upon binding Ca^{2+}. The protein with Ca^{2+} bound perhaps forms a more compact structure and can move through the gel more rapidly.

Although a Ca^{2+} binding site had been localized we had no evidence that this site had any functional significance. To address this issue, we examined the effect

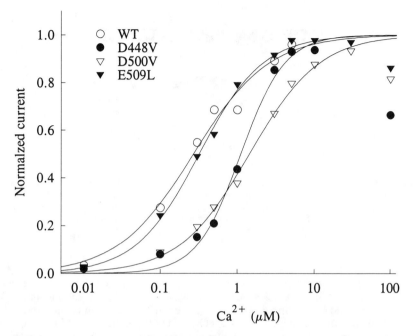

FIGURE 5. Dependencies on cytoplasmic Ca^{2+} of outward exchange currents for wild type and various mutant NCX1 exchangers. In all cases, extracellular (pipette) Ca^{2+} is present and only bath regulatory Ca^{2+} is varied. The two mutants (D448V and D500V) with reduced apparent affinity also have reduced affinity for $^{45}Ca^{2+}$ on Ca^{2+} overlays. (From Matsuoka et al.[10] Reprinted by permission from the *Journal of General Physiology*.)

on Na^+-Ca^{2+} exchange activity of mutations which had been found to alter $^{45}Ca^{2+}$ binding.[10] As shown in FIGURE 5, two mutants (D448V and D500V) with reduced affinity for $^{45}Ca^{2+}$ binding also had reduced apparent affinity for regulatory Ca^{2+}. A mutant (E509L) with normal $^{45}Ca^{2+}$ binding had normal Ca^{2+} regulation. In fact, as shown in TABLE 2, these correlations are maintained for a range of mutants.

TABLE 2. Correlation of Effects of Mutations on $^{45}Ca^{2+}$ Binding and Apparent Ca^{2+} Affinity for Regulation[a]

	Ca^{2+} Binding	Ca^{2+} Regulation
WT	high	high
R441L	high	high
D453V	high	high
E509L	high	high
D447V	low	low
D448V	low	low
D498I	low	low
D498K	low	low

[a] Data are from Refs. 10 and 12.

We conclude that we have localized the functionally important regulatory Ca^{2+} binding site. Although much has been learned about Ca^{2+} regulation of the Na^+-Ca^{2+} exchanger, there are still some surprising aspects which have not yet been fully analyzed. For example, we have found that a Na^+-Ca^{2+} exchanger cloned from *Drosophila* has anomalous Ca^{2+} regulation. In this case, Ca^{2+} inhibits, rather than stimulates, exchange activity.[13] The regulatory properties are almost completely opposite to those observed for NCX1. Further investigation of the *Drosophila* exchanger is the subject of ongoing investigation.

Molecular Regulation: Inactivation

As shown in FIGURE 2, the Na^+-Ca^{2+} exchanger displays a form of intrinsic regulation in addition to secondary Ca^{2+} regulation. Referred to as Na^+-dependent inactivation, this form of regulation is manifested as a slow, partial decline in outward Na^+-Ca^{2+} exchange current upon application of bath Na^+ to initiate exchange.[9] The inactivation can be modeled to originate from a state of the exchanger with Na^+ ions bound at the intracellular surface.[14] That is, upon application of Na^+ to the intracellular surface, three Na^+ ions will bind at the intracellular transport sites. From this Na^+-bound state, the exchanger will translocate Na^+ ions across the membrane to the external surface or will enter an inactivated state. The steady state level of outward exchange current will reflect the distribution of active and inactive states.

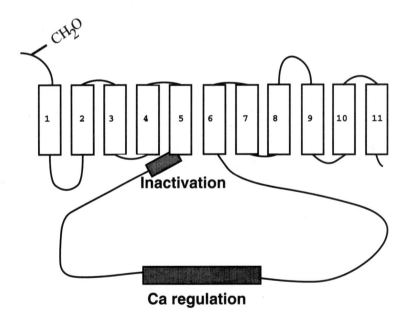

FIGURE 6. Model of the Na^+-Ca^{2+} exchanger. Indicated are regions of the large intracellular loop involved in regulation. The endogenous XIP region is involved in the Na^+-dependent inactivation process, and amino acids 371–508 are involved in Ca^{2+} regulation.

We have engineered mutations which greatly alter the inactivation process. These mutations are in the endogenous XIP region of the exchanger molecule. The endogenous XIP region is that part of the protein with the same amino acid sequence as the exchanger inhibitory peptide (XIP).[15] The XIP region is located at the very beginning (5' end) of the large intracellular loop. We had postulated that this 20-amino acid portion of the exchanger might have an important regulatory role. When a small (4-amino acid) deletion is made in this region, Na^+-dependent inactivation is totally eliminated, whereas Ca^{2+} regulation is still present. Analysis of XIP region mutants is being actively pursued.

Our knowledge of the location of regulatory sites on the exchanger is summarized in FIGURE 6. The endogenous XIP site is associated with inactivation, whereas a site towards the middle of the loop is involved with Ca^{2+} regulation. The two forms of regulation are not completely independent, and we are working to more fully understand these processes.

REFERENCES

1. NICOLL, D. A., S. LONGONI & K. D. PHILIPSON. 1990. Science **250:** 562–565.
2. REILÄNDER, H., A. ACHILLES, U. FRIEDEL, G. MAUL, F. LOTTSPEICH & N. J. COOK. 1992. EMBO J. **11:** 1689–1695.
3. PHILIPSON, K. D. & D. A. NICOLL. 1992. Curr. Opin. Cell Biol. **4:** 678–683.
4. SCHWARZ, E. & S. BENZER. 1993. Expression and evolution of CALX, a sodium-calcium exchanger of *Drosophila melanogaster*. *In* Abstracts of papers presented at the 1993 meeting on Neurobiology of *Drosophila*: Oct. 6–10, 1993. 128. Cold Spring Harbor Laboratory Press. Plainview, NY.
5. LI, Z., S. MATSUOKA, L. V. HRYSHKO, D. A. NICOLL, M. M. BERSOHN, E. P. BURKE, R. P. LIFTON & K. D. PHILIPSON. 1994. J. Biol. Chem. **269:** 17434–17439.
6. CALDIVIA, C., P. KOFUJI, W. J. LEDERER & D. H. SCHULZE. 1995. Biophys. J. **68:** A410.
7. SCHWARZ, E. & S. BENZER. Submitted.
8. DIPOLO, R. 1979. J. Gen. Physiol. **73:** 91–113.
9. HILGEMANN, D. W. 1990. Nature **344:** 242–245.
10. MATSUOKA, S., D. A. NICOLL, L. V. HRYSHKO, D. O. LEVITSKY, J. N. WEISS & K. D. PHILIPSON. 1995. J. Gen. Physiol. **105:** 403–420.
11. MATSUOKA, S., D. A. NICOLL, R. F. REILLY, D. W. HILGEMANN & K. D. PHILIPSON. 1993. Proc. Natl. Acad. Sci. USA **90:** 3870–3874.
12. LEVITSKY, D. O., D. A. NICOLL & K. D. PHILIPSON. 1994. J. Biol. Chem. **269:** 22847–22852.
13. HRYSHKO, L. V., D. A. NICOLL, S. MATSUOKA, J. N. WEISS, E. SCHWARZ, S. BENZER & K. D. PHILIPSON. 1995. Biophys. J. **68:** A410.
14. HILGEMANN, D. W., S. MATSUOKA, G. A. NAGEL & A. COLLINS. 1992. J. Gen. Physiol. **100:** 905–932.
15. LI, Z., D. A. NICOLL, A. COLLINS, D. W. HILGEMANN, A. G. FILOTEO, J. T. PENNISTON, J. N. WEISS, J. M. TOMICH & K. D. PHILIPSON. 1991. J. Biol. Chem. **266:** 1014–1020.

The Structural Basis of Na^+-Ca^{2+} Exchange Activity[a]

HANNAH RAHAMIMOFF, WALTER LOW, ORNA COOK,
IAN FURMAN, JUDITH KASIR, AND RAFI VATASHSKI

Department of Biochemistry
Hebrew University–Hadassah Medical School
Jerusalem, 91120 Israel

Since publication of the first studies that established the existence of a Na^+-dependent Ca^{2+} extrusion mechanism in the heart and the squid giant axon,[1,2] studying the molecular and cellular properties of the Na^+-Ca^{2+} exchanger has been a real challenge for several reasons. First, the protein was of low abundancy[3] in most tissues. Second, there was no specific inhibitor, drug or affinity ligand[4] available to label the protein and manipulate its cellular activity; hence, its physiological importance in maintaining cellular Ca^{2+} homeostasis was often challenged. Moreover, studies in which the exchanger's kinetic parameters were determined and compared to the resting $[Ca^{2+}]$ levels in different cells cast serious doubt on its prime role.[5] In addition, in different mechanistic and molecular studies, considerable heterogeneity has been noted.[6]

During the last five years, since the first Na^+-Ca^{2+} exchanger was cloned,[7] new tools have become available to reexamine old questions. It has become clear that the protein is present in different tissues in different forms[8-12] that are the product of two genes. One of these genes is responsible for the existence of eight different isoforms. These differ in their 5' untranslated region and within the large putative cytoplasmic loop of the protein.

Since cloning the three rat isoforms,[9,10] we have been involved in studies designed to understand the biological rationale in expression of different Na^+-Ca^{2+} exchanger isoforms and have been trying to correlate between the structure of the Na^+-Ca^{2+} exchanger isoforms, their localization and transport activity. The results of some of these experiments are described in this paper.

Preparation of Site-Specific Antipeptide Antibodies Designed to Identify Specific Na^+-Ca^{2+} Exchanger Isoforms

We have successfully prepared two site-specific antipeptide antibodies. FIGURE 1 shows the variable region of the three rat Na^+-Ca^{2+} exchanger isoforms, the protein segments that were chosen to prepare the antibodies and the binding of

---

[a] The experiments described in this research were supported in part by the Basic Research Fund (BRF) of the Israel Academy of Sciences, the National Council for Research and Development of the Ministry of Science MOST, the Israel Ministry of Health and The U.S.A.–Israel Binational Science Foundation, BSF; R. Vatashski is a recipient of the Kornfeld Foundation fellowship.

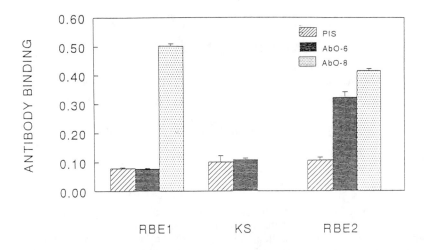

FIGURE 1. Production of site-specific antipeptide antibodies and determination of their specificity. HeLa cells were transfected with the cloned exchanger isoforms RBE-1 and RBE-2 or mock transfected with the plasmid vector pBluescript KS. Cells were disrupted by sonication to obtain crude lysate. Microtiter plated cells were coated with 50 μg lysate protein and the binding of the antibodies AbO-6 and AbO-8 as well as the preimmune sera at a dilution of 1 : 500 was determined by enzyme-linked immunosorbent assay (ELISA).

these antibodies to cell extracts obtained from clone RBE-1 transfected HeLa cells, clone RBE-2 transfected HeLa cells and mock transfected HeLa cells with the plasmid vector pBluescript, into which the exchangers are cloned. Based on the amino acid segment against which antibody AbO-8 was raised, it binds to proteins derived from both clones RBE-1 and RBE-2. AbO-6, an antibody raised against a peptide segment present only in RBE-2, does not bind to HeLa cells transfected with RBE-1. Both antibodies bind also to cell extracts obtained from RHE-1 (the cloned rat heart exchanger) transfected HeLa cells (not shown).

FIGURE 2 shows the immunofluorescence of HeLa cells transfected with RBE-2 using the antipeptide antibody AbO-8. As secondary antibody, rhodamine conjugated goat anti-rabbit immunoglobulin G (IgG) was used. The secondary antibody by itself, in the absence of the primary antibody did not bind to the transfected cells. Panel A shows the transfected HeLa cells as seen in the phase microscope, panel B shows the same field in the fluorescence microscope and in panel C, one of the transfected cells is shown at a 100-fold magnification.

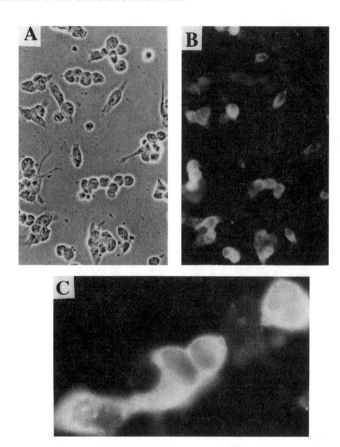

FIGURE 2. Expression of the cloned Na^+-Ca^{2+} exchanger gene RBE-2 in HeLa cells by immunofluorescence. Transfected HeLa cells (with clone RBE-2) were grown on polylysine coated cover slips. 17 hours post transfection the cells were fixed in 4% paraformaldehyde, permeabilized with 0.25% triton ×100 and incubated with AbO-8 at a dilution of 1:1000. Rhodamine conjugated goat anti-rabbit IgG was used as a secondary antibody. **(A)** Transfected HeLa cells, 25 × magnification; phase microscopy; **(B)** same field, fluorescence microscopy; **(C)** 100 × magnification, a transfected cell.

Is There a Functional Difference between the Cloned Rat Heart Na^+-Ca^{2+} Exchanger Gene RHE-1 and the Cloned Rat Brain Exchanger RBE-2?

The rat heart exchanger RHE-1 and its two brain homologues represent two types of Na^+-Ca^{2+} exchanger variants. The heart clone RHE-1 differs from the two brain clones RBE-1 and RBE-2 both in the 5' untranslated region and in the variable region present in cytoplasmic loop (see FIG. 1).

To assess, whether functional differences existed between the two subtypes of these exchangers, we carried out a comparative analysis of the rat heart clone RHE-1 and that of the brain clone RBE-2.

Both isoforms were expressed in the recombinant vaccinia virus VTF-7 infected lipofectin mediated plasmid DNA transfected HeLa cell expression system.[15] Between 16–18 hours post transfection, expression of Na^+ gradient-dependent Ca^{2+} transport was determined.[9,10] This was done both in "whole" cells and following functional reconstitution of the transfected HeLa cell proteins into proteoliposomes.[9,10] To compare the transport activity obtained in the transfected cells, the net Na^+ gradient-dependent Ca^{2+} transport activity, was normalized per total HeLa cell proteins. FIGURE 3 shows that the net transport activity of HeLa cells transfected with the rat heart clone is identical to that of the cells transfected with the rat brain clone RBE-2. However, when the expression of Na^+-Ca^{2+} exchanger protein in the transfected cells has been determined by Western blot analysis it is clearly seen (FIG. 4) that a much higher amount of RBE-2-derived protein (lane 3) is formed in the transfected cells than RHE-1-derived protein (lane 2). Lane 1 shows the immunoblot of the cell extract obtained from HeLa cells that were mock transfected with the plasmid vector pBluescript KS. Our results suggest, that the specific activity of the protein expressed from clone RHE-1 is higher than that of RBE-2.

The Role of the N-Terminal Segment of the Cloned Na^+-Ca^{2+} Exchanger in Functional Expression of the Protein in HeLa Cells

Hydropathy analysis using a window of 20 amino acids of the cloned rat Na^+-Ca^{2+} exchanger isoforms as well as of the other members of this family of proteins

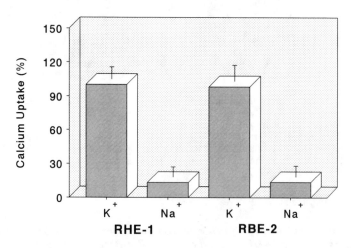

FIGURE 3. Comparison of expression of Na^+ gradient-dependent Ca^{2+} transport activity of the cloned rat exchangers RHE-1 and RBE-2. HeLa cells were transfected in parallel with the cloned rat heart exchanger RHE-1 and the cloned rat brain exchanger RBE-2 as described.[9,10] Seventeen hours post transfection Na^+ gradient-dependent Ca^{2+} uptake was determined.[9,10] Since expression in individual experiments varied, the transport activity of clone RHE-1, which was determined in nmoles Ca^{2+} transported in a Na^+ gradient-dependent manner in 10 minutes in each experiment, was defined as 100% and that of RBE-2 was normalized. The data are compiled from n = 10 different transfection experiments; each data point was done in triplicate.

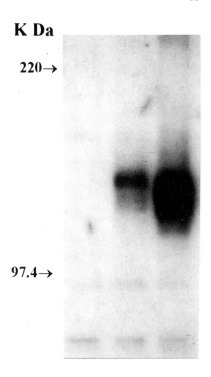

FIGURE 4. Western blot analysis of proteins expressed in RHE-1 and RBE-2 transfected HeLa cells. HeLa cells were transfected as described in FIGURE 3. Seventeen hours post transfection the cells were scraped from the culture dishes and divided into two parts: one part was used to determine transport activity and the other for the protein profile analysis. The cells were clarified by a brief sonication and the proteins in the crude lysate were separated by sodium dodecyl sulfate polyacrylamide gel electrophoresis (SDS/PAGE). Following electrophoretic transfer to nitrocellulose, the blot was treated with a 1:500 dilution of AbO-8. I^{125} Protein A was used as a secondary antibody. *Lane 1*, HeLa cell extract from mock transfected cells with the plasmid vector pBluescript KS; *lane 2*, HeLa cells transfected with RHE-1; *lane 3*, HeLa cells transfected with RBE-2.

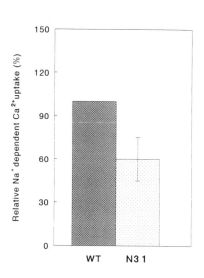

FIGURE 5. Expression of Na^+ gradient-dependent Ca^{2+} uptake in HeLa cells transfected with the N-terminal truncated Na^+-Ca^{2+} exchanger mutant N31. HeLa cells were transfected with the cloned wild type Na^+-Ca^{2+} exchanger gene RBE-1 and its N-terminal truncated mutant N31. The average transport activity of the wild type clone RBE-1 was defined as 100% in each separate experiment and the transport activity of the mutant clone which was always expressed in parallel is presented in relative values.

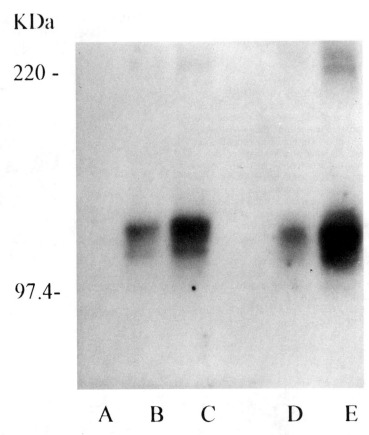

FIGURE 6. Western blot analysis of the proteins synthesized in HeLa cells transfected with the plasmid vector pBluescript KS, the cloned Na^+-Ca^{2+} exchanger RBE-1 and its N-terminal truncated mutant N31. A crude sonicated cell extract obtained from HeLa cells transfected with pBluescript KS (*lane A*), RBE-1 (*lane B*) or N31 (*lane D*) was separated by SDS/PAGE and transferred to nitrocellulose. The protein profile was analyzed to reveal Na^+-Ca^{2+} exchanger-derived proteins by binding of the antipeptide antibody AbO-8 (see FIG. 1). A subcellular fraction enriched in transfected HeLa cell plasma membranes was analyzed as well. In *lane C*, membranes obtained from HeLa cells transfected with RBE-1 are shown and in *lane E*, from HeLa cells transfected with N31. I^{125} Protein A was used as a secondary antibody.

indicated that the protein can be organized into 12 transmembrane α helices.[7,9] Based on microsequencing of the N-terminal amino acids of the purified bovine Na^+-Ca^{2+} exchanger Durkin *et al.*[14] suggested that the initial 32 amino acids of the cloned gene constitute a cleavable signal peptide, since the mature protein starts at amino acid 33 of the cloned gene. We prepared three N-terminal truncated mutants of the rat brain Na^+-Ca^{2+} exchanger gene RBE-1 and tested their functional expression in the recombinant vaccinia virus VTF-7 infected/transfected HeLa cell expression system.[9,10,15]

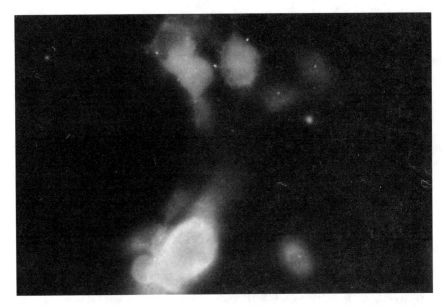

FIGURE 7. Detection of Na^+-Ca^{2+} exchanger protein in astrocytes. A primary culture enriched in astrocytes was grown on polylysine-coated cover slips. Permeabilization, fixation and exposure to antibody were done as in FIG. 2. As a secondary antibody FITC conjugated goat anti-rabbit IgG was used.

FIGURE 5 shows the summary of n = 7 experiments, in which the cells were transfected with the deletion mutant N31 in parallel with the wild type clone RBE-1. The deletion removed all the 31 amino acids between the initiating methionine and the putative N-terminal aspartic acid of the mature RBE-1 protein. This included the entire putative signal peptide, the putative signal peptidase recognition site and the cleavage site. Na^+-Ca^{2+} exchange activity has been determined both in "whole" HeLa cells and in the reconstituted HeLa cell preparation. It can be seen that mutant N31 retained 60% of the Na^+-Ca^{2+} exchange activity relative to the wild type clone. Other N-terminal truncated mutants, in which 21 and 26 amino acids beyond the initiating methionine were deleted, retained 108% and 37% of their transport activity relative to the wild type exchanger. Hence, it seems that the initial 32 N-terminal amino acids that constitute the putative signal peptide are not mandatory for functional expression of the Na^+-Ca^{2+} exchanger.

Western blot analysis (FIG. 6) of the protein profile of transfected HeLa cell extract and that of a plasma membrane enriched fraction using the antipeptide antibody AbO-8 indicates that the signal peptide truncated mutant is both synthesized and inserted into the plasma membrane. In lanes B and D, the immunoblot of the RBE-1 and N31 transfected crude HeLa cell extract is shown. Lanes C and E show the immunoblot of the corresponding protein profile of the plasma membrane enriched fractions, and lane A shows the immunoblot of the protein profile obtained from mock transfected HeLa cells with pBluescript KS.

The Na^+-Ca^{2+} Exchanger in Astrocytes

To identify the Na^+-Ca^{2+} exchanger that is present in astrocytes,[16] we prepared a primary cell culture enriched in astrocytes from newborn (0–24 hours) rats. The astrocytes were characterized by using a primary mouse monoclonal antibody against glial fibrillary acidic protein (GFAP). As a secondary antibody, rhodamine conjugated goat antimouse IgG was used. The presence of Na^+-Ca^{2+} exchanger protein in these astrocytes was tested by using the two polyclonal antipeptide antibodies AbO-6 and AbO-8 (see FIG. 1).

Binding of AbO-6 to the primary culture of astrocytes and detection of the antigen-antibody complexes with fluorescein isothiocyanate (FITC) conjugated goat anti-rabbit IgG showed that many of the cells in the culture were highlighted. No binding was obtained with the secondary antibody alone. AbO-8 bound to the cultures with similar affinity to that of AbO-6. The immunofluorescence detected in astrocytes is shown in FIGURE 7.

REFERENCES

1. REUTER, H. & N. SEITZ. 1968. J. Physiol. (London) **195:** 451–470.
2. BAKER, P. F., M. P. BLAUSTEIN, A. L. HODGKIN & R. A. STEINHARDT. 1969. J. Physiol. (London) **200:** 431–458.
3. ALLEN, T. J. A., D. NOBLE & H. REUTER, Eds. 1989. Sodium-Calcium Exchange. Oxford University Press. Oxford.
4. KACZOROWSKI, G. J., R. S. SLAUGHTER, V. F. KING & M. L. GARCIA. 1989. Biochim. Biophys. Acta **988:** 287–302.
5. BLAUSTEIN, M. P., R. DIPOLO & J. P. REEVES, Eds. 1991. Sodium-Calcium Exchange. Ann. N. Y. Acad. Sci. Vol. 639.
6. RAHAMIMOFF, H. 1990. Curr. Top. Cell. Regul. **31:** 241–271.
7. NICOLL, D. A., S. LONGONI & K. D. PHILIPSON. 1990. Science **250:** 562–565.
8. REILLY, R. F. & C. A. SHUGRUE. 1992. Am. J. Physiol. **262:** F1105–F1109.
9. LOW, W., J. KASIR & H. RAHAMIMOFF. 1992. FEBS Lett. **316:** 63–67.
10. FURMAN, I., O. COOK, J. KASIR & H. RAHAMIMOFF. 1993. FEBS Lett. **319:** 105–109.
11. LEE, S-L., A. S. L. YU & J. LYTTON. 1994. J. Biol. Chem. **269:** 14849–14852.
12. KOFUJI, P., W. J. LEDERER & D. H. SCHULZE. 1994. J. Biol. Chem. **269:** 5145–5149.
13. LI, Z., S. MATSUOKA, L. V. HRYSHKO, D. A. NICOLL, M. M. BERSOHN, E. P. BURKE, R. P. LIFTON & K. D. PHILIPSON. 1994. J. Biol. Chem. **269:** 17434–17439.
14. DURKIN, J. T., D. C. AHRENS, Y-C. E. PAN & J. P. REEVES. 1991. Arch. Biochem. Biophys. **290:** 369–375.
15. BLAKELY, R. D., J. A. CLARK, G. RUDNIK & S. AMARA. 1991. Anal. Biochem. **194:** 302–308.
16. BLAUSTEIN, M. P., W. F. GOLDMAN, G. FONTANA, B. K. KRUEGER, E. M. SANTIAGO, T. D. STEELE, D. N. WEISS & P. J. YAROWSKY. 1991. In Sodium-Calcium Exchange. M. P. BLAUSTEIN, R. DIPOLO & J. P. REEVES, Eds. Ann. N. Y. Acad. Sci. Vol. 639: 254–274.

A New Splicing Variant in the Frog Heart Sarcolemmal Na-Ca Exchanger Creates a Putative ATP-Binding Site[a]

TOMOKO IWATA, ALEXANDER KRAEV,
DANILO GUERINI, AND ERNESTO CARAFOLI[b]

Institute of Biochemistry
Swiss Federal Institute of Technology (ETH)
8092 Zürich, Switzerland

INTRODUCTION

The sarcolemmal Na-Ca exchanger is one of the systems used for the extrusion of Ca from cells.[1,2] The other is the plasma membrane Ca-ATPase.[3] The exchanger takes effect primarily in excitable tissues, *e.g.*, heart. It has high transient capacity (V_{max} = 30–40 nmol) and low affinity for calcium (Km > 10 μM), thus leaving the role of fine regulator of cytosolic Ca to the pump. It transports one Ca ion out of the cell and 3 Na ions in the opposite direction: however, the transport is reversible, *i.e.*, under certain conditions the exchanger may function in the Ca influx mode. Whereas the driving force for Ca transport by the pump is the hydrolysis of adenosine triphosphate (ATP), that for the exchanger is the chemical gradient of Na ions across the cell membrane, together with the membrane potential. That is, the Na-Ca exchanger function is dictated by the concentration of its own substrates (intracellular and extracellular), Ca and Na ions. However, the exchanger activity may be regulated by various extra- and intracellular factors, including ATP, phosphorylation, lipids,[4] cyclic guanosine monophosphate (cGMP),[5] intracellular Ca (allosteric),[6] calmodulin (CaM),[7] and proteinase treatment.[8]

This paper focuses on the possible regulation of the heart sarcolemmal Na-Ca exchanger by ATP and by kinase-promoted phosphorylation. A review on this subjects prefaces recent findings in the authors' laboratory of an ATP binding consensus sequence in a frog heart exchanger clone.

Regulation by Phosphorylation

Kinase-directed phosphorylation is one of the most important means of regulating the protein systems involved in cellular signal transduction. Early biochemical studies on cardiac sarcolemmal vesicles had indicated a regulatory phosphorylation process of the Na-Ca exchanger: in bovine heart sarcolemmal vesicles, a kinase-

---

[a] Supported by a grant from the Swiss National Science Foundation (Grant No. 31-30859.91).
[b] Corresponding author: Dr. E. Carafoli, Laboratory for Biochemistry III, Swiss Federal Institute of Technology (ETH), ETH-Zentrum, Universitätsstr. 16, CH-8092 Zürich, Switzerland.

mediated phosphorylation step activated the Na-Ca exchange, whereas a phosphatase-mediated dephosphorylation inactivated it.[7] The actual product of the phosphorylation reaction, *i.e.*, a phosphorylated band in autoradiograms of sodium dodecyl sulfate (SDS)-polyacrylamide gels, however, has not been seen, since at the time of the observation no clues were available on the molecular nature of the exchanger. In general, the concept of heart exchanger regulation by phosphorylation encountered difficulties and gradually faded away in the literature. It was firmly established, however, for the axonal exchanger by a series of convincing experiments by DiPolo and Beaugé.[9,10] The ATP-dependent "flippase" hypothesis, which explains the ATP regulation of exchanger activity with the asymmetry of membrane phospholipid, was also recently explored in patches of guinea pig myocyte sarcolemma.[11] The cloning of the heart exchanger gene in 1990[12] in principle opened new avenues to the investigation of exchanger phosphorylation (FIG. 1). Unfortunately, isolation and purification of the functionally active exchanger protein, an essential step in studying its phosphorylation, has not yet been achieved.

The application of molecular biological approaches to the exchanger has been made possible by the isolation of dog cardiac Na-Ca exchanger cDNA.[12] The deduced primary structure of the protein has revealed several potential kinase phosphorylation sites. Protein kinase C (PKC), cyclic adenosine monophosphate

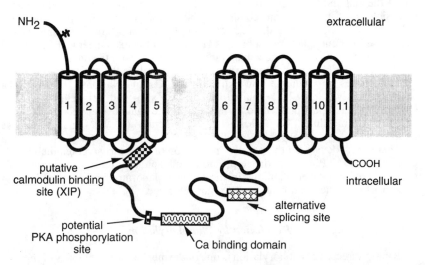

FIGURE 1. A model of the mammalian heart Na-Ca exchanger, based on the work quoted in References 12, 26–31. The deduced sequence predicts 11 putative transmembrane domains (TM) and a long intracellular loop after TM5. The N-terminal signal peptide might be cleaved during the posttranslational processing. Sequence comparison studies[31] have shown that the regions of TM4 and 5 facing the extracellular surface are homologous to those of the Na/K-ATPase and TM2, 3, and 8 to those of the Na-Ca,K exchanger of rod photoreceptors.[32] The main intracellular loop mediates the effects of signals from the cytosolic side of the cell. It contains the XIP (exchanger inhibitory peptide) domain, which has been described as a (pseudo)-calmodulin binding domain,[33] and Ca binding domain, which is responsible for the regulation of the exchanger by intracellular Ca.[6] Asparagine 9 was found to be glycosylated.[34]

(cAMP)-dependent protein kinase (PKA) and Ca/CaM-dependent protein kinase II are among the candidate kinases to phosphorylate the exchanger. The dog heart exchanger contains at least 12 potential PKC phosphorylation sites (3 in the short intracellular stretch between transmembrane domain 1 (TM1) and TM2, one between TM3 and TM4, and 8 in the main intracellular loop), and at least one for PKA or by the Ca/CaM-dependent protein kinase II in the main intracellular loop.

The smooth muscle exchanger has been reported to be phosphorylated by PKC.[13] A phosphorylated band of about 120 kDa was observed in smooth muscle cells in the absence of PKC inducers: phorbol esters increased the intensity of the phosphorylated band, at best, by about 25%. The exchange activity was similarly increased by about 25%. The work also showed that the phosphorylation was on a serine residue. By contrast, the exchange activity was found to decrease when membrane preparations from adrenal medula chromaffin cells were incubated with the catalytic subunit of PKA.[14] In those cells, the exchanger was phosphorylated by PKA and by PKC, the exchange activity decreasing in both cases. It was also reported that the activation of PKC and PKA in renal epithelial cells and in myocytes, respectively, downregulated the exchanger mRNA, with a concomitant decrease in the exchange activity.[15] A recent study found no phosphorylated exchanger protein in Chinese hamster ovary (CHO) cells stably expressing the bovine cardiac isoform, even if the main cytosolic loop contains consensus sequences for PKA or casein kinase II, and can indeed be phosphorylated *in vitro* by either of them.[16] The regulation of the Na-Ca exchanger by phosphorylation may thus be tissue (and species)-specific, and could be (partly) responsible for the variations in the physiological role of the exchanger in different tissues.

Peculiarities of the Frog Heart

Frog cardiac myocytes are much smaller than those of mammalian heart (the surface-volume ratio in the frog cardiac cells is much larger than in most mammalian hearts), and the sarcoplasmic reticulum (SR) is far less developed in frog heart.[17] Since the release of Ca from internal stores is likely to be of lesser importance in frog heart, all Ca that activates the myofibrils must in this case by necessity come from outside; the dehydropyridine (DHP)-sensitive Ca channels, but also the Na-Ca exchanger, are the two systems that can mediate Ca influx. Ca extrusion from frog heart will also be essentially performed by the exchanger as in all other heart types, except that the exchanger will presumably have to be more active, given the essential absence of SR.

β-agonists have profound effects on mammalian heart excitation and contraction:[17] their positive inotropic action is mediated by the increased Ca influx through the DHP-sensitive channel which is phosphorylated by the cAMP-dependent protein kinase (PKA). β-agonists also induce a fuller relaxation mediated by the PKA phosphorylation of the phospholamban, which de-represses the SR-Ca pump, resulting in the decrease of the cytosolic Ca concentration. In spite of the virtual absence of SR, β-adrenergic effects were also observed in frog heart and it has been assumed that the Na-Ca exchange activity was indeed modulated by cAMP in frog cardiomyocytes.[18]

These results indicate that frog heart could be a good subject for molecular level investigations of the Na-Ca exchanger, *i.e.*, to establish whether the PKA effect is indeed mediated by the exchanger itself. Efforts to clone the Na-Ca exchanger from frog heart were thus initiated. The results have led to the discovery of a novel exon which produces an exchanger isoform in which alternative splicing of the primary transcript introduces a 9-aa insertion which completes a P-loop type ATP-binding consensus site.

Cloning of the Frog Heart Na-Ca Exchanger

The screening of a *Xenopus laevis* heart cDNA library was carried out using a probe from the dog heart Na-Ca exchanger cDNA which covered most of the coding sequence (FIG. 2). Two clones, H3 and H6, were obtained; the first spanned a portion of the main intracellular loop, and H6 extended to the 3'-untranslated region. These two clones overlapped in the region where alternative splicing occurs in mammalian clones. Since the library did not contain the N-terminal portion of the exchanger, it was decided to screen a *Xenopus laevis* genomic library using a fragment produced by polymerase chain reaction (PCR) at the N-terminus of the H3 clone as a probe. The stretch coding for the remainder of the region was found in this genomic library.

The molecular features of the frog heart Na-Ca exchanger can be described as follows. a) Its deduced primary structure showed 89% identity (95% similarity) to the dog heart NCX1 isoform (79% identity at the nucleotide level). b) As is the case among mammalian clones, the sequences of the transmembrane domains of the frog clone were very well conserved with respect to the dog heart clone. By contrast, those of the N-terminus and of the main intracellular loop were less well conserved. c) The putative PKA phosphorylation site[12] and the Ca-binding domain[6] seen in mammalian clones were also conserved in the frog exchanger. d) The frog clone showed the "heart type" splicing pattern[19,20] with a novel exon of 27 bp,

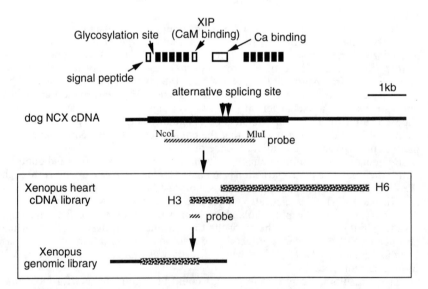

FIGURE 2. Cloning of the frog heart Na-Ca exchanger. The dog heart exchanger cDNA was a gift from Dr. K. D. Philipson, UCLA, Los Angeles, CA. The *Xenopus laevis* heart cDNA library was a gift from Dr. M. Morad, Georgetown University, Washington, DC. The *Xenopus laevis* genomic library was purchased from Stratagene (Zurich, Switzerland).

```
H3      PTITGKILY.........GKPVLRKVQ
H6      PTITGKILYGKSVIQKNTGKPVLRKVQ

dog     PTITGKILY.........GQPVFRKVH
```

Consensus sequence of
P-loop (Walker's A) nucleotide-binding sites:

GXXXXGKS

FIGURE 3. Amino acid sequences of the putative ATP binding region of the frog Na-Ca exchanger isoforms. Clone H3, which does not have the 9-aa insertion described in the text, and clone H6, which has it, are indicated. The sequence of the corresponding region of the dog heart isoform is also shown.

whose insertion resulted in the completion of a P-loop type, ATP-binding consensus sequence (FIG. 3) which is described below.

The splicing pattern of mammalian clones is shown in FIGURE 4, in which only some of the splicing variants are listed. The novel exon in the frog clone is located between exons E and F of the mammalian clones nomenclature.[19,20] Since the composition of the exons is similar throughout the species, that of the novel exon found in the frog Na-Ca exchanger could in principle also be present in mammalian clones, even if not reported so far. If so, however, this spliced isoform would be

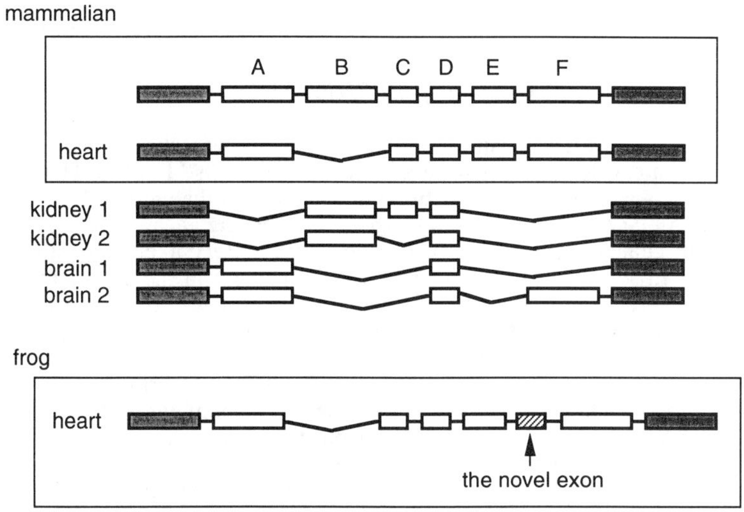

FIGURE 4. Alternative splicing domains of the Na-Ca exchanger. The novel exon identified in frog heart is indicated as a *hatched box*.

of low abundance: sequencing of the 4-kb intron between exons E and F in a human genomic clone of the exchanger has been initiated and will hopefully soon reveal whether the extra exon is present. It is, of course, also possible that this exon is species specific, *i.e.*, only occurring in the frog. The new frog isoforms have been named fncx1a (the clone with the 9-aa insert) and fncx1b (the clone without it).

The ATP-Binding Consensus Site in the Frog Na-Ca Exchanger

The sequence of the frog Na-Ca exchanger encompassing the 9-aa insertion coded by the novel exon is shown in FIGURE 3. As mentioned, the insertion created a canonical nucleotide binding site, most of which (7 out of 8 residues) was already present in the "wild type" clone: the insert "duplicated" the GK pair at the C-terminal of the incomplete P-loop domain of the "wild type" clone, and contributed to it the C-terminal essential serine of the P-loop nucleotide-binding sequence, (GXXXXGKT/S).[21,22] Tests of the ability of the domain with the complete P-loop motif to bind ATP are currently under way using the polypeptides expressed in *E. coli*.

The proteins that bind ATP (or GTP) frequently have a "glycine cluster" in the binding domain.[23] A common sequence for the cluster, as present, for example, in protein kinases (but also in other proteins, for example, the IP_3 receptor[24]) is GXGXXG. The glycine-rich motif known as the P-loop, or Walker's A motif,[21,22] instead has the consensus sequence GXXXXGKS/T. The lysine in position 7 is 100% conserved, since it may interact directly with the β- and γ-phosphates of the bound nucleotide triphosphate (NTP). This motif has been found in other kinases, such as adenylate kinase, in ras oncogene products, in elongation factors, as well as in several ATP-using proteins involved in ion transport, such as the α- and β-subunits of F-type ATPases. ATP bound by P-loop and GXGXXG motifs is in general hydrolyzed, although (*e.g.*, in the inositol 1,4,5-triphosphate (IP_3) receptor) its effect could also only be allosteric.

Regulation of the Exchanger by ATP

MgATP stimulation of Na-Ca exchange currents was observed in squid axons[9,10] and in excised patches from guinea pig heart cells.[8,11] The ATP effects in these systems were similar yet distinct: in squid axons the unhydrolyzable ATP-analogue, ATP-γS activates the exchanger even more significantly than MgATP,[10] whereas in guinea pig myocytes it had no effect.[11] Very likely, the stimulation was indirect, *i.e.*, related to regulatory factors present in the cell or influenced by the membrane phospholipid environment, since it was observed in cardiomyocytes but not in oocytes expressing the exchanger clone.[24b] An axoplasmic factor which potentiated the MgATP stimulatory effect on the squid axon Na-Ca exchange activity was recently reported.[25] The effect described in CHO cells stably expressing the bovine Na-Ca exchanger, *i.e.*, inhibition of the exchange activity by ATP depletion, might have been due to breakdown of the actin cytoskeleton:[16] treatment of the cells with cytochalasin D, which binds to the barbed end of actin microfilament and alters their state of polymerization in intact cells, mimicked the effect of ATP depletion. Other explanations for the effect of ATP are also possible: ATP might, for example, activate chaperons to promote the proper folding of the overexpressed Na-Ca exchanger protein.

CONCLUSIONS

The aim of the work described in this contribution was to establish whether the β-adrenergic effects on frog heart could be mediated by the Na-Ca exchanger in a way that would have been peculiar to this heart species. This would have meant identifying additional consensus sites for PKA phosphorylation in the exchanger sequence: this has not been the case, but a canonical ATP-binding site of the P-loop type has instead been detected. The frog heart Na-Ca exchanger clone, fncx1b, possessed the sequence GXXXGK, *i.e.*, an incomplete P-loop motif, which is not present in mammalian clones. However, the insert of 9 amino acids coded by a novel exon in clone fncx1a has added the essential residue at the end of the above mentioned incomplete GXXXXGK sequence, making it canonical.

Whether the P-loop motif in the frog exchanger binds ATP is not yet known. Although one could cautiously consider the interesting but unorthodox possibility that ATP is not only bound by the motif, but also hydrolyzed, the most plausible hypothesis is that of an allosteric effect of ATP. The physiological meaning of this putative allosteric site is at the moment obscure, but it is tempting to speculate that it could be related to the ATP effect frequently observed in squid axons. cDNA cloning of the squid axon Na-Ca exchanger isoform would presumably answer the question.

REFERENCES

1. PHILIPSON, K. D. & D. A. NICOLL. 1992. Sodium-calcium exchange. Curr. Opin. Cell. Biol. **4:** 678–683.
2. BLAUSTEIN, M. P., R. DIPOLO & J. P. REEVES, Eds. 1991. Sodium-Calcium Exchange: Proceedings of the Second International Conference. New York Academy of Sciences. New York.
3. CARAFOLI, E. 1994. Biosynthesis: plasma membrane calcium ATPase: 15 years of work on the purified enzyme. FASEB. J. **8:** 993–1002.
4. LUCIANI, S., M. ANTOLINI, S. BOVA, G. CARGNIELLI, F. CUSINATO, P. DEBETTO, L. TREVISI & R. VAROTTO. 1995. Inhibition of cardiac sarcolemmal sodium-calcium exchanger by glycerophosphoinositol 4-phosphate and glycerophosphoinositol 4-5-bisphosphate. Biochem. Biophys. Res. Commum. **206:** 674–680.
5. FURUKAWA, K., N. OHSHIMA, Y. TAWADA-IWATA & M. SHIGEKAWA. 1991. Cyclic GMP stimulates Na/Ca exchange in vascular smooth muscle cells in primary culture. J. Biol. Chem. **266:** 12337–12341.
6. LEVITSKY, D. O., D. A. NICOLL & K. D. PHILIPSON. 1994. Identification of the high affinity Ca-binding domain of the cardiac Na-Ca exchanger. J. Biol. Chem. **269:** 22847–22852.
7. CARONI, P. & E. CARAFOLI. 1982. The regulation of the Na-Ca exchanger of heart sarcolemma. Eur. J. Biochem. **132:** 451–460.
8. HILGEMANN, D. W. 1990. Regulation and deregulation of cardiac Na-Ca exchange in giant excised sarcolemmal membrane patches. Nature **344:** 242–245.
9. DIPOLO, R. & L. BEAUGÉ. 1994. Effects of vanadate on MgATP stimulation of Na-Ca exchange support kinase-phosphatase modulation in squid axons. Am. J. Physiol. **266:** C1382–C1391.
10. DIPOLO, R. & L. BEAUGÉ. 1991. Regulation of Na-Ca exchange. An overview. Ann. N. Y. Acad. Sci. **639:** 100–111.
11. HILGEMANN, D. W., A. COLLINS, D. P. CASH & G. A. NAGEL. 1991. Cardic Na-Ca exchange system in giant membrane patches. Ann. N. Y. Acad. Sci. **639:** 126–139.

12. NICOLL, D. A., S. LONGONI & K. D. PHILIPSON. 1990. Molecular cloning and functional expression of the cardiac sarcolemmal Na-Ca exchanger. Science **250**: 562–565.
13. SHIGEKAWA, M., T. IWAMOTO & S. WAKABAYASHI. 1995. Phosphorylation and modulation of Na/Ca exchanger in vascular smooth muscle cells. This volume.
14. LIN, L., L. KAO & E. W. WESTHEAD. 1995. Agents that promote protein phosphorylation increase catecholamine secretion and inhibit the activity of the Na/Ca exchanger in bovine chromaffin cells. This volume.
15. SMITH, J. B., H-W. LEE & L. SMITH. 1995. Regulation of sodium-calcium exchanger expression. This volume.
16. CONDRESCU, M., J. P. GARDNER, G. CHERNAYA, J. F. ACETO, C. KROUPIS & J. P. REEVES. 1995. ATP-dependent regulation of sodium-calcium exchange in Chinese hamster ovary cells transfected with the bovine cardiac sodium-calcium exchanger. J. Biol. Chem. **270**: 9137–9146.
17. SHEU, S-S. & M. P. BLAUSTEIN. 1986. Sodium/calcium exchange and regulation of cell calcium and contractility in cardiac muscle, with a note about vascular smooth muscle. *In* The Heart and Cardiovascular System. H. A. FOZZARD *et al.*, Eds. Chapter 26. 509–535. Raven Press. New York.
18. FAN, J., Y. SHUBA & M. MORAD. 1995. Modulation of sodium-calcium exchanger by beta-adrenergic agonists in frog ventricular myocytes. Biophys. J. **68**: A136.
19. KOFUJI, P., W. J. LEDERER & D. H. SCHULZE. 1994. Mutually exclusive and cassette exons underlie alternatively spliced isoforms of the Na/Ca exchanger. J. Biol. Chem. **269**: 5145–5149.
20. LEE, S-L., A. S. L. YU & J. LYTTON. 1994. Tissue-specific expression of Na-Ca exchanger isoforms. J. Biol. Chem. **269**: 14849–14852.
21. WALKER, J. E., M. SARASTE, M. J. RUNSWICK & N. J. GAY. 1982. Distantly related sequences in the α- and β-subunits of ATP synthase, myosin, kinases and other ATP-requiring enzymes and a common nucleotide binding fold. EMBO J. **8**: 945–951.
22. SARASTE, M., P. R. SOBBALD & A. WITTINGHOFER. 1990. The P-loop—a common motif in ATP- and GTP-binding proteins. Trends Biochem. Sci. **15**: 430–434.
23. SCHULZ, G. E. 1992. Binding of nucleotides by proteins. Curr. Opin. Struct. Biol. **2**: 61–67.
24. KUME, S., A. MUTO, J. ARUGA, T. NAKAGAWA, T. MICHIKAWA, T. FURUICHI, S. NAKADE, H. OKANO & K. MIKOSHIBA. 1993. The Xenopus IP_3 receptor: structure, function, and localization in oocytes and eggs. Cell **73**: 555–570.
24b. MATSUOKA, S., D. A. NICOLL, R. F. REILLY, D. W. HILGEMANN & K. D. PHILIPSON. 1993. Initial localization of regulatory regions of the cardiac sarcolemmal sodium-calcium exchanger. Proc. Natl. Acad. Sci. USA **90**: 3870–3874.
25. BEAUGÉ, L. & R. DIPOLO. 1995. A cytosolic factor is required for a Mg ATP stimulation of Na-dependent ^{45}Ca uptake in plasma membrane vesicles from squid optic nerve. This volume.
26. KOFUJI, P., R. W. HADLEY, R. S. KIEVAL, W. J. LEDERER & D. H. SCHULZE. 1992. Expression of the Na-Ca exchanger in diverse tissues: a study using the cloned human cardiac Na-Ca exchanger. Am. J. Physiol. **263**: C1241–C1249.
27. KOMURO, I., K. E. WENNINGER, K. D. PHILIPSON & S. IZUMO. 1992. Molecular cloning and characterization of the human cardiac Na/Ca exchanger cDNA. Proc. Natl. Acad. Sci. USA **89**: 4769–4773.
28. ACETO, J. F., M. CONDRESCU, C. KROUPIS, H. NELSON, N. NELSON, D. NICOLL, K. D. PHILIPSON & J. P. REEVES. 1992. Cloning and expression of the bovine cardiac sodium-calcium exchanger. Arch. Biochem. Biophys. **298**: 553–560.
29. LOW, W., J. KASIR & H. RAHAMIMOFF. 1993. Cloning of the rat heart Na-Ca exchanger and its functional expression in HeLa cells. FEBS Lett. **316**: 63–67.
30. REILLY, R. F. & C. A. SHUGRUE. 1992. cDNA cloning of a renal Na-Ca exchanger. Am. J. Physiol. **262**: F1105–F1109.
31. NICOLL, D. A. & K. D. PHILIPSON. 1991. Molecular studies of the cardiac sarcolemmal sodium-calcium exchanger. Ann. N. Y. Acad. Sci. **639**: 181–188.

32. REILENDER, H., A. ACHILLES, U. FRIEDEL, G. MAUL, F. LOTTSPEICH & N. J. COOK. 1992. Primary structure and functional expression of the Na/Ca,K-exchanger from bovine rod photoreceptors. EMBO J. **11:** 1689–1695.
33. LI, Z., D. A. NICOLL, A. COLLINS, D. W. HILGEMANN, A. G. FILOTEO, J. T. PENNISTON, J. N. WEISS, J. M. TOMICH & K. D. PHILIPSON. 1991. Identification of a peptide inhibitor of the cardiac sarcolemmal Na-Ca exchanger. J. Biol. Chem. **266:** 1014–1020.
34. HRYSHKO, L. V., D. A. NICOLL, J. N. WEISS & K. D. PHILIPSON. 1993. Biosynthesis and initial processing of the cardiac sarcolemmal Na-Ca exchanger. Biochim. Biophys. Acta **1151:** 35–42.

Alternative Splicing of the Na^+-Ca^{2+} Exchanger Gene, NCX1

D. H. SCHULZE,[a,f] P. KOFUJI,[b] C. VALDIVIA,[a,c] S. HE,[a]
S. LUO,[a] A. RUKNUDIN,[a] S. WISEL,[a] M. S. KIRBY,[c]
W. DuBELL,[c,d] AND W. J. LEDERER[c,e]

[a]Department of Microbiology and Immunology
[b]Department of Pharmacology and Experimental Therapeutics
[c]Department of Physiology
[d]Department of Biological Chemistry
[e]The Medical Biotechnology Center
University of Maryland School of Medicine
655 West Baltimore Street
Baltimore, Maryland 21201

INTRODUCTION

Na^+-Ca^{2+} exchanger activity had been demonstrated functionally as a Na^+-dependent Ca^{2+} flux from such varied sources as the giant squid axon,[1] barnacle[2] and mammalian cardiac cells.[3] The identification of the Na^+-Ca^{2+} exchange protein was initially made using specific antibodies as recently as 1988,[4] followed closely by the cloning of the Na^+-Ca^{2+} exchanger cDNA transcripts by Nicoll et al. using the canine cardiac tissue,[5] which identified the first gene, NCX1. Molecular characterization of the Na^+-Ca^{2+} exchanger suggests that it is a glycosylated[6] multipass membrane protein that contains a cleavable leader peptide[7] followed by 5 putative transmembrane regions at the N-terminus and 6 transmembrane regions at the C-terminus with a large intracellular loop in between.[5] In mammalian systems, cardiac cells contain the largest amount of the Na^+-Ca^{2+} exchanger RNA transcripts, but almost all tissues studied contain hybridizable levels of RNA.[8]

Our analysis of the human cardiac Na^+-Ca^{2+} exchanger[8] demonstrated that there is little variability (98% identity at the deduced amino acid level and 94% at the nucleic acid level) when compared to the canine cardiac Na^+-Ca^{2+} exchanger.[5] This level of conservation across species is comparable to that observed for other mammalian ion transporter systems such as the Na^+ pump[9-11] and for the β_1-subunit of the voltage-dependent calcium channel.[12] Analysis of other mammalian cardiac Na^+-Ca^{2+} exchangers from rat,[13] cow,[14] guinea pig[15] and recently mouse[16] demonstrates similar levels of sequence identity to both the human clone p9-4[8,17] and the canine sequence.[5]

When the human Na^+-Ca^{2+} exchanger was transfected into HEK 293 cells and stained with a polyclonal antibody to the Na^+-Ca^{2+} exchanger (generously provided by Ambesi and Lindenmeyer[18,19]), a bright fluorescence can be seen at the cell membrane, FIGURE 1. Only the transfected cells expressed a protein that was detected by the antibody while untransfected cells and those transfected with the vector alone displayed no fluorescence at an identical exposure (data not shown).

---

[f] To whom reprint requests should be addressed.

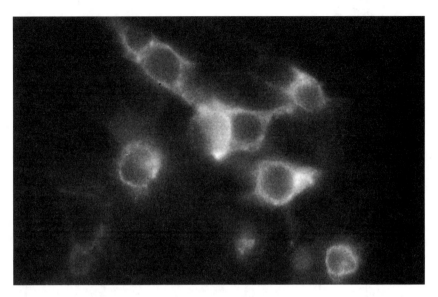

FIGURE 1. Immunofluorescence of cells transfected with the clone human Na^+-Ca^{2+} exchanger. The human Na^+-Ca^{2+} exchanger clone p9-4[8] was subcloned into a mammalian expression vector $pBex_1$, in which transcription is under control of the cytomegalovirus promotor and enhancer. After transfection into 293 HEK cells they were stained with the antibody specific to the exchanger[18] and a secondary FITC-linked antibody.

Transport of Na^+ by the Na^+-Ca^{2+} exchanger is accompanied by the countertransport of Ca^{2+}, and this characteristic has been routinely used to test for the presence and capacity of the exchanger in other studies.[20,21] We used HEK 293 cells transfected with the human cDNA clone and loaded these cells with the Ca^{2+}-sensitive fluorescent indicator fluo-3.[22] Exposure of these cells to a Na^+-free solution produced an increase in the intracellular $[Ca^{2+}]_i$ while the untransfected cells displayed no change (FIG. 2). The Na^{2+}-dependent Ca^{2+} flux was only observed in cells transfected with the human clone 9-4, demonstrating Na^+-Ca^{2+} exchanger activity as was expected.

Na^+-Ca^{2+} Exchanger in Other Tissues

It was not until the sequence of the Na^+-Ca^{2+} exchanger from rabbit kidney was published[23] that potentially relevant differences in exchanger homology were observed. The rabbit kidney Na^+-Ca^{2+} exchanger sequence differed in subtle ways from cardiac Na^+-Ca^{2+} exchanger sequences from other mammalian species. Most of the kidney sequence displayed a similar pattern of hydrophobic regions, but there were two distinguishing characteristics of the sequence in the intracellular loop. First, there was a cluster of deduced amino acid changes in a defined region of the intracellular loop followed by a short deletion when the rabbit kidney sequence was compared to the other mammalian cardiac sequences.[5,10,11] We wanted to determine the nature of these differences and whether they were due

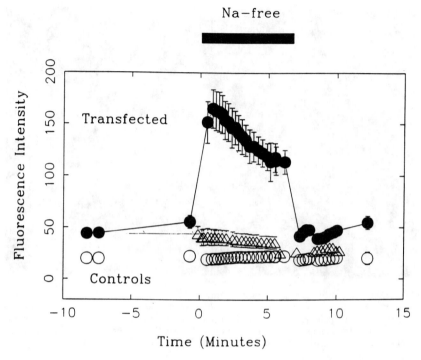

FIGURE 2. Transfected cells demonstrate extracellular Na^+-dependent changes in intracellular Ca^{2+} concentration, $[Ca^{2+}]_i$. Fluo-3 was loaded into cultured HEK 293 cells transfected with the human Na^+-Ca^{2+} exchanger clone p9-4[8] and visualized using confocal microscopy. Time-series confocal images were obtained after loading with fluo-3 and equilibration of the cells in normal extracellular solution (140 mM NaCl). The extracellular Na^+ was replaced by Cs^+ at *time 0*. After 7 min exposure the external solution was returned to 140 mM NaCl. Analysis of the fluorescence signal is presented for the transfected (●), nonresponding cells (○) and untransfected cells (△).

to the species studied (rabbit versus human or dog) or to the tissue (kidney versus cardiac).

Polymerase chain reaction (PCR) primers that flanked the region of diversity (based on the rabbit sequence[23]) were used to amplify cDNA made from various tissues. FIGURE 3 shows that different tissues produce multiple-sized bands when the PCR products are separated on an agarose gel. The single cardiac band and the two distinct kidney bands were subcloned and sequenced. Comparison of the rabbit cardiac sequence, FIGURE 4, with the human cardiac sequence shows that greater than 94% of the amino acids are identical in the portion compared. Sequence comparison of the PCR products from rabbit heart and rabbit kidney demonstrated that all of the sequences are identical at the 5′ and 3′ regions near the primers but the identity falls to only 30% in the region preceding the deletion (FIG. 4). We identified two different clones from the rabbit kidney samples that only differed from each other by the presence or absence of 21 base pairs (bp), *i.e.*, 7 amino acids (FIG. 4). We demonstrated using RNase protection that the shorter form is

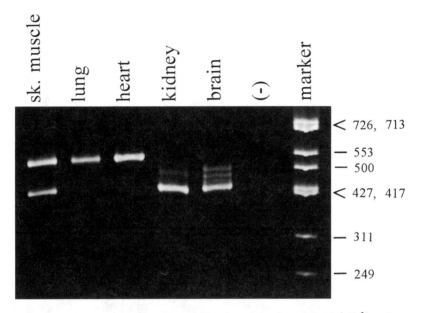

FIGURE 3. Reverse-transcribed PCR amplification of a region of the Na^+-Ca^{2+} exchanger from various tissues. Primers complementary to nucleotides encoding amino acid positions 511–516 and 677–683 of the cardiac exchanger[8] were used in the amplification of cDNA from various tissues. The products are separated on an agarose gel that was stained with ethidium bromide. The lane designated (−) contains all of the components except template DNA. The marker lane is φX174 DNA digested with *Hinf*F1.

the predominate in the kidney RNA.[24] In analyzing the full set of sequences for the heart and kidney from the rabbit, we suggested that alternative splicing of a single gene is the most reasonable explanation for the pattern that we observed.

Rabbit Genomic Clone Analysis

To formally test the possibility that alternative splicing is producing these differences, we needed to determine the germline organization of the Na^+-Ca^{2+} exchanger sequence in this region. A genomic phage clone (L211) that contained the region of the diversity was obtained by standard cloning and screening procedures. Analysis of the genomic clone[25] demonstrated that the region of divergence is represented in the cardiac cDNA by about 76 amino acids is coded for by more than 10 Kb of the rabbit genome (FIG. 5). Sequence analysis of the intron/exon boundaries identified 6 distinct exons labelled A–F, which contained the expected donor/acceptor sites.[25] We demonstrated by sequence analysis that either exon A or B must be used, making these mutually exclusive exons, in that either one but not both must be used to maintain the appropriate reading frame downstream. The next four exons (C–F) that were identified coded for 7, 6, 5 and 24 amino acids respectively are cassette-type exons, in that any combination of these exons could be assembled resulting in a protein that will remain in frame downstream

```
Heart   (NACA1)    EFQNDEIVKTISVKVIDDEEYEKNKTFFLEIGEPRLVEMS    600
Kidney  (NACA2)    EFQNDEIVKIITIRIFDREEYEKECSFSLVLEEPKWI-RR    599
Kidney  (NACA3)    EFQNDEIVKIITIRIFDREEYEKECSFSLVLEEPKWI-RR    599
                   *********  *......* ***** .*.  **..

Heart   (NACA1)    EKKALLNELGGFTITGKYLYGQPVLRKVHARDHPVSTVI     641
Kidney  (NACA2)    GMKALLNELGGFTIT------------------------     615
Kidney  (NACA3)    GMK------GGFTIT------------------------     608
                    *       ******
                           .

Heart   (NACA1)    TIAEEYDDKQPLTSKEEERRIAEMGRPILGEHTKLEVII     681
Kidney  (NACA2)    ---EEYDDKQPLTSKEEERRIAEMGRPILGEHTKLEVII     652
Kidney  (NACA3)    ---EEYDDKQPLTSKEEERRIAEMGRPILGEHTKLEVII     645
                      ************************************
```

FIGURE 4. Sequence comparisons of deduced amino acids for 3 different isoforms of rabbit Na$^+$-Ca^{2+} exchanger. NACA1 rabbit heart,[24] NACA2 kidney form[17,24] and NACA3.[25] *Dashes* (−) indicate that there are no corresponding amino acids in that region. The symbols below the sequence indicate the agreement of the kidney forms with the cardiac isoform where identity (*) or a conservative amino acid replacement (.) occurs.

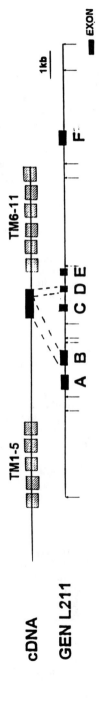

FIGURE 5. Representation of intron/exon arrangement in the rabbit genomic clone for the region that displays alternative splicing and cDNA for the predominant kidney isoform. The exons are represented by the *filled boxes* (not drawn to scale) on the 9 *Eco*RI fragments with sites at *arrow* (↑). The *shaded boxes* on the line representing the cDNA are the putative transmembrane (TM1-11) regions. The *dashed lines* represent the relationship of the germline exons to the predominate kidney cDNA.[24]

from the sites of alternative splicing. This organization explains the results that we observed for the kidney clone in that the kidney form utilizes exon B instead of exon A that the cardiac form utilizes. There are sequence differences between exons A and B explaining the cluster of differences in the comparison of cardiac with kidney in FIGURE 4. Additionally the predominant kidney form (exons B–D) has lost all of the cassette-type exons except D, while the cardiac form is similar to that described for all of the mammalian cardiac sequences (exons A-C-D-E-F).

More recently, alternative splicing has bene described in the 5' untranslated region of the NCX1 gene.[26,27] Analysis of the 5' untranslated region from several tissues identify three unique ends for different messages. These results could occur by the use of a single promotor or multiple independent promotor regions and the subsequent splicing of the 5' untranslated sequence to a conserved region at -34 bp upstream of the start codon.[26] In either case there appears to be rather intricate splicing at the 5' end in the untranslated region of the message, which could be important for message stability or for the efficiency of translation in different tissues.

Cloning the Drosophila Na^+-Ca^{2+} *Exchanger*

With the large extent of sequence identity observed between the various mammalian NCX1 cDNAs that have been characterized, little insight into structure/function relationships for this protein can be obtained. It became important to characterize the Na^+-Ca^{2+} exchanger from a very disparate specie. Genomic Southern analysis demonstrated that there were DNA sequences in *Drosophila* that would cross-hybridize with out human Na^+-Ca^{2+} exchanger cDNA.

We used the human cardiac Na^+-Ca^{2+} exchanger probe and reduced stringency hybridization conditions to identify *Drosophila* cDNA clones. We obtained several cDNA clones from two of these libraries. Sequence analysis revealed an open reading frame of 950 AA with a Kozak consensus initiation site followed by twelve putative hydrophobic regions, separated in the middle by a large hydrophilic region, thought to be an intracellular loop similar in overall structure to the mammalian Na^+-Ca^{2+} exchangers characterized in various mammalian species.[5,8,13–17] The most N-terminal hydrophobic region could be a potentially cleavable leader peptide as has been described for the mammalian exchangers,[8] but this has not been demonstrated to date.

Additionally the sequence and overall structure of the *Drosophila* Na^+-Ca^{2+} exchanger is similar to the single published study of the rat NCX2 cDNA that is found in both brain and skeletal muscle.[28] This second Na^+-Ca^{2+} exchanger gene in the mammalian genome displays 66% amino acid identity and similar functional characteristics to NCX1 sequence even though there is considerable difference in sequence.[28]

The comparison of the deduced amino acid sequence for the *Drosophila* Na^+-Ca^{2+} exchanger with the human Na^+-Ca^{2+} exchange gene NCX1 and the second gene, NCX2, revealed an overall structure of the *Drosophila* protein that was similar with respect to the number, size and position of the transmembrane regions. There are numerous stretches of identity both in the transmembrane and intracellular loop regions with the identity being greater in the transmembrane regions. A few of these areas of conservation correspond to regions of the protein that are thought to have a functional significance for the exchanger. For example, the region of binding to the exchange inhibitory peptide (XIP) that was initially speculated to play a possible role in calmodulin binding[29] and those regions with high affinity for intracellular Ca^{2+} [30] are conserved to a high degree with the *Drosophila* Na^+-Ca^{2+} exchanger sequence.

Does the Drosophila Na^+-Ca^{2+} *Exchanger Undergo Alternative Splicing?*

Interestingly, in the original examination of *Drosophila* cDNAs, we obtained clones from two different libraries (R and S). These libraries differed in that one was prepared from adult male flies (R) and the second from adult female flies (S). In sequence analysis it became clear that subtle differences could be demonstrated between two partial cDNA clones. These two clones were identical in most of the sequence but differed in a 15-bp stretch in one region. It is interesting that the position of these differences corresponds to the region of alternative splicing demonstrated in the rabbit Na^+-Ca^{2+} exchanger gene.[25] Additionally, these differences are short in length, a feature similar to that observed in exons C through E of NCX1.[25] More complete genomic sequence analysis of *Drosophila* may define the extent of potential diversity of this protein by alternative splicing.

Structural Homology of Transporters

As described, NCX1 and NCX2 proteins from mammalian sources have a great deal of similarity in overall structural homology with the *Drosophila* Na^+-Ca^{2+} exchanger, and this is demonstrated in FIGURE 5. The bovine rod Na^+-Ca^{2+} exchanger, which also co-transports K^+, contains little sequence homology to the NCX1 sequence[31] but does contain 6 putative transmembrane regions at the 5' and 3' ends of the molecule (FIG. 6). The Na^+,K^+ ATP-ase a-subunit also has a similar structural design to these Na^+-Ca^{2+} exchangers but lacks a single transmembrane region at the 5' end that is present in the other Na^+-Ca^{2+} exchangers. All the molecules have a large intracellular loop portion and minimal sequences extracellular with the exception of the bovine rod (FIG. 6).

While the general structure of these transporters seems to have been conserved (FIG. 6), there is much less amino acid homology than one might imagine. For example, the Na^+,K^+ ATP-ase, which has a similar general structure to the other exchangers,[32] displays only marginal homology in one transmembrane region to the mammalian Na^+-Ca^{2+} exchanger.[11] Such conservation in overall structure in the absence of amino acid homology could be interpreted either as convergent evolution or as these transport molecules having originated from a common progenitor. If the latter case is true, then sufficient divergence has occurred so that only a small stretch of homologous amino acids have been conserved but the overall structure has not changed.

Phylogeny of Transporter Sequences

When the sequences for the different transporters which share overall structural similarity described above are compared an interesting pattern of relatedness emerges. The mammalian NCX1 Na^+-Ca^{2+} exchangers are dramatically conserved (about 98% identical at the amino acid level, FIG. 7) and most of the diversity present is in the leader peptide region. The different NCX1 exchanger isoforms can vary in the region of alternative splicing but these differences are reduced when the same isoforms are compared between species. The NCX2 gene has only been described in the rat and while almost functionally indistinguishable from the NCX1 gene products, it nevertheless has only marginal sequence homology (65%) to NCX1 (FIG. 7). The *Drosophila* Na^+-Ca^{2+} exchanger is related to an equal extent to both NCX1 or NCX2, with about 44% amino acid identity.

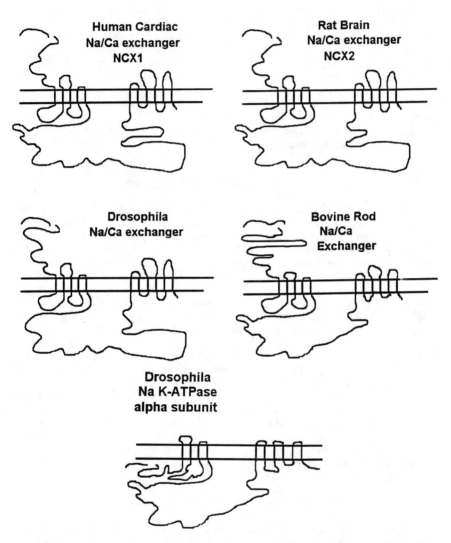

FIGURE 6. Cartoon representation of several transporters and their putative topographical structure. The structures were constructed based on our work and that of others, from human heart (mammalian) NCX1,[5,8] rat brain NCX2,[28] the bovine rod Na^+-Ca^{2+} exchanger[30] and the *Drosophila* $(Na^+ + K^+)$-ATPase.[31] The *solid double horizontal lines* represent the membrane with the *top part* of each cartoon representing the extracellular side of the membrane.

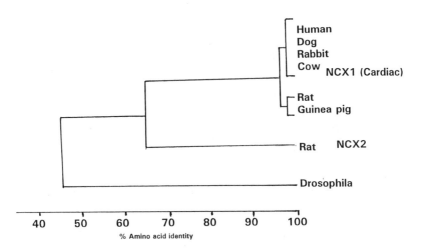

FIGURE 7. Phylogeny of several transport molecules. The rootless tree was constructed from the deduced amino acid identity for sequences from various transporters: NCX1 human,[8,17] NCX1 dog,[5] NCX1 rabbit,[23] NCX1 bovine,[14] NCX1 rat[13] and NCX1 guinea pig,[15] NCX2 rat brain[29] and Drosophila ($Na^+ + K^+$)-ATPase[11] Drosophila (Accession #L39835).

In vertebrate systems there are at least two different loci that code for genes that perform Na^+-Ca^{2+} exchange function. Genomic Southern analysis and *in situ* hybridization of the *Drosophila* Na^+-Ca^{2+} exchanger suggests that there is a single gene sequence. A similar situation exists for the a-subunit of the ($Na^+ + K^+$)-ATPase. In vertebrate genomes the ($Na^+ + K^+$)-ATPase has at least three unlinked a-subunit genes,[9] and the same gene in *Drosophila* is thought to be coded for by a single gene. The extent of similarity for these two transport systems will only be known as more information is learned about the Na^+-Ca^{2+} exchanger structure and function in mammalian and nonmammalian systems.

SUMMARY

We describe an analysis of the NCX1 gene and show that various tissues express different alternatively spliced forms of the gene. Alternative splicing has been confirmed by the genomic analysis of the Na^+-Ca^{2+} exchanger gene. We also describe the *Drosophila* Na^+-Ca^{2+} exchanger as having many of the same structural characteristics of the mammalian exchangers and this locus as possibly undergoing alternative splicing in the same region that has been described in the NCX1 gene. The general structure of the exchangers is similar to that of the a-subunit of the ($Na^+ + K^+$)-ATPase. Finally, sequence comparison of the various molecules demonstrates that structural characteristics of these molecules are more strongly conserved than the primary sequence of these products.

REFERENCES

1. BAKER, P. F., M. P. BLAUSTEIN, A. L. HODGKIN & R. A. STEINHARDT. 1969. The influence of calcium and sodium efflux in squid axons. J. Physiol. (London) **200:** 431–458.
2. LEDERER, W. J. & M. T. NELSON. 1983. Effect of extracellular sodium on calcium efflux and membrane current in single muscle cells from the barnacle. J. Physiol. **341:** 325–339.
3. REUTER, H. & N. SEITZ. 1968. The dependence of calcium efflux from cardiac muscle on temperature and external ion composition. J. Physiol. **195:** 451–470.
4. PHILIPSON, K. D., S. LONGONI & R. WARD. 1988. Purification of the cardiac Na^+-Ca^{2+} exchange protein. Biochim. Biophys. Acta **945:** 298–306.
5. NICOLL, D. A., S. LONGONI & K. D. PHILIPSON. 1990. Molecular cloning and functional expression of the cardiac sarcolemmal Na^+-Ca^{2+} exchanger. Science **250:** 562–565.
6. HRYSHKO, L. V., D. A. NICOLL, J. N. WEISS & K. D. PHILPSON. 1993. Biosynthesis and initial processing of the cardiac sacrolemmal Na^+-Ca^{2+} exchanger. Biochim. Biophys. Acta **1151:** 35–42.
7. DURKIN, J. T., D. C. AHRENS, Y-C. E. PAN & J. P. REEVES. 1991. Purification and amino-terminal sequence of the bovine cardiac sodium-calcium exchanger: evidence for the presence of a signal sequence. Arch. Biochem. Biophys. **290:** 369–375.
8. KOFUJI, P., R. W. HADLEY, R. S. KIEVAL, W. J. LEDERER & D. H. SCHULZE. 1992. Expression of the Na-Ca exchanger in diverse tissues: a study using the cloned human cardiac Na-Ca exchanger. Am. J. Physiol. **263:** C1241–C1249.
9. SHULL, G. E., J. GREEB & J. B. LINGREL. 1986. Molecular cloning of three distinct forms of the Na^+,K^+-ATPase a-subunit from rat brain. Biochemistry **25:** 8125–8132.
10. TAKEYASU, K., M. M. TANKUM, K. J. RENAUD & D. M. FAMBROUGH. 1988. Ouabain-sensitive (Na^+ + K^+)-ATPase activity expressed in mouse L cells by transfection with DNA encoding the a-subunit of an avian sodium pump. J. Biol. Chem. **263:** 4347–4354.
11. LEBOVITZ, R. M., K. TAKEYASU & D. M. FAMBROUGH. 1989. Molecular characterization and expression of the (Na^+ + K^+)-ATPase alpha subunit in *Drosophila melanogaster*. EMBO J. **8:** 193–202.
12. POWERS, P. A., S. LIU, K. HOGAN & R. G. GREGG. 1992. Skeletal muscle and brain isoforms of a β-subunit of human voltage-dependent calcium channels are encoded by a single gene. J. Biol. Chem. **267:** 22967–22972.
13. LOW, W., J. KASIR & H. RAHAMIMOFF. 1993. Cloning of the rat heart Na^+-Ca^{2+} exchanger and its functional expression in HeLa cells. FEBS Lett. **316:** 63–67.
14. ACETO, J. F., M. CONDRESSCU, C. KROUPIS, H. NELSON, D. A. NICOLL, K. D. PHILIPSON & J. P. REEVES. 1992. Cloning and expression of the bovine sodium-calcium exchanger. Arch. Biochem. Biophys. **298:** 553–560.
15. TSURUYA, Y., M. M. BERSOHN, Z. LI, D. A. NICOLL & K. D. PHILIPSON. 1994. Molecular cloning and functional expression of the guinea pig cardiac Na^+-Ca^{2+} exchanger. Biochim. Biophys. Acta **1196:** 97–99.
16. KIM, I. & C. O. LEE. 1995. Cloning of the mouse cardiac Na^+-Ca^{2+} exchanger and functional expression in *Xenopus* oocytes. This volume.
17. KOMURO, I., K. E. WENNINGER, K. D. PHILIPSON & S. IZUMO. 1992. Molecular cloning and characterization of the human cardiac Na^+/Ca^{2+} exchanger cDNA. Proc. Natl. Acad. Sci. USA **89:** 4769–4773.
18. AMBESI, A., E. L. VANALSTYNE, E. E. BAGWELL & G. E. LINDENMAYER. 1991. Effect of polyclonal antibodies on the sodium-calcium exchanger. Ann. N. Y. Acad. Sci. **639:** 245–247.
19. KIEVAL, R. S., R. J. BLOCH, G. E. LINDERNMAYER, A. AMBESI & W. J. LEDERER. 1992. Immunofluorescence localization of the Na-Ca exchanger in heart cells. Am. J. Physiol. **263:** C545–550.
20. ALLEN, T. J. A., D. NOBEL & H. REUTER, Eds. 1989. Sodium-Calcium Exchange. Oxford University Press. Oxford, UK.
21. EISNER, D. A. & W. J. LEDERER. 1985. Na-Ca exchange: stoichiometry and electrogenicity. Am. J. Physiol. **284:** C189–202.

22. MINTA, A., J. R. Y. KAO & R. Y. TSIEN. 1989. Fluorescent indicators for cytosolic calcium based on rhodamine and fluorescein chromophores. J. Biol. Chem. **264:** 8171–8178.
23. REILLY, R. F. & C. A. SHUGRUE. 1992. cDNA cloning of a renal Na^+-Ca^{2+} exchanger. Am. J. Physiol. **262:** F1105–F1109.
24. KOFUJI, P., W. J. LEDERER & D. H. SCHULZE. 1993. Na/Ca exchanger isoforms expressed in the heart and kidney. Am. J. Physiol. **263:** C1241–C1249.
25. KOFUJI, P., W. J. LEDERER & D. H. SCHULZE. 1994. Mutually exclusive and cassette exons underlie alternatively spliced isoforms of the Na/Ca exchanger. J. Biol. Chem. **269:** 14849–14852.
26. LEE, S-L., S. L. YU & J. LYTTON. 1994. Tissue specific expression of Na^+-Ca^{2+} isoforms. J. Biol. Chem. **269:** 14849–14852.
27. MENICK, D. R., K. V. BARNES, U. F. THACKER, M. M. DAWSON, D. E. McDERMOTT, J. D. ROZICH, R. L. KENT & G. COOPER IV. 1995. The exchanger and cardiac hypertrophy. This volume.
28. LI, Z., S. MATSUOKA, L. V. HRYSHKO, D. A. NICOLL, M. M. BERSOHN, E. P. BURKE, R. P. LIFTON & K. D. PHILIPSON. 1994. Cloning of the NCX2 isoform of the plasma membrane Na^+-Ca^{2+} exchanger. J. Biol. Chem. **269:** 17434–17439.
29. LI, Z., D. A. NICOLL, A. COLLINS, D. W. HILGEMANN, A. G. FILOTEO, J. T. PENNISON, J. N. WEISS, J. M. TOMICH & K. D. PHILIPSON. 1991. Identification of a peptide inhibitor of the cardiac sarcolemmal Na^+-Ca^{2+} exchanger. J. Biol. Chem. **266:** 1014–1020.
30. LEVITSKY, D. O., D. A. NICOLL & K. D. PHILIPSON. 1994. Identification of the high affinity Ca^{2+}-binding domain of the cardiac Na^+-Ca^{2+} exchanger. J. Biol. Chem. **269:** 22847–22852.
31. REILANDER, H. A., A. ACHILLES, T. FRIEDEL, G. MAUL, F. LOTTSPEICH & N. J. COOK. 1992. The primary structure and functional expression of the Na/Ca, K exchanger from bovine rod photoreceptors. EMBO J. **11:** 1689–1695.
32. SONG, Y. & D. FAMBROUGH. 1994. Molecular evolution of the calcium-transporting ATPases analyzed by the maximum parsimony method. *In* Molecular Evolution of Physiological Processes. D. M. Fambrough, Ed. 271–283. The Rockefeller University Press. New York.

The Kidney Sodium-Calcium Exchanger[a]

JONATHAN LYTTON,[b,c] SHWU-LUAN LEE,[b]
WEN-SEN LEE,[b] J. VAN BAAL,[d] RENÉ J. M. BINDELS,[d]
RACHEL KILAV,[e] TALLY NAVEH-MANY,[e]
AND JUSTIN SILVER[e]

[c]*Renal Division*
Department of Medicine
Brigham and Women's Hospital
and
Harvard Medical School
Boston, Massachusetts 02115

[d]*Department of Cell Physiology*
University of Nijmegen
The Netherlands

[e]*Minerva Center for Calcium and Bone Metabolism*
Nephrology Services
Hadassah University Hospital
Jerusalem il-91120, Israel

INTRODUCTION

The maintenance of systemic calcium balance is of essential importance to terrestrial vertebrates due to the critical involvement of this ion in a variety of both intracellular signaling and extracellular roles, as well as in the maintenance of the skeletal system. Three organ systems—intestine, bone and kidney—control body calcium under the regulation of hormones such as parathyroid hormone (PTH), 1,25-dihydroxy-vitamin D_3 (1,25$[OH]_2D_3$), and calcitonin, all coordinated by the parathyroid gland. The function of the kidney is essential to this complex network since, of all the organs, it handles by far the most calcium and is the site of conversion of vitamin D to the active metabolite, 1,25$[OH]_2D_3$.[1–3]

In the kidney more than 97% of the free calcium filtered at the glomerulus is reabsorbed as the filtrate passes down the nephron. A large fraction of this calcium reabsorption occurs passively and via paracellular pathways in the proximal tubule and the loop of Henle. The final control of calcium reabsorption and its hormonal regulation, however, takes place in the distal nephron. At this site, calcium transport proceeds against an electrochemical gradient in a transcellular manner. Two transport systems on the basolateral membrane of distal nephron cells provide

---

[a] This work was supported in part by a grant from the National Institutes of Health (DK42789) and a grant-in-aid from the American Heart Association (to JL), and by the Thyssen Foundation (to JS).

[b] An Established Investigator of the American Hearth Association. Address for correspondence: Department of Medical Biochemistry, University of Calgary Health Sciences Centre, 3330 Hospital Drive, NW, Calgary, Alberta, Canada, T2N 4N1.

the driving force for transcellular flux: the plasma membrane calcium-pumping ATPase, and the sodium-calcium exchanger. Several lines of evidence argue persuasively for a substantial involvement of the exchanger in driving calcium reabsorption.[4,5] Nevertheless, due to the lack of specific inhibitors and the potential for secondary artifacts due to the experimental manipulations, the precise relative roles of these two enzymes remain the subject of debate. Indeed, the location of expression of Na-Ca exchanger in the kidney is also controversial. Although several groups find an enrichment in membranes of distal tubules, others have presented arguments for exchanger presence in proximal tubule and collecting duct.[1,6-8]

Several factors are thought to modulate calcium reabsorption in the distal nephron, including 1,25[OH]$_2$D$_3$, PTH, and both thiazide and amiloride diuretics.[1,8] 1,25[OH]$_2$D$_3$ is believed to exert its effect through regulation of the expression of a 28 kDa cytoplasmic calcium binding protein, calbindin-D28k. However, regulation of transporter protein expression by 1,25[OH]$_2$D$_3$, as has been observed in the intestine,[9] has not been carefully examined in kidney. PTH has been suggested to stimulate both the entry of calcium into distal nephron cells,[10] as well as its exit via the Na-Ca exchanger,[11] although the mechanism for these events has not been clearly elucidated. The diuretics, through their inhibition of sodium uptake at the apical membrane, stimulate calcium reabsorption indirectly. Whether this is due to potential changes across the apical membrane that increase calcium entry, or to an increase in the sodium gradient which drives basolateral exit through the Na-Ca exchanger, is unclear.[12-15]

Nevertheless, it is evident that the Na-Ca exchanger plays an important role in the active reabsorption of calcium in the distal nephron. Inasmuch as this action serves to help regulate systemic, *extracellular* calcium rather than *intracellular* calcium, the factors that regulate kidney Na-Ca exchanger activity are likely to be different from those that regulate exchange activity in other cells, such as cardiac myocytes. The investigation of such matters requires a better understanding of the molecule that underlies renal Na-Ca exchange activity, and thus in recent years several groups, including our own, have cloned the kidney Na-Ca exchanger.[16-20] Surprisingly, the molecule turns out to be an alternatively spliced product of the NCX1 gene expressed most abundantly in cardiac tissue. In this paper we describe our work on the expression of the rat Na-Ca exchanger gene in kidney, and other tissues, which suggests that regulation of this molecule is under the control of different promoters in different tissues.

METHODS

In situ hybridization was performed using a partially base-hydrolyzed ^{35}S-labeled riboprobe corresponding to the entire rat kidney F1 clone, as previously described.[21,22] Immunocytochemical procedures were as previously described.[23] The RA-RRCaBP antiserum against calbindin was a generous gift of Sylvia Christakos, University of Medicine and Dentistry of New Jersey, and the C2C12 antibody against the Na-Ca exchanger was a generous gift of Kenneth Philipson, UCLA.

Molecular cloning, RNA isolation, Northern blot hybridization, and 5'-rapid amplification of cDNA ends (RACE) were all as previously described.[17] RNase protection experiments, using ^{32}P-labeled riboprobes corresponding to our genomic clones, or to chimeras between genomic and cDNA clones, were performed according to established procedures.[24]

Models of dietary induced hypo- and hyperparathyroidism, parathyroidectomy, and vitamin-D deficiency were all induced in adult rats as previously described.[25] Following a treatment of two weeks, the animals were sacrificed, the kidneys removed and immediately frozen. Total RNA was isolated from whole kidneys according to standard procedures as previously described.[17]

RESULTS AND DISCUSSION

Our initial studies published three years ago employed the polymerase chain reaction (PCR) to identify a fragment of what proved to be the NCX1 gene that was expressed in kidney.[19] Northern blotting data (see FIG. 4, top panel) revealed that in fact NCX1 was expressed at high abundance, not only in kidney and heart, but also in various brain regions. FIGURE 1 shows the distribution of Na-Ca

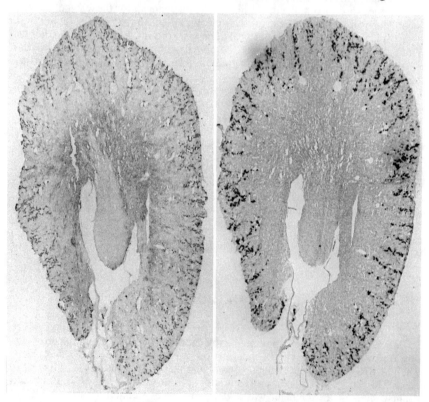

FIGURE 1. Localization of Na-Ca exchanger transcripts in rat kidney. Serial sections of fixed rat kidney were **(A)** stained for calbindin-D28k with RRCaBP antiserum by the immunoperoxidase method or **(B)** hybridized to an antisense riboprobe for Na-Ca exchanger and exposed to emulsion autoradiography. Note the significant overlap of signal for these two probes, which is particularly evident toward the *top left corner* of the figure.

exchanger transcripts in a section of rat kidney determined by *in situ* hybridization, and compared to the expression in a serial section of calbindin-D28k, a protein known to be expressed in the distal nephron, predominantly in distal convoluted tubule and connected segments. The pattern of expression of these two species is clearly similar, and demonstrates that virtually *all* of the abundant Na-Ca exchanger mRNA seen on Northern blots is restricted to only a small fraction of nephron segments—the distal tubule—while essentially no Na-Ca exchanger message is observed elsewhere in the kidney.

The distribution of Na-Ca exchanger protein in rabbit kidney is shown at higher magnification in FIGURE 2, and compared to the expression of calbindin-D28k and the plasma membrane Ca-ATPase in serial sections. The Na-Ca exchanger protein is clearly expressed on the basolateral membrane of what appear to be connecting segment cells. Similar data have also been published by others.[26,27] Interestingly, it had previously been reported that expression of the Na-Ca exchanger was restricted to the connecting segment, while expression of the plasma membrane Ca-ATPase was only in the distal convoluted tubule.[28] Our data show a remarkable overlap of expression of these two proteins in rabbit kidney, together with calbindin-D28k. Further analysis by double-label immunofluorescent staining of rat kidney sections (data not shown) indicates an almost complete overlap of expression of Na-Ca exchanger and plasma membrane Ca-ATPase. The calcium transporter positive cells form a subset of the calbindin-D28k positive cells, and overlap with expression of the thiazide-sensitive Na,Cl-cotransporter protein. These comparisons, together with the morphology of the tubule sections, suggest that Na-Ca exchanger is expressed most strongly in the late portion of the distal convoluted tubule (in true distal convoluted tubule cells), and in essentially all true connecting segment cells throughout the connecting segment, although at levels which diminish as one travels further along this segment. The extremely high level of expression of Na-Ca exchanger mRNA and protein, together with other calcium transport proteins, in precisely those nephron segments associated with active transcellular calcium reabsorption, clearly demonstrates that the exchanger plays an important role in this transport process.

Examination of our rat renal Na-Ca exchanger clone, F1, with other Na-Ca exchanger clones from various species revealed two regions of significant divergence.[17] One region corresponded to the site of alternative splicing within the central cytoplasmic loop of the molecule, which has been well characterized by the studies of Kofuji *et al.*[29] (and presented elsewhere in this volume). The other region corresponded to the 5'-untranslated portion of the transcript. To characterize further the 5'-end of Na-Ca exchanger transcripts expressed in various rat tissues, we used the technique of 5'-RACE. As illustrated in FIGURE 3, a single round of amplification revealed a single major band in RNA from heart, two major bands in kidney and three bands in brain. Analysis by subcloning, sequencing, and restriction mapping, allowed us to identify each of the major bands, indicated by the labeling to the right of the gel and the cartoon above it in FIGURE 3. Three separate species were defined by this analysis, each comprising a unique 5'-end sequence joined to a common sequence at position −34 relative to the initiating methionine codon. One such species, Kc, was isolated from kidney cortex mRNA. Another, Br (previously referred to as Br1), also was identified in brain mRNA. The third species, Ht (previously Br2), also was identified in all three tissues, as bands of different size. In heart the band at ~300 nt corresponds to Ht(s) with 112 nt of 5'-untranslated sequence. Two Ht species were identified in brain, Ht(m) and Ht(l). The latter, which was also found in kidney, contained 220 nt of 5'-untranslated sequence.

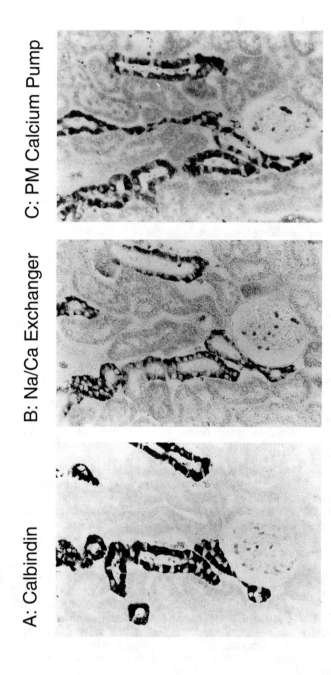

FIGURE 2. Localization of calcium transport proteins in rabbit kidney. Serial sections of fixed rabbit kidney were reacted with **(A)** antibody 0089 against calbindin, **(B)** antibody C2C12 against the Na-Ca exchanger, or **(C)** antibody 5F10 against the plasma membrane Ca-ATPase. Antibody reactivity was visualized with peroxidase staining. Note that the tubule which reacts with the plasma membrane Ca-ATPase antibody, and runs vertically toward the center of (C), cuts obliquely through these sections and only the portion toward the *top* of the figure is present in (A) and (B). Note also the intense basolateral staining with the Na-Ca exchanger antibody.

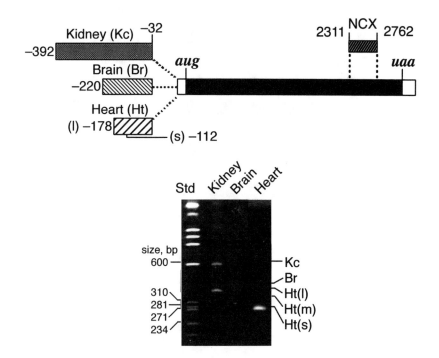

FIGURE 3. 5'-RACE amplification of Na-Ca exchanger transcripts in rat kidney, brain, and heart. The *cartoon* illustrates the coding region of the rat Na-Ca exchanger transcript in *black*, the extent of the three unique 5'-untranslated sequences, and the location of the common coding region probe, NCX-1. The *lower panel* shows an ethidium bromide stained polyacrylamide gel of the products from a typical RACE experiment. The antisense 3' primer used for these experiments anneals beginning at nt 168, while the 5' anchoring primer adds 35 nt to the length of the product. Thus the size of bands corresponding to the indicated Kc, Br, Ht(l) and Ht(s) products are 595, 423, 381 and 315 nt in length, respectively.

The expression of Na-Ca exchanger transcripts possessing each of these unique 5'-end sequences was examined by Northern blot analysis, as shown in FIGURE 4. The Br probe reveals a pattern of Na-Ca exchanger mRNA expression quite similar to the pattern seen with the common coding region probe, NCX-1, except for the conspicuous absence of a band in heart, and a dramatically reduced signal in kidney cortex. As expected, the Ht probe revealed a strong band in RNA from heart, but also a weaker signal in large intestine, and a weak signal in cerebrum RNA (at ~14 kb). The Kc probe revealed a strong band in only kidney cortex mRNA. Thus it is clear that Na-Ca exchanger transcripts with unique 5'-ends are expressed in a tissue-specific fashion, most likely due to the existence of three separate promoters which drive expression in heart, kidney, and elsewhere. Indeed, such a scenario would impart the potential for independent regulation of the Na-Ca exchanger, as anticipated by the different roles the molecule plays, particularly in the kidney compared to other tissues.

Investigation of selective Na-Ca exchanger expression and its regulation clearly requires an analysis of the gene, particularly the promoter regions. Toward that

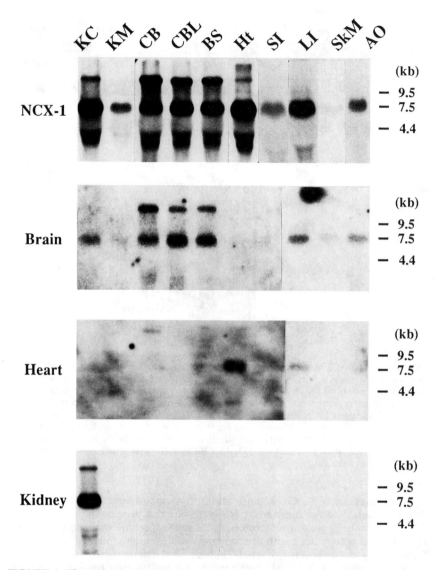

FIGURE 4. Tissue distribution of Na-Ca exchanger isoforms. Four parallel Northern blots were hybridized with probes from the common region (NCX-1) or the unique 5'-end exons corresponding to brain (Br), heart (Ht) and kidney (Kc). The brain and heart probes were short DNA probes, and hence the signal-to-noise ratio is poorer than for the longer riboprobes used for the kidney and NCX-1 probes. Ten micrograms of total RNA was loaded in each lane, isolated from the following rat tissues: KC, kidney cortex; KM, kidney medulla; CB, cerebrum; CBL, cerebellum; BS, brainstem; Ht, heart left ventricle; SI, small intestine; LI, large intestine; SkM, skeletal muscle (diaphragm); AO, aorta.

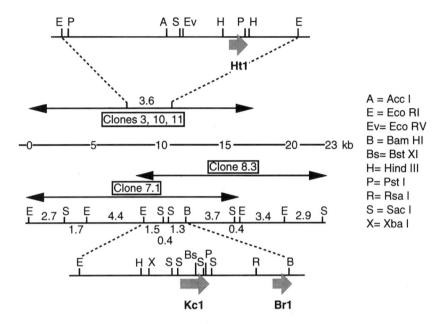

FIGURE 5. Genomic clones encoding the 5'-end exons of the rat Na-Ca exchanger gene. Partial restriction maps of two separate contigs encompassing the Ht exon (contig comprising clones 3, 10 and 11), and the Kc and Br exons (contig comprising clones 7.1 and 8.3). Exons were mapped via a combination of restriction digests, hybridization, and sequencing. The direction of transcription is indicated. Note that the two contigs do not overlap.

end, we have isolated rat genomic clones which contain each of the three unique 5'-untranslated sequences, Ht, Br, and Kc. As illustrated in the cartoon of FIGURE 5, we have located the Kc and Br exons within about 1.5 kb of each other in the center of a ~23 kb genomic contig. Clones containing the Ht exon were also isolated, but do not overlap with the Kc/Br contig, indicating that these exons are separated by at least 10 kb. Moreover, none of these clones contains sequence from the subsequent downstream exon 2, which includes the protein initiation codon. From these data we cannot order the Ht exon with respect to the Kc/Br exons. Data from Don Menick's laboratory on the Na-Ca exchanger gene of the cat (presented elsewhere in this volume) as well as anecdotal evidence based on the two bovine Na-Ca exchanger clones described by John Reeves' laboratory,[30] however, suggest that Ht lies upstream of Kc/Br. The genomic sequences match our cDNA and RACE clones exactly but all end in a common CAG sequence, suggesting that the true splice site which generates the mature Na-Ca exchanger transcripts lies at position −32 in the mRNA.

To determine if our RACE and cDNA clones corresponded to full-length transcripts, we used the genomic clones as templates to perform RNase protection experiments, as illustrated in FIGURE 6. In these experiments, antisense ^{32}P-labeled riboprobes based on genomic sequences are annealed to RNA from various tissues. The annealed complexes are then digested with an RNase that will degrade any single-stranded RNA. Thus, only those genomic sequences that are present in the

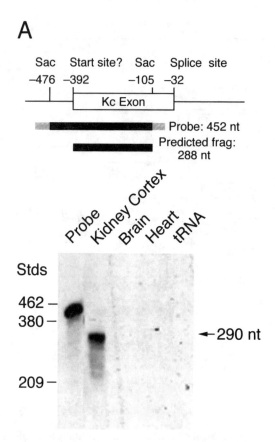

FIGURE 6. RNase protection analysis of the Na-Ca exchanger gene 5'-end. In the *cartoon* illustrations of the probe maps, *open boxes* indicate exons from the genomic clones, while the *shaded box* is part of exon 2 derived from cDNA. The probe constructs themselves are *black* in the regions corresponding to the Na-Ca exchanger gene/cDNA, and *grey* at the ends to denote vector sequences. **(A)** A 452-nt ^{32}P-riboprobe was prepared from the genomic region encompassing the Kc exon. Full length protection of the Kc exon based on cDNA and RACE cloning experiments is expected to result in a 288-nt fragment. **(B)** A chimeric construct was prepared from the genomic clone encompassing the Ht 5'-end exon, and from part of exon 2 from cDNA, to generate a 550-nt ^{32}P-riboprobe. Expression of the Ht(l) exon, as identified in brain and kidney tissue (see FIG. 3) is predicted to result in a species of 274 nt, while protection of the Ht(s) exon identified in heart will yield a 208-nt band. The presence of an exon other than Ht spliced at position −32 will generate a band of 128 nt from the common region of exon 2. If an exon intervenes between the Ht exon and exon 2, then protected bands of 128 nt (exon 2) and 146 nt (Ht(l)) or 80 nt (Ht(s)) are expected. **(C)** A chimeric 600-nt ^{32}P-riboprobe from the Br exon and part of exon 2. Full-length protection of the Br exon, as identified from brain RNA, is expected to result in a fragment of 301 nt. If another exon is spliced in place of the Br exon at position −32, a band of 113 nt is expected. Riboprobes were each annealed to 20 µg of total RNA from the indicated rat tissues, or 20 µg of tRNA (as negative control). Following RNase digestion, the protected products were separated on a 5% acrylamide-50% urea gel, together with an aliquot of the undigested probe, and exposed to X-ray film. The position of size standards (Stds), as well as the approximate size of the protected species (*arrows*), are indicated. Due to the rather high nonspecific background, two representative experiments are illustrated for the Br probe in (C).

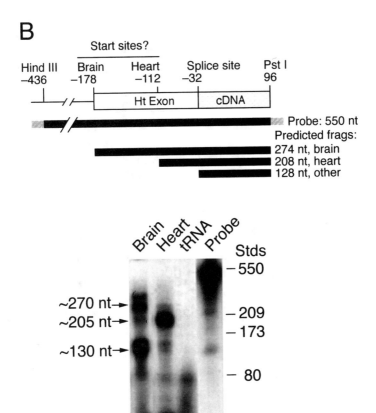

FIGURE 6. (*Continued*)

transcripts will be protected from degradation. The length of these products then defines the start of transcription (if, indeed, the probes derive from the 5'-most exon in the gene).

Panel A of FIGURE 6 demonstrates that RNA from kidney, but not from brain or heart, protects a ~290-nt fragment of the genomic clone which corresponds almost exactly to our predictions for the start site of the kidney Na-Ca exchanger transcript. To examine the heart transcript, we prepared a chimeric genomic/cDNA construct which fuses the Ht exon to the downstream exon 2. This probe results in a longer protected fragment than would be obtained using the Ht exon alone and, thus, a greater signal-to-noise ratio. Panel B shows that RNA from heart protects a band of ~205 nt, which matches very closely the predicted size based on the cDNA and RACE cloning data. No other major bands were observed in heart. Brain RNA, on the other hand, protected an Ht probe band of ~270 nt, which matched the expectations for the length of the cDNA observed. In addition, however, a strong doublet was observed at ~130/140 nt. Since the major Na-Ca exchanger transcript in brain possesses a Br exon end, the Ht probe is expected to produce only a partially protected band of 128 nt. Additionally, if another exon

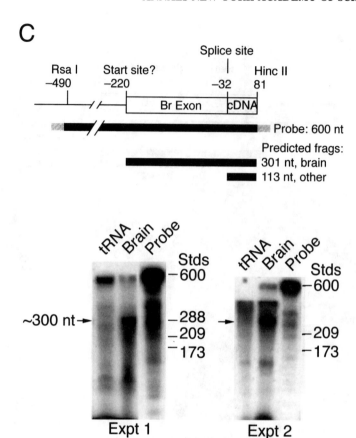

FIGURE 6. (*Continued*)

(Br, Kc?) were to intervene between the Ht exon and exon 2 in brain, then both 128 nt and 146 nt protected fragments would be observed. Finally, as shown in panel C, a chimeric Br construct results in the protection of a ~300 nt fragment as predicted. The Br probe is from a very GC-rich genomic region, however, which has given us trouble with the signal-to-noise ratio. Two representative experiments are shown to help illustrate the specificity of the 300 nt fragment.

These data provide rather good evidence that transcription of the Na-Ca exchanger in kidney, brain and heart derives from different points in the gene. Further experiments are now in progress to determine promoter elements in the 5'-flanking region, and to investigate the nature of both tissue-specific and regulated expression.

Since the renal Na-Ca exchanger is implicated in the control of calcium reabsorption and systemic calcium homeostasis, we have begun experiments to investigate how alterations to body calcium balance influence the expression of the exchanger in kidney. As shown in FIGURE 7A, rats treated with a diet deficient

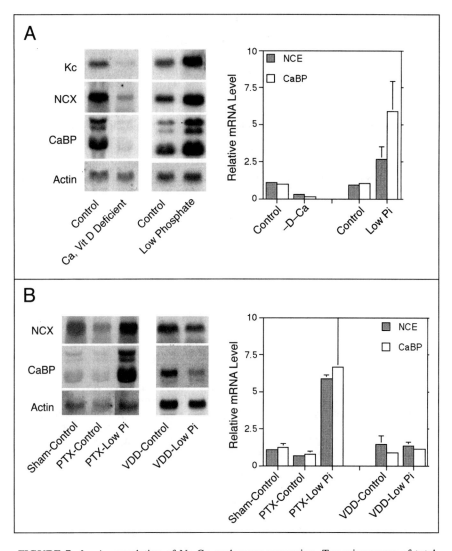

FIGURE 7. *In vivo* regulation of Na-Ca exchanger expression. Ten micrograms of total RNA isolated from kidneys of rats treated as indicated were resolved on formaldehyde-agarose gels, transferred to nylon membranes, and hybridized with probes corresponding to the common coding region of the Na-Ca exchanger (NCX), the kidney 5'-end exon of the Na-Ca exchanger (Kc), calbindin-D28k (CaBP), or β-actin (Actin). Single representative autoradiographs are shown to the *left*. The level of Na-Ca exchanger (NCE) and calbindin-D28k (CaBP) mRNA is shown quantitatively, normalized to the β-actin content with control diet levels set to 1.0 and averaged over a number of experiments, in the graphs to the *right*. *Error bars* indicate the standard error of the mean. (A) Control (n = 11) *versus* calcium and vitamin D-deficient (–D–Ca) (n = 4) and control (n = 9) *versus* low phosphate (Low Pi) (n = 4) diets. (B) Either sham-operated, parathyroidectomizid (PTX), or second-generation vitamin D-deficient (VDD), animals were treated with control or low phosphate (Low Pi) diets (n = 2 for each group, except the sham control, for which n = 4).

in both calcium and vitamin D show a marked reduction in the content of renal Na-Ca exchanger transcript. Conversely, rats treated with a diet deficient in phosphate had elevated levels of Na-Ca exchanger mRNA. In both cases, hybridization with either a common coding region (NCX), or the unique kidney 5'-end (Kc), probe from the Na-Ca exchanger showed similar changes, while levels of calbindin-D28k mRNA responded in parallel with those of the exchanger. Due to the known response of calbindin-D28k to $1,25[OH]_2D_3$, it therefore seemed likely that the renal Na-Ca exchanger was also controlled by levels of this hormone. Nevertheless, the above treatments result in alterations to circulating levels of PTH, as well as ionized calcium and phosphate, all of which are potential mediators of Na-Ca exchanger regulation in these whole animal experiments.

To help distinguish among these alternatives, the experiments illustrated in FIGURE 7B were performed. Parathyroidectomy had little, if any, effect on the levels of mRNA encoding either the Na-Ca exchanger or calbindin-D28k. Moreover, parathyroidectomized animals still responded to low phosphate diet with an increase in the transcripts for these proteins. This rules out PTH as mediator of the observed changes in expression. On the other hand, second-generation vitamin D-deficient rats failed to respond to the low phosphate diet, supporting the suggestion that circulating levels of $1,25[OH]_2D_3$ may control renal Na-Ca exchanger expression.

CONCLUSIONS

Aside from its very important role in controlling intracellular calcium in cardiac myocytes and in neurons, the function of the Na-Ca exchanger is critical for renal reabsorption of calcium, and thus for the control of systemic calcium balance. An alternatively spliced product of the NCX1 gene is expressed very abundantly in kidney cortex, where it is localized to the basolateral membranes of distal nephron cells, primarily in the connecting segment. Na-Ca exchanger expression appears to be under the control of different promoters in kidney, in heart, and in other tissues, imparting the potential for independent regulation of this important transporter. In the kidney, expression of the Na-Ca exchanger responds to altered states of systemic calcium homeostasis, probably through a $1,25[OH]_2D_3$-dependent pathway.

ACKNOWLEDGMENTS

We wish to thank Sylvia Christakos (UMDNJ) and Ken Philipson (UCLA) for the generous gift of antisera, and Matthew Plotkin (Brigham & Women's Hospital) for immunocytochemical analysis of Na-Ca exchanger, plasma membrane Ca-ATPase, and calbindin-D28k expression in rat kidney.

REFERENCES

1. FRIEDMAN, P. A. & F. A. GESEK. 1993. Calcium transport in renal epithelial cells. Am. J. Physiol. **264:** F181–F198.
2. KUROKAWA, K. 1994. The kidney and calcium homeostasis. Kidney Int. **45:** S97–S105.
3. BINDELS, R. J. M., A. HARTOG, J. TIMMERMANS & C. H. VAN OS. 1991. Active

calcium transport in primary cultures of rabbit kidney CCD: stimulation by 1,25-dihydroxyvitamin D_3 and PTH. Am. J. Physiol. **261:** F799–F807.
4. BINDELS, R. J. M., P. L. M. RAMAKERS, J. A. DEMPSTER, A. HARTOG & C. H. VAN OS. 1992. Role of Na^+/Ca^{2+} exchange in transcellular Ca^{2+} transport across primary cultures of rabbit kidney collecting system. Pflügers Arch. **40:** 566–572.
5. SHIMIZU, T., K. YOSHITOMI, M. NAKAMURA & M. IMAI. 1990. Effects of PTH, calcitonin and cAMP on calcium transport in rabbit distal nephron segments. Am. J. Physiol. **259:** F408–F414.
6. RAMACHANDRAN, C. & M. G. BRUNETTE. 1989. The renal Na/Ca exchange system is located exclusively in the distal tubule. Biochem. J. **257:** 259–264.
7. DOMINGUEZ, J. H., M. JUHASZOVA, S. B. KLEIBOEKER, C. C. HALE & H. A. FEISTER. 1992. Na,Ca-exchanger of rat proximal tubule: gene expression and subcellular localization. Am. J. Physiol. **263:** F945–F950.
8. BINDELS, R. J. M. 1993. Calcium handling by the mammalian kidney. J. Exp. Biol. **184:** 89–104.
9. CAI, Q., J. S. CHANDLER, R. H. WASSERMAN, R. KUMAR & J. T. PENNISTON. 1993. Vitamin-D and adaptation to dietary calcium and phosphate deficiencies increase intestinal plasma membrane calcium pump gene expression. Proc. Natl. Acad. Sci. USA **90:** 1345–1349.
10. FRIEDMAN, P. A. & F. A. GESEK. 1994. Hormone-responsive Ca^{2+} entry in distal convoluted tubules. J. Am. Soc. Nephrol. **7:** 1396–1404.
11. BOUHTIAUY, I., D. LAJEUNESSE & M. G. BRUNETTE. 1991. The mechanism of parathyroid hormone action on calcium reabsorption by the distal tubule. Endocrinology **128:** 251–258.
12. GESEK, F. A. & P. A. FRIEDMAN. 1992. Mechanism of calcium transport stimulated by chlorothiazide in mouse distal convoluted tubule cells. J. Clin. Invest. **90:** 429–438.
13. SHIMIZU, T., M. NAKAMURA, K. YOSHITOMI & M. IMAI. 1991. Interaction of trichlormethiazide or amiloride with PTH in stimulating Ca^{2+} absorption in rabbit CNT. Am. J. Physiol. **261:** F36–F43.
14. LAJEUNESSE, D., M. G. BRUNETTE & J. MAILLOUX. 1991. The hypocalciuric effect of thiazides: subcellular localization of the action. Pflügers Arch. **417:** 454–462.
15. COSTANZO, L. S. 1985. Localization of diuretic action in microperfused rat distal tubules: Ca and Na transport. Am. J. Physiol **248:** F527–F535.
16. KOFUJI, P., W. J. LEDERER & D. H. SCHULZE. 1993. Na/Ca exchanger isoforms expressed in kidney. Am. J. Physiol. **265:** F598–F603.
17. LEE, S-L., A. S. L. YU & J. LYTTON. 1994. Tissue-specific expression of Na/Ca exchanger isoforms. J. Biol. Chem. **269:** 14849–14852.
18. REILLY, R. F. & C. A. SHUGRUE. 1992. cDNA cloning of a renal Na^+-Ca^{2+} exchanger. Am. J. Physiol. **262:** F1105–F1109.
19. YU, A. S. L., S. C. HEBERT, S-L. LEE, B. M. BRENNER & J. LYTTON. 1992. Identification and localization of renal Na^+-Ca^{2+} exchanger by polymerase chain reaction. Am. J. Physiol. **263:** F680–F685.
20. LOO, T. W. & D. M. CLARKE. 1994. Functional expression of human renal Na^+/Ca^{2+} exchanger in insect cells. Am. J. Physiol. **267:** F70–F74.
21. WU, K-D., W-S. LEE, J. WEY, D. BUNGARD & J. LYTTON. 1995. Localization and quantification of endoplasmic reticulum Ca^{2+}-ATPase isoform transcripts. Am. J. Physiol. **269:** C775–C784.
22. LEE, W-S., M. J. BERRY, M. A. HEDIGER & P. R. LARSEN. 1993. The type I iodothyronine 5'-deiodinase messenger ribonucleic acid is localized to the S3 segment of the rat kidney proximal tubule. Endocrinology **132:** 2136–2140.
23. BINDELS, R. J. M., J. A. H. TIMMERMANS, A. HARTOG, W. COERS & C. H. VAN OS. 1991. Calbindin-D9k and parvalbumin are exclusively located along basolateral membranes in rat distal nephron. J. Am. Soc. Nephrol. **2:** 1122–1129.
24. AUSUBEL, F. M., R. BRENT, R. E. KINGSTON, D. D. MOORE, J. G. SEIDMAN, J. A. SMITH & K. STRUHL, Eds. 1995. Current Protocols in Molecular Biology. John Wiley & Sons. New York, NY.
25. KILAV, R., J. SILVER, J. BIBER, H. MURER & T. NAVEH-MANY. 1995. Coordinate

regulation of the rat renal parathyroid hormone receptor mRNA and the Na/Pi-cotransporter mRNA and protein. Am. J. Physiol. **268:** F1017–F1022.
26. REILLY, R. F., C. A. SHUGRUE, D. LATTANZI & D. BIEMESDERFER. 1993. Immunolocalization of the Na^+/Ca^{2+} exchanger in rabbit kidney. Am. J. Physiol. **265:** F327–F332.
27. BOURDEAU, J. E., A. N. TAYLOR & A. M. IACOPINO. 1993. Immunocytochemical localization of sodium-calcium exchanger in canine nephron. J. Am. Soc. Nephrol. **4:** 105–110.
28. BORKE, J. L., J. T. PENNISTON & R. KUMAR. 1990. Recent advances in calcium transport by the kidney. Semin. Nephrol. **10:** 15–23.
29. KOFUJI, P., W. J. LEDERER & D. H. SCHULZE. 1994. Mutually exclusive and cassette exons underlie alternatively spliced isoforms of the Na/Ca exchanger. J. Biol. Chem. **269:** 5145–5149.
30. ACETO, J. F., M. CONDRESCU, C. KROUPIS, H. NELSON, N. NELSON, D. NICOLL, K. D. PHILIPSON & J. P. REEVES. 1992. Cloning and expression of the bovine cardiac Na,Ca-exchanger. Arch. Biochem. Biophys. **298:** 553–560.

Sodium-Calcium Exchange and Calcium Homeostasis in Transfected Chinese Hamster Ovary Cells[a]

JOHN P. REEVES, GALINA CHERNAYA, AND
MADALINA CONDRESCU

*Department of Physiology
University of Medicine and Dentistry–
New Jersey Medical School
Newark, New Jersey 07103*

INTRODUCTION

Our laboratory has utilized transfected Chinese hamster ovary (CHO) cells expressing the bovine cardiac Na-Ca exchanger and various mutants created by site-directed mutagenesis to investigate the regulatory behavior of the exchanger. Since the parental CHO cells do not display Na-Ca exchange activity, we reasoned that the transfected cells would provide a versatile experimental system for investigating the role of exchange activity in Ca homeostasis and for clarifying the physiological importance of adenosine triphosphate (ATP)-dependent and Ca-dependent regulation of exchange activity. The results described below have indeed revealed new aspects of exchanger function in its interactions with the actin cytoskeleton and intracellular Ca stores. These findings have led to a better understanding of the impact of Na-Ca exchange activity on Ca homeostatic mechanisms. On the other hand, some of our observations are difficult to explain in conventional terms and reveal the limitations of our current understanding of Ca homeostasis.

Na-Ca Exchange in Transfected CHO Cells

CHO cells were transfected by the Ca phosphate method with a mammalian expression vector (pcDNA; Invitrogen) containing an insert coding for the bovine cardiac Na-Ca exchanger.[1,2] A clone of cells expressing high levels of Na-Ca exchange activity (CK1.4 cells) was used for most of the studies described here. The properties of these cells have been described in detail elsewhere.[2] A similar procedure was followed for obtaining CHO cells expressing altered exchangers created by site-directed mutagenesis. For control cells, either nontransfected parental cells or CHO cells transfected with the expression vector alone (*i.e.*, with no insert), were used; no differences have been observed between the nontransfected and vector-transfected control cells.

The Na-Ca exchanger can mediate net Ca movements either into or out of the cell depending upon the transmembrane gradients of Na and Ca and the membrane

---

[a] This work was supported by NIH Grant HL 49932.

potential. For detecting exchange activity in the transfected CHO cells, we elevate the cytosolic Na concentration ($[Na]_i$) by treating the cells with ouabain and subsequently monitor ^{45}Ca uptake upon reduction of $[Na]_o$. As shown in FIGURE 1, cells expressing the cardiac Na-Ca exchanger accumulate large amounts of Ca whereas vector-transfected control cells take up very little Ca under identical conditions. ^{45}Ca uptake by the transfected cells is greater in a low $[Na]_o$ medium than in a high $[Na]_o$ medium; this reflects the competition between Na and Ca for transport sites on the exchanger. In cells loaded with the Ca indicator dye fura-2, Ca accumulation during Na-Ca exchange is associated with an increase in cytosolic Ca ($[Ca]_i$).[2] In most instances, there is a reasonable correlation between the results of the ^{45}Ca uptake assay and those of fura 2-based assays. However, for some cells which express certain isoforms or mutants of the exchanger, Ca accumulation via Na-Ca exchange activity (measured as ^{45}Ca uptake) is associated with much smaller changes in $[Ca]_i$. These results, which will be presented elsewhere, imply that intracellular Ca often travels along pathways that are secluded from the bulk cytosol. These pathways are difficult to identify and may involve gradients of [Ca] in local regions of the cytosol that are not reflected in bulk cytosolic measurements using fura 2. A theme that emerges from the studies described below is that the local interactions between the exchanger and restricted cytosolic regions or specialized compartments play an important but poorly understood role in cellular Ca homeostasis.

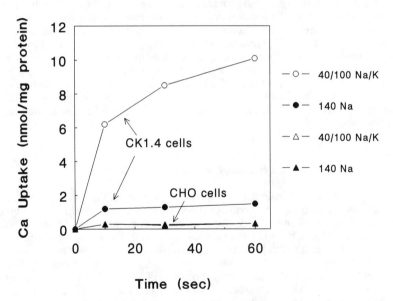

FIGURE 1. ^{45}Ca uptake by transfected and nontransfected CHO cells. Cells were pretreated for 30 min with 0.4 mM ouabain in nominally Ca-free physiological salts solution (PSS) containing (in mM) 140 NaCl, 5 KCl, 1 $MgCl_2$, 10 glucose and 20 Mops, buffered to pH 7.4 (37°C) with Tris. The cells were then assayed for ^{45}Ca uptake in PSS with or without partial substitution of NaCl with 100 mM KCl. The concentration of ^{45}Ca was 1 mM. ^{45}Ca uptake was terminated by washing the cells 3 times with 100 mM $MgCl_2$ containing 10 mM $LaCl_3$ and 5 mM Mops, pH 7.4. Results are shown for transfected cells permanently expressing the bovine cardiac Na-Ca exchanger (CK1.4 cells) and for nontransfected CHO cells.

Regulation of Na-Ca Exchange Activity

The regulation of Na-Ca exchange activity has been extensively studied in dialyzed squid axons and sarcolemmal patches, but its role in cellular physiology remains unclear. Two distinct but interacting regulatory processes have been identified for the cardiac Na-Ca exchanger. The two processes involve different inactive states of the carrier:[3-7] the first inactive state is promoted by the presence of cytosolic Na (Na-dependent inactivation), and the second is promoted by the absence of cytosolic Ca (secondary Ca activation). The presence of cytosolic ATP attenuates Na-dependent inactivation and increases the affinity of the secondary activation site for cytosolic Ca. ATP also alters the kinetics of the Na-Ca exchange activity. In dialyzed squid giant axons, ATP reduces the K_m value for both cytosolic Ca and extracellular Na.[8] In adult heart myocytes[9,10] and cultured vascular smooth muscle cells[11] from the rat, ATP depletion using metabolic inhibitors reduces Ca entry via Na-Ca exchange by more than 80%.

The mechanism by which ATP regulates exchange activity is unknown. DiPolo and Beaugé have suggested that in squid giant axons, the exchanger is regulated by a phosphorylation process, although the kinases/phosphatases involved have not been identified.[12-14] The effects of ATP on exchange currents in cardiac sarcolemmal patches are not consistent with the phosphorylation hypothesis, however; these studies suggest that exchange activity is indirectly regulated by aminophospholipid translocase activity.[4] This ATP-dependent enzyme maintains a high density of phosphatidylserine at the cytosolic surface of the bilayer and it is thought that the asymmetric distribution of this lipid is essential for maximal exchange activity. It seems unlikely that either hypothesis by itself could fully explain the effects of ATP in intact cells, since changes in cellular ATP produce a broad spectrum of effects that could directly or indirectly influence exchange activity.

In transfected CHO cells, ATP depletion with mitochondrial inhibitors reduces both Ca influx and Ca efflux mediated by the exchanger,[15] in agreement with results obtained in other cell types. As discussed in detail in Ref. 15, these effects are not solely explained by changes in cytosolic pH or organellar Ca sequestration. Protein kinase or phosphatase inhibitors had no effect on exchange activity, nor could we detect any phosphorylation of the exchanger by immunoprecipitation techniques; thus, ATP-dependent regulation probably does not involve phosphorylation of the exchanger in these cells. In contrast to these findings, phosphorylation of the Na-Ca exchanger from rat aortic smooth muscle cells was recently reported.[16] Phosphorylation and exchange activity were both stimulated by phorbol ester and thrombin in the smooth muscle cells. The differences between the squid axon, smooth muscle and the cardiac results suggest that the mechanisms of regulation of Na-Ca exchangers may be highly tissue specific.

In our studies with CHO cells, we also observed that ATP depletion affects the competitive interaction between Na and Ca at the extracellular membrane surface. As shown in the left panel of FIGURE 2, Na is more effective in inhibiting the rate of ^{45}Ca uptake in ATP-depleted cells than in control cells; the half-inhibitory concentration of Na is reduced from 90 mM in control cells to 55 mM in ATP-depleted cells. The Hill coefficients for these data are approximately the same (2.0) under both conditions. This effect appears to be mediated by changes in the actin cytoskeleton, since it can be mimicked by treating the cells (under ATP-replete conditions) with cytochalasin D (right panel, FIG. 2). Both ATP depletion and cytochalasin D treatment cause an extensive breakdown of the polymerized actin network in these cells, as determined by staining with fluorescein isothiocyanate (FITC)-labelled phalloidin.[15] The results complement a recent re-

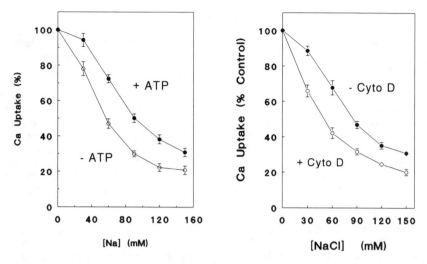

FIGURE 2. Effect of cellular ATP depletion on Na-Ca exchange activity. *Left panel.* CK1.4 cells were pretreated with ouabain in PSS as described in FIGURE 1 and then treated for an additional 10 min with 2.5 μg/ml oligomycin plus 2 μM rotenone. In the absence of glucose, but not in its presence, these agents reduce cellular ATP levels by more than 90% within 10 min. The cells were then assayed for ^{45}Ca uptake (15 sec) as described in FIGURE 1. Results are shown for cells incubated in the absence (*open circles*) or presence (*filled circles*) of 10 mM glucose. *Right panel.* Cells were pretreated for 1 hour with or without 1 μM cytochalasin D and assayed for Na-Ca exchange activity (15 sec ^{45}Ca uptake) as described in FIGURE 1. (From Condrescu et al.[15] Reprinted by permission from the *Journal of Biological Chemistry*.)

port by Philipson and his colleagues indicating that the Na-Ca exchanger interacts with the cytoskeletal protein ankyrin.[17]

It is important to note that ATP depletion and cytochalasin D treatment had no effect on the Na_o-inhibition profile in cells expressing a mutant exchanger in which 440 out of 520 amino acids within the exchanger's central hydrophilic domain had been deleted.[15] The Na_o-inhibition profile of the mutant cells resembled that of ATP-depleted wild-type cells, with a half-inhibitory Na concentration of 55 mM. The effects of ATP depletion and cytochalasin D were also greatly reduced in a second exchanger mutant. In this case, a string of acidic residues in the C-terminal portion of the hydrophilic domain (723-EDDDDDECGEE) were changed to alanines. This region is of particular interest because an even more extensive string of acidic residues is found in the same relative position in the retinal rod Na-Ca + K exchanger.[18] Our results suggest that this region may be involved, directly or indirectly, in the interaction of the exchanger with cytoskeletal elements.

In summary, the actin cytoskeleton appears to influence exchanger function, possibly by means of an interaction between ankyrin and regions of the central hydrophilic domain of the exchanger, as illustrated in FIGURE 3. The actin cytoskeleton is likely to emerge as a physiologically important regulator of Na-Ca exchange activity, as it is for several other carrier-mediated transport processes.[19,20]

Na-Ca Exchange Activity and Ca Mobilization

Agents which promote the production of 1,4,5-inositol trisphosphate (InsP$_3$) give rise to a biphasic increase in [Ca]$_i$. The initial, transient phase is primarily due to release of Ca from intracellular stores while the second phase is a sustained elevation of [Ca]$_i$ due to the influx of Ca from the exterior. The Ca influx pathway involves low conductance Ca channels and is activated, through a poorly understood mechanism, by the loss of Ca from the InsP$_3$-sensitive stores (reviewed in Ref. 21). This process is designated as store-dependent Ca influx, or SDCI. When cells are exposed to a Ca-mobilizing agent in the absence of extracellular Ca, the SDCI pathway remains activated (even after removal of the agent) until Ca is restored and the InsP$_3$-sensitive stores refill with Ca. Prolonged activation of SDCI can be brought about by inhibitors of the SERCA ATPase such as thapsigargin (Tg) which induce a permanent loss of Ca from the InsP$_3$-sensitive Ca stores.[22]

The effects of Tg in transfected CHO cells suggest that the Na-Ca exchanger interacts with the SDCI pathway. When cells are treated with Tg under Ca-free conditions, they exhibit a slowly developing [Ca]$_i$ transient due to release of Ca from intracellular stores. When extracellular Ca is subsequently restored to the Tg-treated cells, they exhibit a rapid rise in [Ca]$_i$ due to Ca entry via the SDCI pathway. FIGURE 4 compares the effects of extracellular Ca addition in Tg-treated

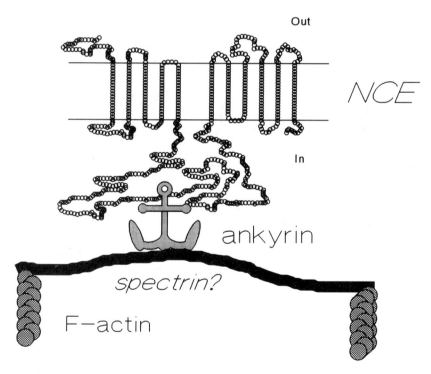

FIGURE 3. Diagram illustrating possible interactions of the Na-Ca exchanger with the actin cytoskeleton.

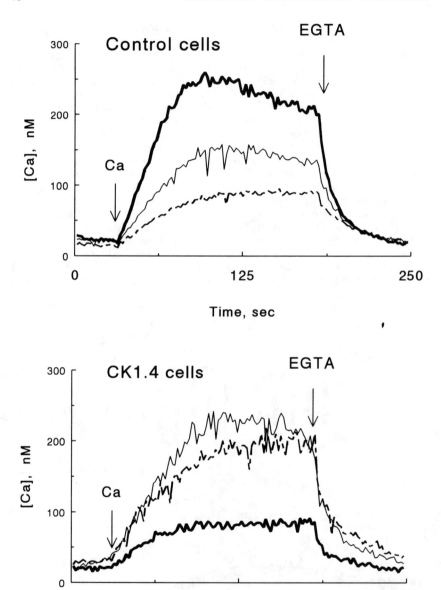

FIGURE 4. Ca entry by SDCI in Tg-treated CHO cells. *Upper panel.* Suspensions of vector-transfected control CHO cells were loaded with fura 2 (30 min; 3 μM fura 2-AM) and pretreated for 1 min with 200 nM Tg in Ca-Free PSS containing 0.3 mM EGTA. The cells were then centrifuged and added to a cuvette containing nominally Ca-free PSS with (*light traces*) or without (*bold traces*) substitution of NaCl with LiCl. $CaCl_2$ (1 mM) was added to the cuvette after 30 sec and 3 mM EGTA was added after 180 sec. $[Ca]_i$ was monitored using standard fura 2 procedures. For the data represented by the *dashed line*, 50 μM SK&F

CHO cells expressing the Na-Ca exchanger and in vector-transfected control cells. In these experiments, fura 2-loaded cells were pretreated for 1 min with 200 nM Tg in a Ca-free medium, and then added to a cuvette containing a nominally Ca-free physiological salts solution. After 30 seconds, 1 mM $CaCl_2$ was added to the cuvette. The data in the top panel of FIGURE 4 depict the results obtained with vector-transfected control cells. In these cells, addition of Ca results in a large increase in $[Ca]_i$ (bold trace) and this increase is markedly inhibited by the SDCI channel blocker SK&F 96365[23] (dashed trace). Removal of extracellular Na (Li replacement) attenuates the rise in $[Ca]_i$ by approximately 50% in these cells.

As shown by the data in the bottom panel of FIGURE 4, cells expressing the exchanger behave quite differently. These cells show a greatly attenuated increase in $[Ca]_i$ at physiological concentrations of Na (bold trace) compared to control cells. Moreover, if extracellular Na is replaced with Li (light trace) the rise in $[Ca]_i$ is dramatically enhanced and is relatively resistant to inhibition by SK&F 96365 (dashed trace). For the experiment shown in FIGURE 4, practically no inhibition by SK&F 96365 was observed, but in several other experiments, SK&F inhibited by approximately 50%. It should be noted that the increases in $[Ca]_i$ observed in these experiments are highly dependent upon pretreatment with Tg; if Tg treatment is omitted, the transfected cells in a Na-free medium show a slightly greater increase in $[Ca]_i$ than in a Na-containing medium, but this difference is small relative to that observed in FIGURE 4.

To explain these results, we suggest that most of the Ca entering the cell by the SDCI channels is transported back out of the cell by Na-Ca exchange, thereby attenuating the rise in $[Ca]_i$ at physiological concentrations of Na. Upon removal of Na_o, exchange-dependent Ca efflux would be blocked and Ca could enter the cell by either the SDCI pathway or the Na-Ca exchanger operating in its "reverse mode" (Na_i-dependent Ca influx). The results with the SDCI Ca channel blocker SK&F 96365 indicate that both pathways for Ca entry are operative under these conditions.

Thus, Ca efflux mediated by the Na-Ca exchanger appears to regulate net Ca entry via SDCI channels in the transfected cells. One might therefore expect to find that the efflux of Ca from the cells expressing the exchanger would be considerably more rapid than in nontransfected cells and would also be highly dependent upon the presence of Na_o. Curiously, however, this is not the case. Thus, when experiments similar to those in FIGURE 4 were carried out at several different concentrations of Na, we found that the rise in $[Ca]_i$ upon addition of Ca_o increased as $[Na]_o$ decreased, but that the rate of decline in $[Ca]_i$ when EGTA was added was approximately the same for the transfected and control cells, and was not markedly affected by the presence of Na.

The effect of Na on Ca efflux was examined more thoroughly in the experiment shown in FIGURE 5. Concentrated suspensions of transfected cells loaded with fura 2 were pretreated with Tg in a Na-free medium containing 1 mM Ca in order to elevate $[Ca]_i$. They were then diluted 30-fold into a cuvette containing a physiological salts solution with or without 140 mM NaCl (Li substitution) and $[Ca]_i$ was monitored continuously thereafter. With this protocol, we could demon-

---

96365, a blocker of SDCI Ca channels, was included in the Na-PSS. *Lower panel.* Transfected CHO cells expressing the bovine cardiac Na-Ca exchanger (CK1.4 cells) were treated as described for the upper panel. For the data indicated by the *dashed line*, 50 μM SK&F 96365 was included in the Li-PSS.

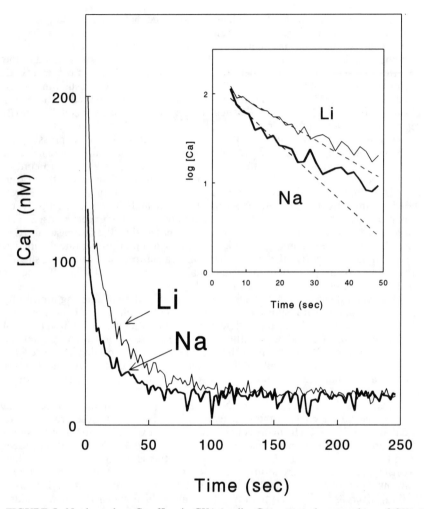

FIGURE 5. Na-dependent Ca efflux in CK1.4 cells. Concentrated suspensions of CK1.4 cells were loaded with fura 2 as described in FIGURE 4 and preincubated for 1 min in 100 μl of Li-PSS containing 1 mM $CaCl_2$ plus 200 nM Tg. The cells were then diluted into a cuvette containing 3 ml of PSS with or without substitution of NaCl with LiCl and $[Ca]_i$ was monitored continuously thereafter. The *inset* depicts the initial decline in $[Ca]_i$ as a first order plot. The half-time for the initial decline of $[Ca]_i$ in Na-PSS is 8.4 sec, compared to 13.4 sec in Li-PSS.

strate that the presence of Na does indeed accelerate the decline of $[Ca]_i$ in the transfected cells. As shown by the inset in FIGURE 5, the first order rate constant for the initial decline in $[Ca]_i$ is approximately 60% higher in the presence of Na than in its absence. When vector-transfected control cells were used in similar experiments, no difference was observed between efflux media with and without Na, and the rate of decline in $[Ca]_i$ was similar to that observed with the transfected cells in Na-free media (data not shown). Thus, Na stimulates Ca efflux in the transfected cells, as expected, but the amount of stimulation is surprisingly small given the dramatic effects of Na on $[Ca]_i$ in the Tg-treated cells.

These observations raise the possibility that the exchanger is more effective as a Ca efflux mechanism when the local concentration of cytosolic Ca is elevated due to Ca entry by SDCI than when Ca entry is blocked. Thus, as indicated in FIGURE 6, if the local concentration of Ca in the vicinity of the exchanger were substantially elevated over that of the bulk cytosol during SDCI, the exchanger could function very efficiently as a Ca extrusion mechanism because of its high turnover rate ($>1,000$ sec^{-1}).[24,25] In contrast, when Ca entry is blocked by the addition of EGTA, the local $[Ca]_i$ would decline rapidly to bulk cytosolic levels, where the exchanger, because of its relatively high K_m (4 μM),[25] would function much less efficiently. This hypothesis implies that the exchanger and the SDCI channels are either very close neighbors, or that they both influence $[Ca]_i$ within a cytosolic domain of restricted Ca diffusion. Focusing on the exchanger's "sphere of influence" at the local level provides new insights into possible mechanisms of regulating Ca homeostasis through Na-Ca exchange; it also suggests a teleological rationale for the kinetic properties of the exchanger, *i.e.*, its relatively low affinity for Ca and its high turnover rate.

Ca Compartmentation in CHO Cells

Transfected CHO cells accumulate 3–6 mmol of Ca per liter of cell water during Na-Ca exchange, assuming an intracellular water content of 6 μl/mg protein.[2] The rise in $[Ca]_i$ under similar conditions, as determined in fura 2-loaded cells, does not exceed 1 μM, indicating that less than 0.03% of the Ca entering the cells appears as free cytosolic Ca ($[Ca]_i$); the remainder must be either buffered by cytosolic constituents or sequestered by intracellular organelles. Recent experiments with bovine chromaffin cells indicate that 1–2% of the Ca entering these

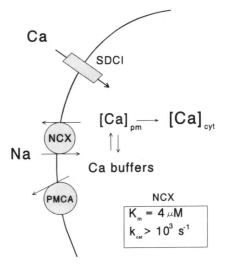

FIGURE 6. Diagram depicting possible relations between Ca entry via SDCI Ca channels and Ca exit via Na-Ca exchange. Diffusion of cytosolic Ca is slowed by interaction with endogenous Ca buffers, thereby promoting a local elevation of [Ca] in the vicinity of the exchanger ($[Ca]_{pm}$), which is substantially greater than [Ca] in the bulk cytosol ($[Ca]_{cyt}$). See text for additional details.

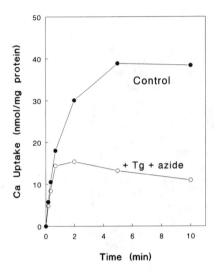

FIGURE 7. ^{45}Ca uptake in CK1.4 cells with and without inhibitors of organellar Ca sequestration. CK1.4 cells were pretreated with ouabain as described in FIGURE 1. During the final 15 min of the preincubation period, some of the cells were exposed to 50 nM Tg. ^{45}Ca uptake was assayed in PSS containing 40 mM NaCl plus 100 mM KCl. For the Tg-treated cells, 10 mM of the NaCl was substituted with 10 mM Na azide.

cells during Ca channel activity appears as free Ca and that, on a millisecond time scale, the rest is rapidly bound to endogenous buffers of low affinity.[26] If we assume a similar buffering ratio in CHO cells, the remaining disparity between the total Ca entering the cell and that appearing as free cytosolic Ca (*i.e.*, more than a factor of 30), must represent Ca that is sequestered in various cellular compartments or organelles.

Which cellular compartments are involved? The two most obvious candidates are the endoplasmic reticulum (ER) and the mitochondria. Inhibitors that block Ca sequestration by these organelles would therefore be expected to inhibit ^{45}Ca uptake by the transfected cells. Indeed, as shown by the data in FIGURE 7, ^{45}Ca uptake by transfected cells is inhibited approximately 60% by the combination of Tg and 10 mM Na azide, agents which inhibit Ca accumulation by the ER and the mitochondria, respectively. However, even in the presence of these inhibitors, cellular ^{45}Ca uptake is equivalent to approximately 2 mmol/l cell water. This figure indicates that there are other compartments within the cell that sequester substantial quantities of Ca under these conditions. The inhibition of ^{45}Ca uptake shown in FIGURE 7 is due entirely to the presence of azide; Tg alone does not inhibit total ^{45}Ca uptake (cf. FIG. 8), and in fact enhances ^{45}Ca uptake under certain conditions.[27] Other mitochondrial inhibitors (*e.g.*, 10 μM Cl-CCP and 2.5 μg/ml oligomycin plus 2 μM rotenone) are as effective as azide in inhibiting ^{45}Ca uptake, suggesting that each of these agents totally blocks mitochondrial Ca accumulation in these cells.

These results imply that Ca buffering and/or Ca sequestration by the ER and the mitochondria account for only a fraction of the Ca accumulated by the cells during Na-Ca exchange. However, the use of inhibitors does not necessarily provide an accurate index of organellar Ca sequestration in intact cells, since they are likely to induce shifts of intracellular Ca to other compartments. For example, Tg actually enhances ^{45}Ca uptake during Na-Ca exchange by shifting Ca to the mitochondria.[27] An even more striking example is described below, which suggests that some of the "intracellular" Ca measured in the presence of mitochondrial inhibitors exchanges very rapidly with the external medium and may in fact be in an extracellular location.

In the presence of Na azide or other mitochondrial inhibitors, approximately half of the ^{45}Ca taken up by Na-Ca exchange in the transfected cells is lost within 5–10 sec of washing the cells. This rapid efflux of ^{45}Ca is blocked by La at concentrations greater than 0.3 mM and can be revealed as the difference in cell-associated ^{45}Ca with and without La in the wash medium. An example of an experiment demonstrating this effect is shown in FIGURE 8. Note that in the absence of 10 mM Na azide, the amount of ^{45}Ca in the "rapid efflux" compartment (given as Δ in FIG. 8) is less than in the presence of azide and represents only a small fraction of the total ^{45}Ca accumulated by the cell. In the presence of mitochondrial inhibitors, this "rapid efflux" compartment contains a substantial amount of Ca (>1 mmol/liter cell water). Given the rapidity of its loss from the cell, it may reflect Ca that has already been extruded from the cell and remains associated with an extracellular compartment that can be "sealed off" by La. The identity of this compartment is completely unknown although several possibilities are suggested below.

We offer the following speculative explanation of these results: Much of the ^{45}Ca that enters the cell via "reverse" Na-Ca exchange is accumulated by mitochondria. Adding mitochondrial inhibitors therefore elevates $[Ca]_i$ and accelerates Ca efflux from the cell, resulting in a partial inhibition of ^{45}Ca uptake (cf. FIG. 7). A portion of the Ca that has been removed from the cell remains transiently associated with the cell surface and can be trapped by the addition of La. La binds tightly to negatively charged groups on the cell surface and might seal off invaginated surface structures such as caveolae, increase the resistance to outward diffusion of Ca through the glycocalyx, or trap Ca within the space between the cell monolayer and the plastic surface of the culture dish. Since our normal assay

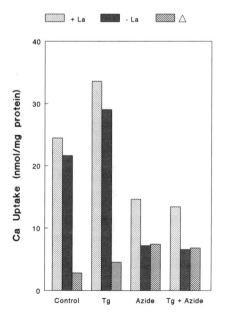

FIGURE 8. ^{45}Ca uptake by CK1.4 cells with and without La in the wash medium. Cells were pretreated with 0.4 mM ouabain and assayed for ^{45}Ca uptake (10 min) in PSS containing 40 mM NaCl plus 100 mM KCl as described in FIGURE 1. Results are shown for cells terminated with the normal termination medium (100 mM $MgCl_2$, 10 mM $LaCl_3$) or termination medium in which the $LaCl_3$ was omitted. The difference in ^{45}Ca uptake with and without La is indicated by the *cross-hatched bars* indicated by Δ at the top of the figure.

protocol uses a termination medium containing 10 mM La, some of the ^{45}Ca uptake that we measure might include Ca that has already been recycled out of the cell. Under normal assay conditions, the fraction of the total cellular Ca in this external compartment is relatively small, but it can increase markedly when intracellular Ca sequestration is blocked.

SUMMARY AND CONCLUSIONS

Our experiments with transfected cells provide new insights into the role of Na-Ca exchange activity in Ca homeostasis and emphasize the role of local interactions in determining exchanger function. Thus, the effects of ATP depletion and cytochalasin D highlight the influence of the actin cytoskeleton in regulating exchange activity. Cytoskeletal interactions could provide a mechanism for modulating exchange activity by mechanical stretch and might constitute a novel feedback mechanism for regulating contractile activity in the heart. The effects of Na on Ca entry during SDCI in the transfected cells suggest that local gradients of $[Ca]_i$ are important determinants of exchanger function. The surface distribution of exchanger proteins in relation to that of Ca channels therefore represents another area in which interactions with the cytoskeleton may be a central element in understanding the physiological function(s) of the exchange activity. At present, it seems likely that the exchanger's central hydrophilic domain mediates the connection between the exchanger and the cytoskeleton. This provides a rationale for understanding the importance of tissue-specific alterations in the exchanger's hydrophilic domain, which appear to have little affect on the kinetic behavior of the exchanger. Future work in our laboratory will be directed toward clarifying the role of cytoskeletal interactions in exchanger function.

ACKNOWLEDGMENTS

We thank Drs. Jeffrey P. Gardner and Abraham Aviv for stimulating discussions and helpful advice during the course of this work.

REFERENCES

1. ACETO, J. F., M. CONDRESCU, C. KROUPIS, H. NELSON, N. NELSON, D. NICOLL, K. D. PHILIPSON & J. P. REEVES. 1992. Cloning and expression of the bovine cardiac sodium-calcium exchanger. Arch Biochem. Biophys. **298**: 553–560.
2. PIJUAN, V., Y. ZHUANG, L. SMITH, C. KROUPIS, M. CONDRESCU, J. F. ACETO, J. P. REEVES & J. B. SMITH. 1993. Stable expression of the cardiac sodium-calcium exchanger in CHO cells. Am. J. Physiol. **246**: C1006–C1074.
3. COLLINS, A., A. V. SOMLYO & D. W. HILGEMANN. 1992. The giant cardiac membrane patch method: stimulation of outward Na/Ca exchange current by MgATP. J. Physiol. **454**: 27–57.
4. HILGEMANN, D. & A. COLLINS. 1992. Mechanism of cardiac Na/Ca exchange current stimulation by MgATP: possible involvement of aminophospholipid translocase. J. Physiol. **454**: 59–82.
5. HILGEMANN, D. W., S. MATSUOKA, G. A. NAGEL & A. COLLINS. 1992. Steady-state and dynamic properties of cardiac sodium-calcium exchange. Sodium-dependent activation. J. Gen. Physiol. **100**: 905–932.
6. HILGEMANN, D. W., A. COLLINS & S. MATSUOKA. 1992. Steady-state and dynamic

properties of cardiac sodium-calcium exchange. Secondary modulation by cytoplasmic calcium and ATP. J. Gen. Physiol. **100:** 933–961.
7. DiPolo, R. & L. Beaugé. 1987. Characterization of the reverse Na/Ca exchange in squid axons and its modulation by Ca_i and ATP. J. Gen. Physiol. **90:** 505–525.
8. Blaustein, M. P. 1977. Effects of internal and external cations and of ATP on sodium-calcium exchange in squid axons. Biophys. J. **20:** 79–111.
9. Haworth, R. A., A. B. Goknur, D. R. Hunter, J. O. Hegge & H. A. Berkoff. 1987. Inhibition of calcium influx in isolated adult rat heart cells by ATP depletion. Circ. Res. **60:** 586–594.
10. Haworth, R. A. & A. B. Goknur. 1992. ATP dependence of calcium uptake by the Na-Ca exchanger of adult heart cells. Circ. Res. **71:** 210–217.
11. Smith, J. B. & L. Smith. 1990. Energy dependence of sodium-calcium exchange in vascular smooth muscle cells. Am. J. Physiol. **252:** C302–C309.
12. DiPolo, R. & L. Beaugé. 1991. Regulation of Na-Ca exchange. An overview. Ann. N. Y. Acad. Sci. **639:** 100–111.
13. DiPolo, R. & L. Beaugé. 1993. Effects of some metal-ATP complexes on Na-Ca exchange in internally dialyzed squid axons. J. Physiol. **462:** 71–86.
14. DiPolo, R. & L. Beaugé. 1994. Effects of vanadate on MgATP stimulation of Na-Ca exchange support kinase-phosphatase modulation in squid axons. Am. J. Physiol. **266:** C1382–C1391.
15. Condrescu, M., J. P. Gardner, G. Chernaya, J. F. Aceto, C. Kroupis & J. P. Reeves. 1995. ATP-dependent regulation of sodium-calcium exchange in CHO cells transfected with the bovine cardiac sodium-calcium exchanger. J. Biol. Chem. **270:** 9137–9146.
16. Iwamoto, T., S. Wakabayashi & M. Shigekawa. 1995. Growth factor-induced phosphorylation and activation of aortic smooth muscle Na/Ca exchanger. J. Biol. Chem. **270:** 8996–9001.
17. Li, Z., E. P. Burke, J. S. Frank, V. Bennett & K. D. Philipson. 1993. The cardiac Na-Ca exchanger binds to the cytoskeletal protein ankyrin. J. Biol. Chem. **268:** 11489–11491.
18. Reiländer, H., A. Achilles, U. Friedel, G. Maul, F. Lottspeich & N. J. Cook. 1992. Primary structure and functional expression of the Na/Ca,K exchanger from bovine rod photoreceptors. EMBO J. **11:** 1689–1695.
19. Mills, J. W. & L. J. Mandel. 1994. Cytoskeletal regulation of membrane transport events. FASEB J. **8:** 1161–1165.
20. Goss, G. G., M. Woodside, S. Wakabayashi, J. Pouysségur, T. Waddell, G. P. Downey & S. Grinstein. 1994. ATP dependence of NHE-1, the ubiquitous isoform of the Na/H antiporter. Analysis of phosphorylation and subcellular localization. J. Biol. Chem. **269:** 8741–8748.
21. Pozzan, T., R. Rizzuto, P. Volpe & J. Meldolesi. 1994. Molecular and cellular physiology and intracellular calcium stores. Physiol. Rev. **74:** 595–636.
22. Takemura, H., A. R. Huges, O. Thastrup & J. W. Putney, Jr. 1989. Activation of calcium entry by the tumor promoter thapsigargin in parotid acinar cells. J. Biol. Chem. **264:** 12266–12271.
23. Merritt, J. E., W. P. Armstrong, C. D. Benham, T. J. Hallan, R. Jacob, A. Taxa-Chamiec, B. K. Leigh, S. A. McCarthy, K. E. Moores & T. J. Rink. 1990. SK&F 96365, a novel inhibitor of receptor-mediated calcium entry. Biochem. J. **271:** 515–522.
24. Cheon, J. & J. P. Reeves. 1988. Site density of the sodium-calcium exchange carrier in reconstituted vesicles from bovine cardiac sarcolemma. J. Biol. Chem. **263:** 2309–2315.
25. Hilgemann, D. W., D. A. Nicoll & K. D. Philipson. 1991. Charge movement during Na translocation by native and cloned cardiac Na/Ca exchanger. Nature **352:** 715–718.
26. Neher, E. & G. J. Augustine. 1992. Calcium gradients and buffers in bovine chromaffin cells. J. Physiol. **450:** 273–301.
27. Reeves, J. P., G. Chernaya, J. R. Aceto, J. P. Gardner & M. Condrescu. 1994. Thapsigargin (Tg) inhibits calcium recycling in CHO cells transfected with the bovine cardiac sodium-calcium exchanger. Biophys. J. **66:** A148.

Mutagenesis Studies of the Cardiac Na$^+$-Ca^{2+} Exchanger

D. A. NICOLL, L. V. HRYSHKO,[a] S. MATSUOKA,[b]
J. S. FRANK, AND K. D. PHILIPSON[c]

*Departments of Physiology and Medicine
and the
Cardiovascular Research Laboratories
University of California, Los Angeles
School of Medicine
Los Angeles, California 90095-1760*

The cardiac Na$^+$-Ca^{2+} exchanger exchanges 3 Na$^+$ ions with 1 Ca^{2+} ion in an electrogenic manner. Three considerations indicate a role for transmembrane amino acid residues in the binding and transport of ions. First, transport of ions through the hydrophobic core of the membrane must involve the hydrophilic portions of transmembrane amino acids. Second, ion binding to the exchanger has been determined to be voltage dependent[1] and hence involving residues in the transmembrane portion of the exchanger. Finally, proteolysis[2] and deletion mutagenesis[3] experiments have demonstrated that exchanger activity can take place in the presence of only the two transmembrane domains (containing five and six transmembrane segments, respectively) with the large hydrophilic domain between transmembrane segments 5 and 6 deleted.

We are attempting to define, by site-directed mutagenesis, portions of the exchanger and specific amino acids in the transmembrane segments which are involved in ion binding and transport. Specific mutations have been made and the mutant exchangers expressed in *Xenopus* oocytes. Exchanger activity was examined by measuring ^{45}Ca fluxes into oocytes and exchanger-associated currents.

We have used a number of criteria in selection of amino acids to mutate, since the exchanger is modelled to contain 234 amino acids in 11 transmembrane segments (FIG. 1). Fortunately, nature has already constructed a number of "mutations" in the transmembrane segments of the exchanger. In addition to NCX1 (the cardiac-type exchanger),[4] a second mammalian exchanger, NCX2 (the brain-type exchanger),[5] has been identified and more recently an exchanger from *Drosophila*, Calx,[6,7] has been cloned and sequenced and shown to be a Na$^+$-Ca^{2+} exchanger[6] with different properties of regulation compared to NCX1 and NCX2.[8] Between NCX1 and NCX2 or Calx there are 98 differences in amino acids in the proposed transmembrane segments. The nonidentical residues are less likely to be involved in ion binding or transport. However, most of the amino acid substitu-

---

[a] Current address: Division of Cardiovascular Sciences, St. Boniface Hospital Research Centre, Winnipeg, Manitoba, Canada R2G 2A6.
[b] Current address: Department of Physiology, Faculty of Medicine, Kyoto University, Kyoto, 606-01 Japan.
[c] Correspondence address: Cardiovascular Research Laboratories, MRL Bldg. 3645, UCLA School of Medicine, Los Angeles, CA 90095-1760.

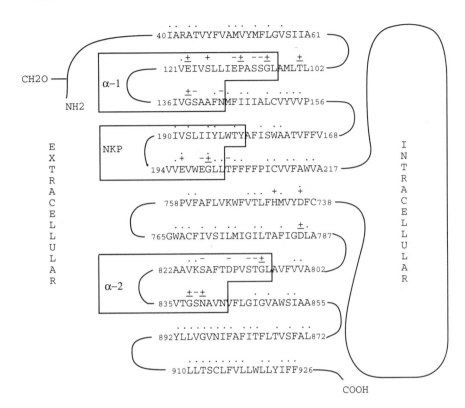

FIGURE 1. A model for the canine cardiac Na^+-Ca^{2+} exchanger. Residues modelled to be transmembrane are shown with the first putative transmembrane segment at the *top* of the figure and the last at the *bottom*. Amino acids which are not conserved between the cadiac,[4] brain[5] and *Drosophila*-type[7] exchangers are indicated with a *dot* above the residue. Residues which have been mutated are indicated with a − for inactive mutants, ± for mutants with reduced activity and + for mutants with wild-type levels of activity. Also shown are regions corresponding to the α-1, α-2 repeats and Na^+,K^+ pump regions.

tions are conservative. Also, the differences between NCX1 and NCX2 or Calx are not isolated, single amino acid changes but a number of changes. In other proteins a single amino acid mutation can destroy activity and a second mutation at another site can restore it.[9]

The natural "mutations" of the exchangers reduce the number of probable residues involved in ion binding and transport from 234 to 135. We have further reduced our search for the functionally important residues of the transmembrane segments by examining residues in regions that are conserved in other transport proteins and acidic and basic residues.

The exchanger transmembrane domains have three regions of sequence similarity to other transport proteins. The first described was a region of similarity to the Na^+,K^+ pump.[4] This region encompasses putative transmembrane segments 4 and 5 (FIG. 1). In the Na^+,K^+ and Ca^{2+} pumps the analogous region has been determined to be involved in ion binding.[10,11] The other two regions of sequence similarity are to two regions of the Na^+-Ca^{2+},K^+ exchanger of the photoreceptor cell (RetX).[12] These regions encompass transmembrane segments 2–3 and 8–9 of NCX1 (FIG. 1, α-1 and α-2). A functional role for the regions in RetX has not been defined. The striking degree of conservation between RetX and NCX1 at these regions, as well as a recent observation by Schwartz and Benzer[6] that the two regions (designated α-1 and α-2) show sequence similarity and are likely to be the result of a gene duplication event, indicate the importance of the regions in exchanger function.

METHODS

Mutations were introduced into cassette portions of the exchanger using Sculptor *In Vitro* Mutagenesis System by Amersham Corporation (Arlington Heights, IL). Cassettes were completely sequenced and subcloned into the full-length exchanger. cRNA was synthesized after linearizing the plasmid DNA at a HindIII site in the 3'-noncoding region. The T3 promotor was used for RNA synthesis with T3 mMessage mMachine from Ambion, Inc. (Austin, TX). cRNA was injected into *Xenopus* oocytes. Three to five days after injection, exchanger activity was analyzed by determining the Na^+-dependent uptake of ^{45}Ca.[13] In some cases oocytes were also examined using the giant excised patch technique.[14]

RESULTS AND DISCUSSION

FIGURE 1 and TABLE 1 summarize the results from mutations in putative transmembrane segments of the canine cardiac Na^+-Ca^{2+} exchanger. Activity was very sensitive to changes in the α-repeat and Na^+,K^+-pump regions. All of the inactive mutants and all but two of the mutants with low activity can be found in these regions. Only one mutated exchanger with an α-repeat mutation (S117A) had wild-type levels of activity. In the Na^+,K^+-pump region, mutations at residues E199 or T203 resulted in inactive exchangers, mutation at G200 resulted in an exchanger with moderate levels of activity and mutation at residue E196 resulted in an exchanger with wild-type levels of activity. On the other hand, mutation of charged residues in other regions of the exchanger all resulted in active exchangers (D740, H744, D785). Two of these mutants, D740N and H744W, had high levels of activity, and two mutants, D785E and D785N, had reduced levels of activity.

All inactive mutants were examined by Western blot and immunofluorescence (not shown). In all cases, exchanger protein was synthesized and expressed at the oocyte cell surface. Therefore, loss of activity following mutagenesis was probably not due to lack of expression or to gross structural changes in the exchanger protein.

Mutations at Serines and Threonines

Half of the inactive exchangers were mutated at the residue serine (six of twelve mutants) and only one of the serines which was mutated (S117) yields a fully functional exchanger. Residues S109 and S110 were mutated to several different amino acids to examine the relative importance of side chain chemical

TABLE 1. Summary of Exchanger Transmembrane Mutations

Extent of Activity	Location[a]	Residue Mutated	Mutant Amino Acid[b]	Activity [%WT ± SE, (n)]
Inactive	α-1	S109	A	2 ± 0 (4)
	α-1	S109	G	6 ± 2 (2)
	α-1	S110	A	2 ± 0 (2)
	α-1	S110	T	3 ± 2 (3)
	α-1	S110	C	6 ± 2 (3)
	α-1, AB	E113	D	5 ± 3 (3)
	α-1, AB	E113	Q	9 ± 0 (2)
	α-1	S139	A	4 ± 1 (5)
	α-1	N143	V	2 ± 0 (10)
	NKP, AB	E199	Q	3 ± 1 (7)
	NKP, AB	E199	D	4 ± 1 (3)
	NKP, AB	T203	V	9 ± 1 (2)
	α-2	T810	A	3 ± 1 (5)
	α-2	S811	T	2 ± 1 (3)
	α-2, AB	D814	N	4 ± 1 (2)
	α-2	S818	A	8 ± 2 (6)
	α-2	S838	A	5 ± 1 (5)
Low	α-1	T103	V	22 ± 5 (8)
	α-1	P112	A	18 ± 3 (9)
	α-1	G138	S	19 ± 4 (5)
	AB	D785	N	11 ± 3 (8)
	AB	D785	E	28 ± 11 (3)
	α-2	G809	A	32 ± 8 (8)
	α-2	N842	V	29 ± 5 (6)
	α-2	N842	D	35 ± 2 (4)
Moderate	α-1	G108	A	52 ± 10 (3)
	α-1, AB	E120	Q	64 ± 28 (2)
	α-1	G138	A	53 ± 13 (2)
	NKP	G200	A	60 ± 31 (5)
	α-2	G837	A	50 ± 6 (3)
High	α-1	S117	A	111 ± 24 (7)
	NKP, AB	E196	Q	96 ± 4 (2)
	AB	D740	N	108 ± 48 (5)
	AB	H744	W	79 ± 14 (4)

[a] α-1, α-1 repeat region. α-2, α-2 repeat region. NKP, Na^+,K^+ pump region. AB, acidic or basic residue.
[b] Single letter amino acid code.

group and size at this location. Mutants at S109 had no activity when serine was substituted with either glycine or alanine, both with side groups which are smaller than serine. This indicates that position 109 has a minimum size requirement. Residue S110 was mutated to alanine, threonine, or cysteine. Neither alanine, with a smaller side group, nor threonine, which has a larger side group but maintains the hydroxyl group, nor cysteine, which is nearly the same size as serine and has a thiol group that shares some of the chemical properties of hydroxyl groups, could substitute for serine. This indicates a stringent requirement for serine at this location in the exchanger.

Mutations at another hydroxyl containing amino acid, threonine, also had large effects. Mutation at two of the three threonines (T203, T810) resulted in loss of activity. When the third threonine (T103) was mutated, the resultant exchanger expressed low apparent levels of activity.

Mutations at Aspartate and Glutamate Residues

Exchanger activity was also sensitive to mutation of acidic amino acids when the mutated residue is modelled to be centrally located in the transmembrane α-helix. Residues E113, E199, and D814 are all modelled to be near the center of their respective helices, and mutations at each of the residues results in inactive exchangers. Two of the residues, E113 and E199, have been conservatively mutated to both a neutral (asparagine) and an acidic (aspartate) amino acid. Neither amino acid substitution could prevent loss of exchanger activity. Therefore, at positions 113 and 199, both the chemical nature and size of the amino acid side group is important in maintaining exchanger activity.

Residue D785 is modelled to be near the cytoplasmic surface of transmembrane segment 7. Mutations at this residue reduce the apparent activity to 28% of wild-type for D785E and 11% for D785N. There is higher apparent activity observed when D785 is mutated to another acidic residue compared to when it is mutated to a neutral residue of nearly the same size. This suggests a greater importance of the chemical nature of the side group as compared to the size of the side group for D785.

Residues E120, E196, and D740 are all modelled to be near the surface of their respective α-helices and each can be mutated to a neutral amino acid and still maintain moderate to high levels of exchanger activity. Also residues E196 and D740 are not conserved among the different exchangers. At the E196 position, there is a glutamine in NCX2[5] and a leucine in Calx.[7] These amino acids (glutamate, glutamine and leucine) have similar accessible surface areas (190, 180 and 170 Å^2, respectively) and may reflect an importance for size at this position. At residue D740, there is a serine residue in Calx.[7] The surface areas of residues which can be located at this position is more varied (asparagine = 160, aspartate = 150, and serine = 115Å^2). This suggests that size is not an important determinant at this site.

Mutation of Glycine Residues

All of the glycine residues which were mutated (G108, G138, G809, G837 and G200) resulted in functional exchangers with low to moderate levels of activity. Each of the glycine residues, except G200, is located at the adjacent, upstream side of a mutation-sensitive serine (or threonine) residue. Mutation at G138 or G837 alters the current-voltage (IV) relationship of the exchanger (FIG. 2). The IV relationship for the wild-type exchanger tends to be linear although on occasion (<25%) a slight upward curvature is observed at high voltage. Mutants G138A or G837A display a pronounced upward curvature of the IV relationship relative to the wild-type exchanger. A double mutant, G138A,G837A displays an even greater curvature. According to the models of Matsuoka and Hilgemann[1] both the Na^+-translocation pathway and, to a much lesser extent, the Ca^{2+}-occlusion step of the exchanger are voltage dependent. The change in IV relationship upon mutating glycine residues indicates that these regions are involved in ion transport.

Residue G200 is a nonconserved residue located in the Na^+,K^+-pump region of the exchanger. Following mutation of G200 to an alanine, the exchanger activity

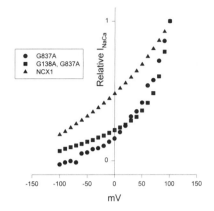

FIGURE 2. Current-voltage relationships for wild-type, mutant G837A, and double mutant G138A,G837A exchangers. Current-voltage relationships, measured as outward currents in giant excised patches of oocyte membranes, are the average for 6 measurements and have been normalized to current at 100 mV. ▲ Wild-type cardiac exchanger. ● Mutant G837A. ■ Double mutant G138A,G837A. The IV curve for mutant G138A was similar to that for mutant G837A (not shown).

is slightly reduced, thereby confirming the relative lack of importance of the residue in exchanger activity.

Importance of α-Repeat Regions in Exchanger Activity

A number of observations point to the importance of the α-repeat regions in exchanger function. The first is the surprising extent of conservation of amino acid sequence both between exchangers of other species or tissues as well as the conservation between the sequences of α-1 and α-2 within a given exchanger. The α-repeats are among the most highly conserved regions between the NCX1, NCX2, and Calx exchangers,[6] and are also the only regions of significant similarity between the Na^+-Ca^{2+} and Na^+-Ca^{2+},K^+ exchangers.[12]

A second observation which points to the importance of the α-repeats is the parallel effects of mutation at corresponding pairs of residues in the α-1, α-2 repeats. For example, the pair G108, G809 occupy similar locations in the α-1 and α-2 repeats and mutations of either of these residues results in an exchanger with reduced apparent activity. Likewise the pair S110, S811 occupy similar locations and are similarly affected by mutation.

Residues G138 and G837, which are also a corresponding pair in the α-repeats, both display altered IV relationships when mutated. The altered IV relationships seen in these mutants and the mutant G138A,G837A are highly suggestive of an important role for these amino acids and the α-repeats in exchanger function.

Importance of the Na^+,K^+-Pump Region in the Exchanger

As with the α-repeats regions, the conservation of this region among both exchangers and pumps indicates that this region is important in exchanger function. Mutagenesis of residues in this region lends further support for this theory. Mutation at the nonconserved residues E196 and G200 has little apparent effect on exchanger activity. Mutation at the conserved residues E199 and T203 eliminates exchanger activity. Also, the E199-equivalent residues in the Na^+,K^+ pump and in the Ca^{2+} pump have been shown to be involved in ion transport.[10,11]

Importance of Acidic Residues

Mutation of acidic residues which are modelled to be near the center of the membrane spanning segments results in inactive exchangers. Mutation of acidic residues modelled to be near the outside of the transmembrane segments has either no effect or decreases the apparent level of exchanger activity. These results suggest that some of the acidic residues may be involved in aspects of exchanger function.

REFERENCES

1. MATSUOKA, S. & D. W. HILGEMANN. 1992. Steady-state and dynamic properties of cardiac sodium-calcium exchange. Ion and voltage dependencies of the transport cycle. J. Gen. Physiol. **100:** 963–1001.
2. HILGEMANN, D. W. 1990. Regulation and deregulation of cardiac Na^+-Ca^{2+} exchange in giant excised sarcolemmal membrane patches. Nature **344:** 242–245.
3. MATSUOKA, S., D. A. NICOLL, R. F. REILLY, D. W. HILGEMANN & K. D. PHILIPSON. 1993. Initial localization of regulatory regions of the cardiac sarcolemmal Na^+-Ca^{2+} exchanger. Proc. Natl. Acad. Sci. USA **90:** 3870–3874.
4. NICOLL, D. A., S. LONGONI & K. D. PHILIPSON. 1990. Molecular cloning and functional expression of the cardiac sarcolemmal Na^+-Ca^{2+} exchanger. Science **250:** 562–565.
5. LI, Z., S. MATSUOKA, L. V. HRYSHKO, D. A. NICOLL, M. M. BERSOHN, E. P. BURKE, R. P. LIFTON & K. D. PHILIPSON. 1994. Cloning of the NCX2 isoform of the plasma membrane Na^+-Ca^{2+} exchanger. J. Biol. Chem. **269:** 17434–17439.
6. SCHWARZ, E. & S. BENZER. 1993. Expression and evolution of *CALX*, a sodium-calcium exchanger of *Drosophila melanogaster*. *In* Abstracts of papers presented at the 1993 meeting on neurobiology of *Drosophila*: Oct. 6–10, 1993:128. Cold Spring Harbor Laboratory Press. Plainview, NY.
7. VALDIVIA, C., P. KOFUJI, M. KIRBY, W. J. LEDERER & D. H. SCHULZE. 1995. Sequence analysis of a *Drosophila* Na/Ca exchanger. Biochim. Biophys. Acta. In press.
8. HRYSHKO, L. V., D. A. NICOLL, S. MATSUOKA, J. N. WEISS, E. SCHWARZ, S. BENZER & K. D. PHILIPSON. 1995. Anomalous regulation of the Na^+-Ca^{2+} exchanger from *Drosophila*. Biophys. J. **68:** A410.
9. DUNTEN, R. L., M. SAHIN-TÓTH & H. R. KABACK. 1993. Role of charge pair aspartic acid-237-lysine-358 in the lactose permease of *Escherichia coli*. Biochemistry **32:** 3139–3145.
10. LINGREL, J. B. & T. KUNTZWEILER. 1994. Na^+,K^+-ATPase. J. Biol. Chem. **269:** 19659–19662.
11. SKERJANC, I. S., T. TOYOFUKU, C. RICHARDSON & D. H. MACLENNAN. 1993. Mutation of glutamate 309 to glutamine alters one Ca^{2+}-binding site in the Ca^{2+}-ATPase of sarcoplasmic reticulum expressed SF9 cells. J. Biol. Chem. **268:** 15944–15950.
12. REILÄNDER, H., A. ACHILLES, U. FRIEDEL, G. MAUL, F. LOTTSPEICH & N. J. COOK. 1992. Primary structure and functional expression of the Na/Ca,K-exchanger from bovine rod photoreceptors. EMBO J. **11:** 1689–1695.
13. LONGONI, S., M. J. COADY, T. IKEDA & K. D. PHILIPSON. 1988. Expression of cardiac sarcolemmal Na^+-Ca^{2+} exchange activity in *Xenopus laevis* oocytes. Am. J. Physiol. **C255:** 870–873.
14. HRYSHKO, L. V., D. A. NICOLL, J. N. WEISS & K. D. PHILIPSON. 1993. Biosynthesis and initial processing of the cardiac sarcolemmal Na^+-Ca^{2+} exchanger. Biochim. Biophys. Acta **1151:** 35–42.

Antisense Oligodeoxynucleotides Directed against the Na-Ca Exchanger mRNA

Promising Tools for Studies on the Cellular and Molecular Level[a]

ERNST NIGGLI,[b,d] BEAT SCHWALLER,[c]
AND PETER LIPP[b]

[b]Department of Physiology
University of Bern
3012 Bern, Switzerland

[c]Department of Histology and General Embryology
University of Fribourg
1705 Fribourg, Switzerland

INTRODUCTION

It is well established that the Na-Ca exchange is essential for the Ca^{2+} homeostasis in many different cell types. However, functional studies of the Na-Ca exchange have been notoriously difficult on the cellular as well as on the molecular level. Particularly in cells exhibiting complex Ca^{2+} signaling systems with several different pathways for Ca^{2+} entry and removal, it has been very demanding if not impossible to sort out the physiological role of the Ca^{2+} fluxes via Na-Ca exchange. In this regard, cardiac muscle preparations are particularly challenging for the experimenter. Although cardiac muscle cells are a rich source of Na-Ca exchanger, studies on the physiological role of this transporter have been complicated by the complexity of cardiac Ca^{2+} signaling. One of the major reasons for these difficulties, if not the most important one, is the lack of a specific inhibitor for the Na-Ca exchange. Despite a significant effort undertaken by several research laboratories and pharmaceutical companies, no specific pharmacological inhibitor for the Na-Ca exchange has been discovered yet. Potential therapeutical benefits of Na-Ca exchange inhibition in man have been discussed, especially with regard to improvements of cardiac muscle function.[1] For experimental *in vitro* studies, a number of inorganic (*e.g.*, Ni^{2+}, La^{3+}) and organic inhibitors (*e.g.*, amiloride, dichlorobenzamil DCB) have been used,[2,3] but all compounds have turned out to be unspecific. It is particularly unfortunate that all presently known inhibitors also block Ca^{2+}-channels and/or K^+-channels.[4] Therefore, these inhibitors cannot be used to separate different Ca^{2+} signal transduction pathways and to elucidate

---

[a] This work was supported by a grant from the Swiss National Science Foundation to E.N. (3100-037417.93/1).
[d] Address for reprint requests and correspondence: Ernst Niggli, Department of Physiology, University of Bern, Bühlplatz 5, 3012 Bern, Switzerland.

the cellular role of the Na-Ca exchange in cardiac muscle cells in a straightforward way.

The lack of a specific inhibitor has not only slowed the research in the field of cellular Ca^{2+} signaling by the Na-Ca exchange, but has also impeded attempts to understand the molecular operations of the Na-Ca exchanger protein. The molecular function of the Na-Ca exchanger is believed to comprise a cycle of several molecular reaction steps leading to Na^+ and Ca^{2+} translocation across the cell membrane. However, it is not clear in which states the Na-Ca exchanger can exist and what general scheme the exchange cycle follows. A number of models and corresponding computer simulations for this biochemical reaction cycle have been developed, all based on available steady-state kinetic data.[5-7] In principle, two fundamentally different transport mechanisms can be distinguished and both have been proposed for the Na-Ca exchange: i) a consecutive exchange cycle in which Na^+ and Ca^{2+} are moved across the membrane in two separate steps and ii) a simultaneous mechanism in which both Na^+ and Ca^{2+} must bind to the transporter molecule before it can undergo a molecular rearrangement involving ion translocation.

In principle, transient state kinetic studies could provide the necessary data to more directly pin down the molecular functioning of the Na-Ca exchanger. Unfortunately, only few studies have revealed transient-state data, and the interpretation of the available information is not very clear.[8,9] There are several reasons for these difficulties: i) apparently, the Na-Ca exchanger has a very fast turnover rate, challenging the temporal resolution of voltage-clamp amplifiers available today;[10] ii) again, no specific inhibitor for the Na-Ca exchange is available; iii) in addition, for the available unspecific blockers the exact mechanism of inhibition is not known on the molecular level, *i.e.,* it is not known which partial reaction of the (unknown) reaction cycle is inhibited by a particular blocker.

After paramount efforts in several laboratories to obtain some structural information about the Na-Ca exchanger, the protein was isolated, sequenced and cloned in 1990.[11] Meanwhile, the Na-Ca exchanger has been functionally expressed in numerous cell systems and various isoforms have been identified in different tissues and organisms.[12-14] With this molecular information available, several studies have been initiated with the goal to understand the structure-function relationship. A variety of mutants have been constructed using molecular biology techniques, some with very interesting functional properties.[15] As far as the tertiary structure of the Na-Ca exchanger protein is concerned, no information is available at present.

Recently, a polypeptide resembling a region of the Na-Ca exchanger was synthesized and was found to significantly inhibit the Na-Ca exchange when applied from the inside of the cell, *i.e.,* the exchange inhibitory peptide or XIP.[16] Experiments carried out with this peptide revealed the important role of the Na-Ca exchange during cardiac Ca^{2+} signaling and suggested a possible role for the exchange during excitation-contraction coupling in cardiac muscle.[17] While this blocker may be more specific than those available previously, the exact molecular inhibitory mechanism is not yet understood, despite the clear structural concept that led to its synthesis.

Using the available cDNA sequence information about the cardiac Na-Ca exchanger of different species we chose an alternative and potentially advantageous approach to "inhibit" the Na-Ca exchange in cultured cardiac myocytes.[18] An antisense oligodeoxynucleotide (AS-ODN) directed against 19 nucleotides in the 3' nontranslated region of the Na-Ca exchanger mRNA was synthesized. A number of biophysical techniques including confocal ratiometric Ca^{2+} measurements, flash photolysis of caged Ca^{2+} and the voltage-clamp technique in the whole-cell mode were used to quantify the Na-Ca exchange function in these cells and to compare control cells with cells exposed to the AS-ODN. We found that exposure of the cells to AS-ODN led to an almost complete suppression of Na-Ca exchange function as determined by measuring Ca^{2+} transport capacity as well as membrane currents generated by electrogenic Na-Ca exchange. The treatment of the cells with antisense ODNs not only reduced the exchange function but is also expected to nearly completely prevent de-novo synthesis of the protein. Therefore, this approach is not only useful to investigate the cellular role of the Na-Ca exchange in various cells but may also be helpful to delineate the molecular functioning of this transporter.

METHODS

Primary cultures of neonatal rat cardiac myocytes were prepared with established methods. Intracellular Ca^{2+} was measured ratiometrically using a confocal microscope with a mixture of two fluorescent Ca^{2+} indicators and an *in vivo* calibration. Cells were dialyzed with fluo-3 and fura-red[19] in the whole-cell recording mode of the patch-clamp technique. Rapid line-scans were performed with a high temporal resolution (up to 500 Hz). Summary results are shown as mean ± standard error. The superfusion solution contained (in mM): NaCl 140, KCl 5, $CaCl_2$ 2, HEPES 10, glucose 10, ryanodine 0.01 and thapsigargin 0.0002, pH 7.4. Temperature 20°–22°C. Li^+ was used in Na^+_o-free solutions. The pipette filling solution contained (in mM): Cs-aspartate 120, NaCl 10, K-ATP 4, tetraethylammonium-Cl 20, HEPES 10, fluo-3 0.033, fura-red 0.066, pH 7.2. For some experiments, Na_4-DM-nitrophen 2 mM, reduced glutathione (GSH) 2 mM and Ca^{2+} 0.5 mM were included in the pipette. The antisense ODN was added to the culture medium at an initial concentration of 3 µM. For details of the target sequence, synthesis and application of the AS-ODN see Ref. 18.

RESULTS AND DISCUSSION

When the experimental strategy involving antisense oligodeoxynucleotides (AS-ODNs) was developed we first had to adopt an experimental protocol to reliably detect and quantitate the inhibitory effects of this treatment. Initially, we expected the onset of an inhibition to be not very pronounced and slow, on the order of several days. Therefore, we decided to statistically compare cells cultured under control conditions with others treated with the AS-ODNs. Despite significant changes in cellular volume and surface of the exposed and control myocytes during the first few days in culture, the rate of Ca^{2+} increase (V_{up}) during rapid superfusion with a Na^+_o-free solution turned out to be a remarkably stable parameter with little cell-to-cell variability. Removal of extracellular Na^+ reverses the electrochemical gradient for Na^+ and forces the Na-Ca exchange to operate in the Ca^{2+}-influx mode (also termed "backward mode"). Applying this technique with a rapid superfusion switcher ($t_{1/2} < 0.5$ s) under voltage-clamp conditions (holding potential −50 mV), we were able to compare the Ca^{2+} transport capacity of the Na-Ca exchange in different cells. During these experiments, ryanodine and thapsigargin were always present to avoid interference from intracellular Ca^{2+} stores with the Ca^{2+} signals generated by the Na-Ca exchange.

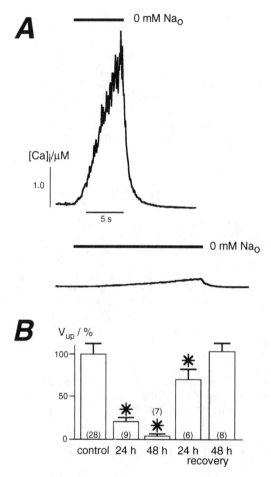

FIGURE 1. Ca^{2+}-transients induced by removal of $[Na^+]_o$. Representative Ca^{2+}-transients elicited by removal of $[Na^+]_o$ in a control cell (*upper trace*) and in a cell treated with 3 μM of the AS-ODN for 24 h (*lower trace*) are shown in (**A**). The initial Ca^{2+} increase rate (V_{up}) is much smaller in the exposed cell than in the control cell. Normalized values for V_{up} are compiled in (**B**) for control cells (pooled data) and cells exposed for 24 or 48 h. The histogram also shows recovery of Na-Ca exchange after removal of the ODN (following 24 h exposure). A * denotes a significant suppression compared to control as assessed by a paired Student t test ($p < 0.05$).

Panel A of FIGURE 1 illustrates such an experiment. For the upper trace $[Ca^{2+}]_i$ was recorded from a control cell exhibiting a rise of $[Ca^{2+}]_i$ of about 3 μM during the Na^+_o-free period. Readdition of Na^+_o lead to a rapid decay of $[Ca^{2+}]_i$ reflecting Ca^{2+} removal via the Na-Ca exchange. The lower trace shows an analogous experiment performed on a cell that had been treated with 3 μM of the antisense ODN for 24 hours. Obviously, the rate of increase of $[Ca^{2+}]_i$ during the first second (V_{up}) was much slower and much less pronounced, despite the longer lasting period of Na^+_o removal. In order to be able to test for statistically significant effects we then determined V_{up} from a large number of cells, either exposed to the AS-ODN or kept under control conditions. On average, V_{up} was 350 nM per second ($\pm$35 nM) in control cells. We normalized all V_{up} measurements with respect to the control cells and used pooled data to compile panel B of FIGURE 1. After 24 hours, the mean Na-Ca exchange activity in the presence of 3 μM AS-ODN dropped significantly to about 20%, and after 48 hours most cells exhibited no detectable increase in $[Ca^{2+}]_i$ upon Na^+_o removal (5 out of 7 cells; 2 still had some residual Na-Ca exchange activity).

We were very much amazed by these results. First, the onset of inhibition was much faster than what we had expected. Second, the inhibition was more profound than what we had hoped for. Unfortunately, both surprising findings could in principle also be explained by various undesirable side-effects of the ODNs. A first possibility would be an unspecific toxic effect essentially killing the cells. In order to rule out this potential problem, each cell was initially tested for viability by eliciting Na^+-currents before continuing the experiment. In addition, a series of recovery experiments was carried out to demonstrate that the cells are not severely damaged and can restore the normal Na-Ca exchange function. For this purpose, cultures were exposed to 3 μM of the AS-ODN for 24 hours. At this time, the ODN was removed by changing the culture medium several times. Recovery of Na-Ca exchange was assayed 24 hours and 48 hours after removing the antisense ODN. Panel B of FIGURE 1 illustrates the recovery in the right two columns. After 24 hours, Na-Ca exchange recovered to 70% of control, after 48 hours V_{up} returned to 100%. Please note that the recovery had a time-course very similar to the onset of inhibition, suggesting that the turnover of the Na-Ca exchanger protein in the membrane is indeed very rapid.

We also carried out several other series of control experiments in an attempt to show that it was a true antisense effect that suppressed the Na-Ca exchange and not an unspecific inhibitory effect of the ODN. One established approach for this control experiment is to test an ODN with a reversed (5' to 3' orientation) but otherwise identical sequence, an oligonucleotide called a "non-sense" ODN.[20,21] Although chemically very similar to the antisense ODN, the non-sense ODN should have no inhibitory effect on the Na-Ca exchange, since it cannot anneal to the target sequence. Similarly, a few "errors" can be introduced into the antisense sequence. These mismatches are known to lower the temperature of annealing between the ODN and the target sequence and thus to destabilize the DNA-RNA complex. A minimal number of mismatches were introduced that prevent the formation of a stable DNA-RNA complex under the experimental conditions and this resulted in the "mismatched" ODN. Obviously, the mismatched ODN should also not interfere with Na-Ca exchange function. In addition to a mismatched ODN with three exchanged nucleotides, we tested a completely unrelated antisense ODN directed against calretinin, a Ca^{2+}-binding protein not expressed in cardiac muscle cells. All three control ODNs described above had no significant inhibitory effect on the Na-Ca exchange function[18] suggesting that the inhibition by the antisense ODN was indeed specific for the particular target sequence.

As already mentioned, the exact mode of inhibition is not known for the presently used blockers of the Na-Ca exchange. There is however some data suggesting that diverse blockers inhibit the Na-Ca exchange with some asymmetries. For example, DCB seems to be more potent when applied from the outside of the cell in whole-cell patch-clamp experiments[9] (but see Ref. 22). Another example is the exchange inhibitory peptide (XIP), which exclusively inhibits from the cytosolic side.[16] It has also been reported that XIP is a more potent inhibitor of Na^+-Na^+ homeo-exchange than of the Ca^{2+}-Ca^{2+} exchange mode.[23] With regard to the symmetry of its inhibitory effect, treatment of the cells with AS-ODN should suppress all modes of the Na-Ca exchange since the protein itself is expected to be removed from the cell membrane. In order to test this prediction we needed to show that Ca^{2+} removal via the Na-Ca exchange (i.e., the forward mode) was also inhibited in cells showing no detectable increase of $[Ca^{2+}]_i$ via backward mode upon removal of Na^+_o. In such cells, another technique was obviously required to elevate $[Ca^{2+}]_i$. This was accomplished by combining the measurements of membrane current and $[Ca^{2+}]_i$ with flash-photolysis of caged Ca^{2+} (DM-nitrophen), and the results are shown in FIGURE 2. Since for this type of experiment recording of fluorescence was required during an extended period of time, we only measured $[Ca^{2+}]_i$ every 2 seconds to minimize bleaching. Panel A shows the Ca^{2+} trace (●) with a prompt increase of $[Ca^{2+}]_i$ during Na^+_o removal in a control cell, an experiment similar to that shown in FIGURE 1A. Readdition of Na^+_o was again followed by rapid Ca^{2+} removal. When Ca^{2+} was uncaged from DM-nitrophen with an intense flash of UV light, $[Ca^{2+}]_i$ rose immediately to a level comparable to the Na^+-free transient. After the rapid rise, Ca^{2+} slowly decayed at a rate similar to the Na^+-free transient. Quite in contrast, the cell exposed to 3 μM AS-ODN for 48 hours exhibited no increase of $[Ca^{2+}]_i$ upon removal of Na^+_o, suggesting that the Ca^{2+}-influx mode of the Na-Ca exchange was not operational. Flash photolysis of caged Ca^{2+} produced a Ca^{2+} concentration jump comparable to the control cell, but the process of Ca^{2+} removal was altered quite dramatically in the exposed cell. Over the time examined (about 1 minute), $[Ca^{2+}]_i$ did not decay appreciably, indicating that the forward mode of the Na-Ca exchange was also completely suppressed in this cell.

These profound changes of Ca^{2+} signaling were complemented by a parallel effect of the AS-ODN treatment on the whole-cell membrane currents. The inset in panel A illustrates changes in membrane current during Na^+_o removal (upper trace) and after a flash photolytic Ca^{2+} concentration jump (lower trace) in the control cell. During the Na^+_o free period, a slowly decaying outward current was observed, probably largely reflecting electrogenic Na-Ca exchange. Upon readdition of Na^+ and thus reversal of the Na-Ca exchange an inward "creep" current was seen corresponding to electrogenic Ca^{2+} removal via Na-Ca exchange. An equivalent inward current was also induced by flash photolytic liberation of intracellular Ca^{2+}.[24] In the cell treated with AS-ODN (inset in panel B of FIG. 2) no significant changes of the membrane currents were detectable with both interventions. The small shift of the holding current during Na^+_o removal probably reflects a somewhat different background conductance for Li^+. In summary, the results on Ca^{2+} transport and Na-Ca exchange currents after 48 hours in 3 μM AS-ODN are consistent with the view that the functional Na-Ca exchanger proteins have disappeared completely from the cell membrane.

Taken together, these first results indicate that our AS-ODN may be a valuable tool to investigate a large range of issues related to the cellular and molecular function of the Na-Ca exchanger. Below we outline a few problems that may be successfully addressed in the near future using this approach.

In the cardiac EC-coupling field, there has been much recent interest in the question whether Ca^{2+} influx via Na-Ca exchange can trigger Ca^{2+} release from the SR. This could, in principle, happen for two reasons. i) Early during the action potential the membrane potential is more positive than the reversal potential of the Na-Ca exchange, and this favors the Ca^{2+} influx mode of the exchange.[17] ii) The massive Na^+ influx during the Na^+ current may lead to an accumulation of

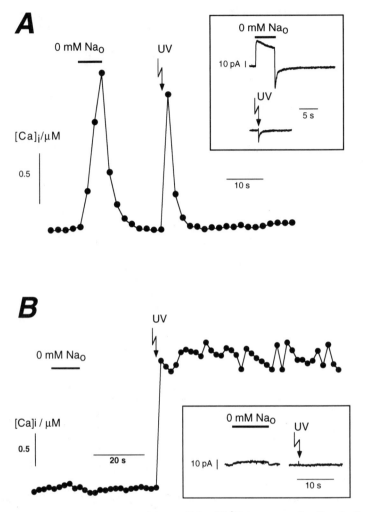

FIGURE 2. Membrane currents and intracellular [Ca^{2+}] in a control cell and after 48-h treatment with antisense ODN. Each data point (●) reflects the cytosolic [Ca^{2+}] determined in a ratiometric confocal image. Frames were acquired at 0.5 Hz. Removing extracellular Na$^+$ for 10 seconds and flash-photolysis of caged Ca^{2+} resulted in transient increases of [Ca^{2+}]$_i$ in a control cell **(A)**. Corresponding changes of Na-Ca exchange outward and inward current are shown in the *inset*. Quite in contrast, Na$^+$ removal did not change [Ca^{2+}]$_i$ significantly in the exposed cell **(B)**, and the flash photolytic Ca^{2+} concentration jump did not decay. Furthermore, no Na-Ca exchange currents could be detected.

Na^+ in a restricted space under the sarcolemma ("fuzzy space"[25]) giving rise to Na^+-induced Ca^{2+} entry via the Na-Ca exchange.[26,27] This Ca^{2+} entry could then trigger Ca^{2+} release from the sarcoplasmic reticulum. Depending on the experimental protocol and probably also under physiological conditions both mechanisms, the membrane potential and the Na^+ accumulation, may affect the Na-Ca exchange cooperatively. However, several laboratories have not been able to reproduce these experimental results[28,29] and there is no obvious reason for this discrepancy, although species differences of the Na-Ca exchange activity have been discussed besides possible experimental pitfalls. An obvious possibility to address this question would be to use the AS-ODNs on adult cardiac myocytes kept in primary short-term culture[30] to assess the role of the Na-Ca exchange during these rapid events of cardiac EC-coupling.

As mentioned above, the treatment with AS-ODN is expected to completely remove the Na-Ca exchanger protein from the cell membrane. This feature could be used to improve the identification of membrane currents attributed to the electrogenic Na-Ca exchange. For example, it has been very difficult to clearly distinguish Ca^{2+}_i-induced membrane currents in cardiac muscle cells and to separate the Na-Ca exchange current I_{NaCa} from a Ca^{2+} activated nonselective cation current that can be observed under some conditions.[31,32]

This approach could also be applied to improve the identification of membrane currents attributed to partial reactions or half-cycles of the Na-Ca exchange.[8,9] The identification of a transient inward current (I_{conf}) that we observed after flash photolytic Ca^{2+} concentration jumps and that we associated with a molecular conformational change of the Na-Ca exchanger immediately after Ca^{2+} binding is still not satisfactory. So far, the identification is based on indirect evidence including the observation that intracellular DCB, an organic inhibitor of the Na-Ca exchange, shifts the voltage dependence of I_{conf} dramatically.[33]

In summary, the AS-ODN synthesized by us completely and rapidly suppresses Na-Ca exchange in cultured cardiac myocytes. The ODN can simply be added to the culture medium and no loading protocol like permeabilization or electroporation is required. The same AS-ODN has also been successfully used with astrocytes and with co-injection experiments in *Xenopus laevis* oocytes (W. J. Lederer, D. H. Schulze, personal communication). Therefore, we anticipate the widespread use of this approach and even this particular antisense ODN to investigate cellular and molecular functions of the Na-Ca exchange in a variety of cell types from different species.

ACKNOWLEDGMENTS

We would like to thank M. Herrenschwand, B. Herrmann and A. Wyss for excellent technical help and Dr. H. Porzig for helpful discussions.

REFERENCES

1. REEVES, J. P. 1989. Sodium-calcium exchange: a possible target for drug development. Drug. Dev. Res. **18:** 295–304.
2. KIMURA, J., S. MIYAMAE & A. NOMA. 1987. Identification of sodium-calcium exchange current in single ventricular cells of guinea-pig. J. Physiol. **384:** 199–222.
3. LIPP, P. & L. POTT. 1988. Voltage dependence of sodium-calcium exchange currents

in guinea-pig atrial myocytes determined by means of an inhibitor. J. Physiol. **403:** 355–366.
4. BIELEFELD, D. R., R. W. HADLEY, P. M. VASSILEV & J. R. HUME. 1986. Membrane electrical properties of vesicular Na-Ca exchange inhibitors in single atrial myocytes. Circ. Res. **59:** 381–389.
5. LÄUGER, P. 1987. Voltage dependence of sodium-calcium exchange: predictions from kinetic models. J. Membrane Biol. **99:** 1–11.
6. JOHNSON, E. A., D. R. LEMIEUX & J. M. KOOTSEY. 1992. Sodium-calcium exchange—derivation of a state diagram and rate constants from experimental data. J. Theor. Biol. **156:** 443–483.
7. HILGEMANN, D. W. 1988. Numerical approximations of sodium-calcium exchange. Prog. Biophys. Mol. Biol. **51:** 1–45.
8. HILGEMANN, D. W., D. A. NICOLL & K. D. PHILIPSON. 1991. Charge movement during Na^+ translocation by native and cloned cardiac Na^+/Ca^{2+} exchanger. Nature **352:** 715–718.
9. NIGGLI, E. & W. J. LEDERER. 1991. Molecular operations of the Na-Ca exchanger revealed by conformation currents. Nature **349:** 621–624.
10. HILGEMANN, D. W., K. D. PHILIPSON & D. A. NICOLL. 1992. Possible charge movement of extracellular Na^+ binding by Na/K pump and cardiac Na/Ca exchanger in giant patches. Biophys. J. **61:** A390.
11. NICOLL, D. A., S. LONGONI & K. D. PHILIPSON. 1990. Molecular cloning and functional expression of the cardiac sarcolemmal Na^+-Ca^{2+} exchanger. Science **250:** 562–565.
12. LI, Z. P., S. MATSUOKA, L. V. HRYSHKO, D. A. NICOLL, M. M. BERSOHN, E. P. BURKE, R. P. LIFTON & K. D. PHILIPSON. 1994. Cloning of the NCX2 isoform of the plasma membrane Na^+-Ca^{2+} exchanger. J. Biol. Chem. **269:** 17434–17439.
13. LEE, S. L., A. S. L. YU & J. LYTTON. 1994. Tissue-specific expression of Na^+-Ca^{2+} exchanger isoforms. J. Biol. Chem. **269:** 14849–14852.
14. KOFUJI, P., W. J. LEDERER & D. H. SCHULZE. 1994. Mutually exclusive and cassette exons underlie alternatively spliced isoforms of the Na/Ca exchanger. J. Biol. Chem. **269:** 5145–5149.
15. MATSUOKA, S., D. A. NICOLL, L. V. HRYSHKO, D. O. LEVITSKY, J. N. WEISS & K. D. PHILIPSON. 1995. Regulation of the cardiac Na^+-Ca^{2+} exchanger by Ca^{2+}—mutational analysis of the Ca^{2+}-binding domain. J. Gen. Physiol. **105:** 403–420.
16. LI, Z., D. A. NICOLL, A. COLLINS, D. W. HILGEMANN, A. G. FILOTE, J. T. PENNISTON, J. N. WEISS, A. M. TOMICH & K. D. PHILIPSON. 1991. Identification of a peptide inhibitor of the cardiac sarcolemmal Na^+-Ca^{2+} exchanger. J. Biol. Chem. **266:** 1014–1020.
17. KOHMOTO, O., A. J. LEVI & J. H. B. BRIDGE. 1994. Relation between reverse sodium-calcium exchange and sarcoplasmic reticulum calcium release in guinea pig ventricular cells. Circ. Res. **74:** 550–554.
18. LIPP P., B. SCHWALLER & E. NIGGLI. 1995. Specific inhibition of Na-Ca exchange function by antisense oligodeoxynucleotides. FEBS Lett. **364:** 198–202.
19. LIPP, P. & E. NIGGLI. 1993. Ratiometric confocal Ca^{2+}-measurements with visible wavelength indicators in isolated cardiac myocytes. Cell Calcium **14:** 359–372.
20. STEIN, C. A. & A. M. KRIEG. 1994. Problems in interpretation of data derived *in vitro* and *in vivo* use of antisense oligodeoxynucleotides. Antisense Res. Dev. **4:** 67–69.
21. WAGNER, R. W. 1994. Gene inhibition using antisense oligodeoxynucleotides. Nature **372:** 333–335.
22. CHERNAYA, G. & J. P. REEVES. 1993. Transmembrane asymmetry of Na^+-Ca^{2+} exchange inhibitors. Biophys. J. **64:** A399.
23. SHANNON T. R., C. C. HALE & M. A. MILANIK. 1994. Potency of exchange inhibitory peptide (XIP) inhibition of the Na-Ca exchanger in alternate transport modes. Biophys. J. **66:** A332.
24. NIGGLI, E. & J. W. LEDERER. 1993. Activation of Na-Ca exchange current by photolysis of caged calcium. Biophys. J. **65:** 882–891.
25. LEDERER, W. J., E. NIGGLI & R. W. HADLEY. 1990. Sodium-calcium exchange in excitable cells: fuzzy space. Science **248:** 283.

26. LEBLANC, N. & J. R. HUME. 1990. Sodium current-induced release of calcium from cardiac sarcoplasmic reticulum. Science **248:** 372–376.
27. LIPP, P. & E. NIGGLI. 1994. Sodium current-induced calcium signals in isolated guinea-pig ventricular myocytes. J. Physiol. **474:** 439–446.
28. SHAM, J. S. K., L. CLEEMANN & M. MORAD. 1992. Gating of the cardiac Ca^{2+} release channel—the role of Na^+ current and Na^+-Ca^{2+} exchange. Science **255:** 850–853.
29. BOUCHARD, R. A., R. B. CLARK & W. R. GILES. 1993. Role of sodium-calcium exchange in activation of contraction in rat ventricle. J. Physiol. **472:** 391–413.
30. LIPP, P., J. HÜSER, L. POTT & E. NIGGLI. 1994. The role of the T-tubules in Ca^{2+} signaling and excitation-contraction coupling in guinea pig cardiac myocytes. J. Physiol. **477:** 18P.
31. COLQUHOUN, D., E. NEHER, H. REUTER & C. F. STEVENS. 1981. Inward current channel activated by intracellular Ca in cultured cardiac cells. Nature **294:** 752–754.
32. EHARA, T., A. NOMA & K. ONO. 1988. Calcium-activated non-selective cation channel in ventricular cells isolated from adult guinea-pig hearts. J. Physiol. **403:** 117–133.
33. NIGGLI, E. & P. LIPP. 1994. Voltage dependence of Na-Ca exchanger conformational currents. Biophys. J. **67:** 1516–1524.

Molecular Biological Studies of the Cardiac Sodium-Calcium Exchanger[a]

ALEXANDER KRAEV, ILYA CHUMAKOV,[b] AND
ERNESTO CARAFOLI[c]

Laboratory of Biochemistry III
Swiss Federal Institute of Technology (ETH)
Universitätsstrasse 16
CH-8092 Zürich, Switzerland

[b]*Centre d'Etude du Polymorphisme Humain*
Paris, France

INTRODUCTION

Two biochemically defined classes of Na-Ca-exchangers, namely, a "cardiac"[1] and a "retinal rod" exchanger,[2] were recently characterized on the molecular level by means of cDNA cloning. The two exchanger types have similar architecture, *i.e.*, 11 transmembrane domains and a large cytoplasmic loop; however, their amino acid sequences have very little similarity. Recently, an exchanger was identified in rat brain, which had 65% sequence homology to the cardiac exchanger.[3] The cardiac exchanger gene was designated NCX1 and was located[4] at the short arm of human chromosome 2. The gene for the recently identified brain exchanger, NCX2, was mapped[3] to human chromosome 14.

The NCX1 gene appears to be ubiquitously expressed in mammals, although the mRNA levels in different tissues may differ by almost two orders of magnitude. Abundant data has been accumulated on the products of the mammalian NCX1 gene,[5–7] which arise from the differential splicing of several small exons.[8] The splicing variants differ in a limited portion of the main cytoplasmic loop, which has been suggested[9] to play a role in the regulation of the protein by cytoplasmic Ca^{2+}. The other two exchangers appear to be more specialized, *i.e.*, one is only expressed in retinal rods and cones[2] and, possibly, platelets;[10] the other in brain and skeletal muscle.[3]

Very little is known on the regulation of expression of the NCX1 gene, although recently it was shown that mRNA leves and exchange activity are regulated in vascular smooth muscle by glucocorticoids and growth factors.[11] As a prerequisite of the studies of gene regulation, we determined the gene structure of the human NCX1 gene and part of the human NCX2 gene. We also identified a putative homolog of the vertebrate NCX gene in *Caenorhabditis elegans*.

METHODS

All DNA manipulations were done essentially by standard procedures.[12] The methodology of gene analysis will be presented in detail elsewhere (Kraev, Chuma-

---

[a] This study was supported by Swiss National Science Foundation Grant No. 3100-30858.91.

[c] Corresponding author.

kov and Carafoli, in preparation). Data on chromosome localization of human genes NCX1, NCX2 and NCKX1, as well as of a putative *C. elegans* homolog of vertebrate NCX (EST yk24h3) are available via computer network from Genome Data Base and ACEDB, respectively, and DNA sequences under Genbank/EMBL Database Accession numbers X92368 and X91803.

RESULTS

Exon Composition of the NCX1 mRNA

The initial studies using the cloned cardiac exchanger cDNA had shown that the same 7.2-kb transcript was found, although in different amounts, in many tissues.[13] A 14-kb transcript was also detected in the brain.[13] To determine transcript boundaries, a combination of cDNA library screening and reverse transcriptase-assisted polymerase chain reaction (RT-PCR) analysis was used in this work. The data obtained account for 6.3 out of the 7.2 kb mRNA length estimated by agarose gel electrophoresis.[13] A human brain cDNA EST94366 represents the missing 5'-end portion of this transcript, and extends our data by 127 bases. Recent studies in rat[14] suggest that the extreme 5'-end portion of this transcript may be subject to differential splicing in a tissue-specific manner. The human expressed sequence tag (EST) shows a limited homology to the "ubiquitous" 5'-end[14] reported for rat NCX1.

The current membrane architectural model for the cardiac exchanger suggests the presence of 11 transmembrane segments with a cytoplasmic loop in the central part of the protein. It was therefore reasonable to assume that the N- and C-terminal portions of the protein might be coded for by single exons, and a series of oligonucleotide primers was thus designed in an attempt to isolate the genomic sequence by polymerase chain reaction. Indeed, a pair of primers, spanning the almost entire N-terminal half of the protein generated the expected 1.7-kb fragment on amplification of human genomic DNA, suggesting that the corresponding portion of the gene, coding for the first 5 transmembrane segments and a substantial fraction of the main cytoplasmic loop, lacked introns. Direct sequencing of this PCR fragment showed that it corresponded exactly to the previously published[15] N-terminal portion of the cDNA. When this PCR fragment was used as a probe in a Southern blot analysis of human DNA, all relevant restriction endonuclease cleavages generated only fragments with the expected sizes (not shown). However, other primers, flanking the central segment of the mRNA, which is different in various tissues due to alternative splicing, failed to generate products on amplification of the human genomic DNA, suggesting the presence of introns. A number of clones from human genomic library in Lambda FIX II was isolated using PCR-generated portions of the cDNA as probes. Analysis of these clones led to the reconstruction of the exon-intron structure of gene NCX1 shown in FIGURE 1: the gene contains 12 exons; however, about 90% of the mature message is contained in two exons only, namely the 2nd and the 12th.

Identification of the Human NCX2 Gene

A subset of the lambda clones, hybridizing to the middle part of the cDNA, was identified as belonging to a second, similar gene, which was identified as

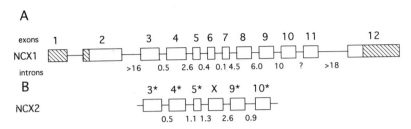

FIGURE 1. Exon-intron organization of the NCX1 gene and of the central part of the NCX2 gene. Exons are shown as *boxes*, introns as *connecting lines*. Neither exons nor introns are drawn to scale. The actual length of introns in kilobases is shown below the connecting line. The portions of the exons, coding for the untranslated regions are *hatched* (**A**) The exons of NCX1 are numbered from 1 to 12, although the evidence for exon 1 was indirect. The length of intron 10 is not known. The length of introns 2 and 11 was estimated from mapping on lambda clones and on genomic DNA and it may actually exceed the value given. (**B**) The exons of NCX2 are shown with the same numbers as those of the homologous exons of NCX1 to simplify the comparison of the two genes.

NCX2 on the basis of the following criteria: a) primers designed for putative exons 3* and 10* (showing substantial homology to NCX1 exons 3 and 10, FIG. 2) amplified a set of fragments specifically from brain mRNA; b) a second set of primers was capable of selecting a yeast artificial chromosome (YAC) clone, mapping to human chromosome 14p24.1; c) sequence analysis of PCR products, amplified from brain RNA, showed that they have a relationship similar to the one known for the corresponding portion of NCX1 mRNA, *i.e.*, display block deletions and insertions. Further analysis of the lambda clones with primers, designed for putative exons revealed that these sequences are indeed interrupted by stretches of DNA, bordered by standard consensus sequences (to be described in detail elsewhere).

Although the complete exons 2*, 11* and 12* of NCX2 have not yet been isolated, location of the available splicing points suggests that also this portion of the two genes is organized in a similar way. Given the otherwise high sequence homology and very similar exon-intron structure in the central portion of NCX1 and NCX2, it is tempting to speculate that they had originated from a common ancestor via duplication and translocation in the course of evolution. Thus, the "cardiac" type Na-Ca-exchanger appears to be coded by a gene family, as is the case for the other plasma membrane Ca^{2+}-transporter, the calcium pump,[16,17] as well as for other mammalian ion-exchangers, *e.g.*, the Na^+-H^+-exchanger[18] and cotransporters, *e.g.*, the Na^+-glucose cotransporter.[19] It is thus conceivable that a substantial diversity of exchanger species is required in some tissues (particularly in brain), if two different mechanisms, transcription and splicing, must be used to achieve it.

A Putative NCX Gene from the Nematode Caenorhabditis elegans

Although mammalian NCX1 and NCX2 genes display overall 61% identity at the nucleotide level, in localized areas identity is much higher, for example, in the area of the first 5 transmembrane domains.[3] We used this sequence for routine

```
Exon 3       KTISVKVIDDEEYEKNKTFFLEIGEPRLVEMSEKK...
Exon 4       KIITIRIFDRREYEKECSFSLVLEEPKWIRRGMK....
Exon 3*      KTIHIKVIDDEAYEKNKNYFIEMMGPRMVDMSFQK...
Exon 4*      KTIRVKIVDEEYERQENFFIALGEPKWMERGIS....
matches      *.*....*..**.......*.

exons 9-10    DEYDDKQPLTSKEEERRIAEMGRPILGEHTKLEVIIEESYEFKSTVDKLIKKTNLALVVGTNSW
exons 9*-10*  ..DVTDRKLTMEEEAKRIAEMGKPVLGEHTKLEVIIEESYEFKTTVDKLIKKTNLALVVGPFLE
matches       **  ***..*******.*.*************************.********************
```

FIGURE 2. Alignment of amino acid sequences encoded in homologous exons. The *stars* below the sequences indicate an exact match; the *dots* indicate conservative aminoacid substitutions. The exons are numbered as in FIGURE 1.

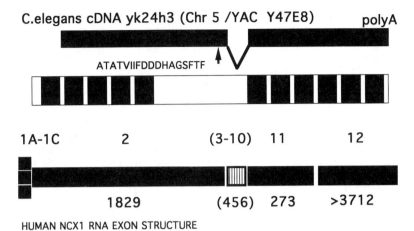

FIGURE 3. Alignment of the nematode cDNA, displaying homology to the mammalian NCX cDNA with the model of protein organization and the exon composition of the mammalian mRNA. The *solid boxes* are transmembrane domains. The *solid bar* shows the mRNA, with exon numbers shown over it, and exon length (in nucleotides) below it. The position of the putative calcium binding site motif is indicated by an *arrow*, with the actual nematode sequence at the left of it.

search in current EST databases and recently detected a cDNA clone in a *C. elegans* database showing 62% nucleotide identity in a 346-bp overlap, spanning transmembrane domains 2-4 of the human NCX1 cDNA. We obtained this clone from the author and sequenced the entire 2.2-kb insert (FIG. 3). Although this sequence at the time of writing still contains some ambiguities, it does display an average 57% identity to human NCX cDNAs. While the sequence data alone are not sufficient to conclude that this protein functions as a sodium-calcium exchanger in the nematode, it is worth noting that it does show several features that identify it as a homolog of the mammalian "cardiac/brain" (NCX) sodium-calcium exchanger. Particularly interesting is the presence of a motive ATATVIIFDDDHAGSFTF, highly resembling the "high affinity calcium-binding domain II"[20] at a similar location relative to the hydrophobic amino acid regions. However, nematode mRNA has a much shorter 3'-untranslated region than the mammalian counterpart. In addition, when 10 regions of similarity, corresponding to hydrophobic amino acid stretches, are aligned with mammalian cDNA, a central portion of the nematode cDNA appears to have a deletion, roughly corresponding to the differentially spliced area of the mammalian cDNA. It is thus tempting to speculate that in the nematode this protein functions only as an exchanger of the primitive nervous system, and thus it lacks the mosaic of the differentially spliced exons coding for the distal part of the cytoplasmic loop. Indeed, comparison of the spliced NCX1 transcripts (see also a paper by Iwata *et al.* in this volume) reveals that NCX1 isoforms have tissue-specific and species/development-specific exons. Splicing variants of NCX2, identified in human and rat brain, are not detected in other tissues (except skeletal muscle) and are not yet characterized in sufficient detail to allow extensive comparison. As more data on the NCX gene structure in lower vertebrates and invertebrates are obtained, and the gaps in the data continuity between the nematode and man are possibly filled, one will be able to evaluate this hypothesis.

SUMMARY

The intron-exon organization of the entire human Na-Ca-exchanger gene NCX1 and of the central part of the related gene NCX2 has been determined. The NCX1 gene is at least 75 kb long and consists of at least 12 exons, the two largest (the 2nd and the 12th) coding for the N-terminal half of the exchanger sequence and for the last three C-terminal transmembrane domains. They also code for the 3.3-kb 3'-untranslated region and account for more than 90% of the length of the mature mRNA. The remainder of the NCX1 (NCX2) gene, coding for a putative cytoplasmic regulatory domain, is split into 9 (7) small exons. In spite of the limited (65%) average homology of the two cDNAs, analogous exons are readily identified within this portion of the two genes based on their high (80–95%) pairwise homology and similar patterns of differential splicing in brain. Human YAC clones have been identified in the CEPH library, which contain the entire NCX1/2 and NCKX1 (retinal rod exchanger) genes, and are used for chromosomal localization of the three genes. A distant homolog of the mammalian NCX genes has been identified in the *C. elegans* EST database and has been completely sequenced. It encodes a 20% shorter protein, which has an average 55% homology to human NCX1, and lacks most of the region that is known to be encoded by multiple differentially spliced exons in vertebrates. Comparison of available data on the gene structure of the NCX homologs in various species suggests that this protein has emerged in the primitive nervous system and has been subsequently adapted to other cellular environments by the use of novel domains, encoded in additional exons.

ACKNOWLEDGMENTS

The authors are indebted to the late Dr. O. W. McBride (NIH, Bethesda, MD) and to T. Scheller for providing unpublished data, as well as to Dr. Yuji Kohara (National Institute of Genetics, Mishima, Japan) and to Dr. Alan Coulson (Sanger Centre, Cambridge, UK) for providing the nematode cDNA and genomic clones.

REFERENCES

1. NICOLL, D. A., S. LONGONI & K. D. PHILIPSON. 1990. Molecular cloning and functional expression of the cardiac sarcolemmal Na^+-Ca^{2+} exchanger. Science **250:** 562–565.
2. REILANDER, H., A. ACHILLES, T. FRIEDEL, G. MAUL, F. LOTTSPEICH & N. J. COOK. 1992. Primary structure and functional expression of the Na/Ca,K-exchanger from bovine rod photoreceptors. EMBO J. **11:** 1689–1695.
3. LI, Z., S. MATSUOKA, L. V. HRYSHKO, D. A. NICOLL, M. M. BERSOHN, E. P. BURKE, R. P. LIFTON & K. D. PHILIPSON. 1994. Cloning of the NCX2 isoform of the plasma membrane Na^+-Ca^{2+} exchanger. J. Biol. Chem. **269:** 17434–17439.
4. SHIEH, B-H., Y. XIA, R. S. SPARKES, I. KLISAK, A. J. LUSIS, D. A. NICOLL & K. D. PHILIPSON. 1992. Mapping of the gene for the cardiac sarcolemmal Na^+-Ca^{2+} exchanger to human chromosome 2p21-p23. Genomics **12:** 616–617.

5. REILLY, R. F. & C. A. SHUGRUE. 1992. cDNA cloning of a renal Na^+-Ca^{2+} exchanger. Am. J. Physiol. **262:** F1105–F1109.
6. KOFUJI, P., W. J. LEDERER & D. H. SCHULZE. 1993. Na/Ca exchanger isoforms expressed in kidney. Am. J. Physiol. **265:** F598–F603.
7. FURMAN, I., O. COOK, J. KAZIR & H. RAHAMIMOFF. 1993. Cloning of two isoforms of the rat brain Na^+-Ca^{2+} exchanger gene and their functional expression in HeLa cells. FEBS Lett. **319:** 105–109.
8. KOFUJI, P., W. J. LEDERER & D. H. SCHULZE. 1994. Mutually exclusive and cassette exons underlie alternatively spliced isoforms of the Na/Ca exchanger. J. Biol. Chem. **269:** 5145–5149.
9. MATSUOKA, S., D. A. NICOLL, R. F. REILLY, D. W. HILGEMANN & K. D. PHILIPSON. 1993. Initial localization of regulatory regions of the cardiac sarcolemmal Na^+-Ca^{2+} exchanger. Proc. Natl. Acad. Sci. USA **90:** 3870–3874.
10. KIMURA, M., A. AVIV & J. P. REEVES. 1993. K^+ dependent Na^+/Ca^{2+} exchange in human platelets. J. Biol. Chem. **268:** 6874–6877.
11. SMITH, L. & J. B. SMITH. 1994. Regulation of sodium-calcium exchanger by glucocorticoids and growth factors in vascular smooth muscle. J. Biol. Chem. **269**(44): 27527–27531.
12. AUSUBEL, F. M. et al., Eds. 1987. Current Protocols in Molecular Biology. Vol. 1 & 2. Greene Publishing Associates, Inc. & John Wiley and Sons, Inc. New York.
13. KOFUJI, P., R. W. HADLEY, R. S. KIEVAL, W. J. LEDERER & D. H. SCHULZE. 1992. Expression of the Na-Ca exchanger in diverse tissues: a study using cloned human cardiac Na-Ca exchanger. Am. J. Physiol. **263:** C1241–C1249.
14. LEE, S-L., A. S. L. YU & J. LYTTON. 1994. Tissue-specific expression of Na-Ca exchanger isoforms. J. Biol. Chem. **269:** 14849–14852.
15. KOMURO, I., K. E. WENNINGER, K. D. PHILIPSON & S. IZUMO. 1992. Molecular cloning and characterization of the human cardiac Na^+/Ca^{2+} exchanger cDNA. Proc. Natl. Acad. Sci. USA **89:** 4769–4773.
16. GREEB, J. & G. E. SHULL. 1989. Molecular cloning of a third isoform of the calmodulin-sensitive plasma-membrane Ca^{2+} transporting ATPase that is expressed predominantly in brain and skeletal muscle. J. Biol. Chem. **264:** 18569–18576.
17. STREHLER, E. E., P. JAMES, R. FISCHER, R. HEIM, T. VORHERR, A. G. FILOTEO, J. T. PENNISTON & E. CARAFOLI. 1990. Peptide sequence analysis and molecular cloning reveal two calcium pump isoforms in the human erythrocyte membrane. J. Biol. Chem. **265:** 2835–2842.
18. WANG, Z., J. ORLOWSKI & G. E. SHULL. 1993. Primary structure and functional expression of a novel gastrointestinal isoform of the rat Na^+/H^+ exchanger. J. Biol. Chem. **268:** 11925–11928.
19. THORENS, B. 1993. Facilitated glucose transporters in epithelial cells. Annu. Rev. Physiol. **55:** 591–608.
20. LEVITSKY, D. O., D. A. NICOLL & K. D. PHILIPSON. 1994. Identification of the high affinity Ca^{2+}-binding domain of the cardiac Na^+-Ca^{2+} exchanger. J. Biol. Chem. **269**(36): 22847–22852.

Expression of Na^+-Ca^{2+} Exchanger with Modified C-Terminal Hydrophobic Domains and Enhanced Activity

N. GABELLINI,[a,c] T. IWATA,[b] AND E. CARAFOLI[a,b]

[a]*Department of Biological Chemistry*
University of Padova
Via Trieste 75
35121 Padova, Italy

[b]*Institute of Biochemistry*
Swiss Federal Institute of Technology (ETH)
8092 Zürich, Switzerland

INTRODUCTION

The first cloning of the Na^+-Ca^{2+} exchanger (NCE) showed that the coding sequence was located in the 5' proximal segment (3 Kb) of a larger transcript (7 Kb).[1] The deduced protein sequence consists of 970 amino acids with a molecular mass of 108 KDa. The primary structure indicates a cleavable NH_2-terminal signal peptide and 11 transmembrane (TM) domains with a large hydrophilic loop between TM V and VI. Several tissue-specific isoforms are produced by alternative splicing of a single gene product.[2,3] A second gene for the Na^+-Ca^{2+} exchanger was found to be preferentially expressed in brain.[4] The alternative splicing mechanism provides a great variability in the region encoding the large hydrophilic loop. In the predicted topology of the protein this region protrudes into the cytoplasm, and includes important regulatory domains, *e.g.*, a calcium regulatory site and the site of the inhibitory peptide XIP.[5] Additional variability is determined by splicing in the C-terminal hydrophobic domains of the exchanger.[6] Expression studies of the cloned cDNA in several systems have indicated that the product of the 3-Kb mRNA is the active protein.[1,7,8] However, expression of a larger segment (6 Kb) of cDNA produced shortened versions of the protein with different C-terminal sequences and with increased Ca^{2+} uptake activity.[6]

A 5' Alternative Splicing Modifies the C-Terminal Hydrophobic Segments

To investigate the function of the "3' untranslated" region present in the exchanger mRNA, different lengths of this region (3.7 Kb and 6 Kb of the canine NCE sequence) were cloned in an expression vector (pcDNAI) under the control of the CMV (cytomegalovirus) promoter.[6] They were used to transiently transfect cultured 293 cells by the calcium phosphate coprecipitation procedure. The expression products that were analyzed in parallel by Northern and Western blotting revealed that equally high levels of mRNA and proteins were expressed by the

[c] Corresponding author.

two constructs. A discrepancy in the expected mRNA length was observed in cells expressing the 6 Kb cDNA: the main transcript was shorter than expected (4 Kb). Moreover, the polypeptide pattern revealed by a specific antibody directed to the 120-KDa NCE polypeptide from heart sarcolemma was similar but not identical in the two types of transfected cells: in the products of the 6-Kb cDNA, an additional polypeptide could be observed in the 120-KDa region. These results suggested that the observed modification in the mRNA could alter the coding sequence.

To explore the molecular basis of this variability, the mRNA was extracted from cells expressing the 6-Kb cDNA, reverse transcribed using oligo dT as primer, and amplified by PCR. Amplification from the SP6 primer located in the vicinity of the 3' end of the transcripts, in combination with a specific primer at position 3043, produced a DNA fragment of smaller than expected size. Using a primer from position 1666 in combination with SP6, three fragments of irregular sizes were amplified. Cloning and sequencing of these amplified fragments revealed the presence of a 5' alternative splicing site in the 6-Kb transcript. The 3' splicing site was located 47 bp upstream of the 6-Kb cDNA, and was found to be alternatively connected to four upstream positions. The spliced products, shown in FIGURE 1, are numbered I to IV according to their length:

I) The 5' splicing site is at G-3198. This cleavage produces the shortest intron excision. This site is the most commonly processed in 293 cells.[6,9] The modification occurs downstream of the coding sequence, which therefore conserves the proposed sequence and topology,[1] as shown in FIGURE 1.

II) The 5' splicing site in this case is at G-2821. The exon sequence links up in frame with Gly-931 in the fifth putative intracellular loop and encodes a hydrophobic segment. The exon sequence is shown in FIGURE 2. Its length and polarity would be adequate to replace the 11th TM segment of the proposed model shown in FIGURE 1. The new exon is not completely included in the 6-Kb cDNA, the translation continuing with an artificial sequence encoded by the vector pcDNAI as shown in FIGURE 2. The predicted protein could conserve the 11 TM domains, with the possible replacement of the last segment.

III) The 5' splicing site is at G-2620 and modifies the NCE protein from Ile-864 in the fourth extracellular loop. The sequence continues with the same exon frame described for alternative splicing II, and could possibly replace TM segment 9. The resulting protein is shorter including only nine TM, the last of which would be replaced (FIG. 1).

IV) This is the shortest splicing product identified. The 5' splicing site is at G-1844 and the predicted isoform is modified after Lys-606. This is linked up in a different frame of the spliced-in exon that encodes one amino acid (Leu) followed by a stop codon. The encoded protein is much shorter, having only the five NH_2-terminal TM segments and terminating in the large intracellular loop.

The complete DNA sequence of the 3' untranslated region of the canine exchanger was determined by Alexander Kraev (ETH, Zurich, Switzerland) and was submitted to the EMBL Nucleotide Sequence Database, accession No. Z49266CFNCXUTR.

The Expression of Modified Exchanger with Increased Ca^{2+} Uptake Activity

Expression of the 3.7-Kb cDNA encoding the originally proposed Na^+-Ca^{2+} exchanger in 293 cells produced a Ca^{2+} uptake activity of 6 nMol/min/mg protein. The Ca^{2+} uptake of 293 cells expressing the 6-Kb cDNA was in the range of 18

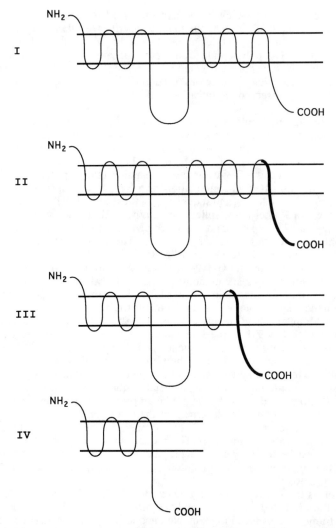

FIGURE 1. A model of the four types of Na^+-Ca^{2+} exchanger expressed from the 6-Kb cDNA. The C-terminal segments of isoforms II and III encoded by the spliced-in exon are in bold.

nmol/min/mg protein.[6] The results indicated that the structural differences in the C-terminal region of the protein produced by the alternative splicing positively influenced the transport function. Although at present it is not possible to trace back the enhancement of activity to any one of the splicing variants in particular, the unexpected possibility that the deletion of two to five C-terminal TM domains is compatible with the maintenance of activity, and even with its stimulation, must be considered. One could speculate that the deletions relieve a negative control on the activity, or that the sequence encoded by the spliced-in exon replacing TM segments (11 and 10) functions as an enhancer.

                    5'exon   vector

pcNCE6Kb   ALGGKCTNCAVPYAAANSCSPGDPLVLEGPIL*

pcNCEΔ17   ALGGKCTNCAVPYAAANSCSPGESL*

pcNCEΔ6   ALGGKCPYSIGT*

pcNCEΔ3   ALGGKSLFYRCHLNARGSL*

FIGURE 2. The C-terminal sequence of isoforms II and III produced by cell transfection with pcNCE6Kb. The aminoacid shown in bold is encoded in the exon sequence partially included at the 3' end of the 6-Kb cDNA. The remainder of the sequence is encoded by the vector. The C-terminal sequence generated by bidirectional deletions in clones Δ17, Δ6 and Δ3 is aligned for comparison.

The splicing products II and III (shown in FIG. 1) included in their C-termini an artificial sequence encoded by the vector. This sequence was deleted to exclude the possibility of its interference with the transport activity. Bidirectional digestion of this region was performed by a short treatment with exonuclease III from the unique BamHI site in the polylinker of the vector located at the 3' end of the 6-Kb cDNA insert.[6] The extent of the deletions was evaluated by restriction cleavage and DNA sequencing. Three clones were selected for functional analysis; their C-terminal sequence is shown in FIGURE 2. The Ca^{2+} uptake activity was measured in whole cells transfected with each type of deletion mutant as previously described.[6] The transport activity of the expressed protein measured 24–48 hours from transfection with pcNCEΔ17 and pcNCEΔ6 was in the range of that measured with the whole 6-Kb construct: 18 nMol/min/mg. This result indicated that the vector sequence did not contribute to the increase in the activity. The recovery of the full activity in the expression products of pcNCEΔ6 also indicated that a portion of the NCE exon sequence was not involved in the enhancement. Interestingly, the Ca^{2+} uptake activity expressed by the deletion mutant pcNCEΔ3 was significantly reduced: 11 nMol/min/mg, as shown in TABLE 1. In this construct the deletion extended to the second Cys of the exon sequence, which was replaced by a Ser. Possibly this Cys residue could be required for the enhancement of the transport activity; alternatively, TM 10 and/or 11, eliminated by the splicing, could contain inhibitory sequences.

TABLE 1. Ca^{2+} Uptake Activity of Cells Transfected with the DNAs Indicated

DNA	% of Uptake Activity
pcNCE6Kb	100
pcNCEΔ17	100
pcNCEΔ6	100
pcNCEΔ3	60

SUMMARY

A 6-Kb canine cDNA fragment complementary to the 5' region of the 7-Kb mRNA encoding the cardiac Na^+-Ca^{2+} exchanger was expressed in human kidney 293 cells. The mRNA products were reverse transcribed and amplified by PCR. The determined DNA sequence of the amplified DNA fragments revealed the presence of an intron that was alternatively spliced. The partial exon sequence, located at the 3' end of the 6-Kb cDNA, was alternatively connected to bases 3198, 2821, 2620 and 1844 in four types of splicing products identified. In the largest product the adjoining exon was located after the putative stop codon of the regular sequence. In a second and third type of shortened transcripts, a hydrophobic sequence encoded by the spliced-in exon was linked with the 4th or the 5th extracellular loops, and could possibly replace transmembrane segments 9 or 11. In the fourth type of spliced transcript the in-frame exon sequence introduced one Leu followed by a stop codon in the large hydrophilic loop. Measurements of Ca^{2+} uptake in 293 cells expressing the modified exchanger indicated a higher activity in comparison with 293 cells expressing the 3.7-Kb cDNA, in which this alternative splicing does not occur. Deletion mutagenesis of the C-terminal region encoded by the spliced-in exon was performed to investigate its role in the enhancement of the transport activity.

REFERENCES

1. NICOLL, D. A., S. LONGONI & K. D. PHILIPSON. 1990. Molecular cloning and functional expression of the cardiac sarcolemmal Na^+-Ca^{2+} exchanger. Science **250:** 562–565.
2. FURMAN, I., O. COOK, J. KASIR & H. RAHAMIMOFF. 1993. Cloning of two isoforms of the rat brain Na^+-Ca^{2+} exchanger gene and their functional expression in HeLa cells. FEBS Lett. **319:** 105–109.
3. KOFUJI, P., W. J. LEDERER & D. H. SCHULZE. 1994. Mutually exclusive and cassette exons underlie alternatively spliced isoforms of the Na/Ca exchanger. J. Biol. Chem. **269:** 5145–5149.
4. LI, Z., S. MATSUOKA, L. V. HRYSHKO, D. A. NICOLL, M. M. BERSOHN, E. P. BURKE, R. P. LIFTON & K. D. PHILIPSON. 1994. Cloning of the NCX2 isoform of the plasma membrane Na^+-Ca^{2+} exchanger. J. Biol. Chem. **269:** 17484–17489.
5. MATSUOKA, S., D. A. NICOLL, R. F. REILLY, D. W. HILGEMANN & K. D. PHILIPSON. 1993. Initial localization of regulatory regions of the cardiac sarcolemmal Na^+-Ca^{2+} exchanger. Proc. Natl. Acad. Sci. USA **90:** 3870–3874.
6. GABELLINI, N., T. IWATA & E. CARAFOLI. 1995. An alternative splicing site modifies the carboxyl-terminal trans-membrane domain of the Na^+/Ca^{2+} exchanger, J. Biol. Chem. **270:** 6917–6924.
7. PIJUAN, V., Y. ZHUANG, L. SMITH, C. KROUPIS, M. CONDRESCU, J. F. ACETO, J. P. REEVES & J. B. SMITH. 1993. Stable expression of the cardiac sodium-calcium exchanger in CHO cells. Am. J. Physiol. **264:** C1066–C1074.
8. KOFUJI, P., R. W. HADLEY, R. S. KIEVAL, W. J. LEDERER & D. H. SCHULZE. 1992. Expression of the Na-Ca exchanger in diverse tissue: a study using the cloned human cardiac Na-Ca exchanger. Am. J. Physiol **263:** C1241–C1249.
9. GREEN, M. R. 1991. Biochemical mechanisms of constitutive and regulated pre-mRNA splicing. Annu. Rev. Cell Biol. **7:** 559–599.

Identification and Antisense Inhibition of Na-Ca Exchange in Renal Epithelial Cells[a]

K. E. WHITE, F. A. GESEK, AND P. A. FRIEDMAN[b]

Department of Pharmacology and Toxicology
Dartmouth Medical School
7650 Remsen
Hanover, New Hampshire 03755-3835

INTRODUCTION

Calcium transport in absorptive epithelial cells is a two-step process. Calcium enters the cell across apical plasma membranes and is extruded across basolateral membranes. Na^+-Ca^{2+} exchange is thought to mediate Ca^{2+} efflux from calcium-absorbing renal epithelial cells.[1] It is proposed that the entry of three sodium ions down their electrochemical gradient into the cell is coupled to the efflux of one calcium ion. The exchanger has been localized to basolateral membranes of distal nephron cells.[2] cDNAs for several renal exchangers have been cloned and contain regions that undergo alternative splicing to produce distinct isoforms.[3-6] The purpose of this study was to determine the presence and isoforms of Na^+-Ca^{2+} exchangers in distal convoluted tubule (DCT) cells, and to test their function through the use of antisense oligonucleotides.

METHODS AND RESULTS

Partial clones encoding a downstream region, amino acids (785–903),[7] of the Na-Ca exchanger were identified by an homology-based cloning strategy. Reverse transcriptase-polymerase chain reaction (RT-PCR) was performed with RNA from a mouse DCT cell line[8] using deoxyoligonucleotide primers specific for the exchanger. The amplified product was subcloned and sequenced and possesses 98% homology to the rat kidney exchanger within this region.

The transcript encoding the putative intracellular loop located between the fifth and sixth transmembrane domains of the Na-Ca exchanger undergoes alternative splicing to generate several isoforms. The portion of the gene encoding the variable region has been identified and contains six exons labeled A–F.[5] We have identified transcripts encoding the kidney isoforms NACA2 (exons B, C, D) and NACA3 (exons B and D), as well as the previously described brain isoform, NACA6 (exons A, C, and D) in the DCT cell line (FIG. 1).

To assess functional Na-Ca exchange, the direction of exchange was reversed by loading the cells with sodium and measuring free intracellular calcium ($[Ca^{2+}]_i$)

---

[a] These studies were supported by National Institutes of Health Grant GM-34399.
[b] Corresponding author.

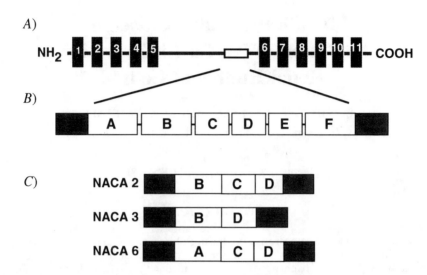

FIGURE 1. (A) Putative protein structure of Na^+-Ca^{2+} exchanger. Transmembrane domains are numbered 1–11. The large intracellular loop contains a region that undergoes tissue-specific alternative splicing (*boxed*). **(B)** Gene structure of the variable region of the intracellular loop. Six exons are spliced to form mature mRNA. Exons A and B are mutually exclusive, while any of exons C–F can be expressed. **(C)** Exchanger isoforms in DCT cells. NACA2 and NACA3 are previously identified kidney isoforms, whereas NACA6 was thought to be exclusively localized to brain.

in single cells with fura-2.[8] DCT cells were plated on glass coverslips, loaded with the Ca^{2+}-sensitive dye fura-2 or in separate experiments, the Na^+-sensitive dye SBFI.[9] Cells were mounted on an inverted microscope and assessed for Na^+-Ca^{2+} exchange activity by monitoring changes of epifluorescence.

Resting $[Ca^{2+}]_i$ and $[Na^+]_i$ averaged 100 nM and 7.5 mM, respectively. $[Na^+]_i$ was increased by inhibiting the Na^+-K^+ ATPase with ouabain, and nifedipine was used to block Ca^{2+} channels. During this treatment, there was a gradual elevation of $[Na^+]_i$ and $[Ca^{2+}]_i$. Isosmotic replacement of extracellular Na^+ with tetramethylammonium reversed the direction of Na-Ca exchange and caused an abrupt increase of $[Ca^{2+}]_i$ by 240% and coincident with a decrease of $[Na^+]_i$ by 40% (FIG. 2A). To determine the source of the rise $[Ca^{2+}]_i$, the experiment was repeated in the absence of extracellular Ca^{2+}. Exposure of DCT cells to Na^+- and Ca^{2+}-free buffer inhibited the rise of $[Ca^{2+}]_i$ by 70%. These findings indicate that the majority of the Na^+-dependent increase of $[Ca^{2+}]_i$ originated from extracellular Ca^{2+} entry and demonstrates the presence of Na^+-Ca^{2+} exchange with Ca^{2+} transport in DCT cells mediated by Na-Ca exchange.

To relate structure and function, antisense oligonucleotides (ODNs) were used to assess exchange activity in DCT cells. Antisense and sense 20 bp ODNs were designed to a downstream region of the cloned DCT NACA2, NACA3, and NACA6 isoforms. ODNs were introduced into DCT cells after brief permeabilization with streptolysin O in the presence of 100 nM ODN.[10] Control cells were permeabilized but not treated with ODN. After 18–24 hr, Na^+-Ca^{2+} exchange activity was analyzed by measuring changes of $[Ca^{2+}]_i$ as described above. The rise of $[Ca^{2+}]_i$ was

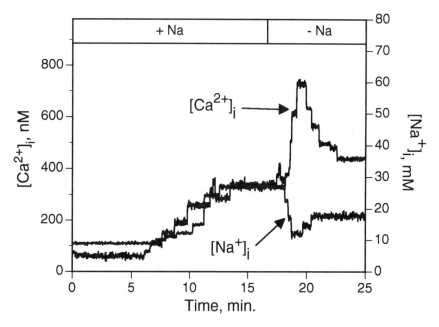

FIGURE 2A. Na^+-Ca^{2+} exchange in DCT cells. Cells were loaded with fura-2 or SBFI to measure free intracellular Ca^{2+} ($[Ca^{2+}]_i$) or Na^+ ($[Na^+]_i$), respectively, in single cells in separate experiments. Cells were treated with ouabain to increase $[Na^+]_i$ and nifedipine to block Ca^{2+} channels. Upon reversal of the Na^+ gradient by isosmotic replacement of extracellular Na^+, $[Ca^{2+}]_i$ increased concomitantly with the decrease of $[Na^+]_i$.

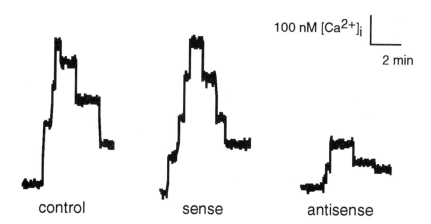

FIGURE 2B. Representative traces of changes of intracellular Ca^{2+} ($[Ca^{2+}]_i$) in DCT cells as a function of time. Cells were permeabilized with streptolysin O in the presence of sense or antisense oligonucleotides to a downstream region of the Na^+-Ca^{2+} exchanger. Control cells were permeabilized but not treated with oligonucleotides.

reduced 60% in antisense ODN-treated cells compared to control cells or cells treated and the sense ODN (FIG. 2B).

SUMMARY AND CONCLUSION

In summary, DCT cells express multiple isoforms of the Na-Ca exchanger and exhibit functional exchange, and antisense oligonucleotides to a downstream region of the exchanger transcript inhibit activity. These experiments provide direct evidence for Na-Ca exchange in DCT cells mediated by NACA2, NACA3, or NACA6.

REFERENCES

1. BOURDEAU, J. E. 1993. Semin. Nephrol. **13:** 191–201.
2. REILLY, R. F., C. A. SHUGRUE, D. LATTANZI & D. BIEMESDERFER. 1993. Am. J. Physiol. **265:** F327–F332.
3. REILLY, R. F. & C. A. SHUGRUE. 1992. Am. J. Physiol. **262:** F1105–F1109.
4. YU, A. S. L., S. C. HEBERT, S-L. LEE, B. M. BRENNER & J. LYTTON. 1992. Am. J. Physiol. **263:** F680–F685.
5. KOFUJI, P., W. J. LEDERER & D. H. SCHULZE. 1994. J. Biol. Chem. **269:** 5145–5149.
6. LEE, S-L., A. S. L. YU & J. LYTTON. 1994. J. Biol. Chem. **269:** 14849–14852.
7. NICOLL, D. A., S. LONGONI & K. D. PHILIPSON. 1990. Science **250:** 562–565.
8. GESEK, F. A. & P. A. FRIEDMAN. 1992. J. Clin. Invest. **90:** 749–758.
9. GESEK, F. A. 1993. Am. J. Physiol. **265:** F561–F568.
10. BARRY, E. L. R., F. A. GESEK & P. A. FRIEDMAN. 1993. BioTechniques **15:** 1016–1020.

Effects of NCX1 Antisense Oligodeoxynucleotides on Cardiac Myocytes and Primary Neurons in Culture[a]

K. S. BLAND, K. TAKAHASHI, S. ISLAM, AND
M. L. MICHAELIS[b]

Department of Pharmacology/Toxicology
and
Center for Neurobiology and Immunology Research
University of Kansas
Lawrence, Kansas 66045

The molecular characterization of the Na^+-Ca^{2+} protein NCX1 has led to a wealth of information about the structural properties of the transporter, the multiplicity of splice variants, and the chromosomal localization of the gene for the exchanger.[1-3] Nevertheless, there is still a great deal of argument and ambiguity regarding the functional properties of the exchanger in various tissues, particularly in excitable cells such as cardiac myocytes and brain neurons. In efforts to probe the contributions of exchanger activity to Ca^{2+}-regulated events in intact cells, we designed antisense (AS) oligodeoxynucleotides for NCX1 to assess the effects of decreasing expression of this exchanger on primary cells in culture.

In one series of studies, we prepared cardiac myocytes from day-18 embryonic rats. In these cultures ~90% of the myocytes contracted spontaneously.[4] Twenty-four hours after plating, a 20-mer AS oligo complementary to a region near the 3' end of NCX1 was added to the cultures at a final concentration of 10 μM. A second addition (5 μM) was made 20 h later, and the assays were carried out 4 h after the second addition. Control cultures were treated with vehicle only (lipofectamine) or with the 20-mer sense (S) oligo. Exchanger activity was assessed in myocytes using fura-2 to monitor Na^+-mediated Ca^{2+} influx. Beating rates of cultured myocardial cells were measured at 23°C with an inverted phase-contrast microscope.[4] Essentially similar conditions were used to treat rat primary cortical neurons (embryonic day 18), and exchanger activity was measured as described previously with neuronal cultures.[5] Neuronal cell viability was monitored via the mitochondrial dehydrogenase assay (MTT).

In cardiac myocytes the AS treatment described above led to an ~30% decrease in exchanger activity in 5 culture preparations. AS-treated cells also consistently exhibited higher beating rates than control or S-treated cells. Mean beating rates in 6 culture preparations were: control cells, 40 ± 2 (n = 142); S-treated cells, 41 ± 1 (n = 285); and AS-treated cells, 69 ± 1 (n = 309). Enhancement of the mean beating rate in AS-treated cells was statistically significant ($p < 0.01$),

---

[a] This work was supported by PHS Grant #AA 04732, the Alzheimer's Association, and the American Heart Association, KS Affiliate.
[b] Corresponding author.

TABLE 1. Effects of NMDA and NCX1 Oligomers on Survival of Primary Cortical Neurons in Culture

Treatment	Neuronal Cell Death (% of Control)[a]
1 mM NMDA	20%
2 mM NMDA	39%
AS Oligo	38%
AS + 1 mM NMDA	52%
AS + 2 mM NMDA	68%

[a] Data represent the mean percentage of cells lost in 6 culture preparations with triplicate determinations for each condition with each culture.

regardless of whether AS cells were compared with control or S-treated cells. Thus reduced expression of NCX1 activity led to changes in $[Ca^{2+}]_i$ regulation that substantially altered the duration of the cardiac cycle.

Treatment of primary neurons with AS oligos produced a dose-dependent decrease in exchanger activity with a maximal reduction to ~50% of the activity present in untreated or S-treated cells. Maximal suppression was observed 12 h after the second AS addition, with either 5 or 10 μM AS oligo. Higher concentrations did not decrease activity further. Since disruption of $[Ca^{2+}]_i$ regulation is quite toxic to neurons, we examined the viability of neuronal cultures treated with the NCX1 AS oligo under resting conditions and under conditions that lead to a potentially toxic influx of Ca^{2+} mediated by activation of N-methyl-D-aspartate (NMDA) receptors. TABLE 1 shows the effect of both AS and NMDA treatments (12 h exposure in normal culture medium) on the percentage of surviving cells in neuronal cultures. Both the NCX1 AS treatment alone and exposure to NMDA led to the loss of a substantial proportion of cells. The combination of AS treatment and exposure to NMDA led to an additive loss of cells. It was somewhat surprising that exposure of AS-treated cells to NMDA did not lead to greater than additive cell loss. It is possible that the remaining exchanger activity is adequate to protect cells from any synergistic effects of a toxic insult. On the other hand, it may be that NCX1 does not play a critical role in the extrusion of Ca^{2+} entering neurons as a result of NMDA receptor activation.

These initial observations with primary neurons and myocardial cells demonstrate that AS oligo technology can be used with intact cells to enhance our understanding of the role the Na^+-Ca^{2+} exchanger plays in Ca^{2+} signaling events. It should be possible to find conditions which lead to an even greater suppression of expression of NCX activity, thereby providing critical new information about the physiological actions of this important Ca^{2+} transporter.

REFERENCES

1. NICOLL, D. A., S. L. LONGONI & K. D. PHILIPSON. 1990. Science **250:** 562–565.
2. KOMURO, I., K. E. WENNINGER, K. D. PHILIPSON & S. IZUMO. 1992. Proc. Natl Acad. Sci. USA **89:** 4769–4773.
3. SHIEH, B. H., Y. XIA, R. S. SPARKES, I. KLISAK, A. J. LUSIS, D. A. NICOLL & K. D. PHILIPSON. 1992. Genomics **12:** 616–617.
4. TAKAHASHI, K., Y. FUJITA, T. MAYUMI, T. HAMA & T. KISHI. 1987. Chem. Pharm. Bull. **35:** 326–334.
5. MICHAELIS, M. L., J. L. WALSH, R. PAL, M. HURLBERT, G. HOEL, K. BLAND, J. FOYE & W. H. KWONG. 1994. Brain Res. **661:** 104–116.

Initial Characterization of the Feline Sodium-Calcium Exchanger Gene[a]

KIMBERLY V. BARNES, MYRA M. DAWSON, AND
DONALD R. MENICK[b]

Cardiology Division
Department of Medicine
and
Gazes Cardiac Research Institute
Medical University of South Carolina
Charleston, South Carolina 29425

The sodium-calcium exchanger plays a key role in calcium efflux and, therefore, in the control and regulation of intracellular calcium in the cardiocyte. We have previously shown the exchanger to be acutely sensitive to changes in pressure-induced load; one hour of pressure overload produces a two-to-fourfold increase in the level of exchanger mRNA. Maintenance of this load for forty-eight hours results in increased exchanger protein.[1] Message half-life analysis implicates a transcriptional mechanism. Presently we have determined the upregulation of the exchanger to be persistent with maintenance of load as long as two weeks.[2] Here we report identification of brain, kidney, and cardiac isoforms of the feline exchanger and outline their patterns of expression in the adult and fetus.

Cardiocyte cDNA library screening and 5′ rapid amplification of cDNA ends (5′RACE) were utilized to clone the 5′ untranslated regions (UTRs) of cardiac exchanger mRNA (FIG. 1). 5′RACE was carried out using a primer specific for bases 250–270 of the open reading frame. Amplification resulted in two products approximately 650 and 375 base pairs in size. The 3′ most 250 bases of both bands were identical to that of the exchanger open reading frame. The smaller band (cardiac isoform 1) extends 122 bases 5′ of the translational start and shares 68% identity with bases −112 to −34 of the rat brain 2 clone[3] and 84% homology with bases −268 to −218 of the bovine p17 clone.[4] The most 5′ 122 bases of the larger band (cardiac isoform 2) are identical to those of isoform 1. Three classes of clones were isolated by library screening with a PCR probe specific for bases +1 to +85 of the open reading frame. Two contained 5′UTR sequence identical to that of either isoform 1 or 2 described above. The third (cardiac isoform 3) was 100% identical to isoform 2 for 117 bases 5′ of the translational start, then diverged for another 106 bases.

The 5′UTRs of the exchanger transcripts expressed in brain and kidney were identified utilizing 5′RACE (FIG. 1). Four unique products were uncovered in brain (brain 1–brain 4) which extended 123, 203, 102, and 66 base pairs 5′ of the translational start, respectively. Cat brain 1 has 60% homology with the rat brain 1 clone.[3] The other brain clones have no homology with any other known sequences. One 78-nucleotide product was found in kidney which was identical to

---

[a] This research is supported in part by National Institutes of Health Grant HL48788.

[b] Corresponding author: Donald R. Menick, Ph.D., Cardiology Division, Medical University of South Carolina, Charleston, SC 29425-2221.

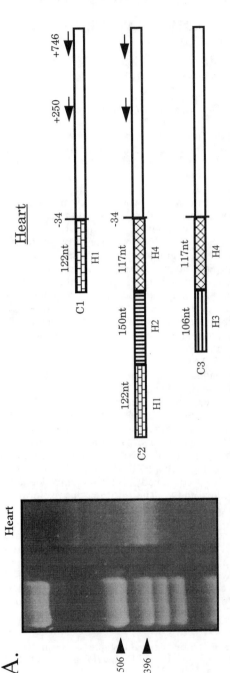

FIGURE 1. 5'RACE products from heart, kidney and brain. A primer 746 nucleotides 5' of the AUG was used to reverse transcribe first strand cDNA using total RNA from heart, brain, and kidney as template. Thirty-five rounds of amplification of each cDNA were carried out using a gene specific primer 250 bases 3' of the AUG. The kidney and brain reactions were then reamplified with a nested primer 165 bases upstream. The 5'RACE products were separated by agarose gel electrophoresis. Two products were seen in heart, one in kidney, and four in brain. The products were subcloned and sequenced. Each is identical from its 3' end up to −34 with the exchanger open reading frame. **(A)** The sizes of the cardiac products, C1 and C2, are 122 and 394 base pairs. Clone C3 was obtained from our feline cardiac cDNA library and contains 223 base pairs of 5'UTR. **(B)** The four brain products, Br 1–Br 4, are 125, 281, 102 and 66 base pairs, respectively. The kidney product, K1, is 78 base pairs in size and is identical to the 3' end of Br 2. The cardiac forms have been mapped to genomic clones and are formed by alternative splicing of four exons, H1, H2, H3, and H4.

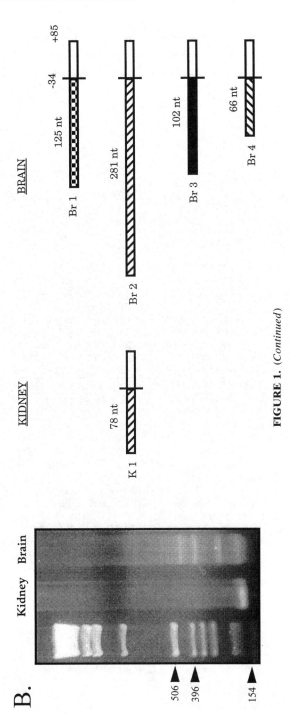

FIGURE 1. (Continued)

TABLE 1. Northern Analysis of the 5' Untranslated Regions[a]

	Probe			
Tissue	H1	H2	H3	Br1
Fetal				
Heart	+++	+++	+++	n.d.
Brain	++		+	n.d.
Kidney		+	+	n.d.
Liver				
Adult				
Heart	+++	+++	+++	+
Brain	+		++	+++
Kidney	+			+
Liver				

[a] 15 µg of total RNA isolated from the adult and fetal tissues were electrophoresed, transferred to Duralon-UV membrane, and hybridized to PCR probes generated from exons H1, H2, H3, or Br1. The number of (+) indicates the relative level of intensity of a 7.2 kb band. (n.d.) denotes hybridizations not yet completed.

the 3' end of the brain 2 isoform. TABLE 1 summarizes the adult and fetal tissue distribution of each isoform.

There is an area of variable sequence in the putative intracellular loop which results from alternative combination of 6 exons.[5] PCR of first strand cDNA was carried out to analyze the loop variation in adult versus fetal heart, brain, and kidney as well as pressure overloaded versus control heart. Feline adult and fetal products were visible which represented those reported previously in rabbit by Kofuji *et al*. Interestingly, in addition to the expected cardiac product, a second band approximately 50 base pairs smaller in size was amplified from both adult and fetal cDNA. A third smaller product was seen in pressure overloaded but not in normal adult heart. Similarly, an additional product approximately 75 base pairs smaller than observed in the rabbit NACA3[5] was seen in fetal but not adult kidney cDNA. The significance of these findings is presently being determined.

The H1, H2, H3, and H4 regions (FIG. 1) of the cardiac 5'UTRs have been used to screen a feline genomic library (Stratagene). Three classes of clones have been identified. The first class (R clones) contains exon H1 (the most 5' 122 bases of isoform 2, identical to all of the 5' untranslated sequence present in isoform 1). The second class (M clones) contains exon H2 (150 bases of isoform 2, sequence 3' of H1). Finally, P clones contain exon H4 (the remaining 117 bases of isoform 2, also found in isoform 3). P clones also contain H3 (the most 5' 106 bases of isoform 3). All sequences of the cardiac isoforms have been mapped to our genomic clones. Southern analysis shows that brain 1 lies on the M clone. We are continuing to map the remaining isoforms.

REFERENCES

1. KENT, R. L., J. D. ROZICH, P. L., MCCOLLAM, D. E. MCDERMOTT, U. F. THACKER, D. R. MENICK, P. J. MCDERMOTT & G. COOPER IV. 1993. Am. J. Physiol. **265** (Heart Circ. Physiol. 34): H1024–H1029.
2. MENICK, D. R., K. V. BARNES, U. F. THACKER, M. M. DAWSON, D. E. MCDERMOTT, J. D. ROZICH, R. L. KENT & G. COOPER IV. This volume.

3. LEE, S., A. S. L. YU & J. LYTTO. 1994. J. Biol. Chem. **269**(21): 14849–14852.
4. ACETO, J. F., M. CONDRESCU, C. KROUPIS, H. NELSON, N. NELSON, D. NICOLL, K. PHILIPSON & J. P. REEVES. 1992. Arch. Biochem. Biophys. **298**(2): 553–560.
5. KOFUJI, P., W. J. LEDERER & D. H. SCHULZE. 1994. J. Biol. Chem. **269**(7): 5145–5149.

Cloning of the Mouse Cardiac Na^+-Ca^{2+} Exchanger and Functional Expression in *Xenopus* Oocytes

INJUNE KIM AND CHIN O. LEE[a]

Department of Life Science
Pohang University of Science and Technology
Pohang 790-784, Korea

The Na^+-Ca^{2+} exchanger is known to regulate intracellular Ca^{2+} concentration using the Na^+ electrochemical gradient in a variety of cells. In cardiac muscle cells, sarcolemmal Na^+-Ca^{2+} exchanger is the major mechanism in returning the myocardial cells to the resting state by extruding Ca^{2+} which enters the cells during excitation.

The mouse is often used as a mammalian system for transgenic animal and for developmental studies. The mouse cardiac Na^+-Ca^{2+} exchanger has not been cloned although other Na^+-Ca^{2+} exchangers have been cloned from several animal species. In the present study, the Na^+-Ca^{2+} exchanger was cloned from the mouse heart to investigate its characteristics using the transgenic mouse. We report the cloning, sequencing, and functional expression in *Xenopus* oocytes of the mouse cardiac Na^+-Ca^{2+} exchanger.

The full length cDNA of the mouse cardiac Na^+-Ca^{2+} exchanger was cloned by connecting four RT-PCR products. The primers for PCR were designed on the basis of the published sequences of other cardiac Na^+-Ca^{2+} exchangers. The sequence showed a single open reading frame of 2910 nucleotides encoding a protein of 970 amino acids (FIG. 1A). Sequence analysis showed that the mouse cardiac Na^+-Ca^{2+} exchanger has 12 putative transmembrane regions and one large cytoplasmic loop like other Na^+-Ca^{2+} exchangers. The deduced amino acid sequence of the mouse clone was about 98% and 95% identical to rat[1] and other cardiac exchangers, including canine,[2] bovine,[3] human,[4,5] and guinea pig,[6] respectively. Although the cardiac exchangers of several species are well conserved in primary structure and hydropathy analysis, the least difference between mouse and rat suggests that the mouse cardiac Na^+-Ca^{2+} exchanger is similar to the rat cardiac exchanger in functions.

cRNA was synthesized *in vitro* and injected into *Xenopus* oocytes to test whether the function of the cloned cDNA could be expressed. After 3-day cultured oocytes were loaded with Na^+ using nystatin, the Na^+-Ca^{2+} exchanger activity was assayed by measuring Na^+ gradient-dependent $^{45}Ca^{2+}$ uptake in either 90 mM choline chloride or 90 mM NaCl (FIG. 2). The Ca^{2+} uptake in cRNA-injected oocytes was hundreds-fold greater than that in water-injected oocytes, confirming that this cloned cDNA encoded the functional Na^+-Ca^{2+} exchanger and that the Na^+-Ca^{2+} exchanger was expressed functionally in *Xenopus* oocytes. Interestingly, the Ca^{2+} uptake of cRNA-injected oocytes in 90 mM NaCl was substantial whereas no Ca^{2+} uptake of water-injected oocytes was observed (FIG. 2). In the

---

[a] Corresponding author.

studies of other Na^+-Ca^{2+} exchangers, Ca^2 uptake of cRNA-injected oocytes was insignificant in the absence of Na^+ gradient across the oocyte membrane. The reason that a substantial Ca^{2+} uptake of cRNA-injected oocytes in the absence of Na^+ gradient was observed in our study is not clear. Further study is required to clarify the reason.

Large variations in sequence can be seen in the cleaved leader peptide[7-9] of the first transmembrane region (FIG. 1B). Only 3 amino acids are different between the leader peptide of mouse Na^+-Ca^{2+} exchanger and that of rat Na^+-Ca^{2+} exchanger. However, there are 10 or 11 different amino acids among the leader peptides of other cardiac Na^+-Ca^{2+} exchangers. In the leader peptides, the nucleotide sequence of the mouse cardiac exchanger is about 97% and 77%–80% identical to rat and other cardiac exchangers, respectively.

We screened the mouse genomic library with the 250-bp mouse specific probe including leader peptide region. One genomic clone was isolated in high stringency

A

```
  1 MLRLSLPPNVSMGFRLVALVALLFSHVDHITADTEAETGGNETTECTGSYYCKKGVILPIWEPQDPSFGDKIARATVYFV
 81 AMVYMFLGVSIIADRFMSSIEVITSQEKEITIKKPNGETTKTTVRIWNETVSNLTLMALGSSAPEILLSVIEVCGHNFTA
161 GDLGPSTIVGSAAFNMFIIIALCVYVVPDGETRKIKHLRVFFVTAAWSIFAYTWLYITLSVSSPGVVEVWEGLLTFFFFP
241 ICVVFAWVADRRLLFYKYVYKRYRAGKQRGMIIEHEGDRPASKTEIEMDGKVVNSHVDNFLDGALVLEVDERDQDDEEAR
321 REMARILKELKQKHPEKEIEQLIELANYQVLSQQQKSRAFYRIQATRLMTGAGNILKRHAADQARKAVSMHEVNMEMAEN
401 DPVSKIFFEQGTYQCLENCGTEALTIMRRGGDLSTTVFVDFRTEDGTANAASDYEFTEGTVIFKPGETQKEIRVGIIDDD
481 IFEEDENFLVHLSNVRVSSDVSEDGILESNHVSSIACLGSPSTATITIFDDDHAGIFTFEEPVTHVSESIGIMEVKVLRT
561 SGARGNVIIPYKTIEGTARGGGEDFEDTCGEPEFQNDEIVKTISVKVIDDEEYEKNKTFFIEIGEPRLVEMSEKKALLLN
641 ELGGFTLTGKEMYGQPIFRKVHARDHPIPSTVITISEEYDDKQPLTSKEEEERRIAEMGRPILGEHTKLEVIIQESYEFK
721 STVDKLIKKTNLALVVGTNSWREQFIEAITVSAGEDDDDDECGEEKLPSCFDYVMHFLTVFWKVLFAFVPPTEYWNGWAC
801 FIVSILMIGLLTAFIGDLASHFGCTIGLKDSVTAVVFVALGTSVPDTFASKVAATQDQYADASIGNVTGSNAVNVFLGIG
881 VAWSIAAIYHAANGEQFKVSPGTLAFSVTLFTIFAFINVGVLLYRRRPEIGGELGGPRTAKLLTSSLFVLLWLLYIFFSS
961 LEAYCHIKGF
```

B

mouse	MLRLSLPPNVSMGFRLVALVALLFSHVDHITA				
rat		T L	T		
canine	Q R L TF	CH L V		L S	
bovine	QF S TL	HVI M		S	
human	R S TF	H LVT S		VI	
guinea pig	S TY L	H L MMT	I		

FIGURE 1. Deduced amino acid sequence of the cardiac Na^+-Ca^{2+} exchangers. **(A)** The deduced amino acid sequence of the mouse cardiac exchanger. The putative transmembrane regions are *underlined*. **(B)** Comparison of the cleaved leader peptide regions of mouse, rat, canine, bovine, human and guinea pig cardiac exchangers. Only different amino acids are shown.

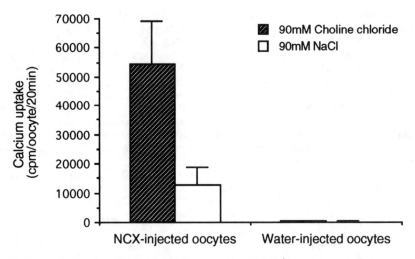

FIGURE 2. Expression of the cloned mouse cardiac Na^+-Ca^{2+} exchanger in *Xenopus* oocytes. Oocytes were injected with 40 nl of water or 40 ng of cRNA. Ca^{2+} uptakes of the Na^+-loaded oocytes were measured in Na^+-free solution and high-Na^+ solution, respectively. The Ca^{2+} uptake solutions contained 100 μM $CaCl_2$ and 5 μCi/ml $^{45}Ca^{2+}$. *Bars* represent means ± SD (n = 4 to 9).

and the first coding exon was mapped on this clone. Using this genomic clone, we are presently preparing the targeting vector for a transgenic mouse line with disrupted cardiac Na^+-Ca^{2+} exchanger gene.

REFERENCES

1. Low, W., J. Kasir & H. Rahamimoff. 1993. FEBS Lett. **316:** 63–67.
2. Nicoll, D. A., S. Longoni & K. D. Philipson. 1990. Science **250:** 562–565.
3. Aceto, J. F., M. Condrescu, C. Kroupis, H. Nelson, N. Nelson, D. A. Nicoll, K. D. Philipson & J. P. Reeves. 1992. Arch. Biochem. Biophys. **298:** 553–560.
4. Komuro, I., K. E. Wenninger, K. D. Philipson & S. Izumo. 1992. Proc. Natl. Acad. Sci. USA **89:** 4769–4773.
5. Kofuji, P., R. W. Hadley, R. S. Kieval, W. J. Lederer & D. H. Schulze. 1992. Am. J. Physiol. **263:** C1241–C1249.
6. Tsuruya, Y., M. M. Bersohn, Z. Li, D. A. Nicoll & K. D. Philipson. 1994. Biochim. Biophys. Acta **1196:** 97–99.
7. Von Hejine, G. 1986. Nucleic Acids Res. **14:** 4683–4690.
8. Durkin, J. T., D. C. Ahrens, Y. E. Pan, & J. P. Reeves. 1991. Arch. Biochem. Biophys. **290:** 369–375.
9. Hryshko, L. V., D. A. Nicoll, J. N. Weiss & K. D. Philipson. 1993. Biochim. Biophys. Acta **1151:** 35–42.

Identification of a Novel Alternatively Spliced Isoform of the Na^+-Ca^{2+} Exchanger (NACA8) in Heart

R. F. REILLY AND D. LATTANZI

Section of Nephrology
Department of Medicine
Yale University School of Medicine
and
WHVA Medical Center
P.O. Box 208047
333 Cedar Street
New Haven, Connecticut 06520-8047

To date three members of the Na^+-Ca^{2+} exchanger gene family have been cloned: NCX1;[1] NCX2;[2] and the retina rod outer segment Na^+-Ca^{2+},K^+ exchanger.[3] NCX1

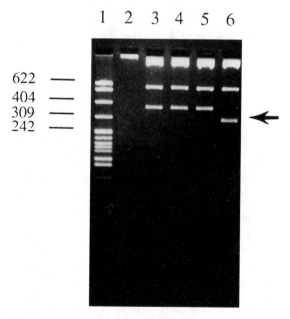

FIGURE 1. Restriction analysis of four minipreparations with *Eco* RI and *Hind* III. Size fractionation was carried out on 1.5% LE-agarose. To the *left* of the gel are listed the DNA size standards. *Lane 1*, DNA size standards (pBR322 DNA-*Msp* I digest); *lane 2*, uncut insert-containing plasmid; and *lanes 3–6*, digests of the four minipreparations. The lower restriction fragment in lane 6 (shown at the *arrowhead*) is approximately 70 base pairs smaller than that of lanes 3, 4, and 5.

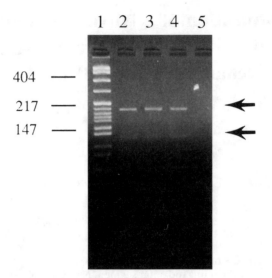

FIGURE 2. Expression of NACA1 and NACA8 examined in the left ventricle, right ventricle, and both atria using RT-PCR. Amplified PCR products were size-fractionated on 3% NuSieve-1% LE-agarose. To the *left* of the gel are listed the DNA size standards. *Lane 1*, DNA size standards (pBR322 DNA-*Msp* I digest); *lanes 2–4*, amplified products from left ventricular, right ventricular, and atrial cDNA, respectively. *Lane 5*, negative control with no template added. Shown at the *arrowheads* are bands of the expected size for NACA1 and NACA8.

is widely expressed in a variety of tissue including heart, brain, kidney, lung, pancreas, placenta, and skeletal muscle.[4] Extensive alternative splicing has been described in three different regions of this transcript.[5–7] The second region corresponds to the C-terminal portion of the large cytoplasmic loop where to date seven distinct alternatively spliced transcripts have been described that are expressed in a tissue-specific fashion. The genomic organization and the mechanism by which splicing occurs in this region was delineated by Kofuji, Lederer, and Schulze.[6] Six exons (A–F) are present in this part of the gene. Exons A and B are mutually exclusive, with exon A expressed primarily in brain and heart, and exon B primarily in kidney. Exons C–F are cassette exons that can be either included or excluded independently. Only one of these seven alternatively spliced gene products, NACA1 (containing exons A, C, D, E, and F), is expressed in heart.

In the present study, we were attempting to PCR across a large region of the cytoplasmic loop using primers that flanked unique Bgl II and Nar I restriction sites at nucleotides 1214 and 2094 (numbered according to the rabbit kidney sequence[8]), respectively, employing cardiac cDNA as template. The region of alternative splicing is located between nucleotides 1800 and 2029. Unique 17-mer oligonucleotides (1201+ and 2143−) were employed. The products of the PCR reaction were size-fractionated on a 1% agarose gel, and a single apparent product of the expected size (942 nucleotides) was amplified and subcloned. Miniprepa-rations were digested with *Eco* RI and *Hind* III and size-fractionated on a 1.5% agarose gel (FIG. 1). The inserted cDNA was known to contain an internal *Eco* RI site, and it was expected that it would be restricted into two fragments. However, in one of the minipreparations (lane E, FIG. 1), the second restriction fragment was approximately 70 base pairs smaller than expected. Minipreparations B and E were sequenced using the dideoxy chain termination method. The DNA sequence of B was identical to the previously described NACA1. Minipreparation E contained exons A, C, D, and E, but in contrast to NACA1 did not contain the 69 nucleotide cassette exon F.

The expression of NACA1 and NACA8 was examined in the left ventricle, right ventricle, and both atria using RT-PCR (FIG. 2). PCR was carried out using the unique oligonucleotides 1883+ located in exon A and 2094− located 3' of exon F. These primers would be expected to amplify cDNAs encoding both NACA1 and NACA8 with expected sizes of 211 nucleotides, and 142 nucleotides, respectively. It would appear qualitatively that the expression of each alternatively spliced isoform is similar in left ventricle, right ventricle and atria, and that NACA1 is more highly expressed than NACA8.

In conclusion, we employed the polymerase chain reaction to amplify a cDNA from rabbit heart that encodes a previously undescribed alternatively spliced isoform of NCX1 (NACA8). NACA8 contains the mutually exclusive exon A and cassette exons C, D, and E. However, in contrast to NACA1, it does not contain the cassette exon F. Using RT-PCR we amplified cDNAs encoding NACA1 and NACA8 from the left ventricle, right ventricle, and both atria. NACA8 was qualitatively present in lower abundance than NACA1.

REFERENCES

1. NICOLL, D. A., S. LONGONI & K. D. PHILIPSON. 1990. Science **250:** 562–565.
2. LI, Z., S. MATSUOKA, L. V. HRYSHKO, D. A. NICOLL, M. M. BERSOHN, E. P. BURKE, R. P. LIFTON & K. D. PHILIPSON. 1994. J. Biol. Chem. **269:** 17434–17439.
3. REILÄNDER, H., A. ACHILLES, U. FRIEDEL, G. MAUL, F. LOTTSPEICH & N. J. COOK. 1992. EMBO J. **11:** 1689–1695.
4. KOFUJI, P., R. W. HADLEY, R. S. KIEVAL, W. J. LEDERER & D. H. SCHULZE. 1992. Am. J. Physiol. **263**(Cell Physiol. 32): C1241–C1249.
5. LEE, S-L., A. S. L. YU & J. LYTTON. 1994. J. Biol. Chem. **269:** 14849–14852.
6. KOFUJI, P., W. J. LEDERER & D. H. SCHULZE. 1994. J. Biol. Chem. **269:** 5145–5149.
7. GABELLINI, N., T. IWATA & E. CARAFOLI. 1995. J. Biol. Chem. **270:** 6917–6924.
8. REILLY, R. F. & C. A. SHUGRUE. 1992. Am. J. Physiol. **262**(Renal Fluid Electrolyte Physiol. 31): F1105–F1109.

Presence of NACA3 and NACA7 Exchanger Isoforms in Insulin-Producing Cells

F. VAN EYLEN,[a] M. SVOBODA,[b] A. BOLLEN,[c] AND
A. HERCHUELZ[a]

[a]*Laboratory of Pharmacology*
[b]*Laboratory of Biochemistry and Nutrition*
[c]*Laboratory of Applied Genetics*
Free University of Brussels
Route de Lennik 808
B-1070 Brussels, Belgium

INTRODUCTION

In the pancreatic B cell, Na-Ca exchange displays a quite high capacity and participates in the control of cytosolic free calcium concentration.[1,2] The exchanger was recently cloned in heart, kidney, brain and vascular smooth muscle.[3,4] Two genes coding for two different exchangers (NCX1 and NCX2) have been identified.[3,5] The coding sequences of these genes present 61% identity and the genes are located on human chromosomes 2 and 14, respectively.[3,5] Several isoforms for NCX1, displaying high identity ($\geq$90%), have been identified and are called NACA1 . . . NACAn. The only structural diversity among the NCX1 isoforms lies in a small region toward the end of the cytoplasmic loop, as a consequence of alternative splicing.[6]

AIM

The aim of the present study was to identify the Na-Ca exchange isoform(s) in insulin-producing cells.

RESULTS AND DISCUSSION

In order to identify the isoform(s) present in the pancreatic B cell, we performed a reverse-transcribed polymerase chain reaction (RT-PCR) analysis on mRNA from rat pancreatic islets, rat purified B cells and rat insulinoma B cells (RINm5F cells).

For NCX1, the primers were designed to anneal to conserved sequences flanking the putative splicing area. For NCX2, the primers corresponded to nucleotides 1760–1782 and 2344–2362, respectively, of the NCX2 brain sequence.

In the three preparations, PCR amplification yielded two bands of 244 bp and 313 bp. The nucleotide and deduced amino acid sequences of the PCR products were identical to the NACA3 and NACA7 isoforms of NCX1 identified in several tissues like kidney, aorta, intestine and thymus.

By contrast, using specific primers to NCX2 cDNA, RT-PCR did not yield the expected fragment of 602 bp either in rat islets or in rat B cells and insulinoma RINm5F cells. Rat brain, where the NCX2 exchanger has been cloned, was used as a positive control.

CONCLUSIONS

In the pancreatic B cell, two Na-Ca exchange isoforms (NACA3 and NACA7) were identified, as in kidney, aorta, intestine and thymus.[6,7] By contrast, there was no evidence of the presence of the NCX2 exchanger in the pancreatic B cell. Further study is required to elucidate the significance of the presence of two different exchanger isoforms in the B cell.

REFERENCES

1. PLASMAN, P-O., P. LEBRUN & A. HERCHUELZ. 1990. Am. J. Physiol. **259:** E844–E850.
2. VAN EYLEN, F., P. LEBRUN & A. HERCHUELZ. 1994. Fund. Clin. Pharmacol. **8:** 425–429.
3. PHILIPSON, K. D. & D. A. NICOLL. 1993. Int. Rev. Cytol. **137C:** 199–227.
4. NAKASAKI, Y., T. IWAMOTO, H. HANADA, T. IMAGAWA & M. SHIGEKAWA. 1993. J. Biochem. **114:** 528–534.
5. LI, Z., S. MATSUOKA, L. V. HRYSHKO, D. A. NICOLL, M. M. BERSOHN, E. P. BURKE, R. P. LIFTON & K. D. PHILIPSON. 1994. J. Biol. Chem. **269**(26): 17434–17439.
6. KOFUJI, P., W. J. LEDERER & D. H. SCHULZE. 1994. J. Biol. Chem. **269**(7): 5145–5149.
7. LEE, S-L., A. S. YU & J. LYTTON. 1994. J. Biol. Chem. **269**(21): 14849–14852.

Part II
Kinetics, Mechanism, and Regulation
Introduction

The study of the *Kinetics, Mechanism, and Regulation* of the Na^+-Ca^{2+} exchanger has benefited from technical advances. The use of the giant excised patch, for example, has allowed measurement of voltage-induced charge movements and capacitive changes associated with exchanger electrogenicity. The cardiac exchanger has a consecutive reaction mechanism though multiple electrogenic steps appear to be involved in the ion translocation process. The exact reaction mechanism and regulation may differ for mammalian exchangers and for exchangers from lower species such as the barnacle and the squid. The Na^+-Ca^{2+},K^+ exchanger of rod photoreceptors has a much lower turnover number than other exchangers and may use a distinctive reaction mechanism.

Direct phosphorylation of a mammalian exchanger molecule has been detected for the first time though detailed information on the physiological and functional significance is still lacking. An initial description of cytosolic factors which may regulate the exchanger and detailed kinetic effects of pH and XIP were presented at the conference.

Molecular biology has begun to contribute to studies on exchanger kinetics and expression. Both kinases and hormones may be major factors in regulating exchanger transcript levels. In the future, analysis of mutated exchangers should enhance our understanding of exchanger function.

The Cardiac Na-Ca Exchanger in Giant Membrane Patches

DONALD W. HILGEMANN

*Department of Physiology
University of Texas Southwestern Medical Center at Dallas
5323 Harry Hines Boulevard
Dallas, Texas 75235-9040*

INTRODUCTION

Many techniques have contributed to progress in Na-Ca exchange research. They include isotope flux methods,[1-3] voltage clamp,[4,5] optical methods,[6] and molecular biology.[7] The giant membrane patch technique[8] was developed to study cardiac Na-Ca exchange currents with improved control of ion concentrations on both membrane sides, free access to the cytoplasmic membrane side, and faster control of membrane potential. At the time of the Second International Meeting on Sodium-Calcium Exchange in 1991, these methodological advantages had already led to several new insights.[9]

First, it was discovered that exchanger activity is modulated by complex autoinhibitory reactions which can move the exchanger out of an active transport cycle.[10,11] These reactions can be analyzed as 'inactivation' processes, analogous to a greater or lesser extent to channel inactivation reactions. Second, in agreement with others,[12,13] the apparent ion affinities of the exchanger depend on ion concentrations on the opposite membrane side as expected for a 'consecutive' exchange cycle (*i.e.*, moving first 3 Na in one direction and then 1 Ca in the other direction across the membrane).[14,15] Third, current transients (or 'charge movements') were identified for sodium translocation in the absence of calcium, but not for calcium translocation in the absence of sodium.[14] These results, together with improved data on the voltage- and ion-dependencies of exchanger currents, suggested that ion binding sites may have a net charge of close to -2. Thus, approximately one net positive charge would be moved across the membrane electrical field during the translocation of three sodium ions, while calcium translocation can be largely electroneutral.

Fourth, the exchange current in cardiac membrane was found to be strongly stimulated by an adenosine triphosphate (ATP)-dependent process[10] which did not 'coexpress' with the exchanger in *Xenopus* oocytes.[16] Many tests for the involvement of protein kinases were negative.[17] At the same time, negatively charged lipids, in particular phosphatidylserine (PS), were found to mimic the effects of ATP from the cytoplasmic side. It was hypothesized that the normal asymmetry of PS in the cardiac membrane (*i.e.*, with most PS being located in the cytoplasmic bilayer leaflet) may be disrupted by procedures to form giant excised membrane patches, and that cytoplasmic ATP fuels an aminophospholipid translocase (or 'flippase') which restores the asymmetry by moving PS from the extracellular to the cytoplasmic membrane leaflet.[11]

As is often the case with a new method, the initial phase of rapid progress has led quickly to a new set of barriers. Four years ago, it was expected that exchanger charge movements would be quickly resolved in voltage jump experiments, thereby allowing detailed studies of individual exchanger reactions. While such electrical signals are indeed isolated, they are substantially smaller and faster than anticipated. They are therefore more difficult to study than expected, and their interpretation is less secure than expected. Second, our data set on exchange current, in particular on current-voltage relations,[15] cannot be accounted for by simple transport models with just one or two voltage-dependent steps. The exchange process is clearly a multi-step process with multiple sources of electrogenicity, and it includes multiple reactions leading out of (or impeding) an active exchange cycle. The resulting complexities make it much more difficult to test definitively specific assumptions about the exchange mechanism. As for the action of ATP, experimental work in the last three years has not definitively supported (or contradicted) our suggested involvement of an aminophospholipid translocase. And during the same time no viable alternative has emerged to replace it.

With this background, the goals of this article will be twofold. First, recent work on the function and the regulation of the cardiac exchanger will be reviewed. Second, new results will be presented which may finally help to overcome our present limitations.

EXPERIMENTAL METHODS

The methods and experimental conditions employed have been described elsewhere in detail.[17] Solution compositions for current measurements, charge movement measurements, and capacitance measurements were also the same as described previously.[11,14] The solutions for noise measurements minimized all ion gradients at 0 mV, except the calcium gradient to drive outward exchange current. Solutions on both sides contained: 10 mM HEPES, 1 mM EGTA, 2 mM $MgCl_2$, and 20 mM TEA-Cl. The extracellular solution contained 4 mM $CaCl_2$ and 30 mM NaCl; the cytoplasmic solution contained 0.7 mM $CaCl_2$, either 30 mM NaCl or (15 mM LiCl + 15 mM CsCl), and 6 mM HCl. pH was adjusted to 7.0 with NMG. Abbreviations are those used elsewhere. The fast charge movements were recorded with a customized Axopatch 200 patch clamp allowing 1 MHz voltage clamp (Axon Instruments). Records were digitized at 8 MHz.

Exchanger Modulation Reactions

FIGURE 1 shows typical records of the outward exchange current. The current is activated by switching solutions on the cytoplasmic side from one with 90 mM cesium to one with 90 mM sodium, whereby the pipette solution contains 4 mM calcium and no sodium. The left panel of FIGURE 1 is the control current response obtained with 1 μM cytoplasmic free calcium (see references for further details). Its current decays exponentially with a time constant of about 4 s to a plateau, steady-state amplitude about 20% of the peak amplitude. This decay reflects the putative sodium-dependent inactivation process. In the steady state, it is interpreted that individual exchangers will be randomly fluctuating between an inactive (inactivated) and an active state. Although the exchanger generates no current in the absence of cytoplasmic sodium, the exchanger is largely in the 'active' active state. It simply lacks substrate for transport.

From several types of experiments we have concluded that inactivation depends on 3 Na ions being bound to transport sites on the cytoplasmic side.[18,19]

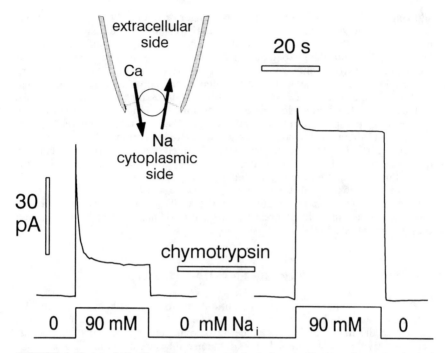

FIGURE 1. Outward Na-Ca exchange current in a giant cardiac membrane patch. *Left record.* Current response under our standard experimental conditions. Note decay of current during application of cytoplasmic sodium. *Right record.* Current response after treatment of the cytoplasmic membrane face with chymotrypsin. Note attenuation of the current decay.

The inactivation does NOT depend on transport activity *per se*. As shown in the right panel of FIGURE 1, the inactivation property is largely destroyed when the cytoplasmic membrane side is treated with chymotrypsin, or other proteases, and the exchange current in steady state is highly stimulated. We have suggested that chymotrypsin does not fundamentally change the actual exchange cycle.[15,18]

Note: The outward exchange current (i.e., the 'calcium influx mode') has been studied more extensively than the inward current (i.e., the 'calcium extrusion mode'). One reason is that the high cytoplasmic calcium concentrations, needed to fully activate the calcium extrusion mode, can destabilize membrane patches. Particularly in the oocyte membrane, high cytoplasmic calcium induces additional current components. With appropriate controls, we have established that the inactivation reactions do indeed modulate the calcium extrusion mode of exchange. Cytoplasmic sodium inhibits the inward exchange current both via a direct mechanism (probably competition with calcium at transport sites) and via sodium-dependent inactivation.

Our interpretation is presented schematically in FIGURE 2. The exchanger transport cycle exchanges three sodium ions for one calcium ion in a consecutive cycle of reactions. As mentioned above, the empty binding sites can be assumed to have a net charge of -2. Regardless of which direction the cycle is operating, the exchanger will occasionally leave the cycle and enter an inactive state when three sodium ions are bound on the cytoplasmic side. The disruption of inactivation by limited proteolysis from the cytoplasmic side is analogous to proteolytic disruption of inactivation in voltage-gated channels. Thus, the inactivation reaction is called 'Transporter Gating' in FIGURE 2.

Note: All of the exchanger modulation reactions can in principle be called 'gating' reactions. However, the term 'gating' has also been used to describe current transients thought to arise from partial transport reactions of the ion transport cycle.[20] This use of the term 'gating' is not straightforward. Channel gating processes are those processes which open and close a channel. In this sense, individual reactions of the transport cycle might be analogous physically to channel gating reactions, but they are not analogous from a functional viewpoint to the switching on and off of transmembrane ion movements.

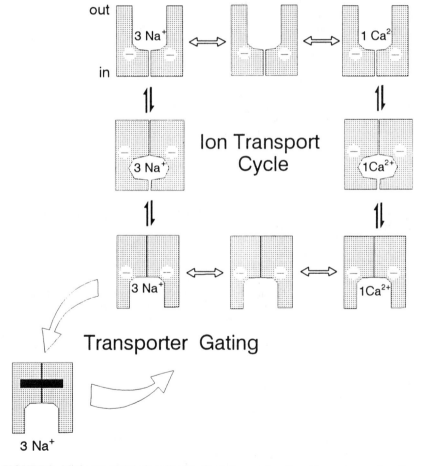

FIGURE 2. Minimum model of cardiac Na-Ca exchanger function. The exchanger functions in a consecutive cycle. Ion translocation takes place in two steps. Binding sites have a net negative charge of -2. The exchanger can enter an inactive state whenever binding sites facing the cytoplasmic side are loaded with 3 sodium ions.

It is not well established whether sodium and calcium ions can freely bind to and unbind from inactive exchanger state(s), but it is a strong possibility. It is well established that cytoplasmic calcium promotes recovery of exchangers from the sodium-induced inactive states, largely by accelerating the recovery back to the active exchange cycle[21] (see also article by Matsuoka et al. in this volume). One attractive explanation is that calcium may bind to the transport site of the inactive exchanger and thereby favor recovery from inactivation, independent of a secondary calcium regulatory site.

The sodium-dependent inactivation, which we also call 'I_1' inactivation, appears to be the end point of several physiological regulatory mechanisms. These include upregulation of exchange activity by high cytoplasmic free calcium, upregulation by high cytoplasmic concentrations of Mg-ATP, upregulation by an increase in negatively charged lipids such as phosphatidylserine (PS) and phosphatidyl-inositolbisphosphate (PIP2), and downregulation by cytoplasmic acidosis.[11,18] In addition, several chemical agents, mostly hydrophobic substances, are known to act on the sodium-dependent (I_1) inactivation process. Agents which favor inactivation, when applied from the cytoplasmic side, include long-chain alkyl amines[11] and cationic phospholipids (Hilgemann, unpublished results). A number of cationic peptides, such as petalysine and neomycin (but not polyamines such as spermine) are strong exchange inhibitors. However, they do not appear to act primarily by modifying the inactivation kinetics (Hilgemann, unpublished results). Inhibition by all these agents is greatly reduced or removed by treatment of the exchanger with chymotrypsin.

FIGURE 3 describes how we have identified a second type of modulation reaction by studying the secondary activation of outward exchange current by cytoplasmic calcium in relation to the direct activation by cytoplasmic sodium. The record #1 is the usual outward exchange current response during application of cytoplasmic sodium in the presence of 2 μM free cytoplasmic calcium. The current inactivates partially, as just described. Record #2 shows the current response to removing and reapplying cytoplasmic calcium in the presence of cytoplasmic sodium. The current turns off and on with a time course very similar to the inactivation time course ($\tau \approx 4$ s), and this probably reflects modification of the sodium-dependent inactivation by cytoplasmic calcium.[21] Record #3 shows the usual absence of any current when sodium is applied in the absence of cytoplasmic calcium. Thus, even in the absence cytoplasmic sodium, the exchanger enters an inactive state in the absence of cytoplasmic calcium. Finally, the record #4 shows the exchange current response for application of sodium plus calcium after a sodium-free, calcium-free incubation period. The exchange current is fully activated within solution switching times. Since no exchange current can be activated in this sequence by application of sodium without calcium, the response in record #4 must reflect recovery from one inactivated state followed by the usual sodium-dependent inactivation.

The inactivation processes have been simulated as independent reactions.[21] However, it seems more likely that one of the reactions will influence the other. In this case, the minimum 'gating' scheme gives rise to four exchanger states, which are shown in the state diagram in the upper right of FIGURE 3; the active transporting state of the exchanger ('A'), the inactive state induced by sodium-dependent (I_1) inactivation, a second inactive state induced by low cytoplasmic calcium per se ('I_2'), and a 'deep' inactive state when both of the inactivation reactions have taken place ('I_3'). The vertical reactions in the diagram correspond to the sodium-dependent inactivation (A to I_1 and I_2 to I_3); the horizontal reactions correspond to the sodium-independent inactivation, which we have called 'I_2' inactivation (A to I_2 and I_1 to I_3).

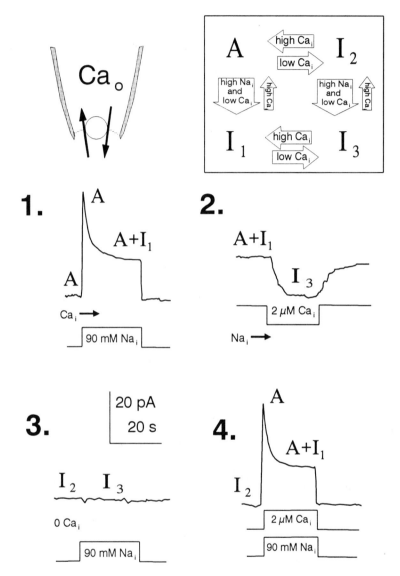

FIGURE 3. Differentiation of two types of exchanger inactivation reactions. *Record 1.* Activation of outward exchange current by 90 mM sodium in the presence of 2 μM free cytoplasmic calcium. *Record 2.* Decay and reactivation of outward exchange current on removal and reapplication of cytoplasmic calcium. *Record 3.* Absence of outward exchange current on application of cytoplasmic sodium in the absence of cytoplasmic calcium. *Record 4.* Immediate activation of outward exchange current after calcium-free period on application of 2 μM free cytoplasmic calcium together with 90 mM sodium. *Upper right panel.* State diagram of minimum exchanger gating reactions to account for the experimental results. Letters in the experimental records indicate the predominant states of the exchanger during the protocols. See text for details of experiments and interpretations.

The relative populations of the exchanger in the different states are indicated above the experimental results. In the presence of cytoplasmic calcium and absence of cytoplasmic sodium, as at the onset of record #1, the exchanger is mostly in the active (A) state. The I_2 inactivation becomes important only when cytoplasmic free calcium falls below 1 μM. In the absence of cytoplasmic sodium and calcium, as at the onset of records #3 and #4, the exchanger will accumulate entirely in the I_2 inactive state. Application of cytoplasmic sodium then drives the exchanger to the I_3 inactive state (record #3). When cytoplasmic calcium is raised simultaneously with cytoplasmic sodium (record #4), however, the outward exchange current is activated with the time course of the recovery from the I_2 to the active (A) exchanger state.

What is the Physiological Significance of the Exchanger Modulation Reactions?

This question is important, but our answers remain speculative. One possibility is that the inactivation mechanisms represent the primary control of resting free calcium. In models of cell calcium homeostasis, a realistic density of exchanger without inactivation reactions can lower free cytoplasmic calcium very close to the expected value for Na-Ca exchange at equilibrium (about 10^{-8} M in resting heart cells). Unrealistically large calcium leaks into the cell must be assumed to achieve a resting free calcium of 10^{-7} M. Thus, inactivation of the exchanger may prevent regulatory and structural problems which might occur if cytoplasmic free calcium falls too low. A second possibility is that the modulation reactions will stimulate exchange activity during periods of increased cell activity and higher average free calcium concentrations. Because the I_2-inactivation reacts to submicromolar free calcium changes, it would control the resting free calcium. Because the I_1 inactivation reacts to free calcium changes in the range of several micromolar it would respond to calcium transients occurring during periods of increased activity. A third consideration is that the exchanger will tend to overload the cell with calcium when cytoplasmic sodium rises pathologically. The sodium-dependent inactivation mechanism may tend to turn off the exchanger when it cannot fulfill its role as a calcium extrusion mechanism for thermodynamic reasons. Finally, the exchanger modulation reactions may be the end point of cell signaling pathways, perhaps involving the ATP-dependent mechanism. But for now, we know nothing concrete about such pathways.

Exchange Mechanism and Kinetics

Minimum Exchange Model

With small modifications, the functional model of the exchanger shown in FIGURE 2 can reproduce a wide range of experimental results.[15,18,19] The binding sites can be assumed to accommodate either 3 Na ions or 1 Ca ion, and the actual binding reactions can be assumed to take place very quickly (instantly) in relation the ion-translocating reactions. The ion binding reactions can be assumed to be entirely voltage-independent. The translocation of 3 Na or 1 Ca must be assumed to take place in at least two steps. A first reaction 'occludes' the ions into the binding sites, and a second reaction opens binding sites to the other side, thereby 'deoccluding' the ions. As already mentioned, the net charge of binding sites must be close to −2. Thus, the movement of 3 sodium ions is electrically equivalent to the movement of about one net positive charge. There may be small net charge movements when calcium is translocated. The reaction which moves the most net charge across membrane field appears to be the reaction occluding and deoccluding sodium from the extracellular side. New evidence for this last assumption is provided in this article using capacitance measurements.

Note: Ion flux experiments in the squid axon[22] and the barnacle muscle (see article by Rasgado-Flores et al. in this volume) do not support these conclusions about exchanger electrogenicity. In those systems, calcium movements appear to be more electrogenic than sodium movements.

Current-Voltage Relations Suggest that Sodium Translocation is Electrogenic

FIGURE 4 shows the first type of experiment which suggested that sodium translocation is electrogenic in the cardiac exchanger, while calcium translocation is largely electroneutral. Under the assumption that the exchange cycle indeed takes place by a consecutive mechanism, one means to identify the electrogenic reactions is to vary the concentrations of transported ions under the 'zero-trans' condition (*i.e.*, with only sodium on one side and calcium on the other). As the ion concentration driving one of the transport reactions is reduced, that reaction (or reactions) will logically take place at a slower rate and ultimately will become rate-limiting in the cycle. If corresponding ion binding and occlusion reactions are voltage-dependent, then the exchange current will remain voltage-dependent (or even become more voltage-dependent) as the driving ion concentration is reduced. This is case when the driving sodium concentration is reduced in both the calcium influx and extrusion exchange modes.[15] If corresponding ion binding and occlusion reactions are electroneutral, then the current-voltage relations will become less steep as the driving ion concentration is reduced.

As shown in FIGURE 4A, the outward exchange current becomes voltage-independent as extracellular calcium is reduced from 10 to 0.1 mM in the absence of extracellular sodium. Note that all results in FIGURE 4 have been normalized to one current value within the data set. This same result is obtained whether or not the exchanger has been 'deregulated' with chymotrypsin. However, the current remains voltage-dependent at low extracellular calcium when high concentrations of monovalent ions (*e.g.*, lithium), other than N-methylglucamine and cesium, are included with calcium on the extracellular side (*e.g.*, 120 mM lithium).[19,23] In the giant patch, we can explain our results by assuming that lithium binds to sodium transport sites from the extracellular side in a voltage-dependent manner, thereby acting as a voltage-dependent exchange current blocker. In the whole-cell configuration, however, lithium stimulates exchange current as well as induces a greater voltage dependence. Presumably, these effects involve lithium binding sites other than transport sites. It is possible that lithium is fundamentally changing the calcium translocation steps, *e.g.* by inducing a change of binding site charge.

As shown in FIGURE 4B, the complementary result is obtained for the inward exchange current when the exchanger is not 'deregulated' by chymotrypsin. The inward exchange current becomes almost voltage-independent as the cytoplasmic free calcium concentration is reduced into the range of 2 μM in the absence of cytoplasmic sodium. As shown in FIGURE 4C, 'deregulation' of the exchanger by chymotrypsin changes these results. The current-voltage relationship of inward exchange current always retains some voltage-dependence as cytoplasmic free

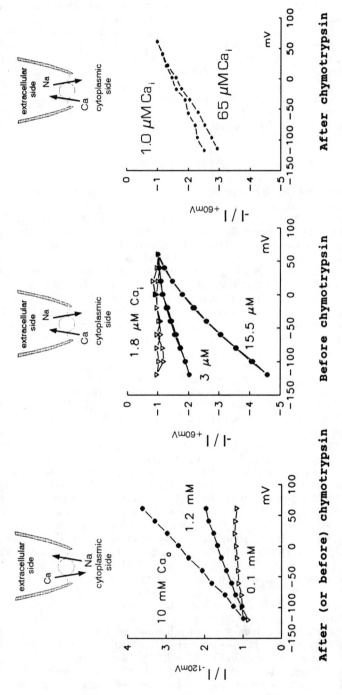

FIGURE 4. Normalized Na-Ca exchange current-voltage relationships. **(A)** Outward exchange current at 10, 1.2 and 0.1 mM extracellular sodium.[15] Chymotrypsin-treated patch. **(B)** Inward exchange current with 15.5, 3, and 1.8 μM free cytoplasmic calcium (120 mM extracellular sodium; 10 mM EGTA). No chymotrypsin treatment. **(C)** Inward exchange current with 65 and 1 μM free cytoplasmic calcium (120 mM extracellular sodium; 10 mM EGTA). The outward currents are normalized to the current magnitude at −120 mV. The inward currents are normalized to the current magnitudes at +60 mV. The 'Kd' for cytoplasmic calcium was similar in (B) and (C).

calcium is reduced. Our tentative explanation is that, after chymotrypsin treatment, a negative charge on the cytoplasmic side of the exchanger can partially enter the membrane electrical field during occlusion of calcium from the cytoplasmic side. This increases the apparent negative charge of the exchanger binding sites from -2 to about -2.2. Less net positive charge is moved during sodium translocation and, as expected, the voltage dependence of the fully activated inward exchange current is less steep after chymotrypsin.

Note: It has become apparent that the exchanger was not fully 'deregulated' by chymotrypsin in one of our published data sets,[14] *which showed complete loss of voltage-dependence of the exchanger at a low free cytoplasmic calcium.*

Results with three further methods support the idea that sodium translocation moves net charge through the membrane electrical field and therefore is electrogenic, while calcium translocation by the cardiac exchanger is less electrogenic. First, an ion concentration jump technique was employed.[14] Briefly, current transients could be induced specifically by application of cytoplasmic sodium under conditions which allow sodium to be transported from the cytoplasmic to the extracellular side and to unbind from the extracellular side. No charge movements were detected for the equivalent experiments to induce calcium translocation. The quantities of charge moved during sodium translocation suggested exchanger densities in cardiac membrane of 300–400 per square μm. Maximum exchanger turnover rates of about 5000 per second would then account for the magnitudes of exchange currents recorded in the same experiments.

FIGURE 5 presents the logic of two other types of experiments designed to identify sources of electrogenicity; fast charge movement measurements and capacitance measurements. *All results are after chymotrypsin treatment so that the modulation reactions do not play a role in results.* Experiments on sodium transport are performed in the absence of calcium on both sides; those on calcium transport are performed in the absence of sodium on both sides. When sodium is present only on the outside (*i.e.*, in the pipette), extracellular sodium can bind and be transported from outside to inside. But without cytoplasmic sodium the transport reactions back to the extracellular side cannot take place, and the exchanger should accumulate entirely in the configuration with empty binding sites open to the cytoplasmic side. This is our putative 'Null condition' (see upper panel of FIG. 5), in which charge movements related to sodium transport should be minimal or absent. When cytoplasmic sodium is added (lower panel; 'Charge movement condition'), the sodium transport reactions can take place in either direction, giving rise to sodium-sodium exchange in ion flux experiments. If the underlying reactions carry a net charge through the membrane electrical field, they will be detected as a 'charge movement' or 'current transient' when voltage pulses are applied in the charge movement condition.

Experimentally, the charge movements of sodium translocation are defined by subtracting records taken in the presence of cytoplasmic sodium from records in the absence of cytoplasmic sodium (see Ref. 24 for equivalent results with the Na/K pump). The patch clamp employed, a modified Axopatch 200, measures directly the amount of charge transfer, rather than current (charge moved per time unit). *Charge is the time integral of membrane current, and membrane current is the first derivative of the charge signals.* The charge signals are less noisy than current signals. At the same time, they display the kinetic properties of interest and provide direct measurement of charge movements. Due to the low resistance of the large pipettes and the fast electronics employed, voltage clamp speeds of 1 to 2 MHz have been achieved in oocyte patches. That the voltage

Null condition

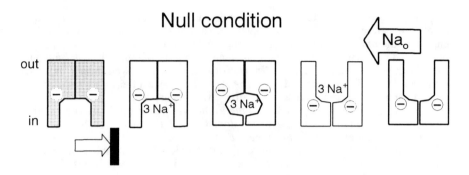

Charge movement / Na-Na exchange condition

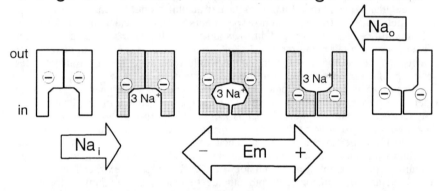

FIGURE 5. Rationale of experimental protocols used in charge movement and capacitance measurements. The sodium transport reactions described in FIGURE 2 are presented horizontally. *Top panel.* In the presence of extracellular sodium, all exchangers should accumulate in the left-most state with binding sites oriented to the cytoplasmic side. *Bottom panel.* In the presence of sodium on both membrane sides, transport can take place in either direction. If these reactions move a net positive charge, they will be driven to the right by depolarization, to the left by hyperpolarization.

clamp is stable within 1 microsecond has been verified by measurements of channel currents such as the calcium-activated chloride current in *Xenopus* oocyte membrane.

Kinetic Resolution of Electrogenic Sodium Transport Steps

FIGURE 6 shows the charge movements of sodium translocation in *Xenopus* oocyte membrane expressing the canine cardiac exchanger. The results are with 40 mM extracellular sodium (no calcium on either side; 10 mM EGTA on both sides; 30°C; for solution compositions see Ref. 14). The voltage pulses are 100 microseconds in duration. The records are a subtraction of results without cytoplasmic sodium from those with 40 mm cytoplasmic sodium. The charge signals return to control values at the end of the voltage pulse. This indicates that the exchanger does not generate a transport current during this protocol, and that good equality of the 'on' and 'off' charge movements is given. For orientation, it is pointed out that exchanger binding sites are expected to be open to the extracellular side at very positive membrane potentials (uppermost record) and open to the cytoplasmic side at very negative membrane potentials (lowermost records).

The charge records from -50 mV to potentials of -200 to 150 mV are all reasonably described by single exponential functions. The fits are included as fine lines. The records for voltage pulses from positive potentials to -50 mV are made up of a fast and slow component. The rate constants of exponential are plotted as a function of membrane potential in the upper right panel. Note that the rates increase with hyperpolarization, similar in principle to results with the Na/K pump. They are fitted to the exponential function shown in the figure. The slowest rate is 80,000 per s, about 8 times faster than predicted. The magnitudes of charge movements, measured at the end of applied voltage

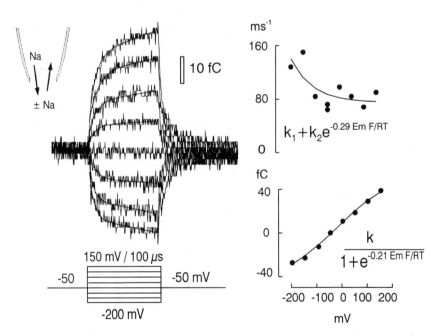

FIGURE 6. Isolated charge movements of sodium transport by cardiac Na-Ca exchange. *Left panel.* Voltage pulses are from -50 mV potentials of -200 mV up to 150 mV in 50-mV steps for 100 μs. 40 mM extracellular sodium. Results in the absence of cytoplasmic sodium are subtracted from those in the presence of 40 mM cytoplasmic sodium. The results were fitted to an exponential function plus a line, corresponding to a constant small conductance (fine lines plotted with the charge traces). *Upper right panel.* Rate constants of the fitted exponential functions at different membrane potentials. *Lower right panel.* Magnitudes of the fitted exponential functions at different membrane potentials.

pulses, were fitted to a Boltzmann equation (see lower right panel). The slope parameter is only 0.21, indicative of a rather weak voltage-dependence. If the charge movements reflected movement of a single charge across the entire membrane potential in a single step, as occurs in the simplest possible exchange models, then the slope of the Boltzmann equation would be 1. The charge movements would come to complete saturation in the voltage range studied.

Why are the charge movements of sodium translocation faster than expected and only weakly voltage-dependent? First, it should be kept in mind that the current-voltage relations of steady state exchange current can also extend over a very large voltage range almost linearly. Thus, the shape of the charge-voltage relations is, in principle, not surprising. A possible explanation for the rapidity of charge movements is that exchanger densities could be less than predicted, while the rates of transport are faster. This does not seem to be the major explanation because the magnitudes of charge movements in the present experiments are at most 50% less than predicted. The second possibility is that sodium translocation is much faster than calcium translocation, and this is consistent with reports that the magnitude of sodium-sodium exchange fluxes can be much larger than those of calcium-calcium exchange in squid axon and barnacle muscle (R. DiPolo, L. Beaugé, and H. Rasgado-Flores, personal communications). This assumption is not consistent with our modeling of exchange currents. Roughly equal rates of sodium and calcium translocation are required to account for current-voltage relations and their changes with ion concentration changes. A third possibility, then, is that a large fraction of the exchanger at any one time can carry out sodium translocation (*i.e.,* sodium-sodium exchange) but cannot carry out calcium translocation. Fast transporter gating would rapidly enable or disable calcium translocation, and on a microsecond time scale exchangers would be 'bursting' between an inactive state and an active state with a very high turnover rate.

Capacitance Measurements of Exchanger Electrogenicity

The charge movements just presented are faster than the ability of most voltage clamp methods to resolve their time course. When voltage is changed in any voltage clamp method, a certain amount of charge must be transferred to change the membrane potential by a certain amount. This is the membrane capacitance. The fast charge movements will equilibrate in the time course of the membrane charging, and they will constitute a small component of the total membrane capacitance. Changes of membrane capacitance can be monitored at very high resolution, and we recently introduced such measurements into the study of transporter electrogenicity. Briefly, transporters are forced into different configurations by changes of solution composition or of the membrane potential. Those conditions which allow fast electrogenic reactions to take place can be identified by an increase of membrane capacitance. Methodologically, a sinusoidal voltage perturbation is applied to the membrane patch (0.5–2 mV in magnitude) at a frequency which is less than the rate at which the expected charge movements take place. Capacitance is measured at an angle of the sine wave where conductance changes or changes of series resistance to the pipette have no influence. The techniques employed are often used to monitor membrane fusion.[25]

FIGURE 7 shows results obtained for the exchanger expressed in *Xenopus* oocyte membrane, thereby allowing the important control that the capacitance changes monitored depend on high expression of the cardiac exchanger. Experiments have been performed to test for electrogenicity in both the sodium and calcium translocation reaction pathways. As first step, it was verified that membrane capacitance increases significantly when cytoplasmic sodium is applied under the conditions of the charge movement measurements. As expected for an ion transport pathway, the cytoplasmic sodium dependence of the capacitance changes was found to shift to higher concentrations with higher extracellular sodium concentrations, similar to results described for ion pulse experiments.[14] Subsequent experiments were designed to test more precisely where in the cycle electrogenicity comes about.

A simplified exchanger reaction cycle is shown in FIGURE 7A. Sodium binding (reaction #1) and subsequent transport to the extracellular side (reaction #2) can be switched on by application of cytoplasmic sodium. The specific experiments to be described were performed with an extracellular sodium concentration (20 mm) substantially less than the half-maximum concentration for fully activated exchange current (>70 mM). There is no extracellular calcium, and 10 mM EGTA were included to guarantee low free calcium. In the absence of cytoplasmic sodium and calcium, it is expected that the translocation of sodium from the extracellular side to the cytoplasmic side (reaction #3) will force all transporters to an orientation with binding sites open to the cytoplasmic side ('E_2' configuration). Because extracellular sodium is low, a high cytoplasmic sodium concentration will shift most of the transporters to the E_2 configuration with binding sites open to the extracellular side. As a further important prediction, application of cytoplasmic calcium is also expected to shift the transporter to the E_2 configuration from which extracellular sodium can bind and unbind (see dotted field). That is because cytoplasmic calcium binding (reaction #7), translocation (reaction #7), and extracellular calcium unbinding (reaction #5) are all expected to be more probable than reaction #3.

As shown in FIGURE 7B, application of sodium to the cytoplasmic side had no effect in patches from control (water-injected) oocytes. In contrast, cytoplasmic application of either sodium or calcium in the presence of 20 mM extracellular sodium increased membrane capacitance in injected oocytes, and the maximum responses to cytoplasmic sodium and calcium were of similar magnitude (FIG. 7C). When cytoplasmic calcium (or sodium) was applied in the absence of extracellular sodium, either with or without calcium, there was no capacitance response (FIG. 7D; 0.2 mM extracellular calcium).

The response to cytoplasmic sodium (FIG. 7C) depends steeply on sodium concentration (n_{Hill} = 3.4), and the half-maximal response occurs at 5.1 mM (FIG. 7E). The half-maximal response to cytoplasmic calcium (FIG. 7C) occurs at 0.9 μM. (FIG. 7F; n_{Hill} = 1.2). These concentration dependencies are both shifted by a factor of about 3 from the corresponding concentration dependencies of transport in the presence of a high extracellular concentration of the counter transported ion (Na^+ or Ca^{2+}). The results fulfill important predictions of the consecutive transport model: When the transport rate for orienting binding sites to the cytoplasmic side (#3) is low, relatively small concentrations of transporter substrates on the cytoplasmic side should suffice to orient binding sites toward the extracellular side (rates #2 and #6). To summarize, capacitance measurements for the canine cardiac Na-Ca exchanger expressed in *Xenopus* oocytes have verified an electrogenicity of extracellular sodium binding. Consistent with the interpretation of current-voltage relations, no evidence has been obtained for electrogenic reac-

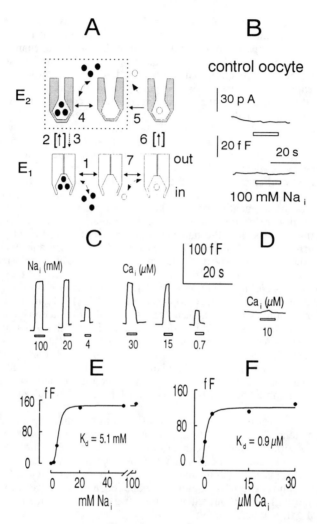

FIGURE 7. Capacitive measurements of cardiac Na-Ca exchanger electrogenicity in *Xenopus* oocyte membrane. **(A)** Simplified state diagram of the cardiac Na-Ca exchange cycle. With low (20 mM) extracellular sodium and no extracellular calcium (10 mM EGTA), the transition from E_2 to E_1 (reaction #3) is slow but significant. In the absence of cytoplasmic sodium and calcium, the exchanger will accumulate in E_1 with empty binding sites. Application of either cytoplasmic sodium (*filled circles*) or calcium (*open circles*) will quantitatively shift the exchanger to the E_2 state, via reactions #2 or #6 respectively, in which extracellular sodium can bind and unbind. **(B)** Lack of effect of cytoplasmic sodium on membrane current and capacitance in a patch from a control (water-injected) egg. **(C–F)** Patches from exchanger-injected eggs. **(C)** Capacitance increases observed on application of the indicated cytoplasmic sodium and free calcium concentrations. **(D)** Lack of effect of cytoplasmic calcium in the presence of 0.2 mM extracellular calcium and no sodium. **(E,F)**. Concentration dependence of sodium-induced (E) and calcium-induced (F) capacitance changes.

tions in the calcium translocation pathway without using high extracellular lithium concentrations.

Possible Exchanger Noise

As outlined above, the sodium-dependent inactivation process is expected to move the exchanger into an inactive state. During steady state current activation, individual exchangers are expected to fluctuate randomly between the inactive and active states, similar to the flickering of ion channels between open and closed states. It is very unlikely that the function of a single exchanger will be resolved, equivalent to single channel recording. However, it may be possible to obtain evidence for this flickering function via the methods of noise analysis. That is to say, by monitoring fluctuations of the membrane current due to random 'flickering' of transporters. If successful, this approach could not only verify the nature of the sodium-dependent inactivation process, it could also provide an independent estimate of single exchanger turnover rates.

FIGURE 8 demonstrates the logic of these experiments by simulating the sodium-dependent inactivation process as an ensemble of transporters whose transitions to and from an inactive state are calculated individually as a random process. For simplicity, it is assumed that all exchangers enter the active state in the absence of sodium (state 'A' in FIG. 3). Accordingly, the peak current amplitude on applying sodium reflects the activity of all exchangers. The probabilities of single channels to enter the inactive state or to recover to the active state were adjusted to give about 60% inactivation. These probabilities determine the kinetics and extent of inactivation, as well as the frequency characteristics of noise observed in the steady state. The simulation results are shown for the assumption that a patch contains 200, 1000, 5000, or 25000 transporters. From these normalized simulations it can be predicted that significant exchange current noise could be resolved in

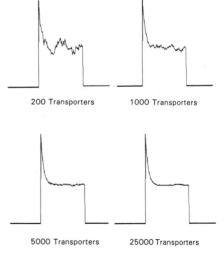

FIGURE 8. Simulated outward exchange current transients. The indicated numbers of transporters were simulated to fluctuate randomly between an active and inactive state. At the peak of the current transient, all transporters were assumed to occupy the active state. Each transporter in each simulation contributes the same unitary current.

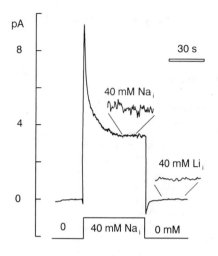

FIGURE 9. Outward exchange current transient in a small patch at 40°C. Short sections of the records with sodium and with lithium are shown at higher resolution. Current noise is clearly increased when the exchange current is activated. See text for further details.

small patches with good seal resistances if exchanger rates are as high as 6000 per s (*i.e.*, 1 fA).

FIGURE 9 shows the corresponding experimental result using a pipette with an inner diameter of about 5 micrometers and a temperature of 40°C to enhance exchanger turnover rates. To minimize leak currents, the low ionic strength solutions described in Methods were employed. There is an evident increase of current noise when the exchanger is activated by 40 mM sodium, and the noise is comparable to that predicted for a patch with 5000 transporters. The peak exchange current in this patch was 9 pA. For 5000 transporters, this would correspond to a single exchanger turnover rate of about 11000 per s. If verified by further work, this will be the first example of noise generated by a transporter carrying out tightly coupled ion transport.

Mechanism of Mg-ATP in Cardiac Membrane

The mechanism by which Mg-ATP stimulates Na-Ca exchange current ties the exchanger to other processes in the cell. Cell regulatory processes would of course be of most interest. In the last three years, I have tried to test three possibilities; the involvement of an aminophospholipid translocase, the involvement of protein kinases and the involvement of ATP-dependent cytoskeletal processes. From this work, the results which seem of most importance will be described.

A number of observations suggest that Mg-ATP is acting by two mechanisms. One observation, shown in FIGURE 10, is that the Mg-ATP effect often decays in two distinct phases after removal of ATP. In the record shown, the outward exchange current was first activated by cytoplasmic sodium. Then, the nonhydrolyzable ATP analogue, 5'-adenylimidodiphosphate (AMP-PNP), was applied at a concentration of 4 mM. It was without effect. However, the exchange current is strongly stimulated by application of 2 mm Mg-ATP over about 30 s. Upon removal of Mg-ATP, there is a delay of a few seconds, and then the current decays to about 50% of the stimulated level over the time course of 30–40 seconds. Subsequently, the current decays only very slowly, over >3 min (not shown), back toward baseline level. Either the response to ATP is heterogeneous in the patch, or else multiple mechanisms are involved in the effect (or its reversal).

No Evidence for Protein Kinase or Cytoskeleton Involvement

The initial rate of the ATP effect has a K_d for ATP of >4 mM.[17] This already makes the involvement of most protein kinases unlikely. In addition to the published experiments (ibid), further experiments have tested for protein kinase involvement. Further protein kinase inhibitors have been tested as possible inhibitors of the ATP effect with no positive result. These include a number of peptide inhibitors, tyrosine kinase inhibitors, and protein kinase C inhibitors. A variety of phosphatases and phosphatase inhibitors (*e.g.*, microstatin, okadaic acid) have also been retested for effects on reversal of the ATP effect. No positive results have been obtained.

The second possible mechanism is the involvement of ATP-dependent processes of the submembrane cytoskeleton. In particular, actin polymerization was an attractive candidate. Cytoskeleton disrupters, such as cytocalasin, were without effect on basal and stimulated exchange current. Enzymes which depolymerase action, such as gelsolin and a 'mini'-gelsolin, were without effect. Enzymes which 'cap' actin filaments were also without effect on reversal of the ATP effect. *These enzymes were all kindly provided by Dr. Helen Yin at this institution.*

EDTA, AMP-PNP and Low Concentrations of Fluoride Block Reversal of the ATP Effect; Low Concentrations of Aluminum Ions Accelerate Reversal of the ATP Effect

Reversal of the ATP effect, after ATP removal, varies remarkably in rate and extent in patches from different myocyte batches. To identify the cause of variability, a large number of experimental factors were systematically changed. After elimination of many possibilities, it is concluded that the presence of trace multivalent metals in our cytoplasmic solutions is probably an important factor. This, in spite of the fact that the cytoplasmic solutions contain 10 mM EGTA.

The results which suggest this are twofold. First, a number of compounds were found to completely block reversal of the ATP effect. They include 2 mM AMP-PNP, 100 μM fluoride, and 1 mM ethylene diaminetetraacetic acid (EDTA). (In the interest of brevity, no results are presented.) All these substances bind multivalent metal ions, and in the case of EDTA many multivalent metals bind

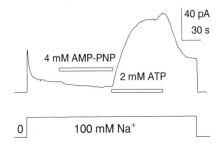

FIGURE 10. Stimulation of outward exchange current by cytoplasmic Mg-ATP. In the first part of the record, 4 mM AMP-PNP was applied and removed with no effect. Thereafter, 2 mM Mg-ATP was applied. The current is stimulated about 10-fold over 80 s. When ATP is removed the current declines in 1 min to an intermediate level.

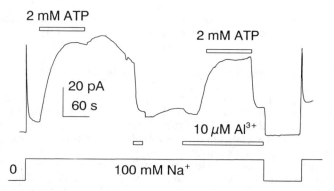

FIGURE 11. Micromolar concentrations of aluminum ions stimulate reversal of the ATP effect in the presence of 10 mM EGTA. Outward exchange current was activated, as indicated below the records by application of 90 mM sodium. 2 mM ATP stimulates the exchange current by about 8-fold, and the effect reverses only 20% in the course of 90 s. After application of 10 μM aluminum chloride, current declines in about 15 s to the prestimulatory magnitude. Removal of aluminum has only a small effect on basal current, and 2 mM ATP restimulates the current in the presence of aluminum. When ATP is removed, the current again returns to baseline in about 15 s.

with a higher affinity than to EGTA. Also for fluoride, a higher affinity of some metals, than to EGTA, can be expected. Second, it was found that aluminum ions (Al^{3+}) drastically accelerate reversal of the ATP effect. This result seems relevant because Al^{3+} is a possible contaminant in some of our chemicals, and the microelectrodes employed represent a possible sink of Al^{3+} in experiments.

The effect of 10 μM Al^{3+} in the presence of 10 mM EGTA and 7 mm calcium is demonstrated in FIGURE 11. In the first part of the record, the exchange current is stimulated by application of 2 mM ATP. After removal of ATP, the current begins to reverse slowly over several minutes. As indicated, 10 μM Al^{3+} was applied, and the current returned within 15 s to the prestimulated level. Next in the record, Al^{3+} was removed and reapplied to test for effects on basal exchange current. As verified in 5 further experiments, the basal current is only slightly inhibited. 100 μM Al^{3+} had only a small further effect, and similar results were obtained on outward current in chymotrypsin-treated patches. Next in the experiment, ATP was applied in the presence of Al^{3+}. The ATP effect is nearly as fast and large as without Al^{3+}. Upon removal of ATP in the continued presence of Al^{3+}, the stimulatory effect of ATP reverses within a few seconds. Thus, Al^{3+} appears to specifically accelerate the reversal of the ATP effect. As in control experiments, a portion of the ATP effect often did not reverse on application of Al.

Do Phospholipid Messengers Regulate Na-Ca Exchange Physiologically?

One possible approach to testing the 'flippase' hypothesis is based on our apparent ability to influence membrane composition by applying exogenous phospholipid to patches.[26] Phospholipid vesicles are prepared in distilled water and are diluted into the experimental solution just before application to the cytoplasmic side of patches, usually at an apparent concentration of 0.3 mM. It is known from experiments with fluorescent lipids that the vesicles bind to the pipette surface. That phospholipids diffuse into (and probably out of) the membrane patch seems evident from the large functional effects obtained by application of different phospholipids to the patches. Membrane capacitance increases only little (2–5%), suggesting that an exchange is taking place between phospholipids on the pipette wall and in the cytoplasmic leaflet of the bilayer. In short, negatively charged lipids, including phosphatidylserine, phosphatidylglycerol, phosphatidic acid, and phosphatidylinositols, stimulate strongly the basal exchange current over 1 to 3 min. Phosphatidycholine abolishes exchange current over 2 to 5 min, and positively charged phospholipid (*e.g.*, used for transfections) abolishes the exchange current in <1 min (for the sake of brevity, no results are presented). The fact that phosphatidylcholine vesicles can completely abolish exchange current over time is consistent with the idea that phospholipid exchange can be very extensive in this type of experiment.

The flippase hypothesis predicts that the ATP effect would be abolished if phosphatidylserine were removed from the membrane. FIGURE 12 describes the relevant experiment. In the experiment presented, ATP was first applied and removed. Then, during reversal of the ATP effect, phosphatidylcholine vesicles were applied. Over the course of 3 min, current declined to below baseline. Thereafter, ATP was again applied. It was nearly without effect, although second and third ATP responses were large and rapid in 4 control patches from the same myocyte batch (not shown). Subsequently, phosphatidylserine vesicles were applied, and the current was strongly stimulated. Thus, phosphatidylserine remains active when ATP is not.

Whether or not changes of membrane lipids are involved in the effects of ATP, the sensitivity of exchange activity to membrane phospholipid environment clearly deserves attention. The large effects of charged lipids are unique in our experience, we having tested these same interventions on several channel currents and the sodium pump current. Regulation might occur through synthesis and degradation of the lipids themselves, through the aminophospholipid translocase, or it might

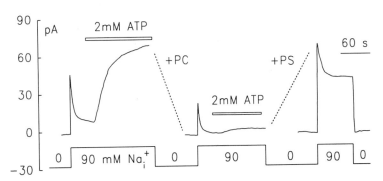

FIGURE 12. Phosphatidylcholine (PC) vesicles abolish a second ATP effect. As indicated, outward exchange current was stimulated by 2 mM Mg-ATP, and ATP was removed. PC vesicles (apparent concentration, 0.5 mM) were applied for 4 min and removed (*dotted line*). The second ATP response is blunted, although control patches responded reproducibly to multiple ATP applications. Finally, phosphatidylserine (PS) vesicles were applied for 2 min. The PS vesicles strongly stimulate the exchange current after PC.

occur through lipid-dependent messengers. An attractive candidate is the positively charged MARCKS protein, which binds to phosphatidylserine in a phosphorylation-dependent fashion, and which interacts with other membrane-associated proteins.[27,28]

CONCLUSIONS

The results presented in this article outline the present status of our work on Na-Ca exchange function and regulation in giant cardiac membrane patches. In summary, our ability to study electrogenic transporter reactions has been greatly improved over the last 3 years. Megahertz resolution of partial exchanger reactions is achieved. However, the task of piecing together details to a coherent picture of molecular exchanger function is only just beginning. The analysis of exchange current noise may provide independent information about both exchanger modulation reactions and single exchanger turnover rates. Capacitance measurements provide a relatively facile 'on-line' monitoring of electrogenic reactions. It should be possible, for example, to study changes in Na-Na exchange or Ca-Ca exchange function in different regulatory states of the exchanger. It is evident that the stimulation of Na-Ca exchange current by Mg-ATP in giant cardiac membrane patches is not taking place by any of the well-established cell regulation pathways. It remains to be seen which experimental avenue ultimately elucidates the underlying mechanisms and how (or if) the ATP effects are tied to cell regulation pathways. In short, we are in a transitional period of Na-Ca exchange work. The complexity of the exchange system continues to surprise and impress us. It is hoped that recent refinements of the giant patch methods will be useful tools in dealing experimentally with this complexity.

ACKNOWLEDGMENTS

I express my gratitude to Dr. Gary Frazier (Texas Instruments) for discussions and for preparing the data acquisition system used in these experiments; to Dr. Rich Lobdill (Axon Instruments) for preparing the voltage clamp unit employed; to Drs. Kenneth D. Philipson and Debrah Nicoll (UCLA) for providing cRNA for the canine cardiac Na-Ca exchanger; to Dr. Helen Yin (UTSW) for providing the microfilament-directed enzymes; to Steve Calloway (UTSW) for excellent technical assistance; to Anatolii Kabakov for performing one of the experiments in FIGURE 4; and to Chin-Chih Lu (UTSW) for helpful discussions and criticism of the manuscript.

REFERENCES

1. BAKER, P. F. & M. P. BLAUSTEIN. 1968. Sodium-dependent uptake of calcium by crab nerve. Biochim. Biophys. Acta **150:** 167–170.
2. REUTER, H. & N. SEITZ. 1968. The dependence of calcium efflux from cardiac muscle on temperature and external ion composition. J. Physiol. **195:** 451–470.
3. REEVES, J. P. & J. L. SUTKO. 1979. Sodium-calcium ion exchange in cardiac membrane vesicles. Proc. Natl. Acad. Sci. USA **76:** 590–594.
4. HODGKIN, A. L., P. A. MCNAUGHTON & B. J. NUNN. 1987. Measurement of sodium-calcium exchange in salamander rods. J. Physiol. **391:** 347–370.

5. KIMURA, J., S. MIYAMAE & A. NOMA. 1987. Identification of sodium-calcium exchange current in single ventricular cells of guinea pig. J. Physiol. **384:** 199–222.
6. BARCENAS-RUIZ, L., D. J. BEUCKELMANN & W. G. WIER. 1987. Sodium-calcium exchange in heart: currents and changes in $[Ca^{2+}]I$. Science **238:** 1720–1722.
7. NICOLL, D. A., S. LONGONI & K. D. PHILIPSON. 1990. Molecular cloning and functional expression of the cardiac sarcolemmal Na^+-Ca^{2+} exchanger. Science **250:** 562–565.
8. HILGEMANN, D. W. 1989. Giant excised cardiac sarcolemmal membrane patches: sodium and sodium-calcium exchange currents. Pflügers Arch. **415:** 247–249.
9. HILGEMANN, D. W., A. COLLINS & D. P. CASH. 1991. Mechanism and regulation of the sodium-calcium exchange system in giant cardiac membrane patches. *In* Sodium-Calcium Exchange: Proceedings of the Second International Conference. Ann. N. Y. Acad. Sci. **639:** 126–139.
10. HILGEMANN, D. W. 1990. Regulation and deregulation of cardiac sodium-calcium exchange in giant excised sarcolemmal patches. Nature **334:** 242–245.
11. HILGEMANN, D. W. & A. COLLINS. 1992. The mechanism of sodium-calcium exchange stimulation by ATP in giant cardiac membrane patches: Possible role of aminophospholipid translocase. J. Physiol. **454:** 59–82.
12. LI, J., & J. KIMURA. 1990. Translocation mechanism of Na-Ca exchange in single cardiac cells of guinea pig. J. Gen. Physiol. **96:** 777–788.
13. KHANANSHVILI. D. 1990. Distinction between two basic mechanisms of cation transport in the cardiac Na^+-Ca^{2+} exchange system. Biochemistry **29:** 2437–2442.
14. HILGEMANN, D. W., D. A. NICOLL & K. D. PHILIPSON. 1991. Charge movement during sodium translocation by native and cloned cardiac Na/Ca exchanger in giant excised membrane patches. Nature **352:** 715–719.
15. MATSUOKA, S. & D. W. HILGEMANN. 1992. Dynamic and steady state properties of cardiac sodium-calcium exchange: Ion- and voltage dependencies of transport cycle. J. Gen. Physiol. **100:** 962–1001.
16. MATSUOKA, S., D. A. NICOLL, R. F. REILLY, D. W. HILGEMANN & K. D. PHILIPSON. 1993. Identification of regulatory regions of the cardiac sarcolemmal Na^+-Ca^{2+} exchanger. Proc. Natl. Acad. Sci. USA **90:** 3870–3874.
17. COLLINS, A., A. SOMLYO & D. W. HILGEMANN. 1992. The giant cardiac membrane patch method: stimulation of outward Na/Ca exchange current by MgATP. J. Physiol. **454:** 37–57.
18. HILGEMANN, D. W., S. MATSUOKA & A. COLLINS. 1992. Dynamic and steady state properties of cardiac sodium-calcium exchange: Sodium-dependent inactivation. J. Gen. Physiol. **100:** 905–932.
19. MATSUOKA, S. & D. W. HILGEMANN. 1994. Inactivation of outward Na/Ca exchange current in guinea pig ventricular myocytes. J. Physiol. **476:** 443–458.
20. NIGGLI, E. & W. J. LEDERER. 1991. Molecular operations of the sodium-calcium exchanger revealed by conformation currents. Nature **349:** 621–624.
21. HILGEMANN, D. W., A. COLLINS, & S. MATSUOKA. 1992. Dynamic and steady state properties of cardiac sodium-calcium exchange: Calcium- and ATP-dependent activation. J. Gen. Physiol. **100:** 933–961.
22. DIPOLO, R. & L. BEAUGÉ. 1990. Asymmetrical properties of the Na-Ca exchange in voltage-clamped, internally dialyzed squid axons under symmetrical ionic conditions. J. Gen. Physiol. **95:** 819–835.
23. GADSBY, D. C., M. NODA, R. N. SHEPHERD & M. NAKAO. 1991. Influence of external monovalent cation on Na-Ca exchange current-voltage relationships in cardiac myocytes. *In* Sodium-Calcium Exchange: Proceedings of the Second International Conference. Ann. N. Y. Acad. Sci. **639:** 140–146.
24. HILGEMANN, D. W. 1994. Channel-like function of the Na/K pump probed at microsecond resolution in giant membrane patches. Science **263:** 1429–1432.
25. FIDDLER, J. & J. FERNANDEZ. 1989. Phase tracking: an improved phase detection technique for cell membrane capacitance measurements. Biophys. J. **56:** 1153–1162.
26. COLLINS, A. & D. W. HILGEMANN. 1993. A novel method for direct application of

phospholipid to giant excised membrane patches in the study of sodium-calcium exchange and sodium channel currents. Pflügers Arch. **423:** 347–355.
27. LI, J. & A. ADEREM. 1992. MacMARCKS, a novel member of the MARCKS family of protein kinase C substrates. Cell **70**(5): 791–801.
28. ADEREM, A. 1995. The MARCKS brothers: A family of protein kinase C substrates. Cell **71**(5): 713–716.

Multiple Functional States of the Cardiac Na^+-Ca^{2+} Exchanger

Whole-Cell, Native-Excised, and Cloned-Excised Properties

SATOSHI MATSUOKA,[a,d] KENNETH D. PHILIPSON,[b] AND DONALD W. HILGEMANN[c]

[a]Department of Physiology
Faculty of Medicine
Kyoto University
Kyoto 606, Japan

[b]Departments of Medicine and Physiology
and
The Cardiovascular Research Laboratories
University of California at Los Angeles
School of Medicine
Los Angeles, California 90095

[c]Department of Physiology
University of Texas
Southwestern Medical Center at Dallas
Dallas, Texas 75235

INTRODUCTION

Modulation properties of the cardiac Na^+-Ca^{2+} exchanger have been studied electrophysiologically in giant membrane patches from cardiac myocytes,[1-4] in giant patches from *Xenopus* oocytes expressing the canine cardiac exchanger (NCX1),[5] and in single guinea-pig ventricular myocytes.[6] A comparison of exchanger function between these different systems may help to define exchanger properties which are inherent to the exchanger protein itself, as opposed to properties which depend on exchanger environment (*e.g.*, cell regulatory processes, auxiliary proteins, and/or a different processing of the exchanger in different cell types). As first important example, it was found that MgATP is without effect on the exchange current expressed in *Xenopus* oocyte membrane,[6] thereby suggesting that proteins other than the exchanger are involved in the ATP effect.

Qualitatively similar exchanger modulation reactions have been identified in each of the three experimental models.[1,2,5,6] Many results can be well explained by assuming that, in analogy to ion channels, the exchanger fluctuates between

---

[d] Address for correspondence: Satoshi Matsuoka, Department of Physiology, Faculty of Medicine, Kyoto University, Sakyo-ku, Kyoto 606, Japan.

one active state and multiple inactive states (see SCHEME 1). The outward current activated by cytoplasmic Na^+ (reverse mode of exchange) decays during Na^+ application on a multisecond time scale (Na^+-dependent inactivation (I_1)). The transition to the I_1-inactive state is assumed to take place from a fully-Na^+ loaded exchanger with a cytoplasmic orientation of the binding site[1] ($E_1 3Na_i$). Cytoplasmic Ca^{2+} reduces the extent of Na^+-dependent inactivation, and also increases peak amplitude of the outward current by relieving exchangers from another inactive state (Na^+-independent Ca^{2+}-dependent inactivation (I_2)).

However some different properties have been observed between the exchanger in these three systems, suggesting a complex functional state of the exchanger. In this study, modulation of the cardiac Na^+-Ca^{2+} exchange current will be demonstrated, and possible mechanisms will be discussed.

MATERIALS AND METHODS

Inside-out giant membrane patch recording was carried out using guinea-pig single ventricular cells and *Xenopus* oocytes expressing canine cardiac exchanger (NCX1).[7] Whole-cell Na^+-Ca^{2+} exchange currents were recorded from single guinea-pig ventricular cells. Holding membrane potential was 0 mV in all experiments. Experiments were carried out at 35°C in whole-cell and giant patch experiments from guinea-pig myocytes, and at 30°C in experiments of *Xenopus* oocytes.

Methods for isolation of single guinea-pig myocytes, oocyte preparation, and the giant membrane patch are described elsewhere.[8,1,5] Standard solutions used are listed in TABLE 1. Modification of the solution will be described in the text or in figure legends.

RESULTS

Na^+-Dependent Inactivation

SCHEME 1 is our model of the transport cycle and inactivation of Na^+-Ca^{2+} exchange. The ion binding site alternatively faces the intracellular or extracellular medium in the E_1 and E_2 states, respectively. The exchanger is assumed to enter into a Na^+-dependent inactive state ($I_1 3Na_i$) from a fully-Na^+ loaded exchanger with a cytoplasmic orientation of the binding site ($E_1 3Na_i$)[1] and to enter an I_2-inactive state (I_2) from the E_1 state.[6]

FIGURE 1A demonstrates representative Na^+-Ca^{2+} exchange currents from a giant membrane patch expressing NCX1. With 8 mM Ca^{2+} in the pipette, outward currents were activated by replacing 100 mM Cs^+ with Na^+ in the superfusate. While an increase of Na^+ concentration augmented current amplitude, the current decayed in a Na^+-dependent manner (Na^+-dependent inactivation (I_1-inactivation)). Current amplitudes at peak (open circles) and at steady state (open squares) are plotted against Na^+ concentration in FIGURE 1B. Decaying phases of the current transients were fit by single exponentials, and time constants (τ) and ratios of steady state current to peak current (Fss) are plotted in FIGURE 1C. As Na^+ increased, the speed and extent of the inactivation increased and saturated at a cytoplasmic Na^+ concentration of 25–50 mM, whereas current magnitudes continued to increase somewhat at higher Na^+ concentrations.

TABLE 1. Composition of Standard Solution for Outward Na$^+$-Ca^{2+} Exchange Current Measurement (in mM)

	Giant Membrane Patches from Cardiac Myocytes		Giant Membrane Patches from *Xenopus* Oocytes		Whole-Cell Currents	
	Pipette Solution	Perfusion Solution	Pipette Solution	Perfusion Solution	Pipette Solution	Perfusion Solution
EGTA	—	10	—	10	50	0 or 0.5
CaCl$_2$	0.1–6	0–10	8	0–10	6.7	0 or 1
CaCO$_3$	—	—	—	—	—	—
MgCl$_2$	2	1	2	—	1.1	2
Mg(OH)$_2$	—	—	—	1–1.5	—	—
BaCl$_2$	0.5	—	2	—	—	2
Ba(OH)$_2$	—	—	—	—	—	—
NaOH	—	—	—	—	100	—
(Na + Li)-Cl	—	—	—	—	—	140
CsOH	—	—	—	—	20	—
NMG-MES	100	—	100	—	—	—
(Na + Cs)-MES	—	100	—	100	—	—
Li-MES	10	—	—	—	—	—
TEA-Cl	—	—	—	—	20	10
TEA-MES	20	20	20	20	—	—
Cs-MES	20	20	20	20	50	10
HEPES	10	20	20	20	—	—
Ouabain	0.2	—	0.25	—	—	0.05
Verapamil	0.002	—	—	—	2	0.002
MgATP	—	—	—	—	—	—
Tris-creatine phosphate	—	—	—	—	2	—
pH	7.0	7.0	7.0	7.0	7.0	7.0

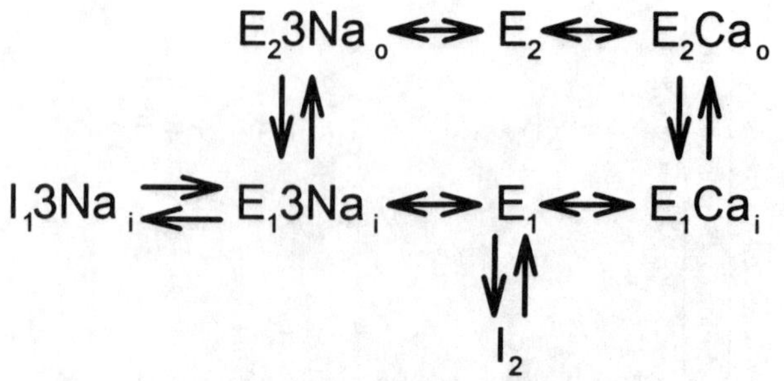

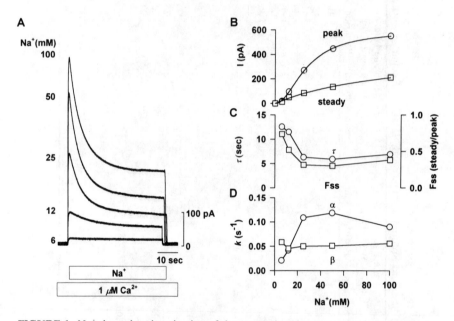

SCHEME 1. Model of transport cycle and inactivation of Na^+-Ca^{2+} exchange.

FIGURE 1. Na^+-dependent inactivation of the outward exchange current in a giant membrane patch from a *Xenopus* oocyte expressing NCX1. Pipette solution contains 8 mM Ca^{2+} and 0 mM Na^+. **(A)** Outward currents activated by cytoplasmic Na^+. Free cytoplasmic Ca^{2+} concentration is 1 μM and Na^+ concentrations are denoted at the *left side* of each trace. **(B)** Na^+ dependence of the exchange current. Peak (*circles*) and steady state (*squares*) current amplitudes are plotted against Na^+ concentration. Curves are fitted Hill equations. Half maximal concentrations (K_h) are 28 mM and 102 mM; Hill coefficients are 2.0 and 1.0 at peak and steady state, respectively. **(C)** Na^+-dependence of the inactivation. Time constants (τ) to fit data in (A) and ratio of steady state current to peak current (Fss) are plotted against Na^+ concentration. **(D)** Na^+-dependence of rate constants of the inactivation. Forward (α) and backward (β) rates are plotted as a function of Na^+ concentration.

With an assumption of a simple inactivation reaction as follows,

$$\text{Active} \underset{\beta}{\overset{\alpha}{\rightleftarrows}} \text{Inactive}$$

the apparent inactivation and recovery rate constants, α and β, can be evaluated assuming that the time constant (τ) of the observed reaction is equal to $1/(\alpha + \beta)$, and that the fraction of current which does not inactivate (Fss: steady state current/peak current) is equal to $\beta/(\alpha + \beta)$.

In FIGURE 1D, calculated α and β are plotted as a function of Na$^+$ concentration. With an increase of Na$^+$ concentration, α increases and saturates, while β remains unchanged. Essentially the same results were obtained in three other patches. They are closely consistent with our proposal that the exchanger enters into the I_1 inactive state from a conformation loaded with 3 cytoplasmic Na$^+$.

Ca^{2+} Regulation

Micromolar concentrations of free cytoplasmic Ca^{2+} stimulate the outward Na$^+$-Ca^{2+} exchange current.[2] FIGURE 2 demonstrates stimulation of the outward current of NCX1 by cytoplasmic Ca^{2+}. Currents were activated by 100 mM cyto-

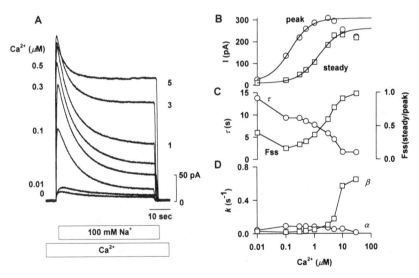

FIGURE 2. (A) Ca^{2+} stimulation of the outward exchange current of NCX1. Pipette solution is the same as in FIGURE 1. Outward currents were activated by 100 mM cytoplasmic Na$^+$ at various Ca^{2+} concentrations as denoted. (B) Ca^{2+} dependence of the current. Peak (*circles*) and steady state (*squares*) current amplitudes are plotted against Ca^{2+} concentration. Curves are fitted Hill equations. K_h values are 0.1 μM and 1.4 μM and Hill coefficients are 1.2 and 1.1 at peak and steady state, respectively. (C) Ca^{2+} dependence of I_1-inactivation τ (*circles*) and Fss (*squares*) values are plotted against Ca^{2+} concentration. (D) Ca^{2+} dependence of inactivation rate constants. Rate constants α and β are plotted as a function of free Ca^{2+} concentration.

plasmic Na$^+$ at various Ca^{2+} concentrations (0–5 μM). As with native cardiac patches, Ca^{2+} appears to have two effects. The peak current is augmented in a low free Ca^{2+} concentration range, and the Na$^+$-dependent inactivation is attenuated in a higher free Ca^{2+} concentration range. The Ca^{2+} dependence of the peak current (k_h = 0.1 μM) is shifted one log unit to lower Ca^{2+} concentrations from the steady state current (k_h = 1.4 μM; FIG. 2B). Ca^{2+} augmentation of peak current can be explained by an increase of the number of exchangers relieved from I$_2$-inactive state.[2]

The time constant (τ) of the I$_1$-inactivation and the fraction of current which does not inactivate (Fss) are plotted in FIGURE 2C. The Ca^{2+} dependence of steady state exchange current parallels that of the Fss. This indicates that reduction of I$_1$-inactivation is the primary basis for the rise of steady state current. In FIGURE 2D, the calculated rate constants α and β are plotted. While α changes only little, β increases remarkably at Ca^{2+} concentration over 1 μM. Thus, it would appear that in the expressed exchanger cytoplasmic Ca^{2+} is acting primarily to facilitate the recovery from I$_1$-inactivation, rather than inhibiting entrance to inactivation.

These stimulatory effects of Ca^{2+} on NCX1 are very similar to those described for the native exchanger in excised patches. However, one notable difference is the existence of a Ca^{2+}-insensitive current component in the cloned exchanger. There is usually no measurable current without cytoplasmic Ca^{2+} in the native exchanger in excised patches.[2]

FIGURE 3A shows outward currents in the presence and absence of 1 μM cytoplasmic Ca^{2+}. Without Ca^{2+}, current activation by 100 mM Na$^+$ was not immediate and the current decay (I$_1$-inactivation) was slower. In FIGURE 3B, effects of Ca^{2+} on Na$^+$ dependence of the peak outward current were examined. With an increase of cytoplasmic Ca^{2+}, the apparent K_h value decreased, approaching the K_h value after chymotrypsin treatment. Since proteolysis by chymotrypsin eliminates both I$_1$- and I$_2$-inactivations,[9] the K_h value after chymotrypsin treatment should reflect Na$^+$ affinity of the transport cycle itself. Therefore, this finding indicates that cytoplasmic Ca^{2+} modifies transport properties by changing apparent Na$^+$ affinity and not only controls the fraction of exchangers in an active state.[10]

As mentioned in the Introduction, work in giant cardiac patches and in whole myocytes strongly suggests that I$_1$-inactivation takes place from the E$_1$ state with 3 Na$^+$ bound (SCHEME 1). As just pointed out, and as discussed elsewhere,[6] some data also suggest that the I$_2$-inactivation takes place from the E$_1$ state of the exchanger. If this is correct, activation of the exchange current by cytoplasmic Ca^{2+} should depend on the distribution of exchangers in the transport cycle. The more exchangers are present in the E$_1$ state, the stronger should be the inactivation. In FIGURE 4, outward currents of the native exchanger in giant membrane patches were activated by 100 mM cytoplasmic Na$^+$ in the presence of 0.1 mM (circles) and 6 mM (squares) extracellular Ca^{2+}. Peak currents are plotted as a function of cytoplasmic Ca^{2+}. The K_h for Ca^{2+} decreased from 0.24 to 0.08 μM Ca^{2+} when the extracellular Ca^{2+} was reduced from 6 to 0.1 mM. With a high concentration of Ca^{2+} on the extracellular side, and without either Na$^+$ or Ca^{2+} on the cytoplasmic side, many exchangers are predicted to locate in the E$_1$ and I$_2$ states. Reduction of extracellular Ca^{2+} promotes exchangers to locate into the E$_2$ state, and thereby promotes recovery from inactivation. Accordingly, less cytoplasmic free Ca^{2+} can activate the exchange current when extracellular free Ca^{2+} is relatively low. As a parallel finding in whole-cell experiments,[6] it was demonstrated that current inactivation by cytoplasmic Ca^{2+} depends on extracellular Na$^+$, which affects the

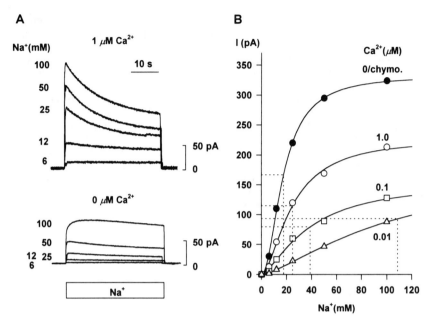

FIGURE 3. Modification of apparent Na$^+$ affinity by cytoplasmic Ca^{2+}. **(A)** Outward currents from NCX1 with or without cytoplasmic Ca^{2+}. The current was activated by 6–100 mM Na$^+$ in the presence of 1 μM Ca^{2+} (*upper panel*) and in the absence of Ca^{2+} (*lower panel*) in the same patch. **(B)** Na$^+$ dependence of peak current. Peak current amplitudes with 0.01 μM (*open triangles*), 0.1 μM (*open squares*), 1.0 μM (*open circles*), and 0 μM Ca^{2+} after treatment with 1 mg/ml α-chymotrypsin (*filled circles*) are plotted as a function of Na$^+$ concentration. *Solid curves* are fitted Hill equations and *dotted lines* indicate K_h values. K_h values are 109, 39, 25, 18 mM at 0.01, 0.1, 1.0 μM Ca^{2+} and 0 μM Ca^{2+} after α-chymotrypsin treatment, respectively. (Modified from Matsuoka *et al.*[10])

fraction of exchangers in the transport cycle. The model of SCHEME 1 could explain all of these findings.

Inactivation of Whole-Cell Exchange Current

In order to test whether the inactivation process exists in the exchanger of intact myocytes, whole-cell voltage clamp experiments were carried out. In FIGURE 5A, outward Na$^+$-Ca^{2+} exchange current was activated by applying 1 mM extracellular Ca^{2+} in the entire absence of extracellular Na$^+$. Pipette solution contains a saturating concentration of Na$^+$ (100 mM) and 0.1 μM Ca^{2+} adjusted with 50 mM EGTA. The current decayed during Ca^{2+} application. The decay could be fit to two exponentials (FIG. 5B). Several lines of evidence support that this decay reflects in part the inactivation process which exists in both native exchanger and NCX1 in giant patches.[6] One evidence is the effect of trypsin. Trypsin eliminates both I_1- and I_2-inactivations of the exchanger in giant membrane patch, like chymotrypsin. In FIGURE 5C, the effect of trypsin on whole-cell currents is described.

Outward currents were activated by applying 1 mM extracellular Ca^{2+}. 0.5 mg/ml trypsin was superfused into myocytes by using a pipette perfusion device.[11] Trypsin attenuated the current decay without marked change of peak amplitude.

A consecutive exchange cycle model with inactivation processes can successfully simulate characteristics of the exchange current in giant membrane patches,[1] and also can explain many of the complex properties of whole-cell current.[6] Therefore, we are confident that the inactivation processes also exist in intact myocytes and are relevant to physiological exchanger function.

The whole-cell exchange current appears to have some different properties from the current of both the native exchanger in patches and NCX1. FIGURE 6A shows current trace and current-voltage (I-V) relations of whole-cell exchange current. The current was activated by 1 mM extracellular Ca^{2+} in the absence of extracellular Na^+, replaced with Li^+ (upper panel), and I-V relations were measured. The lower panel shows I-V relations during the current decay. Control current was subtracted as denoted. The I-V relation became flatter during the process of inactivation, and at steady state, the I-V relation was almost voltage

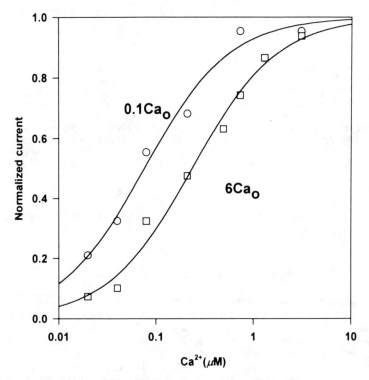

FIGURE 4. The effect of extracellular Ca^{2+} on I_2-inactivation of the native exchanger in giant membrane patches. Outward currents were activated by 100 mM Na^+ at various Ca^{2+} concentrations with 0.1 and 6 mM Ca^{2+} in pipette. Currents were normalized to fitted maximal concentration. Curves are Hill equations. K_h values are 0.08 and 0.24 μM Ca^{2+} at 0.1 and 6 mM extracellular Ca^{2+}, respectively, and Hill equations are 1.0 in both cases.

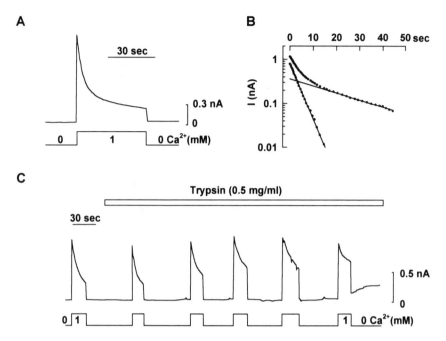

FIGURE 5. The inactivation of whole-cell exchange current. **(A)** Outward current activated by 1 mM extracellular Ca^{2+}. Pipette solution contains 100 mM Na^+ and 0.1 μM Ca^{2+} adjusted with 50 mM EGTA. **(B)** Semilogarithmic plot of the current decay. The decaying phase and the fast component of the current are plotted in a semilogarithmic fashion (*dotted curves*). An asymptote was subtracted from data. *Continuous lines* are fitted exponentials. **(C)** Effect of trypsin on the inactivation. Outward currents were activated with the same protocol as (A), and 0.5 mg/ml trypsin was superfused into a cell by a pipette perfusion device. Note that trypsin reduced the extent of inactivation. (Modified from Matsuoka & Hilgemann.[6])

independent. Possible leak changes caused by 1 mM Ca^{2+} did not account for this alternation.[6] Importantly, this flattening of the I-V relation could *not* be correlated with the inactivation process.[6]

The voltage-independent components of the outward Na^+-Ca^{2+} exchange current have been reported in the absence of extracellular monovalent cations in single myocytes.[6,12] As shown in FIGURE 6B, removal of extracellular monovalent cations resulted in a decrease of current size, but the inactivation still persisted. The lower panel of FIGURE 6B shows the effects of monovalent cations on the I-V relation of outward Na^+-Ca^{2+} exchange current. The exchange current in the absence of extracellular monovalent cations was almost voltage independent, and the I-V relation became flatter during the current decay. This pattern has never been observed in excised membrane patches. Since voltage-dependency of the exchanger originates from the transport cycle itself, and not from inactivation properties, it would appear that some factor or factors in the whole cell can importantly modulate individual steps of the transport cycles.

DISCUSSION

Modulation of the cardiac Na^+-Ca^{2+} exchange current by cytoplasmic Na^+ and Ca^{2+} has been studied in giant membrane patches from cardiac myocytes.[1,2] The mechanism of the modulation could be explained in approximation by assuming a model with one active state and two types of inactivation reactions.[2]

Qualitatively, similar inactivation reactions were demonstrated also in patches from *Xenopus* oocytes expressing the canine cardiac exchanger (NCX1),[5] and in single guinea-pig ventricular myocytes.[6] However, as outlined in Results, some different properties have been observed between the exchanger in these systems.

I_1-inactivation properties of NCX1 expressed in *Xenopus* oocytes are similar to those of the native exchanger in guinea-pig giant membrane patches.[1] However, there is a tendency that the speed of I_1-inactivation is slower and the extent of the inactivation is smaller in NCX1 than the native exchanger in giant membrane

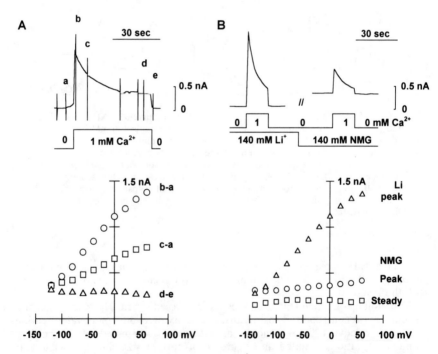

FIGURE 6. I-V relations of the outward Na^+-Ca^{2+} exchange current in single myocytes. **(A)** I-V relations during the decay of outward current. The current was induced with the same protocol as in FIGURE 5. Difference currents between currents during Ca^{2+} application (b, c and d) and control current (a or e) are shown in *lower panel*. **(B)** Effects of monovalent cations on the outward Na^+-Ca^{2+} exchange current and its voltage dependence. Outward currents were activated by 1 mM extracellular Ca^{2+} with (140 mM Li^+) or without (140 mM NMG) monovalent cations (*upper panel*). I-V relations at current peak at 140 mM Li^+ (*triangles*), and at peak (*circles*) and steady state (*squares*) in the absence of monovalent cations are shown in *lower panel*. Control currents are subtracted. (Modified from Matsuoka & Hilgemann.[6])

patches. I_2-inactivation properties of NCX1 are also similar to those of the native exchanger. However, there is a Ca^{2+}-insensitive current component in NCX1. Although the current size is variable, the Ca^{2+}-insensitivity is not due to deregulation (i.e., proteolysis), because the apparent Na^+ affinity of the current was different from that of a patch deregulated by chymotrypsin (FIG. 3). Furthermore, the apparent affinity for cytoplasmic Na^+ changed depending on cytoplasmic Ca^{2+}. Therefore, it is likely that cytoplasmic Ca^{2+} controls the exchange activity by modulating transport properties (Na^+ affinity) of the exchanger, in addition to modulating the major inactivation reactions.[10]

It was demonstrated that inactivation exists in intact exchangers from single guinea-pig myocytes using whole-cell voltage clamp.[6] The current decay during extracellular Ca^{2+} application may reflect both I_1- and I_2-inactivation, since the great majority of exchangers are predicted to locate in the E_2 state without extracellular Na^+ and Ca^{2+}, being relieved from both inactive states. Applying extracellular Ca^{2+} causes exchangers to move into the E_1 state, and then into both the I_1 and I_2 states.

Ca^{2+} regulation of the cardiac exchanger was first reported in whole-cell experiments.[13] While I_2-modulation in both native and cloned exchangers from giant membrane patches is usually in the range of hundreds of nanomolar cytoplasmic free Ca^{2+}, the activation of outward current in intact myocytes by cytoplasmic Ca^{2+} takes place in a severalfold lower concentration range.[14,15] This difference may suggest additional regulatory systems under physiological conditions.

The inactivation processes do not affect the voltage dependence of the outward exchange current of either native exchanger[1] or NCX1[5] in giant membrane patches. This seems also to be true in the exchanger of single myocytes.[6] However, an almost voltage-independent outward current is observed under two conditions in intact myocytes, but not in excised patches: after the process of inactivation has come to equilibrium and in the absence of extracellular monovalent cations. For several reasons, these myocyte-specific phenomena cannot be explained by a decrease of current size or by contamination of other membrane currents.[6] Accordingly, additional regulatory factors which affect voltage dependence of the exchanger appear to exist under physiological conditions in intact myocytes.

The difference between cloned canine exchanger and guinea-pig exchanger could be due to molecular differences of the exchanger protein, as an exchanger isoform NCX2 has different properties of outward exchange current.[16] Guinea-pig cardiac Na^+-Ca^{2+} exchanger was cloned and functionally expressed in *Xenopus* oocytes.[17] The deduced sequence of the guinea-pig exchanger protein is 98% identical to the canine cardiac exchanger, NCX1. They appear to be structurally and functionally very similar, because differences of amino acids between the two species are very conservative. However, electrophysiological studies on the function of the cloned guinea-pig exchanger have not been carried out.

Kinetic studies on I_1- and I_2-modulation are limited, because cytoplasmic Ca^{2+} affects both inactivation processes. However, the two inactivation processes do not seem independent, because mutation of the regulatory Ca^{2+} binding domain altered both I_1- and I_2-inactivation properties.[10] Mutational analysis of exchangers which lack I_1- or I_2-inactivation may help understanding the interaction between I_1- and I_2-inactivation, and also the interaction between the transport cycle and the inactivation processes. The model in SCHEME 1 is our present working hypothesis for inactivation processes, although it is not completely proved by experiments. Especially, it is not certain from which substates in the E_1 state the exchanger enters the I_2 state. Modeling and simulation of the NCX1 current are currently being carried out.

REFERENCES

1. HILGEMANN, D. W., S. MATSUOKA, G. A. NAGEL & A. COLLINS. 1992. Steady-state and dynamic properties of cardiac sodium-calcium exchange: sodium-dependent inactivation. J. Gen. Physiol. **100**: 905–932.
2. HILGEMANN, D. W., A. COLLINS & S. MATSUOKA. 1992. Steady-state and dynamic properties of cardiac sodium-calcium exchange: secondary modulation by cytoplasmic calcium and ATP. J. Gen. Physiol. **100**: 933–961.
3. DOERING, A. E. & W. J. LEDERER. 1993. The mechanism by which cytoplasmic protons inhibit the sodium-calcium exchanger in guinea-pig heart cells. J. Physiol. **466**: 481–499.
4. DOERING, A. E. & W. J. LEDERER. 1994. The action of Na^+ as a cofactor in the inhibition by cytoplasmic protons of the cardiac Na^+-Ca^{2+} exchanger in the guinea-pig. J. Physiol. **480**: 9–20.
5. MATSUOKA, S., D. A. NICOLL, R. F. REILLY, D. W. HILGEMANN & K. D. PHILIPSON. 1993. Initial localization of regulatory regions of the cardiac sarcolemmal Na^+-Ca^{2+} exchanger. Proc. Natl. Acad. Sci. USA **90**: 3870–3874.
6. MATSUOKA, S. & D. W. HILGEMANN. 1994. Inactivation of outward Na^+-Ca^{2+} exchange current in guinea-pig ventricular myocytes. J. Physiol. **476**: 443–458.
7. NICOLL, D. A., S. LONGONI & K. D. PHILIPSON. 1990. Molecular cloning and functional expression of the cardiac sarcolemmal Na^+-Ca^{2+} exchanger. Science **250**: 562–565.
8. COLLINS, A., A. V. SOMLYO & D. W. HILGEMANN. 1992. The giant cardiac membrane patch method: stimulation of outward Na^+-Ca^{2+} exchange current by MgATP. J. Physiol. **454**: 27–57.
9. HILGEMANN, D. W. 1990. Regulation and deregulation of cardiac Na^+-Ca^{2+} exchange in giant excised sarcolemmal membrane patches. Nature **344**: 242–245.
10. MATSUOKA, S., D. A. NICOLL, L. V. HRYSHKO, D. O. LEVITSKY, J. N. WEISS & K. D. PHILIPSON. 1995. Regulation of the cardiac Na^+-Ca^{2+} exchanger by Ca^{2+}. Mutational analysis of the Ca^{2+}-binding domain. J. Gen. Physiol. **105**: 403–420.
11. SOEJIMA, M. & A. NOMA. 1984. Mode of regulation of the Ach-sensitive K-channel by the muscarinic receptor in rabbit atrial cells. Pflügers Arch. **400**: 424–431.
12. GADSBY, D. C., M. NODA, R. N. SHEPHERD & M. NAKAO. 1991. Influence of external monovalent cations on Na-Ca exchange current-voltage relationships in cardiac myocytes. Ann. N. Y. Acad. Sci. **639**: 140–146.
13. KIMURA, J., A. NOMA & H. IRISAWA. 1986. Na-Ca exchange current in mammalian heart cells. Nature **319**: 596–597.
14. NODA, M., R. N. SHEPHERD & D. C. GADSBY. 1988. Activation by $[Ca]_i$ and block by $3'4'$-dichlorobenzamil of outward Na/Ca exchange current in guinea-pig ventricular myocytes. Biophys. J. **53**: 342a.
15. MIURA, Y. & J. KIMURA. 1989. Sodium-calcium exchange current. Dependence on internal Ca and Na and competitive binding of external Na and Ca. J. Gen. Physiol. **93**: 1129–1145.
16. LI, Z., S. MATSUOKA, L. V. HRYSHKO, D. A. NICOLL, M. M. BERSOHN, E. P. BURKE, R. P. LIFTON & K. D. PHILIPSON. 1994. Cloning of the NCX2 isoform of the plasma membrane Na^+-Ca^{2+} exchanger. J. Biol. Chem. **269**(26): 17434–17439.
17. TSURUYA, Y., M. M. BERSOHN, Z. LI, D. A. NICOLL & K. D. PHILIPSON. 1994. Molecular cloning and functional expression of the guinea pig cardiac Na^+-Ca^{2+} exchanger. Biochim. Biophys. Acta **1196**: 97–99.

Mechanism of XIP in Cardiac Sarcolemmal Vesicles[a]

CALVIN C. HALE

Department of Veterinary Biomedical Sciences
and
The Dalton Cardiovascular Research Center
University of Missouri
Columbia, Missouri 65211

INTRODUCTION

The Na-Ca exchange inhibitory peptide known as XIP is a 20 amino acid peptide (RRLLF YKYVY KRYRA GKQRG) that corresponds to residues 219–238 of the cardiac Na-Ca exchange protein. It has been postulated that the endogenous XIP domain may play a regulatory role in the Na-Ca exchange process. Our working hypothesis is that the XIP domain interacts with either the membrane phospholipid bilayer or another domain of the transporter thus upregulating or downregulating the exchange process. We have tested this hypothesis by examining the interactions of XIP and phospholipids on Na-Ca exchange activity in cardiac sarcolemmal vesicles. This report summarizes those observations which have led us to develop a model on how the XIP domain may regulate the exchange process.

BACKGROUND

The Na-Ca exchange protein primary sequence contains a positively charged region rich in arginine and lysine. Hydropathy plot analysis places this 20 amino acid region, which extends from residues 219 through 238, at the N-terminal side of cytoplasmic loop f (FIG. 1). Nicoll et al.[1] noted that because of its positive nature, this area of loop f resembles a calmodulin binding site. Li et al.[2] demonstrated that a synthetic peptide corresponding to residues 219–238 (FIG. 2) binds calmodulin and is a potent inhibitor of the Na-Ca exchange process. Despite this observation, it has not been conclusively demonstrated that cardiac Na-Ca exchange is modulated by calmodulin. Because of its inhibitory properties, this peptide is referred to as XIP for exchange inhibitory peptide. XIP is not the only positively charged peptide capable of inhibiting Na-Ca exchange. Truncated XIP analogs, C28R2 (analogous to the Ca pump calmodulin

---

[a] This work was supported by the American Heart Association and the American Heart Association—Missouri Affiliate.

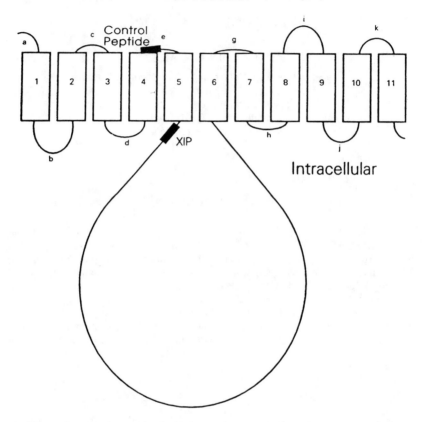

FIGURE 1. Structural model of Na-Ca exchange protein. Membrane spanning regions are *numbered* and extramembranal regions are designated by *letters*. The XIP domain and control peptide region (used in peptide-liposome binding experiments) are shown. (From Shannon *et al.*[3] Reprinted by permission from the American Physiological Society.)

binding domain), and pentalysine (described later) also inhibit Na-Ca exchange although with less potency than XIP.[2,3]

Interest in XIP and the XIP domain in the Na-Ca exchange protein has centered around its possible role as an autoregulatory (inhibitory) region.[2,4] The assumption implicit with an autoregulatory function is that the native XIP domain in the exchange protein must be capable of interacting with another domain of the exchanger in a manner similar to that observed with calmodulin's modulation of Ca pump activity. We present here evidence that Na-Ca exchange activity and XIP are modulated by and interact with negatively charged phospholipids. The notion that the membrane phospholipid environment can modulate Na-Ca exchange activity was previously suggested by Vemuri and Philipson.[5]

```
XIP            RRLLFYKYVYKRYRAGKQRG

Control        YTWLYIILSVI
Peptide

Pentalysine    KKKKK
```

FIGURE 2. Amino sequence of XIP, control peptide, and pentalysine.

MATERIALS AND METHODS

Cardiac Sarcolemmal Vesicle Preparation and Na-Ca Exchange Measurements

Bovine cardiac sarcolemmal vesicles were prepared and Na-Ca exchange activity was assayed as previously described.[6] Briefly, bovine ventricular tissue was obtained fresh from a local abattoir and trimmed to remove epicardium and endocardium before homogenizing. Sarcolemmal vesicles were isolated by differential and gradient centrifugation. Na-Ca exchange activity was determined as the Na_i-

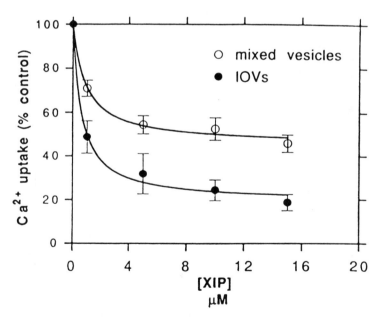

FIGURE 3. Effect of XIP on Na_i-dependent Ca uptake in cardiac sarcolemmal vesicles. Inhibition of the initial rate (2 sec) of Na_i-dependent Ca uptake is shown as a function of the XIP concentration. XIP was present in the uptake solution only during the transport period (2 sec). IOV = inside-out vesicles. (From Kleiboeker et al.[4] Reprinted by permission from the American Society for Biochemistry and Molecular Biology.)

dependent incorporation of $^{45}Ca^{2+}$ into the vesicles. In peptide inhibition experiments, peptides were present only in the uptake solution and thus were exposed to the Na-Ca exchanger only for the period of transport (usually 2 sec).

Synthesis of Peptides

Peptides were synthesized at the Peptide Synthesis Laboratory at the University of Kentucky. Peptides were sequenced to assure purity. Radiolabeling and cross-linking of peptides were performed as previously described.[4]

RESULTS

XIP Inhibition of Na-Ca Exchange in Cardiac Sarcolemmal Vesicles

Na_i-dependent Ca uptake in cardiac sarcolemmal vesicles is inhibited by XIP as shown in FIGURE 3. In these experiments, we were able to distinguish right-side-out from inside-out vesicles by Na-loading inside-out vesicles via the Na-K ATPase as previously described.[7] For a mixed population of vesicles (right-side-

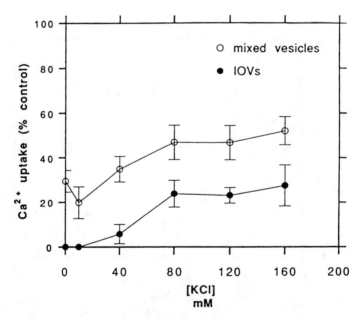

FIGURE 4. Effect of ionic strength of the uptake solution on XIP inhibition. Inhibition of the initial rate (3 sec) of Na_i-dependent Ca uptake is shown as a function of the ionic strength of the external uptake solution. Mannitol was added at appropriate concentrations so that the total osmotic concentration of KCl and mannitol was 320 mosm. Each point contained 5 µM XIP. IOV = inside-out vesicles. (From Kleiboeker et al.[4] Reprinted by permission from the American Society for Biochemistry and Molecular Biology.)

TABLE 1. Effects of XIP on Na-Ca Exchange in Cardiac Sarcolemmal Vesicles[a]

Condition	% Control (±SE)
Na_i-dependent Ca uptake	54.2 (± 4.2) (n = 5)
Na_o-dependent Ca efflux	82.5 (± 7.5) (n = 3)
Pronase digestion	45.2 (n = 1)
Chymotrypsin digestion	46.3 (± 0.8) (n = 2)
Reconstitution (asolectin)	104.3 (± 1.4) (n = 4)
Reconstitution (1 : 1 PC/PS)	66.2 (± 0.3) (n = 2)

[a] After Kleiboeker et al.[4]

out and inside-out), maximal inhibition by XIP was about 46% of control with an IC_{50} of 0.9 μM XIP. For inside-out vesicles, maximal inhibition was 20% of control with an IC_{50} of approximately 0.61 μM XIP. In other experiments, we examined the effect of solution ionic strength on XIP inhibition. As the ionic strength of the external uptake solution decreased, XIP inhibition was enchanced (FIG. 4). In inside-out vesicles, XIP inhibition of Na-Ca exchange reached 100% at low ionic strengths (approximately 20 mM KCL and below). The results shown in FIGURES 3 and 4 suggest that XIP interaction/inhibition of Na-Ca exchange is charge related and occurs on the cytoplasmic surface of the sarcolemma.

TABLE 1 summarizes some of the characteristics of XIP inhibition in cardiac sarcolemmal vesicles observed in our laboratory. XIP inhibits both Na-driven Ca uptake and efflux in vesicle preparations. Mild proteolysis of the vesicles with either pronase or chymotrypsin did not eliminate XIP inhibition. Interestingly, however, following detergent solubilization and reconstitution of sarcolemmal vesicle proteins with soy bean phospholipids (asolectin), XIP did not inhibit activity. When reconstituted into a 1 : 1 mixture of phosphatidylcholine (PC) and phosphatidylserine (PS), XIP inhibited Na-Ca exchange activity. Reconstitution into PC (only) liposomes resulted in a complete loss of Na-Ca exchange activity although re-extraction by detergent and re-reconstitution into 1 : 1 PC/PS restored activity (data not shown).

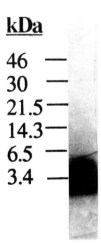

FIGURE 5. Radiolabeling of XIP with $Na^{125}I$. XIP was iodinated as previously described[4] with iodobeads (Pearce). Shown is an autoradiogram of radiolabeled XIP following fractionation on a 17.5% SDS-polyacrylamide gel.

Cross-Linking XIP to Na-Ca Exchange Protein

XIP is readily iodinatable (FIG. 5). The commercially available cross-linking reagent DTSSP (3,3'-dithiobis(sulfosuccinimidylpropionate); Pearce) was used to covalently cross-link ^{125}I-XIP to sarcolemmal vesicle proteins. The results of these experiments are shown in FIGURE 6. Proteins at approximately 75, 120 and 220 kDa were identified by autoradiograms following SDS-PAGE. A fourth band at 48 kDa, which may be an exchanger breakdown product, also was observed. In the presence of 100-fold excess XIP, cross-linking of ^{125}I-XIP was not observed. Also shown in FIGURE 6 is evidence that the cross-linking reagent DTSSP did alter the Coomassie blue stainable protein pattern.

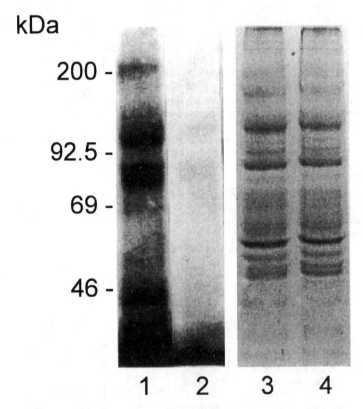

FIGURE 6. Specificity of XIP cross-linking to Na-Ca exchange in cardiac sarcolemmal vesicles. 125-Iodine-XIP was covalently linked to Na-Ca exchange protein in bovine cardiac sarcolemmal vesicles. Following cross-linking, vesicles were washed by centrifugation to remove unbound XIP and subjected to 6.5% SDS-PAGE. *Lane 1,* autoradiogram of 1 μM ^{125}I-XIP cross-linked to 30 μg of sarcolemmal vesicle protein. *Lane 2,* as in lane 1 except cross-linking was performed in the presence of 100 μM XIP (unlabeled). *Lanes 3 and 4,* Coomassie blue staining of experiments shown in lanes 1 and 2, respectively. (After Kleiboeker *et al.*[4])

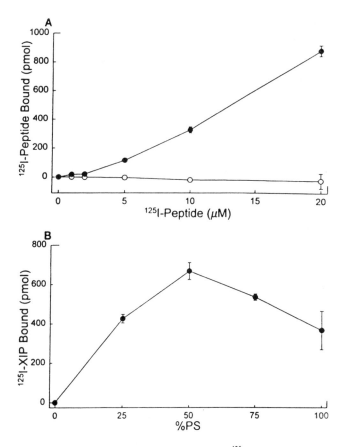

FIGURE 7. XIP binding to liposomes. The binding of ^{125}I-XIP or control peptide was measured. **(A)** The dose-dependent relationship of XIP binding to phosphatidylserine (PS) vesicles is shown. Control peptide binding was not higher than background. Binding of neither XIP nor control peptide to phosphatidylcholine vesicles was higher than background (not shown). **(B)** XIP binding to liposomes as a function of PS content. (From Shannon et al.[3] Reprinted by permission from the American Physiological Society.)

XIP and the Phospholipid Environment

The phospholipid environment can determine the sensitivity of Na-Ca exchange inhibition by XIP. Reconstitution experiments such as those shown in TABLE 1 and reports from other laboratories suggest a potential relationship of negative phospholipids on Na-Ca exchange activity. We hypothesized that the endogenous XIP domain may act as a regulatory region that is affected by negatively charged phospholipids. The behavior of synthetic XIP was used to predict how the endogenous XIP domain may act in the intact exchange protein. Experiments such as those shown in FIGURE 7A indicate that XIP binds to vesicles composed of PS in a dose-dependent manner. The optimal liposome phospholipid composition in these experiments was 1:1 PC/PS (FIG. 7B).

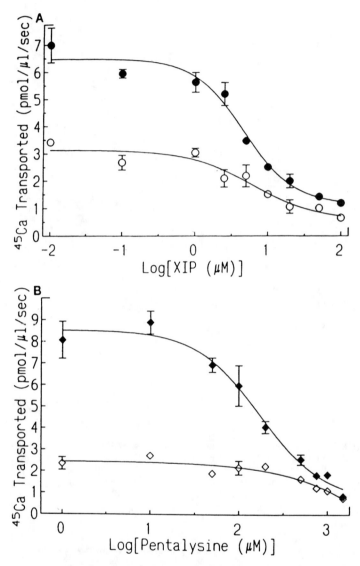

FIGURE 8. Inhibition of Na-Ca exchange by XIP or pentalysine in native and proteolyzed sarcolemmal vesicles. Na-Ca exchange activity was measured in native or mildly proteolyzed cardiac sarcolemmal vesicles (0.1 mg/ml chymotrypsin for 10 min at 37°C) in the indicated concentrations of either XIP or pentalysine. **(A)** Native $IC_{50} = 4.3 \pm 1.5\ \mu M$; proteolyzed $IC_{50} = 3.0 \pm 1.0\ \mu M$. **(B)** Native $IC_{50} = 500\ \mu M$; proteolyzed $IC_{50} = 150 \pm 20\ \mu M$. (From Shannon et al.[3] Reprinted by permission from the American Physiological Society.)

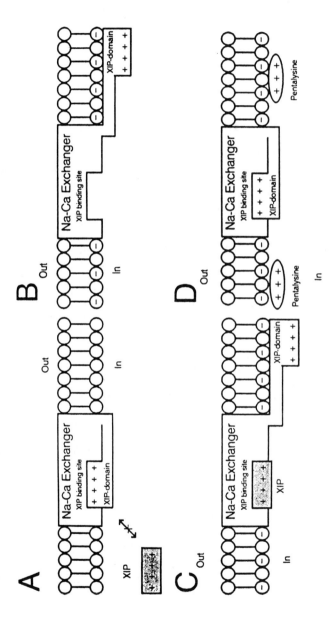

FIGURE 9. A model for the interaction of XIP and the XIP domain with membrane phospholipids and the Na-Ca exchange protein. (**A**) Downregulated exchanger, no negative phospholipids present. (**B**) Negative phospholipids upregulate the exchanger. (**C**) Upregulated exchanger can be inhibited by XIP. (**D**) Downregulated exchanger due to pentalysine occupation of negatively charged phospholipid head groups. (From Shannon et al.[3] Reprinted by permission from the American Physiological Society.)

When Na-Ca exchange activity was reconstituted into 1:1 PC/PS, transport was inhibited by both XIP (TABLE 1) and pentalysine (not shown). In other experiments, we found that both XIP and pentalysine inhibited Na-Ca exchange in both native and mildly proteolyzed native cardiac sarcolemmal vesicles (FIG. 8). In these experiments, proteolysis did not alter the IC_{50} for XIP (4.3 ± 1.5 μM for native and 3.0 ± 1.0 μM after proteolysis). The IC_{50} for pentalysine inhibition in native vesicles (500 μM) was over 100 times higher than the XIP IC_{50} and over 3-fold greater than the pentalysine IC_{50} in proteolyzed vesicles (150 μM). These results suggest that XIP and pentalysine may not bind at the same site.

DISCUSSION

While the physiological relevance of the XIP domain in regulation of the Na-Ca exchange process has not been proved, this region of the transporter has provided some interesting insight regarding *in vitro* manipulation of exchange activity. The initial observation that the XIP domain resembles a calmodulin binding region led to experiments demonstrating inhibition of Na-Ca exchange by synthetic XIP and truncated XIP analogs.[2] Although less potent, Na-Ca exchange also is inhibited by C28R2 (a peptide analogous to the plasma membrane Ca pump calmodulin binding site). M13, a positively charged peptide analogous to the calmodulin binding site of skeletal muscle light chain kinase, is at best a weak inhibitor[2]. It therefore appears that these positively charged peptides can act as cross-inhibitors although their potency may be diminished. It may be reasonable to consider these positive domains a motif as they are found in a variety of proteins and are usually associated with a particular function, which in this case appears to be regulation of transport.

Based upon the results presented above, we have proposed a model for a possible mechanism for XIP and XIP domain interaction with the Na-Ca exchanger (FIG. 9). We have formulated this model as a mechanism to explain how the phospholipid membrane environment may in part control cardiac Na-Ca exchange activity. FIGURE 9A shows an inhibited form of the Na-Ca exchanger in the absence of negatively charged phospholipids. FIGURE 9B depicts an active exchanger due to the presence of negatively charged phospholipids which interact with the XIP domain. This form (FIG. 9B) can be inhibited by the exogenously added XIP peptide (FIG. 9C). FIGURE 9D represents a possible mechanism to explain the lower potency of pentalysine and the observation that it may not interact at the "XIP binding site."

REFERENCES

1. NICOLL, D. A., S. LONGONI & K. D. PHILIPSON. 1990. Molecular cloning and functional expression of the cardiac Na^+-Ca^{2+} exchanger. Science **250:** 562–565.
2. LI, Z., D. NICOLL, A. COLLINS, D. W. HILGEMANN, A. G. FILOTEO, J. T. PENNISTON, J. N. WEST, J. M. TOMICH & K. D. PHILIPSON. 1991. Identification of a peptide inhibitor of the cardiac sarcolemmal Na^+-Ca^{2+} exchanger. J. Biol. Chem. **266:** 1014–1020.
3. SHANNON, T. R., C. C. HALE & M. A. MILANICK. 1994. Interaction of the Na-Ca exchanger and exchange inhibitory peptide with membrane phospholipids. Am. J. Physiol. **266:** C1350–C1356.
4. KLEIBOEKER, S. B., M. A. MILANICK & C. C. HALE. 1992. Interactions of the exchange

inhibitory peptide (XIP) with Na-Ca exchange in bovine cardiac sarcolemmal vesicles and ferret red cells. J. Biol. Chem. **267:** 17836–17841.
5. VEMURI, R. & K. D. PHILIPSON. 1988. Phospholipid composition modulates the Na^+-Ca^{2+} exchange activity of cardiac sarcolemma in reconstituted vesicles. Biochim. Biophys. Acta **937:** 258–268.
6. SLAUGHTER, R. S., J. L. SUTKO & J. P. REEVES. 1983. Equilibrium calcium-calcium exchange in cardiac sarcolemmal vesicles. J. Biol. Chem. **258:** 3183–3190.
7. PHILIPSON, K. D. & A. Y. NISHIMOTO. 1982. Na^+-Ca^{2+} exchange in inside-out cardiac sarcolemmal vesicles. J. Biol. Chem. **257:** 5111–5117.

Cardiac Na-Ca Exchange and pH

ANDREA E. DOERING,[a] DAVID A. EISNER,[b]
AND W. JON LEDERER[c]

[a]*Cardiovascular Research Laboratories
University of California, Los Angeles School of Medicine
MRL Building, Room 3645
Los Angeles, California 90024-1760*

[b]*Department of Veterinary Preclinical Sciences
University of Liverpool
Liverpool L69 3BX, United Kingdom*

[c]*Department of Physiology
University of Maryland School of Medicine
Baltimore, Maryland 21201*

INTRODUCTION

Protons have a powerful modulatory effect on the Na-Ca exchanger which may have important implications for its function. Although pH is tightly regulated in the environment of the cell membrane, it is known to change, for example, in disease states: hypoxia results in cytoplasmic acidification.[1] In the cardiac cell there are constant fluctuations of cytoplasmic calcium concentration, which may be accompanied by significant fluctuations in pH since protons compete with calcium for intracellular buffers.[2]

Modulation of Na-Ca exchange by protons was first reported in 1968 by Baker and Blaustein, who observed that extracellular alkalinization from pH 6.0 to 8.5 increased sodium-dependent calcium efflux from crab nerve by a factor of three.[3] This showed not only that protons could inhibit Na-Ca exchange, but that at physiological pH around 7, the exchanger would be partially inhibited. In 1977, Baker and McNaughton looked at the effect of cytoplasmic protons in giant squid axon and showed that acidification inhibited calcium-dependent sodium efflux.[4] Wakabayashi and Goshima showed that extracellular acidification inhibited sodium-dependent calcium uptake in mouse heart in 1981,[5] and in 1982, Philipson, Bersohn, and Nishimoto did an extensive study of the effects of pH on Na-Ca exchange measured in canine cardiac sarcolemmal vesicles.[6] This study showed that changing pH from 6.0 to 10.0 increased sodium-dependent calcium uptake by a factor of 15. Presumably the measured effect was mostly exerted from the cytoplasmic side of the membrane. Protons inhibited by competing at the calcium translocation site, and also possibly at the sodium translocation site.

In order to further investigate the mechanism or mechanisms of the proton inhibitory effect, we measured Na-Ca exchange current under voltage clamp in the whole cell and the inside-out excised patch configurations from ventricular myocytes. Protons appear to modulate cardiac Na-Ca exchange in a complex manner, acting at more than one site on the molecule. An understanding of these regulatory effects is important to predict the physiological role of the exchanger.

METHODS

Giant excised patch experiments: adult guinea pig ventricular cells were prepared as previously described.[7] The extracellular (pipette) solution contained (in mM) 160 NMGCl, 7.5 CaCl$_2$, 0.02 EGTA, 20 HEPES, 10 TEACl, 0.025 ouabain, and 0.0025 D600. The cytoplasmic (bath) solution contained 140 NMGCl, 1.5 MgCl$_2$, 20 BAPTA, 20 HEPES, 5 MgATP, and 0.0002 free calcium. The intracellular sodium was raised by substituting NaCl for NMGCl. Changes in cytoplasmic pH had no effect on the background membrane current with 0 mM cytoplasmic sodium.

Whole cell experiments: acutely dissociated adult rat ventricular myocytes were loaded with the fluorescent pH indicator carboxy-SNARF. SNARF was incorporated into the cells as the acetomethoxy ester. The fluorescence signal

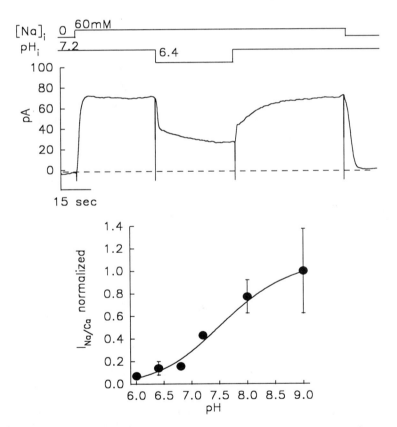

FIGURE 1. Cytoplasmic protons inhibit Na-Ca exchange current. *Top panel*, outward Na-Ca exchange current is activated in the giant excised patch by raising [Na]$_i$ from 0 to 60 mM. When pH$_i$ is lowered from 7.2 to 6.4, the current is partially and reversibly inhibited. *Bottom panel*, current amplitude is plotted versus pH$_i$. The points are averages of up to 17 measurements. (From Doering & Lederer.[7] Reprinted by permission from the *Journal of Physiology*.)

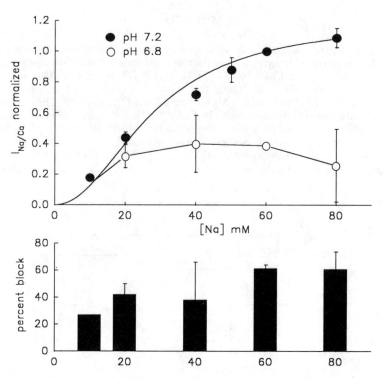

FIGURE 2. Protons do not compete at the cytoplasmic sodium translocation site. *Top panel*, Na-Ca exchange current amplitude is plotted versus [Na]$_i$. *Filled circles*, pH 7.2. *Open circles*, pH 6.8. *Bottom panel*, the per cent block when pH is lowered from 7.2 to 6.8 is plotted versus [Na]$_i$. There is no relief of proton block at high [Na]$_i$.

was calibrated using the free acid form of the dye in solution in the cell bath. The extracellular (bath) solution contained (in mM) 140 NaCl, 2 MgCl$_2$, 2 BaCl$_2$, 0.05 EGTA, 10 HEPES, 10 TEACl, 0.05 ouabain, and 0.004 verapamil. In the cytoplasmic (pipette) solution was 100 NaOH, 20 CsCl, 20 TEACl, 0.0003 free Ca, 1.1 MgCl$_2$, 50 EGTA, 2 MgATP, 2 Tris phosphocreatine, and 0.05 dimethylamiloride. The calcium-activated Na-Ca exchange current was completely blocked by 5 mM nickel. Intracellular pH was altered by isosmotic substitution of the weak acid butyric acid (BA) or the weak base tetramethylammonium (TMA) into the bath solution. The undissociated form of the acid or base crosses the cell membrane and then dissociates in the cytoplasm, where it can act as a proton donor or acceptor, respectively. The background current measured with 0 mM external calcium was not altered by nickel, TMA, or butyric acid.

The membrane potential was held at 0 mV. All experiments were carried out at 33–35°C.

RESULTS

The giant excised patch preparation, described by Hilgemann in 1990,[8] is ideal for measuring Na-Ca exchange current and manipulating cytoplasmic pH.

In FIGURE 1, the top panel shows the outward Na-Ca exchange current activated when cytoplasmic sodium is raised from 0 to 60 mM. When cytoplasmic pH is reduced from 7.2 to 6.4, the current is substantially inhibited, and this inhibition is completely reversible when pH is raised to 7.2 again. The bottom panel shows Na-Ca exchange current plotted versus cytoplasmic pH. The points are the mean values from up to seventeen membrane patches, and the error bars show standard error. Na-Ca exchange is extremely sensitive to cytoplasmic pH, which agrees with work done by other investigators. It is partially inhibited at physiological pH of 7.2, and is particularly sensitive to pH in the physiological range. The Na-Ca exchange current is completely blocked below pH 6.0.

To characterize the site of proton interaction with the Na-Ca exchange molecule, we first investigated the possibility that protons compete at the sodium translocation site. The top panel of FIGURE 2 shows the dependence of the Na/Ca exchange current amplitude on cytoplasmic sodium concentration, measured at pH 7.2 and at pH 6.8. At pH 6.8, the current amplitude is decreased, and per cent inhibition is plotted in the bottom panel versus sodium concentration. If protons inhibited via simple competition at the sodium translocation site, increased

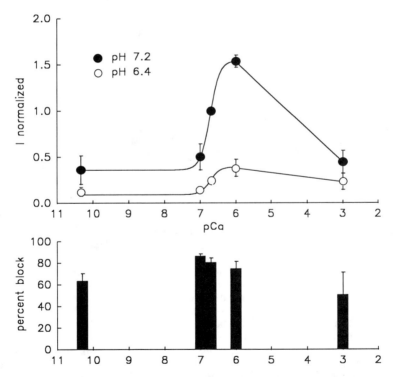

FIGURE 3. Protons may compete at a low-affinity calcium binding site. *Top panel*, Na-Ca exchange current amplitude is plotted versus pCa_i. *Filled circles*, pH 7.2. *Open circles*, pH 6.4. *Bottom panel*, the per cent block when pH is lowered from 7.2 to 6.4 is plotted versus pCa_i. There is slight relief of proton block at 1 mM $[Ca]_i$. (From Doering & Lederer.[7] Reprinted by permission from the *Journal of Physiology*.)

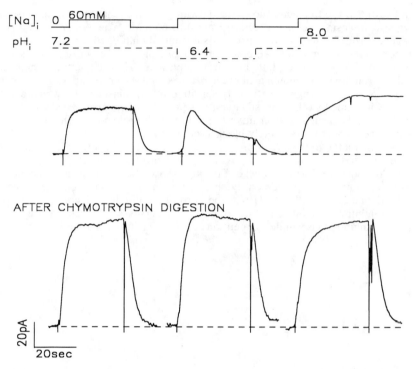

FIGURE 4. Chymotrypsin digestion relieves proton inhibition. Na-Ca exchange current is activated by raising [Na]$_i$ at pH$_i$ 7.2, 6.4, and 8.0. *Top panel,* control current is inhibited by acidification and stimulated by alkalinization. *Bottom panel,* after 1 minute exposure to cytoplasmic chymotrypsin, pH changes have no effect on the current amplitude. (From Doering & Lederer.[7] Reprinted by permission from the *Journal of Physiology*.)

sodium concentration would be expected to reduce proton inhibition. This does not happen; in fact per cent inhibition increases slightly with increased sodium concentration. We concluded that protons do not exert their inhibitory effect at the sodium translocation site.

Protons compete at the calcium translocation site to inhibit Na-Ca exchange,[6] but that would not affect our measurements because we looked only at the effect of cytoplasmic protons on outward current, which is produced by transport of calcium from the external to the cytoplasmic side. It is possible, however, that the proton inhibition we observe could be due to protons competing at the cytoplasmic calcium stimulatory site. In FIGURE 3, we show the dependence of the Na-Ca exchange current amplitude on cytoplasmic calcium. In the top panel, current is plotted versus pCa, and the current amplitude increases up to 1 μM calcium, and then is strongly inhibited at 1 mM calcium due to a decrease in the driving force for Na-Ca exchange. The calcium dependence has the same shape at pH 7.2 and at pH 6.4, and the percent inhibition is plotted versus pCa in the bottom panel. There is no relief of proton block in the range of calcium concentration at which calcium has its stimulatory effect. There is some reduction in inhibition at 1 mM calcium, although the significance of the difference is difficult to evaluate because the current amplitude is small. Some competition between protons and calcium may be expected at any site of proton interaction on the Na-Ca exchange molecule, since protons and calcium tend to compete for binding sites. This could be evidence of proton competition at the calcium stimulatory site, if the site has an extremely high affinity for protons relative to calcium, but the data are inconclusive.

The calcium stimulatory site, along with modulatory sites for cytoplasmic sodium and ATP, is disrupted or removed by partial proteolysis of the Na-Ca exchanger, so that the modulatory effects are lost.[8] We found that the proton inhibitory effect was also removed by partial proteolysis. The top panel of FIGURE 4 shows control current records, in which Na-Ca exchange current is activated at pH 7.2, then again as pH is lowered to 6.4, and again as pH is raised to 8.0. The current is inhibited at pH 6.4 and stimulated at pH 8.0. After one minute of exposure to chymotrypsin in the bath solution, the protocol is repeated. The current amplitude is doubled at pH 7.2, and is unchanged at 6.4 and 8.0. The results are summarized in FIGURE 5. Na-Ca exchange current is plotted versus cytoplasmic pH. The filled circles show a steep sensitivity before digestion, and the open circles show that after digestion there is no change in current amplitude with pH changes above 6.8. Because the Na-Ca exchanger is strongly inhibited

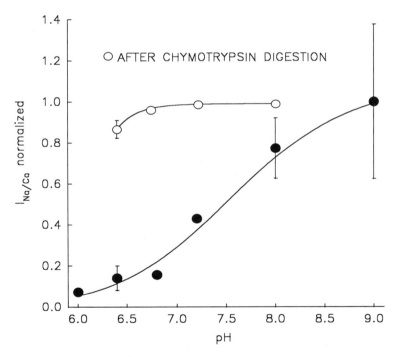

FIGURE 5. Chymotrypsin digestion relieves proton inhibition. Na-Ca exchange current is plotted versus pH_i. *Filled circles,* before chymotrypsin digestion. *Open circles,* after chymotrypsin digestion. (From Doering & Lederer.[7] Reprinted by permission from the *Journal of Physiology*.)

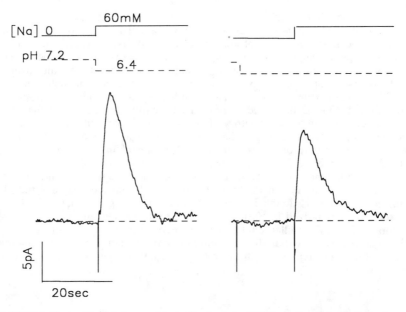

FIGURE 6. Proton block is incomplete in the absence of [Na]$_i$. *Left panel*, Na-Ca exchange current is activated by raising [Na]$_i$, and pH$_i$ is simultaneously lowered. An outward current declines to a low steady-state level as proton block develops in 15 sec. *Right panel*, pH$_i$ is lowered to inhibit the Na-Ca exchanger 15 sec before [Na]$_i$ is raised. The peak current amplitude is reduced by 30%. The current then declines to the same steady-state level as proton block develops to completion 15 sec after [Na]$_i$ has been raised. (From Doering & Lederer.[13] Reprinted by permission from the *Journal of Physiology*.)

by protons at physiological pH, this removal of proton inhibition offers a useful explanation for the well-known stimulation of Na-Ca exchange activity by partial proteolysis.[9] There is a slight decrease in current amplitude at pH 6.4, after digestion. This may be due to a decrease in the affinity of the inhibitory site for protons, which would shift the pH dependence to the left, or it may be that there are multiple mechanisms of proton inhibition, and a powerful one is removed by proteolysis, leaving a weaker one intact.

From the data presented so far, we concluded that protons inhibit the Na-Ca exchanger, not by competing at the sodium translocation site, but by interacting at a low affinity calcium binding site which is located near the cytoplasmic sodium, calcium, and ATP modulatory sites. The calcium modulatory site is known to be located in the cytoplasmic loop of the molecule.[10] We went on to investigate the characteristics of proton block in terms of its rate of development.

FIGURE 6 shows an attempt to measure the rate of proton block in the absence of exchanger activity. The first current record is Na-Ca exchange current activated when sodium is raised from 0 to 60 mM and pH is simultaneously lowered from 7.2 to 6.4. Because Na-Ca exchange current activates more quickly than proton block develops, the current is transient, declining in 15 seconds to a steady-state blocked level. The second current record shows what happened when pH was lowered from 7.2 to 6.4 15 seconds before Na-Ca exchange current was activated by raising sodium. Surprisingly, there is still a transient current, which suggests that proton inhibition does not develop to its steady-state level in the absence of Na-Ca exchange activity.

There seem to be two components to proton inhibition. The first develops in the absence of Na-Ca exchange activity and is relatively fast, as the following observation indicates. When the time of preexposure to acidic pH was varied between one second and 30 seconds, the peak amplitude of the transient Na-Ca exchange current did not change, indicating that this component of proton block reached a steady-state level in less than one second. The first component of proton block decreases peak current by about 30%. The second component only develops after the Na-Ca exchanger is activated by raising cytoplasmic sodium, and is relatively slow, with a half-time of 6 seconds. Since complete block is about 90%, the second component is responsible for decreasing peak current by about 60%. This second component could somehow require cycling of the Na-Ca exchanger, or it could require cytoplasmic sodium. We discovered that the latter is true.

FIGURE 7 shows that proton inhibition can be partially relieved by removing cytoplasmic sodium. Na-Ca exchange current is activated by raising cytoplasmic sodium from 0 mM to 60 mM while pH is simultaneously lowered from 7.2 to 6.4. The current declines to a steady-state blocked level. Sodium is then decreased to 0 for increasing intervals, while pH is held acidic at 6.4. Na-Ca exchange current

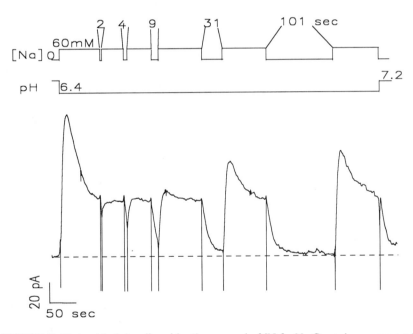

FIGURE 7. Proton block is relieved by the removal of $[Na]_i$. Na-Ca exchange current is activated by 60 mM $[Na]_i$, pH_i 6.4. $[Na]_i$ is repeatedly reduced to 0 for increasing durations. After 9.3 sec in 0 mM $[Na]_i$, current amplitude increases above steady-state proton-blocked level, indicating partial relief of proton block, although pH_i is still 6.4. Peak current increases for increased durations of exposure to 0 mM $[Na]_i$. (From Doering & Lederer.[13] Reprinted by permission from the *Journal of Physiology*.)

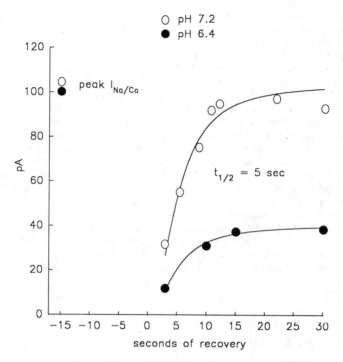

FIGURE 8. Relief of proton block in 0 [Na]$_i$ at pH$_i$ 7.2 and 6.4. Similar experimental protocol to FIGURE 7. Na-Ca exchange current is activated by raising [Na]$_i$ from 0 to 60 mM and simultaneously lowering pH$_i$ from 7.2 to 6.4. [Na]$_i$ is then lowered to 0 mM for increasing intervals and then returned to 60 mM, ph$_i$ 6.4. The peak current is plotted versus the duration in 0 mM [Na]$_i$. The increase in peak current with time reflects the rate of relief of proton block. *Open circles,* pH$_i$ is raised to 7.2 for the intervals in 0 mM [Na]$_i$, so the peak currents represent the rate of relief of proton block. *Filled circles,* pH$_i$ was held at 6.4 for the intervals in 0 mM [Na]$_i$, so the peak currents represent the rate of relief of [Na]$_i$-dependent proton block.

at first deactivates and reactivates to the same steady-state blocked level as sodium is decreased to 0 and raised again to 60 mM. After 9 seconds in 0 mM sodium, however, a current transient reappears, and the peak amplitude of the transient increases with increased time in 0 mM sodium. This shows that there is a sodium-dependent component to proton inhibition which can be relieved by removing sodium, even though pH is unchanged.

The two components of proton inhibition are distinguished in a different way in FIGURE 8. The protocol shown in FIGURE 7 is carried out and then repeated, with pH raised from 6.4 to 7.2 for the intervals in 0 mM sodium. In the first protocol, sodium-dependent block only is relieved, whereas in the second protocol, both components of proton block are relieved. Peak currents are plotted versus the interval in 0 mM sodium. The points plotted at −15 seconds represent the peak current measured when Na-Ca exchange current is activated and pH is simultaneously lowered from 7.2 to 6.4. The filled circles then show the time course of relief of proton block when sodium is removed for increasing intervals but pH is held acidic at 6.4. There is about 40% relief of block. The open circles show relief of block when pH is returned to 7.2 as sodium is removed, and there is 100% relief. There is an early fast component to relief in pH 7.2 which takes place in under three seconds. This is presumably relief from the sodium-independent component of block. After that, both plots can be fitted with a curve which gives a half-time for relief of 5 seconds. This represents the rate of relief from the sodium-dependent secondary component in both cases. We can thus separate proton inhibition into two processes. The first develops and is relieved too quickly for the rate to be resolved by our methods, and the second develops and is relieved with a half-time of around 5 seconds and requires cytoplasmic sodium.

We constructed a mathematical model to help interpret these results. The Na-Ca exchange cycle, diagrammed in FIGURE 9, was modeled as a series of first- and second-order reactions, with the goal of reproducing four of our experimental results: the dependence of outward Na-Ca exchange current on i) cytoplasmic sodium concentration and ii) cytoplasmic pH, and the phenomena of iii) partial proton block in 0 mM sodium and iv) partial recovery from proton block in 0 mM sodium. We used a consecutive transport scheme, based on the evidence for partial reactions, and we did not take any modulatory effects other than that of pH into consideration. The exchanger molecule (E) is oriented with sodium- and calcium-translocation sites facing either the external or cytoplasmic side of the membrane, and when it is bound to a cytoplasmic proton, it cannot participate in the exchange cycle. We gave the cytoplasmic sodium-bound forms of the exchanger a higher affinity for protons than the sodium-free forms in order to reproduce two components to proton block, one of which requires sodium. This model was successful in reproducing the experimental results, as shown in FIGURE 10.

The top panel in FIGURE 10 is simulated Na-Ca exchange current. An outward current is activated when cytoplasmic sodium concentration is raised from 0 to 60 mM. The current has a transient component, which slowly declines to a steady-state level with a half-time of about 30 seconds. We have inadvertently reproduced a modulatory effect which we did not consider in

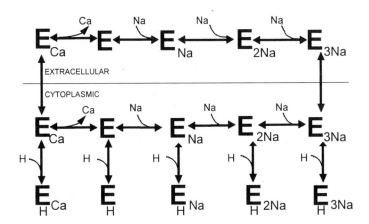

FIGURE 9. Model of the Na-Ca exchange cycle and inhibition by cytoplasmic protons.

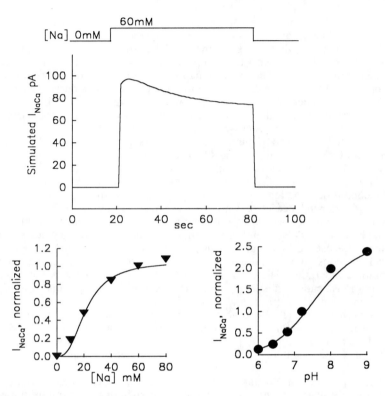

FIGURE 10. Simulated Na-Ca exchange current. *Top panel,* simulated current produced by the model of the proton-sensitive Na-Ca exchanger when $[Na]_i$ is raised from 0 to 60 mM at pH_i 7.2. There is a transient component to the current due to the development of $[Na]_i$-dependent proton block, even though pH_i is unchanged. *Bottom panel,* $[Na]_i$- and pH_i-dependence of simulated Na-Ca exchange current. The *solid line* is a fit to *experimental* data. (From Doering & Lederer.[13] Reprinted by permission from the *Journal of Physiology.*)

designing the model. In the absence of any changes in pH, the Na-Ca exchanger is inhibited by cytoplasmic sodium. This phenomenon has been described by Hilgemann and co-workers in both the giant excised patch[11] and the whole cell.[12] It is seen when outward Na-Ca exchange current is activated by raising cytoplasmic sodium concentration, as a slow decline of the current amplitude over seconds. This intrinsic sodium inhibition is variable from membrane patch to patch, and is removed by partial proteolysis. We have reproduced such an effect by modeling proton inhibition as a partially sodium-dependent process. Therefore, intrinsic sodium-dependent inactivation of Na-Ca exchange current may be at least in part due to sodium enhancing the proton inhibition which is already quite strong around physiological pH.

In addition, our own experimental results were reproduced by the model. Sodium dependence and pH dependence of the simulated current are plotted in the bottom panel of FIGURE 10. The points are steady-state simulated current amplitudes, and the curves are fits to the experimental data. FIGURE 11 shows that the phenomena of partial proton inhibition and partial relief of inhibition in 0 mM sodium are also reproduced by the model. The top panel simulates the experimental protocol illustrated in FIGURE 6, and the bottom panel simulates the protocol illustrated in FIGURE 7. We did not attempt to reproduce the rates of proton block or relief due to variations in the experimental values. This model is a simplification which cannot prove the mechanism of proton inhibition. Its success does show that the proposal that the sodium-bound forms of the Na-Ca exchange molecule may have a higher affinity for inhibitory protons than the sodium-free forms is consistent with the experimental results.

The results are summarized in TABLE 1. We have measured the inhibitory effect of cytoplasmic protons on the outward Na-Ca exchange current and found a strong sensitivity to changes of cytoplasmic pH in the physiological range. This effect does not appear to be due to protons competing at the sodium translocation site or at the calcium modulatory site, although the proton

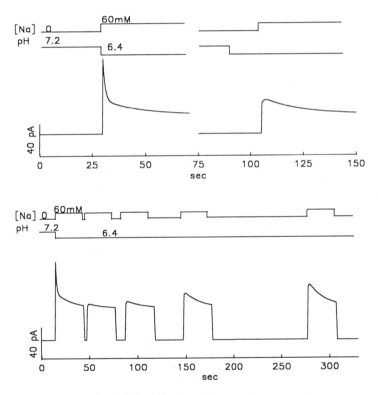

FIGURE 11. $[Na]_i$-sensitive proton inhibition of simulated Na-Ca exchange current. *Top panel*, the experimental protocol of FIGURE 6 is reproduced by the model. Preexposure to cytoplasmic protons results in partial block of the Na-Ca exchanger, but block is not complete until $[Na]_i$ is raised. *Bottom panel*, the experimental protocol of FIGURE 7 is reproduced by the model. After proton inhibition has developed to completion in the presence of 60 mM $[Na]_i$, there is partial relief of proton block when $[Na]_i$ is removed, even though pH_i remains acidic. (From Doering & Lederer.[13] Reprinted by permission from the *Journal of Physiology*.)

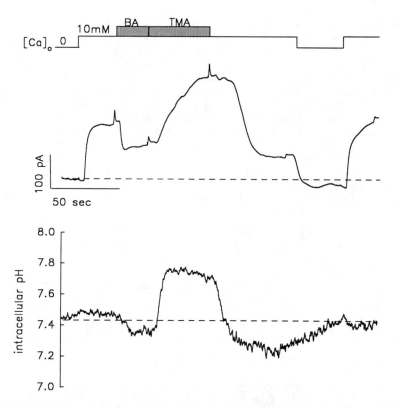

FIGURE 12. Effect of changes in pH_i on whole-cell Na-Ca exchange current. *Top panel,* outward Na-Ca exchange current activated by raising $[Ca]_0$ from 0 to 10 mM. *Bottom panel,* pH_i 10 mM butyric acid and then 10 mM TMA were added to the extracellular solution, and pH_i decreased, then increased. Na-Ca exchange current was inhibited by intracellular acidification and stimulated by intracellular alkalinization.

interaction site may have a low affinity for calcium. Partial proteolysis effectively removes the proton inhibitory effect, indicating that the site or sites of proton interaction with the Na-Ca exchange molecule may be in the cytoplasmic loop near the calcium modulatory site and may be located near the sodium and ATP modulatory sites, all of which are also sensitive to proteolysis.

There are two components to proton inhibition, one of which is fast and one of which is slow and requires cytoplasmic sodium. When sodium is 60 mM and pH is lowered from 7.2 to 6.4, the Na-Ca exchange current is inhibited about 98%. This inhibition develops with a half-time of about 6 seconds. When sodium is 0 mM and pH is lowered from 7.2 to 6.4, the Na-Ca exchange current is inhibited about 34% in under 1 second. After inhibition has developed in 60 mM sodium at pH 6.4, if sodium is lowered to 0 mM, inhibition is relieved about 55%. Presumably, this is because exchanger molecules which are bound to sodium and protons lose their sodium and in consequence lose their protons and are freed from inhibition.

TABLE 1. Characteristics of Proton Inhibition of the Outward Na-Ca Exchange Current

Competition with $[Na]_i$	Insignificant
Competition with $[Ca]_i$	Insignificant
Chymotrypsin digestion	Strongly decreased
Block in 60 mM Na, pH 6.4	$98 \pm 5\%$
Rate of block	$t_{1/2} = 5.7 \pm 0.6$ seconds
($I_{Na/Ca}$ activation rate)	$t_{1/2} = 1.3 \pm 0.2$ seconds
Block in 0 mM Na, pH 6.4	$34 \pm 11\%$
Rate of block	$t_{1/2} < 1$ second
Recovery in 0 mM Na, pH 6.4	$55 \pm 34\%$
Rate of recovery	$t_{1/2} = 5.0 \pm 1.4$ seconds

Finally, we measured the effects of pH changes on Na-Ca exchange current measured in the whole cell. The outward current was activated by raising external calcium from 0 to 10 mM, and cytoplasmic pH was monitored using the fluorescent indicator SNARF. Due to the limitations of the experimental technique, cytoplasmic pH cannot be exactly quantified, but deviations from the resting pH are estimated. The results are shown in FIGURE 12. When butyric acid is added to the external solution, the cytoplasmic pH is acidified, and Na-Ca exchange current amplitude decreases. Butyric acid is replaced with tetramethylammonium, the cytoplasmic pH goes alkaline, and Na-Ca exchange current amplitude increases. When TMA is removed, the cytoplasm is acidified in a rebound effect, for reasons which are not well understood, and then returns to resting pH level. The current activated by 10 mM calcium returns to control level. Results from seven cells are summarized in FIGURE 13. Whole-cell Na-Ca exchange current is plotted versus cytoplasmic pH, and the result agrees qualitatively with the giant patch results. The pH dependence is considerably steeper than that measured in the excised patch. This may be a species difference since the whole-cell experiments were done in rat, or it may indicate that the giant patch measurements underestimate the sensitivity of the Na-Ca exchanger to pH changes.

The effect of changes in extracellular pH on Na-Ca exchange current was also measured, as shown in FIGURE 14. The pH of the external solution was

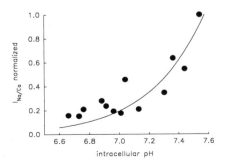

FIGURE 13. pH_i-dependence of whole-cell Na-Ca exchange current. Current amplitudes measured in 6 cells are plotted versus pH_i. Na-Ca exchange current is extremely sensitive to a pH_i change of only one pH unit.

monitored using SNARF, and the cytoplasmic pH was buffered with 50 mM HEPES and did not change when external pH was changed (data not shown). When external pH was acidified, the outward current was, unexpectedly, stimulated. External alkalinization strongly inhibited the Na-Ca exchanger. The current record shown here illustrates an extreme degree of current rundown, as the current does not appear to recover from inhibition by external alkalinization. In five other cells, the current did recover, but this record illustrates that the stimulatory effect of external acidification may be underestimated in these experiments. The results from six cells are summarized in FIGURE 15. We do not have any evidence as to the site of action of extracellular protons on the Na-Ca exchanger, and cannot speculate as to why protons should have opposing modulatory effects on opposite sides of the membrane. It is clear that the Na-Ca exchanger is about three times more sensitive to changes in cytoplasmic pH than to changes in extracellular pH, so under physiological conditions where extracellular pH changes may result in parallel cytoplasmic pH changes, the inhibitory effect of protons may be expected to dominate.

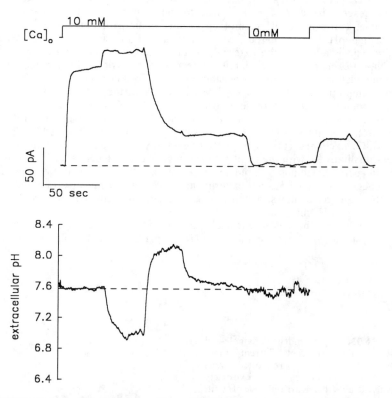

FIGURE 14. Effect of changes in pH_o on Na-Ca exchange current. *Top panel,* outward current activated by raising $[Ca]_o$ from 0 to 10 mM. *Bottom panel,* pH_o. Extracellular pH was increased and then decreased, and the Na-Ca exchange current was inhibited by extracellular alkalinization and stimulated by extracellular acidification.

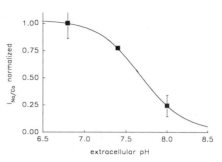

FIGURE 15. pH_o-dependence of whole-cell Na-Ca exchange current. Current amplitudes measured in 5 cells are plotted versus pH_o. The modulatory effect of pH_o on Na-Ca exchange current is opposite to the modulatory effect of pH_i.

CONCLUSION

Protons affect the cardiac Na-Ca exchanger at multiple sites via multiple mechanisms. We have distinguished two separate components of the inhibitory effect of cytoplasmic protons, one of which requires cytoplasmic sodium, and have shown a stimulatory effect of extracellular protons. The cytoplasmic site or sites of proton interaction are disrupted by partial proteolysis and thus may be located in the intracellular loop of the Na-Ca exchange molecule. Relief of proton inhibition may be largely responsible for the stimulatory effect of proteolysis on the Na-Ca exchanger. In addition, the sodium-dependent component of proton inhibition may also underlie intrinsic sodium-dependent inactivation of the Na-Ca exchanger. These contributions should be useful for understanding the structure-function relationship of the Na-Ca exchanger and predicting its role in cardiac function.

REFERENCES

1. MURPHY, E., M. PERLMAN, R. E. LONDON & C. STEENBERGEN. 1991. Amiloride delays the ischemia-induced rise in cytosolic free calcium. Circ. Res. **68:** 1250–1258.
2. BERS, D. M. & D. ELLIS. 1982. Intracellular calcium and sodium activity in sheep heart Purkinje fibers. Pflugers Arch. **393:** 171–178.
3. BAKER, P. F. & M. P. BLAUSTEIN. 1968. Sodium-dependent uptake of calcium by crab nerve. Biochem. Biophys. Res. Commun. **150:** 167–170.
4. BAKER, P. F. & P. A. MCNAUGHTON. 1977. Selective inhibition of the Ca-dependent Na efflux from intact squid axons by a fall in intracellular pH. J. Physiol. **269:** 78p–79p.
5. WAKABAYASHI, S. & K. GOSHIMA. 1981. Kinetic studies on sodium-dependent calcium uptake by myocardial cells and neuroblastoma cells in culture. Biochim. Biophys. Acta **642:** 158–172.
6. PHILIPSON, K. D., M. M. BERSOHN & A. Y. NISHIMOTO. 1982. Effects of pH on Na^+-Ca^{2+} exchange in canine cardiac sarcolemmal vesicles. Circ. Res. **50:** 287–293.
7. DOERING, A. E. & W. J. LEDERER. 1993. The mechanism by which cytoplasmic protons inhibit the sodium-calcium exchanger in guinea-pig heart cells. J. Physiol. (London) **466:** 481–499.
8. HILGEMANN, D. W. 1990. Regulation and deregulation of cardiac Na^+-Ca^{2+} exchange in giant excised sarcolemmal membrane patches. Nature **344:** 242–245.
9. PHILIPSON, K. D. & A. Y. NISHIMOTO. 1982. Stimulation of Na-Ca exchange in cardiac sarcolemmal vesicles by proteinase pretreatment. Am. J. Physiol. **243:** 16–19.
10. MATSUOKA, S., D. A. NICOLL, R. F. REILLY, D. W. HILGEMANN & K. D. PHILIPSON.

1993. Initial localization of regulatory regions of the cardiac Na/Ca exchanger. Proc. Natl. Acad. Sci. USA **90:** 3870–3874.
11. HILGEMANN, D. W., S. MATSUOKA, G. A. NAGEL & A. COLLINS. 1992. Steady-state and dynamic properties of cardiac Na/Ca exchange. Sodium-dependent inactivation. J. Gen. Physiol. **100:** 905–932.
12. MATSUOKA, S. & D. W. HILGEMANN. 1994. Inactivation of outward Na/Ca exchange current in guinea-pig ventricular myocytes. J. Physiol. **476:** 443–458.
13. DOERING, A. E. & W. J. LEDERER. 1994. The action of Na^+ as a cofactor in the inhibition by cytoplasmic protons of the cardiac Na^+-Ca^{2+} exchanger in the guinea-pig. J. Physiol. **480**(1): 9–20.

In Squid Axons Phosphoarginine Plays a Key Role in Modulating Na-Ca Exchange Fluxes at Micromolar $[Ca^{2+}]_i$[a]

REINALDO DiPOLO[b] AND LUIS BEAUGÉ[c]

[b]*Centro de Biofísica y Bioquímica IVIC*
Apartado 21827
Caracas 1020A, Venezuela

[c]*Instituto de Investigación Médica*
"Mercedes y Martín Ferreyra"
Casilla de Correo 389
5000 Córdoba, Argentina
and
Marine Biological Laboratory
Woods Hole, Massachusetts 02543

INTRODUCTION

In several biological preparations including cardiac muscle and nerve cells, metabolic regulation of the Na-Ca exchanger has been directly related to the MgATP effect.[1,2] This crucial mechanism, which promotes Na-Ca exchange by increasing the affinity of both the intracellular regulatory Ca_i site and the extracellular Na^+ transport site, also diminishes the Na^+-dependent inactivation.[1] In squid axons, more than 90% of the Na_o-dependent Ca fluxes under nearly physiological conditions (0.1–0.3 μM Ca^{2+}_i) are dependent on MgATP.[3] One of the remarkable characteristics of this regulation is its high degree of specificity, since no other high energy rich compound is able to replace adenosine triphosphate (ATP). In fact, only the hydrolyzable analogs of ATP (2-deoxy-ATP, AMP-CPP and AMP-NPP) as well as ATP-gamma-S can serve as substrates.[3] In squid axons the mechanism by which MgATP regulates the exchanger is best explained by the involvement of a kinase/phosphatase system through a phosphorylation process.[4] Nevertheless, the kinase(s) and phosphatase(s) involved have not yet been identified (however, see Beaugé *et al.*, this volume).

As is the case for ATP, in both invertebrates and vertebrates the most common high energy-rich compounds present in millimolar quantities are the phosphagens: N-phosphoarginine (Pa) in invertebrates and phosphocreatine (Pc) in vertebrates.

---

[a] This work was supported by grants from: Consejo Nacional de Investigaciones Cientificas y Tecnologicas (CONICIT—S1-2651), Consejo Nacional de Investigaciones Cientificas y Tecnicas (CONICET—Argentina) and the National Science Foundation USA (BNS-9120177).

In squid axons Pa is present in millimolar concentrations (up to 5–10 mM).[5] Phosphoarginine and ATP concentrations are practically at equilibrium in the squid axoplasm due to a powerful arginine kinase. The fast turnover of this enzyme (53 μmol · mg^{-1} · min^{-1}; which is 100 times more active than ATPases and adenylate kinase) contributes to quickly buffer the cytosolic [ATP][5] (see FIG. 5). In fact, the Pa system has traditionally been used in squid axon to control ATP concentration during ion transport experiments.[4–6]

Beside their involvement in the energy buffer of the cell, neither Pa, nor Pc, its counterpart in vertebrates, has been reported to be involved directly in modulating ion transport processes. In this work we demonstrate for the first time that the phosphagen N-phosphoarginine is indeed a modulator of the Na-Ca exchanger different from the metabolic regulation caused by MgATP.

METHODS

We used two different types of squid: *Loligo pealei* from the Marine Biological Laboratory, Woods Hole, MA; and *Loligo plei* from the Instituto Venezolano de Investigaciones Cientificas, Caracas, Venezuela. The experimental procedures for internal dialyzing giant squid axons have been described elsewhere.[7] In order to add or remove compounds from the axoplasm as quickly as possible we used regenerated cellulose fibers with large diameters (230 μm OD; 220 ID Spectra/ Por #132225, Spectrum, Houston, TX) and high molecular weight cut off (18,000 Da). These capillaries have permeability coefficients which are on the average 4.2 times greater than those of cellulose acetate used before (from Fabric Research,

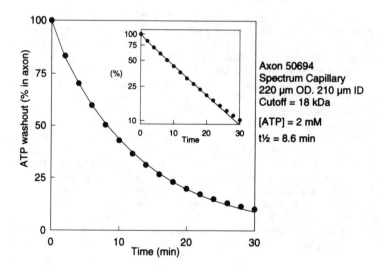

FIGURE 1. ATP wash out curve from a 520-μm giant axon using a regenerated cellulose acetate capillary of 18-kDa cut off. *Ordinate*: percentage (%) of [ATP] in the axon. *Abscissa*: Time in minutes. The *line* through the experimental points represents the best fit to a single exponential function. The t$_{\frac{1}{2}}$ for this particular experiment was 8.6 min. See Methods for details. Temperature: 17°C.

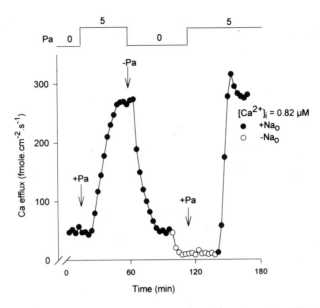

FIGURE 2. Effect of Pa on Na_o-dependent (forward Na-Ca exchange) and Na_o-independent (Ca pump) Ca efflux. *Ordinate*: Ca efflux in fmole · cm^{-2} · s^{-1}. *Abscissa*: time in minutes. Unless specified, all concentrations are in millimolar (mM). (●) Ca^{2+} efflux in Na-containing medium ($+Na^+_o$, $+Ca^{2+}_o$). (○) Ca^{2+} efflux in the absence of Na^+_o and Ca^{2+}_o. The *arrows* indicate the addition or the removal of Pa. Notice the lack of effect of Pa on the Ca pump.

MA). The standard dialysis medium had the following composition (mM): Tris-MOPS (3-(N-morpholino) propanesulphonic acid), 385; NaCl, 45; $MgCl_2$, 4; glycine, 285; Tris-EGTA (ethylene glycol-bis-β(aminoethyl ether)-N,N,N′,N′-tetraacetic acid), 1–3. The pH was 7.3 and the temperature between 17–18°C. The osmolarity of all solutions was adjusted to 940 mosmols. The external medium had the following composition (mM): Na^+, 440; Ca^{2+}, 0.5; Mg^{2+}, 60; Cl^-, 570. The pH was 7.6. Removal of external sodium was compensated with Li^+. Since in these experiments it is critical to have a complete control of the ATP concentration during an experiment, we re-evaluated the ATP washout curve using the new generation of dialysis capillaries. FIGURE 1 shows a washout curve of a 520 μm axon using [^{32}P]γ-ATP to follow the ATP levels in the axon after the onset of the dialysis (see legend to FIG. 1). With these capillaries, ATP is removed from the axoplasm with a $t_{\frac{1}{2}}$ of 7.2 ± 1.4 minutes (n = 3) following a single exponential decay. In all the experiments to be shown 1 mM NaCN was always present in the external medium to lower the [ATP] prior to the onset of the dialysis ([ATP] is close to 100 μM in cyanide poisoned axons[8]). Considering that the axons were predialyzed for about 45 minutes without ATP before the addition of the ^{45}Ca dialysis medium, the [ATP] at the beginning of that period ought to be less than 2 μM. This has been confirmed experimentally by analyzing the [ATP] of axons dialyzed for 1 hr without ATP (experiments not shown). Phosphoarginine was obtained either from Sigma Chemical Co., St. Louis, MO or from Fluka Chemical Co., New York. Purity assays reported by these companies indicate no major contaminants, with less than 1% inorganic phosphate and arginine. Protein kinase and phosphatase inhibitors were purchased from LC Laboratories, MA. All other chemicals were of analytical grade.

RESULTS AND DISCUSSION

To examine the effect of Pa on the Na_0-dependent Ca efflux, axons were depleted of ATP by prolonged dialysis before addition of the phosphagen (5 mM; which is a physiological concentration). FIGURE 2 shows an experiment in which after more than one hour of dialysis without ATP, 5 mM Pa induces a large activation in the calcium efflux from a base line of 46 fmole $\cdot$ cm$^{-2} \cdot$ s^{-1} to about 280 fmole $\cdot$ cm$^{-2} \cdot$ s^{-1}. The effect is totally reversible, for upon removal of Pa from the internal medium Ca efflux rapidly drops to its base line level.

Although in these experiments the concentration of ATP and adenosine diphosphate (ADP) prior to the addition of Pa must be very small (<1 μM), it could still be argued that Pa activation is secondary to the formation of ATP from ADP in a hypothetical compartment close to the membrane. In order to test this possibility we took advantage of the ten times difference in the apparent $K_{\frac{1}{2}}$ for ATP between the Ca pump[2] (Na_0-independent Ca efflux; $K_{\frac{1}{2}}$ for ATP = 25 μM) and the Na-Ca exchange[2] (Na_0-dependent Ca efflux. $K_{\frac{1}{2}}$ for ATP = 250 μM). In principle, if ATP is produced from ADP during the addition of Pa, one should see an activation of the Na_0-independent Ca efflux component (MgATP-dependent Ca pump). FIGURE

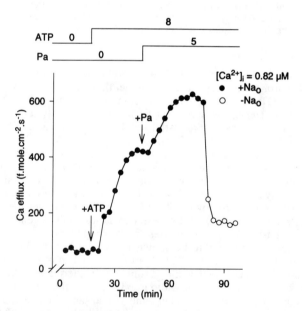

FIGURE 3. Effect of PA on Na_0-dependent Ca efflux in the presence of a saturating (8-mM) ATP concentration. *Ordinate*: Ca efflux in fmole $\cdot$ cm$^{-2} \cdot$ s^{-1}. *Abscissa*: time in minutes. Unless specified, all concentrations are in millimolar (mM). (●) Ca^{2+} efflux in Na^+-containing medium (+Na^+_0, +Ca^{2+}_0). (○) Ca^{2+} efflux in the absence of Na^+_0 and Ca^{2+}_0. The *arrows* indicate the additions or removal of ATP and PA from the dialysis medium.

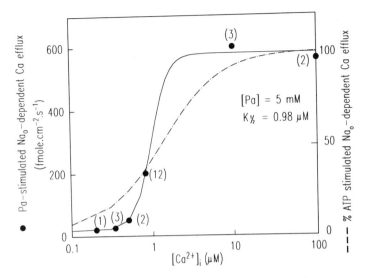

FIGURE 4. Ca^{2+}_i dependence of the Pa and ATP stimulation of the Na-Ca exchange. *Left ordinate*: Pa-stimulated Na_0-dependent Ca efflux in $fmole \cdot cm^{-2} \cdot s^{-1}$. *Abscissa*: $[Ca^{2+}]_i$ in μM. *Numbers in parenthesis* represent different axons. The *line* through the points (*continuous line*) is the best fit to the Hill equation. $V_{max} = 579\ fmole \cdot cm^{-2} \cdot s^{-1}$, $K_{\frac{1}{2}} = 0.98\ \mu M$ and $n = 4.23$. *Right ordinate*: Ca^{2+}_i dependence of ATP effect (from Blaustein[9]). *Discontinuous line*: best fit to a Hill equation with $n = 1.4$. (From DiPolo.[15] Reprinted by permission from the *Journal of Physiology*.)

2 shows that this is not the case, since in the absence of external sodium (open circles), 5 mM Pa has no effect on Ca efflux. Therefore, Pa activation of the exchanger cannot be ascribed to a secondary production of ATP. The end of FIGURE 2 shows that Pa activation of Ca efflux requires the presence of external Na^+ (closed circles), since replacing Li_0 by Na_0 brings Ca efflux to a steady level of around 300 $fmole \cdot cm^{-2} \cdot s^{-1}$, a value similar to the Ca efflux induced by Pa at the beginning of the experiment.

A further strong argument in favor of a genuine Pa effect is shown in FIGURE 3. In this axon, after completely depleting all nucleotides, a saturating ATP concentration (8 mM; thirty times its apparent K_m) was added to the dialysis medium. Ca efflux raises from about 30 $fmole \cdot cm^{-2} \cdot s^{-1}$ to a steady level of 410 $fmole \cdot cm^{-2} \cdot s^{-1}$. Nevertheless, addition of 5 mM Pa (in the presence of 8 mM ATP) induces a further increase in Ca efflux to about 610 $fmole \cdot cm^{-2} \cdot s^{-1}$. This increment is of the same magnitude as that seen in the absence of ATP (see FIG. 2). The removal of external Na^+ at the end of the experiment allows one to dissect the three components of the Ca efflux at 0.82 μM Ca^{2+}_i in the presence of both ATP and Pa: i) MgATP-dependent Na_0-dependent Ca efflux (243 $fmole \cdot cm^{-2} \cdot s^{-1}$), ii) Pa-dependent Na_0-dependent Ca efflux (216 $fmole \cdot cm^{-2} \cdot s^{-1}$), and iii) MgATP-dependent Na_0-independent Ca efflux (Ca pump) (167 $fmole \cdot cm^{-2} \cdot s^{-1}$).

In summary, three findings demonstrate that the Pa effect is indeed different from that of ATP: i) The presence of very low levels (<2 μM) of ATP and ADP prior to the addition of Pa, ii) the absence of activation of the Ca pump by Pa, and iii) the additional activation of the Na-Ca exchange by Pa even at saturating [ATP].

Experiments aimed to determine the affinity for the Pa effect indicate that the phosphagen stimulates the exchanger with low apparent affinity ($K_{\frac{1}{2}}$ of 7.7 mM). This value is almost thirty times larger than that for ATP and close to the Pa physiological concentration; hence variations in that concentration may induce important changes in the magnitude of the Na-Ca exchange fluxes. The effect of Pa seems to be quite specific, since phosphocreatine, a very similar molecule differing in a methyl group, is without effect. Both Pa and Pc are phosphorylated at the guanidine moiety with the difference that it is methylated in phosphocreatine, thus indicating that the place at which Pa operates requires an N-phosphorylated guanidine structure without methyl residues.

Since the activation of the Na-Ca exchange by ATP has an absolute requirement for intracellular Mg^{2+}, the next aspect we explored was the dependence of the Pa effect on this divalent cation. In axons previously depleted of internal Mg^{2+} by dialyzing with an EDTA-containing medium, addition of 5 mM Pa causes a very small (<10%) activation of the Na_0-dependent Ca efflux. Readdition of Mg^{2+}_i

TABLE 1. Some Features of ATP- and Pa-Dependent Stimulation of Na-Ca Exchange in Squid Giant Axons[a]

	Na_0-Dependent Ca Efflux (Forward Na-Ca Exchange)	
	ATP-Dependent	Pa-Dependent
Sizes of Na_0-dependent fluxes at nearly resting conditions (0.1–0.3 μM Ca^{2+}_i)	>90%	very small (<10%)
Size of Na_0-dependent fluxes at 0.82 μM Ca^{2+}_i	55%	45%
Apparent affinity	K^{ATP} = 0.25 mM	K^{Pa} = 7.7 mM
Mg^{2+}_i requirement	absolute	partial but strong
Ca^{2+}_i dependence	slight sigmoidal activation Hill coeff. = 1.4 $K_{\frac{1}{2}}$ = 1–3 μM	strong sigmoidal activation Hill coeff. = 4.23 $K_{\frac{1}{2}}$ = 0.9–1.0 μM
Kinase inhibitors:		
Staurosporine (several kinases)	negative	negative
H_7 (PKC)	negative	negative
Genistein (TyK)	negative	negative
Lavendusting A (TyK)	negative	negative
Tyrphostin (TyK)	negative	negative
Calmidazolium (Calm)	negative	negative
Phosphatase inhibitors:		
Vanadate	activates ATP-stimulated Na-dependent Ca fluxes[4]	variable slight activation or no effect
Okadaic acid (PP-1 and PP-2A)	negative	negative
Microcystin (PP-1 and PP-2A)	negative	negative

[a] f/cs = fmole · cm^{-2} · s^{-1}; TyK = tyrosine kinase; PKC = protein kinase C; Calm = calmodulin. All concentrations of inhibitors were 5–10 × IC_{50}.

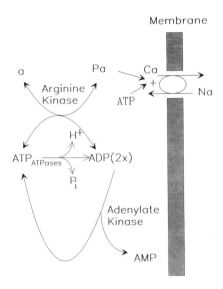

FIGURE 5. Schematic representation of the "energy buffer system" for the control of the cellular ATP/ADP ratio and the direct stimulatory effect of ATP and Pa on the plasma membrane Na-Ca exchanger.

(4 mM at a constant $[Ca^{2+}]_i$ of 0.82 μM) induces a large increment in the Pa effect indicating that, as with ATP, the Pa effect is largely stimulated by internal Mg^{2+}.

During preliminary experiments carried out at low submicromolar $[Ca^{2+}]_i$ (0.2–0.4 μM), we found that the Pa effect was strongly diminished. Nevertheless, the ATP effect on both the Na_0-dependent and Na_0-independent (Ca pump) fluxes were of the expected magnitude. We then decided to investigate the Ca^{2+}_i dependence of the phosphoarginine effect. FIGURE 4 summarizes the results of 23 different axons in which Pa stimulation (at [Pa] of 5 mM) of the Na_0-dependent Ca efflux was measured at different $[Ca^{2+}]_i$. One of the striking features of the Pa effect is its steep dependence on Ca^{2+}_i. In fact, below 0.5 μM, the Pa effect is negligible, becoming sizable between 0.8 and 2 μM. The data of FIGURE 4 were fitted (continuous line) to the Hill equation (see legend of FIG. 4); the values from the fit were: $K_{\frac{1}{2}}$ of 0.98 μM, V_{max} of 579 fmole $\cdot$ cm$^{-2} \cdot$ s^{-1} and n = 4.23. This is in marked contrast with the Ca^{2+}_i activation of the MgATP effect (broken line, taken from Blaustein[9]) which has an n = 1.4. The Ca^{2+}_i dependence of the Pa effect could have important physiological implications, since although at low $[Ca^{2+}]_i$ (<0.5 μM) its contribution to the total Na-Ca exchange flux is small, an important extra activation of the Na-Ca exchanger, above that of MgATP, will occur at micromolar Ca^{2+}_i.

At present, the mechanism by which Pa activates the Na-Ca exchanger remains unknown. In eukaryotic cells it has been shown that many proteins undergo changes through phosphorylation in the side chain of basic N-amino acids like arginine,[10] histidine[11] and lysine.[12] There is no experimental evidence to attribute a basic phosphorylated residue as responsible for the Pa effect on the Na-Ca exchanger; however, recent data demonstrate the presence of arginine-specific protein kinases.[13] Whether these proteins are involved in the Pa effect is unknown.

TABLE 1 shows some of the basic properties and differences between MgATP-dependent and the Pa-dependent stimulation of the Na-Ca exchanger. It is clear from experiments in squid axons (see TABLE 1) that neither the ATP- nor the Pa-requiring mechanism is affected by specific inhibitors of the classical serine-threonine or tyrosine kinases. On the other hand, vanadate, a potent inhibitor of phosphatases, does affect the MgATP effect in a way consistent with a phosphorylation/dephosphorylation process mediated by a kinase/phosphatase system.[4] In the case of Pa, the effect of vanadate is variable: in some axons we observed a slight increase in the exchange fluxes although much smaller than that seen with ATP, while the exchanger remained activated after removal of Pa. In other axons vanadate shows no effect.

In conclusion, we have discovered that N-phosphoarginine, a physiological substrate present in millimolar concentrations in the squid cytosol, is able to activate the Na-Ca exchanger in a Ca^{2+}_i-dependent way (FIG. 5). This new metabolic regulation of the Na-Ca exchanger might play a physiological role in extruding calcium ions from regions of the neuron in which the $[Ca^{2+}]_i$ can reach high levels of ionized Ca_i (>50 μM) such as the synaptic terminal.[14] This finding opens the possibility that other phosphagens present in other species such as phosphocreatine in vertebrates may also modulate the activity of the Na-Ca exchanger *in vivo*.

REFERENCES

1. HILGEMANN, D. W. & A. COLLINS. 1993. Mechanism of cardiac Na^+-Ca^{2+} exchange current stimulated by MgATP: possible involvement of amino-phospholipid translocase. J. Physiol. **454:** 59–73.
2. DIPOLO, R. & L. BEAUGÉ. 1991. Regulation of Na-Ca exchange. An overview. Ann. N.Y. Acad. Sci. **639:** 100–111.
3. DIPOLO, R. & L. BEAUGÉ. 1979. Physiological role of ATP-driven Ca pump in squid axons. Nature **278:** 271–273.
4. DIPOLO, R. & L. BEAUGÉ. 1994. Effects of vanadate on MgATP stimulation of Na-Ca exchange support kinase-phosphatase modulation in squid axons. Am. J. Physiol. **266:** C1382–C1391.
5. MULLINS, L. & F. J. BRINLEY. 1967. Some factors influencing sodium extrusion by internally dialyzed squid axons. J. Gen. Physiol. **50:** 2333–2345.
6. DE WEER, P. 1970. Effects of intracellular adenose-5'-diphosphate and orthophosphate on the sensitivity of the sodium efflux from squid axons to external sodium and potassium. J. Gen. Physiol. **56:** 583–620.
7. DIPOLO, R., F. BEZANILLA, C. CAPUTO & H. ROJAS. 1985. Voltage dependence of the Na/Ca exchange in voltage-clamped, dialyzed squid axons. J. Gen. Physiol. **86:** 457–478.
8. CALDWELL, P. C. & H. SCHIMER. 1965. The free energy available to the sodium pump of squid axons and changes in the sodium efflux on removal of the extracelluar potassium. J. Physiol. **181:** 25P.
9. BLAUSTEIN, M. P., R. DIPOLO & J. REEVES. 1991. Sodium-calcium exchange: proceedings of the second international conference. Ann. N.Y. Acad. Sci. **639:** 1–671.
10. SCOTT, L. S., C. W. KERN, R. M. HALPERN & R. A. SMITH. 1976. Phosphorylation on basic amino acids in myelin basic protein. Biochem. Biophys. Res. Commun. **71:** 459–465.
11. FUJITAKI, J. M., G. FUNG, E. L. OH & R. A. SMITH. 1981. Characterization of chemical and enzymatic acid-labile phosphorylation of histone H4 using phosphorus-31 nuclear magnetic resonance. Biochemistry **20:** 3658–3664.
12. CHEN, C. C., B. B. BRUEGGER, C. W. KERN, L. C. LIN, R. M. HALPERN & R. A. SMITH. 1977. Phosphorylation of nuclear proteins in rat regenerated liver. Biochemistry **16:** 4852–4855.

13. LEVY-FAVATIER, F., M. DELPECH & J. KRUH. 1987. Characterization of an arginine-specific protein kinase tightly bound to rat liver DNA. Eur. J. Biochem. **166:** 617–621.
14. LLINAS, R. R., M. SUGIMORI & R. B. SIVER. 1994. Localization of calcium concentration microdomains at the active zone in the squid giant synapsis. Adv. Second Messenger Phosphoprotein Res. **29:** 133–137.
15. DIPOLO, R. 1995. Phosphoarginine stimulation of Na-Ca exchange in squid axons. J. Physiol. **487:** 57–66.

A Nerve Cytosolic Factor Is Required for MgATP Stimulation of a Na$^+$ Gradient-Dependent Ca^{2+} Uptake in Plasma Membrane Vesicles from Squid Optic Nerve[a]

LUIS BEAUGÉ,[b] DANIEL DELGADO,[c] HÉCTOR ROJAS,[c]
GRACIELA BERBERIÁN,[b] AND REINALDO DiPOLO[c]

[b]*Instituto de Investigación Médica*
"Mercedes y Martín Ferreyra"
Casilla de Correo 389
5000 Córdoba, Argentina
and
Marine Biological Laboratory
Woods Hole, Massachusetts 02543

[c]*Centro de Biofísica y Bioquímica, IVIC*
Apartado Postal 21827
Caracas 1020-A, Venezuela

INTRODUCTION

As early as 1979[1] and 1982[2] we collected evidence that in dialyzed axons Ca$^{2+}_i$ was required for Ca^{2+} influx through the Na$^+$-Ca^{2+} exchanger. This need for Ca^{2+} binding to nontransport internal sites was unambiguously shown in 1986 by following Na$^+$ efflux through the reversal exchange; we named that effect Ca$^{2+}_i$ regulation, and the intracellular site was called the Ca^{2+} regulatory site.[3] In addition, and in the same paper, we presented data indicating that (i) ATP also stimulated reversal exchange and (ii) there seemed to be a relationship (or interaction) between Ca$^{2+}_i$ and ATP stimulation. Later work demonstrated that ATP acted precisely by increasing the apparent affinity of the intracellular regulatory site for calcium ions.[4]

In the work that followed (see Ref. 4 for references), always in dialyzed squid giant axons, we accumulated evidence that consistently suggested the effect of ATP was a consequence of a phosphorylation process that involved the interplay of a coupled kinase(s)-phosphatase(s) system(s). The main features of the ATP stimulation of the Na$^+$-Ca^{2+} exchanger that led to this conclusion are:

- Requires Mg^{2+}.
- Only hydrolyzable ATP analogues can mimic ATP.
- ATP-gamma-S, a kinase but not an ATPase substrate, is even a better activator than ATP.

---

[a] This work was supported by grants from the National Science Foundation USA (BNS 9120177), CONICET, Argentina (PID-BID 1053) and CONICIT, Venezuela (S1-2651).

- Is increased by Pi, a phosphatase product inhibitor.
- Is increased by pNPP, a phosphatase substrate, likely by pulling away phosphatases (unpublished work).
- Is inhibited by ADP, a product kinase inhibitor.
- Is increased by micromolar vanadate concentrations (K_i around 5 μM), at which this compound acts as phosphatase inhibitor and likely as kinase activator.
- Is inhibited by millimolar concentrations of vanadate (K_i about 5 mM), at which this compound acts as kinase inhibitor.
- Is inhibited by Cr(III)ATP, a powerful kinase inhibitor.
- Is not affected by Co(NH4)3ATP, a compound inert to most kinases.

In other cell preparations the effect of MgATP was different both qualitatively and quantitatively, and in some cases was totally absent.[5-8] Particularly interesting, we systematically faced negative results when we explored the MgATP effect in isolated membrane vesicles from squid optic nerve.

In recent unpublished work, we followed the phosphorylation patterns of squid giant axons under different experimental conditions. In injected axons (with their axoplasm intact), the general level of phosphorylation was dramatically increased when 100 μM vanadate was added. Likewise, in axons longitudinally opened and treated with 100 μM vanadate, the membrane phosphorylation pattern was noticeably increased when the axoplasm was present as compared with those axons that had the axoplasm removed. As mentioned above, in dialyzed axons, which preserve their axoplasm, micromolar vanadate concentration markedly enhances MgATP stimulation of the exchanger. It is possible that the lack of MgATP stimulation in isolated nerve vesicles was due to the lack of an essential compound present in the cytosol and lost during vesicle preparation. In addition, the usual [Ca^{2+}] used in most vesicle experiments, including cardiac sarcolemmal, was saturating; since ATP acts by increasing the affinity of the regulatory Ca^{2+}_i site the conditions used would mask the nucleotide effect.[8] We therefore decided to look into these possibilities by exploring the effects of axoplasm and MgATP on Ca^{2+} uptake via the Na^+-Ca^{2+} exchanger in vesicles from squid optic nerve at submicromolar [Ca^{2+}].

METHODS

Membrane vesicles from squid optic nerve were prepared by differential centrifugation[9] and loaded with 300 mM NaCl, 0.1 mM EDTA and 30 mM MOPS-Tris (pH 7.4 at 20°C). Axoplasm of giant axons taken from fresh squid were homogenized (1 : 1 w/v ratio) in 20 mM MOPS-Tris (pH 7.3 at 20°C), 1 mM DDT, 0.1 mM EDTA, 0.2 mM EGTA and an antiprotease cocktail (0.5 mM PMFS and 10 μg/ml of aprotinine, leupeptin and pepstatin A). The mixture was then homogenized in small glass tubes with a teflon homogenizer and centrifuged at 12,000 × g for 10 minutes. All these procedures were done in the cold (0–4°C). In the follow-up work, that supernatant was further centrifuged for 30 minutes at 40,000 rpm in a T50 Beckman rotor (100,000 × g). The supernatant of that centrifugation was then filtered through Amicon Centricon tubes of different molecular weight cut off. Assays on Ca^{2+} influx were performed with both supernatant and the different filtrates. [^{45}Ca]Ca^{2+} uptake was measured at room temperature by incubating the vesicles for 10 seconds in media with high (300 mM) and low (30 mM) Na^+. In addition, all extravesicular solutions contained 1 mM $MgCl_2$, 0.1 mM vanadate and 20 mM MOPS-Tris (pH 7.3 at 20°C) and a final $[Ca^{2+}]$ of 0.6 μM; in low Na^+ media osmolarity was preserved with NMG-Cl. The reaction was stopped with 0.8 ml of an ice cold solution containing 20 mM MOPS-Tris (pH 7.3 at 20°C), 300 mM KCl and 1 mM EGTA-Tris and filtered through Whatman F/GF glass fiber filters. The filters were washed with 5 ml of the same solution, immersed into 5 ml of scintillator fluid and counted in a liquid scintillation counter. Other procedures are described in the Results and Discussion section.

RESULTS AND DISCUSSION

In the first experiments supernatant of the 12,000 × g centrifuged axoplasm extruded from squid giant axons was used for transport assays; the volume added was 10 μl into a final volume of 200 μl. For an average total protein concentration of 20–25 mg/ml, that meant an addition of about 200–250 μg protein to the assay media. As TABLE 1 illustrates this promoted a MgATP stimulation of a Na^+ gradient-dependent Ca^{2+} uptake in the vesicles. In this particular experiment MgATP was present under all conditions, with and without Na^+ gradient or axoplasm. Other features of this stimulation (experiments not shown here) are: (i) ATP stimulation fails to appear in the absence of Mg^{2+}. (ii) Axoplasm by itself is ineffective when there is no added MgATP, indicating that the effect is not due simply to some compound extracted form the cytosol of the axons. (iii) Albumin, at the same concentration as the total axoplasmic protein, has no effect in the absence or presence of MgATP; therefore the stimulation is not a consequence of an unspecific protein protection against the eventual presence of inhibitors as occurs with "*in vitro*" enzyme systems.[10] (iv) Heat denaturation (15 minutes at 70°C) and trypsin digestion rendered axoplasm ineffective. (v) Likewise, and similar to what happens in dialyzed squid axons, there was no stimulation by

TABLE 1. MgATP Stimulation of Na^+ Gradient-Dependent Ca^{2+} Uptake in Membrane Vesicles of Squid Optic Nerve Due to the Addition of the 12,000 × g Supernatant of Axoplasm from Squid Giant Axons[a]

Main Solution Composition			
$[Na^+]_i$ (mM)	$[Na^+]_o$ (mM)	Axoplasm Supernatant (μl)	Ca^{2+} Uptake (nmoles/mg Prot. 10 s)
300	300	none	0.026 ± 0.012
300	30	none	0.020 ± 0.009
300	300	axp.	0.022 ± 0.013
300	30	axp.	0.257 ± 0.039**

[a] (i) The composition of solutions was as follows: intravesicular (extracellular): 20 mM MOPS-Tris (pH 7.4 at 20°C) and 0.1 mM EDTA; extravesicular (intracellular): 0.6 μM Ca^{2+}], 100 μM EGTA, 1 mM ATP, 1 mM $MgCl_2$ and 100 μM vanadate; removal of Na^+ (NaCl) was osmotically balanced with NMG-Cl. (ii) Axp. means that 10 μl (150 μg total protein) of 12,000 × g supernatant of centrifuged axoplasm were present. (iii) The total volume was 0.2 ml. (iv) No Na^+ gradient-dependent Ca^{2+} uptake was observed when axoplasm was added in the absence of membrane vesicles. (v) Each entry is the mean ± SEM of triplicate determinations. (vi) ** $p < 0.001$ for the difference between means. (vii) (i) means intravesicular and (o) extravesicular.

TABLE 2. Preliminary Fractionation Data of a Cytosolic Factor from Squid Optic Ganglia That Reconstitutes the MgATP Stimulation of Na^+-Ca^{2+} Exchange in Membrane Vesicles from Squid Optic Nerves[a]

Treatment	Cytosolic Sample Added		Na^+-Dependent Ca^{2+} Uptake (nmoles/mg Prot. 10 s)
	[Protein] (mg/ml)	Volume (μl)	
Control	—	—	0.030 ± 0.011
12,000 × g supernat.	23.5	10	0.334 ± 0.020*
100,000 × g supernat.	14.6	10	0.357 ± 0.017*
100,000 × g supernat. 30 kD filtrated	0.5	10	0.333 ± 0.015*
100,000 × g supernat. 30 kD filtrated	0.5	5	0.317 ± 0.018*
100,000 × g supernat. 10 kD filtrated	n.d.	10	0.054 ± 0.014**

[a] (i) The total incubation volume was 200 μl; (ii) Controls had no additions other than 10 μl buffer. (ii) Each entry is the mean ± SEM of quadruplicate determinations. (iii) * The difference with control is statistically significant ($p < 0.001$). (iv) ** The difference with control is not statistically significant.

MgATP + axoplasm when extravesicular [Ca^{2+}] was increased to 100 μM. Taken altogether, these results indicate that an axoplasmic factor, likely a protein, is required for the MgATP stimulation of Na^+-Ca^{2+} exchange in nerve.

One of the problems of using axoplasm from giant axons is the small amount of material obtained. This difficulty becomes critical when purifying the compound or compounds responsible for the MgATP effect. Therefore, we decided to investigate whether a similar effect could be obtained with cytosolic material extracted from squid optic ganglia, a much more abundant tissue (compare 20 mg of axoplasm against 1 gram of ganglia per squid). To that end, optic ganglia were homogenized and centrifuged using the same procedure as that for axoplasm and then assayed for activity. Fortunately, the same MgATP stimulatory effect was observed. Consequently, material from optical ganglia was used in the remainder of the work.

A partial purification of the responsible factor was initiated by a further centrifugation and subsequent filtration through filters of different molecular weight cut off. Following the procedures of Gould et al.[11] We removed all particulate fractions by centrifuging the original supernatant at 100,000 × g for 30 minutes. The 100,000 × g supernatant (post-microsomal fraction) was then subjected to filtration through 100 kD, 50 kD, 30 kD and 10 kD MW cut off filters and the activity of the filtrate assayed in each of these fractions. The results of one of these experiments (TABLE 2) show that, although the total protein is reduced by the successive centrifugation and filtrations, the filtrate through filters down to 30,000 MW cut off remained fully active, while activity was completely lost when a 10,000 MW filter was used; i.e., the unknown factor (or factors) must have a molecular weight between 10 kD and 30 kD. In addition, and considering the sample volumes used and the total concentration of protein in each case, TABLE 2 shows that we could reduce the total protein in the assay media from 235 μg down to 2.5 μg maintaining full effect.

A 15 percent PAGE analysis of the 30-kD fraction stained with Coomasie blue displayed only three net bands (not shown). We also ran an HPLC analysis of this fraction, and FIGURE 1 is the result of one of them. A distinct large peak is seen at 30.5 minutes elution which is preceded by five and followed by one much smaller peaks. The peak corresponding to 26.86 minutes' elution has a mobility almost identical to that of calmodulin. We therefore performed some experiments where the effect of calmodulin, in the absence of axoplasm, was investigated. Even at concentrations as high as 5 μg/200 μl calmodulin showed no effect of the Na^+ gradient dependent and independent Ca^{2+} uptake either in the absence or presence of MgATP. Hence, we can rule out any involvement of calmodulin as the responsible factor for restoring the MgATP stimulation of the Na^+-Ca^{2+} exchanger in optic squid nerve vesicles.

In the past few years we have been using a dialysis capillary (Spectrum, Houston, TX) with a molecular weight cut off around 18 kD. With the usual 2–3-

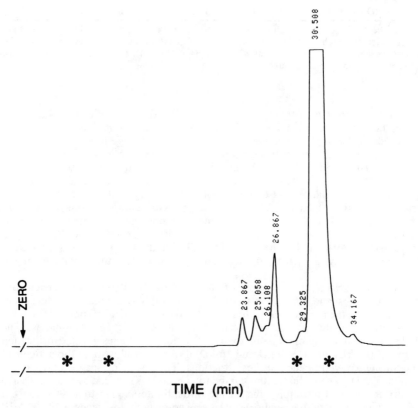

FIGURE 1. HPLC analysis of the 30,000-kD filtrate from squid optic ganglia. Homogenized and 100,000 × g centrifuged optic ganglia were filtrated through 30,000 molecular weight cut off Centricon filters. An aliquot of 20 μl (about 30 μg total protein) suspended in 150 mM MOPS-Tris (pH 7.4 at 20°C) was injected at time = 0 in the HPLC system using a Protein-Pack-125 Millipore column and a 280 nm detector. The dilution was performed with the same buffer at a 0.4-ml/min flow rate. The *stars* indicate the molecular weight standards, which from left to right correspond to bovine albumin (67 kD), egg albumin (43 kD), trypsinogen (25.5 kD) and ribonuclease (14.4 kD).

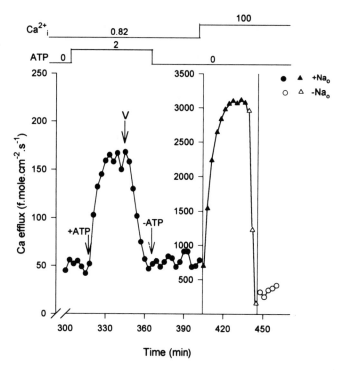

FIGURE 2. Loss of MgATP stimulation of Na^+-Ca^{2+} exchange in squid giant axon by prolonged dialysis with an 18,000 molecular weight cut off capillary (Spectrum, Houston, TX). A 540-μm diameter axon was predialyzed for 5 hours with a standard dialysis solution containing 0.8 μM Ca^{2+} and no ATP. At the indicated times different ligands (2 mM ATP, 100 μM vanadate or 100 μM Ca^{2+}) were added to the intracellular medium. *Ordinate*: Ca efflux in fmole $\cdot$ cm$^{-2}\cdot$ s^{-1}. *Abscissa*: time in minutes. (●) Ca efflux in full Na_o; (o) Ca efflux in the absence of Na_o. Notice that while ATP does not stimulate the exchanger, the rate of Na^+-Ca^{2+} exchange at nonlimiting (100 μM) $[Ca^{2+}]_i$ is similar to that usually found in axons subjected to short dialysis periods.

hour dialysis periods we had not seen any significant variation in the MgATP stimulation of the Na^+-Ca^{2+} exchanger. In view of the findings described above we decided to explore whether an axoplasmic factor with a molecular weight close to 18 kD could be washed away by prolonged dialysis time. FIGURE 2 is one of the experiments set up to check this hypothesis. As the figure shows, after about five hours dialysis with 0.82 μM Ca^{2+} and no ATP, the efflux of Ca^{2+} in the presence of external Na^+ (NaSW) has the usual value of around 40 fmole $\cdot$ cm$^{-2}\cdot$ s^{-1}. Addition of 2 mM ATP in the presence of 4 mM Mg^{2+} resulted in a stimulation of a Ca^{2+} efflux which went only through the Ca^{2+} pump for it was completely abolished by 100 μM vanadate; *i.e.*, the MgATP stimulation of the exchanger was lost. This disappearance of the MgATP effect is not due to an inactivation of the exchanger since, as shown in the figure, raising the $[Ca^{2+}]$ from 0.8 μM to 100 μM increases Ca^{2+} efflux to values seen in normal axons. Furthermore, this increase in Ca^{2+} efflux was totally dependent on $Na^+{}_o$. Although the expectation based on the removal of a soluble factor with a molecular weight below 18–20 kD was fulfilled, it must be emphasized that other explanations are possible. For instance, we may be removing a compound or compounds required for the MgATP effect but different from that isolated from squid axoplasm and optic ganglia. Alternatively, it might be that prolonged dialysis without any substrate produced a generalized run down of several metabolic pathways, including the MgATP regulation of the exchanger (for instance, the Ca^{2+} flux in this particular axon was reduced from the usual 130–150 fmole·cm^{-2}·s^{-1} to about 80 fmole·cm^{-2}·s^{-1}). Future experiments might help to answer the question, particularly if we are able to restore the MgATP stimulation in axons subjected to prolonged dialysis by injecting the cytosolic factor or fraction that promotes that MgATP effect in nerve vesicles.

The results presented here show that MgATP stimulation of the Na^+-Ca^{2+} exchange can be observed in isolated membrane vesicles from nerve provided a cytosolic factor, present in a fraction between 10 kD and 30 kD, is included in the assay media. Taken together both the above data and previous observations in dialyzed squid axons, the most plausible explanation is that the reconstitution of the MgATP effect is a result of a phosphorylation of a structure(s) that otherwise does not take place in this membrane preparation. Which are the actual structures that undergo phosphorylation or the enzymatic pathways involved in the nucleotide regulation of the Na^+-Ca^{2+} exchanger are not known. FIGURE 3 illustrates two

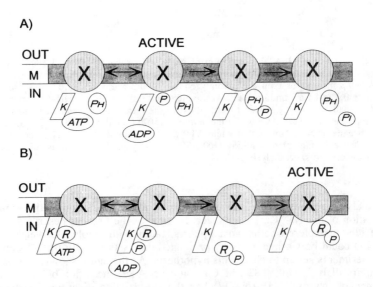

FIGURE 3. Two possible mechanisms of metabolic regulation of the Na^+-Ca^{2+} exchange through phosphorylation from MgATP. **(A)** The final target for phosphorylation is the exchange carrier itself. The low active state of the exchanger corresponds to the unphosphorylated form; upon phosphorylation the exchanger is transferred to a high active state. **(B)** The final target for phosphorylation is a cytosolic soluble protein (type of response regulator). The activity level of the carrier is high when bound to the regulator and low when detached from it. The binding affinity of the regulator for the exchanger increases dramatically when the regulator is phosphorylated by a membrane-bound kinase.

possible general types of mechanisms, which may act alone or in combination. In one of them (top), stimulation results from phosphorylation of the exchange carrier, either directly or through a cascade of kinase reactions. In this case the high active state of the carrier is in the phosphorylated form and the level of its phosphorylation results from the balance of kinase and phosphatase activities. A direct phosphorylation of the exchanger, which correlates with the transport rate, has recently been shown in smooth muscle though a direct or indirect action of protein kinase C.[12] On the other hand (bottom of FIG. 3), it is possible that the metabolic regulation of the exchanger occurs via a system similar to the so-called "response regulators," a well-known system present in prokariotes, which is now being found also in eukariotes.[13] In this case, regulation of the exchanger depends on the binding of a protein to it. When dephosphorylated, that protein has low affinity for the exchanger, but binds with high affinity to a kinase. After autophosphorylation of the kinase, the response regulator transfer the phosphoryl group into its own domain, becomes detached from the kinase and attaches itself to the exchanger bringing it to a high active state. Upon dephosphorylation of the "regulator" (by its own or other phosphatase activity) the carrier is brought back to a low active state. Note that in this model while the kinase is membrane bound the "response regulator" is a soluble protein located into the cytosol. Finally, other mechanisms could be implicated in the MgATP effect. For instance, in a recent paper on cardiac exchanger expressed in COS cells, Condrescu et al.[14] showed an ATP regulation on Na^+-Ca^{2+} exchange without phosphorylation of the carrier; according to their interpretation an ATP-dependent assembly of cytoskeleton might be involved. However, in dialyzed axons (DiPolo and Beaugé, unpublished) and cultured glial cells (Holgado and Beaugé, unpublished) cytochalasin D, up to 5 μM concentration, failed to show any effect of the ATP-stimulated Na^+-Ca^{2+} exchange fluxes.

ACKNOWLEDGMENTS

We wish to thank the Director and staff of the Marine Station at Mochima, IVIC, Venezuela, for the facilities put at our disposal.

REFERENCES

1. DIPOLO, R. 1979. J. Gen. Physiol. **73:** 91–113.
2. DIPOLO, R., H. ROJAS & L. BEAUGÉ. 1982. Cell Calcium **3:** 19–41.
3. DIPOLO, R. & L. BEAUGÉ. 1986. Biochim. Biophys. Acta **854:** 298–306.
4. DIPOLO, R. & L. BEAUGÉ. 1991. Ann. N.Y. Acad. Sci. **639:** 100–111.
5. CARONI, P. & E. CARAFOLI. 1983. Eur. J. Biochem. **132:** 451–460.
6. COLLINS, A., A. V. SOMLYO & D. W. HILGEMANN. 1992. J. Physiol. **454:** 27–57.
7. REEVES, J. P. & K. D. PHILIPSON. 1989. In Na-Ca Exchange. T. Allen, D. Noble & H. Reuter, Eds. 27–53. Oxford University Press. Oxford.
8. BERBERIÁN, G. & L. BEAUGÉ. This volume.
9. CONDRESCU, M., L. OSSES & R. DIPOLO. 1984. Biochim. Biophys. Acta **769:** 261–269.
10. PEDEMONTE, C. H. & J. KAPLAN. 1988. In The Na^+,K^+-Pump. Part A: Molecular Aspects. J. C. Skou, J. G. Norby, A. B. Maunsbach & M. Esmann, Eds. 324–327. Alan R. Liss. New York.
11. GOULD, R. M., W. D. SPIVACK, D. ROBERTSON & M. J. POZNANSKY. 1983. J. Neurochem. **40:** 1300–1306.

12. IWAMOTO, T., S. WAKABAYASHI & M. SHIGEKAWA. 1955. J. Biol. Chem. **270:** 8896–9001.
13. PARKINSON, J. S. & E. C. KOFOID. 1992. Annu. Rev. Genet. **26:** 71–112.
14. CONDRESCU, M., J. P. GARDNER, G. CHERNAYA, J. F. ACETO, C. KROUPIS & J. P. REEVES. 1995. J. Biol. Chem. **270:** 9137–9146.

Kinetics and Mechanism: Modulation of Ion Transport in the Cardiac Sarcolemma Sodium-Calcium Exchanger by Protons, Monovalent Ions, and Temperature[a,c]

DANIEL KHANANSHVILI,[b]
EVELYNE WEIL-MASLANSKY, AND DAVID BAAZOV

Department of Physiology and Pharmacology
Sackler School of Medicine
Tel-Aviv University
Ramat-Aviv 69978, Israel

INTRODUCTION

The cardiac sarcolemma Na^+-Ca^{2+} exchange cycle and its partial reactions (Na^+-Na^+ and Ca^{2+}-Ca^{2+} exchanges) can be described as separate movements of Na^+ and Ca^{2+} ions through the exchanger (consecutive or ping-pong mechanism).[1-6] Although the electrogenic stoichiometry ($3Na^+ : Ca^{2+}$) of ion exchange is electrogenic (one positive charge is translocated per one cycle),[7-9] it is not clear how the net charge transfer is realized at various stages of the exchange cycle. It was suggested by several opportunities that the charged ion-protein intermediates can determine a response of the Na^+-Ca^{2+} exchange rate to voltage,[5,10-15] but these putative intermediates have not yet been identified and characterized. Whatever the exact mechanism is, three major factors can determine the overall rate and voltage response: a) the fractional concentration of rate-limiting intermediate(s); b) the magnitude of the rate constant that limits the overall exchange rate; c) the charge of rate-limiting intermediate(s). The rate-limiting pathway actually determines the identity of the predominant partial reaction and intermediates for controlling the rate and voltage-sensitive properties of exchange modes.

According to the consecutive mechanism (SCHEME 1) the Na^+-Ca^{2+} exchange and its partial reactions (Na^+-Na^+ and Ca^{2+}-Ca^{2+} exchanges) must involve common reaction routes (*e.g.*, the Ca^{2+} influx for Na^+-Ca^{2+} and Ca^{2+}-Ca^{2+}

---

[a] This work is supported by the USA-Israel Binational Foundation (#9300096), the Israeli Science Foundation (#196/93-1) and the Ministry of Science and the Arts. D. K. holds the Igal Alon Career Development Fellowship.
[b] To whom correspondence should be addressed.
[c] Abbreviations: **Bis-Tris Propane**, 1,3-bis(tris[hydroxymethyl]methylamino)propane; **Caps**, 3-(cyclohexylamino)-1-propanesulfonic acid; **Ches**, 2-(N-cyclohexylamino)ethanesulfonic acid; **Mes**, 2-(N-morpholino)ethanesulfonic acid; **Mops**, 3-(N-morpholino)propanesulfonic acid; **Tris**, tris(hydroxymethyl)aminomethane; **EGTA**, ethylene glycol bis (β-aminoethyl ether)-N,N,N',N'-tetraacetic acid.

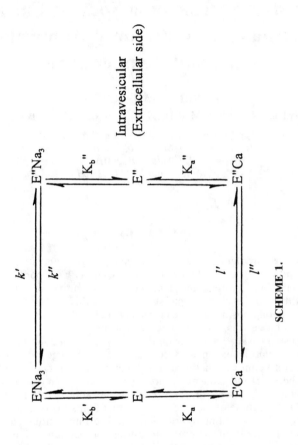

SCHEME 1.

exchanges), although the rate-limiting pathways of various exchange modes could be different. For example, altering the common reaction pathway (*e.g.*, the Ca^{2+} influx) may cause different effects on the steady-state rates of Na^+-Ca^{2+} and Ca^{2+}-Ca^{2+} exchanges: a) Both Na^+-Ca^{2+} and Ca^{2+}-Ca^{2+} exchanges can be accelerated or decelerated to a similar extent if the Ca^{2+} influx is rate limiting for both exchange modes; b) No effects can be seen on either exchange rate if the Ca^{2+} influx is not rate limiting for both exchanges; c) Divers effects can be expected for Na^+-Ca^{2+} and Ca^{2+}-Ca^{2+} exchanges if the distinct reactions (the Ca^{2+} influx, Na^+ efflux or Ca^{2+} efflux) limit each exchange mode.

Recent studies suggest that the binding of either Na^+ or Ca^{2+} to the exchanger is a weekly voltage-sensitive process, while a primary response of Na^+-Ca^{2+} exchange to membrane potential is determined by rate-limiting transport of a charge-carrying intermediate (presumably, the Na^+-bound species).[2-6,15,18] A voltage-curve analysis suggests a model, in which the $E.Na_3$ species are positively charged, while the $E.Ca$ species are "electroneutral" (*e.g.*, the unloaded cation-binding domain may contain -2 charges).[5,14] According to this model the voltage-sensitive and rate-limiting properties of Na^+-Ca^{2+} exchange are determined by charge-carrying intermediates (*e.g.*, the $E.Na_3$ species), while the "uncharged" intermediates (*e.g.*, the $E.Ca$ species) contribute to the rapid reaction(s). A similar model was described before for the Na^+,K^+-ATPase,[19] although the Na^+-transport step is not rate limiting in this system. Interestingly, both the Na^+,K^+-ATPase and the Na^+-Ca^{2+} exchanger are the only known proteins that bind $3Na^+$ ions and have a homologous sequence of 23 amino-acids (a putative domain for ion binding/transport).[21] Complete deletion of the intracellular loop (520 amino acids) does not alter the voltage response of the Na^+-Ca^{2+} exchange currents, suggesting that the voltage-dependent modulation of the exchange cycle is determined by an ion movement pathway through the transmembrane portion of the protein.[22] Recent studies show that among the negatively charged residues of putative transmembrane segments only two, the Glu-113 and Glu-199, are essential for Na^+-Ca^{2+} exchange activity.[23] Interestingly, the Glu-199 is highly conserved in the Na^+,K^+-ATPase, Ca^{2+}-ATPase and H^+,K^+-ATPase,[21-24] suggesting that the Glu-199 might be an integral part of a putative ion binding/transport site(s).

The electrogenic stoichiometry of Na^+-Ca^{2+} exchange (which is a thermodynamic property of the system) cannot always certify the voltage-sensitive response of the exchange rate (which is a kinetic property of the system).[10,13,14] Even if the exchange has the electrogenic stoichiometry, it may show no response to voltage change. For example, no response of the Na^+-Ca^{2+} exchange rate to membrane potential is expected if the partial reactions are being modified in such a way that the electroneutral step (*e.g.*, the Ca^{2+} transport) has become rate limiting in reply to some regulatory setup. This kind of mechanism was proposed recently indicating that high concentrations of protons reduce drastically the voltage sensitivity of Na^+-Ca^{2+} exchange as well as lower severalfold the ratios (R values) of Na^+-Ca^{2+}/Ca^{2+}-Ca^{2+} exchange.[15]

In this work the combined effects of temperature, pH, diffusion potential (potassium-valinomycin) and K^+ have been analyzed on Na^+-Ca^{2+} exchange and its partial reaction, the Ca^{2+}-Ca^{2+} exchange under conditions in which the ion binding is not rate limiting. The preparation of isolated cardiac sarcolemmal vesicles has been used, in which the inside-out vesicles contribute to most, if not all of the Na^+-Ca^{2+} exchange activity.[16-18] The semirapid mixing techniques[1,5,14,18] have been explored for measuring the initial rates of ^{45}Ca-uptake.

This experimental setup allows to avoid the irreversible inactivation, when the exchanger exposes to high temperature or extreme pH during the assay (t = 1 s). It is concluded that under physiologically related conditions (pH 7.4, 37°C and 0–100 mM KCl), the voltage-sensitive Na^+ efflux limits the Na^+-Ca^{2+} exchange, while the electroneutral Ca^{2+} efflux controls the rate of Ca^{2+}-Ca^{2+} exchange. Therefore, under standard conditions the voltage-insensitive Ca^{2+} influx is not rate limiting for both exchange modes. Varying pH, temperature and potasisum modify the rate-limiting pathways, thereby affecting the exchange rate and/or voltage response. The present findings and kinetic analysis could be relevant for resolving the underlying mechanisms of some pathophysiological conditions when the intracellular pH drops (*e.g.*, acidosis, ischemia) or when low temperature alters cardiac functions (cold-induced heart contractures, cardioplegia during open heart surgery, etc.).

The Protonation-Deprotonation State Can Determine the Rate-Limiting Pathway and Voltage Response

A number of studies suggest that protons can strongly modulate the cardiac sarcolemma Na^+-Ca^{2+} exchange.[15,18,22,30] Since intracellular pH in cardiac and other cells can be affected under pathological conditions (*e.g.*, within 1 min after the onset of myocardial ischemia intracellular pH decreases by 0.5–0.8 unit), the interplay and structure-function relationships between specific Ca^{2+} transport systems may have considerable significance from mechanistic and physiological points of view.[15,22,30,31] It became quite obvious that the exchanger can undergo multiple steps of protonation-deprotonation involving a number of regulatory and transport sites.[15,18,22,30] Although the application of molecular biology approaches seems to be very promising in dissecting functional domains of the exchanger,[22] there is no substitution for complementary kinetic studies. Here we describe underlying mechanisms of proton-dependent regulation of the exchanger in isolated cardiac sarcolemma vesicles.

The effect of extravesicular pH 5.0–10.0 was examined on Na^+-Ca^{2+} and Ca^{2+}-Ca^{2+} exchanges in the absence of monovalent cations in the assay medium. The initial rates (t = 1 s) of Na_i-dependent ^{45}Ca uptake (Na^+-Ca^{2+} exchange) and Ca_i-dependent ^{45}Ca uptake (Ca^{2+}-Ca^{2+} exchange) were measured under conditions in which the ion binding is not rate limiting at both sides of the membrane.[15,18] The vesicles were prepared and preequilibrated before the experiment in the presence of 20 mM Mops/Tris pH 7.4. Therefore, it is assumed that the intravesicular pH is not altered significantly during the assay. The pH-titration curve of Ca^{2+}-Ca^{2+} exchange (FIG. 1) shows a bell shape in the acidic range (pK_{a1} = 5.1 ± 0.1 and pK_{a2} = 6.3 ± 0.1) followed by activation of the exchange in the alkaline range (pK_{a3} = 8.6 ± 0.2). In contrast, the Na^+-Ca^{2+} exchange shows a monotonic increase from pH 5.0 to 10.0, but exhibits very similar pK_a values (pK_{a1} = 5.2 ± 0.1, pK_{a2} = 6.4 ± 0.1 and pK_{a3} = 8.7 ± 0.1) as observed for Ca^{2+}-Ca^{2+} exchange. Although the absolute values of pK_a can be varied by 10–15% in different experiments, the typical pH titration curves were obtained in 18 independent experiments by using 7 different preparations of sarcolemma vesicles. Similar pH titration curves were obtained under higher ionic conditions for Ca^{2+}-Ca^{2+} exchange with $[^{45}Ca]_o$ = $[Ca]_i$ = 500 μM.

Although the observed pK_a values do not necessarily represent the pK_a values of specific functional groups involving ion transport and/or modulatory activities, three apparent phases (SCHEME 2) of the exchanger deprotonation can be de-

FIGURE 1. Effect of pH on Na^+-Ca^{2+} and Ca^{2+}-Ca^{2+} exchanges in the absence of monovalent cations in the assay medium. The Na-loaded (160 mM) or Ca-loaded (250 μM) vesicles were diluted 50-fold in the assay medium (20 mM buffer Mes/Tris, Mops/Tris, Tris/Ches or Caps/Tris) with different pH's (5.0–10.0), 0.25 M sucrose and 250 μM ^{25}Ca. The Na^+-Ca^{2+} (●) and Ca^{2+}-Ca^{2+} (○) exchange reactions were measured at 37°C and the reaction of ^{45}Ca-uptake was quenched after 1 s by the addition of cold EGTA buffer (pH 7.4) in the semi-rapid-mixer.[14,15,18] Each point represents the average value of duplicate measurements. The lines were computed to give an optimal fit to three pK_a values.[15] The pK_a values were estimated as pK_{a1} = 5.2 ± 0.1, pK_{a2} = 6.4 ± 0.1 and pK_{a3} = 8.7 ± 0.1 for Na^+-Ca^{2+} exchange and pK_{a1} = 5.1 ± 0.1, pK_{a2} = 6.3 ± 0.1 and pK_{a3} = 8.6 ± 0.2 for Ca^{2+}-Ca^{2+} exchange.

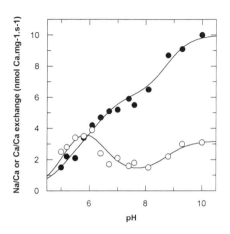

scribed: a) At pH <6.1 both exchange modes have a very similar rate while a deprotonation with an apparent pK_{a1} ≈ 5.1 accelerates both Na^+-Ca^{2+} and Ca^{2+}-Ca^{2+} exchanges in a similar manner (FIG. 1). This phase of deprotonation may involve functionally important carboxyl residues (Glu-113 and/or Glu-199) for ion transport. b) Further deprotonation (pK_{a2} ≈ 6.3) of the exchanger accelerates the Na^+-Ca^{2+} exchange but decelerates the Ca^{2+}-Ca^{2+} exchange (FIG. 1), suggesting that this stage of deprotonation has opposing effects on rate-limiting Na^+ and Ca^{2+} transport. A histidine and/or carboxyl group(s) could be a good candidate for accounting for these diverse effects. c) Deprotonation in the alkaline range (pK_{a3} ≈ 8.7) accelerates both Ca^{2+}-Ca^{2+} and Na^+-Ca^{2+} exchanges, suggesting that a deprotonation of some functional hydroxyl and/or amino groups may result in accelerating the Na^+ and Ca^{2+} movements.

The initial rates (t = 1s) of Na^+-Ca^{2+} exchange were measured by varying the extravesicular pH 5.5–10 at fixed inside positive potential (0 mV or 200 mV) (FIG. 2A) or by varying inside potential (from −80 mV to +140 mV) at fixed pH (5.6 or 7.4) (FIG. 2B). Control experiments show that the potassium itself has no effect on the pH-titration curves of Na^+-Ca^{2+} exchange (not shown). Interestingly, the positive inside potential has no effect on Na^+-Ca^{2+} exchange at pH <6.1, while with increasing pH the membrane potential activates the rate of the exchange reaction (FIG. 2A). Under similar conditions, the Ca^{2+}-Ca^{2+} exchange is still voltage insensitive in a wide range of voltage (from −80 mV to +200 mV) and pH 5.0–10 (not shown). A typical voltage flux relationship was obtained at fixed pH 7.6 and varying voltage (FIG. 2B). With pH 5.5 the voltage response of Na^+-Ca^{2+} exchange is nearly lost in the same range of voltage (FIG. 2B). Therefore, the rate of Na^+-Ca^{2+} exchange becomes nearly voltage insensitive at pH <6.1, presumably, because the voltage-insensitive Ca^{2+} influx becomes rate limiting (FIG. 3). Similar results were obtained in 12 independent experiments by using five different preparations of cardiac sarcolemma vesicles.

As can be seen from FIGURE 3, the ^{45}Ca influx is a common pathway for

$$H_3E \underset{}{\overset{pK_{a1} = 5.1}{\rightleftharpoons}} H_2E \underset{}{\overset{pK_{a2} = 6.3}{\rightleftharpoons}} H_1E \underset{}{\overset{pK_{a3} = 8.7}{\rightleftharpoons}} E$$

SCHEME 2.

FIGURE 2. Effect of pH and inside positive potential on the initial rate of Na^+-Ca^{2+} exchange. **(A)** Effect of varying pH at fixed inside positive potential. Before the experiment, the Na-loaded vesicles were treated with (*open histograms*) or without (*dark histograms*) 1 μM valinomycin. By using the semirapid mixing device the vesicles were diluted in the assay medium containing 20 mM buffer (pH = 5.5–10.0), 160 mM KCl, and 250 μM ^{45}Ca and the reaction of ^{45}Ca uptake was quenched after 1 s. The Na^+-Ca^{2+} exchange activities of valinomycin-untreated vesicles (*open histograms*) represent the control activity (100%) of Na^+-Ca^{2+} exchange at a given pH. *Dark histograms* show the Na^+-Ca^{2+} exchange activity of valinomycin-treated vesicles at various pH. **(B)**. Effect of varying inside potentials at fixed pH 5.6 and 7.4. The sarcolemma vesicles were preequilibrated with 20 mM Mops/Tris, pH 7.4, 160 mM NaCl, and 0.1 mM KCl at 4°C overnight and pretreated with 1 μM valinomycin before the experiment. Different values of the diffusion potential were clamped by 25-fold dilution of vesicles in the assay medium containing various concentrations of KCl and LiCl ($[K]_o + [Li]_o = 160$ mM), 250 μM $^{45}CaCl_2$, and either 20 mM Mops/Tris, pH 7.4 (○), or Mes/Tris, pH 5.6 (●). Each point represents duplicate measurements of the initial rate (t = 1 s) of Na_i-dependent ^{45}Ca uptake. The membrane potential was calculated according to $\Delta\psi = 61.6$ mV log ($[K]_o/[K]_i$).

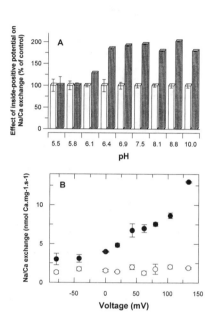

Na^+-Ca^{2+} and Ca^{2+}-Ca^{2+} exchanges. The fact that the deprotonation (pH >6.5) has opposing effects on the two exchange modes (accelerates the Na^+-Ca^{2+}, but decelerates the Ca^{2+}-Ca^{2+} exchange) (FIG. 1), suggests that different partial reactions, the Na^+ efflux and Ca^{2+} efflux, might limit the Na^+-Ca^{2+} and Ca^{2+}-Ca^{2+} exchanges, respectively (see FIG. 3). On the other hand, at pH <6.1 the rates of Na^+-Ca^{2+} and Ca^{2+}-Ca^{2+} exchanges are very similar (FIG. 1), suggesting that the same partial reaction, the Ca^{2+} influx may limit both Na^+-Ca^{2+} and Ca^{2+}-Ca^{2+} exchanges (see FIG. 3). This proposal is confirmed by the fact that at pH >6.5 the inside positive potential accelerates the Na^+-Ca^{2+} exchange by 180–220% (the voltage-sensitive Na^+ efflux limits the exchange rate). Therefore, the acidic pH can switch the rate-limiting pathway to the voltage-insensitive Ca^{2+} influx, causing a loss of voltage-sensitivity of Na^+-Ca^{2+} exchange (see FIGS. 1–3). Two principal mechanisms may account for proton-dependent modulation of voltage response of Na^+-Ca^{2+} exchange: a) Protonation (<6.1) of the exchanger does not alter the stoichiometry ($3Na^+ : Ca^{2+}$), but slows down the electroneutral movement of Ca^{2+} efflux in such a way that it becomes rate limiting. Therefore, the protonated species have a potency to "mask" the electrogenic transport of Na^+ efflux (see FIG. 3), because this step is not rate limiting. This is a kinetic mechanism in nature,

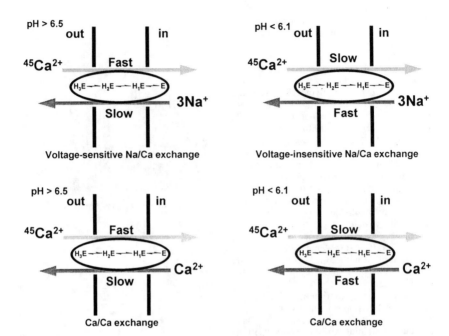

FIGURE 3. Modulation of rate-limiting pathways of Na^+-Ca^{2+} and Ca^{2+}-Ca^{2+} exchanges by protons. A schematic presentation indicates that at pH >6.5 the Na^+ efflux and Ca^{2+} efflux limit the Na^+-Ca^{2+} and Ca^{2+}-Ca^{2+} exchanges, respectively. At pH <6.1, the same partial reaction, the ^{45}Ca influx determines the rates of both Na^+-Ca^{2+} and Ca^{2+}-Ca^{2+} exchanges. At pH >6.5 the Na^+-Ca^{2+} exchange is voltage sensitive because the rate-limiting Na^+ efflux involves the charge carrying species. At pH <6.1 the Na^+-Ca^{2+} exchange becomes voltage insensitive, because the electroneutral ^{45}Ca influx determines the exchange rate. "In" and "out" correspond to the intravesicular (extracellular) and extravesicular (intracellular) spaces in cardiac sarcolemma vesicles.

because the rate-limiting pathway is modified without altering the stoichiometry of exchange. b) Alternatively, high concentrations of protons may switch the electrogenic stoichiometry ($3Na^+:Ca^{2+}$) to the electroneutral one (*e.g.*, $2Na^+:Ca^{2+}$ or $3Na^+:Ca^{2+},H^+$)]. In this case the overall thermodynamics of ion exchange must be altered between affecting the equilibrium between the two bulk phases. Further systematic investigation is necessary to distinguish between these two major mechanisms.

Potassium Interacts with the Deprotonated Species of the Exchanger, Resulting in Accelerating Ca Movements

Although there is solid evidence that K^+ is not cotransported in the cardiac Na^+-Ca^{2+} exchanger,[25] both the voltage sensitivity and the magnitude of outward Na^+-Ca^{2+} current depend critically on the presence of the extracellular monovalent cations.[26] Likewise, the cardiac and neuronal Ca^{2+}-Ca^{2+} exchange can be accelerated severalfold by K^+ or by other monovalent ions, including Na^+.[27-29] The intracellular loop of the cardiac exchanger contains a regulatory domain for Na^+ binding,[22] the occupation of which causes a fast inactivation of the Na^+-Ca^{2+} exchanger. It is not clear whether the regulatory Na^+, K^+ and H^+ compete for the same regulatory and/or transport site(s) or these ions interact with distinct domains causing divers regulatory effects.[15,18,22,30]

In order to investigate the interaction of monovalent cations with different protonated-deprotonated forms of the exchanger, the effect of varying pH 4.8–9.7 was examined on Ca^{2+}-Ca^{2+} exchange at 37°C either with 100 KCl or choline-Cl in the assay medium. Increase of pH from 6.0 to 7.4 decelerates the Ca^{2+}-Ca^{2+} exchange (FIG. 4). In the range of pH 5.0–6.0 potassium has little (if any) effect on Ca^{2+}-Ca^{2+} exchange, while it accelerates the exchange rate 2.0–2.5-fold in the range of pH 7.0–8.1 (FIG. 4). These results were reproducible in 15 independent experiments using five different preparations of vesicles. These data suggest that the accelerating effect of potassium opposes the inhibitory effect of deprotonation at pH 6.0–8.1. This is consistent with our previous finding that potassium affects the pK_{a2} and pK_{a3} components of the pH-titration curve of Ca^{2+}-Ca^{2+} exchange (there is very little, if any, effect of potassium on the pK_{a1}).[18] It is reasonable to assume that K^+ interacts with deprotonated species of the exchanger (pH 6.0–8.1), thereby accelerating the rate-limiting reaction (the Ca^{2+} efflux) of Ca^+-Ca^{2+} exchange (FIG. 5). In contrast to Ca^+-Ca^{2+} exchange, potassium has no detectable effect on Na^+-Ca^{2+} exchange in a wide range of pH 5.0–10 (not shown). This is expected if K^+ does not affect the rate-limiting step (the Na^+ efflux) of Na^+-Ca^{2+} exchange (see FIG. 5). It is possible that potasisum also affects the Ca^{2+} influx, but this effect cannot be seen under steady-state conditions, presumably because this partial reaction is not rate limiting at pH >6.5 for both Na^+-Ca^{2+} and Ca^{2+}-Ca^{2+} exchanges. Therefore, potassium-induced effects can be considered as independent evidence for rate-limiting Ca^{2+} efflux during the Ca^{2+}-Ca^{2+} exchange and rate-limiting Na^+ efflux during the Na^+-Ca^{2+} exchange. It was suggested above that the Ca^{2+} influx may limit both Ca^{2+}-Ca^{2+} and Na^+-Ca^{2+} exchanges at pH <6.1. If so, one may ask: why does K^+ have no effect either on Ca^{2+}-Ca^{2+} or Na^+-Ca^{2+} exchange at low pH? (FIG. 4) The reason could be simple: Potassium-binding site(s) may become functional at pH >6.5 (e.g., deprotonation of histidine and/or carboxyl may permit potassium access).

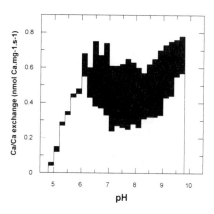

FIGURE 4. Effect of varying pH and fixed K^+ on Ca^{2+}-Ca^{2+} exchange. The effect of K^+ on Ca^{2+}-Ca^{2+} exchange was tested at various pH. The vesicles were loaded with calcium as described in FIGURE 1. The Ca-loaded vesicles were mixed (t = 1 s) with the assay medium containing 20 mM buffer with pH 4.8–9.7, 0.2 M sucrose, 250 μM $^{45}CaCl_2$ plus 100 mM of either KCl (*dark histogram*) or choline-Cl (*open histogram*). Each step of histogram represents a mean value of duplicate measurements.

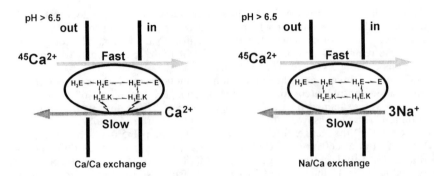

FIGURE 5. Interaction of potassium with the deprotonated species of the exchanger and modulation of rate-limiting steps of Na^+-Ca^{2+} and Ca^{2+}-Ca^{2+} exchanges. The Na^+-Ca^{2+} exchanger is described as a system, which can undergo three steps of protonation-deprotonation ($H_3E \rightleftharpoons H_2E \rightleftharpoons H_1E \rightleftharpoons E$). Potassium can interact with a deprotonated species of the exchanger (e.g., H_2E and H_1E species), yielding the $H_2E.K$ and $H_1E.K$ species. Potassium-bound species have no effect on the rate-limiting and voltage-sensitive Na^+ efflux during the Na^+-Ca^{2+} exchange, but they can speed up the rate-limiting Ca^{2+} efflux during the Ca^{2+}-Ca^{2+} exchange.

Bell-Shaped Temperature Curve of Ca-Ca Exchange Suggests that the Ca-Transport Step Involves More Than Two Species

The Na^+-Ca^{2+} exchange activities of various species and tissues exhibit strikingly alike temperature-dependent curves.[32–34,37] It is possible that the different temperature curves reflect distinct rate-limiting pathways and/or membrane environment. Comparison of temperature dependence of Na^+-Ca^{2+} exchange in reconstituted proteoliposomes and natural membranes lead to the conclusion that the temperature-induced effects reflect the intrinsic nature of the Na^+-Ca^{2+} exchanger protein(s) rather than the properties of lipid environment.[32–34,37] In the present study the effect of temperature was studied on the exchange modes with a goal of identifying and characterizing the rate-limiting pathways during the Na^+-Ca^{2+} and Ca^{2+}-Ca^{2+} exchanges.

The effect of varying temperature (6–45°C) was tested on the initial rates of Na^+-Ca^{2+} and Ca^{2+}-Ca^{2+} exchanges at fixed pH 7.4 with either 100 mM Choline-Cl (FIG. 6A) or KCl (FIG. 6B). By increasing temperature (6–45°C) the Na^+-Ca^{2+} exchange is accelerated 15–20-fold either in the presence (FIG. 6B) or absence of potassium (FIG. 6A). The Na^+-Ca^{2+} exchange is apparently insensitive to potassium in the range of 6–35°C, suggesting that in a wide range of temperature potassium has no affect on the rate-limiting Na^+ efflux. A modest activation (10–20%) by potassium is observed (n = 5) at >37°C, suggesting that at higher temperatures the Ca^{2+} influx may become partially rate limiting. In the absence of potassium the Ca^{2+}-Ca^{2+} exchange exhibits a bell-shaped curve with a broad maximum at 26–33°C (FIG. 6A). The bell-shaped temperature-dependence of Ca^{2+}-Ca^{2+} exchange is even more prominent in the presence of K^+, exhibiting a maximum at 27–29°C (FIG. 6B). Similar results were obtained when the ionic concentrations were increased to $[^{45}Ca]_o = [Ca]_i = 500\ \mu M$ (not shown). The bell-shaped temperature-curve of the Ca^{2+}-Ca^{2+} exchange was observed in 12 independent experiments with five different preparations of sarcolemma vesicles. Control experiments show

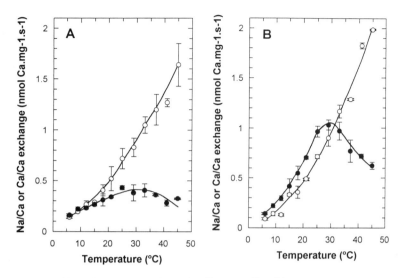

FIGURE 6. Effect of temperature in Na^+-Ca^{2+} and Ca^{2+}-Ca^{2+} exchanges in the presence or absence of K^+. The Na^+-Ca^{2+} (○) and Ca^{2+}-Ca^{2+} (●) exchanges were measured in the absence (**A**) or presence (**B**) of potassium in the assay medium. The vesicles were loaded with sodium or calcium as described in FIGURE 1. The Ca-loaded vesicles were mixed (t = 1 s) at 6–45°C with the assay medium (20 mM Mops/Tris, pH 7.4, 0.2 M sucrose, 250 μM $^{45}CaCl_2$) containing either 100 mM choline-Cl (A) or 100 mK KCl (B). Each point represents a mean value of four independent measurements (*bars* indicate ± SD of the mean). The lines were calculated with Eq. 4 and 6 (see Appendix). For simplicity the total concentration of the exchanger was fixed at E_t = 0.002 mg/mL. The kinetic parameters, derived from the data fitting, are summarized in TABLE 1.

that a short-time exposure (10–15 s) of vesicles to 29–45°C does not cause irreversible inactivation of Ca^{2+}-Ca^{2+} exchange (not shown). This means that the descendic shoulder of Ca^{2+}-Ca^{2+} exchange (FIG. 6) cannot be caused by 'thermal inactivation' of the exchanger.

The significance of the bell-shaped temperature-curve of Ca^{2+}-Ca^{2+} exchange is that it cannot be reconciled with a simple reversible reaction involving two species and two elementary rate constants (l' and l'') (SCHEME 3). Temperature-dependence of $V_{max}(Ca/Ca) = l'l''/(l' + l'')$ can be rewritten as $V_{max}(Ca/Ca) = [ab/(ab + 1)]l'_o Q_{l'}^{(\Delta T/10)}$, in which $a = l''_o/l'_o$, $b = Q_{l''}/Q_{l'}$ and the rate-constants are temperature-dependent, $l'_o Q_{l'}^{(\Delta T/10)}$, and $l'' = l''_o Q_{l''}^{(\Delta T/10)}$ (see Appendix). A simple mathematical analysis show that for a simple reversible reaction (SCHEME

$$E'Ca \underset{l''}{\overset{l'}{\rightleftharpoons}} E''Ca$$

SCHEME 3.

Mechanism A

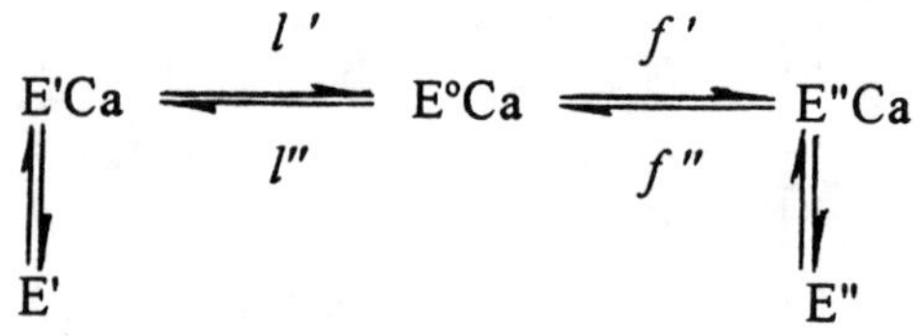

$$E'Ca \underset{l''}{\overset{l'}{\rightleftharpoons}} E°Ca \underset{f''}{\overset{f'}{\rightleftharpoons}} E''Ca$$
$$E' \qquad\qquad\qquad\qquad E''$$

Mechanism B

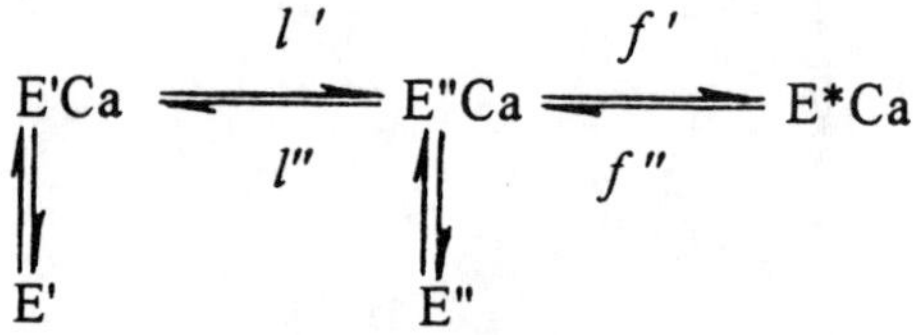

$$E'Ca \underset{l''}{\overset{l'}{\rightleftharpoons}} E''Ca \underset{f''}{\overset{f'}{\rightleftharpoons}} E*Ca$$
$$E' \qquad\qquad E''$$

Mechanism C

$$E'_A Ca \underset{l''}{\overset{l'}{\rightleftharpoons}} E''_A Ca$$
$$E' \qquad\qquad\qquad E''$$
$$E'_B Ca \underset{f''}{\overset{f'}{\rightleftharpoons}} E''_B Ca$$

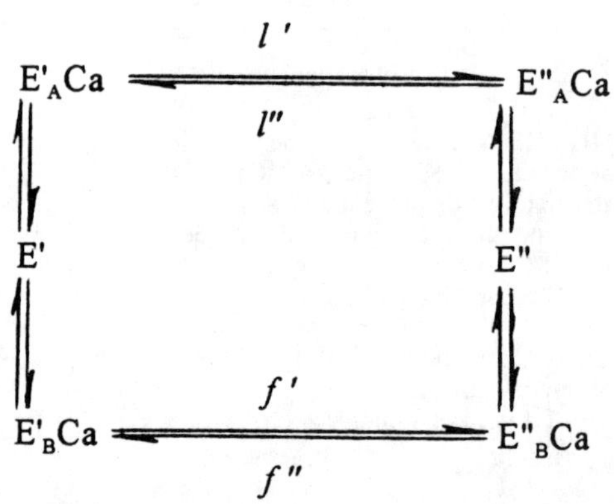

MECHANISMS A–C.

3) the temperature-dependence of V_{max}(Ca/Ca) cannot show a maximum. Therefore, in the frame of the present formalism the Ca^{2+}-Ca^{2+} exchange cannot exhibit a bell-shaped temperature curve.

In order to describe the bell-shaped temperature-curve of Ca^{2+}-Ca^{2+} exchange, it is essential to assume that at least two reactions and three species are involved in the Ca^{2+}-transport. Three possible mechanisms can be considered: (a) A ground-state intermediate (E°Ca) exists on the Ca^{2+} transport pathway (MECHANISM A). (b) Functionally "inactive" species (E*Ca) are in equilibrium with the Ca^{2+}-transporting species (MECHANISM B). (c) Two (or more) intercovertable conformers (e.g., E_A and E_B) operate in parallel (MECHANISM C).

Under conditions in which the calcium concentrations are high at both sides of the membrane, the fractional concentrations of ligand free species (E' and E'') become negligible. Therefore, the Mechanisms A and B cannot be distinguished from each other, yielding a similar mechanism with two reversible reactions (four rate constants) and three species (SCHEME 4). By using a simple kinetic formalism (see derivation of Eq. 4 in Appendix) a reasonable fit can be obtained to the experimental data describing the bell-shaped temperature curve of Ca^{2+}-Ca^{2+} exchange (FIG. 6). According to this formalism the bell-shaped temperature curve

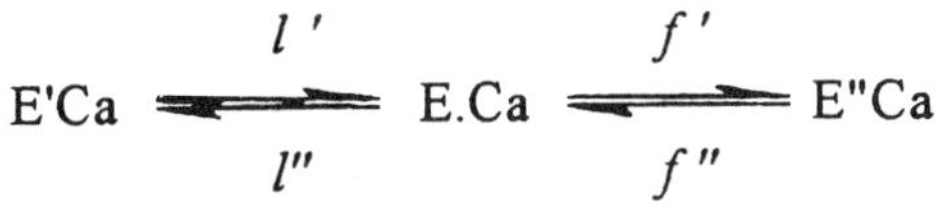

$$E'Ca \underset{l''}{\overset{l'}{\rightleftharpoons}} E.Ca \underset{f''}{\overset{f'}{\rightleftharpoons}} E''Ca$$

SCHEME 4.

is observed because the forward rate constants, f' and l', have much lower Q_{10} values ($Q_{l'} = 1.4$–1.5 and $Q_{f'} = 1.3$–1.4) than the reverse rate constants, f'' and l'' ($Q_{l''} = 2.7$–2.8 and $Q_{f''} = 2.8$–3.7) (see TABLE 1). Therefore, the forward and reverse reactions may involve conformational transitions with very different energy barriers. These provide independent evidence for the functional asymmetry of the exchanger. Despite the fact that potassium has a dramatic effect on the temperature curve of Ca^{2+}-Ca^{2+} exchange (compare FIG. 6A and 6B), potassium does not affect significantly the Q_{10} values, rather it affects a basal rate constant. This is consistent with a proposal that potassium affects the asymmetry of bidirectional calcium movements (see TABLE 1).

The bell-shaped curves can also be computed for Na^+-Ca^{2+} exchange, but in this case the curve maximum shows up at much higher temperatures (>50°C). Unfortunately, the calculated profiles cannot be tested experimentally, because at >45°C the irreversible inactivation of the exchanger may become significantly damaging and thereby prevents reaching meaningful conclusions. As a matter of fact, the bell-shaped temperature curve was observed before in the synaptic Na^+-Ca^{2+} exchanger, exhibiting the curve maximum at rather low temperatures.[33,37] This may suggest that the basic mechanisms and rate-limiting pathways could be the same in the cardiac and synaptic isoforms, although the relative rates of partial reactions might be different.

If indeed the electrogenic Na^+ efflux is rate limiting in a wide range of tempera-

TABLE 1. Rate Constants (at 6°C) and Their Q_{10} Values, Derived from the Temperature Curves of Na^+-Ca^{2+} and Ca^{2+}-Ca^{2+} Exchanges

	Ca^{2+}-Ca^{2+} Exchange		Na^+-Ca^{2+} Exchange	
	$-K^+$	$+K^+$	$-K^+$	$+K^+$
Rate constant at 6°C		s^{-1}		
l'_o	696 ± 304[a]	5960 ± 2470	301 ± 57	674 ± 147
l''_o	284 ± 131	333 ± 150	ND[b]	ND
f'_o	641 ± 250	1154 ± 351	790 ± 347	648 ± 192
f''_o	522 ± 285	1438 ± 300	18 ± 9	7 ± 3
k''_o	ND	ND	98 ± 18	62 ± 10
Q_{10} values				
$Q_{l'}$	1.48 ± 0.23	1.45 ± 0.19	1.35 ± 0.03	1.44 ± 0.25
$Q_{l''}$	2.80 ± 0.40	2.77 ± 0.50	ND	ND
$Q_{f'}$	1.30 ± 0.20	1.30 ± 0.15	1.83 ± 0.44	1.68 ± 0.35
$Q_{f''}$	2.82 ± 0.43	3.71 ± 0.48	4.00 ± 0.35	4.00 ± 0.40
$Q_{k''}$	ND	ND	3.85 ± 0.35	3.41 ± 0.37

[a] ±SE.
[b] Not determined, because by definition k''_o and $Q_{k''}$ exist only for Na^+-Ca^{2+} exchange and l''_o and $Q_{l''}$ only for Ca^{2+}-Ca^{2+} exchange (compare Eqs. 1 and 4 in Appendix). Temperature dependence of Na^+-Ca^{2+} and Ca^{2+}-Ca^{2+} exchange reactions were measured and the lines were computed to fit the data (see FIG. 6 and Appendix). The total concentration of the exchanger was fixed at $E_t = 0.002$ mg/mL.

ture (as was claimed above), one may predict that the net Na^+-Ca^{2+} exchange must also be voltage sensitive in a wide range of temperature. In order to test this proposal the effect of inside positive potential was examined on the initial rates of Na^+-Ca^{2+} exchange by varying temperature from 6°C to 45°C. Different voltage was clamped by exposing the valinomycin-untreated ($\Delta\psi = 0$ mV) or valinomycin-treated vesicles ($\Delta\psi \geq +200$ mV) to the potassium-containing assay medium (FIG. 7). As can be seen from FIG. 7, the inside positive potential accelerates (150–200%) the Na^+-Ca^{2+} exchange at each fixed temperature. This means that despite the rate of Na^+-Ca^{2+} exchange is accelerated 15–20-fold by increasing temperature (FIG. 6), the magnitude of fractional acceleration, induced by inside positive potential, is still comparable at various temperatures (FIG. 7). Therefore, in the range of 6–45°C and pH 6.5–9.7 (either with or without potassium) the electrogenic Na^+-transport is still rate limiting for Na^+-Ca^{2+} exchange.

The Ratio of Na-Ca/Ca-Ca Exchanges as an Estimate for the Asymmetry of Bidirectional Ca Movements

The theoretical and experimental analysis shows that the ratio of Na^+-Ca^{2+}/Ca^{2+}-Ca^{2+} exchange (signed as R value) may reflect a degree of asymmetry of bidirectional Ca^{2+} movements during the Ca^{2+}-Ca^{2+} exchange.[14,15,18] Careful examination of Eq. 8 shows that if the R values exceed the unity ($R \gg 1$) it can be concluded that: a) the Ca^{2+} influx is faster than the Ca^{2+} efflux ($l' > l''$) and b) the

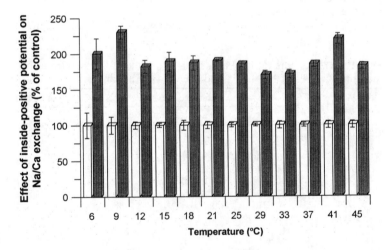

FIGURE 7. Effect of fixed membrane potential on Na^+-Ca^{2+} exchange at various temperatures. The Na-loaded vesicles were obtained as described in FIGURE 1. Before the experiment, the Na-loaded vesicles were treated with (*dark histograms*) or without (*open histograms*) valinomycin. The valinomycin-treated or -untreated vesicles were diluted 25-fold in the assay medium (20 mM Mops/Tris pH 7.4, 0.2 M sucrose, 100 mM KCl, and 250 μM ^{45}Ca) at 6–45°C. The reaction of ^{45}Ca uptake was quenched after t = 1 s. Each point represents a mean of duplicate measurements (*bars* indicate ± SD of the mean). 100% represents the control activities of Na^+-Ca^{2+} exchange at various temperatures at "0 mV."

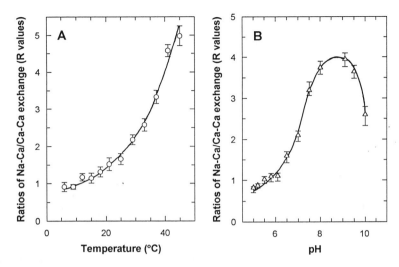

FIGURE 8. Effect of temperature and pH on the ratio of Na^+-Ca^{2+}/Ca^{2+}-Ca^{2+} exchanges. The initial rates (t = 1 s) of Na^+-Ca^{2+} and Ca^{2+}-Ca^{2+} exchanges were measured either at various temperatures (6–45°C) and fixed pH 7.4 (**A**) or at varying pH and fixed 37°C (**B**). The exchange reactions were done in the standard assay medium containing 20 mM buffer, 0.2 M sucrose and 250 µM $^{45}CaCl_2$ (see FIG. 1). The Na^+-Ca^{2+}/Ca^{2+}-Ca^{2+} exchange ratios (R values) represent a mean of four independent measurements of Na^+-Ca^{2+} and Ca^{2+}-Ca^{2+} exchanges and *bars* indicate ± SD of the mean.

Ca^{2+} efflux is slower than the Na^+ efflux ($l'' < k''$). However, if the Na^+-Ca^{2+} and Ca^{2+}-Ca^{2+} exchange rates

$$R = \frac{K_m(Na/Ca)}{K_m(Ca/Ca)} = \frac{V_{max}(Na/Ca)}{V_{max}(Ca/Ca)} = \frac{1 + (l'/l'')}{1 + (l'/k'')} \quad (8)$$

are similar (R ~ 1), no conclusions can be drawn regarding the ratio of Ca^{2+} influx/Ca^{2+} efflux (bidirectional Ca^{2+} movements may be or may not be asymmetric).

The effect of varying temperature (6–45°C) at fixed pH 7.4 (FIG. 8A) and of varying pH (5.0–10.0) at fixed 37°C (FIG. 8B) were examined on the R values. The initial rates (t = 1 s) of Na^+-Ca^{2+} and Ca^{2+}-Ca^{2+} exchanges were measured in the standard assay medium with sucrose, in the absence of potassium (these measurements were done on the same day of experiment by using the same preparation of vesicles). Typical results were obtained in 15 independent experiments by using seven different preparations of vesicles. As can be seen from FIG. 8, the rates of Na^+-Ca^{2+} and Ca^{2+}-Ca^{2+} exchanges are very similar (R ≈ 1) either a low temperature (6–12°C) (FIG. 8A) or low pH (5.0–6.0) (FIG. 8B). Nevertheless, the different mechanisms must be involved in modulating the exchange rates at low temperature and acidic pH. This because the Na^+-Ca^{2+} exchange is voltage-insensitive at the acidic pH (FIG. 2A), while at low temperatures the Na^+-Ca^{2+} exchange is still voltage-sensitive (FIG. 7). Therefore, at low temperatures the rate-limiting reactions, the Na^+ efflux (for Na^+-Ca^{2+} exchange) and Ca^{2+} efflux (for Ca^{2+}-Ca^{2+} exchange), may have similar rate values, while at low pH the same partial reaction, the Ca^{2+} influx might limit the rates of both Na^+-Ca^{2+} and Ca^{2+}-Ca^{2+} exchanges. By increasing temperature from 12°C to 45°C (FIG. 8A) or pH from 6.0 to 9.0 (FIG. 8B) enhances the R values from 0.8–1.0 to 4–5, suggesting that the bidirectional Ca^{2+} movements of Ca^{2+}-Ca^{2+} exchange become considerably asymmetric at pH >6.0 and T° >12°C. It was found recently that addition of potassium to the medium decreases the R values at fixed temperature of pH,[15,18] suggesting a possibility that potassium decreases the degree of asymmetry of bidirectional Ca^{2+} movements. Further investigation is necessary to confirm this proposal.

CONCLUSIONS

1. All the observed effects of temperature, pH, potassium and voltage on Na^+-Ca^{2+} and Ca^{2+}-Ca^{2+} exchanges can be described in the frame of consecutive (ping-pong) mechanism, although the existence of some new intermediates is definitely suggested.

2. Under a wide range of tested conditions (pH >6.5, 6–37°C, 0–100 mM KCl), the voltage-sensitive Na^+ efflux limits the Na^+-Ca^{2+} exchange, while the electroneutral Ca^{2+} efflux controls the Ca^{2+}-Ca^{2+} exchange (the voltage-insensitive Ca^{2+} influx is not rate limiting for both exchange modes).

3. The pH titration curves of both Na^+-Ca^{2+} and Ca^{2+}-Ca^{2+} exchanges can be described by three apparent phases of the exchanger deprotonation ($pK_{a1} \approx 5.1$, $pK_{a2} \approx 6.3$ and $pK_{a3} \approx 8.7$). The first and third steps of deprotonation accelerate the movement of both ions, while the second step of deprotonation has opposite effects of the Na^+ and Ca^{2+} movements: it accelerates the rate-limiting Na^+ efflux for Na^+-Ca^{2+} exchange, but decelerates the rate-limiting Ca^{2+} efflux for Ca^{2+}-Ca^{2+} exchange.

4. The rates of Na^+-Ca^{2+} and Ca^{2+}-Ca^{2+} exchanges are very similar ($R \approx 1$) at both low temperature (6–12°C) or low pH (5.0–6.0). The Na^+-Ca^{2+} exchange becomes voltage-insensitive at pH <6.1, and the Na^+-Ca^{2+} exchange is still voltage sensitive at low temperature. Therefore, at low temperature the Na^+ efflux and Ca^{2+} efflux may have similar rates, while at low pH with the same partial reaction, the Ca^{2+} influx may limit the rates of both Na^+-Ca^{2+} and Ca^{2+}-Ca^{2+} exchanges.

5. Potassium can interact with the deprotonated species of the exchanger when pH >6.1 and thereby accelerates the rate-limiting Ca^{2+} efflux of Ca^{2+}-Ca^{2+} exchange. Potassium has no considerable effect on the rate-limiting Na^+ efflux during Na^+-Ca^{2+} exchange. Potassium exhibits the characteristic shifts of pK_{a2} and pK_{a3} components on the pH titration curve of Ca^{2+}-Ca^{2+} exchange, while it has little (if any) affect on the exchange rate in the acidic range.

6. The Ca^{2+}-Ca^{2+} exchange shows a bell-shaped temperature curve, either in the presence or absence of potassium. This behavior is not caused by irreversible inactivation of the exchanger or heterogeneous orientation of vesicular preparations. This bell-shaped temperature curve cannot be reconciled with a simple reversible reaction involving only two interconvertible species with two rate constants. For successful description of the data it is necessary to assume that at least three species with two reversible reactions are involved in Ca^{2+} transport.

7. Increase of temperature (from 12°C to 45°C) or pH (from 6.0 to 9.0) magnifies the R values from 0.8–1.0 to 4–5, suggesting that the bidirectional Ca^{2+} movements become asymmetric during the Ca^{2+}-Ca^{2+} exchange by increasing temperature and pH.

Although the relevance of our *in vitro* findings to cellular and cardiac physiology needs a systematic and careful examination, the present analytical approaches could be useful for resolving the underlying mechanisms under conditions in which the intracellular pH drops dramatically (*e.g.*, acidosis, ischemia, etc.) or when low temperature alters cardiac functions (cold-induced heart contractures, cardioplegia during open heart surgery, etc.).

APPENDIX

Rate Equations of Ca^{2+}-Ca^{2+} and Na^+-Ca^{2+} Exchanges (Two-Step Ca^{2+} Transport Model)

Equilibrium Ca^{2+}-Ca^{2+} Exchange

Under equilibrium conditions ($[*Ca_o] = [Ca_i]$) the Ca^{2+}-Ca^{2+} exchange can be described as a reversible two-step reaction:

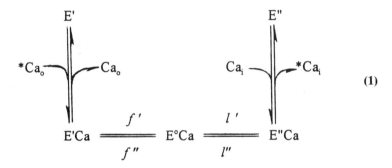

(1)

Since under conditions tested the ion binding is not rate limiting on both sides of the membrane and it is assumed that rate constants for ion binding dissociation are faster than the ion transport, the maximal rate of $*Ca_o$-Ca_i exchange can be written as:

$$V_{max}(Ca/Ca) = k_{cat}[E_t] = [E'Ca]f'l'/(f'' + l') \text{ or } k_{cat} = ([E'Ca]/E_t)f'l'/(f'' + l')$$
(2)

Since the system is at equilibrium (no net flux or charge transfer occurs), one can derive from Eq. 1 that $K_1 = [E°Ca]/[E'Ca] = f'/f''$, $K_2 = [E''Ca]/[E°Ca] = l'/l''$, $K_1K_2 = [E''Ca]/[E'Ca] = l'f'/l''f''$. Therefore, the fractional concentration of $E'Ca$ can be represented as:

$$\frac{[E'Ca]}{[E_t]} = \frac{[E'Ca]}{[E'Ca] + [E°Ca] + [E''Ca]} = \frac{1}{1 + K_1 + K_1K_2}$$

$$= \frac{1}{1 + (f'/f'') + (f'l'/f''l'')} = \frac{f''/f'}{1 + (f''/f') + (l'/l'')}$$
(3)

By substituting $[E'Ca]/[E_t]$ into Eq. 2 one can obtain:

$$V_{max}(Ca/Ca) = \frac{[E_t][f'l'/(f''+l')](f''/f')}{1+(f''/f')+(l'/l'')}$$

$$= \frac{[E_t]l'}{[1+(l'/f'')][1+(f''/f')+(l'/l'')]} \quad (4)$$

Unidirectional Na^+-Ca^{2+} Exchange

With saturating *Ca_o and Na_i (the ion binding is not rate limiting) the unidirectional *Ca_o-Na_i exchange with two-step Ca^{2+} transport can be described as:

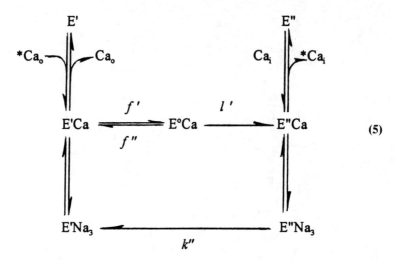

By using the rules of partition analysis and concept of net rate constants,[35,36] the V_{max}(Na/Ca) of *Ca_o = Na_i exchange can be written from Eq. 5 as:

$$V_{max}(Na/Ca) = \frac{E_t}{[1/\{f'l'/(f''+l')\}]+(1/l')+(1/k'')}$$

$$= \frac{E_t l'}{(f''/f')+(l'/f')+(l'/k'')+1} \quad (6)$$

Temperature Dependence of Individual Rate Constants

For the sake of simplicity and data fitting limitations the temperature dependence of individual rate constants were calculated according to the following equation:

$$k^i = k_o^i Q_i^{(\Delta T/10)} \quad (7)$$

in which k^i is a specific rate constant (l', l'', f', f'' or k'') at a given temperature, k_o^i depicts the rate constant (l'_o, l''_o, f'_o, f''_o or k''_o) at the reference temperature (6°C) and Q_i represents the Q_{10} values ($Q_{l'}$, $Q_{l''}$, $Q_{f'}$, $Q_{f''}$ or $Q_{k''}$) of appropriate rate constants. ΔT is a difference between a given temperature and the reference temperature (6°C). The term $\Delta T/10°C$ represents the increase of temperature with respect to the reference temperature (6°C).

REFERENCES

1. KHANANSHVILI, D. 1990. Biochemistry **29**: 2437–2442.
2. NIGGLI, E. & W. J. LEDERER. 1991. Nature **349**: 621–624.
3. HILGEMANN, D. W., D. A. NICOLL & K. D. PHILIPSON. 1991. Nature **352**: 715–718.
4. LI, J. & J. KIMURA. 1991. Ann. N. Y. Acad. Sci. **639**: 48–60.
5. KHANANSHVILI, D. 1991a. J. Biol. Chem. **266**: 13764–13769.
6. MATSUOKA, S. & D. W. HILGEMANN. 1992. J. Gen. Physiol. **100**: 963–1001.
7. REEVES, J. P. & C. C. HALE. 1984. J. Biol. Chem. **259**: 7733–7739.
8. KIMURA, J., S. MIYAMAE & A. NOMA. 1987. J. Physiol. **384**: 199–222.
9. CRESPO, L. N., C. J. GRANTHAM & M. B. CANNEL. 1990. Nature **345**: 616–621.
10. STEIN, W. D. 1986. Transport and Diffusion across Cell Membranes. Academic Press, New York.
11. LAUGER, P. 1987. J. Membr. Biol. **99**: 1–11.
12. KHANANSHVILI, D. 1990. Curr. Opin. Cell Biol. **2**: 731–734.
13. LAUGER, P. 1991. Electrogenic Ion Pumps. 3–135. Sinauer Assoc. Inc. Sunderland, MA.
14. KHANANSHVILI, D. 1991. Ann. N. Y. Acad. Sci. **639**: 85–98.
15. KHANANSHVILI, D. & E. WEIL-MASLANSKY. 1994. Biochemistry **33**: 312–319.
16. LI, Z., D. A. NICOLL, A. COLLINS, D. W. HILGEMANN, A. G. FILOTEO, J. T. PENNISTON, J. M. TOMICH & K. D. PHILIPSON. 1990. J. Biol. Chem. **266**: 1014–1020.
17. KHANANSHVILI, D., D. C. PRICE, M. J. GREENBERG & Y. SARNE. 1993. J. Biol. Chem. **268**: 200–205.
18. KHANANSHVILI, D., G. SHAULOV & E. WEIL-MASLANSKY. 1995. Biochemistry **34**: 10290–10297.
19. GOLDSHLEGGER, R., S. J. D. KARLISH, A. REPHAELI & W. D. STEIN. 1987. J. Physiol. (London) **387**: 331–355.
20. GADSBY, D. C. & M. NAKAO. 1989. J. Gen. Physiol. **94**: 511–537.
21. NICOLL, D. A., S. LONGONI & K. D. PHILIPSON. 1990. Science **250**: 562–564.
22. MATSUOKA, S., D. A. NICOLL, R. F. REILLY, D. W. HILGEMANN & K. D. PHILIPSON. 1993. Proc. Natl. Acad. Sci. USA **90**: 3870–3874.
23. PHILIPSON, K. D., D. A. NICOLL, Z. LI, J. S. FRANK & D. W. HILGEMANN. 1992. 8th Int. Symp: Calcium-Binding Proteins and Calcium Function in Health and Diseases. S23, 23–27. Davos, Switzerland
24. REEVES, J. P. 1992. Arch. Biochem. Biophys. **292**: 329–334.
25. YASUI, K. & J. KIMURA. 1990. Pflügers Arch. **415**: 513–515.
26. GADSBY, D. C., M. NODA, R. N. SHEPHERD & M. NAKAO. 1991. Ann. N. Y. Acad. Sci. **639**: 140–146.
27. REEVES, J. P. 1985. Curr. Top. Membr. Transp. **25**: 77–127.
28. DIPOLO, R. & L. BEAUGÉ. 1991. Ann. N. Y. Acad. Sci. **639**: 100–111.
29. PHILIPSON, K. D. & D. A. NICOLL. 1993. Int. Rev. Cytol. **137C**: 199–226.
30. DOERING, A. E. & W. J. LEDERER. 1993. J. Physiol. **466**: 481–499.
31. DIXON, D. A. & D. H. HAYNES. 1990. Biochim. Biophys. Acta **1029**: 274–284.
32. BERSOHN, M. M., R. VAMURI, D. W. SHUIL, R. S. WEISS & K. D. PHILIPSON. 1991. Biochim. Biophys. Acta **1062**: 19–23.
33. TESSARI, T. & H. RAHAMIMOFF. 1991. Biochim. Biophys. Acta **1066**: 208–218.
34. TIBBITS, G. F., K. D. PHILIPSON & H. KASHIHARA. 1992. Am. J. Physiol. **261**: C411–C417.
35. CLELAND, W. W. 1975. Biochemistry **14**: 3220–3224.
36. HUANG, C. Y. 1979. Methods Enzymol. **63**: 54–84.
37. RAHAMIMOFF, H., D. DAHAN, I. FURMAN, R. SPANIER & M. TESSARI. 1991. Ann. N. Y. Acad. Sci. **639**: 210–221.

Voltage Dependence of Na-Ca Exchange in Barnacle Muscle Cells

I. Na-Na Exchange Activated by α-Chymotrypsin[a]

H. RASGADO-FLORES, R. ESPINOSA-TANGUMA,[b]
J. TIE, AND J. DeSANTIAGO

Department of Physiology and Biophysics
Finch University of Health Sciences/
The Chicago Medical School
3333 Green Bay Road
North Chicago, IL 60064

INTRODUCTION

The Na-Ca exchanger is able to operate in four modes of exchange: it can mediate the efflux of Ca^{2+} in exchange for extracellular Na^+ (*i.e.*, the Ca^{2+} efflux mode);[1] it can also promote the influx of Ca^{2+} in exchange for intracellular Na^+ (*i.e.*, the Ca^{2+} influx mode).[2] In addition, the exchanger can also operate as Na-Na[3] and Ca-Ca[1] modes of exchange. The direction of net Ca^{2+} flux is governed by the electrochemical gradients of Na^+ and Ca^{2+} and the membrane potential (V_M). Interestingly, performance of any of the modes of exchange mediated by the exchanger has an absolute requirement for intracellular Ca^{2+} (Ca_i),[4,5] and they are all activated by intracellular ATP.[4] On the other hand, a mild treatment with the protease α-chymotrypsin removes the Ca_i-requirement of the exchanger as well as its ATP activation rendering a fully activated (*i.e.*, deregulated) exchanger.[6-8]

Given the fact that the stoichiometry of exchange is 3 Na^+ : 1 Ca^{2+},[9] exchange of Na^+ for Ca^{2+} either in the Ca^{2+} influx or efflux modes of exchange are electrogenic[10] and voltage sensitive.[11] On the other hand, Na-Na and Ca-Ca exchange are likely to operate with stoichiometries of 3 Na^+ : 3 Na^+ and 1 Ca^{2+} : 1 Ca^{2+}. Therefore, operation of these modes of exchange is electroneutral. However, electroneutral ionic transporters or pumps may still be voltage sensitive for at least two main reasons. First, in order to gain access to the intra- or extracellular ionic binding site at the exchanger, the ion to be translocated may need to pass through a narrow access channel. In this case, the ion senses the change in voltage across the membrane field and alterations in V_M affect the access of the ion to its binding site.[12,13] The voltage sensitivity becomes more prevalent at low concentrations of the transported ion at the membrane side (*i.e.*, intra- or extracellular) where the access channel is present. Second, voltage sensitivity can also be

---

[a] H. R-F. is an established investigator of the American Heart Association. R. E-T. and J. D. are research fellows of the American Heart Association, Metropolitan Chicago. This work was supported by NIH Grant RO1-NS28563.

[b] Present address: Dept. Physiology and Pharmacology, Univ. Autonoma de San Luis Potosí, School of Medicine, 78210 San Luis Potosí, S.L.P. México.

conferred to the electroneutral exchange if the ionic binding sites at the exchanger are charged and binding of the transported ions only produces a partial screening of these charges. In this case, if the ionic translocation step from the intra- to the extracellular space, and vice versa, is accompanied by displacement of these charges across the membrane voltage field, changes in V_M will affect the translocation process. Membrane polarization will increase or decrease the exchange rate depending on whether the ionic influx or efflux is the rate limiting step.

There is considerable controversy regarding the voltage dependence of the electroneutral Na-Na and Ca-Ca exchange modes. In squid axons and guinea pig myocytes, for example, voltage sensitivity has been reported for Ca-Ca but not for Na-Na exchange.[14,15] However, the opposite results have also been reported for guinea pig myocytes.[16] Therefore, a systematic study of the voltage dependence of these electroneutral modes of exchange is needed using a preparation where ionic fluxes can be measured under voltage clamp conditions. The present work was undertaken to characterize the voltage dependence of the Na-Na exchange mediated by the Na-Ca exchanger. Internally perfused, voltage-clamped barnacle muscle cells were used as models to carry out these studies. Na-Na exchange was measured radiometrically as the extracellular Na^+ (Na_o)-dependent Na^+ efflux.

MATERIALS AND METHODS

Experimental Set-Up

Isolated, internally perfused, voltage-clamped, single barnacle (*Balanus nubilus*) muscle cells were used. The basic methodology followed to carry out these experiments has been published.[5,8,17]

External Solutions

In general, external solutions were free of Ca^{2+} to prevent Ca_o-Na_i exchange. The standard Na^+-sea water Ca^{2+}-free (NaSW0Ca) solution contained (in mM): 456 NaCl, 36 $MgCl_2$, 10 KCl and a mixture of Tris(hydroxy-methyl)aminomethane (Tris)-base/Tris-HCl to adjust the pH (7.8) and osmolality (1000 mOsm/KgH_2O). In one instance (see FIG. 3) the external solution contained 36 mM $CaCl_2$; this was obtained by replacing $MgCl_2$ for $CaCl_2$. For experiments in which the cells were voltage clamped, the solutions were free of Cl^- and K^+ to reduce the membrane conductance. This solution (Na-methanesulfonate 0Ca^{2+}) contained (in mM): 456 Na-methanesulfonate; 36 $MgSO_4$; 3 Trizma Base and a mixture of methanesulfonate/Tris-base to adjust the pH and osmolality. Na_o-free conditions were attained by replacing (mole-for-mole) Na_o with either Tris-base/Tris-HCl (Tris 0Ca solution) or Tris-methanesulfonate (Tris-methanesulfonate 0Ca^{2+} solution) for Cl^--free solutions. Various Na_o concentrations were attained by mixing, in appropriate proportions, either the NaSW0Ca and TrisSW0Ca or the Na-methanesulfonate 0Ca^{2+} and Tris-methanesulfonate 0Ca^{2+}.

Barnacle muscle cells lack Na^+ channels. However, in the absence of Ca_o, Na^+ can permeate through voltage-gated Ca^{2+} channels.[18] Consequently, solutions included a blocker of voltage-gated Ca^{2+} channels as well as blockers of other Na^+ efflux pathways. All solutions contained 0.1 mM ouabain to inhibit Na^+ efflux through the Na/K pump;[19] 0.01 mM bumetanide to block the Na/K/Cl cotransporter;[20] 0.01 mM verapamil to prevent Na^+ efflux through Ca^{2+} channels;[18] and 0.2 mM 4,4'-diisothiocyanostilbene-2,2'-disulfonic acid (DIDS) to inhibit the $Na^+ + HCO_3^- \text{-} H^+ + Cl^-$ countertransport.[21]

Internal Solutions

All perfusion solutions contained (in mM, unless otherwise specified): 46 N-2-hydroxyethyl-piperazine-N'-2-ethanesulfonic acid (HEPES); 3.5 caffeine and 0.025 carbonyl cyanide p-trifluoromethoxyphenylhydrazone (FCCP), to inhibit Ca^{2+} sequestration by the sarcoplasmic reticulum and the mitochondria, respectively; 1.5 phosphoenol pyruvate and 0.08 mg/ml pyruvate kinase, to regenerate ATP; 4 ATP-Mg; 60 HEPES; 8 ethyleneglycol-bis-(β-aminoethylether)-N,N'-tetraacetic acid (EGTA) or 10 N-hydroxyethylethylene-diaminetriacetic acid (HEEDTA); and 200 glycine. For experiments in which Na-Na exchange was stimulated by raising the intracellular free Ca^{2+} concentration ($[Ca^{2+}]_i$) the solutions (NaCl perfusate) also contained (in mM): 38 KCl; 162 K-aspartate; 7 $MgCl_2$ and various amounts of $CaCl_2$ (see below). For experiments in which the exchanger was activated with α-chymotrypsin and the cells were voltage clamped, the perfusate (Na-methanesulfonate perfusate) was free of Cl^- and K^+ to reduce the membrane conductance. This solution also contained (in mM): 7 Mg-methanesulfonate; 8 EGTA; and 0.561 Ca-methanesulfonate.

The osmolality of the internal solutions was 1000 ± 10 mOsm/KgH_2O (adjusted with sucrose), and the pH was 7.3.

Various $[Ca^{2+}]_i$ was obtained by using either a Ca-EGTA or Ca-HEEDTA buffer systems. 8 mM EGTA was used to buffer $[Ca^{2+}]_i$ at 1×10^{-8} M (using 0.56 mM Ca^{2+}) or 5×10^{-6} (with 7.8 mM Ca^{2+}); 10 mM HEEDTA was used to buffer $[Ca^{2+}]_i$ at 2×10^{-5} M (using 8 mM Ca^{2+}). EQCAL software (Biosoft; Cambridge, UK) was utilized to calculate the values of $[Ca^{2+}]_i$. All the pertinent stability constants for the combinations of H^+, Ca^{2+}, Mg^{2+} with EGTA, HEEDTA, aspartate, ATP and methanesulfonate were obtained from the literature.[22,24] Given the fact that methanesulfonate binds Ca^{2+} with high affinity, $[Ca^{2+}]$ was verified and adjusted by adding $CaCl_2$ as needed, using a Ca^{2+}-selective electrode (Quickcal; WPI, Sarasota, FL). Temperature of all experiments was 16°C.

α-Chymotrypsin Treatment

In some experiments the Na-Ca exchanger was activated by exposing the cell to a mild intracellular treatment (perfusion) with the protease α-chymotrypsin (Type I-S from bovine pancreas, Sigma, St. Louis, MO). The methodology to attain this activation has been described.[8]

Statistics

Statistical analysis was performed on a PC using the SigmaStat program (Jandel Scientific; San Rafael, CA). Comparisons of values were made using the Student t analysis.

RESULTS

Na-Na Exchange Is Activated by Increasing $[Ca^{2+}]_i$

FIGURE 1 shows a representative experiment from a series of four where the effect of Na_o removal was assessed in cells perfused with either low (10^{-8} M) or high (2×10^{-5} M) $[Ca^{2+}]_i$ in the presence of ouabain, bumetanide, DIDS and verapamil. The initial internal solution used was the NaCl perfusate containing 10^{-8} M $[Ca^{2+}]$ and the external solution was the NaSW0Ca containing DIDS, verapamil and bumetanide. The figure shows that addition of ouabain (at **a**) induced a reduction in Na^+ efflux from 45 to ~30 pmoles · cm^{-2} · sec^{-1}. Under these conditions, in the presence of low $[Ca^{2+}]_i$, replacement of Na_o by Tris (TrisSW0Ca) (from **b** to **c**) produced no effect on Na^{2+} efflux. At **d**, $[Ca^{2+}]_i$ was raised to 2×10^{-5} M. This manipulation produced a large increase in Na^+ efflux (to ~120 pmoles · cm^{-2} · sec^{-1} in 40 min). This increase was due to activation of Ca_i-activated nonselective cation channels[25] and Na-Na exchange. Evidence for activation of this latter mechanism is provided by the fact that when Na_o was removed (from **e** to **f**), Na^+ efflux was reversibly reduced by ~90 pmoles · cm^{-2} · sec^{-1}. This demonstrates that in barnacle muscle cells, as in squid axons,[4] an increase in $[Ca^{2+}]_i$ is necessary to activate Na-Na exchange.

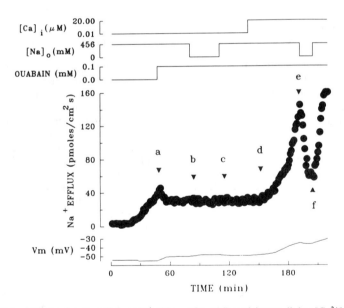

FIGURE 1. Effect of extracellular Na^+ (Na_o, 456 mM) and intracellular $[Ca^{2+}]$ on Na^+ efflux in an internally perfused barnacle muscle cell. The initial internal solution was the NaCl perfusate containing 10^{-8} M $[Ca^{2+}]$, at **d** $[Ca^{2+}]_i$ was raised to 2×10^{-5} M. The external solutions used were the NaSW0Ca and TrisSW0Ca containing DIDS, verapamil and bumetanide. At **a**, ouabain was added to the external solution. See text for further details.

Activation of Na-Na Exchange by External Na^+

To further characterize the Ca_i-activated Na_o-dependent Na^+ efflux, the magnitude of Na-Na exchange was measured at various Na_o concentrations. FIGURE 2A shows data from a representative experiment from a series of four in which step changes in Na_o concentrations were assessed on the efflux of Na^+. To prevent large activation of Na^+ efflux *via* nonselective cation channels, the $[Ca^{2+}]_i$ used was 5×10^{-6} M; the external solution was the NaSW0Ca solution containing standard inhibitors. The figure shows that step reductions in Na_o concentration produced reversible decreases in Na^+ efflux. FIGURE 2B is a summary of four experiments similar to the one shown in FIGURE 2A. The figure depicts the Na_o-dependent Na^+ efflux as a function of various external Na^+ concentrations. The continuous line represents the activation kinetics of the Na-Na exchange assuming that 3 Na_o activate this exchange (*i.e.*, Hill coefficient = 3). The calculated kinetic parameters from the experimental data are $K_{Nao} = 250 \pm 11$ mM, $J_{Na(Max)} = 17.67 \pm 0.8$ pmoles $\cdot$ cm^{-2} $\cdot$ sec^{-1}, and Hill coefficient = 2.8 ± 0.23. These data suggest that 3 Na_o activate Na^+ efflux.

Na-Na Exchange Is Activated by α-Chymotrypsin in the Presence of Low $[Ca^{2+}]_i$

In barnacle muscle cells, activation of the Na-Ca exchanger is complicated by the fact that the increase in $[Ca^{2+}]_i$, besides activating the exchanger, also opens nonselective cation channels. Activation of these channels increases the membrane conductance and makes it difficult to voltage-clamp the cells and study the voltage dependence of the Na-Ca exchanger. Alternatively, a mild treatment with α-chymotrypsin renders a deregulated, fully activated exchanger[6-8] under conditions of low $[Ca^{2+}]_i$ and therefore, in the absence of activation of nonselective cation channels. FIGURE 3 shows a representative example from a series of four similar experiments where the effect of treatment with α-chymotrypsin on Na-Na exchange was assessed. Cells were perfused with the NaCl perfusate containing 10^{-8} M $[Ca^{2+}]$. The external solution was the NaSW0Ca containing the standard blockers. Removal of Na_o before addition of α-chymotrypsin (from **a** and **b**) produced no effect on Na^+ efflux. This indicates that the exchanger was not activated and was unable to engage in Na-Na exchange. However, after exposure to the protease (from **c** to **d**), in the presence of Na_o there was a large increase (~125 pmoles $\cdot$ cm^{-2} $\cdot$ sec^{-1}) in the Na^+ efflux (from **d** to **e**). That this increase was due to activation of Na-Na exchange was demonstrated by the fact that Na_o replacement by Tris (from **e** to **f**) produced a reduction in Na^+ efflux to basal levels. Readmission of normal Na_o (at **f**) induced a larger increase in Na^+ efflux. The greater magnitude of this flux is attributed to the fact that maximal activation of the exchanger by α-chymotrypsin takes 90–120 min after the beginning of the perfusion with the protease.[8] In any event, at this point in the experiment, in the presence of Na_o, addition of Ca_o (from 0 to 36 mM, at **g**), produced a partial and slow inhibition of Na^+ efflux (by 50 pmoles $\cdot$ cm^{-2} $\cdot$ sec^{-1} in 50 min). Simultaneous removal of Na_o and Ca_o (at **h**) produced a fast reduction of Na^+ efflux to about the baseline that slowly increased with time during the experiment. The observation that addition of Ca_o inhibited Na^+ efflux in the presence of Na_o is explained by the fact that Ca_o has a higher affinity than

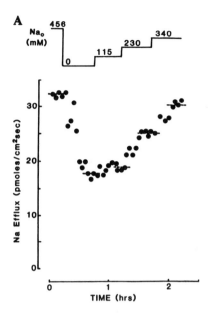
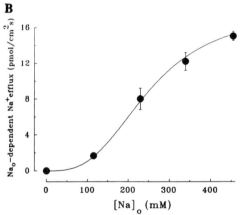

FIGURE 2. Effect of various concentrations of Na_o on Na^+ efflux on barnacle muscle cells perfused with the NaCl perfusate containing 5×10^{-6} M $[Ca^{2+}]_i$ and superfused with either NaSW0Ca or TrisSW0Ca. In this and all subsequent experiments all the external solutions contained the standard inhibitors. **(A)** Example of a cell where Na^+ efflux was measured in response to step changes in the concentration of Na_o from 456, to 0, 115, 230 and 340 mM. **(B)** Summary of four experiments where the effect of various concentrations of Na_o was measured on Na^+ efflux. See text for further details.

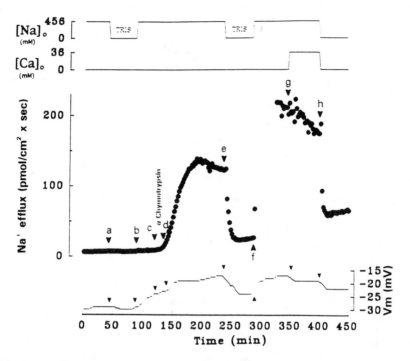

FIGURE 3. Effect of α-chymotrypsin and extracellular Na$^+$ or Na$^+$ efflux and on membrane potential (mV) in an internally perfused barnacle muscle cell. The internal solution was the Na-methanesulfonate perfusate containing 10^{-8} [Ca^{2+}]; the superfusion solution was either the Na-methanesulfonate 0Ca or Tris-methanesulfonate 0Ca except for a portion of the experiment (from **g** to **h**) when the Na-methanesulfonate solution containing 36 mM Ca^{2+} was used. See text for further details. (From Espinosa-Tanguma et al.[8] Reprinted by permission from the *American Journal of Physiology, Cell Physiology*.)

Na$_o$ for the extracellular binding site of the Na-Ca exchanger and because Na-Na exchange operates at a faster rate than Ca$_o$-Na$_i$ exchange.[8]

Voltage Dependence of Na-Na Exchange Mediated by the Na-Ca Exchanger Activated by α-Chymotrypsin Treatment

Two experimental strategies were followed to study the effect of V_M on Na-Na exchange mediated by the Na-Ca exchanger in barnacle muscle cells. The first one consisted of assessing the effect of changes in V_M on Na$^+$ efflux both in the absence and presence of Na$_o$. The second strategy consisted of comparing the magnitude of the Na$_o$-dependent Na$^+$ efflux at various levels of V_M. In addition, both types of experiments were carried out in cells whose Na-Ca exchanger was activated by exposing the cells to α-chymotrypsin. FIGURE 4 illustrates a representative example (from a series of 8) where the effect of V_M was studied on Na-Na exchange in cells perfused with the Na$^+$ efflux solution containing 1 ×

10^{-8} M $[Ca^{2+}]_i$ and exposed to α-chymotrypsin 90 min before starting the collection of the superfusate (time 0 in FIG. 4). FIGURE 4 shows that, in the absence of Na_o (Tris-methanesulfonate $0Ca^{2+}$ solution), successive voltage steps of 20 mV from −40 to +20 mV (at **a**, **b**, and **c**) produced no appreciable change in Na^+ efflux at the scale presented. In fact, a hyperpolarization of 40 mV (from 0 to −40 mV) produced a small voltage-dependent passive (electrophoretic) Na^+ efflux of ~15 pmoles · cm^{-2} · sec^{-1} (data not shown). Simultaneous addition of 456 mM Na_o (Na-methanesulfonate $0Ca^{2+}$ solution) and hyperpolarization to 0 mV, at **d**, produced a large and transient increase in Na^+ efflux which reached a steady value at ~490 pmoles · cm^{-2} · sec^{-1}. The nature of this transient increase in Na^+ efflux before reaching the steady flux is explained below (see FIG. 5). In any event, once the flux reached a steady value in the presence of Na_o, hyperpolarization to −40 mV (at **e**) produced a small reduction in Na^+ efflux of ~40 pmoles · cm^{-2} · sec^{-1}. At this stage, three subsequent depolarization steps of 20 mV (at **f**, **g** and **h**) induced increases in Na^+ efflux of ~30, 10 and 0 pmoles · cm^{-2} · sec^{-1}, respectively.

The example presented in FIGURE 4 suggest that Na-Na exchange activated by α-chymotrypsin is only minimally responsive to changes in V_M (10% change in Na^+ efflux for a change in 40 mV). To further investigate the voltage dependence of this mode of exchange, a comparison was made of the magnitude of the Na_o-

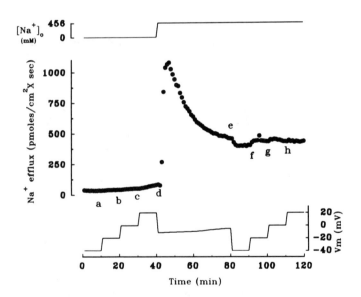

FIGURE 4. Effect of extracellular Na^+ (Na_o, 456 mM) and membrane potential (V_M) on Na^+ efflux in an internally perfused, voltage-clamped barnacle muscle cell. The perfusion fluid was the Na-methanesulfonate perfusate containing 10^{-8} M $[Ca^{2+}]$. The cell was pretreated with α-chymotrypsin 90 min before starting the collection of the superfusate. Initially, the external solution was the tris-methanesulfonate $0Ca^{2+}$. In the absence of Na_o, from **a** to **c** V_M was depolarized from −40 to +20 mV in steps of 20 mV each. At **d** the external solution was replaced for Na-methanesulfonate $0Ca^{2+}$ and V_M was hyperpolarized to 0 mV. In the presence of Na_o, at **e**, V_M was hyperpolarized to −40 mV. From **f** to **h**, V_M was depolarized from −40 to +20 mV in steps of 20 mV each. See text for further details.

dependent Na^+ efflux at -40 and 0 mV. FIGURE 5 illustrates an example of this kind of measurement. FIGURE 5A shows the effect of transiently (40 min) adding Na_o (from **a** to **b**) on Na^+ efflux while the V_M was maintained constant at -40 mV. FIGURE 5B shows a similar experiment but in this instance V_M was held at 0 mV. The Na^+ efflux stimulated by Na_o once the flux reached a steady value at -40 and 0 mV was 150 and 120 pmoles/cm^2 sec, respectively. The average of 13 independent measurements (26 cells) of Na_o-dependent Na^+ efflux at 0 and -40 mV was 204 ± 27 and 178 ± 43 pmoles $\cdot$ cm$^{-2} \cdot$ sec^{-1}, respectively, and is shown in FIGURE 6. The figure demonstrates that Na-Na exchange was not significantly different at this range of voltages.

The simplest interpretation of the transient increase in Na^+ efflux observed in response to the addition of Na_o (FIGS. 4 and 5), is that, as ^{22}Na exits the cell, a portion remains trapped (loosely bound) at the sarcolemmal surface and invaginations. Therefore, addition of a high concentration of nonradioactive Na_o promotes the release of bound ^{22}Na leading to a large, transient release of bound ^{22}Na giving the appearance of an increase in Na^+ efflux. A similar observation has been described for nonradioactive Ca_o-induced release of plasmalemmal-bound ^{45}Ca in squid giant axons.[26] To test this hypothesis, the effect of an increase in Na_o to a lower value (*i.e.*, 30 mM) than the one previously used (*i.e.*, 456 mM) was tested on Na^+ efflux. If the hypothesis is correct, the expected result is that under this condition the transient Na^+ efflux should be greatly diminished. In addition, this kind of experiment offers the opportunity to assess the presence of an extracellular access channel for Na^+ in the exchanger[13] by testing for the voltage dependence of Na-Na exchange at this lower Na_o concentration (see Discussion). FIGURE 7 shows a representative (from a series of 4) experiment of this kind. The figure shows that after having exposed the cell to a mild treatment with α-chymotrypsin, as expected, addition of 30 mM Na_o (from **a** to **b** and from **d** to **e**) produced an increase in Na^+ efflux, but in this instance the flux reached a steady value without going through the large apparent transient increase in Na^+ efflux observed previously (FIGS. 4 and 5). In addition, FIGURE 7 shows that the Na_o-dependnet Na^+ efflux is similar in magnitude ($\sim$35 pmoles $\cdot$ cm$^{-2} \cdot$ sec^{-1}) at 0 mV (from **a** to **b**) and -40 mV (from **d** to **e**). Finally, the figure illustrates that in the absence of Ca_o and Na_o, a hyperpolarization from 0 to -40 mV (at **c**) produced a small decrease in Na^+ efflux of $\sim$15 pmoles $\cdot$ cm$^{-2} \cdot$ sec^{-1} indicating the presence of a small electrophoretic voltage-dependent Na^+ efflux.

DISCUSSION

Reliability of the Na-Na Exchange Measurements Activated by either $[Ca^{2+}]_i$ or α-Chymotrypsin

Internally perfused, voltage-clamped barnacle muscle cells offer the advantage of measuring Na-Na exchange by determining unidirectional efflux of radiolabelled Na^+ by the presence of Na_o. These fluxes cannot be attributed to changes in V_M associated with the changes in the extracellular concentration of Na^+, because in these experiments, this parameter was controlled. Likewise, they are not due to the addition or removal of the extracellular ions used to replace Na_o (*i.e.*, Tris$^+$) since these ions do not affect the efflux of Na_+. Therefore, they represent genuine Na-Na exchange. However, identification of these fluxes as modes of exchange mediated by the Na-Ca exchanger demands justification. In barnacle muscle cells, unidirectional Na^+ efflux can be mediated by the Na-Ca exchanger, the Na/K

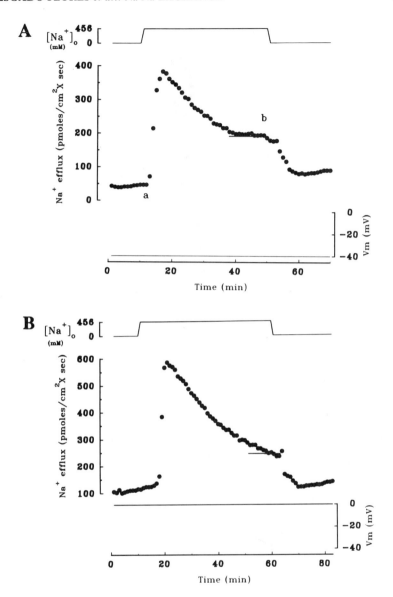

FIGURE 5. Effect of addition of Na_o (456 mM, for **a** to **b**) on Na^+ efflux in two cells clamped at different V_M. The cells were perfused with the Na^+ efflux solution containing 10^{-8} M $[Ca^{2+}]$ and were pretreated with α-chymotrypsin 90 min before starting the collection of the superfusate. **(A)** Na_o-dependent Na^+ efflux in a cell whose V_M was clamped at -40 mV. **(B)** Na-Na exchange measured at 0 mV. The values of Na^+ efflux that were considered for analysis are the ones which reached a steady value (*horizontal bars* in the figures). See text for details.

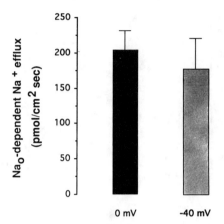

FIGURE 6. Comparison of the Na_o-dependent Na^+ efflux at 0 and -40 mV in cells perfused with the Na-methanesulfonate solution containing 10^{-8} M $[Ca^{2+}]$ and pretreated with α-chymotrypsin. The values of Na_o-stimulated Na^+ efflux used for the comparison were the steady values reached after the transient increase in efflux following the addition of Na_o. The *dark- and light-shaded bars* represent the average Na-Na exchange of 13 cells clamped at 0 and -40 mV, respectively. See text for details.

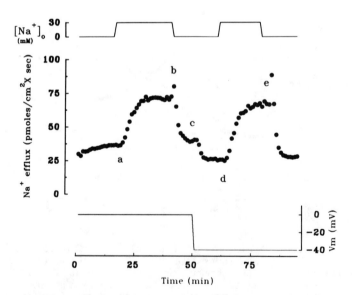

FIGURE 7. Effect of Na_o (30 mM) and V_M on the efflux of Na^+ in a voltage-clamped cell perfused with the Na-methanesulfonate solution containing 10^{-8} M $[Ca^{2+}]$ and pretreated with α-chymotrypsin. The Tris-methanesulfonate $0Ca^{2+}$ external solution was replaced by a Na-methanesulfonate $0Ca^{2+}$ solution containing 30 mM Na^+ from **a** to **b** and from **d** to **e**. V_M was hyperpolarized from 0 to -40 mV at **c**. See text for details.

pump, or the Na-K-Cl cotransporter or *via* voltage-gated Ca^{2+} channels. Three main evidences indicate that in these experiments Na-Na exchange is mediated by the Na-Ca exchanger: i) all the other pathways of Na^+ efflux are inhibited by presence of ouabain, bumetanide, DIDS and verapamil; and ii) all the factors that activate (*e.g.*, absolute requirement of Ca_i, ATP, α-chymotrypsin) or inhibit (*e.g.*, intracellular acidification) the Na-Ca exchanger similarly affect Na-Na exchange; and iii) Ca_o inhibits Na-Na exchange as it promotes the slower Ca_o-Na_i exchange.[8]

Controversy Regarding the Voltage Dependence of Electroneutral Na-Na Exchange

Our results agree with those of DiPolo and Beaugé[14] indicating that, at the range of V_M tested (*i.e.*, -60 to $+20$ mV) and at the two Na_o concentrations tested (456 and 30 mM), Na-Na exchange is voltage insensitive. Both studies were carried out measuring tracer fluxes. These results, however, are in contradiction with those carried out using patch-clamp techniques supporting voltage sensitivity for Na-Na exchange.[16] A plausible explanation for these contradictory results may lie in the use of different techniques to carry out the studies. At present, tracer flux measurement is the most reliable technique to study electroneutral, voltage-insensitive exchanges.

Activation of the Exchanger by α-Chymotrypsin

The mechanism by which a mild treatment with α-chymotrypsin renders a highly activated Na-Ca exchanger is unknown. Given the fact that the intracellular hydrophilic loop between membrane-spanning segments 5 and 6 of the exchanger possesses the regulatory sites for Ca^{2+} and ATP,[27] it is speculated this site could be the target for the protease,[28] but this remains to be demonstrated. In any event, it is clear that α-chymotrypsin activates all the modes of exchange mediated by the Na-Ca exchanger.[8] However, it is still unknown whether treatment with this protease modifies the voltage dependence of the various modes of exchange mediated by the Na-Ca exchanger.

In conclusion, in barnacle muscle cells, Na-Na exchange can be activated by either an increase in $[Ca^{2+}]_i$ or a mild treatment with α-chymotrypsin in the presence of low $[Ca^{2+}]_i$. The activation curve of Na-Na exchange by Na_o indicates that 3 external Na^+ activate Na-Na exchange. Finally, Na-Na exchange in the deregulated exchanger showed insensitivity to changes in V_M from -60 to $+20$ mV. Furthermore, reduction of the Na_o concentration (from 456 to 30 mM) to values below the K_m for Na_o (*i.e.*, 250 mM for barnacle muscle cells, see FIG. 2B) did not modify the voltage insensitivity of the exchanger, suggesting the absence of a narrow access channel for Na_o at the exchanger. These results are consistent with the voltage insensitivity for this mode of exchange reported for squid axons[14] where the exchanger was activated with $[Ca^{2+}]_i$. However, it is still unknown whether Na-Na exchange in barnacle muscle cells is voltage insensitive when the exchanger is activated with $[Ca^{2+}]_i$. Likewise, the voltage dependence of Ca-Ca exchange still needs to be assessed in these cells. Such studies are currently being carried out.

ACKNOWLEDGMENTS

We wish to thank Dr. V. A. Kimler for helpful comments on the manuscript and Mrs. C. Peña-Rasgado for help in the preparation of some of the figures.

REFERENCES

1. BLAUSTEIN, M. P. 1977. Biophys. J. **20:** 79–111.
2. BLAUSTEIN, M. P. & J. M. RUSSELL. 1975. J. Membr. Biol. **22:** 285–312.
3. DIPOLO, R., L. BEAUGÉ & H. ROJAS. 1989. Biochim. Biophys. Acta **978:** 328–332.
4. DIPOLO, R. & L. BEAUGÉ. 1987. J. Gen. Physiol. **90:** 505–525.
5. RASGADO-FLORES, H., E. M. SANTIAGO & M. P. BLAUSTEIN. 1989. J. Gen. Physiol. **93:** 1219–1241.
6. PHILIPSON, K. D. & A. Y. NISHIMOTO. 1982. Am. J. Physiol. Cell Physiol. **243:** C191–C195.
7. HILGEMANN, D. W. 1990. Nature **344:** 242–245.
8. ESPINOSA-TANGUMA, R., J. DESANTIAGO & H. RASGADO-FLORES. 1993. Am. J. Physiol. **265** (Cell Physiol. **34**): C1128–C1137.
9. RASGADO-FLORES, H. & M. P. BLAUSTEIN. 1987. Am. J. Physiol. **252** (Cell Physiol. **21**): C499–C504.
10. MULLINS, L. J. 1984. *In* Electrogenic Transport: Fundamental. M. P. Blaustein & M. Lieberman, Eds. 161–179. Raven Press. New York.
11. ALLEN, T. J. A. & P. F. BAKER. 1986. J. Physiol. (London) **378:** 77–96.
12. LAUGER, P. 1991. Electrogenic Ion Pumps. Sinauer. Sunderland, MA.
13. GADSBY, D. C., R. F. RAKOWSKI & P. DEWEER. 1993. Science **260:** 100–103.
14. DIPOLO, R. & L. BEAUGÉ. 1990. J. Gen Physiol. **95:** 819–835.
15. NIGGLI, E. & W. J. LEDERER. 1991. Nature **349:** 621–624.
16. HILGEMANN, D. W., D. A. NICOLL & K. D. PHILIPSON. 1991. Nature **352:** 715–718.
17. RASGADO-FLORES, H., J. DESANTIAGO & R. ESPINOSA-TANGUMA. 1991. Ann. N. Y. Acad. Sci. **639:** 22–33.
18. PEÑA-RASGADO, C., K. D. MCGRUDER, J. C. SUMMERS & H. RASGADO-FLORES. 1994. Am. J. Physiol. **267** (Cell Physiol. **36**): C768–C775.
19. NELSON, M. T. & M. P. BLAUSTEIN. 1980. J. Gen. Physiol. **75:** 183–206.
20. ALTAMIRANO, A. A. & J. M. RUSSELL. 1987. J. Gen. Physiol. **89:** 669–686.
21. RUSSELL, J. M., W. F. BORON & M. S. BRODWICK. 1983. J. Gen. Physiol. **82:** 47–78.
22. SILLEN, L. G. & A. E. MARTELL. 1964. Stability Constants of Metal-Ion Complexes. Burlington House. London.
23. BLINKS, J. R., W. G. WIER, P. HESS & F. G. PRENDERGAST. 1982. Prog. Biophys. Mol. Biol. **40:** 1–114.
24. DIPOLO, R., J. REQUENA, F. J. BRINLEY, JR., L. J. MULLINS, A. SCARPA & T. TIFFERT. 1976. J. Gen. Physiol. **67:** 433–467.
25. SHEU, S-S. & M. P. BLAUSTEIN. 1983. Am. J. Physiol. **244** (Cell Physiol. **13**), C297–C302.
26. BAKER, P. F. & P. A. MCNAUGHTON. 1990. J. Physiol. (London) **276:** 127–150.
27. PHILIPSON, K. D. & D. A. NICOLL. 1992. Curr. Opin. Cell Biol. **4:** 678–683.
28. HILGEMANN, D. W., A. COLLINS, D. P. CASH & G. A. NAGEL. 1991. Ann. N. Y. Acad. Sci. **639:** 126–139.

Phosphorylation and Modulation of the Na^+-Ca^{2+} Exchanger in Vascular Smooth Muscle Cells[a]

MUNEKAZU SHIGEKAWA, TAKAHIRO IWAMOTO, AND SHIGEO WAKABAYASHI

Department of Molecular Physiology
National Cardiovascular Center Research Institute
5-7-1 Fujishiro-dai
Suita, Osaka 565, Japan

INTRODUCTION

The Na^+-Ca^{2+} exchanger of the plasma membrane is primarily responsible for the removal of excess Ca^{2+} from many mammalian cells during agonist or electrical stimulation. The regulation of the Na^+-Ca^{2+} exchanger (NCX1) has most extensively been studied in cardiac myocytes and squid axons.[1] The cardiac exchanger activity, which has been characterized using giant excised membrane patches from cardiomyocytes or *Xenopus laevis* oocytes expressing the cardiac exchanger cRNA, is modulated positively by cytoplasmic Ca^{2+} and adenosine triphosphate (ATP) and negatively by cytoplasmic Na^+ and exchanger inhibitory peptide (XIP).[2,3] For this cardiac Na^+-Ca^{2+} exchanger, there is currently no strong evidence that protein phosphorylation is involved in its regulation. In squid giant axons, however, there is much evidence suggesting that protein kinase-dependent phosphorylation is responsible for stimulation of the exchanger activity induced by MgATP.[4] On the other hand, the Na^+-Ca^{2+} exchanger in vascular smooth muscle cells (VSMCs) was reported to be stimulated by phorbol esters,[5] 8-bromocyclic guanosine monophosphate (cGMP),[6] norepinephrine and high K^+,[7] and platelet-derived growth factor (PDGF)-BB,[8] suggesting the role of protein phosphorylation in the exchanger activation in this tissue. In the present study, we provide direct evidence that the Na^+-Ca^{2+} exchanger in rat aortic VSMCs is phosphorylated and concomitantly activated in response to growth factors.

METHODS

VSMCs were isolated from the thoracic aorta of male Wistar rats (200–300 g) by enzymatic dispersion.[6] After reaching confluency, cells were cultured in serum-free medium for an additional 24–48 h to enhance redifferentiation. Growth factor-induced phosphorylation of the Na^+-Ca^{2+} exchanger was investigated by immuno-

---

[a] This work was supported by a Grant-in-Aid for Scientific Research on Priority Areas (321) from the Ministry of Education, Science and Culture of Japan and by Special Coordination Funds Promoting Science and Technology (Encouragement System of COE) from the Science and Technology Agency of Japan.

precipitation from 1% octaethylene glycol mono-n-dodecyl ether ($C_{12}E_8$)-solubilized VSMCs metabolically labeled with [^{32}P]orthophosphate using a polyclonal antibody raised against a maltose binding protein fusion protein containing the cytoplasmic domain (amino acids 273–769) of the dog cardiac Na^+-Ca^{2+} exchanger.[9] Tryptic phosphopeptide mapping and phosphoamino acid analysis were carried out as described previously.[9] Measurements of intracellular Ca^{2+} concentration ($[Ca^{2+}]_i$), cell Na^+, cell 1,2-diacylglycerol, and Na^+-dependent $^{45}Ca^{2+}$ uptake were carried out also as described.[9] Experimental results are expressed as means ± SD. Significant differences were assessed with a one-way analysis of variance followed by Dunnett's test. A p value of <0.05 was considered significant.

RESULTS AND DISCUSSION

Phosphorylation of the Na^+-Ca^{2+} Exchanger by PDGF-BB and Other Agents

We investigated growth factor-induced phosphorylation of the Na^+-Ca^{2+} exchanger by immunoprecipitation from serum-depleted rat aortic VSMCs metabolically labeled with [^{32}P]orthophosphate. A polyclonal antibody against the Na^+-Ca^{2+} exchanger recognized a relatively broad band covering the expected molecular weight for the Na^+-Ca^{2+} exchanger (~125 kDa), when the immunoprecipitated material was analyzed by sodium dodecyl sulfate polyacrylamide gel electrophoresis (SDS-PAGE) and protein staining (FIG. 1A). This protein band was phosphorylated in unstimulated cells (FIG. 1B). When these cells were stimulated with 10 or 20 ng/ml PDGF-BB for 10 min, the extent of phosphorylation increased to 146 ± 9% or 166 ± 15% (n = 3) of that for the untreated cells (FIG. 1B). Treatment of cells with 10 nM α-thrombin or 100 nM phorbol 12-myristate 13-acetate (PMA) for 10 min also increased the phosphorylation to 140 ± 9% or 145 ± 11% (n = 3), respectively (FIG. 1B and 1C). However, 100 nM angiotensin II did not increase the phosphorylation significantly (118 ± 6%, n = 3) under the equivalent experimental conditions (FIG. 1C). Thus PDGF-BB, α-thrombin, and PMA induced Na^+-Ca^{2+} exchanger phosphorylation in serum-depleted aortic VSMCs. We found that removal of extracellular Ca^{2+} decreased the basal and PDGF-BB-induced exchanger phosphorylation slightly; in 4 mM ethylene glycol-bis-(β-aminoethyl ether)-N,N,N',N',-tetraacetic acid (EGTA), the basal phosphorylation was 83% (n = 2) of the control value, whereas PDGF-BB (10 ng/ml, 10 min) increased the exchanger phosphorylation to 132% (n = 2) under the same conditions.

FIGURE 2 shows a time course of PDGF-BB-induced phosphorylation of the Na^+-Ca^{2+} exchanger as analyzed densitometrically, as well as those of PDGF-BB-induced changes in $[Ca^{2+}]_i$ and a 1,2-diacylglycerol level. The PDGF-BB-induced phosphorylation increased with time, reaching 169 ± 17% (n = 3) of the control level at 20 min. Under the same conditions, $[Ca^{2+}]_i$ reached a peak (450 ± 40 nM, n = 4) within 2 min followed by a slow decline. The formation of 1,2-diacylglycerol, on the other hand, increased from a basal level of 91 ± 8 pmol/10^6 cells (n = 4) to a high level of 191 ± 20 pmol/10^6 cells at 20 min. In contrast to PDGF-BB, 100 nM angiotensin II, which did not enhance the exchanger phosphorylation significantly (FIG. 1C), increased the formation of 1,2-diacylglycerol to a lower level (126 ± 16 pmol/10^6 cells (n = 4) at 10 min), as compared with PDGF-BB (189 ± 30 pmol/10^6 cells (n = 4) at 10 min).

We performed two-dimensional tryptic phosphopeptide mapping to characterize sites for the basal and stimuli-enhanced phosphorylation in the Na^+-Ca^{2+}

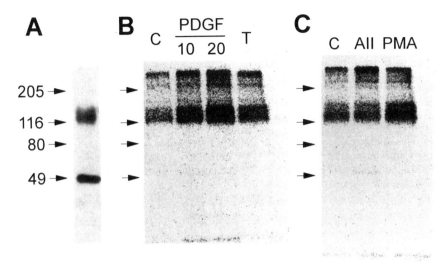

FIGURE 1. Immunological detection of phosphorylated Na^+-Ca^{2+} exchanger in unstimulated and growth factor-stimulated VSMCs. Rat aortic VSMCs labeled with (panels B and C) or without (panel A) [^{32}P]orthophosphate (10 MBq/ml) were treated with growth factors for 10 min and then solubilized with 1% $C_{12}E_8$ lysis buffer. Cell lysates were incubated with a polyclonal antibody against the dog cardiac Na^+-Ca^{2+} exchanger (1/100 dilution) and the resultant immunoprecipitates were subjected to SDS-PAGE. In panel A, a nitrocellulose transfer of gel was probed with the antibody, which visualized the Na^+-Ca^{2+} exchanger (*upper band*) and immunoglobulin derived from the immunoprecipitate (*lower band*). In panels B and C, phosphorylated proteins in the gels were visualized by Bioimage analyzer. In each panel, cells were stimulated with no agent (C), 10 or 20 ng/ml PDGF-BB (PDGF 10 and 20), 10 nM α-thrombin (T), 100 nM angiotensin II (AII), or 100 nM PMA, respectively. Molecular mass markers (in kDa) are shown on the *left*. (From Iwamoto et al.[9] Reprinted by permission from the *Journal of Biological Chemistry*.)

exchanger (FIG. 3). Tryptic digestion of the ^{32}P-labeled Na^+-Ca^{2+} exchanger in unstimulated cells generated four major phosphopeptides (P1, P2, P3, and P4). Occasionally one minor spot (P5), which may represent a partially digested phosphopeptide, was observed. Stimulation of cells with PDGF-BB (20 ng/ml) and PMA (100 nM) for 20 min resulted in a significant increase in the amounts of ^{32}P-label incorporated into P1 and, to a lesser extent, into P2. In unstimulated cells, P1 was weakly phosphorylated, representing less than 15% of the total ^{32}P-label incorporated. In contrast, it accounted for 40–50% of the total ^{32}P-label in PDGF-BB- and PMA-stimulated cells. These results indicate that phosphorylation of the Na^+-Ca^{2+} exchanger occurs at multiple sites and that PDGF-BB and PMA enhance phosphorylation of the same phosphopeptides. By phosphoamino acid analysis, we found that the Na^+-Ca^{2+} exchanger phosphorylation in the basal, and PDGF-BB- and PMA-stimulated cells occurred exclusively on serine residues, not on threonine or tyrosine residues.

These results indicate the involvement of serine kinase(s) in the Na^+-Ca^{2+} exchanger phosphorylation. Since the phosphopeptide maps were essentially the same for PDGF-BB- or PMA-treated cells, we conclude that the aortic Na^+-Ca^{2+} exchanger is phosphorylated in response to PDGF-BB or α-thrombin via a protein kinase C-dependnet pathway. We do not know, however, whether protein kinase C directly phosphorylates the exchanger or whether it phosphorylates and activates a secondary serine/threonine kinase. The involvement of protein kinase C is consistent with the well-known facts that signaling by PDGF-BB and α-thrombin occurs through activation of phospholipase C-γl and phospholipase C-β1, respectively, that produce inositol 1,4,5-triphosphate and 1,2-diacylglycerol from phosphatidylinositol 4,5-bisphosphate.[10] Indeed, we observed elevation of 1,2-diacylglycerol in parallel to the exchanger phosphorylation in PDGF-BB-stimulated cells (FIG. 2). At present, however, we cannot rule out the possibility that other protein kinases such as Ca^{2+}/calmodulin kinase II were also involved in the exchanger phosphorylation, because PDGF-BB induced a prolonged increase in $[Ca^{2+}]_i$ (FIG. 2) and because increased $[Ca^{2+}]_i$ seemed to be able to enhance the exchanger phosphorylation (see above).

Effect of PDGF-BB and Other Agents on Na^+-Ca^{2+} Exchanger Activity

We determined the effect of treatment with PDGF-BB (10 or 20 ng/ml), α-thrombin (10 nM), angiotensin II (100 nM), or PMA (100 nM) for 10 min on Na^+_i-dependent $^{45}Ca^{2+}$ uptake (reverse mode of the Na^+-Ca^{2+} exchanger) in Na^+-loaded rat aortic VSMCs (FIG. 4). Because these agents except for PMA increased $[Ca^{2+}]_i$ from basal levels of 100–150 nM to peak levels of 400–550 nM and because Na^+-Ca^{2+} exchange depends on $[Ca^{2+}]_i$, cells had been exposed to a Ca^{2+}-free solution containing these agents during the last 3 min of the 10-min treatment in order to reduce $[Ca^{2+}]_i$ to a basal level. Treatment of cells with PDGF-BB or α-thrombin caused a significant increase (120–130% of the control value) in Na^+_i-

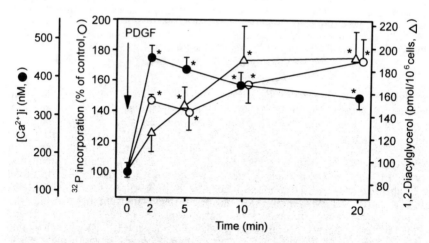

FIGURE 2. Time courses of PDGF-BB-induced changes in Na^+-Ca^{2+} exchanger phosphorylation, $[Ca^{2+}]_i$, and 1,2-diacylglycerol level. In cells stimulated with 10 ng/ml PDGF-BB for indicated periods of time, changes in the Na^+-Ca^{2+} exchanger phosphorylation, $[Ca^{2+}]_i$, and the 1,2-diacylglycerol content were determined. ^{32}P incorporation into the Na^+-Ca^{2+} exchanger was quantified by densitometry. Results are presented as the mean ± SD (n = 3 or 4). * Significant difference from cells at 0 min. (From Iwamoto *et al.*[9] Reprinted by permission from the *Journal of Biological Chemistry*.)

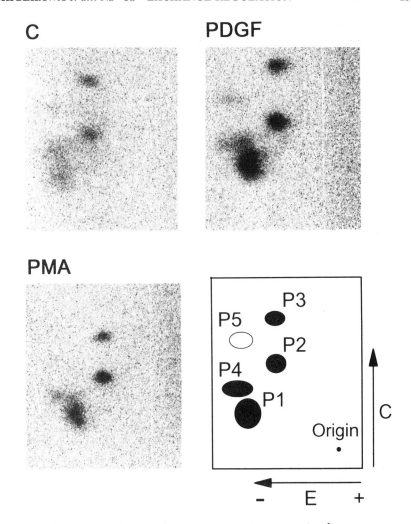

FIGURE 3. Two-dimensional tryptic phosphopeptide maps of Na^+/Ca^{2+} exchanger. Cells labeled with [^{32}P]orthophosphate were stimulated with no agent (C), PDGF-BB (20 ng/ml), or PMA (100 nM) for 20 min. The phosphorylated Na^+-Ca^{2+} exchanger isolated by immunoprecipitation and subsequent SDS-PAGE was digested with tosylamide-2-phenylethylchloromethyl ketone-treated trypsin and subjected to two-dimensional phosphopeptide mapping.[9] The *right bottom panel* schematically represents the location of phosphopeptides and direction of peptide migration (*arrow*). (From Iwamoto *et al.*[9] Reprinted by permission from the *Journal of Biological Chemistry*.)

dependent $^{45}Ca^{2+}$ uptake. Cell Na^+ loading itself was not affected by prior treatment with PDGF-BB. The extent of such PDGF-BB-induced activation was similar, when cells had been loaded with different levels of Na^+ (30 to 146 mM).

We also examined whether PDGF-BB was able to enhance Na^+_o-dependent decline in $[Ca^{2+}]_i$ via the forward mode of Na^+-Ca^{2+} exchanger in aortic VSMCs

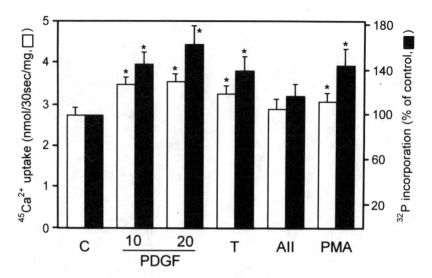

FIGURE 4. Correlation between phosphorylation and activity of Na$^+$-Ca^{2+} exchanger. Na^+_i-dependent ^{45}Ca^{2+} uptake and ^{32}P incorporation into the immunoprecipitated Na$^+$-Ca^{2+} exchanger were determined in cells pretreated with no agent (C), 10 or 20 ng/ml PDGF-BB (PDGF 10 or 20), 10 nM α-thrombin (T), 100 nM angiotensin II (AII), or 100 nM PMA for 10 min. Results are presented as the mean ± SD (n = 3 or 4). * Significant difference from unstimulated cells. (From Iwamoto et al.[9] Reprinted by permission from the *Journal of Biological Chemistry*.)

(FIG. 5). We treated cells with 50 μM cyclopiazonic acid in Ca^{2+}- and Na$^+$-free, high pH (pH 8.8) BSS containing 20 mM Mg^{2+}, which inhibits Ca^{2+} extrusion via the Na$^+$-Ca^{2+} exchanger and the ATP-dependent Ca^{2+} pump.[11] After [Ca^{2+}]$_i$ reached a plateau level of about 300 nM (about 2 min later), Na$^+$ was added extracellularly to a final concentration of 50 mM. In unstimulated cells, Na^+_o evoked a decline in [Ca^{2+}]$_i$ (initial rate of decline, 33 ± 6 nM/10 s (n = 3)) (FIG. 5A), although addition of choline chloride, in place of NaCl, did not affect [Ca^{2+}]$_i$ (data not shown). When cells were stimulated with 10 ng/ml PDGF-BB for 20 min, the initial rate of Na^+_o-induced [Ca^{2+}]$_i$ decline significantly increased to 89 ± 12 nM/10 s (n = 3), and [Ca^{2+}]$_i$ reached a lower steady state level (FIG. 5A). Such stimulation of Na^+_o-dependent [Ca^{2+}]$_i$ decline by PDGF-BB was abolished (initial rate of decline, 36 ± 8 nM/10 s (n = 3)) by prolonged (24 h) pretreatment of cells with PMA (FIG. 5B), a procedure that downregulates the protein kinase C content in the plasma membrane.[12] All these results suggest that activity of the Na$^+$-Ca^{2+} exchanger is positively regulated by growth factors and that the PDGF-BB-induced increase in Na$^+$-Ca^{2+} exchange activity, like the phosphorylation of the exchanger, is mediated via a protein kinase C-dependent pathway.

In contrast to other growth factors tested, angiotensin II did not significantly increase both Na$^+$-Ca^{2+} exchange and phosphorylation of the exchanger (FIG. 1C and FIG. 4). α-Thrombin and PMA induced intermediate levels of activation of these parameters (FIG. 4). Thus, there is apparently a good correlation between the extents of stimulation of Na$^+$-Ca^{2+} exchange and phosphorylation of the exchanger induced by different growth factors. We found that angiotensin II increased cell

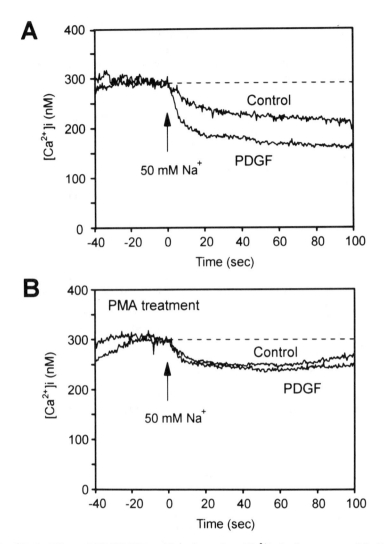

FIGURE 5. Effect of PDGF-BB on Na^+_o-dependent $[Ca^{2+}]_i$ decline measured in PMA-pretreated cells in Na^+- and Ca^{2+}-free high pH/high Mg^{2+} medium. Cells were treated with or without 10 ng/ml PDGF-BB for 20 min in BSS after they had been incubated **(B)** with or **(A)** without 100 nM PMA for 24 h. Then they were transferred to a Ca^{2+}- and Na^+-free, high pH medium (pH 8.8) containing 20 mM Mg^{2+} and 50 μM cyclopiazonic acid. After $[Ca^{2+}]_i$ reached a plateau level (about 300 nM), NaCl was applied to the medium to a final concentration of 50 mM to induce Na^+_o-dependent $[Ca^{2+}]_i$ decline. (From Iwamoto et al.[9] Reprinted by permission from the *Journal of Biological Chemistry*.)

1,2-diacylglycerol to a low level under our experimental conditions when compared with PDGF-BB (see above). Therefore, the inability of angiotensin II to enhance Na^+-Ca^{2+} exchange and the exchanger phosphorylation seems to be due to a low level of 1,2-diacylglycerol produced. Production of different levels of 1,2-diacylglycerol in response to PDGF-BB and angiotensin II seems to indicate that these two growth factors differ in their capacity to activate protein kinase C in the primary cultured VSMCs used in this study.

The Na^+-Ca^{2+} exchanger (NCX1) has been cloned from heart,[13] aorta,[14] kidney,[15] and brain.[16] By hydropathy analysis, the cardiac clone is modeled to have an amino-terminal cleaved signal sequence, 11 transmembrane segments, and a large hydrophilic cytoplasmic domain between the fifth and sixth transmembrane segments.[13] There is much evidence showing that the cytoplasmic domain is involved in the regulation of the exchanger activity.[1,2] Primary structure of the rat aortic Na^+-Ca^{2+} exchanger is highly homologous to the canine cardiac exchanger, except for part (amino acid residues 570–621) of the cytoplasmic domain.[14] In the cytoplasmic domain of the aortic smooth muscle Na^+-Ca^{2+} exchanger, we note three candidate serine residue-containing sequences phosphorylatable by protein kinase C or Ca^{2+}/calmodulin kinase II, which are also conserved in the cardiac exchanger.

PDGF and α-thrombin are powerful mitogens for cells of mesenchymal origin including VSMCs. They increase $[Ca^{2+}]_i$ transiently and induce cascades of events such as activation of ion transporters, phosphorylation of proteins, and expression of a number of genes that are associated with cell proliferation. Transient exposure of growth-arrested VSMCs or other types of cells to PDGF, α-thrombin, or phorbol esters result in activation of the Na^+-H^+ exchanger and Na^+/K^+/Cl^- cotransporter.[17,18] In addition, phorbol esters are able to stimulate the plasma membrane Ca^{2+} pump in VSMCs and other cell types.[19,20] Protein kinase C directly phosphorylates the plasma membrane Ca^{2+}-ATPase.[21,22] The results presented here indicate that the Na^+-Ca^{2+} exchanger also is one of the target ion transporters whose activation was induced by growth factors to help cellular proliferative cycle proceed properly.

CONCLUSION

The Na^+-Ca^{2+} exchanger in rat aortic VSMCs is phosphorylated in serine residues in response to physiological ligands such as PDGF-BB and α-thrombin or PMA. These agents also enhance Na^+-Ca^{2+} exchange activity. In contrast, angiotensin II does not enhance the phosphorylation and the exchange activity significantly. We provide evidence that activation of both is mediated via a protein kinase C-dependent pathway. Thus regulation of Na^+-Ca^{2+} exchange by phosphorylation currently seems to be a property shared by squid axon and smooth muscle exchangers.

REFERENCES

1. PHILIPSON, K. D. & D. A. NICOLL. 1993. Int. Rev. Cytol. **137C:** 199–227.
2. HILGEMANN, D. W., A. COLLINS, D. P. CASH & G. A. NAGEL. 1991. Ann. N. Y. Acad. Sci. **639:** 126–139.
3. LI, Z., D. A. NICOLL, A. COLLINS, D. W. HILGEMANN, A. G. FILOTEO, J. T. PENNIS-

TON, J. N. WEISS, J. M. TOMICH & K. D. PHILIPSON. 1991. J. Biol. Chem. **266:** 1014–1020.
4. DIPOLO, R. & L. BEAUGÉ. 1991. Ann. N. Y. Acad. Sci. **639:** 100–111.
5. VIGNE, P., J-P. BREITTMAYER, D. DUVAL, C. FRELIN & M. LAZDUNSKI. 1988. J. Biol. Chem. **263:** 8078–8083.
6. FURUKAWA, K-I., N. OHSHIMA, Y. TAWADA-IWATA & M. SHIGEKAWA. 1991. J. Biol. Chem. **266:** 12337–12341.
7. KHOYI, M. A., R. A. BJUR & D. P. WESTFALL. 1991. Am. J. Physiol. **261:** C685–C690.
8. CIRILLO, M., S. J. QUINN, J. R. ROMERO & M. L. CANESSA. 1993. Circ. Res. **72:** 847–856.
9. IWAMOTO, T., S. WAKABAYASHI & M. SHIGEKAWA. 1995. J. Biol. Chem. **270:** 8996–9001.
10. RHEE, S. G. & K. D. CHOI. 1992. J. Biol. Chem. **267:** 12393–12396.
11. FURUKAWA, K-I., Y. TAWADA & M. SHIGEKAWA. 1988. J. Biol. Chem. **263:** 8058–8065.
12. KIKKAWA, U. & Y. NISHIZUKA. 1986. Annu. Rev. Cell Biol. **2:** 149–178.
13. NICOLL, D. A., S. LONGONI & K. D. PHILIPSON. 1990. Science **250:** 562–565.
14. NAKASAKI, Y., T. IWAMOTO, H. HANADA, T. IMAGAWA & M. SHIGEKAWA. 1993. J. Biochem. (Tokyo) **114:** 528–534.
15. REILLY, R. & C. A. SHUGRUE. 1992. Am. J. Physiol. **262:** F1105–F1109.
16. FURMAN, I., O. COOK, J. KASIR & H. RAHAMIMOFF. 1993. FEBS Lett. **319:** 105–109.
17. MOOLENAAR, W. H. 1986. Annu. Rev. Physiol. **48:** 363–376.
18. PANET, R. & H. ATLAN. 1991. J. Cell Biol. **114:** 337–342.
19. SMALLWOOD, J. I., B. GUGI & H. RASMUSSEN. 1988. J. Biol. Chem. **263:** 2195–2202.
20. FURUKAWA, K-I., Y. TAWADA & M. SHIGEKAWA. 1989. J. Biol. Chem. **264:** 4844–4849.
21. FUKUDA, T., T. OGURUSU, K-I. FURUKAWA & M. SHIGEKAWA. 1990. J. Biochem. (Tokyo) **108:** 629–634.
22. OGURUSU, T., S. WAKABAYASHI, K-I. FURUKAWA, Y. TAWADA-IWATA, T. IMAGAWA & M. SHIGEKAWA. 1990. J. Biochem. (Tokyo) **108:** 222–229.

Regulation of Expression of Sodium-Calcium Exchanger and Plasma Membrane Calcium ATPase by Protein Kinases, Glucocorticoids, and Growth Factors[a]

JEFFREY BINGHAM SMITH, HYEON-WOO LEE, AND LUCINDA SMITH

Department of Pharmacology and Toxicology
Schools of Medicine and Dentistry
University of Alabama at Birmingham
Birmingham, Alabama 35294-0019

INTRODUCTION

The plasma membrane of mammalian cells has two distinctly different mechanisms for ejecting Ca^{2+}: the sodium-calcium exchanger (NCX) and the plasma membrane calcium ATPase (PMCA). NCX is a carrier mechanism, which is readily reversible, with a turnover number of several thousand Ca^{2+} per s.[1] PMCA is an adenosine triphosphate (ATP)-driven pump, which is not readily reversible, with a turnover number of 50 Ca^{2+} per s.[2,3] The stoichiometry of NCX is 3 Na^+ per Ca^{2+} and that of PMCA is one Ca^{2+} per ATP.[3,4] Humans have at least two Na^+-Ca^{2+} exchanger genes, NCX1 and NCX2, and four PMCA (1–4) genes.[3,5] Different NCX and PMCA genes and isoforms, which are produced by alternative splicing, are expressed in a tissue-specific manner.[6–9] PMCA appears to be expressed in most, if not all mammalian cells, whereas NCX is expressed in some cell types, but not others.[3,10] While much is known about the structure and posttranslational regulation of PMCA and NCX1,[11–14] little is known about the regulation of PMCA and NCX expression. During cardiac development, NCX expression becomes maximal at birth and declines postnatally.[15] Pressure overload of feline heart increased NCX expression and accelerated general and contractile protein synthesis.[16] Recently we observed that glucocorticoids, growth factors, and activation of protein kinase A markedly influence NCX expression in vascular smooth muscle cells.[17,18] Activation of protein kinase C in kidney epithelial cells produces a substantial depletion of mRNA and protein.[19] In contrast to the depletion of NCX in kidney cells, activation of protein kinase C in endothelial cells strikingly induced PMCA.[20] In aortic myocytes, thrombin rapidly increased PMCA transcripts.[21] The data presented here indicate that NCX and PMCA expression by mammalian cells is remarkably sensitive to external stimuli.

---

[a] This work was supported by grants from the American Heart Association, Alabama Affiliate, Inc. (AL-G-940014), and the National Institutes of Health (HL44408).

MATERIALS AND METHODS

Cell Culture

Rat aortic myocytes immortalized by simian virus 40 large T antigen and the LLC-MK$_2$ line (ATCC CCL 7.2) of renal epithelial cells were grown in Dulbecco's modified Eagle's medium (DMEM) containing 5% fetal bovine serum as described.[22,23]

^{3}H-Monoclonal Antibody Binding

Monoclonal antibody (MAb) R3F1 to canine NCX expressed in Sf9 cells was purified and labeled with N-succinimidyl [2,3-^{3}H]propionate as previously described.[24] ^{3}H-MAb binding was done essentially as described.[24] Briefly, confluent cultures were rinsed twice with ice-cold physiological salts solution (PSS), scraped with PSS, and sonicated for 10 s with a 1/8 inch microtip with output control at 4 and 40% maximal output (continuous) with a Sonifier cell disrupter 350 (Branson Ultrasonics Corp., Danbury, CT). PSS contained (in mM) 120 NaCl, 5 KCl, 1 CaCl$_2$, 1 MgCl$_2$, and 20 HEPES-Tris, pH 7.4. Sonicate (0.25 ml) containing 0.6 mg protein (or as indicated) was diluted to 1 ml with PSS containing 1 mg/ml saponin and 1 mg/ml BSA, and 5 µl of ^{3}H-MAb (~200,000 cpm) was added to start the binding reaction. The specific activity of the ^{3}H-MAb was estimated to be 570 cpm/ng. Sonicate protein was measured with the BCA protein assay kit (Pierce Chemical Company, Rockford, IL). After 1 h at 37°C bound ^{3}H-MAb was collected by vacuum filtration on Whatman GF/F filters that had been wetted with a solution of 1% (w/v) polyethylenimine (Sigma Chemical Co.). Filters were rinsed three times with saline containing 10 mM HEPES/Tris, pH 7.5, dried, and put in a vial with 10 ml Budget-Solve (Research Products International Corp., Mount Prospect, IL) for β counting.

Northern Analysis of NCX, PMCA and c-Myc Transcripts

Poly(A)$^+$ RNA was prepared by oligo(dT)-cellulose chromatography from total RNA which was extracted with acidified guanidinium thiocyanate-phenol-chloroform (Ultraspec RNA, Biotecx Laboratories, Inc., Houston, TX). Poly(A)$^+$ RNA was quantified by absorbance at 260 nm and size-fractionated by electrophoresis in 1% (w/v) agarose gels (Bio-Rad Laboratories, Hercules, CA). Gels were prepared in electrophoresis running buffer which contained (in mM): 20 3-(N-morpholino)propane sulfonate (MOPS), pH 7.4, 1 ethylenediaminetetraacetic acid (EDTA), 5 sodium acetate, and 400 formaldehyde. RNA was transferred to Duralon UV membranes (Stratagene, La Jolla, CA) by the downward capillary method and cross-linked by UV irradiation. Membranes (11 × 14 cm) were prehybridized with 6 ml QuikHyb (Stratagene, La Jolla, CA) for 10 min at 68°C in a roller-bottle oven. ^{32}P-labeled cDNA probe (3–5 µCi/ml) was mixed with 100 µl denatured salmon sperm DNA (10 mg/ml) and added to the roller-bottle. After 2 h at 68°C the membrane was washed twice at room temperature for 15 min with twice concentrated sodium chloride sodium citrate (SSC) containing 0.1% sodium dodecylsulfate and twice at 60°C for 15 min with SSC containing 0.1% sodium dodecylsulfate. SSC contained 8.8 g NaCl and 4.4 g sodium citrate per liter and was adjusted to pH 7.0 with HCl. Probes were synthesized with 2 units DNA polymerase I, Klenow fragment (sequencing grade, Boehringer Mannheim, Indianapolis, IN) using random hexamer primers, [α^{32}P]dCTP and a cDNA template that was purified by agarose gel electrophoresis following plasmid digestion with the indicated restriction enzymes. The following cDNAs were used: 1.5 kb cDNA to the 3'-end of guinea pig cardiac NCX, 1.4 kbp Clal-EcoRI fragment of pHSR-1 of human c-Myc (ATCC 41010), 1.4 kbp Aval fragment of rat brain plasma membrane Ca^{2+} ATPase (PMCA), a 735-kb β-actin cDNA probe derived from a PC12 cDNA clone, and 1.5 kbp EcoRI fragment of human 18S rRNA (ATCC 77242). The specific activity of the ^{32}P-labeled cDNA probes ranged from $2-8 \times 10^8$ cpm/μg. Transcripts were quantified by autoradiography for ≤ 3 days at $-70°C$ with Konica PPB film and an intensifying screen. Membranes were stripped and hybridized sequentially with each cDNA probe. Autoradiograms were scanned and analyzed with a model GS-670 Imaging Densitometer using Molecular Analyst 2.0 software (Bio-Rad Laboratories, Hercules, CA).

RESULTS AND DISCUSSION

Activation of Protein Kinase C Downregulates NCX Expression in Renal Epithelial Cells

Determination of NCX Protein by ^{3}H-MAb Binding

In order to understand the regulation of NCX expression it is necessary to quantify NCX message, protein, and activity. We have used a ^{3}H-labeled MAb to cardiac NCX1 to quantify NCX protein.[19] FIGURE 1 shows ^{3}H-MAb binding to sonicates of LLC-MK$_2$ cells and CK1.4 cells, a Chinese hamster ovary (CHO) line that stably expresses cardiac NCX. The amount of ^{3}H-MAb bound increased linearly as a function of the amount of CK1.4 sonicate added to the assay. By contrast to CK1.4 cells, there was no significant binding of ^{3}H-MAb to nontransfected CHO cells. Nontransfected CHO cells lacked detectable NCX activity or transcript. Note that the filter blank (no cell sonicate) was plotted on the vertical axis, which shows total ^{3}H-MAb binding. ^{3}H-MAb binding to LLC-MK$_2$ cells

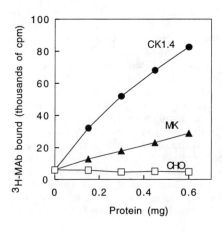

FIGURE 1. Dependence of amount of ^{3}H-MAb binding on the amount of CK1.4, LLC-MK$_2$ (MK), or CHO cell sonicate added to the assay. Binding was assayed at 0.35 ng/ml [^{3}H]MAb and the indicated amount of cell sonicate. Sonicates were prepared from nontransfected CHO cells which lack NCX protein, mRNA, and activity; CK1.4 cells, which stably express cardiac NCX; and LLC-MK$_2$ cells. Values are means of two experiments. (From Smith *et al.*[19] Reprinted by permission from the *American Journal of Physiology*.)

increased linearly with amount of sonicate added to the assay. The slope of the line for binding to LLC-MK$_2$ cells was approximately a third that for binding to CK1.4 cells. Therefore, CK1.4 cells expressed 3 times more NCX protein per cell than LLC-MK$_2$ cells. It is noteworthy that the CK1.4 cells used for this experiment were subconfluent, rapidly growing cultures.

Downregulation of NCX Protein by Phorbol Myristate Acetate

Next we determined the B_{max} and K_d for ^{3}H-MAb binding to confluent, stationary phase CK1.4 or LLC-MK$_2$ cell sonicates. Binding curves were well fitted by nonlinear regression assuming a single class of binding sites.[19] The K_d values for binding to CK1.4 and LLC-MK$_2$ cells did not differ significantly. K_d values were 0.17 ± 0.04 µg/ml (n = 5) and 0.13 ± 0.01 (n = 4) for CK1.4 and LLC-MK$_2$ cells, respectively.[19] B_{max} values indicated that CK1.4 cells contained 56% more NCX protein than LLC-MK$_2$ cells. B_{max} for stationary cultures of CK1.4 cells was 204 ± 65 ng/mg protein (n = 5) compared to 131 ± 10 ng/mg protein for LLC-MK$_2$ cells (n = 4).[19] There are ~250,000 NCX per LLC-MK$_2$ cell based the B_{max} assuming 2 million cells per mg protein. The smaller difference in the B_{max} values of CK1.4 and LLC-MK$_2$ cells compared to larger difference in ^{3}H-MAb binding to these two cell types shown in FIGURE 1 appears to be due to the fact that subconfluent, log phase CK1.4 cells contain substantially more NCX than confluent, stationary phase cells.

Phorbol myristate acetate (PMA), which activates protein kinase C (PKC), markedly decreased ^{3}H-MAb binding to LLC-MK$_2$ cells. The cells were incubated in serum-free DMEM for 24 h with or without 0.1 µM PMA. PMA treatment decreased ^{3}H-MAb binding to $58 \pm 6\%$ control.[19]

Downregulation of NCX mRNA by Phorbol Ester

Because PMA decreased NCX protein, we performed northern analysis to determine whether PMA affected the level of NCX transcript in LLC-MK$_2$ cells (FIG. 2). FIGURE 2A shows the effect of a 24-h incubation of LLC-MK$_2$ cells in serum-free DMEM in the presence or absence of 0.1 µM PMA. Poly(A)$^+$ RNA (1, 2, or 3 µg), which was quantified by O.D. at 260 nm, was size fractionated on a formaldehyde agarose gel, transferred to a nylon membrane and hybridized with a cDNA probe to cardiac NCX1. PMA treatment markedly decreased the level of the NCX transcript. This blot also shows that our northern analysis conditions accurately quantify NCX mRNA because volume integration (O.D. × area of autoradiogram) of NCX bands gave relative values that were directly proportional to the amount of RNA electrophoresed (FIG. 2A). For example, the relative volumes of the NCX band at 1, 2, and 3 µg of poly(A)$^+$ RNA from PMA-treated cells were 1.0, 2.1, and 2.9, respectively. The relative volumes of the NCX bands from the control cells were 4.0, 10.1, and 12.6 for 1, 2, and 3 µg poly(A)$^+$ RNA electrophoresed, respectively. Therefore, in this experiment PMA decreased NCX mRNA by 75, 79, and 77% at 1, 2, and 3 µg poly(A)$^+$, respectively. After probing the membrane for NCX, we probed it with a β-actin cDNA as an internal standard for the quality of poly(A)$^+$ RNA from the PMA-treated and control cells.

FIGURE 2B shows that a 4-h treatment of the cells with PMA markedly decreased NCX mRNA. A 2-h incubation with PMA had no significant effect on NCX mRNA (FIG. 2B). Additional experiments showed that 10 nM PMA produced

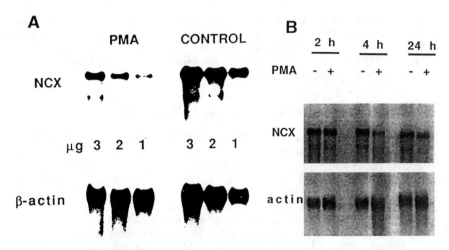

FIGURE 2. Downregulation of NCX mRNA in renal epithelial cells by PMA. **(A)** LLC-MK$_2$ cells were incubated in serum-free DMEM with or without 0.1 μM PMA. Poly(A)$^+$ (1, 2, or 3 μg) was size fractionated by electrophoresis in 1% agarose gel containing formaldehyde. After hybridization with a ^{32}P-labeled cDNA probe to cardiac NCX1, the membrane was stripped and probed for β-actin. **(B)** LLC-MK$_2$ cells were incubated in serum-free DMEM with or without 0.1 μM PMA for the indicated interval and RNA was extracted. Poly(A)$^+$ enriched RNA (2 μg) was size fractionated and subjected to Northern analysis.

almost the maximal decrease in NCX mRNA in 24 h.[19] The activation of PKC appears to cause the decrease in NCX mRNA because a 24-h incubation of the cells with 4αPMA, a stereoisomer that neither binds nor activates PKC,[25] had no effect on NCX mRNA.[19] These findings suggest that activation of PKC downregulates expression of NCX protein by decreasing the level of the NCX transcript.

PMA Produces a Prolonged Activation of PKC

Experiments reported elsewhere determined that PKCα was the predominant PMA-sensitive isoform of PKC expressed in LLC-MK$_2$ cells.[19] A 24-h treatment of the cells with 0.1 μM PMA decreased total PKCα protein by 65% as determined by Western analysis. Despite the partial depletion of PKCα protein by PMA, there was substantially greater PKC activity in the cells treated with PMA for 24 h compared to control cells.[19] First, particulate PKCα was approximately 2-fold greater in the PMA-treated compared to the control cells. Second, *in vivo* ^{32}P-labeling of MARCKS (myristoylated, alanine-rich C kinase substrate), a prominent and selective PKC substrate,[26] provided further evidence that the activation of PKC by PMA was sustained for 24 h. Treatment with 0.1 μM PMA for 10 min or 24 h approximately doubled the ^{32}P-labeling of MARCKS as determined by immunoprecipitation from cells that were labeled with [^{32}P]orthophosphate.[19]

Downregulation of NCX Activity by PMA in Renal Epithelial Cells

NCX activity was assayed as ^{45}Ca^{2+} influx that depended on inverting the Na$^+$ gradient, *i.e.*, raising intracellular Na$^+$ and replacing external Na$^+$ with K$^+$.

Treatment of LLC-MK$_2$ cells with PMA decreased NCX activity after a 4-h lag. Between 4 and 48 h PMA decreased NCX activity exponentially with a half-time of ~19 h.[19] 4αPMA, which does not activate PKC, had no effect on NCX activity.[19] PMA markedly (≥50%) decreased NCX mRNA in 4 h, as already indicated (FIG. 2B). Therefore the decrease in NCX mRNA preceded the decrease in NCX activity. These findings suggest that the activation of PKC downregulates NCX activity by decreasing the number of NCX transcripts per cell. The 42% decrease in NCX protein produced by a 24-h 0.1-μM PMA treatment largely accounts for the ~60% decrease in NCX activity. The presence of nonfunctional NCX protein in the PMA-treated cells may account for the slightly larger decrease in NCX activity compared to protein. PMA may downregulate NCX mRNA by increasing its degradation and/or decreasing transcription. Additionally PMA may affect the degradation of NCX protein.

Cyclic AMP Downregulates NCX Expression in Vascular Smooth Muscle

Chronic Elevation of Cyclic AMP Decreases NCX Activity

Recently we used forskolin, which raises cellular 3',5'-cyclic adenosine monophosphate (cAMP) by activating adenylyl cyclase, to further investigate the possibility that NCX expression is sensitive to stimuli that inhibit or evoke a mitogenic response. Chronic elevation of cAMP by hormones, or activation of protein kinase A by pharmacologic agents, inhibits the proliferation of arterial myocytes from various species and blood vessels.[27-30] We found that forskolin decreases NCX mRNA and activity, which were restored by fetal bovine serum. Addition of forskolin plus methylisobutyl xanthine (MIX) to rat aortic myocytes, which produces a prolonged elevation of cAMP, decreased NCX activity after a 6-h lag.[18] From 6 to 40 h NCX activity decreased almost linearly. Forskolin plus MIX decreased NCX activity by 50% at 24 h and 85% at 40 h. Forskolin alone decreased NCX activity by 40% in 20 h, whereas MIX alone had no effect.[18] MIX at the concentration used (50 μM) only slightly increases cAMP in aortic myocytes. In the presence or absence of MIX, 1,9-dideoxyforskolin, which fails to activate adenylyl cyclase, had no effect on NCX activity. These findings suggested that the prolonged elevation of cAMP decreased NCX expression because posttranslational regulation of NCX activity by a cAMP-dependent protein kinase would be expected to occur within minutes of raising cAMP.

Forskolin Decreases NCX mRNA

Forskolin, but not dideoxyforskolin, decreased NCX mRNA. NCX mRNA decreased by 60% in 6 h and 85% in 24 h.[18] A 24-h incubation with MIX or dideoxyforskolin had no effect on NCX mRNA. These findings indicate that the decrease in NCX mRNA preceded the decrease in NCX protein, which is consistent with the view that activation of a cAMP-dependent protein kinase downregulated NCX expression at the transcript level.

Forskolin Downregulates c-Myc, but not PMCA

Expression of c-Myc is known to be low in quiescent rat aortic myocytes and elevated throughout the cycle in proliferating myocytes.[28,31] Downregulation of c-Myc appears to be a prerequisite for growth arrest of rat aortic myocytes.[27,31] Forskolin plus MIX decreased the steady state level of c-Myc mRNA by ~70% in 24 h.[18] Dideoxyforskolin had no significant effect on c-Myc mRNA. Rat aortic myocytes express two PMCA genes, PMCA1 and PMCA4, which encode transcripts of 5.5 and 8.5 kb, respectively.[32] A 24-h incubation with forskolin plus MIX had no effect on either of the PMCA transcripts. The PMCA data are important because they indicate that raising cAMP selectively affected NCX expression.

Restoration of NCX and c-Myc mRNA by Serum

Myocytes were incubated for 48 h with forskolin plus MIX to downregulate NCX mRNA. Then the myocytes were rinsed and incubated in DMEM containing 10% fetal bovine serum, with or without cycloheximide or actinomycin D. Removal of forskolin plus MIX and addition of serum increased NCX mRNA ~8-fold in 6 h.[18] Cycloheximide or actinomycin D prevented the increase in NCX mRNA indicating that it depended on protein and RNA synthesis.[18] Removal of forskolin plus MIX increased c-Myc mRNA ~7-fold in 6 h similarly to NCX mRNA.[18] Interestingly, addition of cycloheximide together with serum increased c-Myc mRNA ~22-fold.[18] Inhibitors of protein synthesis are known to superinduce immediate early genes by prolonging the activation of transcription and stabilizing the transcripts.[33] By contrast to cycloheximide, actinomycin D prevented c-Myc induction by serum and decreased c-Myc mRNA by 75%.[18] The decrease in c-Myc RNA by actinomycin D is consistent with the short half-life of the transcript.[33]

Restoration of NCX Activity by Serum

Myocytes were treated with forskolin and MIX for 48 h under the conditions used to quantify NCX mRNA as described above. Then they were rinsed and incubated with DMEM in the presence or absence of fetal bovine serum. Removal of forskolin plus MIX and addition of serum restored full NCX activity in 24 h.[18] Restoration of NCX activity required serum addition.

The following observations suggest that forskolin decreased NCX activity by decreasing the NCX protein. First, if raising cAMP inhibited NCX activity postranslationally, for example, by phosphorylation of NCX by protein kinase A, then forskolin would be expected to affect NCX activity within minutes rather than hours as was observed. Second, forskolin markedly and selectively decreased NCX mRNA. Third, restoration of NCX activity required a prolonged incubation of the myocytes with fetal bovine serum. Fourth, the changes in NCX mRNA preceded the changes in NCX activity. Activation of adenylyl cyclase decreased the c-Myc transcript, an immediate early gene whose expression directly correlates with cell proliferation.[28,31] Serum rapidly increased c-Myc mRNA. Transcription factors that regulate smooth muscle growth may mediate the opposing influences of serum and forskolin on NCX mRNA and activity.

Glucocorticoids Downregulate NCX Expression in Vascular Smooth Muscle: Reversal by Growth Factors

Glucocorticoids Decrease NCX Activity in Aortic Myocytes

Glucocorticoids produce hypertension in part by potentiating the pressor responses to vasoconstrictors through a direct action on vascular smooth muscle (reviewed in Ref. 17). Glucocorticoids apparently enhance vascular responsiveness by potentiating the free Ca^{2+} responses to several vasoconstrictors. NCX is largely responsible for the net Ca^{2+} efflux provoked by vasoconstrictors in aortic myocytes.[34,35] Decreased Ca^{2+} efflux via NCX would be expected to prolong the Ca^{2+} transient produced by vasoconstrictors.

Dexamethasone (DEX), cortisol, or aldosterone decreased NCX activity by ~55% in 24 h.[17] Other steroids such as testosterone, progesterone, and β-estradiol, had no effect on NCX activity. Activation of a glucocorticoid receptor is probably responsible for the downregulation of NCX activity because DEX was >100 times more potent than aldosterone.[17] DEX half-maximally decreased NCX activity in 12 h and maximally decreased NCX activity in 24 h.[17] There was no further decrease in NCX activity between 24 and 48 h. Glucocorticoids are known to downregulate expression of their own receptor, which may have limited the effect of DEX on NCX activity. The downregulation of NCX in vascular smooth muscle may contribute to the increased vascular responsiveness that occurs in hypertension produced by glucocorticoid excess.

Dexamethasone Decreases NCX and c-Myc, but not PMCA, Transcripts

FIGURE 3 shows of the effects of a 24-h incubation with 10 or 100 nM DEX on NCX mRNA. Both concentrations of DEX decreased NCX mRNA ~90%. A 5-h incubation with 100 nM DEX decreased NCX mRNA by 60%. This experiment agrees well with those reported previously. In addition to decreasing NCX mRNA,

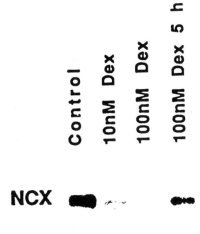

FIGURE 3. Effect of DEX concentration on NCX mRNA in aortic myocytes. Immortalized rat aortic myocytes were incubated in serum-free DMEM without or with the indicated concentration of DEX for 24 (*first three lanes*) or last 5 h of the 24-h incubation with DMEM (*fourth lane*). RNA was extracted, and 1 μg of poly(A)$^+$ enriched RNA was subjected to Northern analysis.

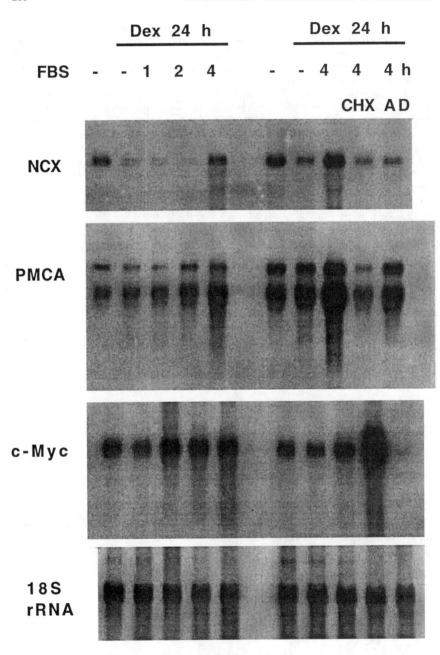

FIGURE 4. Serum increases NCX, PMCA, and c-Myc mRNA in DEX-treated myocytes: dependence on RNA and protein synthesis. **(A)** Aortic myocyte cultures were incubated for 24 h in serum-free DMEM with or without 0.1 µM DEX as indicated. Fetal bovine serum (FBS, 20%) was added to DEX-containing DMEM, and the incubation continued for 1, 2, or 4 h as indicated. **(B)** Cycloheximide (CHX, 30 µg/ml) or actinomycin D (AD, 2.5 µg/ml) was added to DEX-containing DMEM as indicated. After 1 h, 20% FBS was added. Four h later, RNA was extracted and 3 µg poly(A)$^+$ enriched RNA from (A) and (B) were size fractionated on the same agarose gel and subjected to Northern analysis. The membrane was sequentially hybridized with ^{32}P-labeled cDNA to NCX, PMCA, c-Myc, and 18S rRNA.

DEX decreased c-Myc mRNA by ~38% in 24 h (FIG. 4). The decreases in c-Myc and NCX mRNA produced by DEX are similar to those produced by forskolin as described above. DEX had no effect on the level of the PMCA transcripts (FIG. 4A,B).

Restoration of NCX mRNA and Activity by Growth Factors

Following a 24-h treatment with DEX, addition of serum markedly increased NCX mRNA and activity. The peak increase in NCX mRNA occurred 6 h after the addition of serum and was ~10-fold greater than the level in the DEX-treated cells.[17] Serum fully restored NCX activity to DEX-treated myocytes, which required ~16 h.[17] Thrombin, which is strongly mitogenic for aortic myocytes, partially restored NCX mRNA and activity.[17] Cycloheximide or actinomycin D abolished the increase in NCX mRNA produced by serum indicating that it depended on protein and RNA synthesis (FIG. 4B).

Upregulation of PMCA and c-Myc Transcripts in Vascular Smooth Muscle

A 4-h incubation with fetal bovine serum markedly increased PMCA1 and PMCA4 transcripts in DEX-treated myocytes (FIG. 4A,B), as well as quiescent myocytes that were not DEX-treated.[21] The increases in PMCA transcripts produced by serum were evident at 2 h and preceded NCX induction (FIG. 4A). Thrombin increased PMCA transcripts similarly to serum. The peak increase, which occurred 3 h after the addition of thrombin, was 7–15-fold greater than control.[21] Thrombin and serum also markedly increased c-Myc mRNA in quiescent myocytes. The increase in c-Myc mRNA by thrombin preceded the increase in PMCA transcripts. Thrombin substantially increased c-Myc in 1 h, at which time there was no detectable increase in PMCA transcripts.[21] c-Myc mRNA was substantially increased 8 h after addition of thrombin. Induction of PMCA by thrombin depended on protein kinase C, as reported previously for endothelial cells,[20] and also tyrosine phosphorylation.[21]

Cycloheximide or actinomycin D abolished the effects of serum on PMCA transcripts in DEX-treated myocytes (FIG. 4B). Cycloheximide markedly potentiated the effect of serum on c-Myc mRNA (FIG. 4B), similarly to c-Myc superinduction described above in relation to the restoration of NCX mRNA in forskolin-treated myocytes. Induction of PMCA by thrombin in quiescent myocytes also depended on protein and RNA synthesis, and cycloheximide plus thrombin superinduced c-Myc.[18]

Coordinate Upregulation of Na-Ca Exchanger and Plasma Membrane Ca ATPase

The data presented and reviewed here are the initial studies of the regulation of NCX and PMCA expression. These findings suggest that hormones, growth factors, and protein kinases tightly regulate expression of NCX and PMCA. TABLE 1 summarizes the stimuli that regulate NCX and PMCA expression. The regulation of NCX and PMCA mRNA and activity exhibited stimulus and cell type specificity. Activation of protein kinase C downregulated NCX mRNA and activity in kidney epithelial cells, but had no effect on NCX mRNA or activity in vascular smooth muscle cells (TABLE 1).[19] Activation of protein kinase C upregulated PMCA mRNA in aortic myocytes and endothelial cells, but had no effect on PMCA mRNA in kidney epithelial cells.[20,21] In aortic myocytes dexamethasone downregulated NCX mRNA and activity without affecting PMCA mRNA.[17] Dexamethasone had no effect on NCX mRNA or activity in kidney epithelial cells.[19] The diversity of the stimuli that influence NCX expression is consistent with a model proposed by Lytton and co-workers.[8] Expression of the same NCX protein appears to be driven by at least three different tissue-specific promoters, upstream of a unique 5′-end exon that is spliced to a common acceptor (nt-34).[8] Tissue or cell type specific promoter elements may at least partially account for the distinct regulation of NCX expression in response to different demands placed on its

TABLE 1. Summary of Stimuli that Regulate the Expression of NCX and PMCA[a]

	NCX			PMCA
	mRNA	Protein	Activity	mRNA
Aortic myocytes[18-20]				
PMA	→	ND	→	↑
cAMP	↓	ND	↓	→
DEX	↓	ND	↓	→
Thrombin	↑	ND	↑	↑
FBS	↑	ND	↑	↑
Kidney epithelial cells[21]				
PMA	↓	↓	↓	ND
cAMP[b]	→	ND	→	ND
DEX[b]	→	ND	→	ND
Aortic endothelial cells[17]				
PMA	ND	ND	ND	↑
cAMP	ND	ND	ND	→
Angiotensin	ND	ND	ND	↑

[a] ↑ = increased, ↓ = decreased, → = unchanged, ND = not done. The indicated cell type was incubated for several hours with the indicated stimulus as described.[17-21] Upregulation of PMCA mRNA peaked in 3 to 4 h, whereas NCX mRNA peaked in 4 to 6 h.[20,21] Downregulation of NCX mRNA was maximal in 24 h.[17-19] PMA, phorbol myristate acetate; cyclic AMP (cAMP) was elevated by incubation with forskolin plus methylisobutyl xanthine; DEX, dexamethasone; FBS, fetal bovine serum.

[b] Smith, J. B., unpublished data.

function in different environments. The biochemical pathways by which protein kinases and external stimuli regulate NCX and PMCA expression remain to be elucidated.

Accelerated Ca^{2+} cycling between the cell and the environment is a hallmark of cells stimulated by neurohormones and growth factors.[36] Serum and purified growth factors such as thrombin increased NCX mRNA and activity in aortic myocytes following their downregulation by glucocorticoid or cAMP.[19,20] Glucocorticoid or elevated cAMP inhibits myocyte growth and decreases expression of immediate early genes such as c-Myc.[27-31] Serum and thrombin also markedly increased PMCA and c-Myc transcripts in quiescent aortic myocytes.[21] Induction or repression of transcription factors that regulate myocyte proliferation, such as c-Myc, may be responsible for the opposing influences of mitogenic stimuli and inhibitors on NCX and PMCA expression. Greater NCX and/or PMCA expression would enhance the Ca^{2+} efflux capacity of the stimulated cell. In contrast to the marked decreases in NCX transcript and activity produced by glucocorticoid or raising cAMP with forskolin, neither agent affected the level of the PMCA transcripts.[17,18] The retention of PMCA under conditions that downregulate NCX may help preserve cell viability, because NCX and PMCA are the only known mechanisms by which mammalian cells expel Ca^{2+}. Therefore, coordinate negative regulation of NCX and PMCA is potentially lethal. Coordinate positive regulation of NCX and PMCA preserves Ca^{2+} homeostasis and apparently fosters cell growth.

ACKNOWLEDGMENTS

We thank Dr. K. D. Philipson for providing the pC8-1,E-1 vector containing guinea pig cardiac exchanger cDNA and Dr. G. E. Shull for the vector containing PMCA1 cDNA (RB-11-1).

REFERENCES

1. HILGEMANN, D. W., D. A. NICOLL & K. D. PHILIPSON. 1991. Charge movement during Na^+/Ca^{2+} exchanger. Nature **352**: 715–718.
2. REGA, A. F. & P. J. GARRAHAN. 1985. The Calcium Pump of Plasma Membranes. 62–63. CRC Press, Inc. Boca Raton, FL.
3. CARAFOLI, E. 1994. Biogenesis: plasma membrane calcium ATPase: 15 years of work on the purified enzyme. FASEB J. **8**: 993–1002.
4. REEVES, J. P. & C. C. HALE. 1984. The stoichiometry of the cardiac sodium-calcium exchange system. J. Biol. Chem. **259**: 7733–7739.
5. LI, Z., S. MATSOKA, L. V. HRYSHKO, D. A. NICOLL, M. M. BERSOHN, E. P. BURKE, R. P. LIFTON & K. D. PHILIPSON. 1994. Cloning of the NCX2 isoform of the plasma membrane Na^+-Ca^{2+} exchanger. J. Biol. Chem. **269**: 17434–17439.
6. REILLY, R. F. & C. A. SHUGRUE. 1992. cDNA cloning of a renal Na^+-Ca^{2+} exchanger. Am. J. Physiol. **262**: F1105–F1109.
7. NAKASAKI, Y., T. IWAMOTO, H. HANADA, T. IMAGAWA & M. SHIGEKAWA. 1993. Cloning of the rat aortic smooth muscle Na^+/Ca^{2+} exchanger and tissue-specific expression of isoforms. J. Biochem. (Tokyo) **114**: 528–534.
8. LEE, S-L., A. S. L. YU & J. LYTTON. 1994. Tissue-specific expression of Na^+-Ca^{2+} exchanger isoforms. J. Biol. Chem. **269**: 14849–14852.
9. KOFUJI, P., J. W. LEDERER & D. H. SCHULZE. 1994. Mutually exclusive and cassette exons underlie alternatively spliced isoforms of the Na/Ca exchanger. J. Biol. Chem. **269**: 5145–5149.
10. PIJUAN, V., Y. ZHUANG, L. SMITH, C. KROUPIS, M. CONDRESCU, J. F. ACETO, J. P.

Reeves & J. B. Smith. 1993. Stable expression of the cardiac sodium-calcium exchanger in CHO cells. Am. J. Physiol. **264:** C1066–C1074.
11. Nicoll, D. A., S. Longoni & K. D. Philipson. 1990. Molecular cloning and functional expression of the cardiac sarcolemmal Na^+-Ca^{2+} exchanger. Science **250:** 562–565.
12. Matsuoka, S., D. A. Nicoll, R. F. Reilly, D. W. Hilgemann & K. D. Philipson. 1993. Initial localization of regulatory regions of the cardiac sarcolemmal Na^+-Ca^{2+} exchanger. Proc. Natl. Acad. Sci. USA **90:** 3870–3874.
13. Li, Z., D. A. Nicoll, A. Collins, D. W. Hilgemann, A. G. Filoteo, J. T. Penniston, J. N. Weiss, J. M. Tomich & K. D. Philipson. 1991. Identification of a peptide inhibitor of the cardiac sarcolemmal Na^+-Ca^{2+} exchanger. J. Biol. Chem. **266:** 1014–1020.
14. Levitsky, D. O., D. A. Nicoll & K. D. Philipson. 1994. Identification of the high affinity Ca^{2+}-binding domain of the cardiac Na^+-Ca^{2+} exchanger. J. Biol. Chem. **269:** 22847–22852.
15. Boerth, S. R., D. B. Zimmer & M. Artman. 1994. Steady-state mRNA levels of the sarcolemmal Na^+-Ca^{2+} exchanger peak near birth in developing rabbit and rat hearts. Circ. Res. **74:** 354–359.
16. Kent, L., J. D. Rozich, P. L. McCollam, D. E. McDermott, U. F. Thacker, D. R. Menick, P. J. McDermott & G. Cooper, IV. 1993. Rapid expression of the Na^+-Ca^{2+} exchanger in response to cardiac pressure overload. Am. J. Physiol. **265** (Heart Circ. Physiol. 34): H1024–H1029.
17. Smith, L. & J. B. Smith. 1994. Regulation of sodium-calcium exchanger by glucocorticoids and growth factors in vascular smooth muscle. J. Biol. Chem. **269:** 27527–27531.
18. Smith, L. & J. B. Smith. 1995. Activation of adenylyl cyclase down-regulates sodium-calcium exchanger of arterial myocytes. Am. J. Physiol. In press.
19. Smith, L., H. Porzig, II-W. Lee & J. B. Smith. 1995. Phorbol esters down-regulate expression of the sodium-calcium exchanger in renal epithelial cells. Am. J. Physiol. **269** (Cell Physiol. 38): C457–C463.
20. Kuo, T. H., B-F. Liu, C. Diglio & W. Tsang. 1993. Regulation of the plasma membrane calcium pump gene expression by two signal transduction pathways. Arch. Biochem. Biophys. **305:** 428–433.
21. Smith, L., H-W. Lee & J. B. Smith. 1995. Induction of plasma membrane calcium ATPase (PMCA) by thrombin in vascular smooth muscle cells. J. Mol. Cell. Cardiol. **27:** A51.
22. Lyu, R-M., L. Smith & J. B. Smith. 1993. Ca^{2+} influx via Na^+-Ca^{2+} exchange in immortalized aortic myocytes. I. Dependence on $[Na^+]_i$ and inhibition by external Na^+. Am. J. Physiol. **263:** C628–C634.
23. Lyu, R-M., L. Smith & J. B. Smith. 1991. Sodium-calcium exchange in renal epithelial cells; dependence on cell sodium and competitive inhibition by magnesium. J. Membr. Biol. **124:** 73–83.
24. Porzig, H., Z. Li, D. A. Nicoll & K. D. Philipson. 1993. Mapping of the cardiac sodium-calcium exchanger with monoclonal antibodies. Am. J. Physiol. **265:** C748–C756.
25. Blumberg, P. M. 1980. *In vitro* studies on the mode of action of phorbol esters, potent tumor promoters: Part 1. Crit. Rev. Toxicol. **8:** 153–198.
26. Blackshear, P. J. 1993. The MARCKS family of cellular protein kinase C substrates. J. Biol. Chem. **268:** 1501–1504.
27. Assender, J. W., K. M. Southgate, M. B. Hallett & A. C. Newby. 1992. Inhibition of proliferation, but not Ca^{2+} mobilization, by cyclic AMP and GMP in rabbit aortic smooth-muscle cells. Biochem. J. **288:** 527–532.
28. Bennet, M. R., G. I. Evan & A. C. Newby. 1994. Deregulated expression of the c-myc oncogene abolishes inhibition of proliferation of rat vascular smooth muscle cells by serum reduction, interferon-γ, heparin, and cyclic nucleotide analogues and induces apoptosis. Circ. Res. **74:** 525–536.
29. Jonzon, B., J. Nilsson & B. B. Fredholm. 1985. Adenosine receptor-mediated changes in cyclic AMP production and DNA synthesis in cultured arterial smooth muscle cells. J. Cell. Physiol. **124:** 451–456.

30. OREKHOV, A. N., V. V. TERTOV, S. A. KUDRYASHOV, KH. A. KHASHIMOV & V. N. SMIRNOV. 1986. Primary culture of human aortic intima cells as a model for testing antiatherosclerotic drugs. Effects of cyclic AMP, prostaglandins, calcium antagonists, antioxidants, and lipid-lowering agents. Atherosclerosis **60:** 101–110.
31. BIRO, S., Y. M. FU, Z-X. YU & S. E. EPSTEIN. 1993. Inhibitory effects of antisense oligodeoxynucleotides targeting c-myc mRNA on smooth muscle cell proliferation and migration. Proc. Natl. Acad. Sci. USA **90:** 654–658.
32. HAMMES, A., S. OBERDORF, E. E. STREHLER, T. STAUFFER, E. CARAFOLI, H. VETTER & L. NEYES. 1994. Differentiation-specific isoform mRNA expression of the calmodulin-dependent plasma membrane Ca^{2+} ATPase. FASEB J. **8:** 428–435.
33. LAU, L. F. & D. NATHANS. 1987. Expression of a set of growth-regulated immediate early genes in BALB/c 3T3 cells: coordinate regulation of c-fos and c-myc. Proc. Natl. Acad. Sci. USA **84:** 1182–1186.
34. SMITH, L. & J. B. SMITH. 1987. Extracellular Na^+ dependence of changes in free Ca^{2+}, $^{45}Ca^{2+}$ efflux, and total cell Ca^{2+} produced by angiotensin II in cultured arterial muscle cells. J. Biol. Chem. **262:** 17455–17460.
35. SMITH, J. B., R-M. LYU & L. SMITH. 1991. Inhibition of sodium-calcium and sodium-proton exchangers by amiloride congeners in arterial muscle cells. Biochem. Pharmacol. **41:** 601–609.
36. VILLEREAL, M. L. & K. L. BYRON. 1992. Calcium signals in growth factor signal transduction. Rev. Physiol. Biochem. Pharmacol. **119:** 68–121.

Enzyme Kinetics

Thermodynamic Constraints on Assignment of Rate Coefficients to Kinetic Models[a]

JONATHAN WAGG AND PETER H. SELLERS[b]

The Laboratory of Cardiac/Membrane Physiology
[b]*The Laboratory of Mathematics in Molecular Biology*
The Rockefeller University
1230 York Avenue
New York, New York 10021

INTRODUCTION

Enzyme-catalyzed reactions are complex in that they proceed from an initial set of reactants to a final set of products via a mechanism comprising more than one elementary reaction. This is because catalysis not only involves the binding and dissociation of reactant/s and product/s (terminal species) of the overall catalyzed reaction to and from the enzyme, but also transitions between enzyme states. In general, this does not involve the net consumption of enzyme because it is "cycled" among the various intermediate states. Indeed, there may be many distinct ways of cycling around a given mechanism or partial mechanism. Cycles associated with no net chemical change are called null cycles. All other cycles, by definition, involve net chemical change. For a null cycle, the product of all the equilibrium constants for the individual reaction steps in the cycle should equal unity. However, for any cycle associated with an overall reaction this product should equal the equilibrium constant for that reaction. This, the principle of detailed balance, follows from The Second Law of Thermodynamics and it provides an important constraint on the number of degrees of freedom permissible in assigning rate coefficients to the steps of a mechanism comprising one or more cycles.[1] In this paper, we present a general method for explicit identification of the constraints imposed on the set of rate coefficients for a kinetic model by the requirement to satisfy detailed balance for all cycles in the model. The approach is exemplified by considering a hypothetical mechanism for a second order carrier-mediated transport process.

A Second Order Carrier-Mediated Transport Process

Consider the following second order transport reaction: $2S_o \leftrightarrow 2S_i$, where S_o and S_i denote a chemical species, S, as present in the extracellular and intracellular compartments, respectively. Forward and reverse reaction are defined as occurring from left to right and right to left, respectively. The overall reaction involves the net movement of S across a biological membrane from the extracellular compart-

---

[a] Supported by HHMI and NIH HL-36783. Jonathan Wagg is a Howard Hughes Medical Institute Postdoctoral Physician.

ment to the intracellular compartment. The equilibrium constant for this reaction is unity since it is a transport reaction that is not coupled to any chemical reaction (such as ATP hydrolysis). Carriers are intrinsic membrane proteins that catalyze transport reactions. A hypothetical mechanism for the catalytic action of a carrier (denoted E) mediating the above reaction is presented in FIGURE 1A (see figure legend for details).

The minimum constraints that must be satisfied by the kinetic parameters for this mechanism to ensure detailed balance for all its cycles may be derived by the methods of Happel and Sellers.[2] This involves mapping of the mechanism onto a matrix which is then transformed. Inspection of the transformed matrix allows derivation of the relevant constraints. These procedures are outlined below.

Mapping the Mechanism onto a Matrix

Any reaction may be written in vector notation. This involves listing the names of all chemical species in the reaction and preceding each name with its respective stoichiometric coefficient, negative or positive according to whether it is a reactant or product. For example, step 1 of FIGURE 1A, $ES_2 \leftrightarrow E\_S + S_i$, may be written as: $-ES_2 + E\_S + S_i$. The negative stoichiometric coefficient of ES_2 indicates that it is a reactant, while the positive coefficients of E_S and S_i indicate that these are products. Identification of reactants and products is arbitrary and depends on the definition of forward and reverse directions for the reaction. The ability to map any single reaction onto a vector means that any set of steps comprising a mechanism may be mapped onto a set of vectors. For example, each of the ten steps of the mechanism of FIGURE 1A may be mapped onto its respective vector and the resulting set of ten vectors tabulated to define a Step by Species matrix, which is the vector space representation of this mechanism (FIG. 1B). Each horizontal row of this matrix corresponds to a specific step, the nth row corresponding to the nth step, and the absolute value of each vertical column element gives the stoichiometric coefficient of the respective species in the correspondig reaction step. Thus, in FIGURE 1B, the nth column $[1 \Leftarrow n \Leftarrow 8]$ corresponds to the nth intermediate state of the carrier, and the 9th and 10th columns correspond to the terminal species, S_i and S_o, respectively. Zeros indicate that the corresponding species do not participate in that step. Note that the species are divided into two groups: intermediate species, which are states of the carrier, and terminal species, which are reactants and products of the overall reaction. This separation provides the basis for defining a Step by Intermediate Species part of the matrix (first 8 columns) and a Step by Terminal Species part of the matrix (last 2 columns).

Transforming the Matrix Mapping of the Mechanism

Once a mechanism has been mapped onto a matrix the methods of linear algebra can be used to identify a set of linearly independent cycles: by definition, any given cycle within such a set cannot be obtained by addition or subtraction of any linear combination of the other cycles in this set. This involves performing a set of mathematical operations on the Step by Intermediate species part of this matrix to transform it to its row reduced form. This set of mathematical operations, the process of row reduction, simply involves adding or subtracting one or more rows of the original matrix from one or more other rows of this matrix in accordance with a set of clearly defined rules. As part of the present application, one extends

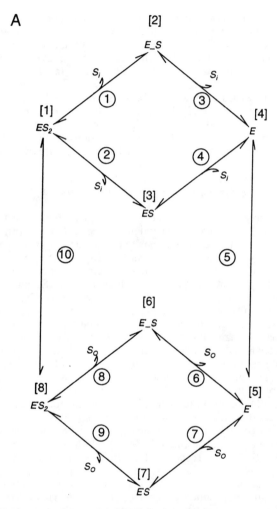

FIGURE 1. (A) Hypothetical mechanism for a second order uniport involving the binding of a chemical species (S) to a membrane bound enzyme (E) and a subsequent translocation of this species from one side of the membrane to the other two particles at a time. Numbers in *square brackets* denote intermediate states of the protein (often referred to as a carrier or porter). Two major conformational states of the carrier are identified, E and E', according to whether the species binding sites are free accessible from the intracellular compartment or form the extracellular compartment, respectively. *Encircled numbers* denote reaction steps. Forward reaction for each step is defined as that direction that advances the overall reaction ($2S_o \leftrightarrow 2S_i$) in the forward direction. Two distinct binding sites are assumed and thus four distinct forms of the partially loaded carrier are identified, states 2 and 6, and 3 and 7, according to which of the sites is loaded. Steps 5 and 10, the *vertical arrows*, represent conformational transitions of the free and loaded forms of the carrier, respectively. All other reaction steps involve the binding and dissociation of substrate. **(B)** The Step by Species matrix mapping of the mechanism of (A). Every step of this mechanism was mapped onto a respective vector (using the method outlined in the text) and the resulting set of ten vectors (one for each step) tabulated to define the Step by Species matrix.

B

		[1]	[2]	[3]	[4]	[5]	[6]	[7]	[8]	S_i	S_o
step	①	-1	1	0	0	0	0	0	0	1	0
	②	-1	0	1	0	0	0	0	0	1	0
	③	0	-1	0	1	0	0	0	0	1	0
	④	0	0	-1	1	0	0	0	0	1	0
	⑤	0	0	0	-1	1	0	0	0	0	0
	⑥	0	0	0	0	-1	1	0	0	0	-1
	⑦	0	0	0	0	-1	0	1	0	0	-1
	⑧	0	0	0	0	0	-1	0	1	0	-1
	⑨	0	0	0	0	0	0	-1	1	0	-1
	⑩	1	0	0	0	0	0	0	-1	0	0

FIGURE 1 (*Continued*)

these row operations (defined by reference to the Step by Intermediate species part of the matrix) to the rest of the matrix. The resulting matrix is presented in FIGURE 2A. Each row of this new matrix is a linear combination of rows from the original matrix obtained by adding together, and/or subtracting, rows from the original matrix. The combinations of original matrix rows corresponding to the rows of the new matrix are indicated at the left.

Inspection of the row reduced matrix (FIG. 2A) yields the number of linearly independent cycles within the mechanism, which equals the number of null rows (containing only zeros) in the Step by Intermediate species part of the new matrix. Rows 8, 9 and 10 are the three null rows in this part of the matrix in FIGURE 2A. Each of these rows is associated with a unique cycle whose nature may be determined by inspection of the sum given to the left of the row. For this purpose, addition or subtraction of original matrix rows may be thought of as accounting for reaction in the forward or reverse direction across the corresponding step. For example, the first null row, row 8, results from adding together rows 2 and 4 from the matrix of FIGURE 1B and subtracting rows 1 and 3 from this sum. The cycle associated with this row (cycle 1) thus comprises forward reaction across steps 2 and 4 and reverse reaction across steps 1 and 3 (FIG. 2B). Likewise, null rows 9 and 10 are associated with reverse reaction across steps 7 and 9 and forward reaction across steps 6 and 8 (cycle 2), and forward reaction across steps 1, 3, 5, 7, 9 and 10 (cycle 3), respectively (FIG. 2B, right).

The overall reaction associated with a given cycle may be deduced by direct inspection of the corresponding row of the Step by Terminal Species part of the

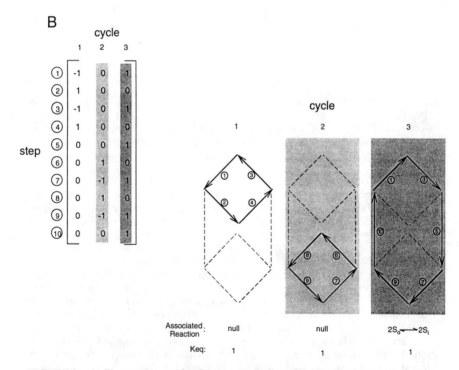

FIGURE 2. (A) The matrix resulting from row reduction of the Step by Intermediate species partition of the matrix presented in FIGURE 1B. Each row of this matrix is a linear combination of rows from the matrix of FIGURE 1B. The linear combinations of rows of the matrix of

row reduced matrix (last 2 columns, FIG. 2A). For example, for null rows 8 and 9, these last entries are both zero, meaning that each of the corresponding cycles is null. On the other hand, the bottom row has entries of 2 and -2 in the 9th and 10th columns, respectively, indicating that the cycle represented by this row is associated with the overall transport reaction: $2S_o \leftrightarrow 2S_i$. The steps of the mechanisms comprising each of the three linearly independent cycles, with their relevant directionality, are summarized in the Step by Cycle matrix presented in FIGURE 2B (left). Each column of this matrix represents a cycle derived from a null row of the row reduced matrix (FIG. 2A). Below the illustrations of these cycles (FIG. 2B, right) are statements of their overall reactions and their associated equilibrium constants.

Although the set of three linearly independent cycles identified above does not include every possible cycle within the mechanism, it provides the basis for enumerating all such cycles.[3] In the present example, there are a total of six possible cycles throughout the mechanism: these reflect reaction across steps (a) 2, 4, -3 and -1, (b) 6, 8, -9 and -7, (c) 1, 3, 5, 7, 9 and 10, (d) 1, 3, 5, 6, 8 and 10, (e) 2, 4, 5, 7, 9 and 10, and (f) 2, 4, 5, 6, 8 and 10. The first three cycles correspond to those already identified above, while the others correspond to their linear combinations. Thus, cycles d, e and f may be obtained by addition of cycles b and c, a and c, and a, b and c, respectively.

Deriving the Thermodynamic Constraints on Assignment of Rate Coefficients to the Mechanism

If a set of kinetic parameters for the mechanism is such that detailed balance is satisfied for the first three cycles (linearly independent cycles), then it can be shown that all other cycles through the mechanism (necessarily linear combinations of these cycles) will also satisfy this principle.[3] Accordingly, the constraints these parameters must fulfill can be explicitly identified by applying the principle of detailed balance to each of the first three cycles:

$$(K_2 . K_4)/(K_1 . K_3) = 1 \qquad \text{(Cycle 1)}$$

$$(K_6 . K_8)/(K_7 . K_9) = 1 \qquad \text{(Cycle 2)}$$

$$K_1 . K_3 . K_5 . K_7 . K_9 . K_{10} = 1 \qquad \text{(Cycle 3)}$$

where K_i denotes the equilibrium constant for the ith step (the ratio of the forward rate coefficient to the reverse rate coefficient for the ith step), defined according to the directionality specified for each step. The number on the right hand side corresponds to the equilibrium constant for the overall reaction mediated by the given cycle. The number of terms on the left hand side corresponds to the number

---

FIGURE 1B corresponding to the rows of this matrix are presented to its *left*. Mathematically, the new matrix contains the same information that was in the original Step by Species matrix but in a more informative format. **(B)** The Step by Cycle matrix tabulating the steps of the mechanism of FIGURE 1A that comprise a set of three linearly independent cycles through this mechanism. Each of these three cycles is illustrated to the *right* of the Step by Cycle matrix and below each illustration is a statement of the overall reaction mediated by the cycle and the associated equilibrium constant.

of steps in the corresponding cycle. If reaction occurs across a given step in a forward or reverse direction as part of a cycle, then the term for that step (its equilibrium constant) appears as part of the numerator or denominator, respectively, of the corresponding equation. If a set of kinetic parameters assigned to the mechanism of FIGURE 1A satisfies these three equations, then detailed balance has been achieved for all six cycles through the mechanism.

The approach described above is completely general. Any mechanism can be mapped onto a vector space as outlined, whereupon row reduction allows identification of a set of linearly independent cycles through this mechanism. It can be shown that this set is maximal in the sense that all possible cycles through the mechanism may be derived as linear combinations of the cycles within this subset.[3] This means that the problem of satisfying detailed balance for the mechanism may be reduced to that of ensuring that the linearly independent cycles satisfy this principle. This is because, if the linearly independent cycles satisfy this principle then all other possible cycles (which necessarily correspond to linear combinations of these cycles) must also satisfy it. The constraints that must be satisfied for the linearly independent cycles to satisfy detailed balance may be derived by application of this principle to each such cycle. In this way, a set of mathematical equations is obtained, one for each independent cycle. These equations are the thermodynamic constraints that apply to assignment of rate coefficients in the development of any kinetic model based on the particular mechanism. Although the method was outlined within the context of a set of reactions for a given enzyme catalyzed process, it is applicable to any set of reactions. For example, it is applicable to sets of elementary or complex reactions comprising overlapping metabolic pathways that are mediated by the action of a number of distinct enzymes. In addition, these methods are readily amenable to computer based implementation and are currently being incorporated into an existing software package (SCoP; distributed by Simulation Resources of Berrien Springs, MI[4]) to serve as a tool to ensure that quantitatively explicit formulations of kinetic models for biological processes are consistent with The Second Law of Thermodynamics.

ACKNOWLEDGMENTS

The authors want to thank David C. Gadsby for his support and encouragement and Mari Kuwabara for assistance in the preparation of figures.

REFERENCES

1. HILL, T. L. 1977. Free Energy Transduction in Biology. Academic Press. New York.
2. HAPPEL, J. & P. H. SELLERS. 1992. J. Phys. Chem. **96:** 2593–2597.
3. WAGG, J. & P. H. SELLERS. 1996. To be submitted.
4. KOOTSEY, J. M. 1995. This volume.

Effects of External Monovalent Cations on Na^+-Ca^{2+} Exchange in Cultured Rat Glial Cells[a]

ANDREA HOLGADO AND LUIS BEAUGÉ

Instituto de Investigación Médica "M. y M. Ferreyra"
Casilla de Correo 389
5000 Córdoba, Argentina

Only recently has attention been paid to Na^+-Ca^{2+} exchange in glia.[1,2] In this work we looked into the effects of external monovalent cations on the Na^+-Ca^{2+} exchanger in primary glial cultures and pure type I astrocytes. Cells were isolated from the brain cortex of 1–2-day-old rats according to Ref. 3 for primary cultures and to Refs. 3, 4 for type I astrocytes. Na^+-Ca^{+2} exchange was estimated as the Na_i^+-stimulated $[^{45}Ca]Ca^{2+}$ uptake (see Ref. 5). A threefold increase in $[Na^+]_i$ was obtained by adding 1 mM ouabain to the culture media 2 hours before the $[^{45}Ca]Ca^{2+}$ uptake. Calcium uptake solutions had 1 mM $MgCl_2$, variable $[^{45}Ca]CaCl_2$ and enough Bis-Tris.Propane.HCl + NaCl, KCl or LiCl to reach 330 mosmol. $[^{86}Rb]K$ uptake was measured in 40 mM KCl, 130 mM Bis-Tris.Propane.HCl, 0.1 mM EGTA and 1 mM $MgCl_2$ with or without added calcium. The solutions also contained 1 mM ouabain, 400 nM tetrodotoxin and 100 μM Bumetanide.[6] Temperature was 37°C and pH 7.4.

The results showed that glial Na^+-Ca^{+2} exchanger has an external site where K^+, Li^+ and Na^+ stimulate the reversal mode. The mechanism of that stimulation can be explained by an increase in the apparent affinity of the carrier for $[Ca^{2+}]_o$ without effect on the maximal translocation rate. The data fit two models equally well: (i) the formation of ECa_o is essential for the binding of the monovalent cation, and (ii) the activating cation can bind even when the carrier is free of $[Ca^{2+}]_o$ (see curve fitting to data points in FIG. 1). K^+ and Li^+ produced only stimulation, although that of K^+ seemed to require actions other than the chemical effect (FIG. 1). The response to Na^+ was biphasic (FIG. 1); this can be accounted for by considering that at low concentrations $[Na^+]_o$ binds mainly to the activating monovalent site while at high concentrations it displaces Ca^{2+} from its external transporting site. The monovalent cations are not transported during the exchange cycle (FIG. 2). Therefore, glia exchanger resembles those of cardiac myocytes and squid nerves and differs from those seen in retinal rod and platelet (see Refs. 7 and 8 for references).

REFERENCES

1. GOLDMAN, W. F., P. J. YAROWSKY, M. JUHASZOVA, B. K. KRUEGER & M. P. BLAUSTEIN. 1994. J. Neurosci. **14:** 5834–5843.
2. TAKUMA, K., T. MATSUDA, H. HASHIMOTO, S. ASANO & A. BABA. 1994. Glia **12:** 336–342.

---

[a] Supported by Grants PID-BID 1053 from CONICET AND 3102/94 from CONICOR.

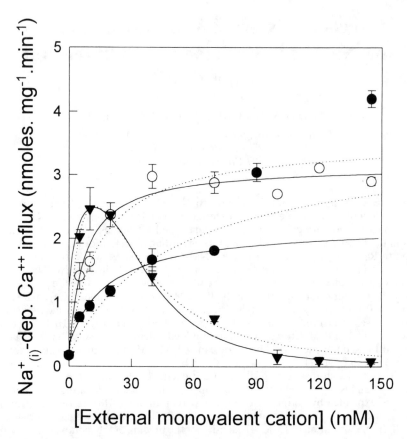

FIGURE 1. $[Na^+]_i$-dependent $[^{45}Ca]Ca^{2+}$ influx in primary glial cultures as a function of the concentration of external K^+ (*filled circles*), Li^+ (*open circles*) and Na^+ (*filled triangles*) at a limiting (0.2 mM) external Ca^{2+} concentration. The osmolarity of the solutions was kept constant at 330 mosmol with Bis-Tris.Propane.HCl. All values correspond to the influx of Ca^{2+} in high Na^+ glia minus that observed in glia with normal Na^+ content. *Data points* are the means of two to four different experiments; *vertical bars* are the SEM which corresponds. The *lines* through the points are the best fits to a model where the formation of ECa_o is essential (*solid lines*) or irrelevant (*broken lines*) for the binding of the monovalent cation to the carrier. Identical results were obtained with pure type I astrocytes.

3. MCCARTHY, K. D. & J. DE VELLIS. 1980. J. Cell Biol. **85:** 890–902.
4. BANKER, G. & K. GOSLIN. 1991. Astroglia in culture. *In* Culturing Nerve Cells. G. Banker & K. Goslin, Eds. 309–336. MIT Press. Cambridge, MA.
5. SMITH, J. B., E. J. CRAGOE & L. SMITH. 1987. J. Biol. Chem. **262:** 11988–11994.
6. LONGO, N., L. D. GRIFFIN & L. J. ELSAS. 1991. Am. J. Physiol. **260:** 1341–1346.
7. CONDRESCU, M., H. ROJAS, A. GERARDI, R. DIPOLO & L. BEAUGÉ. 1990. Biochim. Biophys. Acta **1024:** 198–202.
8. KIMURA, M., A. AVIV & J. P. REEVES. 1993. J. Biol. Chem. **268:** 6874–6877.

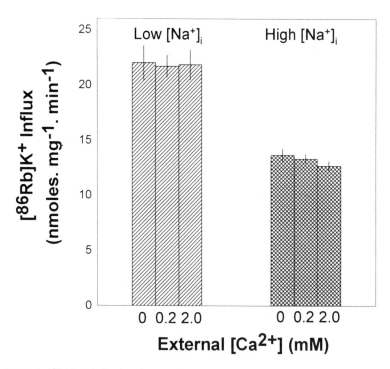

FIGURE 2. [^{86}Rb]K$^+$ influx in primary cultured glia with normal (*hatched bars*) and elevated (*crosshatched bars*) Na$^+$ content incubated in Ca^{2+}-free media and media containing 0.2 mM and 2 mM Ca^{2+} concentrations. Note the lack of effect of external Ca^{2+} on the [^{86}Rb]K$^+$. Identical results were obtained with pure type I astrocytes.

ATP Stimulation of a Na$^+$ Gradient-Dependent Ca^{2+} Uptake in Cardiac Sarcolemmal Vesicles

GRACIELA BERBERIÁN AND LUIS BEAUGÉ

Instituto de Investigación Médica "M. y M. Ferreyra"
Casilla de Correo 389
5000 Córdoba, Argentina

In intact cells and dialyzed squid axons Na$^+$-Ca^{2+} exchange is modulated (stimulated) by ATP (see Refs. 1 and 2). A MgATP stimulation was also shown in giant membrane patches excised from cardiac myocytes.[3] In isolated cardiac membrane vesicles, one reported MgATP activation[4] was not seen by others (see Ref. 5). We therefore decided to reinvestigate the MgATP actions in the cardiac preparation. Cardiac membrane vesicles prepared by differential centrifugation[6] were loaded with 160 mM NaCl, 0.1 mM EDTA and 20 mM MOPS.Tris (pH 7.4 at 37°C). Usually there were around 38% inside out, 36% rightside out and 26% leaky vesicles. ^{45}Ca uptake was measured in media with high (160 mM) and low (10 mM) Na$^+$, 20 mM MOPS.Tris (pH 7.4 at 37°C) and 0.1 mM digitoxigenin. Low Na$^+$ had the equivalent to 260 mosmols of NMG.Cl, Bis.Tris.Propane.Cl (BTP) or KCl.

In the presence of 1 μM Ca^{2+}, 0.5 mM vanadate (V) and 5 mM Mg^{2+}, 1 mM ATP increased two-to-fourfold a Na$^+$ gradient-dependent Ca^{2+} influx; in the absence of only gradient a 20–40% increase was observed. This ATP stimulation had the following characteristics (TABLES 1 and 2): (i) it was seen in NMG and BTP but not in K$^+$ media; (ii) 1 mM ATP-τ-S also stimulated whereas 2 mM AMP-PCP did not; (iii) it was not observed in the presence of 20 μM eosine (E) without V but was reinstated when E and V were together; ATP-τ-S stimulation did not require V. In addition we were able to demonstrate (results not shown here) that: (i) the ATP effect was due to an increased apparent affinity for Ca^{2+}; a similar K_m^{Ca} shift (from 1–2 μM to 0.1–0.2 μM) was caused by K$^+$ with and without ATP; (ii) the $K_{0.5}^{ATP}$ was about 0.5 mM; (iii) the vesicles lost all Ca^{2+} when the ionofore A23187 was added.

Though not definite, it is likely that these fluxes go through the Na$^+$-Ca^{2+} exchanger. In the event they do, there are two main questions: why it was not always detected, and the mechanisms underlying this regulation. On the first, it is quite possible that even minimal differences in the experimental conditions are critical. Also, it must be noted that our results do not coincide completely with others who found an ATP effect in heart preparations. Thus, comparing with excised patch data we observed (i) a much lower K_m^{ATP}, (ii) that ATP-τ-S also stimulates and (iii) that vanadate is required. In contrast with the report in Ref. 4 we do not see an ATP effect in the presence of high [K$^+$] where the K_m^{Ca} is 10 to 20 times lower. Regarding the mechanism, the fact that in the absence of vanadate (with the Ca^{2+} pump inhibited by eosine[7]) there is no ATP stimulation suggests a mechanism similar to that operating in dialyzed squid axons,[1] involving perhaps associated kinase(s)-phosphatase(s) enzymes as proposed earlier.[1,4]

TABLE 1. Effect of Cation Composition in the Extravesicular Medium on the Ca^{2+} Uptake by Bovine Heart Sarcolemmal Vesicles in the Absence and Presence of ATP[a]

i/o	^{45}Ca Uptake (nmoles · mg^{-1} · s^{-1})	
	− ATP	+ ATP
Na$^+$/Na$^+$	0.22 ± 0.04	0.36 ± 0.07
Na$^+$/NMG	1.15 ± 0.16	2.61 ± 0.12
Na$^+$/BTP	0.93 ± 0.13	2.73 ± 0.23
Na$^+$/K$^+$	2.09 ± 0.10	2.18 ± 0.30

[a] ^{45}Ca uptake was measured in media with high (160 mM) and low (10 mM) Na$^+$, 20 mM MOPS.Tris (pH 7.4 at 37°C), 5 mM Mg^{2+}, 0.1 mM EGTA, 0.5 mM vanadate, 0.1 mM digitoxigenin with or without 1 mM ATP. In KCl media there was also 1 μM valinomycin. [Ca^{2+}] was 1 μM. Low Na$^+$ had the equivalent to 260 mosmols of NMG.Cl or Bis.Tris.Propane.Cl or KCl. All entries are the mean ± SD of triplicate determinations. Note that i/o indicates the principal monovalent cation in the intra (i) or extra (o) vesicular media.

TABLE 2. Effects of Eosin and Vanadate, Separate or in Combination, on the Ca^{2+} Uptake of Sarcolemmal Vesicles in the Presence of High (Gradient-Independent) and Low (Gradient-Dependent) Extravesicular [Na$^+$] and of 1 mM ATP[a]

Assay Conditions	160 mM [Na$^+$]$_o$	10 mM [Na$^+$]$_o$
	Ca^{2+} Uptake (nmoles · mg^{-1} · s^{-1})	
Control	4.00 ± 0.12	n.d.
20 μM eosin	0.32 ± 0.05	0.39 ± 0.17
0.5 mM vanadate	0.18 ± 0.01	2.91 ± 0.32
20 μM eosin + 0.5 mM vanadate	0.15 ± 0.01	2.33 ± 0.35
1 mM ATP-τ-S without vanadate	0.45 ± 0.08	2.43 ± 0.22

[a] (i) [Ca^{2+}] was 1 μM and [Mg^{2+}] was 5 mM; (ii) in the absence of ATP the Na$^+$ gradient dependent was not statistically different in eosine, vanadate and eosine plus vanadate (not shown here); (iii) the last row shows that ATP-τ-S stimulation does not require vanadate; (iv) (n.d.) means not determined.

REFERENCES

1. DiPolo, R. & L. Beaugé. 1994. Am. J. Physiol. **266:** C1382–C1391.
2. Reeves, J. P., M. Condrescu, G. Chernaya & J. P. Gardner. 1994. J. Exp. Biol. **196:** 375–388.
3. Collins, A., A. V. Somlyo & D. W. Hilgemann. 1992. J. Physiol. **454:** 27–57.
4. Caroni, P. & E. Carafoli. 1983. Eur. J. Biochem. **132:** 451–460.
5. Reeves, J. P. & K. D. Philipson. 1989. *In* Na-Ca Exchange. T. Allen, D. Noble & H. Reuter, Eds. 27–53. Oxford University Press. Oxford.
6. Reeves, J. P. 1991. *In* Cellular Calcium. J. G. McCormack & P. H. Cobbold, Eds. 283–297. Oxford University Press. Oxford.
7. Gatto, C., C. C. Hale & M. A. Milanick. 1994. Biophys. J. **66:** 331 (Abstract).

Modifications of XIP, the Autoinhibitory Region of the Na-Ca Exchanger, Alter Its Ability to Inhibit the Na-Ca Exchanger in Bovine Sarcolemmal Vesicles[a]

C. GATTO,[b] W-Y. XU,[b] H. A. DENISON,[b] C. C. HALE,[c,d]
AND M. A. MILANICK[b,d,e]

Department of [b]Physiology
[c]Department of Veterinary Biomedical Sciences
[d]Dalton Cardiovascular Research Center
University of Missouri
Columbia, Missouri 65212

Li et al.[1] reported that a synthetic 20-amino acid peptide (RRLLFYKYVYKRYR-AGKQRG) corresponding to residues 251–270 of the Na-Ca exchanger[2] bound calmodulin and, most interestingly, the peptide also inhibited Na-Ca exchange, hence the name, XIP for exchange inhibitory peptide. Li et al.[1] described some initial structural constraints on the peptide that were necessary for inhibition. XIP inhibits less well at high ionic strength.[3] This suggests that the placement of one or more of the positive charges is important. It is possible to make an educated guess as to which of the 8 basic residues are most important by examining the result of Li et al.[1] Removal of the first 5 residues of XIP (XIP 15) only slightly reduced the inhition. The calmodulin-binding domain of the PM Ca pump (C28R2) is also a fairly potent inhibitor of Na-Ca exchange. Any 15mer peptide made from C28R2 has at most 4 positive charges. There are three different alignments in which 3 positive charges are in register. In one alignment, all three charges are lysines; in the other two alignments, two of the charges are lysines. Lysines were modified to determine if one or more of them was most important.

We have chemically modified the lysines in XIP with sulfosuccinimidyl acetate (SNA) by incubation at pH 9 at 37°C for 1 hour (TABLE 1). We previously reported[4] that when XIP was incubated with a 10-fold excess of SNA (per mole of XIP) the IC_{50} for the modified peptide was only slightly greater than for the native peptide. In contrast, when XIP was incubated with a 30-fold excess of SNA, the IC_{50} increased >10-fold; thus at least one of the positive charges at positions 7, 11, and 17 is important. SNA did not alter the IC_{50} of Arg XIP in which all 8 positive charges are Arg, suggesting that SNA does not react with Arg under these conditions. (Note that we previously found that Lys-XIP, Arg-XIP, and XIP all inhibit similarly.[4]) The arginines could also be important; however, we have not found an arginine-specific reagent. Hydroxyphenylglyoxal increased the IC_{50} for Lys-

---

[a] Supported by NIH DK37512 (MAM), a RCDA from NIH-DK (MAM), and AHA (CCH).
[e] Address for correspondence: Mark Milanick, Dept. of Physiology, Room MA415 Med. Sci. Bldg., University of Missouri, Columbia, MO 65212.

TABLE 1. Summary of Results of Chemical Modification of XIP and Analogs[a]

Position	Status	Result
1	not critical	XIP 15 inhibits[b]
2	not critical	XIP 15 inhibits[b]
12	not critical	peptide 1 inhibits[c]
14	not critical	peptide 1 inhibits[c]
19	not critical	SNA-peptide 1 inhibits
7, 11, or 17	critical	SNA-XIP does not inhibit

[a] Details in text.
[b] See Reference 1.
[c] But compare Reference 5.

XIP by >5-fold; in lys-XIP the arginines have been replaced by lysines so that all 8 positive charges are from lysine.

Peptide 1 (RRLLFYRYVYRCYCAGRQKG) inhibits the Na-Ca exchanger; in five different experiments the ratio of the IC_{50} for peptide 1 to the IC_{50} for XIP was 1, 1.1., 2, 5 and 6. In addition, peptide 1 was reacted with IAA, to introduce negative charges at positions 12 and 14. This modified peptide had a similar IC_{50}. This suggests that positive charges at positions 12 and 14 are not very important. We have also found that peptide 1 modified with SNA still inhibits. Peptide 1 has only 1 lysine, at position 19. Thus it appears that the positive charges at 1, 2, 12, 14, and 19 are not as critical. This is consistent with the SNA XIP data, which suggested that at least one of the positive charges at 7, 11 or 17 was critical. These results are in contrast to the results of He and Philipson.[5] This difference may reflect a species difference (bovine vs canine), a difference in assay conditions or other unidentified reasons.

REFERENCES

1. Li, Z., D. A. Nicoll, A. Collins, D. W. Hilgemann, A. G. Filoteo, J. T. Penniston, J. N. Weiss, J. M. Tomich & K. D. Philipson. 1991. Identification of a peptide inhibitor of the cardiac sarcolemmal Na-Ca exchanger. J. Biol. Chem. **266:** 1014–1020.
2. Nicoll, D. A., S. Longoni & K. D. Philipson. 1990. Molecular cloning and functional expression of the cardiac sarcolemmal Na^+-Ca^{2+} exchanger. Science **250:** 562–565.
3. Kleiboeker, S. B., M. A. Milanick & C. C. Hale. 1992. Interactions of the exchange inhibitory peptide (XIP) with Na-Ca exchange in bovine cardiac sarcolemmal vesicles and ferret red cells. J. Biol. Chem. **267:** 17836–17841.
4. Denison, H. A., B. J. WIlson, C. C. Hale & M. A. Milanick. 1994. Modification of the positive charges of the exchange inhibitory peptide (XIP) decreases its inhibition of Na-Ca exchange activity. Biophys. J. **66:** A331.
5. He, Z. & K. D. Philipson. 1995. Identification of important amino acid residues of exchanger inhibitor peptide XIP. Biophys. J. **68:** A410.

Use of Cysteine Replacements and Chemical Modification to Alter XIP, the Autoinhibitory Region of the Na-Ca Exchanger

Inhibition of the Activated Plasma Membrane Ca Pump[a]

W-Y. XU, C. GATTO, C. J. ALLEN,
AND M. A. MILANICK[b]

Department of Physiology
University of Missouri
Room MA415 Medical Science Building
Columbia, Missouri 65212

The plasma membrane (PM) calcium pump and the sodium-calcium exchanger have homologous autoinhibitory domains, called C28 and XIP,[1] respectively. Gentle proteolysis activates the plasma membrane calcium pump, mimicking many features of the activation observed with calmodulin (see Ref. in Ref. 2). Both XIP and C28 contain several positively charged amino acid residues which are believed to be involved in the binding of these peptides to the PM Ca pump, the Na-Ca exchanger, and calmodulin. To gain further insight into the interactions of XIP and C28 with the calcium pump and calmodulin we studied other similarly charged peptides (VIP, PACAP, H9685), synthetic analogs of XIP, and chemical modification of these peptides (TABLE 1). Ca pump activity was assayed by ^{45}Ca uptake into inside-out vesicles from human red blood cells. Calmodulin binding was determined by changes in dansylated calmodulin fluorescence.

XIP inhibits the activated plasma membrane Ca pump, confirming the work of Enyedi and Penniston.[2] In addition, PACAP (1-38) inhibits the Ca pump with an IC_{50} in the range of 10 to 20 μM. In contrast, several fragments of PACAP (1-27, 16-38, 28-38) did not inhibit the pump (IC_{50} values >50 μM). Also, VIP (10-28), which is analogous to PACAP (10-28) and the positive peptide, Bachem peptide #H9685 (QRRQRKSRRTI), do not inhibit.

Sulfo-NHS-acetate (sulfosuccinimidyl acetate, SNA) modified XIP (RRLLFY-KYVYKRYRAGKQRG) no longer inhibits the Ca pump. As SNA reacts preferentially with lysines, this suggests that at least one of the positive charges at positions 7, 11, or 17 is important. SNA did not react with Arg XIP in which all 8 positive charges are Arg, confirming that SNA does not react with Arg under these conditions. The IC_{50} for peptide 1 (RRLLFYRYVYRCYCAGRQKG), an analog of XIP, was similar to the IC_{50} for XIP; this suggests that positive charges at residues 12 and 14 are not critical to binding. SNA increased the IC_{50} for peptide 1 from

---

[a] Supported by NIH DK37512 and a RCDA from NIH-DK to MAM.
[b] Corresponding author.

TABLE 1. Summary of Results of Chemical Modification of XIP and Analogs[a]

Position	Status	Result
1	not test	
2	not tested	
12	not critical, but may contact	peptide 1 inhibits[b]
14	not critical, but may contact	peptide 1 inhibits[b]
19	moderately important	SNA-peptide 1 inhibits weakly
7, 11, or 17	critical	SNA-XIP does not inhibit

[a] Details in text.
[b] IAA appears to prevent inhibition.

8 μM to 30 μM, 10% of the free energy for binding, assuming that IC_{50} = Kd. SNA would modify lysine 19 in peptide 1; thus the positive charge at position 19 is not critical, but may be of mild importance. Iodoacetic acid (IAA) increased the IC_{50} for peptide 1. IAA introduces negative charges at 12 and 14; thus at least one of these residues lies close enough to the protein that a negative charge can prevent inhibition, even though a positive charge is not required. The effects of chemical modification may also be due to changes in peptide folding.

PACAP (1–27 and 16–38), peptide 1 and SNA modified XIP altered dansylcalmodulin fluorescence. Thus these peptides bind calmodulin but do not inhibit the Ca pump, confirming that the specificity for binding to calmodulin and the PM Ca pump are different. Because peptide 1 binds to calmodulin and has cysteines, it may be possible to introduce fluorescent reporter groups onto the peptide. Also, bulky groups can be added. If peptide 1 binds inside the calmodulin hydrophobic channel, bulky groups might prevent entry into the channel and therefore not only decrease the affinity for calmodulin but prevent calmodulin binding altogether. Such a modified peptide might inhibit the Ca pump but not bind to calmodulin.

We are currently testing three new peptides in which we have placed alanine, cysteine and lysine at positions 7, 11, and 17 (*e.g.*, in the sequence ACK, CKA, and KAC). Positions 1, 2, 12, 14 and 19 are arginines. By testing these three peptides and the peptides modified with SNA (to neutralize the positive charge on lysine) or MSTEA (will convert cys to a positive side chain) we can determine if 1 or 2 positive charges are required and which positions are most important.

REFERENCES

1. LI, Z., D. A. NICOLL, A. COLLINS, D. W. HILGEMANN, A. G. FILOTEO, J. T. PENNISTON, J. N. WEISS, J. M. TOMICH & K. D. PHILIPSON. 1991. Identification of a peptide inhibitor of the cardiac sarcolemmal Na-Ca exchanger. J. Biol. Chem. **266:** 1014–1020.
2. ENYEDI, A. & J. T. PENNISTON. 1993. Autoinhibitory domains of various Ca^{2+} transporters cross-react. J. Biol. Chem. **268:** 17120–17125.
3. C. J. ALLEN, S. L. JONES, C. GATTO, & M. A. MILANICK. 1995. Pituitary adenylate cyclase activating peptide (PACAP) inhibits the human erythrocyte calcium pump. Biophys. J. **68:** A371.

Effects of Phe-Met-Arg-Phe-NH$_2$ (FMRFa)-Related Peptides on Na-Ca Exchange and Ionic Fluxes in Rat Pancreatic B Cells

F. VAN EYLEN,[a] P. GOURLET,[b] A. VANDERMEERS,[b]
P. LEBRUN,[a] AND A. HERCHUELZ[a]

[a]Laboratory of Pharmacology
and
[b]Laboratory of Biochemistry and Nutrition
Free University of Brussels
Route de Lennik 808
B-1070 Brussels, Belgium

INTRODUCTION

FMRFamide-related peptides were recently shown to inhibit Na-Ca exchange in cardiac sarcolemmal vesicles.[1] These small (tetra)peptides known as molluscan cardioexcitatory peptides display naloxone-like activity and are widely distributed in mammalian tissues.[2,3] Because these peptides may represent interesting tools to modulate Na-Ca exchange, we examined their effects on ionic fluxes in intact rat pancreatic B cells.

RESULTS AND DISCUSSION

Na-Ca exchange was measured as Na$^+$-dependent Ca^{2+} uptake.[4] In the presence of 2.8 mM glucose, FMRFa almost failed to inhibit Na-Ca exchange. Related peptides such as His-Met-Arg-Phe-NH$_2$ (HMRFa), Val-Met-Arg-Phe-NH$_2$ (VMRFa) and Phe-Leu-Arg-Phe-NH$_2$ (FLRFa) were more potent than FMRFa but their inhibitory effects were far less pronounced in pancreatic B cells (36% at 100 μM) than in cardiac sarcolemmal vesicles (IC$_{50}$ of HMRFa and VMRFa ≈ 2 μM).

Since membrane potential is known to affect Na-Ca exchange activity, the effects of the two most active peptides (HMRFa and VMRFa) were examined under depolarizing conditions (16.7 mM glucose + 20 mM tetraethylammonium (TEA) or glucose + TEA + 1 μM of the Ca^{2+} channel activator BAY K 8644). Exposure of the islet cells to the peptides under depolarizing conditions greatly enhanced the inhibitory activity on Na-Ca exchange of the two most potent peptides, HMRFa and VMRFa (≈25% at 1 μM, FIG. 1). HMRFa and VMRFa failed to affect ^{86}Rb uptake and ^{45}Ca uptake, but decreased, in the nM range, ^{86}Rb outflow from glucose-exposed islet cells.

FIGURE 1. Effect of HMRFa on Na^+_i-dependent $^{45}Ca_o$ uptake in intact islet cells (reverse Na-Ca exchange) exposed to glucose (16.7 mM), TEA (20 mM) and BAY K 8644 (1 μM). Data are expressed in % of control values found at 0 mM extracellular Na^+ in the absence of the peptides, after subtraction of basal uptake measured at 139 mM extracellular Na^+. Mean ± SEM refer to at least 22 individual samples. Basal uptake and control uptake averaged 183 ± 12 (n = 22) fmol Ca/1000 cells and 310 ± 6 (n = 24) fmol Ca/1000 cells, respectively. (From Van Eylen et al.[5] Reprinted by permission from *Molecular and Cellular Endocrinology*.)

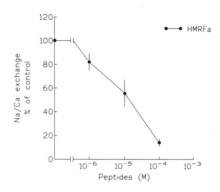

CONCLUSIONS

The present data show that FMRFa-related peptides may inhibit Na-Ca exchange in intact cells under appropriate experimental conditions but that the peptides may also alter other ionic fluxes.

REFERENCES

1. KHANANSHVILI, D., D. C. PRICE, M. J. GREENBERG & Y. SARNE. 1993. Phe-Met-Arg-Phe-NH$_2$ (FMRFa)-related peptides inhibit Na^+-Ca^{2+} exchange in cardiac sarcolemmal vesicles. J. Biol. Chem. **268:** 200–205.
2. PRICE, D. C. & M. J. GREENBERG. 1989. The hunting of the FaRPs: the distribution of FMRFamide-related peptides. Biol. Bull. **177:** 198–205.
3. RAFFA, R. 1988. The action of FMRFamide (Phe-Met-Arg-Phe-NH$_2$) and related peptides on mammals. Peptides **9:** 915–922.
4. PLASMAN, P-O., P. LEBRUN & A. HERCHUELZ. 1990. Characterization of the process of sodium/calcium exchange in pancreatic islet cells. Am. J. Physiol. **259:** E844–E850.
5. VAN EYLEN, F., P. GOURLET, A. VANDERMEERS, P. LEBRUN & A. HERCHUELZ. 1994. Inhibition of Na-Ca exchange by Phe-Met-Arg-Phe-NH$_2$ (FMRFa)-related peptides in intact rat pancreatic B-cells. Mol. Cell. Endocrinol. **106:** 1–5.

Kinetics of Na-Ca Exchange Current after a Ca^{2+} Concentration Jump

M. KAPPL AND K. HARTUNG

Max Planck Institute for Biophysics
Kennedyallee 70
60596 Frankfurt/Main, Germany

METHODS

Single ventricular myocytes from rat and guinea pig were prepared as described before.[1,2] The composition of electrolytes was chosen to minimize currents other than Na-Ca exchange current (see figure legends). The Ca^{2+} activity of all solutions was measured with a Ca^{2+}-selective electrode. Na-Ca exchange currents were measured using the giant excised patch clamp technique.[3] Bath solutions could be exchanged by an eight channel perfusion system. Photolysis of DM-nitrophen (DMN), a Ca^{2+} chelator that changes its affinity from 5 nM to 3 mM within 30 μs upon illumination with UV light,[4,5] was achieved by laser flashes (10 ns, 308 nm, 0.6 mJ), which were transmitted to the experimental chamber by an optical silica fiber. One flash photolysed about 10% of the DMN, increasing the Ca^{2+} concentration to approximately 50 μM. The pipette with the membrane was centered in front of the fiber at a distance of about 100 μm. Experiments were performed at a holding potential of 0 mV and 24°C. Current signals were recorded with a bandwidth of 3 kHz and stored on magnetic disc.

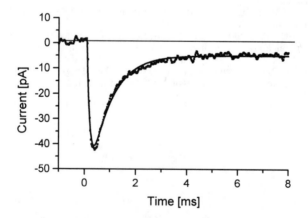

FIGURE 1. Current signal from a guinea pig membrane patch, recorded after a Ca^{2+} concentration jump to approximately 50 μM. The pipette contained (in mM) 100 NaCl, 10 EGTA, 20 CsCl, 20 TEA-Cl, 0.02 verapamil, 10 mM Hepes, pH 7.1 with N-methyl-gluconate (NMG). Bath solutions: 100 LiCl, 0.50 DMN, 0.49 $CaCl_2$, 20 CsCl, 20 TEA-Cl, 10 Hepes, pH 7.1 with NMG. The laser flash occurred at 0 ms. *Continuous line*: fit with two exponential functions and a constant. Time constants of the transient current are $\tau_1 = 0.13$ ms for the rise and $\tau_2 = 0.77$ ms for the decay to the stationary level.

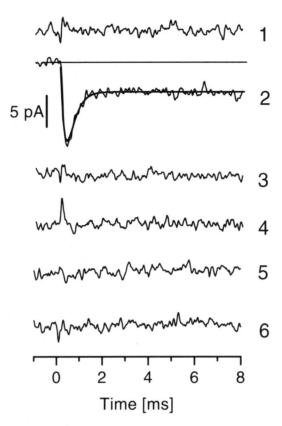

FIGURE 2. Currents recorded from a guinea pig membrane patch. Solutions as in FIGURE 1, but different concentrations of DMN and Ca^{2+} as indicated below. The moment of the laser flash is taken as origin of the time axis. All traces are from the same membrane patch. *Trace 1*: No DMN and Ca^{2+} added. Only a small artifact due to the laser flash is observed. *Trace 2*: 0.5 mM DMN and 0.49 mM Ca^{2+} added. Photolysis causes activation of transient and stationary current. Time constants of a fit with two exponential functions are $\tau_1 = 0.23$ ms for the rise and $\tau_2 = 0.25$ ms for the decay to the plateau level. *Traces 3–4*: Two further flashes, following immediately after trace 2 (exchanger fully activated), do not cause any change. *Trace 5*: 0.5 mM DMN without Ca^{2+} and 1 mM EGTA were added. No activation of transient or stationary currents is observed after the flash. *Trace 6*: 0.5 mM DMN and 0.6 mM Ca added. The exchanger current is already fully activated before the laser flash. The photolysis does not produce a transient signal and does not influence the activity of the exchanger.

RESULTS AND CONCLUSIONS

In this study we have combined the giant patch clamp technique with a Ca^{2+} concentration jump generated by the photolysis of DMN. FIGURE 1 shows the current signal generated by the photolytic release of Ca^{2+}. The signal starts with a peak rising with $\tau = 0.13$ ms and falling to a plateau with $\tau = 0.77$ ms (cf.

Ref. 6). The time constant of the rising phase is relatively constant in different experiments, but could be limited by the recording bandwidth and the release of Ca^{2+}: 0.15 ± 0.09 ms (n = 128, 32 patches), while the time constant of the decay spans a large range: 0.7 ± 0.5 ms (range 0.2 to 1.9 ms). Also the ratio between the amplitude of peak and plateau varies considerably. The kinetic behavior is not influenced by deregulation of the exchanger with chymotrypsin. There is no significant difference between rat and guinea pig with respect to the kinetics of the transient, but the current density at the plateau is approximately 2 times larger in guinea pig membranes. Because the fraction of free DMN is only 2% it can be excluded that the current transient is due to a transient rise of the Ca^{2+} concentration, which occurs if DMN is only partially loaded with Ca^{2+}.[7] A variety of control experiments has been performed to exclude that the current signal or parts of it are generated by reactive photoproducts of DMN or unspecific binding of Ca^{2+} to surface charges of the membrane. FIGURE 2 shows that in the absence of DMN (trace 1) or in the presence of DMN without Ca^{2+} (trace 5) or with $[Ca^{2+}]$ > [DMN] (trace 6) no transient current signal was induced by a laser flash. In the presence of 5 mM Ni^{2+} in the pipette neither a transient nor a stationary current was induced by the photolytic release of Ca^{2+}. Superfusion with a Ca^{2+}-containing solution also did not elicit a stationary current. Correct patch configuration was demonstrated by the activation of the Na-K ATPase current. These control experiments show that under conditions where no activation of Na-Ca exchange current is expected, the photolysis of DMN does not generate a current signal. It seems also not very likely that the transient current is due to gating charges of ion channels, because this should be an outward current if the membrane is depolarized by the screening of negative surface charges on the cytoplasmatic side.[8] If the current transient is indeed related to the Na-Ca exchange current its decay is probably determined by the rate limiting step of the reaction cycle which determines the turnover number. Thus we conclude that under our conditions the turnover number is >1000/s in accordance with earlier reports. This conclusion is even valid if the transient is completely unrelated to the Na-Ca exchanger, because in any case the steady state of the Na-Ca exchanger is reached in a millisecond or less.

REFERENCES

1. ISENBERG, G. & U. KLÖCKNER. 1982. Pflügers Arch. **395:** 6–18.
2. YAZAWA, K., M. KAIBARA, M. OHARA & M. KAMEYAMA. 1990. Japanese J. Physiol. **40:** 157–163.
3. HILGEMANN, D. W. 1989. Pflügers Arch. **415:** 247–249.
4. KAPLAN, J. & G. C. R. ELLIS-DAVIS. 1988. Proc. Natl. Acad. Sci. USA **85:** 6571–6575.
5. VERGARA, J. & A. ESCOBAR. 1993. Biophys. J. **64:** A34.
6. NIGGLI, E. & W. J. LEDERER. 1991. Nature **349:** 621–624.
7. ZUCKER, R. S. 1993. Cell Calcium **14:** 87–100.
8. BEAN, P. B., & E. RIOS. 1989. J. Gen. Physiol. **94:** 65–93.

Calcemic Hormones Regulate the Level of Sodium-Calcium Exchange Protein in Osteoblastic Cells

NANCY S. KRIEGER[a]

Departments of Medicine and Pharmacology
University of Rochester School of Medicine
Rochester, New York 14642

We previously proposed a role for Na^+-Ca^{2+} exchange in hormonally regulated bone resorption[1] and subsequently characterized Na^+-Ca^{2+} exchange activity in osteoblast-like rat osteosarcoma cells (UMR-106) using the Ca-sensitive dye fura-2.[2] We recently demonstrated that Na-dependent Ca influx in these cells was inhibited by 24 h treatment with parathyroid hormone (PTH), prostaglandin E_2 (PGE_2), or 1,25(OH)$_2$ vitamin D_3 (1,25(OH)$_2D_3$) in a dose-dependent manner.[3] To further characterize this inhibition of Na^+-Ca^{2+} exchange transport, we examined the effect of PTH on the level of Na^+-Ca^{2+} exchange protein in the osteoblasts using immunoblot analysis and demonstrated that PTH decreased the level of Na^+-Ca^{2+} exchange protein in these bone cells.[4] To determine whether inhibition of Na^+-Ca^{2+} exchange activity by other calcemic agents was also associated with decreased exchanger protein levels, we have examined the effect of 24 h treatment of UMR-106 cells with PGE_2 or 1,25(OH)$_2D_3$.

UMR-106 cells, a clonal rat cell line that has differential properties of mature osteoblasts,[5] were cultured in the absence or presence of calcemic agents. The cells were collected by trypsinization, homogenized in a hypotonic buffer (25 mM Tris-HCl, pH 7.5, 2 mM $MgCl_2$, 0.1 mM TLCM, 0.1 mM PMSF) and a 7500-g membrane fraction (primarily plasma membrane, based on relative alkaline phosphatase activity) was obtained from each treatment group. Proteins were separated by SDS-PAGE, transferred to nylon membranes (Immobilon-P, Millipore) and immunoblotted with a polyclonal rabbit antibody to canine cardiac Na^+-Ca^{2+} exchanger (generously provided by K. D. Philipson). The detection system utilized a horseradish peroxidase-coupled secondary antibody and chemiluminescence (ECL, Amersham). The developed autoradiographs were quantitated using an LKB Ultroscan XL laser densitometer.

In UMR cell membranes, a specific protein band was detected at ~90 kD that cross-reacted with antibody directed against the cardiac Na^+-Ca^{2+} exchanger. The level of this protein was decreased 36% and 37% after 24 h treatment of cells with 10^{-6} M PGE_2 or 10^{-8} M 1,25(OH)$_2D_3$, respectively (FIG. 1). This was a significant decrease, although not so great as the 64% inhibition observed with a maximally effective concentration of PTH.

Although the size of the protein identified in this study is not identical to that reported for the denatured exchanger in cardiac tissue,[6] the antibody reaction is

---

[a] Address correspondence to: Nancy S. Krieger, Ph.D., Department of Medicine, Box 675, University of Rochester School of Medicine and Dentistry, 601 Elmwood Avenue, Rochester, New York 14642.

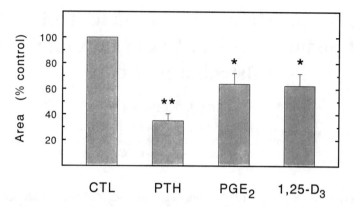

FIGURE 1. The effect of 24 h treatment in culture with calcemic hormones on the level of Na^+-Ca^{2+} exchange protein in osteosarcoma cells. Confluent flasks of UMR-106 cells were incubated in the absence or presence of 10^{-7} M PTH, 10^{-6} M PGE_2 or 10^{-8} M $1,25(OH)_2D_3$ for 24 h. Membranes were then isolated, and 8 μg of protein from the 7500-g pellet from each treatment group were separated on a 7.5% acrylamide gel and immunoblotted using an antibody to the canine cardiac Na^+-Ca^{2+} exchanger. Autoradiographic films obtained after chemiluminescent immunoblot detection were scanned using a laser densitometer. Data presented are the combined results of several experiments. Scans of the protein band at 90 kD were normalized by defining the area of the band from control cells in each experiment as 100%. The density of the band from each treated group was then calculated as a percent of its concurrently run control. Data are the mean ± SE (n = 5–12/group). * $p < 0.05$; ** $p < 0.001$ compared to control.

specific and reproducible in this system. Thus, the previously reported inhibition of Na^+-Ca^{2+} exchange activity by calcemic agents in rat osteoblastic cells appears to be due to regulation of synthesis of the exchanger protein. We have not yet determined whether there is also regulation at the level of transcription. This Na^+-Ca^{2+} exchange transport process may play an integral role in the long-term response of the osteoblast to these calcemic agents.

REFERENCES

1. KRIEGER, N. S. & A. H. TASHJIAN, JR. 1980. Nature **287**: 843–845.
2. KRIEGER, N. S. 1992. J. Bone Min. Res. **8**: 1105–1111.
3. SHORT, C. L., R. D. MONK, D. A. BUSHINSKY & N. S. KRIEGER. 1994. J. Bone Min. Res. **9**: 1159–1166.
4. KRIEGER, N. S. 1994. J. Bone Min. Res. **9**: S214.
5. PARTRIDGE, N. C., D. ALCORN, V. P. MICHELANGELI, G. RYAN & T. J. MARTIN. 1983. Cancer Res. **43**: 4308–4314.
6. VEMURI, R., M. E. HABERLAND, D. FONG & K. D. PHILIPSON. 1990. J. Membr. Biol. **118**: 279–283.

A Novel Approach for Imaging the Influx of Ca^{2+}, Na^+, and K^+ in the Same Cell at Subcellular Resolution

Ion Microscopy Imaging of Stable Tracer Isotopes

SUBHASH CHANDRA[a] AND GEORGE H. MORRISON

Department of Chemistry
Baker Laboratory
Cornell University
Ithaca, New York 14853

Ion fluxes of Ca^{2+}, K^+, and Na^+ are tightly regulated at the plasma membrane at the cost of enormous amounts of energy. These ions are intimately involved in signal transduction pathways for physiological events. Consequently, their measurements in single cells at subcellular resolution may help unravel their roles under physiological and pathological conditions. In this work we show the feasibility of using stable $^{44}Ca^{2+}$, Cs^+, and Rb^+ tracers in the extracellular solution for imaging the influx of Ca^{2+}, Na^+, and K^+, respectively, with ion microscopy. Ion microscopy is based on secondary ion mass spectrometry (SIMS), and the technique is capable of providing visual images of the distribution of any element (isotope) from H to U in relation to cell morphology.[1] The optics of the ion microscope are such that there is a one-to-one correspondence between the relative position of atoms in the sample and the final position of ions as they strike the detector. The ion images represent the surface chemistry of the sample. Since the technique distinguishes between isotopes on the basis of their mass-to-charge (m/z) ratio and the analysis is made by continuously eroding (sputtering) the cell surface, it is possible to image, sequentially, intracellular distributions of several different isotopes from the same cell. An added advantage of isotopic detection is that one can use stable isotopes as tracers for ion transport. Therefore, by using $^{44}Ca^{2+}$ in the extracellular solution, the ion microscope can image its transport in the cell interior at mass 44 (^{44}Ca), and endogenous calcium at mass 40 (^{40}Ca) from the same cell. Similarly, one can also image the transport of other tracer ions such as Rb^+ and Cs^+ from the same cell. With this approach, we have observed that the mitotic metaphase cells strictly regulate the influx of Ca^{2+}, but not Cs^+ and Rb^+, as compared to their neighboring interphase cells.

The pig kidney LLC-PK1 cells were treated with Tyrode solution containing 1.8 mM $^{44}Ca^{2+}$ ($^{44}CaCl_2$, 98.78% ^{44}Ca enrichment) rather than naturally abundant $^{40}CaCl_2$. In addition, the solution also contained 2 mM CsCl, and 2 mM RbCl. After 10 min of exposure of cells to this solution, they were cryogenically prepared with a simple sandwich fracture technique.[2] The frozen freeze-dried cells were imaged first with reflected light Nomarski optics for identifying various stages of mitosis, and then with ion microscopy for imaging the distributions of ^{87}Rb, ^{40}Ca, ^{44}Ca, ^{39}K, ^{23}Na, and ^{133}Cs from the same cells. A Cameca IMS-3f ion microscope

---

[a] Corresponding author.

was used for the study. This instrument has a spatial resolution of 0.5 μm in the imaging mode. Isotope images were recorded with a Photometrics charge-coupled-device (CCD) camera and digitally processed for photographing. FIGURE 1a shows an optical image of several frozen freeze-dried LLC-PK1 cells (top left) and the ion microscopic analysis of the same cells revealing intracellular distributions of ^{87}Rb, ^{40}Ca, and ^{44}Ca. FIGURE 1b shows the further analysis of the same cells for

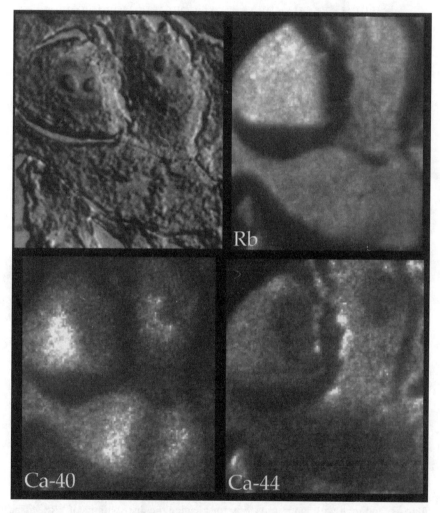

FIGURE 1a. Reflected light Nomarski image of several frozen-fractured freeze-dried LLC-PK1 cells (*top left*). The cell in the metaphase is clearly recognized by the alignment of its chromosomes. Ion microscopy images of the same cells revealing the distribution of ^{87}Rb, ^{40}Ca and ^{44}Ca. Image integration times on the CCD camera were 10 seconds for the ^{87}Rb image, and 120 seconds for the ^{40}Ca and ^{44}Ca images.

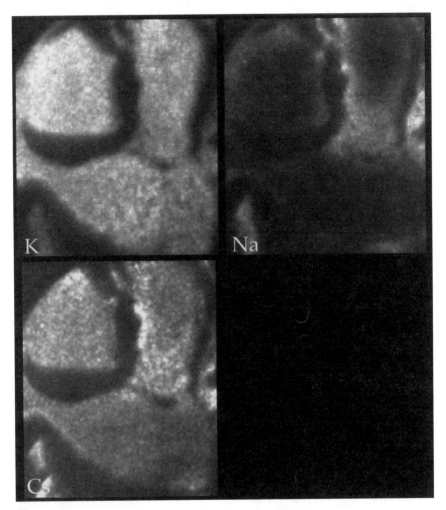

FIGURE 1b. Ion microscopy images of ^{39}K, ^{23}Na and ^{133}Cs from the cells shown in FIGURE 1a. Image integration times on the CCD camera were 0.2 seconds for the ^{39}K and ^{23}Na images, and 10 seconds for the ^{133}Cs image.

^{39}K, ^{23}Na, and ^{133}Cs. Brightness indicates higher signal intensities within an image. The metaphase cell and the neighboring interphase cells can be easily recognized between the optical and the ion microscopy images. Rb has entered the cells and is distributed rather evenly in the metaphase and interphase cells. The ^{40}Ca image represents the distribution of endogenous calcium. The distribution of ^{40}Ca is heterogeneous; the cell nuclei in interphase cells and the region occupied by the chromosomes in metaphase cells reveal lower intensities. The ^{44}Ca image represents the influx of calcium within 10 min. The metaphase cell appears dark in the ^{44}Ca image as compared to the neighboring interphase cells, which indicates the restriction of ^{44}Ca influx in the metaphase cell. This pattern is not followed by the influx of Rb and Cs in the same cells (see Cs image in FIG. 1b). The $^{39}K^+$ and $^{23}Na^+$ images from the same cells are also shown in FIG. 1b. The K^+/Na^+ ratio in these cells was about 12. The $^{23}Na^+$ image was normalized to the $^{39}K^+$ image, and, therefore, the $^{23}Na^+$ image appears dark. The details of this study will be discussed elsewhere.

REFERENCES

1. CHANDRA, S. & G. H. MORRISON. 1988. Methods Enzymol. **158:** 157–179.
2. CHANDRA, S., G. H. MORRISON & C. C. WOLCOTT. 1986. J. Microsc. (Oxford) **144:** 15–37.

Part III

Sodium-Calcium Exchange in the Neural System

Introduction

New detailed information on *Sodium-Calcium Exchange in the Neural System* was presented at the conference. The power of Na^+-Ca^{2+} exchange to promote rapid fluxes across the synaptic membrane was emphasized. Increased density of the exchanger molecule was found at nerve terminals consistent with an important functional role. Likewise, the Na^+-Ca^{2+},K^+ exchange system of the outer segments of rod photoreceptors can mediate the transport of substantial fluxes of Ca^{2+} in brief time periods. Reminiscent of other exchangers, cytosolic Ca^{2+} regulates the rod exchanger, though in a complex manner.

Research has been initiated in some new areas: Na^+-Ca^{2+} exchange may be the Ca^{2+} influx mechanism responsible for irreversible injury in anoxic CNS axons. Both the secretory vesicles and the plasma membranes of adrenal chromaffin cells appear to possess Na^+-Ca^{2+} exchangers. Initial evidence for a role for Na^+-Ca^{2+} exchange in auditory processes was described.

The Na^+-Ca^{2+} Exchanger in Rat Brain Synaptosomes

Kinetics and Regulation[a]

MORDECAI P. BLAUSTEIN,[b] GIOVANNI FONTANA,[c]
AND ROBERT S. ROGOWSKI

Department of Physiology
University of Maryland School of Medicine
655 West Baltimore Street
Baltimore, Maryland 21201

INTRODUCTION

Calcium ions have numerous functions in neurons, including a key role in triggering neurotransmitter release at nerve terminals.[1] Much of this trigger Ca^{2+} enters the terminals from the extracellular fluid via voltage-gated Ca^{2+} channels during depolarization.[2] During repolarization and recovery, Ca^{2+} must, therefore, be extruded in order to maintain steady Ca^{2+}. Several mechanisms participate in the removal of Ca^{2+} from the cytosol, and in the regulation of the cytosolic free Ca^{2+} concentration, $[Ca^{2+}]_i$. These include: i) Ca^{2+} buffering by cytoplasmic proteins;[3] ii) Ca^{2+} sequestration in the endoplasmic reticulum (via a calmodulin-insensitive, adenosine triphosphate (ATP)-driven Ca^{2+} pump[4]), and in mitochondria[5] (but see Ref. 6); and iii) Ca^{2+} extrusion across the plasmalemma (PM) via a calmodulin-sensitive ATP-driven Ca^{2+} pump,[7] and via a Na^+-Ca^{2+} exchanger.[8,9] Ca^{2+} buffering and sequestration may be the main mechanisms for rapidly removing Ca^{2+} from the transmitter releasing sites.[9] The relatives roles of the Na^+-Ca^{2+} exchanger and the PM Ca^{2+} pump in extruding Ca^{2+} and restoring Ca^{2+} homeostasis are not yet clear.

Presynaptic nerve terminals have a high density of Na^+-Ca^{2+} exchanger molecules.[10] These molecules cross-react with antibodies raised against dog heart Na^+Ca^{2+} exchanger, and appear to be similar to the cardiac exchanger molecules in molecular structure.[11,12] The kinetics and voltage-sensitivity of the mammalian heart Na^+-Ca^{2+} exchanger have been extensively investigated.[13–18] Here, we summarize recent studies[19] and unpublished data (Blaustein and Rogowski) on the kinetic properties and regulation of the Na^+-Ca^{2+} exchanger in isolated nerve terminals (synaptosomes) from rat brain. We also compare the properties of the brain and heart Na^+-Ca^{2+} exchangers. To this end, we measured internal

---

[a] These studies were supported by a research grant from the National Institutes of Health (NS-16106). GF was funded, in part, by a stipend from the Ministero dell Università e dell Ricerca Scientifica e Technologica (M.U.R.S.T.), Rome, Italy.

[b] Corresponding author.

[c] Dr. G. Fontana was a visiting scientist from the Istituto di Farmacologia e Farmacognosia, Università di Genova, Viale Cembrano 4, Genova 16148, Italy.

Na- (Na_i-)* dependent ^{45}Ca influxes and external Na- (Na_o-)* dependent ^{45}Ca effluxes in synaptosomes during 1 sec incubations.

METHODS

Preparation of Synaptosomes

Synaptosomes were prepared from the forebrains of adult female Sprague-Dawley rats (175–200 g) as described.[20]

Solution Composition

The standard physiological salt solution (Na-PSS) contained (mM): 145 NaCl, 2.6 KCl, 2.4 KH_2PO_4, 0.2 $CaCl_2$, 1.2 $MgCl_2$, 10 glucose and 10 HEPES; pH was adjusted to 7.4 with NaOH. The ionic composition of the solutions was varied in most experiments. For example, in low Na^+ solutions, Na^+ was replaced isosmotically by Li^+ (Li-PSS) buffered with LiOH to pH 7.4, or by N-methyl-D-glucamine (NMG-PSS) buffered with HCl to pH 7.4. K-free solutions contained 145 mM NaCl and no KCl. In high K^+ solutions, some of the NaCl was replaced by equimolar KCl. Details are given in Results and the figure legends.

^{45}Ca Influx and Efflux

Details of the protocols used for measurements of ^{45}Ca influx and efflux are published.[20] Further information is provided in Results and in the figure legends.

Statistical Analysis

All measurements were made on replicate samples as described in Results. Data are presented as mean values ± SE. Comparisons between means were made using Student's t test for unpaired data. Where indicated, data were fitted to theoretical curves, or least square linear regressions were calculated with SigmaPlot 4.1 software (Jandel Scientific, Corte Madera, CA).

Standard errors for the Na_i-dependent (ΔNa_i) fluxes ($SE\Delta Na_i$) were calculated according to the equation:[21]

$$SE\Delta Na_i = \sqrt{[SEJ(Na_i)]^2 + [SEJ(Li_i)]^2} \tag{1}$$

where [$SEJ(Na_i)$] is the standard error of the Ca^{2+} uptake by Na-loaded synaptosomes, and [$SEJ(Li_i)$] is the standard error of the Ca^{2+} uptake by Li-loaded

---

* Abbreviations used: $[Ca^{2+}]_o$ = external Ca^{2+} concentration; $[Ca^{2+}]_i$ = cytosolic free Ca^{2+} concentration; FCCP = carbonyl cyanide p-trifluoromethoxyphenylhydrazone; $[K^+]_i$ = internal K^+ concentration; $[K^+]_o$ = external K^+ concentration; $[Na^+]_i$ = internal Na^+ concentration; Na_i = internal Na^+ (or intracellular Na^+); $[Na^+]_o$ = external Na^+ concentration; Na_o = external Na^+ (or extracellular Na^+); V_M = membrane potential.

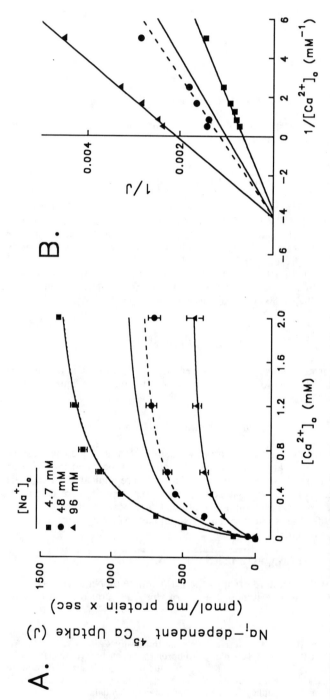

FIGURE 1. Activation of Na_i-dependent Ca^{2+} uptake (J) by $[Ca^{2+}]_o$. Data are shown for $[Na^+]_o = 4.7$ mM (■), 48 mM (●), and 96 mM (▲); the Na_i-independent Ca^{2+} uptake has been subtracted. In the low $[Na^+]_o$ media, Na-PSS was replaced by NMG-PSS during incubation with ^{45}Ca. For each curve in each experiment, 4 replicates were measured for each condition (± internal Na^+). (**A**) Averaged data from three experiments each with $[Na^+]_o = 48$ mM and 96 mM, and five experiments with $[Na^+]_o = 4.7$ mM, all at 30°C. $K_{Ca(o)}$ (the $[Ca^{2+}]_o$ at half-maximal activation) was obtained by fitting each curve in each experiment to the Michaelis-Menten equation using SigmaPlot. The mean $K_{Ca(o)}$ values are: 0.24 ± 0.03, n = 5 ($[Na^+]_o = 4.7$ mM); 0.23 ± 0.04, n = 3 ($[Na^+]_o = 48$ mM); and 0.22 ± 0.03, n = 3 ($[Na^+]_o = 96$ mM). The overall mean $K_{Ca(o)} = 0.23 \pm 0.01$, n = 11. The smooth Na_i-dependent Ca^{2+} influx (= J) curves were calculated fits to the product of (I) a two substrate (Ca^{2+} and M^+)

ordered bi bi activation process,[25] where activation is proportional to:

$$A = \left(\frac{([Ca^{2+}]_o \times [M^+]_o)}{(K_{Ca} \times K_M) + (K_M \times [M^+]_o) + ([Ca_{2+}]_o \times [M^+]_o)} \right) \quad (L1)$$

and (2) noncompetitive inhibition by the cooperative action of two Na^+ ions[25] (also see FIG. 2B and text), where inhibition is proportional to:

$$B = \left(\frac{1}{1 + \left(\frac{([Na^+]_o)^2}{(\overline{K}_{I(Na)})^2} \right)} \right) \quad (L2)$$

$\overline{K}_{I(Na)}$ is the calculated mean $[Na^+]_o$ required for half-maximal inhibition of the Ca^{2+} influx. K_m is the alkali metal ion concentration, $[M^+]_o$, required for half-maximal activation of Ca^{2+} influx by alkali metal ions (see FIGS. 2 and 3, and related text). The Na_i-dependent Ca^{2+} uptake is then given by (see text equation 2):

$$J = J_{max} \times A \times B \quad (L3)$$

where J_{max} is the apparent maximal Ca^{2+} influx at saturating $[Ca^{2+}]_o$, and saturating $[M^+]_o$ (the total alkali metal ion concentration), in the absence of inhibition by Na^+. The solid line curves in (A) were fitted to text equation 2 with the following kinetic constants: $K_{Ca} = 0.23$ mM, $K_M = 1$ mM, $K_{I(Na)} = 60$ mM, and $J_{max} = 1530$ pmol/mg protein × sec. **(B)** Double reciprocal (Lineweaver-Burke) plot of the data from (A). The least squares regression lines intersect the abscissa at nearly the same point; they do not intersect at the ordinate intercept. This indicates that external Na^+ is a noncompetitive inhibitor of Ca^{2+} influx. The abscissa intercepts ($= -1/K_{Ca}$) correspond to $K_{Ca} = 0.25$, 0.29 and 0.25 mM for $[Na^+]_o = 4.7$, 48 and 96 mM, respectively. $\overline{K}_{I(Na)}$ can be calculated from the ordinate intercept ($= 1/J_{max(app)}$) (cf. text equation 2 with $[Ca^{2+}]_o \gg K_{Ca(o)}$ and $[M^+]_o \gg K_M$):

$$J = J_{max(app)} \left(1 + \frac{([Na^+]_o)^2}{(\overline{K}_{I(Na)})^2} \right) \quad (L4)$$

With $J_{max} = 1530$ pmol/mg protein × sec, $\overline{K}_{I(Na)} = 56$ and 65 mM at $[Na^+]_o = 48$ and 96 mM, respectively. (From Fontana et al.[19] Reprinted by permission from the *Journal of Physiology*.)

synaptosomes. Similar formulae were used to calculate the standard errors of the Na_o-inhibitable Ca^{2+} uptake and the Na_o-dependent Ca^{2+} efflux.

RESULTS

Dependence of ^{45}Ca Uptake on $[Ca^{2+}]_o$; the Influence of External Na^+

Preliminary studies established that ^{45}Ca uptake by the synaptosomes was linear for at least 3 sec: i) Whether the synaptosomes were incubated in standard Na-PSS (with 145 mM Na^+ and 1.2 mM Ca^{2+}), or in media with a reduced Na^+ concentration and, ii) Whether the synaptosomes had their normal complement of internal Na^+, or were depleted of internal Na^+ by preincubation for 15 min in low Ca^{2+} Li-PSS. Moreover, the Na_i-dependent Ca^{2+} uptake (influx in Na-loaded minus influx in Li-loaded synaptosomes) was quantitatively equal to the uptake activated by reduction of $[Na^+]_o$ from 145 mM to 5 mM (in Na-loaded synaptosomes).

Activation of the Na_i-dependent (and Na_o-inhibitable) Ca^{2+} uptake (J) by $[Ca^{2+}]_o$ was measured at three different $[Na^+]_o$ at 30°C (FIG. 1). These data demonstrate that increasing $[Na^+]_o$ decreased the apparent maximum rate of Ca^{2+} uptake ($J_{max(app)}$), but did not affect the apparent affinity for Ca^{2+}. Half-maximal activation by external Ca^{2+} (= $K_{Ca(o)}$), determined by fitting the Michaelis-Menten equation, was constant when $[Na^+]_o$ was increased from 4.7 mM ($K_{Ca(o)}$ = 0.24) to 96 mM ($K_{Ca(o)}$ = 0.22) (see FIG. 1 legend). The external Na^+-dependent decrease in $J_{max(app)}$ with negligible effect on $K_{Ca(o)}$ is consistent with noncompetitive inhibition by Na^+. This is illustrated by the double reciprocal (Lineweaver-Burk) plot (FIG. 1B): the ordinate intercepts equal the $1/J_{max(app)}$'s corresponding to the respective $[Na^+]_o$'s. The mean $K_{Ca(o)}$, determined from Michaelis-Menten equation curve fits for all three $[Na^+]_o$, was 0.23 ± 0.03 mM (n = 11). Then, taking the maximal J in the absence of inhibition by external Na^+, J_{max} = 1530 pmol Ca^{2+}/mg × sec, and assuming that 2 Na^+ ions cooperatively inhibit the uptake of 1 Ca^{2+} (see below), the calculated apparent mean inhibitory constant for Na_o (= $\overline{K}_{I(Na)}$) was ≈60 mM for $[Na^+]_o$ = 48–96 mM (FIG. 1 legend).

Inhibition of Na_i-dependent Ca^{2+} uptake by external Na^+ was explored further by examining the effect of varying $[Na^+]_o$ at three different $[Ca^{2+}]_o$ (FIG. 2). The results (especially the Dixon plot in FIG. 2B) confirm that the interaction between external Na^+ and Ca^{2+} is noncompetitive. Furthermore, $\overline{K}_{I(Na)}$ was ≈50 mM, and was independent of $[Ca^{2+}]_o$ for $[Ca^{2+}]_o$ = 0.1–1.2 mM; these $\overline{K}_{I(Na)}$ values are similar to those obtained from the Ca^{2+} activation curves (FIG. 1). [Indeed, a slightly low $\overline{K}_{I(Na)}$ is expected because the Ca^{2+} uptake is not completely inhibited by 145 mM Na^+, the reference used for these calculations (see FIG. 2B legend).] Hill plots of these data for $[Na^+]_o \geq$ 10 mM yielded Hill coefficients of 1.9–2.0; the implication is that 2 Na^+ ions act cooperatively to inhibit the uptake of 1 Ca^{2+}. This is consistent with the fact that the Dixon plot of 1/J versus $([Na^+]_o)^2$ produces straight lines that meet at the abscissa intercept (FIG. 2B). In summary, these data indicate that external Ca^{2+} does not affect the affinity for Na^+ (FIG. 2B) and, conversely, external Na^+ does not affect the affinity for Ca^{2+} (FIG. 1B).

Activation of Na_i-Dependent Ca^{2+} Uptake by Alkali Metal Ions

The curves in FIGURE 2A are biphasic; low external Na^+ concentrations (<10 mM) activate the Ca^{2+} uptake. Similar activation of the Na^+-Ca^{2+} exchanger

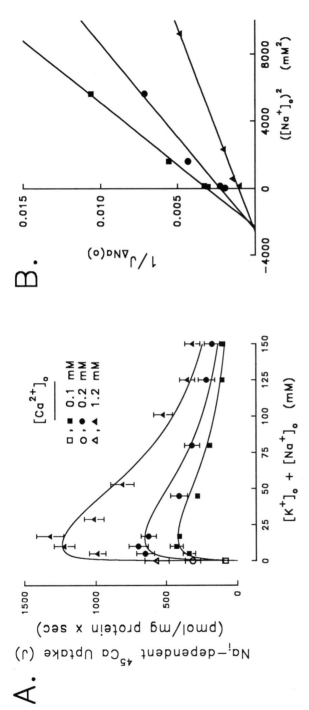

FIGURE 2. Activation of Na_i-dependent Ca^{2+} uptake by external Na^+ and K^+, and inhibition by external Na^+. Synaptosomes were preincubated in Na-PSS or Li-PSS; ^{45}Ca uptake was then measured in mixtures of Na-PSS and NMG-PSS at 30°C with $[Ca^{2+}]_o = 1.2$ mM (▲), 0.2 mM (●), or 0.1 mM (■), and the $[Na^+]_o + [K^+]_o$ indicated on the abscissa (*open symbols*: $[K^+]_o = 0$ mM; *solid symbols*: $[K^+]_o = 5$ mM). The Na_i-independent Ca^{2+} uptake (in Li-loaded synaptosomes) has been subtracted. The symbols are the means of data from 3 to 5 experiments (4 replicates for each data point in each experiment). (**A**) The smooth curves were calculated from text equation 2 using the following kinetic constants: $K_{Ca} = 0.23$ mM (see FIG. 4 legend), $K_M = 2$ mM, $\overline{K}_{I(Na)} = 70$ mM, and $J_{max} = 1530$ pmol/mg × sec. (**B**) Dixon plot[25] of the data in (A) for values of $[Na^+]_o \geq 10$ mM. The Na_o-dependent Ca^{2+} uptake from Na-PSS (with 145 mM Na^+) has been subtracted to obtain the Na_o-inhibitable flux ($J_{\Delta Na(o)}$). The abscissa units are given in terms of $([Na^+]_o)^2$; the resulting regression curves are linear, which indicates that two $\underline{Na^+}$ ions act cooperatively to inhibit the uptake of one Ca^{2+}. The linear least squares regression lines intersect at the abscissa (intercept = $-(\overline{K}_{I(Na)})^2$), which indicates that $\overline{K}_{I(Na)}$ is constant (mean = 50 ± 1 mM for the three curves), and that the interaction between external Na^+ and Ca^{2+} is noncompetitive. (From Fontana *et al.*[19] Reprinted by permission from the *Journal of Physiology*.)

by alkali metal ions, including Na^+, has been observed in squid axons[22,23] and mammalian cardiac muscle.[14,24] This activation of the Na_i-dependent Ca^{2+} uptake is supported by low concentrations (5–10 mM) of all alkali metal ions (not shown). The activation of Ca^{2+} uptake by $[Na^+]_o \leq 10$ mM and $[K^+]_o$, and inhibition by higher $[Na^+]_o$, accounts for the biphasic nature of the J vs $[Na^+]_o$ curves in FIGURE 2A. The apparent $[Na^+]_o$ and $[K^+]_o$ required for half-maximal activation ($K_{M(Na)}$ and $K_{M(K)}$, respectively) were 0.12 mM and 0.10 mM (FIG. 3A); these are almost certainly underestimates because of incomplete removal of extracellular K^+ and Na^+.

Increasing $[K^+]_o$ tended to reduce $K_{Ca(o)}$ and increase $J_{max(app)}$ (FIG. 3B). This suggests that alkali metal ions increase the affinity of the Na^+-Ca^{2+} exchanger for external Ca^{2+}. These kinetics may therefore correspond to an "ordered bi bi" reaction sequence[25] in which the alkali metal ion must bind before Ca^{2+}, and binding is much more rapid than Ca^{2+} translocation. Accordingly, the Ca^{2+} activation data (FIG. 1A) and the biphasic J versus $[Na^+]_o$ data (FIG. 2a) were fitted to the following equation (see FIG. 1 legend, equation L3):

$$J = J_{max} \times \left(\frac{([Ca^{2+}]_o \times [M^+]_o)}{(K_{Ca} \times K_M) + (K_M \times [M^+]_o) + ([Ca^{2+}]_o \times [M^+]_o)}\right)$$
$$\times \left(\frac{1}{1 + \frac{([Na^+]_o)^2}{(\overline{K}_{I(Na)})^2}}\right) \qquad (2)$$

where $[M^+]_o$ is the concentration of the activating external monovalent cation. This equation corresponds to the product of an ordered bi bi activation process involving alkali metal ions (M^+) and Ca^{2+} (FIG. 1 legend, equation L1), and noncompetitive inhibition by the cooperative action of 2 Na^+ (FIG. 1 legend, equation L2). The calculated curves are a good fit for the experimental data.

Dependence of Ca^{2+} Uptake upon Internal Na^+

To determine the relationship between $[Na^+]_i$ and Ca^{2+} influx, we used synaptosomes that were preincubated for 15 min in Na-PSS/Li-PSS mixtures. To promote equilibration of $[Na^+]_i$ with $[Na^+]_o$, these mixtures contained 1 mM ouabain (to inhibit Na^+ extrusion), and 10 μM monensin, 10 μM nigericin, and 5 μM gramicidin D (to increase Na^+, H^+, and K^+ permeability, respectively). We therefore assume that $[Na^+]_i \approx$ preincubation $[Na^+]_o$.

FIGURE 4 illustrates the relationship between preincubation $[Na^+]_o$ ($\approx [Na^+]_i$) and the ^{45}Ca uptake from NMG-PSS. With $[Ca^{2+}]_o = 1.2$ mM (FIG. 4A), the apparent mean $\overline{K}_{Na(i)}$ was about 28 mM, where $\overline{K}_{Na(i)}$ is the $[Na^+]_i$ required for half-maximal activation of Ca^{2+} influx. The $[Na^+]_i$ activation curve was well fit by the Hill equation with $n_H = 3.0$; this is consistent with the view that 3 Na^+ ions are required to activate the entry of 1 Ca^{2+}. The maximum rate of Ca^{2+} uptake [*i.e.*, with a high Na^+ concentration in the preincubation (= intracellular) medium] was, however, significantly lower than observed in the experiments of FIGURES 1 and 2; perhaps this was due to a large decline in $[K^+]_i$, and a consequently large change in V_M.

Similar results were obtained with $[Ca^{2+}]_o = 0.2$ mM, although the apparent $\overline{K}_{Na(i)}$ was then only ≈ 16 mM (FIG. 4B). The higher $\overline{K}_{Na(i)}$ at $[Ca^{2+}]_o = 1.2$ mM

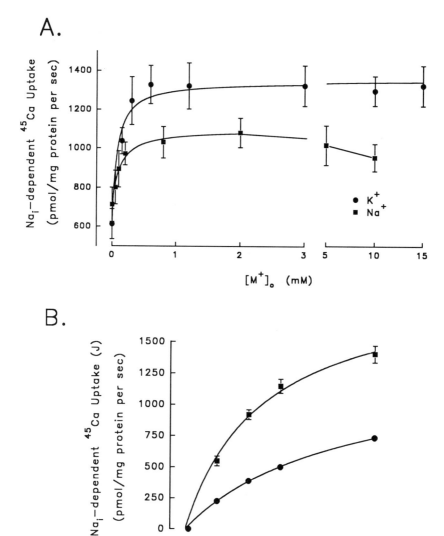

FIGURE 3. Alkali metal ion activation of Na_i-dependent Ca^{2+} uptake. **(A)** Activation of Na_i-dependent Ca^{2+} uptake by external Na^+ (■) and K^+ (●). The nominally alkali metal ion-free medium was K-free NMG-PSS. The calculated half-maximal activation concentrations (based on Michaelis-Menten kinetics), $K_{M(K)}$ and $K_{M(Na)}$, are 0.10 and 0.12 mM, respectively. These K_M values do not take into account the alkali metal ion concentrations in the K-free NMG-PSS. Temp = 30°C; four replicate samples for each condition. **(B)** Ca_o-dependence of the external K^+-activated, Na_i-dependent Ca^{2+} uptake from NMG-PSS. The Ca^{2+} uptake from nominally K-free media has been subtracted. Data for $[K^+]_o = 3$ (and 15) mM (■) and $[K^+]_o = 0.08$ mM (●) are shown; results for $[K^+]_o = 3$ and 15 mM were virtually identical, and have therefore been combined. External K^+ appears to increase the affinity for Ca^{2+} (*i.e.*, reduce K_{Ca}) as well as increase the apparent J_{max} for Ca^{2+} uptake. Temp = 30°C; four replicate samples for each condition in each experiment. (From Fontana *et al.*[19] Reprinted by permission from the *Journal of Physiology*.)

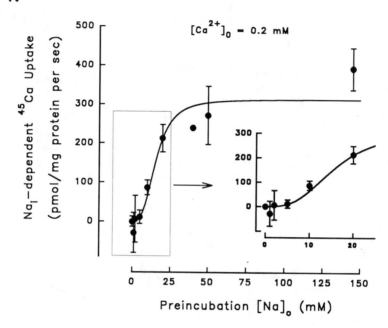

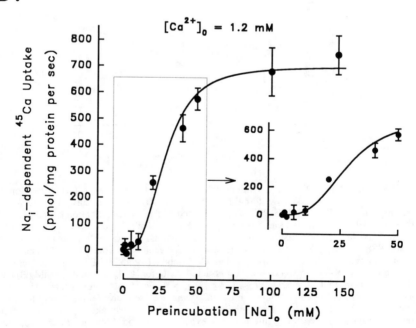

FIGURE 4. Dependence of Ca^{2+} uptake upon $[Na^+]_i$. Synaptosomes were preincubated for 15 min in mixtures of Na-PSS and Li-PSS containing the $[Na^+]_o$ shown on the abscissa. The preincubation media also contained 10 μM monensin, 1 mM ouabain 5 μM gramicidin D and 10 μM nigericin. ^{45}Ca uptake was then measured in either Na-PSS or NMG-PSS containing 0.2 mM Ca^{2+} **(A)** or 1.2 mM Ca^{2+} **(B)** at 30°C. The symbols indicate the differences (*i.e.*, the Na_o-inhibitable = Na_i-dependent Ca^{2+} uptake). Data from two representative experiments are shown; 4 replicates were obtained for each condition in each experiment; error bars indicate ± SE. The Ca^{2+} uptake (J) curves were fitted to the Hill equation:[25]

$$J = J_{max(app)} \left(\frac{([Na^+]_i)^n}{([Na^+]_i + \overline{K}_{Na(i)})^n} \right) \quad (L5)$$

where n is the Hill coefficient. The calculated parameters for the fitted curves are:

$[Ca^{2+}]_o$ (mM)	n	$\overline{K}_{Na(i)}$ (mM)	$J_{max(app)}$ (pmol/mg protein × sec)
1.2 (B)	3.0	28	702
0.2 (C)	3.0	16	314

(From Fontana *et al.*[19] Reprinted by permission from the *Journal of Physiology*.)

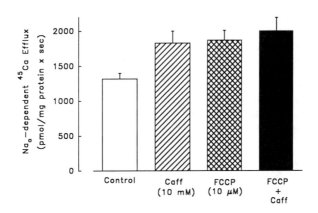

FIGURE 5. Effect of caffeine and FCCP on the Na_o-dependent Ca^{2+} efflux from synaptosomes. The synaptosomes were loaded with ^{45}Ca via voltage-gated Ca^{2+} channels[21] by depolarizing the terminals for 9 sec in 100 mM K-PSS containing 1.2 mM Ca^{2+}. After washing out the extracellular ^{45}Ca, tracer efflux was measured for 1 sec in either Ca-free NMG-PSS ($[Na^+]_o = 0$ mM) or Ca-free Na-PSS, either without or with 10 mM caffeine (Caff) and/or 10 μM FCCP. The bars correspond to the Na_o-dependent Ca^{2+} efflux (*i.e.*, efflux into Ca-free NMG-PSS has been subtracted). Temp = 37°C. Symbols indicate the means ± SE of data from 4–8 separate experiments, with 5 replicate determinations for each condition in each experiment. (Data from Fontana *et al.*[19])

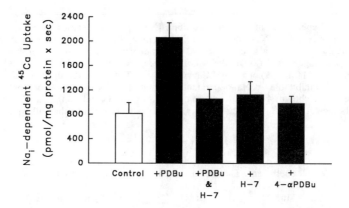

FIGURE 6. Effect of phorbol esters on the Na_i-dependent Ca^{2+} uptake in synaptosomes. Synaptosomes were preincubated for 15 min at 30°C in Na-PSS or Li-PSS without or with phorbol ester, without or with the C kinase inhibitor, H-7. The 1 sec ^{45}Ca uptake was measured in NMG-PSS without or with 100 nM PDBu, 400 nM 4-αPDBu and/or 50 μM H-7. The ordinate shows the Na_i-dependent ^{45}Ca uptake. Error bars indicate the means ± SE of 4 replicate determinations for each condition.

may be an overestimate if the larger influx of Ca^{2+} under these circumstances interferes with the activation by internal Na^+.

"Initial Rates" of Na_o-Dependent Ca^{2+} Efflux from Synaptosomes

Na_o-dependent Ca^{2+} efflux "initial rates" were measured in synaptosomes loaded with ^{45}Ca via voltage-gated Ca^{2+} channels by briefly depolarizing the terminals with 100K-PSS. The mean total Ca^{2+} load was large: ≈7200 pmol Ca^{2+} (as ^{45}Ca) per mg protein. The bar graph in FIGURE 5 shows the 1 sec Na_o-dependent Ca^{2+} efflux under control conditions, and in the presence of 10 mM caffeine (caff), 10 μM FCCP, or both. Caffeine unloads endoplasmic reticulum Ca^{2+} stores, and FCCP uncouples oxidative phosphorylation and releases Ca^{2+} from mitochondria. The caffeine and FCCP were used to raise $[Ca^{2+}]_i$ and saturate the Ca^{2+} transport sites at the cytoplasmic face of the plasmalemma in order to obtain J_{max} values for the Ca^{2+} efflux. These data indicate that the J_{max} for the Na^+-Ca^{2+} exchanger-mediated Ca^{2+} efflux (i.e., the Na_o-dependent efflux component) is ≈2000 pmol/mg × sec at 37°C (assuming a linear rate of efflux during the first 1 sec). This is much larger than the Na_o-independent component (≈ 600–800 pmol/mg sec), which may correspond primarily to the plasmalemmal ATP-driven Ca^{2+} pump and Ca^{2+} "leak."

Effects of Phorbol Esters on Ca^{2+} Influx and Efflux

The regulation of the Na^+-Ca^{2+} exchanger has been studied in several types of cells. The exchanger in vascular smooth muscle cells is, for example, reported to be modulated by a number of agents, including growth factors, cyclic nucleotides, and phorbol esters.[26,27] Very little is known, however, about the regulation of the neuronal Na^+-Ca^{2+} exchanger. FIGURES 6 and 7 show the effects of 100 μM phorbol 12,13-dibutyrate (PDBu) on Na^+-dependent Ca^{2+} fluxes in synaptosomes. When synaptosomes were pretreated with PDBu for 15 min before initiating the flux experiments, this agent markedly stimulated both the Na_i-dependent Ca^{2+} influx (FIG. 6) and the Na_o-dependent Ca^{2+} efflux (FIG. 7); PDBu had no significant effects on the Na-independent Ca^{2+} fluxes (not shown). The data in FIGURE 6 indicate that these effects of PDBu are due to the activation of protein kinase C (PKC): The effect on Ca^{2+} uptake was blocked by the protein kinase inhibitor, H-7, and was not observed with 4-αPDBu, an analogue of PDBu that does not activate PKC. PDBu increased the J_{max} values for both Ca^{2+} influx and efflux, but had negligible effects on the apparent affinities for either external Ca^{2+} or internal Na^+ (not shown). Preliminary studies indicate that PDBu promotes the phosphorylation of the Na^+-Ca^{2+} exchanger in synaptosomes.

Effects of Membrane Potential on Ca^{2+} Influx and Efflux

The effect of membrane potential (V_M) on Na^+-Ca^{2+} exchanger-mediated Ca^{2+} uptake was determined by measuring the influence of raised $[K^+]_o$ on the Na_i-dependent Ca^{2+} uptake from low-Na^+ media. Conversely, the voltage-sensitivity of the Na^+-Ca^{2+} exchanger operating in the Ca^{2+} efflux mode was determined by examining the effect of varying $[K^+]_o$ at constant $[Na^+]_o$ on the Na_o-dependent component of Ca^{2+} efflux. The data are graphed as a function of log $[K^+]_o$ in FIGURE 8; the upper abscissa indicates the calculated V_M.[28] The Na_i-dependent Ca^{2+} uptake (circles) increased by a factor of about 2 with a 60-mV depolarization; the Na_o-dependent Ca^{2+} efflux (diamonds) decreased by a factor of about 2 with a 60-mV depolarization. These results are expected if depolarization promotes the exit (and inhibits the entry) of net positive charge associated with an exchange of 3 or more Na^+ ions for one Ca^{2+}. The curves, based on an Eyring rate theory model[18,29] (FIG. 8), suggest that about 0.58 net elementary charges move through the membrane field during each transport cycle or, alternatively, that the rate-limiting step is about 58% of the distance across the membrane field.

DISCUSSION

The Na_i-dependent Ca^{2+} influx and Na_o-dependent Ca^{2+} efflux in rat brain synaptosomes were used as functional measures of Na^+-Ca^{2+} exchange activity.

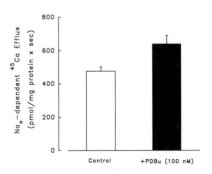

FIGURE 7. Effect of phorbol esters on the Na_o-dependent Ca^{2+} efflux from synaptosomes. The synaptosomes were incubated for 15 min at 30°C without or with 100 nM PDBu. They were then loaded with ^{45}Ca by depolarizing them for 9 sec in 100 mM K-PSS containing 1.2 mM Ca^{2+}, without or with PDBu. After washing out the extracellular ^{45}Ca, tracer efflux was measured for 1 sec in either NMG-PSS or Na-PSS, either without or with 10 μM FCCP and/or 10 mM caffeine. Bars indicate the means ± SE of data from 2 experiments with 4 replicate determinations for each condition in each experiment.

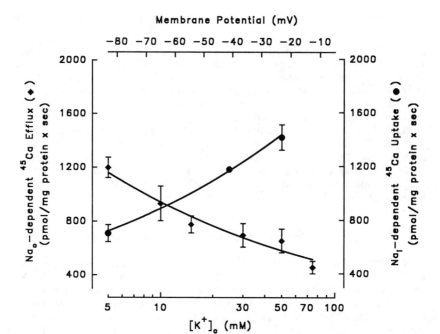

FIGURE 8. Effect of $[K^+]_o$ (and membrane potential) on Na_i-dependent Ca^{2+} uptake and Na_o-dependent Ca^{2+} efflux. Ca^{2+} *uptake* (●). Synaptosomes were loaded with Na^+ or Li^+ by incubation for 15 min at 37°C in Na-PSS or Li-PSS, respectively. They were then pre-depolarized by diluting 50 μl of synaptosome suspension with 50 μl Ca-free (and Na-free) NMG-PSS containing 100 mM K^+, 10 μM verapamil and 0.2 mM EGTA. Seven sec later, ^{45}Ca uptake was measured by adding 1.4 ml of mixtures of NMG-PSS and K-PSS containing 0.2 mM tracer-labelled Ca^{2+}; the final $[K^+]_o$ is indicated on the lower abscissa. Ordinate (right-hand scale) shows the Na_i-dependent Ca^{2+} uptake (uptake in synaptosomes preincubated in Na-PSS minus uptake in synaptosomes incubated in Li-PSS). Upper abscissa scale shows the calculated membrane potential.[28] The symbols are the means of data from 4 experiments, with 4 replicate determinations for each condition in each experiment. The solid line has a slope equivalent to about a 2-fold increase in Na_i-dependent Ca^{2+} uptake per 60 mV depolarization. Temp = 37°C. Ca^{2+} *efflux* (◆). Synaptosomes were loaded with ^{45}Ca via voltage-gated Ca^{2+} channels by depolarizing the terminals for 9 sec in 100 mM K-PSS containing 0.2 mM Ca^{2+}. After washing out extracellular ^{45}Ca, tracer efflux was measured for 1 sec at 37°C in mixtures of Na-PSS, NMG-PSS, and K-PSS. For each $[K^+]_o$, Ca^{2+} efflux was measured in media with $[Na^+]_o = 75$ mM and $[Na^+]_o = 5$ mM; the difference between these two fluxes, the Na_o-dependent Ca^{2+} efflux (◆), is plotted on the ordinate (left-hand sale). Symbols for the efflux experiments correspond to the means of data from 4 experiments, with 5 replicate determinations for each condition in each experiment. The solid lines for both the Ca^{2+} influx and efflux curves were calculated to fit an Eyring rate model:[18,29]

$$J = J_o \exp\left(\frac{z F V_M}{2RT}\right) \quad (L6)$$

J is the Na_i-dependent Ca^{2+} influx or Na_o-dependent Ca^{2+} efflux (which correspond to the net outward or inward current, respectively) at any V_M (in mV). J_0 = the Ca^{2+} influx at $V_M = 0$ mV. R, T and F have their usual meanings, and z is the fraction of the membrane electric field through which an elementary positive charge moves during the rate-limiting step during each complete transport cycle. The fitted lines correspond to 1 elementary positive charge moving through about 58% of the membrane field (or, alternatively, 0.58 elementary charges moving through the entire electric field). (From Fontana et al.[19] Reprinted by permission from the *Journal of Physiology*.)

To approximate initial rates, 1 sec ^{45}Ca fluxes were employed for kinetic analysis. Most experiments were performed at 30°C, rather than at 37°C, to slow the kinetics so that some parameters could be better resolved.

The Large Capacity (Maximum Velocity, J_{max}) of the Synaptosome Na^+-Ca^{2+} Exchanger

The large maximum rate of Ca^{2+} transport (J_{max}) mediated by the Na^+-Ca^{2+} exchanger in rat brain synaptosomes is about 2000 pmol/mg × sec at 37°C (see FIG. 5). This corresponds to a Na^+-Ca^{2+} exchanger-mediated flux of about 500 μmoles of Ca^{2+} per liter cell water per sec (since intra-synaptosome volume is about 3.5 μl/mg protein[30]). This large Ca^{2+} flux is equivalent to a turnover of about ¼ of the total synaptosome Ca^{2+} (about 2 mmoles/liter cell water[31]) in 1 sec.

Interactions between Na^+ and Ca^{2+}

The interference with Ca^{2+} influx by Na^+ at the external face of the synaptosome plasmalemma (FIGS. 1 and 2) appears to be noncompetitive. This contrasts with conclusions drawn from some studies on cardiac muscle which indicate that external Na^+ is a competitive inhibitor at the exchanger's external Ca^{2+} binding site.[13,14] In synaptosomes, as in the heart, the interaction between external Na^+ and Ca^{2+} apparently involves 2 Na^+ ions and 1 Ca^{2+}. It is difficult to imagine how two Na^+ ions (each with an ionic radius ≈ 1 Å) could bind to the identical site to which one Ca^{2+} (ionic radius ≈ 1 Å) binds. Thus, at least superficially, a noncompetitive interaction seems more plausible.

Activation of Ca^{2+} Influx by Alkali Metal Ions

The Na^+-Ca^{2+} exchanger in rod outer segments has an absolute requirement for K^+ and a coupling ratio of 4 Na^+ : (1 Ca^{2+} + 1 K^+).[32] In contrast, the exchanger from mammalian heart does not require K^+, and has a coupling ratio of 3 Na^+ : 1 Ca^{2+}.[33] The brain exchanger, which is structurally similar to the heart exchanger,[11,12] also does not have an absolute requirement for K^+. We found that a low concentration (1–10 mM) of any alkali metal ion (including Na^+ as well as K^+) can activate the Na_i-dependent Ca^{2+} influx. This is consistent with observations in squid axons[22] and cardiac muscle.[24] Thus, the activation of Na_i-dependent Ca^{2+} influx by low $[K^+]_o$ (<2 mM),[34] is likely attributable to an action at the alkali metal ion activation site. The activating ion in the brain/heart exchanger is apparently not transported because Na^+ can serve as an activating cation (FIGS. 2 and 3), even though the stoichiometry of the exchange appears to be 3 Na^+ : 1 Ca^{2+}.

Coupling Ratio and Voltage-Sensitivity

Several observations support the view that the Na^+ : Ca^{2+} coupling ratio in the rat brain exchanger is 3 Na^+ : 1 Ca^{2+}: i) The dependence of Ca^{2+} efflux on external Na^+ is sigmoid, with a Hill coefficient of 2.5 (Ref. 8). ii) The Hill coefficient for activation of Ca^{2+} influx by internal Na^+ is ≈3.0 (FIG. 4). These data suggest that Ca^{2+} transport requires the cooperative action of 3 Na^+ on the opposite side of the membrane. iii) The rat brain synaptic membrane exchanger is immunologically and structurally similar to the dog cardiac exchanger,[11,12] which has a coupling ratio (or stoichiometry) of 3 Na^+ : Ca^{2+}.[33]

An exchanger with this coupling ratio should be voltage sensitive if one or more of the intermediate steps in the transport cycle is voltage sensitive.[29] Indeed, the nerve terminal Na^+-Ca^{2+} exchanger exhibits about a 2-fold change in flux per 60-mV change in V_M; depolarization increases Ca^{2+} influx and decreases Ca^{2+} efflux (FIG. 8). This voltage-sensitivity is comparable to that of the exchanger in cardiac myocytes, where an e-fold change in Ca^{2+} flux per 77-mV change in V_M is observed.[14]

Regulation of Na^+-Ca^{2+} Exchanger Activity

The neuronal exchanger is regulated by intracellular Ca^{2+} and ATP.[23,35] Activation by intracellular Ca^{2+} is required even for Ca^{2+} entry mediated by the exchanger, and phosphorylation by ATP increases the affinity for intracellular Ca^{2+} at the transport site. Here, we have shown that the synaptosome exchanger is influenced by phorbol esters. The data suggest that phosphorylation of the exchanger by PKC can increase the maximum velocity of exchanger-mediated Ca^{2+} influx and efflux by nearly 50%. One possibility is that some nonphosphorylated exchangers are latent, and that phosphorylation increases the number of available exchanger molecules rather than the rate of turnover of individual exchanger molecules.

A Kinetic Model for the Na^+-Ca^{2+} Exchanger

The ping pong (consecutive)[16] bi bi transport model shown in FIGURE 9 is compatible with Equation 2. The reactions in the upper portion of the model are those that take place at the external face of the plasmalemma; those in the lower portion take place at the cytoplasmic face. The clockwise sequence corresponds to the Ca^{2+} influx mode of exchange, and the counterclockwise sequence corresponds to the Ca^{2+} efflux mode of exchange. Activation of the Ca^{2+} influx by external alkali metal ions (M^+) is shown as the ordered binding of M^+ first, and then (transported) Ca^{2+}. The noncompetitive inhibition of Ca^{2+} influx by external Na^+ may then be a manifestation of the tendency for Na^+ to bind and promote Ca^{2+} exit mode exchange (*i.e.*, drive the reaction counterclockwise). Possible internal alkali metal ion activation of Ca^{2+} efflux[23] has been ignored in this model, which is limited to the data from synaptosomes; however, data from squid axons[23] suggests that efflux of Ca^{2+} may require the binding of an internal alkali metal ion.

Physiological Role(s) of the Na^+-Ca^{2+} Exchanger in Nerve Terminals

The prevalence[10] and the consequent large capacity (J_{max}), of the Na^+-Ca^{2+} exchanger at nerve terminals imply that the exchanger plays an important physiological role. The Na^+-Ca^{2+} exchanger has a relatively low affinity for cytoplasmic Ca^{2+}: the $K_{Ca(i)}$ is on the order of 10^{-6} M,[36] in contrast to the plasmalemmal ATP-driven Ca^{2+} pump with a $K_{Ca(i)}$ for Ca^{2+} on the order of 10^{-7} M.[37] Thus, exchanger turnover may be relatively slow under "resting" conditions when $[Ca^{2+}]_i$ is about 10^{-7} M.[38] Nevertheless, the exchanger may modulate resting $[Ca^{2+}]_i$, even if this

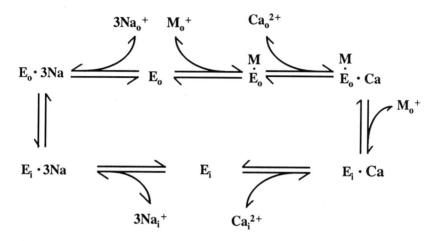

FIGURE 9. Kinetic model of the synaptosome Na^+-Ca^{2+} exchanger. The model corresponds to an ordered ping pong (consecutive) bi bi mechanism.[16,25] E_o is the exchanger with its ion binding sites facing the extracellular fluid; E_i is the exchanger with its ion binding sites facing the cytoplasm. $E_o \cdot 3Na$ and $E_i \cdot Ca$ are examples of ion-bound forms of the exchanger. Possible internal alkali metal ion activation of Ca^{2+} efflux[23] has been ignored in this model, which is limited to the data from synaptosomes (data from squid axons[23] suggests that efflux of Ca^{2+} may require the binding of an internal alkali metal ion). Clockwise cycling of the exchanger corresponds to Ca^{2+} influx mode exchange;[36] counterclockwise cycling corresponds to Ca^{2+} efflux mode exchange.[36]

level is controlled primarily by the ATP-driven Ca^{2+} pump. More importantly, even small changes in $[Ca^{2+}]_i$ modulate the amount of Ca^{2+} in the intracellular stores in the endoplasmic reticulum[38] because more than 99.9% of the intracellular Ca^{2+} is sequestered.[20,39] The exchanger may, thus, influence the numerous processes that depend upon release of Ca^{2+} from the intracellular stores.

The Na^+-Ca^{2+} exchanger also may mediate net Ca^{2+} influx during nerve terminal depolarization. The depolarization (see FIG. 8) and increase in driving force, as well as the transient rise in $[Ca^{2+}]_i$ that activates Ca^{2+} influx mode exchange,[35,36] should enhance exchanger-mediated Ca^{2+} entry. We have no direct information about the relative contributions of Ca^{2+} entry via voltage-gated Ca^{2+} channels, and via the exchanger, to depolarization-induced Ca^{2+} transients. Operating at maximal rate, however, the exchanger could contribute only about 1–2 pmol of Ca^{2+} to the total influx of about 6 pmol/mg msec.[40]

When the plasma membrane repolarizes, during the falling phase of the action potential, the driving force on the Na^+-Ca^{2+} exchanger will favor Ca^{2+} extrusion. Ca^{2+} transport sites at the cytoplasmic face of the membrane will then be saturated by the high $[Ca^{2+}]_i$ (perhaps $\approx 10^{-4}$ M).[41] This will favor rapid Ca^{2+} extrusion via the exchanger. Thus, the exchanger may even help to terminate evoked transmitter release following cell activation. Most of the Ca^{2+} that enters during depolarization will, however, diffuse away from the plasmalemma, and will be rapidly buffered and sequestered. This Ca^{2+} will then be extruded from the terminals over the next 1–2 sec.[9] The fact that most of this Ca^{2+} efflux is Na_o-dependent (FIG. 8)[8] suggests that the Na^+-Ca^{2+} exchanger plays a major role in Ca^{2+} extrusion following nerve terminal activation.

ACKNOWLEDGMENTS

We thank Prof. M. Raiteri for support and encouragement.

REFERENCES

1. AUGUSTINE, G. J., H. BETZ, K. BOMMERT, M. P. CHARLTON, W. M. DEBELLO, M. HANS & D. SWANDULLA. 1994. Molecular pathways for presynaptic calcium signaling. In Molecular and Cellular Mechanisms of Neurotransmitter Release. L. Stjärne, P. Greengard, S. Grillner, T. Hökfelt & D. Ottoson, Eds. 139–153. Raven Press. New York.
2. KATZ, B. 1969. The Release of Neural Transmitter Substances. Thomas. Springfield, IL.
3. ANDRESSEN, C., I. BLÜMCKE & M. R. CELIO. 1993. Calcium-binding proteins: selective markers of nerve cells. Cell Tissue Res. **271:** 181–208.
4. BLAUSTEIN, M. P., R. W. RATZLAFF, N. C. KENDRICK & E. S. SCHWEITZER. 1978. Calcium buffering in presynaptic nerve terminals. I. Evidence for involvement of a nonmitochondrial ATP-dependent sequestration mechanism. J. Gen. Physiol. **72:** 15–41.
5. WERTH, J. L. & S. A. THAYER. 1994. Mitochondria buffer physiological calcium loads in cultured rat dorsal root ganglion neurons. J. Neurosci. **14:** 348–356.
6. RASGADO-FLORES, H. & M. P. BLAUSTEIN. 1987. ATP-dependent regulation of cytoplasmic free calcium in nerve terminals. Am. J. Physiol. **252:** C588–C594.
7. GILL, D. L., E. F. GROLLMAN & L. D. KOHN. 1981. Calcium transport mechanisms in membrane vesicles from guinea pig brain synaptosomes. J. Biol. Chem. **256:** 184–192.
8. SANCHEZ-ARMASS, S. & M. P. BLAUSTEIN. 1987. Role of Na/Ca exchange in the regulation of intracellular Ca^{2+} in nerve terminals. Am. J. Physiol. **252:** C595–C603.
9. BLAUSTEIN, M. P. 1988. Calcium transport and buffering in neurons. Trends Neurosci. **11:** 438–443.
10. LUTHER, P. W., R. K. YIP, R. J. BLOCH, A. AMBESI, G. E. LINDENMAYER & M. P. BLAUSTEIN. 1992. Presynaptic localization of sodium/calcium exchangers in neuromuscular preparations. J. Neurosci. **12:** 4898–4904.
11. YIP, R. K., M. P. BLAUSTEIN & K. D. PHILIPSON. 1992. Immunologic identification of Na/Ca exchange protein in rat brain synaptic plasma membrane. Neurosci. Lett. **135:** 123–126.
12. FURMAN, I., O. COOK, J. KASIR & H. RAHAMIMOFF. 1993. Cloning of two isoforms of the rat brain Na^+-Ca^{2+} exchanger gene and their functional expression in HeLa cells. FEBS Lett. **319:** 105–109.
13. REEVES, J. P. & J. L. SUTKO. 1983. Competitive interactions of sodium and calcium with the sodium-calcium exchange system of cardiac sarcolemmal vesicles. J. Biol. Chem. **258:** 3178–3182.
14. MIURA, Y. & J. KIMURA. 1989. Sodium-calcium exchange current. Dependence on internal Ca and Na and competitive binding of external Na and Ca. J. Gen. Physiol. **93:** 1129–1145.
15. CRESPO, L. M., C. J. GRANTHAM & M. B. CANNELL. 1990. Kinetics, stoichiometry and role of the Na-Ca exchange mechanism in isolated cardiac myocytes. Nature **345:** 618–621.
16. KHANANSHVILI, D. 1990. Distinction between the two basic mechanisms of cation transport in the cardiac Na^+-Ca^{2+} exchange system. Biochemistry **29:** 2437–2442.
17. MATSUOKA, S. & D. W. HILGEMANN. 1992. Steady-state and dynamic properties of cardiac sodium-calcium exchange. Ion and voltage dependencies of the transport cycle. J. Gen. Physiol. **100:** 963–1001.
18. NIGGLI, E. & W. J. LEDERER. 1993. Activation of Na-Ca exchange current by photolysis of "caged calcium." Biophys. J. **65:** 882–891.
19. FONTANA, G., R. S. ROGOWSKI & M. P. BLAUSTEIN. 1995. Kinetic properties of the sodium/calcium exchanger in rat brain synaptosomes. J. Physiol. (London) **485:** 349–364.

20. FONTANA, G. & M. P. BLAUSTEIN. 1993. Calcium buffering and free Ca^{2+} in rat brain synaptosomes. J. Neurochem. **60:** 843–850.
21. NACHSHEN, D. A. & M. P. BLAUSTEIN. 1979. The effects of some organic "calcium antagonists" on calcium influx in presynaptic nerve terminals. Mol. Pharmacol. **16:** 579–586.
22. BAKER, P. F., M. P. BLAUSTEIN, A. L. HODGKIN & R. A. STEINARDT. 1969. The influence of calcium on sodium efflux in squid axons. J. Physiol. (London) **200:** 431–458.
23. BLAUSTEIN, M. P. 1977. Effects of internal and external cations and of ATP on sodium-calcium and calcium-calcium exchange in squid axons. Biophys. J. **20:** 79–111.
24. GADSBY, D. C., M. NODA, R. N. SHEPHERD & M. NAKAO. 1991. Influence of external monovalent cations on Na-Ca exchange current-voltage relationships in cardiac myocytes. Ann. N.Y. Acad. Sci. **639:** 140–146.
25. SEGEL, I. H. 1976. Biochemical Calculations. 2nd Edit. John Wiley and Sons. New York.
26. VIGNE, P., J-P. BREITTMAYER, D. DUVAL, C. FRELIN & M. LAZDUNSKI. 1988. The Na^+-Ca^{2+} antiporter in aortic smooth muscle cells. Characterization and demonstration of an activation by phorbol esters. J. Biol. Chem. **263:** 8078–8083.
27. IWAMOTO, T., S. WAKABAYASHI & M. SHIGEKAWA. 1995. Growth factor-induced phosphorylation and activation of aortic smooth muscle Na^+-Ca^{2+} exchanger. J. Biol. Chem. **270:** 8996–9001.
28. BLAUSTEIN, M. P. & J. M. GOLDRING. 1975. Membrane potentials in pinched-off presynaptic nerve terminals monitored with a fluorescent probe: evidence that synaptosomes have potassium diffusion potentials. J. Physiol. (London) **247:** 589–615.
29. LÄUGER, P. 1991. Electrogenic Ion Pumps. Sinauer Associates. Sunderland, MA.
30. BLAUSTEIN, M. P. 1975. Effects of potassium, veratridine and scorpion venom on calcium accumulation and transmitter release by nerve terminals *in vitro*. J. Physiol. (London) **247:** 617–644.
31. SCHWEITZER, E. S. & M. P. BLAUSTEIN. 1980. Calcium buffering in presynaptic nerve terminals. Free calcium levels measured with arsenazo III. Biochim. Biophys. Acta **600:** 912–921.
32. CERVETTO, L., L. LAGNADO, R. J. PERRY, D. W. ROBINSON & P. A. MCNAUGHTON. 1989. Extrusion of calcium from rod outer segments is driven by both sodium and potassium gradients. Nature **337:** 740–743.
33. YASUI, K. & J. KIMURA. 1990. Is potassium co-transported by the cardiac Na-Ca exchange? Pflügers Arch. **415:** 513–515.
34. DAHAN, D., R. SPANIER & H. RAHAMIMOFF. 1991. The modulation of rat brain Na^+-Ca^{2+} exchange by K^+. J. Biol. Chem. **266:** 2067–2075.
35. DIPOLO, R. & L. BEAUGÉ. 1987. Characterization of the reverse Na/Ca exchange in squid axons and its modulation by Ca_i and ATP. Ca_o-dependent Na_i/Ca_o and Na_i/Na_o exchange modes. J. Gen. Physiol. **90:** 505–525.
36. RASGADO-FLORES, H., E. M. SANTIAGO & M. P. BLAUSTEIN. 1989. Kinetics and stoichiometry of coupled Na efflux and Ca influx (Na/Ca exchange) in barnacle muscle cells. J. Gen. Physiol. **93:** 1219–1241.
37. DI POLO, R. & L. BEAUGÉ. 1979. Physiological role of ATP-driven calcium pump in squid axons. Nature **278:** 271–273.
38. BLAUSTEIN, M. P., W. F. GOLDMAN, G. FONTANA, B. K. KRUEGER, E. M. SANTIAGO, T. D. STEELE, D. N. WEISS & P. J. YAROWSKY. 1991. Physiological roles of the sodium-calcium exchanger in nerve and muscle. Ann. N.Y. Acad. Sci. **639:** 254–274.
39. DUARTE, C. B., C. A. M. CARVALHO, I. L. FERREIRA & A. P. CARVALHO. 1991. Synaptosomal $[Ca^{2+}]_i$ as influenced by Na^+/Ca^{2+} exchange and K^+ depolarization. Cell Calcium **12:** 623–633.
40. NACHSHEN, D. A. 1985. The early time course of potassium-stimulated calcium uptake in presynaptic nerve terminals from rat brains. J. Physiol. **361:** 251–268.
41. SMITH, S. J. & G. J. AUGUSTINE. 1988. Calcium ions, active zones and synaptic transmitter release. Trends Neurosci. **11:** 458–464.

Localization of the Na^+-Ca^{2+} Exchanger in Vascular Smooth Muscle, and in Neurons and Astrocytes[a]

MAGDALENA JUHASZOVA,[b,e] HIROSHI SHIMIZU,[b]
MIKHAIL L. BORIN,[b] RICK K. YIP[b,e]
ELIGIO M. SANTIAGO,[b] GEORGE E. LINDENMAYER,[d]
AND MORDECAI P. BLAUSTEIN[b,c]

Departments of [b]Physiology and [c]Medicine
University of Maryland School of Medicine
Baltimore, Maryland 21201

[d]*Department of Pharmacology*
Medical University of South Carolina
Charleston, South Carolina 29425

INTRODUCTION

The Na^+-Ca^{2+} exchanger was first identified in the late nineteen sixties in squid giant axons,[1] and in cardiac muscle.[2] In 1973, the Na^+-Ca^{2+} exchanger was identified in vascular smooth muscle (VSM) on the basis of ^{45}Ca fluxes and contraction experiments.[3] Nevertheless, the physiological significance and even the presence of the exchanger in smooth muscle cells was questioned for many years,[4,5] although such views have begun to change.[6] The early skepticism was based on the observation that even large changes of the Na^+ electrochemical gradient across the plasma membrane did not significantly alter the resting cytosolic free Ca^{2+} concentration ($[Ca^{2+}]_{cyt}$) in VSM cells. A logical explanation, however, is that even large Na^+-Ca^{2+} exchanger-mediated influxes of Ca^{2+} may be very efficiently buffered by the sarcoplasmic reticulum (SR).

The presence of a high capacity (*i.e.*, large maximal velocity) Na^+-Ca^{2+} exchanger in the mammalian brain is well established.[7] The brain exchanger was cloned and sequenced,[8] and the presence of at least three different brain isoforms was predicted from molecular studies.[9] The Na^+-Ca^{2+} exchanger in neurons apparently plays important roles in extruding Ca^{2+} following cell activation, and in controlling $[Ca^{2+}]_{cyt}$. However, about half of the brain's volume consists of nonexcitable cells, and the most populous among them are the astrocytes, which also possess a Na^+-Ca^{2+} exchanger.

It is now clear that astrocytes play key roles in the normal physiology, pathology

---

[a] Supported by National Institutes of Health Grants NS-16106 and HL-45215 to MPB, HL-42040 to GEL and HL-50700 to MLB, and American Heart Association-Maryland Affiliate Grants-in-Aid to MJ and MLB. HS is on leave of absence from the Sumitomo Pharmaceutical Company, Osaka, Japan.

[e] Address for correspondence: Dr. Magdalena Juhaszova, Department of Physiology, University of Maryland School of Medicine, 655 West Baltimore Street, Baltimore, MD 21201.

and development of the nervous system. For example, astrocytes have receptors for most neurotransmitters and they participate in the metabolism of a variety of neurotransmitters, including glutamate and gamma-aminobutyric acid (GABA). Astrocytes can also release a number of neuronal growth factors. Especially important for the present study is the fact that astrocytes may help to maintain a proper ionic environment for neurons.[10] Despite this wealth of evidence about the importance of astrocytes in brain function, there is only limited information available about Ca^{2+} metabolism in astrocytes. In the past few years, several groups have provided evidence that there is a physiologically significant Na^+-Ca^{2+} exchanger in mammalian astrocytes.[11–13]

MATERIALS AND METHODS

Primary Culture of Rat Arterial Smooth Muscle Cells and Rat Astrocytes

Arterial smooth muscle (ASM) cells were dissociated from the media of mesenteric arteries from adult rats, and were cultured as described.[14]

Astrocytes were cultured using a modification of the method of Booher and Sensenbrenner[15] (see Ref. 13). The astrocytes were prepared from the brains of one-day-old rats by mechanical disruption and filtration of the cell suspension without enzymatic digestion.

Preparation of Freshly Isolated Arterial Myocytes

Rat mesenteric artery or aorta was dissected in Ca^{2+}-free physiological salt solution (PSS) to prevent contraction. After removing the adipose tissue and adventitia, the arteries were incubated with collagenase, washed in PSS, and plated on coverslips coated with Cell-Tak adhesive (Collaborative Biomedical Products, Bedford, MA). The coverslips were immobilized for 45 min to permit time for the cells to settle and stick. The myocytes were then fixed in 2% formaldehyde, permeabilized with 0.5% Brij 58, and labeled with antibodies or $DiOC_6(3)$ (DiOC). In the latter case, the cells were exposed to phosphate buffered saline (PBS) containing 0.5 µg/ml DiOC for 5 min; the extracellular DiOC was then washed away with fresh PBS before mounting the coverslips for examination in the microscope.

Immunofluorescence Microscopy

Cultures were grown on 2.5 cm diameter coverslips. Primary cultured mesenteric artery cells and astrocytes were studied 7–10 days after plating. The freshly isolated cells were labeled immediately after immobilization. The coverslips were washed with PBS and immunolabeled as described.[16] Rabbit polyclonal antibodies were raised against the purified Na^+-Ca^{2+} exchanger from dog heart sarcolemma.[17] Polyclonal antibodies against the sarco-endoplasmic reticulum Ca^{2+} pump (SERCA2b) were kindly provided by Dr. F. Wuytack (Katholieke Universitiet, Leuven, Belgium). Polyclonal antibodies raised against the plasmalemmal (PL) Ca^{2+}-ATPase were the generous gift of Dr. E. Carafoli (Swiss Federal Institute of Technology, Zurich, Switzerland). Cy3-conjugated donkey anti-rabbit IgG (Jackson ImmunoResearch) were used to visualize the primary antibodies.

Frozen hippocampal slices, 25 μm thick, were obtained from the brains of rats anesthetized with Nembutal (the brain slices were kindly provided by Dr. P. M. Wise; University of Kentucky, Lexington, KY). The slices were crossreacted with antibodies raised against the Na^+-Ca^{2+} exchanger and conterstained with immunoperoxidase to visualize the primary antibody.

The specimens were examined with a Nikon Diaphot fluorescence microscope (40× or 100× N.A. 1.3 Nikon UV-Fluor objectives or an Olympus 60× N.A. 1.4 objective). In some instances, images were photographed directly from the video monitor onto Kodak Ektachrome Professional (EPN) film.[18] In other cases, a CELLscan (Scanalytics, Billerica, MA) high resolution fluorescence imaging system was employed to remove out-of-focus fluorescence.

The CELLscan system deconvolution algorithm uses an experimentally generated point spread function to vector out-of-focus fluorescence back to its point of origin in each specimen image plane.[19] Each individual 2-D image generated by this system corresponds to a single deconvolved and restored optical image (*i.e.*, the in-focus fluorescence in that image plane). The Image-1/MetaMorph Imaging System (Universal Imaging Corporation, West Chester, PA) was employed for statistical analysis of the co-localization of the plasma membrane ion transporters with the SR.

Calcium and Sodium Imaging Experiments

To measure $[Ca^{2+}]_{cyt}$, the cultured cells were loaded with the membrane permeable (acetoxymethyl ester) Ca^{2+} sensitive fluorescent dye, Fura 2-AM. The dye was excited alternately with 380 and 360 nm light, and fluorescent emission at 510 nm was measured. The Fura-2 emission ratio evoked by 380/360 nm excitation was calculated and calibrated to express $[Ca^{2+}]_{cyt}$.[18] The Na^+ sensitive dye, sodium-binding benzofuran isophthalate (SBFI) was used to monitor changes in the cytoplasmic Na^+ concentration ($[Na^+]_{cyt}$). The SBFI emission ratio evoked by 340/380 nm excitation was calculated and calibrated to express $[Na^+]_{cyt}$.[20] The fluorescence was imaged using a Nikon Diaphot microscope (Nikon UV-Fluor objective ×40; N.A. 1.3). The imaging system has been described.[18] Small cytoplasmic areas of interest ($\approx 4 \times 6$ pixels), one area per cell, in 8–10 cells per field of view on a single coverslip, were used for the calculations of $[Na^+]_{cyt}$ and $[Ca^{2+}]_{cyt}$.

RESULTS AND DISCUSSION

The Na^+-Ca^{2+} Exchanger as a Regulator of SR and ER Ca^{2+}

Digital imaging experiments on VSM cells were employed to determine the role of the Na^+-Ca^{2+} exchanger in modulation of the intracellular (SR) Ca^{2+} stores. In most types of cells, inhibition of the Na^+ pump with ouabain increases $[Na^+]_{cyt}$ and decreases the Na^+ electrochemical gradient across the plasma membrane. As

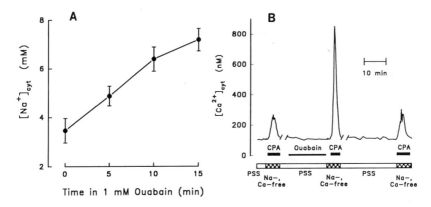

FIGURE 1. Effects of ouabain on $[Na^+]_{cyt}$ and $[Ca^{2+}]_{cyt}$ in rat aortic myocytes. **(A)** SBFI-loaded cells were incubated in PSS with 1 mM ouabain to measure the time course of changes in $[Na^+]_{cyt}$. The data correspond to the mean values for 19 cells. **(B)** Fura-2-loaded cells were incubated in PSS in the absence or presence of 1 mM ouabain, as indicated. Ca^{2+} release was evoked by 5 μM CPA (in Ca^{2+}-free PSS). Data correspond to mean values for 14 cells from a representative experiment.

illustrated in FIGURE 1A, we employed SBFI to show, directly, that 1 mM ouabain increases $[Na^+]_{cyt}$ in primary cultured rat aortic myocytes in a time-dependent manner.

This increase in $[Na^+]_{cyt}$ can be expected to enhance Ca^{2+} entry and reduce Ca^{2+} exit via the Na^+-Ca^{2+} exchanger.[21,22] Under these circumstances the Ca^{2+} entering the cells is very efficiently buffered by the SR.[23,24] To estimate relative changes in the amount of Ca^{2+} stored in the SR, we measured the cyclopiazonic acid (CPA)-evoked rise in $[Ca^{2+}]_{cyt}$ (FIG. 1B). The aortic myocytes were loaded with Fura-2 to determine resting $[Ca^{2+}]_{cyt}$ and the amplitudes of the CPA-evoked Ca^{2+} transients. CPA is a reversible blocker of the sarco-endoplasmic reticulum Ca^{2+} (SERCA) pump. Like thapsigargin, another SERCA blocker, CPA promotes leak of Ca^{2+} from the SR and a transient rise in cytosolic Ca^{2+} (*i.e.*, a "Ca^{2+} transient"); the effects of CPA are, however, reversible.

FIGURE 1B shows data from a representative Ca^{2+} experiment. Under resting conditions $[Ca^{2+}]_{cyt}$ was about 100 nM. External Ca^{2+} was removed just before CPA was added in order to block Ca^{2+} influx (including so-called "capacitative calcium entry"); external Na^+ was removed to prevent Ca^{2+} efflux via Na^+-Ca^{2+} exchange. The cells were then exposed briefly to 5 μM CPA to induce a Ca^{2+} transient. After recovery in normal PSS, the cells were exposed to 1 mM ouabain for varying periods of time. Note that after 12 min (FIG. 1B), there was only a very small rise in the resting $[Ca^{2+}]_{cyt}$; however, the CPA-evoked Ca^{2+} transient was much greater than before ouabain. This effect was reversible, and the Ca^{2+} transient recovered when the ouabain was washed out (FIG. 1B). The implication is that ouabain induced a substantial increase in cell Ca^{2+}, but most of this Ca^{2+} was buffered and stored in the SR. This is presented quantitatively in the graph in FIGURE 2. These $[Na^+]_{cyt}$ and $[Ca^{2+}]_{cyt}$ data were obtained by exposing the aortic myocytes to 1 mM ouabain for 0, 5, 10 or 15 min. The increase in resting $[Ca^{2+}]_{cyt}$ (*open circles*), and in the peak of the CPA-induced Ca^{2+} transient (*solid circles*) obtained during the second exposure to CPA, relative to the values ob-

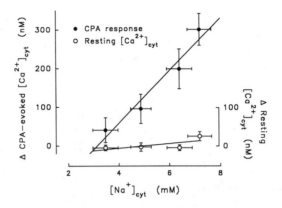

FIGURE 2. Graph of increases in resting $[Ca^{2+}]_{cyt}$ and CPA-evoked Ca^{2+} transients graphed as a function of $[Na^+]_{cyt}$. The *abscissa* values were obtained from FIGURE 1A; *ordinate* values were obtained from experiments similar to that of FIGURE 1B. Δ Resting $[Ca^{2+}]_{cyt}$ corresponds to the increase in $[Ca^{2+}]_{cyt}$ after a 0-, 5-, 10- or 15-min exposure to ouabain, relative to the $[Ca^{2+}]_{cyt}$ before the first (control) response to CPA. Δ CPA-evoked $[Ca^{2+}]_{cyt}$ corresponds to the increase in the amplitude of the second Ca^{2+} transient (after 0, 5, 10 or 15 min in ouabain) relative to the first (control) response to CPA.

tained in the control period, are graphed as a function of the $[Na^+]_{cyt}$ obtained in parallel experiments (FIG. 1A). As expected from the data in FIGURE 1, the relationship between the increase in resting $[Ca^{2+}]_{cyt}$ and $[Na^+]_{cyt}$ has a very shallow slope. In contrast, the relationship between the increase in the peak of the CPA-induced Ca^{2+} transient and $[Na^+]_{cyt}$ has a very steep slope. Very similar results have also been obtained in primary cultured rat astrocytes[25] and neurons (Golovina & Blaustein, unpublished). Data such as these strongly support the view that an important role of the Na^+-Ca^{2+} exchanger in these cells is to help modulate the stores of Ca^{2+} in the SR and ER of cells such as arterial myocytes and brain astrocytes.

Immunolabeling of the Na^+-Ca^{2+} Exchanger in VSM Cells and Astrocytes

Standard immunochemical and molecular biological techniques were used to demonstrate that Na^+-Ca^{2+} exchanger message and protein are expressed in rat VSM cells,[26] and in astrocytes and neurons.[27,13] The Northern and Western blot analyses confirm that the exchanger in these cells is similar to the cardiac Na^+-Ca^{2+} exchanger in terms of molecular size and antigenic properties. Furthermore, the data suggest that the Na^+-Ca^{2+} exchanger protein is present at low density in arterial myocyte membranes,[26] at intermediate density in astrocyte membranes,[13] and at high density in cardiac myocyte and synaptic plasma membranes.[27,28]

In the present study we focused on the immunolocalization of this transporter in the plasma membranes of these cells. The spatial relationship between the plasmalemmal exchanger and underlying SR was also examined. The immunofluorescent localization of the Na^+-Ca^{2+} exchanger in cultured mesenteric artery myocytes is presented in FIGURE 3. The cells were incubated with anti-Na^+-Ca^{2+} exchanger antiserum; these antibodies were affinity purified to increase the specificity and decrease the background. The labeling was punctate, and we attribute each fluorescent spot to a cluster of Na^+-Ca^{2+} exchanger molecules. Specific labeling of the Na^+-Ca^{2+} exchanger was intense at cell edges (arrowheads); labeling on cell surfaces formed regular reticular patterns in many areas (double arrowheads). Labeling of nuclei (N) was nonspecific (FIG. 3B). The reticular labeling pattern of the Na^+-Ca^{2+} exchanger suggests that the exchanger is distributed in an organized manner in the sarcolemma.

A similar distribution pattern of immunoreactive sites was observed when primary cultured rat cortical astrocytes were incubated with anti-Na^+-Ca^{2+} exchanger antiserum (FIG. 4). All cells in the culture were labeled specifically (FIG. 4A); no immunofluorescent labeling was observed when the cells were incubated with preimmune serum (FIG. 4B). Punctate immunofluorescent foci were distributed over entire cells, and the labeling was arrayed in a distinct reticular pattern in many areas (arrowheads in FIG. 4A).

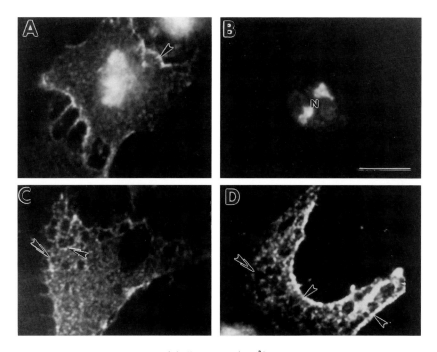

FIGURE 3. Immunofluorescent labeling of Na^+-Ca^{2+} exchanger in cultured mesenteric artery myocytes. Cells in **(A)**, **(C)** and **(D)** were incubated with affinity-purified anti-Na^+-Ca^{2+} exchanger antiserum; cells in **(B)** were incubated with preimmune serum processed in same way as affinity-purified antiserum. Fluoresceinated goat anti-rabbit immunoglobulin G (IgG) was used as the secondary antibody. Labeling of Na^+-Ca^{2+} exchanger is very intense at cell edges (*arrowheads*); labeling on the cell surface forms a regular reticular pattern (*double arrowheads*) in many areas. Labeling of nuclei (N) is nonspecific. *Bar*: 8 μm. (Data modified from Ref. 26.)

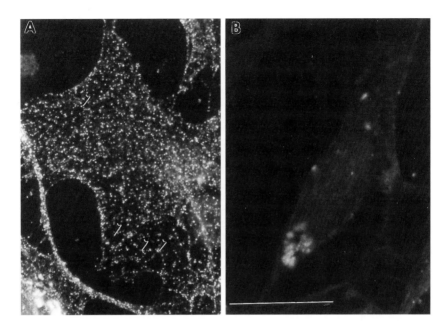

FIGURE 4. Immunofluorescent labeling of Na^+-Ca^{2+} exchangers in cultured rat brain astrocytes. The cell in **(A)** was probed with affinity-purified antibodies raised against canine cardiac sarcolemmal Na^+-Ca^{2+} exchanger. The cell in **(B)** was probed with preimmune serum. The *arrowheads* in (A) point to regions of the cell in which the label has a distinct reticular pattern. *Bar*: 20 μm. (Data reproduced from Ref. 13, with permission.)

Localization of the Na^+-Ca^{2+} Exchanger in VSM Cells and Astrocytes

These data lead to two obvious questions. First: Is the label on the cell surface, or is it in the cell interior? Second: What is the significance of these reticular distribution patterns?

To answer the first question, antibody labeling was studied in freshly-isolated rat mesenteric artery myocytes. The CELLscan imaging system was employed for these experiments. Multiple image planes were collected, out-of-focus fluorescence was reassigned with the deblurring algorithms, and single focal plane images were reconstructed.[19]

FIGURE 5 shows data from such experiments. The myocyte in FIGURE 5A was labeled with affinity purified anti-Na^+-Ca^{2+} exchanger antiserum; the antibody fluorescence was confined to the plasmalemma. The same cell was also stained with DiOC (FIG. 5B). DiOC is a lipophilic, cationic, dicarbocyanine dye, that can be used to identify SR.[29] This dye stains organelles in the cell interior, and its labeling pattern is, thus, distinct from that of the Na^+-Ca^{2+} exchanger antibodies. In freshly isolated arterial myocytes, both labels appear to be co-localized close to the cell surface where the junctional SR membranes are located. The DiOC probably also stains the plasmalemma, but the stain in a single bilayer membrane is undetectable with our imaging methods.

The freshly-isolated myocyte in FIGURE 5C was labeled with antibodies raised against the plasma membrane (PM) Ca^{2+}-ATPase, whereas the myocyte in FIGURE 5D was labeled with antibodies raised against the SR Ca^{2+}-ATPase. The PM Ca^{2+}-ATPase label was confined to the cell surface (FIG. 5C), whereas the SR Ca^{2+}-ATPase label was present only in the cell interior (FIG. 5D).

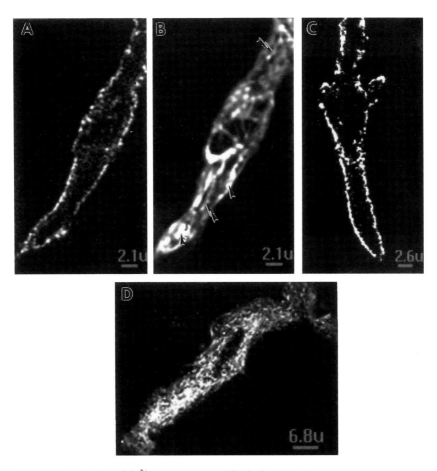

FIGURE 5. Labeling of Ca^{2+} transporters and SR in freshly isolated rat mesenteric artery myocytes. Each image shows a single restored image plane (obtained with the CELLscan System) in the middle of the cell. **(A)** Myocyte probed with affinity-purified anti-Na^+-Ca^{2+} exchanger antiserum; the label is present in punctate foci on the cell surface, and is absent from the cell interior. **(B)** The same cell was stained with DiOC, which labels SR (*double arrowheads*) and mitochondria (*arrowheads*); this label is distributed throughout cell interior. **(C)** Another myocyte was probed with anti-PM Ca^{2+}-ATPase antiserum. This label, too, is confined to the cell surface and is absent from the cytoplasm. This label appears to be more densely distributed on the cell surface than is the anti-Na^+-Ca^{2+} exchanger antibody (contrast FIG. 5B). **(D)** A different myocyte was probed with anti-SERCA 2b antiserum. This label is distributed throughout the cytoplasm.

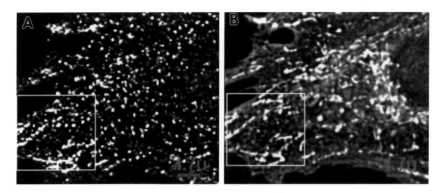

FIGURE 6. Immunocytochemical localization of Na^+-Ca^{2+} exchanger **(A)** and subsequent DiOC staining of SR **(B)** in a cultured rat mesenteric artery myocyte (single focal plane, deconvolved and restored images). The cell was immunolabeled with affinity-purified anti-Na^+-Ca^{2+} exchanger antiserum, and then stained with 500 ng/ml DiOC. The areas within the *boxes* are enlarged in FIGURE 7.

To address the second question, we determined the spatial correlation between the surface Na^+-Ca^{2+} exchanger and the underlying SR by double-labeling cultured cells with antibodies raised against the Na^+-Ca^{2+} exchanger and with DiOC. The CELLscan system was used for all double-labeling experiments.

FIGURE 6 shows data from a double-labeled, cultured mesenteric artery

TABLE 1. Statistical Analysis of the Co-localization of Plasma Membrane Transporters with the Sarcoplasmic/Endoplasmic Reticulum

Labeled Pair[a]	Observed Overlap[b]	Random Overlap[c]	Probability Observed Overlap Is Due to Chance
Mesenteric artery myocyte (FIGS. 6 and 7)			
NCX with SR	30.3 ± 2.7	18.5 ± 2.3	$p < 0.05$
SR with NCX	21.0 ± 1.8	12.6 ± 1.1	$p < 0.05$
Astrocyte (FIGS. 8 and 9)			
NCX with ER	39.6 ± 3.4	7.6 ± 1.2	$p < 0.005$
ER with NCX	24.8 ± 2.5	4.9 ± 0.8	$p < 0.005$
Mesenteric Artery Myocyte (FIGS. 10 and 11)			
PL Ca^{2+}-ATPase with SR	15.9 ± 2.8	16.4 ± 3.5	NS
SR with PL Ca^{2+}-ATPase	7.2 ± 1.1	7.4 ± 1.3	NS

[a] Cells labeled with DiOC and with antibodies raised against either the Na^+-Ca^{2+} exchanger (NCX) or PL Ca^{2+}-ATPase were studied (FIGS. 6–11). Four areas (each ≈3750 pixels or 84.4 μm^2) from each labeled cell (FIGS. 6, 8 and 10) were analyzed.

[b] The "observed overlap" is the percentage of pixels (mean ± SE) containing the label listed first, which also contained the second label.

[c] Random overlap was determined after one of the paired images was shifted 10 pixels along the x axis and 10 pixels along the y axis (1.5 μm).

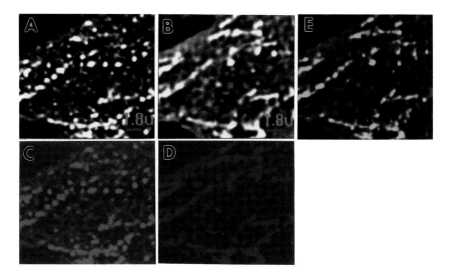

FIGURE 7. Co-localization of the Na^+-Ca^{2+} exchanger and SR in the same myocyte as in FIGURE 6. The *boxed area* of the cell from FIGURE 6 was enlarged **((A) and (B))**. The Na^+-Ca^{2+} exchanger label (A) was colored green **(C)**, the DiOC (B) was colored red **(D)**, and the two colored images were superimposed **(E)**. The yellow corresponds to regions in which the two labels overlapped. TABLE 1 shows that there was significant co-localization of the two labels.

myocyte. In some areas, the fluorescent clusters of Na^+-Ca^{2+} exchanger molecules (FIG. 6A) exhibited the same reticular pattern as did the underlying SR visualized with DiOC (FIG. 6B). The boxed areas from FIGURE 6 were enlarged to show details (FIG. 7A,B). To determine whether the two labels are co-localized, the Na^+-Ca^{2+} exchanger label was colored red (FIG. 7C), and the DiOC fluorescence was colored green (FIG. 7D). The two colored images were superimposed in FIGURE 7E. In the superimposed image, the yellow areas correspond to the regions in which the Na^+-Ca^{2+} exchanger and the SR were co-localized. The overlap (co-localization) between the two labels was quantitated (see Methods and footnotes to TABLE 1). The results (TABLE 1) indicate that the observed overlap is significant, and that the Na^+-Ca^{2+} exchanger is co-localized with a portion of the SR.

Comparable co-localization studies were carried out on cultured astrocytes (FIGS. 8 and 9). The astrocyte in FIGURE 8 was immunolabeled with anti-Na^+-Ca^{2+} exchanger antibodies (A) and DiOC (B). Here the Na^+-Ca^{2+} exchanger labeling appears to form strings of beads arranged in a reticular network (FIG. 8A) very similar to DiOC labeling of the ER in the same cell (FIG. 8B). The regions in the boxes in FIGURE 8 were enlarged (FIGS. 9A and 9B). The exchanger label was then colored green (C), and the DiOC label was colored red (D). The two colored images were superimposed (E); again, yellow areas indicate the regions of overlap of the two labels. The data in TABLE 1 show that in astrocytes, too, the Na^+-Ca^{2+} exchanger is co-localized with some of the SR.

Not all plasma membrane ion transporters are distributed in an organized, reticular pattern over the cell surface. In contrast to the arrangement of the

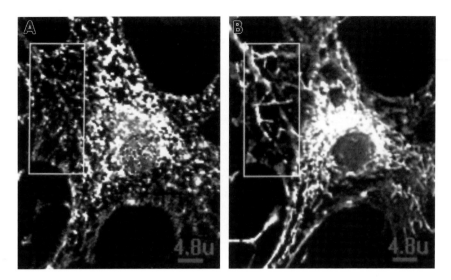

FIGURE 8. Immunocytochemical localization of Na^+-Ca^{2+} exchanger **(A)** and subsequent DiOC staining of ER **(B)** in a cultured rat brain astrocyte (single focal plane deconvolved, restored images). The cell was immunolabeled with affinity-purified anti-Na^+-Ca^{2+} exchanger antiserum, and then stained with 500 ng/ml DiOC. The *areas within the boxes* are enlarged in FIGURE 9.

Na^+-Ca^{2+} exchanger, PM Ca^{2+}-ATPase is much more diffusely distributed: FIGURE 10A shows that immunofluorescently labeled PM Ca^{2+}-ATPase is randomly distributed in a mesenteric artery myocyte. This differs from the reticular pattern of DiOC stain in the same cell (FIG. 10B). Indeed, there was no significant overlap of the PM Ca^{2+}-ATPase and underlying SR structures (FIG. 11 and TABLE 1).

The co-localization of the Na^+-Ca^{2+} exchanger with SR/ER (FIGS. 7E and 9E) provides a structural basis for the functional observations in FIGURE 1: as a consequence of its close apposition to the SR/ER, the plasmalemmal Na^+-Ca^{2+} exchanger may play a key role in the modulation of SR/ER Ca^{2+} stores. In accordance with the "buffer barrier" hypothesis for vascular smooth muscle,[24] substantial movements of Ca^{2+} into the cells may therefore be rapidly buffered, and may have little effect on bulk $[Ca^{2+}]_{cyt}$. Recently, Moore *et al.*[30] found that immunofluorescent patches of calsequestrin (a marker for the SR) were closely associated with plasma membrane regions rich in Na^+/K^+-pump and Na^+-Ca^{2+} exchanger in toad stomach smooth muscle cells. Consequently, they, too, suggested that the Na^+-Ca^{2+} exchanger may have preferred access to the Ca^{2+} in the SR. Our results in astrocytes now imply that this is a much more general phenomenon, and is not limited to smooth muscle cells.

Distribution of the Na^+-Ca^{2+} Exchanger in Neurons

Previous immunocytochemical studies[28] indicate that cultured neurons cross-react strongly with anti-Na^+-Ca^{2+} exchanger antibodies, and that the Na^+-Ca^{2+}

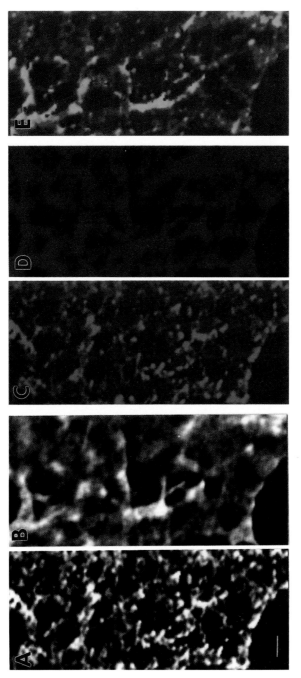

FIGURE 9. Co-localization of the Na^+-Ca^{2+} exchanger and SR in the same astrocyte as in FIGURE 8. The *boxed area* of the cell from FIGURE 8 was enlarged (**A**) and (**B**). The Na^+-Ca^{2+} exchanger label (A) was colored green (**C**), the DiOC (B) was colored red (**D**), and the two colored images were superimposed (**E**). The yellow corresponds to regions in which the two labels were co-localized; the overlap was significant (TABLE 1). *Bar*: 2 μm.

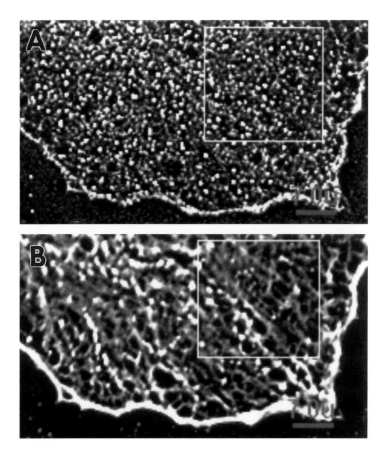

FIGURE 10. Immunocytochemical localization of plasma membrane (PM) Ca^{2+} pump **(A)** and subsequent DiOC staining of SR **(B)** in a cultured mesenteric artery myocyte (single focal plane deconvolved, restored images). The cell was immunolabeled with antibodies raised against the rat PM Ca^{2+} pump and then stained with 500 ng/ml DiOC.

exchanger may be concentrated at presynaptic nerve terminals. FIGURE 12A shows the distribution of anti-Na^+-Ca^{2+} exchanger label in a coronal section of the rat brain hippocampus. The immunolabeling was specific: no labeling was observed in brain slices treated with preimmune serum (FIG. 12B). Note that the synaptic fields in FIGURE 12A are particularly heavily labeled. In contrast, the cell body layers, especially the pyramidal cell layer (pc) and granule cell layer (gc), are very sparsely labeled. It is not clear whether the low level of labeling in the cell body layers is due to a lower surface : volume ratio in these areas, or to a real difference in the distribution of exchanger molecules between the cell body layers and the synaptic fields.

In situ hybridization studies revealed a complementary labeling pattern in the hippocampus. Na^+-Ca^{2+} exchanger mRNA was identified by intense staining in the pyramidal cell and granule cell layers, while the synaptic fields were spared (M. Juhaszova, unpublished data; and see Ref. 31).

Results comparable to those in the hippocampus were obtained in the cerebellum. Here, too, the synaptic fields were heavily labeled with antibodies raised against the Na^+-Ca^{2+} exchanger, while layers enriched with cell bodies were very sparsely labeled (not shown; see Marlier et al., 1993,[31] for complementary in situ hybridization studies of Na^+-Ca^{2+} exchanger mRNA distribution).

Additional detailed information about the distribution of the Na^+-Ca^{2+} exchanger in neurons was required for comparison with the data from astrocytes and VSM cells shown above. Cultured hippocampal neurons were therefore labeled with anti Na^+-Ca^{2+} exchanger antibodies (FIGS. 13 and 14). More immunolabeling was observed in neurons than in astrocytes or arterial myocytes, and a more dilute antibody solution was needed to label the neurons. This is consistent with our immunoblot evidence (above) that the exchanger is much more prevalent (i.e., there is a higher density of molecules) in neurons than in astrocytes or VSM cells. The labeling was discontinuous, and consisted of punctate immunofluorescent patches on cell bodies as well as along axons and dendrites (FIG. 13), and especially on growth cones and nerve terminals. A similar distribution of Na^+-Ca^{2+} exchanger has been observed in cultured Xenopus neurons.[28]

FIGURE 13 shows a montage of reconstructed single focal plane images of the cell body of a hippocampal pyramidal cell; the images were obtained with

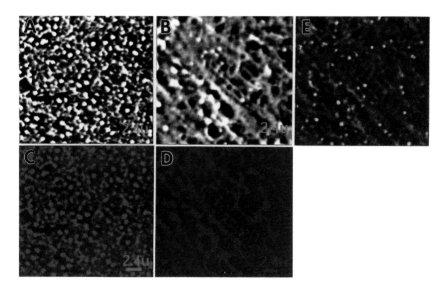

FIGURE 11. Co-localization of the plasmalemmal Ca^{2+} pump and SR in the same myocyte as in FIGURE 10. The *boxed area* of the cell from FIGURE 10 was enlarged (**(A)** and **(B)**). The Na^+-Ca^{2+} exchanger label (A) was colored green **(C)**, the DiOC (B) was colored red **(D)**, and the two colored images were superimposed **(E)**. The yellow corresponds to regions in which the two labels overlapped. As indicated in TABLE 1, there is no significant co-localization of the two labels.

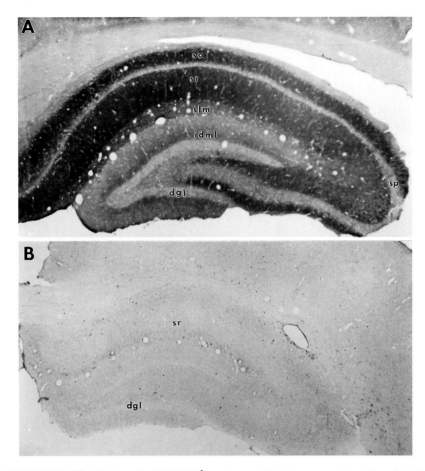

FIGURE 12. Distribution of anti-Na^+-Ca^{2+} exchanger label in a coronal section of adult rat hippocampus. **(A)** The section was incubated with antiserum and stained with immunoperoxidase. The regions containing most of the neuronal cell bodies are only very sparsely labeled: the stratum pyramidale (sp) containing most of the pyramidal cell bodies, and the dentate granule cell layer (dgl). The antiserum strongly labeled all synaptic regions: the strata oriens (so) and radiatum (sr) were most intensely labeled, followed by the stratum lacunosum-moleculare (slm) and the inner third of the dentate molecular layer (idml). **(B)** A parallel section incubated with preimmune serum shows no labeling.

the CELLscan system. Note that the immunolabeling is punctate, and is restricted to the plasma membrane, as in the VSM cell in FIGURE 5A. The single plane images from the cell in FIGURE 13 are stacked and shown as a stereo image pair in FIGURE 14. At the top surface of the cell body, much of the label appears to be arranged in a reticular pattern. This raises the possibility (not yet tested) that here, too, the Na^+-Ca^{2+} exchanger may be organized into

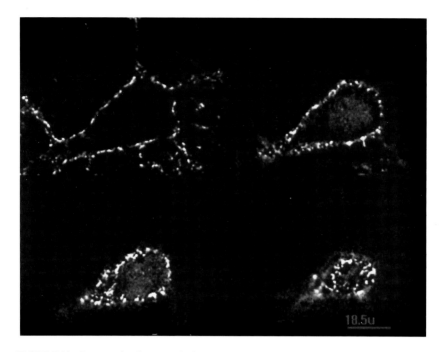

FIGURE 13. Deconvolved, restored (single focal plane) images of a cultured rat hippocampal neuron labeled with antibodies raised against the Na^+-Ca^{2+} exchanger. The figure shows a montage of images (2.5 μm apart) in the X-Y plane. Note that the label is restricted to the cell surface.

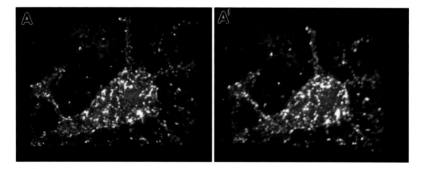

FIGURE 14. Stereo image view of the hippocampal neuron from FIGURE 13 labeled with antibodies raised against Na^+-Ca^{2+} exchanger. The two volume views **(A)** and **(A')** were reconstructed from deconvolved, restored single focal plane images. The volume view consists of a stack of 55 single plane images, 0.25 μm apart. **(A')** is rotated 6° to create the 3-dimensional image.

microdomains, and may be co-localized with underlying cytoplasmic structures—possibly the ER.

CONCLUSIONS

Mammalian VSM, neurons and astrocytes all possess a high capacity, cardiac-type Na^+-Ca^{2+} exchanger in their plasma membranes. Immunoblot and Northern blot as well as immunolocalization data suggest that VSM cells and astrocytes possess much lower densities of Na^+-Ca^{2+} exchanger molecules than do the plasma membranes of cardiac myocytes and neurons. Functional data indicate that one role of the Na^+-Ca^{2+} exchanger is the indirect regulation of the intracellular Ca^{2+} stores in the SR and ER. The immunolocalization data provide evidence for a morphological correlate of this functional relationship. The Na^+-Ca^{2+} exchanger molecules are distributed in organized patterns across the plasmalemma of VSM cells and astrocytes, and probably of neurons as well. The punctate foci of immunostaining appear to be organized in reticular patterns over the cell surfaces. In many areas the exchanger is co-localized with underlying SR or ER structures. This localization is consistent with the view that the exchanger may have preferred access to the Ca^{2+} in the SR/ER.

ACKNOWLEDGMENTS

We thank Dr. P. M. Wise for providing frozen brain slices, Dr. F. S. Fay for very helpful comments on the statistical analysis of co-localized labels, and Drs. E. Carafoli and F. Wuytack for gifts of antisera.

REFERENCES

1. BAKER, P. F., M. P. BLAUSTEIN, A. L. HODGKIN & R. A. STEINHARDT. 1969. The influence of calcium ions on sodium efflux in squid axons. J. Physiol. (London) **200:** 431–458.
2. REUTER, H. & H. SEITZ. 1968. The dependence of calcium efflux from cardiac muscle on temperature and external ion composition. J. Physiol. (London) **95:** 451–470.
3. REUTER, H., M. P. BLAUSTEIN & G. HAUSLER. 1973. Na/Ca exchange and tension development in arterial smooth muscle. Philos. Trans. R. Soc. (London) **B265:** 87–94.
4. SOMLYO, A. P., R. BRODERICK & A. V. SOMLYO. 1986. Calcium and sodium in vascular smooth muscle. Ann. N. Y. Acad. Sci. **488:** 228–239.
5. VAN BREEMEN, C. &. C. SAIDA. 1989. Cellular mechanisms regulating $[Ca^{2+}]_i$ in smooth muscle. Annu. Rev. Physiol. **51:** 315–329.
6. SOMLYO, A. P. & A. V. SOMLYO. 1994. Signal transduction and regulation in smooth muscle. Nature **372:** 231–236.
7. BLAUSTEIN, M. P., G. FONTANA & R. S. ROGOWSKI. 1995. The Na^+Ca^{2+} exchanger in rat brain synaptosomes: kinetics and regulation. This volume.
8. FURMAN, I., O. COOK, J. KASIR & H. RAHAMIMOFF. 1993. Cloning of two isoforms of rat brain Na^+-Ca^{2+} exchanger gene and their functional expression in HeLa cells. FEBS Lett. **319:** 105–109.
9. KOFUJI, P., W. J. LEDERER & D. H. SCHULZE. 1994. Mutually exclusive and cassette exons underlie alternatively spliced isoforms of the Na/Ca exchanger. J. Biol. Chem. **269:** 5145–5149.
10. KIMELBERG, H. K. & M. D. NORENBERG. 1989. Astrocytes. Sci. Am. April: 66–76.

11. BLAUSTEIN, M. P., W. F. GOLDMAN, G. FONTANA, B. K. KRUEGER, E. M. SANTIAGO, T. D. STEELE, D. N. WEISS & P. J. YAROWSKY. 1991. Physiological roles of the sodium-calcium exchanger in nerve and muscle. Ann. N. Y. Acad. Sci. **639:** 254–274.
12. FINKBEINER, S. M. 1993. Glial calcium. Glia **9:** 83–104.
13. GOLDMAN, W. F., P. J. YAROWSKY, M. JUHASZOVA, B. K. KRUEGER & M. P. BLAUSTEIN. 1994. Sodium/calcium exchange in rat cortical astrocytes. J. Neurosci. **14:** 5834–5843.
14. GUNTHER, S., R. W. ALEXANDER, W. J. ATKINSON & M. A. GIMBRONE, JR. 1982. Functional angiotensin II receptors in cultured vascular smooth muscle cells. J. Cell Biol. **92:** 289–298.
15. BOOHER, J. & M. SENSENBRENNER. 1972. Growth and cultivation of dissociated neurons and glial cells from embryonic chick, rat and human brain in flask cultures. Neurobiology **2:** 97–105.
16. LUTHER, P. W. & R. J. BLOCH. 1989. Formaldehyde-amine fixatives for immunocytochemistry of cultured *Xenopus* myocytes. J. Histochem. Cytochem. **37:** 75–82.
17. AMBESI, A., E. E. BAGWELL & G. E. LINDENMAYER. 1991. Purification and identification of the cardiac sarcolemmal Na/Ca exchanger (Abstract). Biophys. J. **59:** 138a.
18. GOLDMAN, W. F., S. BOVA & M. P. BLAUSTEIN. 1990. Measurement of intracellular Ca^{2+} in cultured arterial smooth muscle cells using fura-2 and digital imaging microscopy. Cell Calcium **11:** 221–231.
19. CARRINGTON, W. A., K. E. FOGARTY & F. S. FAY. 1990. 3D Fluorescence imaging of single cells using image restoration. *In* Noninvasive Techniques in Cell Biology. 53–72. Wiley-Liss, Inc.
20. BORIN, M. L., W. F. GOLDMAN & M. P. BLAUSTEIN. 1993. Intracellular free Na^+ in resting and activated cultured vascular smooth muscle cells. Am. J. Physiol. **264:** C1513–C1524.
21. BORIN, M. L., R. M. TRIBE & M. P. BLAUSTEIN. 1994. Increased intracellular Na^+ augments mobilization of Ca^{2+} from SR in vascular smooth muscle cells. Am. J. Physiol. (Cell Physiol.) **266:** C311–C317.
22. BOVA, S., W. F. GOLDMAN, X-J. YUAN & M. P. BLAUSTEIN. 1990. Influence of the Na^+ gradient on Ca^{2+} transients and contraction in vascular smooth muscle. Am. J. Physiol. **259:** H409–H423.
23. BLAUSTEIN, M. P. 1993. The pathophysiological effects of endogenous ouabain: control of stored Ca^{2+} and cell responsiveness. Am. J. Physiol. **264:** C1367–C1387.
24. CHEN, Q. & C. VAN BREEMEN. 1992. Function of smooth muscle sarcoplasmic reticulum. *In* Advances in Second Messenger and Phosphoprotein Research. J. W. Putney, Ed. Vol. **26:** 335–350. Raven Press. New York.
25. GOLOVINA, M., L. L. BAMBRICK, B. K. KRUEGER, P. J. YAROWSKY & M. P. BLAUSTEIN. 1994. Na-Ca exchange regulates intracellular stores of Ca^{2+} in mouse cortical astrocytes. Biophys. J. **66:** A255.
26. JUHASZOVA, M., A. AMBESI, G. E. LINDENMAYER, R. J. BLOCH & M. P. BLAUSTEIN. 1994. The Na/Ca exchanger in arteries: identification by immunoblotting and immunofluorescence microscopy. Am. J. Physiol. **266:** C234–C242.
27. YIP, R. K., M. P. BLAUSTEIN & K. D. PHILIPSON. 1992. Immunologic identification of Na/Ca exchange protein in rat synaptic plasma membrane. Neurosci. Lett. **136:** 123–126.
28. LUTHER, P. W., R. K. YIP, R. J. BLOCH, A. AMBESI, G. E. LINDENMAYER & M. P. BLAUSTEIN. 1992. Presynaptic localization of sodium/calcium exchangers in neuromuscular preparations. J. Neurosci. **12:** 4898–4904.
29. TERASAKI, M. 1989. Fluorescent labeling of endoplasmic reticulum. Methods Cell Biol. **29:** 125–135.
30. MOORE, E. D. W., E. F. ERTTE, K. D. PHILIPSON, W. F. CARRINGTON, K. E. FOGARTY, L. M. LIFSHITZ & F. S. FAY. 1993. Coupling of Na^+/Ca^{2+} exchanger, Na^+/K^+ pump and sarcoplasmic reticulum in smooth muscle. Nature **365:** 657–660.
31. MARLIER, L. N. J-L., T. ZHENG, J. TANG & D. R. GRAYSON. 1993. Regional distribution in the rat central nervous system of a mRNA encoding a portion of the cardiac sodium/calcium exchanger isolated from cerebellar granule neurons. Mol. Brain Res. **20:** 21–39.

Regulation of the Bovine Retinal Rod Na-Ca+K Exchanger[a]

PAUL P. M. SCHNETKAMP, JOSEPH E. TUCKER,
AND ROBERT T. SZERENCSEI

Department of Medical Biochemistry
University of Calgary
Health Science Centre
3330 Hospital Drive N.W.
Calgary, Alberta, T2N 4N1, Canada

INTRODUCTION

The cytosolic free Ca^{2+} concentration in the outer segments of retinal rod photoreceptors (ROS) is thought to arise from a kinetic equilibrium between Ca^{2+} influx via the light-sensitive and cGMP-gated channels and Ca^{2+} efflux via the Na-Ca+K exchanger.[1] Both cGMP-gated channels[2] and Na-Ca+K exchangers[3] are thought to be located exclusively in the ROS plasma membrane and they are the only two Ca^{2+} transport proteins that have been identified with certainty in isolated ROS.[4,5] In a recent study on tiger salamander ROS, used by electrophysiologists for their large size, about 10% of the dark current was attributed to Ca^{2+} and carried 8 pA of Ca^{2+} current into ROS in darkness giving rise to a dark cytosolic free Ca^{2+} concentration of 550 nM.[6] It is currently thought that no other Ca^{2+} transport proteins and no intracellular Ca^{2+} storage and release contribute to Ca^{2+} homeostasis in ROS. The flat pancake-shaped vesicles called disks constitute the only membraneous organelle within ROS.

Since its discovery in squid giant axons and in mammalian cardiac muscle in the sixties, Na-Ca exchange has been described in a number of tissues including the heart, photoreceptors, kidney, synaptic nerve terminals and smooth muscle.[7,8] Perhaps the most detailed studies have recently been conducted on Na-Ca exchange in cardiac myocytes and retinal ROS in view of the high flux density observed in these cells. About six years ago, our laboratory and independently McNaughton's group discovered that Na-Ca exchange in ROS differs fundamentally from that observed in the heart and in most other tissues: Ca^{2+} extrusion from ROS is driven by both transmembrane Na^+ and K^+ gradients and the exchanger operates at an electrogenic 4 $Na^+:(1\ Ca^{2+} + 1\ K^+)$ stoichiometry[9-11] as opposed to the electrogenic 3 $Na^+:1\ Ca^{2+}$ stoichiometry observed in the heart.[12,13] Very similar functional characteristics have been described for the Na-Ca+K exchanger in either the small ($1 \times 20\ \mu m$) bovine ROS or in the large ($6-10 \times 60\ \mu m$) amphibian ROS.[4,5] Cloning and sequencing of the dog heart Na-Ca exchanger[14] and the bovine ROS Na-Ca+K exchanger[15]

---

[a] This research was funded through an operating grant from the Canadian Medical Research Council. PPMS is a scholar of the Alberta Heritage Foundation of Medical Research and recipient of a Roy Allen Investigatorship in Visual Science.

revealed surprisingly little sequence similarity, although the proposed topology for both exchangers is very similar; the two sequences show little or no homology with other ion transporters.

In the above cited studies Na-Ca exchange was measured as Na^+-dependent Ca^{2+} fluxes and Na-Ca exchange currents, often using large nonphysiological Ca^{2+} loads with the exchanger operating far from equilibrium. In the past few years we have taken advantage of the unique situation that the plasma membrane in isolated bleached ROS contains the Na-Ca+K exchanger as the only one functional cation transporter[16] and of the fact that ROS are readily loaded with fluorescent Ca^{2+}-indicating dyes such as fluo-3.[17] With this preparation we have examined the kinetics of changes in cytosolic free Ca^{2+} when the exchanger operates far from its maximal capacity observed with saturating concentrations of Na^+ and Ca^{2+} on the opposite sides of the membrane. Furthermore, several regulatory features were observed, whereas previous flux and current measurements had suggested little regulation with the direction of Ca^{2+} flux only depending on the direction of the Na^+ gradient. Here, we will review our studies on regulation of Na-Ca+K exchange in isolated intact bovine ROS.

RESULTS

Changes in Cytosolic Free Ca^{2+} upon Changes in Transmembrane Na^+ and K^+ Gradients

In all our experiments we used Ca^{2+}-depleted bovine ROS loaded with fluo-3 as described;[17] Ca^{2+}-depleted ROS were obtained by isolating and purifying bovine ROS in a Ca^{2+}-free solution containing 50 mM NaCl.[18] The initial free cytosolic Ca^{2+} concentration was <10 nM when ROS were maintained in a buffered sucrose solution containing EDTA.

To illustrate the coupling of Ca^{2+} fluxes to transmembrane Na^+ and K^+ gradients, Ca^{2+}-depleted ROS were equilibrated with external Ca^{2+} for ten minutes with low concentrations of NaCl (2 mM) and KCl (1 mM) added to the external medium (FIG. 1). Because of the use of Na^+-rich ROS, Na_{in}-dependent Ca^{2+} influx allowed the cytosolic free Ca^{2+} concentration to rise to values close to 3 μM. Subsequent additions of external Na^+ changed the transmembrane Na^+ gradient and led to a rapid lowering of cytosolic free Ca^{2+} until a new equilibrium value was reached; a subsequent fiftyfold increase in external K^+ concentration caused a large rise in cytosolic free Ca^{2+} (FIG. 1). The direction of Na^+- and K^+-induced changes in cytosolic free Ca^{2+} are in accordance with the proposed direction of coupled Na^+ and Ca^{2+} countertransport and K^+ and Ca^{2+} cotransport, respectively. Four features were noted in such experiments[17,19] and they can be noted in FIGURE 1 as well. First, lowering of cytosolic free Ca^{2+} was very rapid at high Na^+ concentrations despite the fact that the Na^+-dependent Ca^{2+} efflux rate was <10% of the maximal rate observed in ROS with very high Ca^{2+} loads. Second, Ca^{2+} influx via Na_{in}-dependent transport could readily be observed at submicromolar external free Ca^{2+} concentrations. Third, the dependence of the equilibrium free Ca^{2+} concentration on changes in the transmembrane Na^+ gradient was not nearly so steep as predicted from a 4:1 $Na^+ : Ca^{2+}$ coupling ratio. Fourth, at high external Na^+ concentrations an inward Na^+ gradient was expected to drive cytosolic free Ca^{2+} well below the external free Ca^{2+} concentration, but this was not observed.

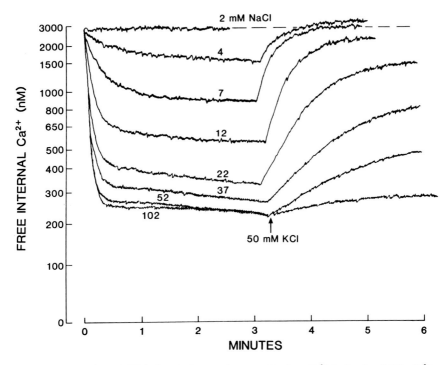

FIGURE 1. Na$^+$- and K$^+$-induced changes in cytosolic free Ca^{2+} in bovine ROS. Ca^{2+}-depleted and fluo-loaded bovine ROS were incubated for ten minutes in 600 mM sucrose, 20 mM Hepes (adjusted to pH 7.4 with arginine), 2 mM NaCl, 1 mM KCl, 500 nM FCCP, 0.5 mM BAPTA and 0.25 mM CaBAPTA. Superimposed traces are illustrated representing separate cuvettes in which Na$_{out}$-dependent Ca^{2+} extrusion was initiated at time zero by addition of NaCl to the indicated final concentration. After about three minutes 50 mM KCl was added to all cuvettes and this caused the indicated rise in free cytosolic Ca^{2+}. Temperature: 25°C.

Potassium Ions Modulate the Competitive Interactions between Ca^{2+} and Na$^+$ Ions

We proposed a three-site model for the ROS Na-Ca+K exchanger in which a single set of sites can accommodate either four sodium ions or one calcium plus one potassium.[20] Two sodium ions or a single calcium ion occupy the A-site, to which neither Li$^+$ nor K$^+$ can bind; the B-site represents a nonselective alkali cation site, and the K-site represents the K$^+$-selective site responsible for K$^+$ transport when Ca^{2+} occupies the A-site. When the A-site is occupied by Na$^+$, both B- and K-site become Na$^+$-selective sites as well. We have reported that the cation dissociation constants for the outward-facing binding sites and the inward-facing cation binding sites, respectively, are quite similar for Ca^{2+},[21] K$^+$ [10] and

Na$^+$.[22] Symmetrical properties of inward- and outward-facing cation binding sites and the occurrence of (K+Ca)-(K+Ca)[20] and Na-Na[5] self-exchange fluxes corroborate our model of a consecutive or ping-pong mechanism of Na-Ca+K exchange, in which a single set of binding sites is alternately exposed to the cytosol and extracellular milieu, respectively. Such a model poses one significant kinetic problem. A common Na$^+$ and Ca^{2+} binding site should balance its relative affinities for Na$^+$ and Ca^{2+} in such a way as to favor Ca^{2+} extrusion at submicromolar intracellular Ca^{2+} concentrations and intracellular Na$^+$ concentrations of up to 10–15 mM. This appears to be reflected by the observed Ca^{2+} and Na$^+$ dissociation constants of about 1 μM and 35 mM, respectively.[21,22] However, at an extracellular Ca^{2+} concentration of >1 mM, the outward-facing A-site would be predominantly occupied by Ca^{2+} and this would result in a futile Ca-Ca exchange cycle rather than the desired Na$_{out}$-dependent Ca^{2+} extrusion. Since two of the proposed sites can accommodate Na$^+$ and K$^+$ at one point in the transport cycle, allosteric interactions between sites may be anticipated, *i.e.*, occupancy of the A-site by Na$^+$ is assumed to render the K-site Na$^+$ selective. We obtained evidence for an allosteric effect of occupancy of the alkali cation sites by K$^+$ on the properties of the outward-facing A-site by examining competition of Na$_{in}$-dependent Ca^{2+} influx by external Na$^+$.[17] At an external K$^+$ concentration of 2 mM, external Na$^+$ was a potent competitor for Na$_{in}$-dependent Ca^{2+} influx (FIG. 2); when K$^+$ was increased from 2 to 100 mM, external Na$^+$ became a significantly less potent inhibitor (notice the difference in external free Ca^{2+} concentration between the two sets of traces). The data illustrated in FIGURE 2 were transformed into a Dixon plot (FIG. 3), which yielded straight lines with a common point of intersection in the second quadrant well above the abscissa indicative of competition between a single Ca^{2+} and two sodium ions (note the scale on the ordinate) for a common site. The apparent K$_i$'s with which Na$^+$ inhibited Ca^{2+} transport were 0.5 and 5.7 mM for K$^+$ concentrations of 2 and 100 mM, respectively. This suggests a kinetic regulation of exchanger function that is directly related to properties of the cation binding sites. High intracellular K$^+$ would result in an exchanger that kinetically favors Ca^{2+} extrusion over Ca^{2+} influx if the properties of the inward-facing sites mirror those of the outward-facing sites.

Inactivation of Na$_{out}$-Dependent Ca^{2+} Extrusion

In the experiment illustrated in FIGURE 1 it was observed that high external Na$^+$ concentrations failed to lower cytosolic free Ca^{2+} below 200 nM, even though the inward Na$^+$ gradient and the 4 Na^{2+} : 1 Ca^{2+} coupling ratio of the exchanger would predict a much lower value. This behavior was consistently observed, Na$_{in}$-dependent Ca^{2+} influx readily increased cytosolic free Ca^{2+} from values of less then 10 nM to micromolar levels, but Na$_{out}$-dependent Ca^{2+} extrusion never lowered cytosolic free Ca^{2+} below 50–100 nM.[19] In two subsequent studies the kinetics of Na$_{out}$-dependent Ca^{2+} extrusion were examined under nonequilibrium conditions, *i.e.*, with zero external Ca^{2+}.[23,24] The major observations and conclusions reported in these two studies will be summarized below. In all these experiments, Ca^{2+}-depleted ROS were loaded with fluo-3 and Ca^{2+} to a moderate total Ca^{2+} load of 300–600 μM (in some cases labelled with ^{45}Ca). ROS were kept in a Na$^+$- and Ca^{2+}-free buffered sucrose solution until Ca^{2+} extrusion was initiated by addition of Na$^+$ to the external medium. Fluo-3 was used as an exclusive indicator of free cytosolic Ca^{2+}, and ^{45}Ca was used as an indicator of total ROS Ca^{2+}. The intracellular free Ca^{2+} recordings of Na$_{out}$-dependent Ca^{2+} extrusion at high extra-

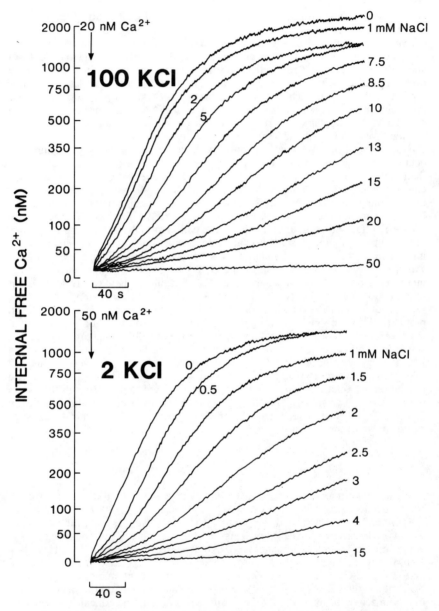

FIGURE 2. Inhibition of Na_{in}-dependent Ca^{2+} influx by external Na^+. Ca^{2+}-depleted and fluo-loaded ROS were incubated in 600 mM sucrose, 20 mM Hepes (pH 7.4), 1 µM FCCP, 1 mM BAPTA and 2 mM KCl or 100 mM KCl as indicated. NaCl was present as indicated. Superimposed traces represent separate cuvettes in which Ca^{2+} influx was initiated at time zero by addition of CaBAPTA to the indicated final free Ca^{2+} concentration. Temperature: 25°C (From Schnetkamp et al.[17] Reprinted by permission from the *Journal of Biological Chemistry*.)

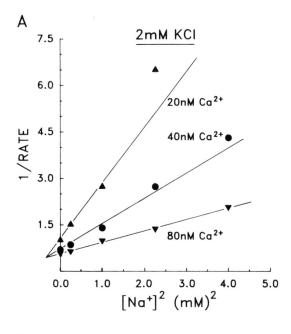
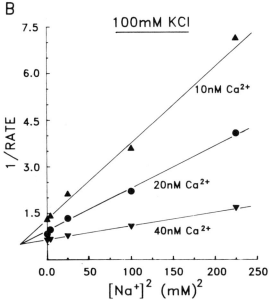

FIGURE 3. Dixon analysis of inhibition of Ca^{2+} influx by external Na^+. Data obtained from experiments like that illustrated in FIGURE 2 were converted into a Dixon plot. (From Schnetkamp et al.[17] Reprinted by permission from the *Journal of Biological Chemistry*.)

cellular Na^+ concentrations of 50–100 mM looked very much like that illustrated here in FIGURE 1 in the presence of a low concentration of extracellular Ca^{2+}: a rapid Ca^{2+} extrusion phase lowered free cytosolic Ca^{2+} from an initial value in the low micromolar range to values between 100 and 200 nM within 20 seconds. The rapid phase came to a rather abrupt end (*i.e.*, the rate of change in free cytosolic Ca^{2+} decreased dramatically over either a very short period of time or over a very narrow range of free Ca^{2+}) and was followed by a sustained (up to forty minutes) period of a steady cytosolic free Ca^{2+} concentration, well above a value of <1 nM predicted from the coupling stoichiometry. Two factors were shown to contribute to this persistent high cytosolic free Ca^{2+} concentration. First, the Ca^{2+} extrusion mode of the Na-Ca+K exchanger inactivated, apparently by lowering the maximal velocity of Ca^{2+} extrusion by at least twentyfold. Second, the sustained steady cytosolic free Ca^{2+} concentration observed arose from an equilibrium between Ca^{2+} release from disks and subsequent Na_{out}-dependent Ca^{2+} removal across the plasma membrane by the exchanger operating at the above noted greatly reduced rate.

Mechanism of Inactivation of Na-Ca+K Exchange

The molecular mechanism of inactivation of the Ca^{2+} extrusion mode of the rod Na-Ca+K exchanger has not been elucidated yet. Our observations have suggested three possibilities:

First, the high-velocity Na_{out}-dependent Ca^{2+} extrusion mode lasted in most cases 20–30 seconds suggesting perhaps a time-dependent inactivation.

Second, pretreatment of ROS with low doses of A23187 prior to addition of Na^+ abolished some of the characteristics of inactivation suggesting a role of intradiskal Ca^{2+} in the inactivation process.[23] Pretreatment with A23187 releases Ca^{2+} from disks that is then removed by the high-velocity mode of the subsequent Na_{out}-dependent Ca^{2+} extrusion. This observation suggests strongly that intradiskal Ca^{2+} or release of intradiskal Ca^{2+} is instrumental in maintaining the sustained high cytosolic Ca^{2+} levels observed.

Third, we suggested that unbinding of Ca^{2+} from multiple allosteric Ca^{2+} binding sites located on the large cytosolic domain of the Na-Ca+K exchanger may act as a sensor that controls inactivation of Ca^{2+} extrusion.[24]

Experiments are under way to evaluate the above proposed mechanisms for inactivation of Na-Ca+K exchange.

Ca^{2+} Sequestration and Release within ROS

The above cited studies demonstrate that ROS contain a membraneous compartment (*i.e.*, the intradiskal space) from which Ca^{2+} can be released into the cytosol. Until recently, little was known about potential Ca^{2+} sequestration within ROS disks; Ca^{2+} fluxes across disks were generally thought not to occur and not to play a role in Ca^{2+} homeostasis in ROS. Recently, we reported that Na_{in}-dependent Ca^{2+} influx into the cytosol of Ca^{2+}-depleted and Na^+-rich ROS was accompanied by Ca^{2+} sequestration into disks. Here we illustrate that Ca^{2+}-depleted ROS initially contain no internal Ca^{2+} store from which Ca^{2+} can be

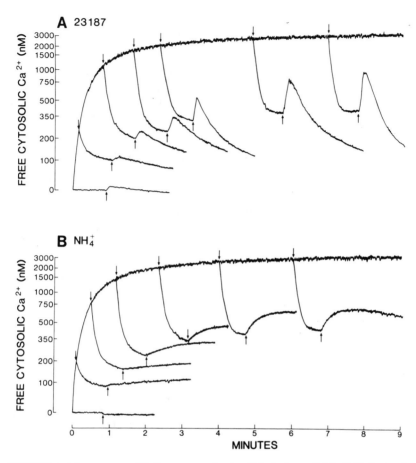

FIGURE 4. Time course of Ca^{2+} sequestration by disks in intact ROS. Ca^{2+}-depleted and fluo-loaded ROS were incubated in 600 mM sucrose, 20 mM Hepes (pH 7.4), 1 μM FCCP, 0.5 mM BAPTA and 2 mM KCl. Superimposed traces represent separate cuvettes in which Ca^{2+} influx was initiated at time zero by addition of 0.5 mM CaBAPTA. At different time points (as indicated by the *downward pointing arrows*), Ca^{2+} influx was stopped and efflux initiated by addition of 1 mM EDTA and 20 mM NaCl, which caused a rapid lowering of cytosolic free Ca^{2+}. Ca^{2+} release from disks was initiated by addition of **(A)** 200 nM A23187 or **(B)** 100 mM ammonium acetate as indicated by the *upward pointing arrows*. Temperature: 25°C.

released by low doses of A23187. Panel A in FIGURE 4 illustrates a series of superimposed Ca^{2+} influx and Ca^{2+} efflux traces. Na_{in}-dependent Ca^{2+} influx was measured at a free external Ca^{2+} concentration of 100 nM. Ca^{2+} influx was stopped and Na_{out}- dependent Ca^{2+} efflux was initiated at different time points (as indicated by the downward-pointing arrows) by addition of 1 mM EDTA and 20 mM NaCl. Na_{out}-dependent Ca^{2+} extrusion was followed for 50 seconds after which 200 nM A23187 was added to monitor Ca^{2+} release from disks. Initially, very little A23187-

induced Ca^{2+} release was observed. As Na_{in}-dependent Ca^{2+} influx proceeded, the A23187-induced Ca^{2+} release signals grew larger on a time scale of minutes illustrating progressive Ca^{2+} sequestration by disks. Changes in intracellular cation concentrations mediated by the alkali cation channel ionophore gramicidin could cause significant changes in the steady cytosolic free Ca^{2+} concentration that followed inactivation of Na_{out}-dependent Ca^{2+} extrusion. In particular, addition of gramicidin in the presence of external Na^+ caused a persistent rise in free cytosolic Ca^{2+} that required Ca^{2+} release from disks.[24] When the intrinsically permeant ammonium acetate (due to permeation of the neutral species ammonia and acetic acid) was added instead of A23187, an increase in cytosolic free Ca^{2+} was observed that became larger on a very similar time scale as observed for the A23187-induced Ca^{2+} release (FIG. 4, panel B). Addition of ammonium acetate is of course a nonphysiological intervention, but it serves to illustrate the principle that Ca^{2+} release from disks can contribute significantly to cytosolic free Ca^{2+} in ROS. Therefore, Ca^{2+} sequestration by disks and Ca^{2+} release from disks may contribute significantly to Ca^{2+} homeostasis in ROS. This may become particularly pronounced in continuous bright light when Ca^{2+} influx through the light-sensitive and cGMP-gated channels is stopped and when inactivation of Ca^{2+} efflux via the Na-Ca+K exchanger greatly reduces Ca^{2+} efflux.

REFERENCES

1. YAU, K-W. & K. NAKATANI. 1985. Nature **313:** 579–582.
2. COOK, N. J., L. L. MOLDAY, D. M. REID, U. B. KAUPP & R. S. MOLDAY. 1989. J. Biol. Chem. **264:** 6996–6999.
3. REID, D. M., U. FRIEDEL, R. S. MOLDAY & N. J. COOK. 1990. Biochemistry **29:** 1601–1607.
4. LAGNADO, L. & P. A. MCNAUGHTON. 1990. J. Membr. Biol. **113:** 177–191.
5. SCHNETKAMP, P. P. M. 1989. Prog. Biophys. Mol. Biol. **54:** 1–29.
6. GRAY-KELLER, M. P. & P. B. DETWILER. 1994. Neuron **13:** 849–861.
7. NICOL, G. D., P. P. M. SCHNETKAMP, Y. SAIMI, E. J. CRAGOE, JR. & M. D. BOWNDS. 1987. J. Gen. Physiol. **90:** 651–669.
8. BLAUSTEIN, M. P., R. DIPOLO & J. P. REEVES, Eds. 1991. Sodium-Calcium Exchange. Ann. N. Y. Acad. Sci. **639.**
9. CERVETTO, L., L. LAGNADO, R. J. PERRY, D. W. ROBINSON & P. A. MCNAUGHTON. 1989. Nature **337:** 740–743.
10. SCHNETKAMP, P. P. M., D. K. BASU & R. T. SZERENCSEI. 1989. Am. J. Physiol. **257:** C153–C157.
11. SCHNETKAMP, P. P. M., R. T. SZERENCSEI & D. K. BASU. 1988. Biophys. J. **53:** 389a.
12. REEVES, J. P. & C. C. HALE. 1984. J. Biol. Chem. **259:** 7733–7739.
13. HILGEMANN, D. W., D. A. NICOLL & K. D. PHILIPSON. 1991. Nature **352:** 715–718.
14. NICOLL, D. A., S. LONGONI & K. D. PHILIPSON. 1990. Science **250:** 562–565.
15. REILANDER, H., A. ACHILLES, U. FRIEDEL, G. MAUL, F. LOTTSPEICH & N. J. COOK. 1992. EMBO J. **11:** 1689–1695.
16. SCHNETKAMP, P. P. M., R. T. SZERENCSEI & D. K. BASU. 1991. J. Biol. Chem. **266:** 198–206.
17. SCHNETKAMP, P. P. M., X-B. LI, D. K. BASU & R. T. SZERENCSEI. 1991. J. Biol. Chem. **266:** 22975–22982.
18. SCHNETKAMP, P. P. M. 1986. J. Physiol. **373:** 25–45.
19. SCHNETKAMP, P. P. M., D. K. BASU, X-B. LI & R. T. SZERENCSEI. 1991. J. Biol. Chem. **266:** 22983–22990.

20. SCHNETKAMP, P. P. M. & R. T. SZERENCSEI. 1991. J. Biol. Chem. **266:** 189–197.
21. SCHNETKAMP, P. P. M. 1991. J. Gen. Physiol. **98:** 555–573.
22. SCHNETKAMP, P. P. M., J. E. TUCKER & R. T. SZERENCSEI. 1995. Am. J. Physiol. In press.
23. SCHNETKAMP, P. P. M. & R. T. SZERENCSEI. 1993. J. Biol. Chem. **268:** 12449–12457.
24. SCHNETKAMP, P. P. M. 1995. J. Biol. Chem. **270:** 13231–13239.

Turnover Rate and Number of Na$^+$-Ca^{2+}, K$^+$ Exchange Sites in Retinal Photoreceptors[a]

GIORGIO RISPOLI,[b] ANACLETO NAVANGIONE, AND VITTORIO VELLANI

Section of General Physiology
Department of Biology
National Institute for the Physics of Matter (INFM)
Via Borsari 46
Ferrara I-44100, Italy

INTRODUCTION

The vertebrate photoreceptor exchanger[1,2] plays an important role in phototransduction, since the sustained exchanger activity in the light induces [Ca^{2+}]$_i$ fall, which in turn triggers several mechanisms taking part in dark state recovery and light adaptation of the photoreceptor.[3-7] It has been found that the exchanger imports 4 Na$^+$ ions for every Ca^{2+} and K$^+$ ion extruded[8,9] (forward mode of exchange), thereby accounting for one net positive charge imported per exchange cycle.[1,2] The Ca^{2+} extrusion in vertebrate photoreceptors relies entirely on the density of exchange sites and on the number of Ca^{2+} ions transported per molecule per second. Several studies have attempted to quantify the density and the turnover number τ of the photoreceptor exchanger, but large discrepancies were found in these estimates:[2,10-12] the former ranged from 200–450 molecules/μm^2 to $3.5 \cdot 10^4$ molecules/μm^2 while the latter ranged from 10 sec^{-1} to 115 sec^{-1}. τ can be estimated reliably by measuring the effect of a jump in the concentration of one or more of the transported ion(s) on the exchange current (recorded in situ). This approach would be particularly feasible if the ion transport was as slow as suggested by the above studies, and in particular if the entire reaction cycle was rate limited by one or more steps that can be forced to occur by manipulating the extracellular solution. This would allow to measure the step(s) kinetics (or τ), no matter if the step is electrogenic or not, and to calculate the number of the exchanger molecules N by the equation:

$$N = \frac{I_E}{e_0 \cdot \tau}$$

where I_E is the saturating exchange current and e_0 the elementary charge. If the ions are not translocated simultaneously (that is, the exchanger reaction cycle is consecutive, as exemplified in FIG. 5), then one would suspect that the reaction(s)

---

[a] This work was supported by grants from the Consiglio Nazionale delle Ricerche (CNR), the Ministero dell'Università e della Ricerca Scientifica e Tecnologica (MURST), and the Istituto Nazionale di Fisica della Materia (INFM).

[b] To whom correspondence should be addressed.

implicated in the Ca^{2+} transport are slower than the ones implicated in the transport of Na^+ and K^+. In fact, the exchanger affinity for Ca^{2+} must change more than 3 orders of magnitude in order to bind Ca^{2+} intracellularly (where its concentration is 500 nM at the most[13,14]) and release it extracellularly (where its concentration is about 1 mM). It is reasonable to expect that this affinity change would require some time to be accomplished. Alternatively, the affinity for extracellular Ca^{2+} could remain elevated, but then the Ca^{2+} release at the extracellular side would take time to be completed. In the present work, the Ca^{2+} dependence of the exchanger reaction cycle was studied by recording the rod outer segment (OS) membrane current under whole-cell or excised-patch voltage-clamp conditions, while performing fast solution changes of the ions transported by the exchanger.[15-18] The peculiar architecture of the OS, in which the entire transduction machinery, the cGMP channels and the exchanger are segregated, makes the patch clamp recording from mechanically isolated OS a powerful technique to study photoreceptor physiology.[8,14,19-21] It was found that the turnover of photoreceptor exchanger is indeed rate limited by the Ca^{2+} transport but not by the Na^+ or K^+ transport and resulted in the extrusion of about 2–3 Ca^{2+} ions/s, which yields a density of about 10^4 molecules/μm^2.

METHODS

OS were mechanically isolated from rods of the nocturnal Tokay Gekko lizard (*Gekko gecko*). The methods are described in detail elsewhere.[18,21] Electrical recordings were carried out using the "whole-cell" or the "outside-out" configuration of the patch recording technique. Pipettes were fabricated from 100 μl Drummond glass capillaries in the conventional manner[22] and fire polished to a pipette resistance of 7–11 MΩ. The OS was perfused intracellularly and extracellularly with solutions containing the ions indicated in the text and in the figure legend. The 0 Ca^{2+} solutions were buffered with 2 mM ethylene glycol-bis(β-aminoethyl ether)-N,N,N',N'-tetraacetic acid (EGTA); all solutions were buffered to pH $\approx$ 7.4 with N-2-hydroxyethylpiperazine-N'-2-ethanesulphonic acid (Hepes) and NaOH, KOH or LiOH depending upon whether the solution contained high Na^+, high K^+ or 0 Na^+ and 0 K^+, respectively. The composition of Ringer solution was: 160 mM Na^+, 3.3 mM K^+, 1 mM Ca^{2+}, 1.7 mM Mg^{2+}, 165 mM Cl^-, 1.7 mM SO_4^-, 2.8 mM Hepes and 10 mM dextrose; 200 μM Ca^{2+} and 0 Ca^{2+} Ringers had the same ion composition of normal Ringer except that they contained 200 μM Ca^{2+} or 0 Ca^{2+} + 2 mM EGTA, respectively, instead of 1 mM Ca^{2+}. Isotonic solutions contained 166 mM of the monovalent cation indicated in the text and 2 mM EGTA; the osmolality of the solutions that contained non-Isotonic concentrations of the cations transported by the exchanger was adjusted to Ringer osmolality with Li^+ and checked with a microosmometer (13/13 DR Roebling, Berlin, Germany). Current was recorded with an Axopatch 1D amplifier (Axon Instruments, Burlingame, CA). Seal resistance measured before rupturing the membrane patch ranged between 10 and 40 GΩ, while series resistance ranged from 20 to 40 MΩ. The external solution was changed rapidly (typically in less than 50 msec) by moving a multibarrelled perfusion pipette on an horizontal plane in front of the OS or excised patch with a stepping motor (FIG. 2, *inset*). Outside-out excised patches were obtained at the end of a whole-cell recording session by blowing away the OS upon abruptly increasing the flow speed; then all the solution changes were repeated on the excised patch.

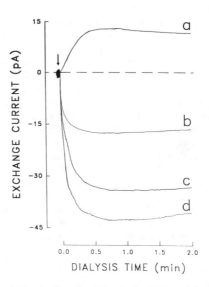

FIGURE 1. Chart record (obtained with a Linseis L 6514 recorder) of exchange currents under various internal and external ionic conditions. Forward exchange currents were obtained in the presence of Ca_i^{2+}, K_i^+ and Na_o^+ while the reverse mode ones in the presence of Na_i^+, K_o^+ and Ca_o^{2+}. Traces (mean current ± SEM at the steady-state, n = number of OS averaged): **(a)** 15.0 ± 1.4 pA, n = 12; **(b)** −15.5 ± 0.7 pA, n = 14; **(c)** −37.0 ± 1.6 pA, n = 5, 139 Na_o^+ + 20 mM K_o^+ solution; **(d)** $I_E \approx$ −48.0 ± 1.9 pA, n = 14. 30 mM Ca_i^{2+} + 124 mM K_i^+ solution in traces (b), (c) and (d). Holding potential: 0 mV in all traces. The breakthrough from cell-attached to whole-cell recording occurred at the *arrow*.

RESULTS

The search for a step that limits the overall exchange cycle was carried out according to the following strategy. The exchange current was recorded in the presence of saturating concentrations of Na^+ at one side of the membrane and Ca^{2+} and K^+ at the other side. Under these conditions, the recorded current is due exclusively to the electrogenic activity of the exchanger,[8,15-18,20,21,23] whose cycle consists in binding ions at one side of the membrane and releasing them at the other side after translocation. The effect on the exchange current of adding high concentrations of one of the transported ions at the side of the membrane where it must be released was then investigated. A large decrease in the current size would indicate that the transport of that particular ion was a rate limiting step in the exchange process. FIGURE 1 illustrates a family of exchange currents recorded from OS using different intracellular and extracellular media. The traces selected represent the typical behavior of an OS in each of the conditions indicated (see legend; the breakthrough occurs at zero time). In the forward mode of operation the current was inward, and attained maximum amplitude I_E using Isotonic Na_o^+ and 124 mM K_i^+ + 30 mM Ca_i^{2+} (FIG. 1, trace *d*). This current was modestly affected by an increase of $[K^+]_o$ (FIG. 1, trace *c*) or $[Na^+]_i$ (FIG. 3, traces *a* and *b*) by tens of mM, while it was strongly reduced when the OS was bathed in Ringer, that is in the presence of 1 mM Ca_o^{2+} (FIG. 1, trace *b*). Indeed, a switch to a Ca_o^{2+}-free Ringer (FIG. 2, lower trace) caused a current increase to an amplitude close to the steady-state value of trace d in Fig. 1 (after correcting for the different holding potential of the two recordings). A reduction of Ca_o^{2+} from 1 mM to 200 μM still caused a marked current increase, while switching from Isotonic Na_o^+ to 159 mM Na_o^+ + 5 mM Ca_o^{2+} caused a 7.5 ± 0.7 (7 OS)-fold decrease of exchange current (data not shown). The effect induced by changes in Ca^{2+} can be accounted for by assuming an apparent dissociation constant for Ca_o^{2+}, K_m^{app} $[Ca_o^{2+}]$, of 1 mM,[21] *i.e.*, about 3 orders of magnitude larger than $K_m^{app}[Ca_i^{2+}]$.[21,24] A similar value for $K_m^{app}[Ca_o^{2+}]$ was estimated from measurements on outside-out excised patches, performed on the same OS from which whole-cell recording was previously made (FIG. 2, upper trace). The effect on reverse exchange (Ca^{2+} and K^+ imported per Na^+ extruded) of adding one ion species at the membrane side where it must be released cannot be accomplished with the strategy of FIG. 1. In fact, Ca^{2+} entering the OS during reverse operation accumulates intracellularly and inhibits reverse exchange itself,[18,23] as follows. Recording a in FIGURE 1 was obtained using the same solutions as in d but placed on opposite sides of the membrane. The exchange current was outward and attained a smaller amplitude compared to the forward mode with corresponding electrochemical gradients. However, if an OS perfused with Isotonic Na_i^+ was switched from Isotonic Na_o^+ to 159 mM K_o^+ + 1 mM Ca_o^{2+}, a large outward current was elicited, which declined rapidly after attaining a peak (FIG. 3, trace a). During the reverse exchange

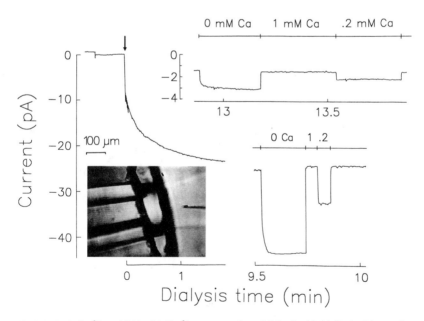

FIGURE 2. 0 Ca_o^{2+} and 200 μM Ca_o^{2+} response in a OS bathed initially in Ringer (*lower trace*; 30 mM Ca_i^{2+} + 124 mM K_i^+ solution). Mean steady-state current ± SEM: Ringer, −24.1 ± 1.6 pA, n = 23; 0 Ca_o^{2+}, −44.1 ± 1.2 pA, n = 6; 200 μM Ca_o^{2+}, −30.6 ± 0.7 pA, n = 4. *Upper trace*, 0 Ca_o^{2+} and 200 μM Ca_o^{2+} response in an outside-out patch excised from the same OS as described in Methods. Holding potential: −20 mV. *Inset*: video recording of an actual experiment. Solution change occurs once the OS crosses the boundary separating two adjacent streams, visible on the *right*.

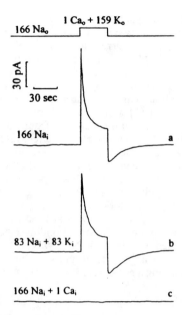

FIGURE 3. Effect of adding K_i^+ or Ca_i^{2+} on reverse exchange. Current peak amplitude: (a) 98 ± 11 pA (n = 9, 13 solution changes); (b) 65 ± 13 pA, (n = 3, 5 solution changes). Internal solution is indicated near each trace, extracellular solution change is indicated on *top*. Holding potential: 0 mV.

Ca^{2+} and K^+ accumulated intracellularly, since it was possible to elicit forward exchange operation upon switching back to Isotonic Na_o^+. The inward current then declined back to 0 as the Ca_i^{2+} and K_i^+ load was extruded by forward exchange (FIG. 3, trace *a*). Since outward currents as large as 70 pA were measured in 159 mM K_o^+ + 1 mM Ca_o^{2+} in the presence of 83 mM K_i^+ (FIG. 3, trace *b*), while no exchange current was recorded in the presence of 1 mM Ca_i^{2+} (FIG. 3, trace *c*), it is concluded that the Ca_i^{2+} (and not K_i^+) accumulation during reverse operation inhibits reverse exchange itself. The inward current peak in Isotonic Na_o^+, elicited by the Ca_i^{2+} and K_i^+ accumulated during reverse exchange operation (FIG. 3, traces *a* and *b*), had a size comparable to trace *d* of FIGURE 1, indicating that the presence of large concentrations of Na_i^+ did not affect forward exchange. Similarly, the size of the reverse exchange peak in 159 mM K_o^+ + 1 mM Ca_o^{2+} was little affected by the presence of tens of mM of K_i^+ (FIG. 3, trace *b*), or Na_o^+ (data not shown). In conclusion, the turnover of both forward and reverse mode of exchange under saturating conditions was little affected by adding concentrations of tens of mM of Na^+ or K^+ at the membrane side where they were released, while it was strongly inhibited by adding much smaller amounts of Ca^{2+}. This suggests that Ca^{2+} transport is a rate limiting step in both forward and reverse mode of exchanger operation, while Na^+ and K^+ transport are not. In order to further investigate the kinetics of Ca^{2+} transport, fast changes of $[Ca^{2+}]_o$ were then performed. Forward exchange current decayed rapidly to zero (<50 msec) upon switching to Isotonic Li_o^+ or Isotonic K_o^+ or 159 mM Li_o^+ + 5 mM Ca_o^{2+} (FIG. 4A and B). At this point, switching back to Isotonic Na_o^+, the current

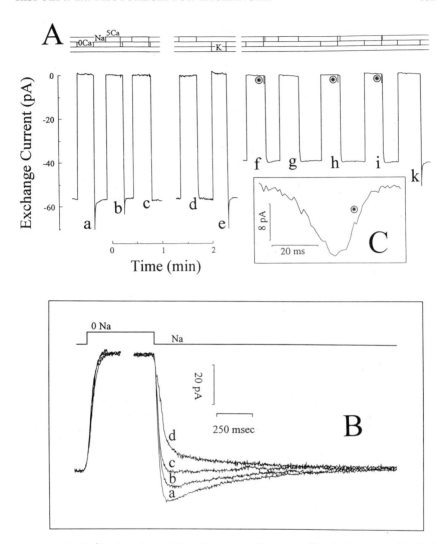

FIGURE 4. Ca_o^{2+}-induced transitions from state X to state Y and vice-versa. **(A)** *Left panel*, chart record of 0 Ca_o^{2+} exposures for different times after 5 mM Ca_o^{2+} preapplication and effect on the exchange current reactivation in Isotonic Na_o^+; *right panel*, 0 Na_o^+ + 5 mM Ca_o^{2+} exposures for different times after 0 Ca_o^{2+} preapplication and effect on the exchange current reactivation in Isotonic Na_o^+. **(B)** Exchange current decays and reactivations of (A), *left panel*, on a faster time scale low pass filtered with an 8-poles Bessel filter at 500 Hz and sampled at 1-msec intervals: a, 19 sec of 0 Na_o^+ + 0 Ca_o^{2+} perfusion; b, 2.5 sec of 0 Na_o^+ + 0 Ca_o^{2+} perfusion after 0 Na_o^+ + 5 mM Ca_o^{2+} preapplication; c, 0.7 sec of 0 Na_o^+ + 0 Ca_o^{2+} perfusion after 0 Na_o^+ + 5 mM Ca_o^{2+} preapplication; d, 21 sec of 0 Na_o^+ + 5 mM Ca_o^{2+} perfusion. **(C)** Average of the 3 current transients indicated with the *asterisk* in (A), *right panel*, recorded upon switching from 0 Ca_o^{2+} to 5 mM Ca_o^{2+} sampled as in (B). The average time integral of the transient resulted of $\approx 0.4 \pm 0.03 \cdot 10^{-12}$ C (18 transients averaged in 3 OS). Holding potential: 0 mV; 30 mM Ca_i^{2+} + 124 mM K_i^+ solution.

recovered I_E amplitude in two phases: a first one, of amplitude I_F and roughly as fast as the solution change, followed by a slower one with an exponential time course. The largest I_F (-57.5 ± 2.7 pA, 16 traces averaged in 9 OS) was recorded after perfusing with 0 Ca_o^{2+} (*i.e.*, after Isotonic Li^+_o or Isotonic K_o^+ perfusion; FIG. 4A trace a, e and k; FIG. 4B, trace a) and I_E was recovered with a time constant of 514 ± 20 msec (16 traces averaged in 9 OS). After 5 mM Ca_o^{2+} perfusion, I_F was instead -37.5 ± 1.4 pA (20 traces averaged in 7 OS) and I_E was recovered with a time constant of 397 ± 16 msec (20 traces averaged in 7 OS; FIG. 4A trace d, f, g, h and i and FIG. 4B, trace d). This indicates that once the exchanger is blocked in a certain state, it takes hundreds of msec to have full transport cycles restored. The waveform of current reactivation depended entirely on $[Ca^{2+}]_o$ in the 0 Na_o^+ solution that was used to block the exchanger. This indicates that 0 Ca_o^{2+} perfusion recruits an exchanger state (call it X) different than the one (Y) recruited by 5 mM Ca_o^{2+} perfusion. In order to investigate the kinetics with which the exchanger can be switched from state X to state Y and vice-versa the following strategy was employed. A first solution was preapplied, containing 0 Na_o^+ + 0 Ca_o^{2+} or 0 Na_o^+ + 5 mM Ca_o^{2+}, for a sufficiently long time (>15 sec) to have all the exchangers distributed at equilibrium between the states. The OS was then switched for different times to a second 0 Na_o^+ solution containing either 5 mM Ca_o^{2+} or 0 Ca^{2+}, respectively, and finally to Isotonic Na_o^+. It was then measured for how long it was necessary to perfuse with the second solution to have a waveform of current reactivation in Na_o^+ identical to the one obtained after a long perfusion with the second solution. It was necessary to perfuse in 0 Ca_o^{2+} for at least 3 sec, after the 5 mM Ca_o^{2+} preapplication, to maximize I_F (FIG. 4A, left panel and FIG. 4B); however, as soon as Ca_o^{2+} was presented after the 0 Ca_o^{2+} preapplication, I_F was reduced (FIG. 4A, right panel). This shows that the Ca^{2+}-induced transition from state X to Y is much faster than the transition from Y to X. These transitions might be electrogenic: an inward current transient was in fact recorded upon switching the OS from 166 mM Li^+_o to 159 mM Li^+_o + 5 mM Ca_o^{2+} (FIG. 4A, right panel and FIG. 4C) while no discernible transient was recorded upon switching from 159 mM Li^+_o + 5 mM Ca_o^{2+} to 166 mM Li^+_o (FIG. 4A, left panel). The time integral of the transient yields the charge transported in the transition from X to Y by N exchangers: from the estimate of N given in the Discussion section, this partial cycle corresponds to a charge movement of ≈ 0.1 e_0 per exchange site, where e_0 is the elementary charge. This indicates that the transition from X to Y is weakly electrogenic and faster than the transition from Y to X.

DISCUSSION

This section will attempt to characterize further the states X and Y. In the exchanger model illustrated in FIGURE 5, in which ionic translocation is "consecutive,"[16,17,23] when in the forward mode of operation, the exchanger binds 4 Na_o^+ in the state $E1$; the Na^+ ions are then occluded and translocated after a conformational change of the antiporter to the state $E2$. After the release of Na^+ inside, the exchanger binds one Ca_i^{2+}, then occludes and translocates it upon returning to the $E1$ state at the expense of the energy stored during the previous translocation; the Ca^{2+} release at the external site terminates the cycle. K^+ translocation is omitted for simplicity; it is not necessary, in the present discussion, to assume that Ca^{2+} and Na^+ binding sites are distinct, as depicted in FIGURE 5. The overall

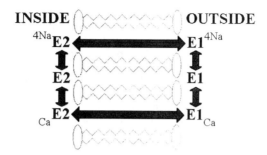

FIGURE 5. Consecutive scheme of exchange cycle. The notation $_{ma}B_c^{nd}$ indicates the exchanger in the state B, where it binds m ions "a" at an intracellular site, one ion "c" at an extracellular site and n ions "d" at a second extracellular site.

exchange cycle must occur with the transport of a charge e_0 in order to account for the observed stoichiometry: let us assume that most of this electrogenicity lies in the reaction/s that are involved in the binding and translocation of Na^+.[23] The similarity between exchange activity in OS and excised patches (FIG. 2) indicates that the exchange process is entirely regulated by ionic gradients and not by intracellular factors. This view is also supported by experiments showing that intracellular applications of adenosine triphosphate (ATP) and/or guanosine triphosphate (GTP) do not change the saturating exchange current level[8,18,21] (whereas ATP regulates the Na^+-Ca^{2+} exchange[25]), thus excluding that the exchange process is regulated by phosphorylation or by G-protein coupled pathway. Thus, it should be possible to force the occurrence of one of the presumed partial cycles by simply changing the electrochemical gradient of the ions transported in that cycle. The perfusion with 0 Na_o^+ + 5 mM Ca_o^{2+} would then recruit some exchanger molecules in the state El_{Ca}, where sites can bind Ca_o^{2+}, and some in the state $_{Ca}E2$, if Ca_o^{2+} can be translocated in the absence of K_o^+. Upon switching the OS from 0 Na_o^+ + 5 mM Ca_o^{2+} to Isotonic Na_o^+, the transition $El_{Ca} \rightarrow El$ must be completed (together with the transition $_{Ca}E2 \rightarrow El_{Ca}$ if the transition $El_{Ca} \rightarrow {}_{Ca}E2$ occurred during the 5 mM Ca_o^{2+} perfusion) before the transitions $El \rightarrow El^{4Na}$ and $El^{4Na} \rightarrow {}^{4Na}E2$ can occur. Let us assume that one or more of the reactions involved in the Ca^{2+} transport occur on a time scale on the order of hundreds of msec and are much slower than the one(s) involved in the Na^+ translocation. The overall current reactivation kinetics in Isotonic Na_o^+ will then have the timing of the Ca^{2+} transport. The fast component of current reactivation in the solution change from 5 mM Ca_o^{2+} to Isotonic Na_o^+ would then result from the exchangers in the state El that performed the transitions $El \rightarrow El^{4Na}$ and $El^{4Na} \rightarrow {}^{4Na}E2$, as soon as Na_o^+ was reintegrated. The slow component would instead originate from the exchangers that performed the transition(s) $El_{Ca} \rightarrow El$ and/or $_{Ca}E2 \rightarrow El_{Ca}$. Similarly, the perfusion with 0 Na_o^+ + 0 Ca_o^{2+} recruited a large fraction of the exchangers in the state El: this would maximize the amplitude of the fast component of current reactivation, thus generating the observed peak overshooting the steady-state amplitude. The amplitude of current reactivation would eventually decrease, as observed, since the rate at which the El states are formed by the transitions $_{Ca}E2 \rightarrow El_{Ca}$ and $El_{Ca} \rightarrow El$ is much slower than the rate of the transitions $El \rightarrow El^{4Na}$ and $El^{4Na} \rightarrow {}^{4Na}E2$. In the above framework, the experiments of FIGURE 4 indicate that it was necessary to perfuse in 0 Ca_o^{2+}

+ 0 Na_o^+ solution for at least 3 sec to maximize the number of exchangers in state $E1$, if most of them were previously recruited in the state $E1_{Ca}$ (and possibly $_{Ca}E2$) by the 0 Na_o^+ + 5 mM Ca_o^{2+} preapplication (FIG. 4A, left panel). By contrast, if the states $E1$ were recruited by the 0 Na_o^+ + 0 Ca_o^{2+} preapplication, then the states $E1_{Ca}$ (and possibly $_{Ca}E2$) were formed upon switching to 0 Na_o^+ + 5 mM Ca_o^{2+} with the kinetics of the solution change. As pointed out in the Introduction, the slow reaction(s) involved in the Ca^{2+} transport might be at the level of the Ca^{2+} translocation step ($_{Ca}E2 \to E1_{Ca}$), given the extremely large change in the affinity for Ca^{2+} that the exchanger must undergo during its operation. Alternatively, the kinetics of Ca^{2+} release at the extracellular side ($E1_{Ca} \to E1$) might be particularly slow, if the exchanger affinity for Ca^{2+} remains elevated after having translocated it across the membrane. If Ca_o^{2+} cannot be imported in the absence of K_o^+, then the data of FIGURE 4C indicate that the Ca_o^{2+} release is the rate limiting step in the exchange cycle. Either way, both the above conjectures predict that the smaller the $[Ca^{2+}]_i$ at which the exchanger operates, the slower the Ca^{2+} transport. The photoreceptor exchanger, being the only Ca^{2+} extrusion system in the OS, is designed to bind Ca_i^{2+} in the nM range, which is one order of magnitude smaller than the $[Ca^{2+}]_i$ at which the muscle exchanger operates. So this exchanger is expected to have a smaller turnover number than the one of cardiac Na^+-Ca^{2+} exchanger, as it was indeed found.[26] It can be tentatively concluded that the turnover of all transporters is rate limited by the reaction(s) which involve a large change in the affinity of binding site(s) for the transported substance(s). Thus, Ca^{2+} transport in the photoreceptor exchanger yields a turnover rate of about 2–3 Ca^{2+} ions/s extruded and a density of about 10^4 molecules/μm^2 (the surface area of an OS^{14} is ≈ 3000 μm^2), as was suggested by Hodgkin et al.[2] This large density may be designed to prevent a nonuniform distribution of $[Ca^{2+}]_i$ that may dramatically affect the photoreceptor light response. The general scheme for a Ca^{2+} exchanger may be the one of FIG. 5, in which the exchanger bears a negative charge ($-3 \cdot e_0$ in the case of the photoreceptor exchanger and $-2 \cdot e_0$ in the case of the Na^+-Ca^{2+} exchanger). This design would minimize the energy expense in the extrusion of Ca^{2+} by (almost) neutralizing its positive charge (the membrane potential is usually negative), while it maximizes the energy acquired in the Na^+ translocation by concentrating most of the exchanger electrogenicity in the latter step.

REFERENCES

1. YAU, K-W. & K. NAKATANI. 1984. Nature **311**: 661–663.
2. HODGKIN, A. L., P. A. MCNAUGHTON & B. J. NUNN. 1987. J. Physiol. **391**: 347–370.
3. TORRE, V., H. R. MATTHEWS & T. D. LAMB. 1986. Proc. Natl. Acad. Sci. USA **83**: 7109–7113.
4. RISPOLI, G., W. A. SATHER & P. B. DETWILER. 1988. Biophys. J. **53**: 390a.
5. HSU, Y-T. & R. S. MOLDAY. 1993. Nature **361**: 76–79.
6. KAWAMURA, S. 1993. Nature **362**: 855–857.
7. LAGNADO, L. & D. A. BAYLOR. 1994. Nature **367**: 273–277.
8. CERVETTO, L., L. LAGNADO, R. J. PERRY, D. W. ROBINSON & P. A. MCNAUGHTON. 1989. Nature **337**: 740–743.
9. SCHNETKAMP, P. P. M., D. K. BASU & R. T. SZERENCSEI. 1989. Am. J. Physiol. **257**: C153–C157.
10. COOK, N. J. & U. B. KAUPP. 1988. J. Biol. Chem. **263**: 11382–11388.
11. FRIEDEL, U., G. WOLBRING, P. WOHLFART & N. J. COOK. 1991. Biochim. Biophys. Acta **1061**: 247–252.

12. NICOLL, D. A. & M. L. APPLEBURY. 1989. J. Biol. Chem. **264:** 16207–16213.
13. GREY-KELLER, M. P. & P. B. DETWILER. 1994. Neuron **13:** 849–861.
14. RISPOLI, G., A. T. FINEBERG & P. B. DETWILER. 1990. Biophys. J. **57:** 368a.
15. RISPOLI, G. & A. NAVANGIONE. 1992. Pflügers Arch. **421:** R22.
16. RISPOLI, G. & A. NAVANGIONE. 1993. Pflügers Arch. **424:** R35.
17. NAVANGIONE, A. 1994. Ph.D. thesis. University of Ferrara. Ferrara, Italy.
18. RISPOLI, G., A. NAVANGIONE & V. VELLANI. 1995. Biophys. J. **69:** 74–83.
19. RISPOLI, G. & P. B. DETWILER. 1990. Biophys. J. **57:** 368a.
20. RISPOLI, G. & P. B. DETWILER. 1991. Biophys. J. **59:** 434a.
21. RISPOLI, G., W. A. SATHER & P. B. DETWILER. 1993. J. Physiol. **465:** 513–537.
22. HAMILL, O. P., A. MARTY, E. NEHER, B. SAKMANN & F. J. SIGWORTH. 1981. Pflügers Arch. **391:** 85–100.
23. PERRY, R. J. & P. A. MCNAUGHTON. 1993. J. Physiol. **466:** 443–480.
24. LAGNADO, L., L. CERVETTO, P. A. MCNAUGHTON. 1988. Proc. Natl. Acad. Sci. USA **85:** 4548–4552.
25. HILGEMANN, D. W. 1990. Nature **344:** 242–245.
26. HILGEMANN, D. W., D. A. NICOLL & K. D. PHILIPSON. 1991. Nature **352:** 715–718.

Na-Ca Exchange in Ca^{2+} Signaling and Neurohormone Secretion

Secretory Vesicle Contributions in Adrenal Chromaffin Cells

ALLAN S. SCHNEIDER AND CHUNG-REN JAN

Department of Pharmacology and Neuroscience, A136
Albany Medical College
Albany, New York 12208

INTRODUCTION

Although calcium signaling plays an essential role in hormone and neurotransmitter secretion, the mechanisms regulating the calcium signal, particularly its buffering and decay, are not well defined. The adrenal chromaffin cell is a prototypic neurosecretory cell in which calcium signaling has been extensively studied and for which evidence for a role of Na-Ca exchange in calcium signaling has begun to accumulate.[1-5] Chromaffin cells are developmentally homologous with sympathetic neurons and have as their main functions the synthesis, storage and release of catecholamine neurohormones during stress. The chromaffin secretory vesicles (chromaffin granules) have similar properties to adrenergic synaptic vesicles. There is also evidence for the existence of a secretory vesicle Na-Ca exchanger[6-8] in addition to a plasma membrane exchanger. The chromaffin secretory vesicle exchanger may contribute to calcium signal decay following exocytotic secretion,[1] as well as to intracellular calcium mobilization during conditions of cell Na^+-loading, as occurs during exposure to cardiac glycosides. We now present results on Na-Ca exchange contributions to calcium signal rise and decay in chromaffin cells and the possible role of exchange proteins on secretory vesicle membranes. Na^+-dependent Ca^{2+} transport in chromaffin cells is demonstrated, including effects on ^{45}Ca uptake and efflux, cytosolic Ca^{2+} signals and catecholamine secretion. Contributions from intracellular organelles are shown to produce cytosolic Ca^{2+} oscillations. The extent of incorporation of secretory vesicle membrane into the plasma membrane during exocytotic secretion is shown to influence the magnitude of Na^+-dependent ^{45}Ca efflux from chromaffin cells.

RESULTS

FIGURE 1 shows the effect of extracellular Na^+ deprivation on 75 mM K^+-stimulated ^{45}Ca uptake and efflux and catecholamine secretion in primary cultures of bovine adrenal chromaffin cells.[1] Elevated K^+ depolarizes chromaffin cells and opens voltage-gated calcium channels leading to calcium uptake. The resulting rise in cytosolic calcium triggers enhanced calcium efflux and exocytotic secretion as shown in the open bars in FIGURE 1. If the cells are depolarized in the absence of extracellular Na^+, ^{45}Ca efflux is markedly reduced while ^{45}Ca uptake is unaffected. There is an accompanying increase in secretion, presumably related to the inhibi-

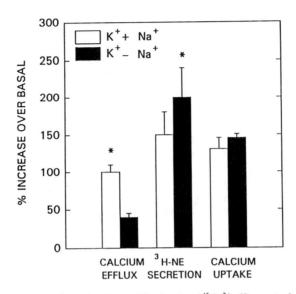

FIGURE 1. Effects of Na^+ deprivation on K^+-stimulated $^{45}Ca^{2+}$ efflux, catecholamine secretion and $^{45}Ca^{2+}$ uptake in cultured bovine adrenal chromaffin cells. Stimulating solutions: $K^+ + Na^+$, high K^+ depolarizing medium containing 75 mM K^+ and 75 mM Na^+. K^+-Na^+, Na^+-free, high K^+ medium containing 75 mM K^+ and 75 mM N-methylglucamine. The stimulation is 4 min at 25°C for all experiments. Catecholamine secretion is measured as ^{3}H-norepinephrine (^{3}H-NE) release. Data are expressed as % increase over basal control, which is the basal response in Locke's medium, and represent the mean ± SEM of 3–5 independent experiments, each done in triplicate. Basal values were: $^{45}Ca^{2+}$ efflux = 15–25% of preloaded ^{45}Ca; catecholamine secretion = 5% of total catecholamines; and $^{45}Ca^{2+}$ uptake = 0.6 nmole/10^6 cells/4 min. * indicates significant difference at $p < 0.05$ by paired student t test, comparing the paired *solid* and *blank* bars for each response. (From Jan & Schneider.[1] Reprinted by permission from the *Journal of Biological Chemistry*.)

tion of Ca efflux and a slowing of the decay of the cytosolic Ca^{2+} signal for release. We have directly measured the effect of Na^+ deprivation on the kinetics of decay of cytosolic Ca signals using dual excitation fura 2 ratio fluorescence as shown in FIGURE 2. The upper curve shows a transient cytosolic Ca^{2+} signal evoked by depolarization of the chromaffin cells with 75 mM K^+ in the presence of extracellular Na^+. The lower curve shows the effects on the Ca^{2+} transient of replacement of Na^+ in the medium with an equivalent amount of N-methylglucamine. The rate of decay of the Ca^{2+} signal is markedly slowed by the removal of extracellular Na^+, consistent with Na^+-dependent Ca^{2+} efflux contributing to Ca^{2+} signal decay.

Previous studies have indicated the presence of a Na-Ca exchange system on chromaffin secretory granules.[6–8] FIGURE 3 shows evidence for Na^+-dependent Ca^{2+} transport across isolated chromaffin granule membranes.[6] The upper curve demonstrates a significant ^{45}Ca uptake within 5 min into chromaffin secretory vesicle ghosts loaded with either NaCl or sodium isothionate. When Na^+ was replaced with various other cations (K^+, Li^+, choline) in the vesicle loading medium, the subsequent ^{45}Ca uptake was substantially reduced.[6]

Since the chromaffin granules are the dominant organelle in chromaffin cells in terms of numbers (approximately 25,000–30,000/cell) and membrane surface

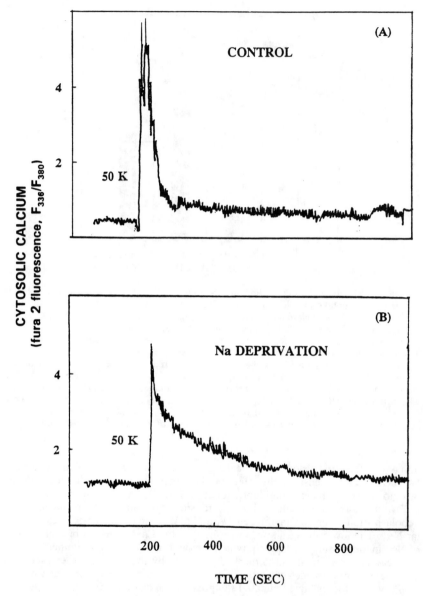

FIGURE 2. Na^+ dependence of Ca^{2+} signal decay. Cytosolic Ca^{2+} was monitored by dual excitation ratio microspectrofluorometry with fura 2 as calcium-sensitive probe in a PT1 Delta Scan microspectrofluorometer. The rise in cytosolic Ca^{2+} was stimulated by depolarizing the cell with 75 mM KCl in **(A)** the presence or **(B)** absence of 75 mM Na^+. In (B), Na^+ was replaced with an equivalent amount of N-methylglucamine. Cytosolic Ca^{2+} is represented by the fluorescence ratio measured for excitations at 336 nm/380 nm with emission measured at 500 nm.

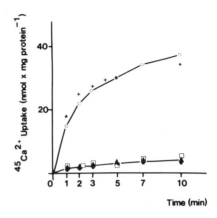

FIGURE 3. Na-dependent ^{45}Ca uptake by chromaffin secretory vesicle ghosts. Secretory vesicle ghosts were loaded in hypotonic medium containing 50 mM of various monovalent cations and washed and resuspended in 100 mM of the same monovalent cation. NaCl-loaded ghosts (○), Na$^+$ isothionate-loaded ghosts (+), KCl-loaded ghosts (□), LiCl-loaded ghosts (●) and choline chloride-loaded ghosts (▲) were incubated with 1.4 × 10^{-6} M free ^{45}Ca^{2+} (sp.act. 10 mCi/mmole Ca^{2+}) in 100 mM KCl. The incubation media was supplemented with 0.5 mM EGTA and 20 mM MOPS (pH 7.3). (From Krieger-Bauer & Gratzl.[6] Reprinted by permission from the *Journal of Neurochemistry*.)

area (tenfold that of the plasma membrane), it was of interest to determine whether Na$^+$-dependent Ca^{2+} mobilization could be demonstrated across intracellular organelle membranes in situ within the cell. To explore this possibility, we loaded chromaffin cells with Na$^+$ in a Ca^{2+}-free medium and monitored cytosolic calcium levels during the Na$^+$ loading. Incubation with ouabain and veratridine in a Na$^+$-containing, Ca^{2+}-free medium was used to load Na$^+$ into the cells, and cytosolic Ca^{2+} was measured in a PTI-Deltascan microspectrofluorometer using dual-excitation ratio fluorescence methods with the Ca^{2+}-sensitive dye, fura 2, . The results are shown in FIGURE 4. Following a brief period (minutes) of Na$^+$-loading in Ca^{2+}-free medium there occurred a spontaneous rise and oscillation of cytosolic Ca^{2+}. Since the experiment was done in Ca^{2+}-free medium, the source of the Na$^+$ loading-induced cytosolic Ca^{2+} rise must be intracellular, suggesting Na-Ca exchange across intracellular organelle membranes. This would be consistent with the known presence of a Na-Ca exchange system on chromaffin secretory vesicle membranes and possibly mitochondrial membranes. These results also suggest an additional secondary action of cardiac glycosides: upon inhibition of the Na/K ATPase by ouabain, the resulting elevation of cytosolic Na$^+$ can mobilize intracellular Ca^{2+} by enhancing Na-Ca exchange across intracellular organelle membranes.

In order to further test for the presence of Na-Ca exchange proteins on chromaffin secretory vesicles we have employed fluorescence immunocytochemistry of intact chromaffin cells using a monoclonal antibody (mAb R3F1) to the cardiac Na-Ca exchanger. Serial optical sections through a labeled chromaffin cell were observed by confocal fluorescence microscopy and the results are shown in FIGURE 5 (A–D). Starting in an upper section through the cell (A) a fine punctate fluorescence can be seen distributed throughout the cell cross section. This presumably includes both plasma membrane and labeled exchangers on organelles. Progressing through sections deeper into the cell, the punctate fluorescence is distributed more brightly in a region of the cytosol outside the nucleus and becomes weaker in the darkening area in which the nucleus begins to appear (panels B,C). These results provide evidence for antibody recognition of a nonplasma membrane source of Na-Ca exchange proteins deep within the chromaffin cell that are in the plane of sections containing the nucleus and which stain with greater intensity than the plasma membrane ring around the edge of the cell. This would be consistent with the presence of Na-Ca exchange proteins on chromaffin secretory vesicles which

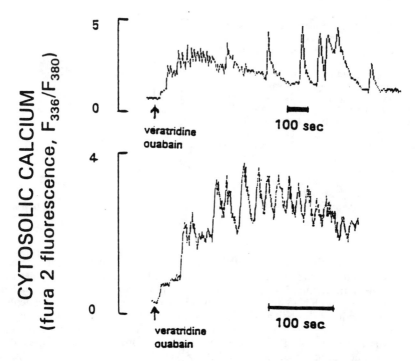

FIGURE 4. Intracellular Na^+-dependent cytosolic Ca^{2+} oscillations stimulated by Na^+-loading in Ca^{2+}-free medium in cultured bovine adrenal chromaffin cells. Na^+-loading was initiated by incubating cells with veratridine (100 μM) and ouabain (200 μM) in Ca^{2+}-free buffer, at the indicated time (*arrow*). The experiments were done at room temperature in a PTI Deltascan microspectrofluorometer as in FIGURE 2. Cytosolic Ca^{2+} is reported in terms of fura 2 fluorescence excitation ratio.

have a distribution throughout the cell like that of the punctate fluorescence in FIGURE 5.

The role of chromaffin secretory vesicle Na-Ca exchange proteins in regulation of Ca^{2+} signaling and catecholamine release is not fully understood. Previous measurements by electron probe microanalysis of the ion concentrations in various subcellular compartments of the chromaffin cell indicate a lack of detectable Na^+ in situ in mature chromaffin secretory vesicles in resting cells.[9] Thus, although there may be a Na-Ca exchange capability of the chromaffin vesicles, Ca^{2+} uptake via Na-Ca exchange into the secretory vesicle of resting chromaffin cells would seem unlikely considering the absence of detectable Na^+ in the vesicle. However, during exocytotic secretion, one face of the secretory vesicle would be open to the high Na^+ of the extracellular medium and the other face would see the elevated cytosolic Ca^{2+} which triggers exocytosis. Thus, there is the distinct possibility that during exocytosis, vesicle Na-Ca exchange would be activated and cause Ca^{2+} efflux from the cell. Following exocytosis, secretory vesicles are retrieved back into the cell, having captured the high Na^+ of the extracellular medium. The retrieved vesicle could continue to lower cytosolic Ca^{2+} in exchange for its captured Na^+. We have begun to test the potential role of Na-Ca exchange across exocytotic secretory vesicles as a contributor to Ca^{2+} signal decay, *i.e.*, as a negative feed back to the Ca^{2+} signal for secretion. One approach has been to either suppress or enhance secretory vesicle membrane incorporation into the plasma membrane and determine whether there are corresponding effects on $^{45}Ca^{2+}$ efflux.[1] It is necessary to use methods of changing vesicle membrane incorporation that do not change Ca^{2+} uptake, since the latter could alter cytosolic Ca levels and affect measured efflux. Several methods of altering exocytotic vesicle membrane insertion into the plasma membrane were tried.[1] Catecholamine secretion is measured to monitor the extent of exocytosis.

To suppress exocytotic vesicle incorporation we used either hypertonic media or low temperature.[1] Hypertonic media is known to inhibit exocytosis in chromaffin cells,[10,11] and the effects of such media on K^+-evoked catecholamine secretion and $^{45}Ca^{2+}$ uptake and efflux are shown in FIGURE 6. Both catecholamine secretion and Ca^{2+} efflux were inhibited by about 50%, while Ca^{2+} uptake was not affected. A similar result was found upon lowering temperature from 25°C to 12°C.[1] Secretion

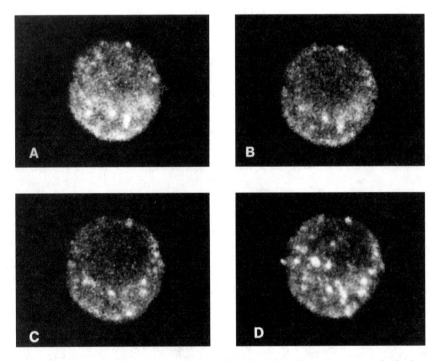

FIGURE 5. Fluorescent immunocytochemical labeling of Na-Ca exchange proteins in chromaffin cells with monoclonal antibody mAb R3F1. Optical sectioning through chromaffin cell (**A–D**) with fluorescence confocal microscope. Punctate fluorescence from interior of cell (**B,C,D**) would be consistent with labeling of Na-Ca exchange proteins on chromaffin secretory vesicles. Scale: chromaffin cells are 16–18 micron in diameter. Measurements were made in the laboratory of Dr. Joy Frank of the UCLA School of Medicine. Cells were fixed in 2% formaldehyde, washed and permeabilized with 0.1 % Triton X-100 for 10 min. Primary antibody 1/50, mAb R3F1 for 1.5 hr followed by secondary fluorescein isothiocyanate (FITC) labeled goat anti-mouse antibody, 1/50, for 1 hr.

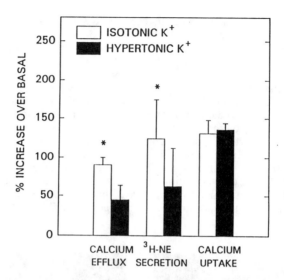

FIGURE 6. Effects of hypertonic medium on K^+-evoked $^{45}Ca^{2+}$ efflux, catecholamine secretion and $^{45}Ca^{2+}$ uptake in cultured bovine adrenal chromaffin cells. Stimulating solutions: Isotonic K^+: high K^+ depolarizing medium containing 75 mM K^+ and 75 mM Na^+; Hypertonic K^+: medium containing 75 mM K^+, 75 mM Na^+ and 300 mM sucrose; total osmolarity, 620 mOsm. Basal: Locke's medium with or without 300 mM sucrose added. The stimulation time for $^{45}Ca^{2+}$ uptake and $^{45}Ca^{2+}$ efflux was 4 min, and for catecholamine secretion was 4–6 min, all at 25°C. Catecholamine secretion is measured as ^3H-norepinephrine (^3H-NE) release. Data are expressed as % increase over basal and represent the mean ± SEM of 4–6 independent experiments each done in triplicate. * indicates significant difference at $p < 0.05$ by paired Student t test, as in FIGURE 1. (From Jan & Schneider. Reprinted by permission from the *Journal of Biological Chemistry*.)

stimulated by elevated K^+ (75 mM) was blocked at the lower temperature as was the stimulated Ca efflux. However, Ca^{2+} uptake was unaffected since the voltage-gated Ca^{2+} channels are relatively insensitive to temperature.

We have also used pertussis toxin (PTX) to enhance exocytotic secretion to see whether there would be a corresponding enhancement of Ca^{2+} efflux.[1] PTX has been reported to enhance catecholamine secretion from chromaffin cells without affecting Ca^{2+} uptake or cytosolic Ca levels.[12,13] FIGURE 7 shows the effects of PTX on basal and 75 mM K^+-evoked secretion (7A) and ^{45}Ca efflux (7B), in the presence and absence of extracellular Na^+ (75 mM). PTX is seen in FIGURE 7A to significantly enhance both basal and stimulated secretion in either the presence or absence of extracellular Na^+. There is a parallel PTX-induced enhancement of basal and stimulated ^{45}Ca efflux in the presence of 75 mM extracellular Na^+ shown in FIGURE 7B. This enhanced Ca^{2+} efflux is Na^+-dependent, since upon replacement of extracellular Na^+ with an equivalent amount of N-methylglucamine (75 mM), the enhanced ^{45}Ca efflux is blocked, including that due to both the K- and PTX-induced increases. Thus, either suppression or enhancement of exocytotic insertion of secretory vesicle membrane into the plasma membrane results in a parallel suppression or enhancement of Ca^{2+} efflux. Furthermore, the enhanced Ca^{2+} efflux is Na^+-dependent. These results would be consistent with a role for

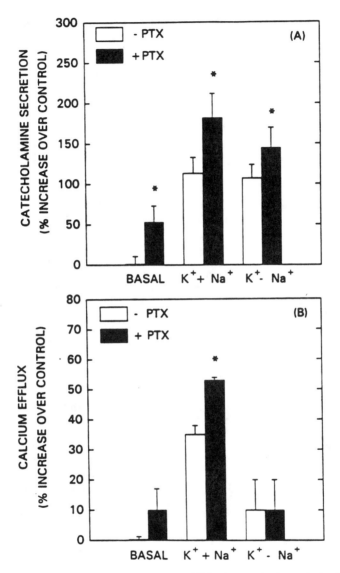

FIGURE 7. Effects of pertussis toxin on basal and K^+-stimulated (A) catecholamine secretion and (B) $^{45}Ca^{2+}$ efflux from cultured bovine adrenal chromaffin cells. Cells were preincubated with or without pertussis toxin (PTX, 100 ng/ml) in Dulbecco's modified Eagle's medium (DMEM) for 20 hours at 37°C. Catecholamine secretion is measured as ^{3}H-norepinephrine (^{3}H-NE) release. All responses were measured for 4 min at 37°C. Stimulating solutions: $K^+ + Na^+$: 75 mM K^+, 75 mM Na^+; K^+-Na^+: 75 mM K^+, 75 mM N-methylglucamine, Na^+-free medium. Data are expressed as % increase over control which is the basal response in the absence of pertussis toxin pretreatment and are mean ± SEM of 3 independent experiments each done in triplicate. *: $p < 0.05$, comparing the *solid* and the *blank bars* in individual pairs. (From Jan & Schneider.[1] Reprinted by permission from the *Journal of Biological Chemistry*.)

Na-Ca exchange across exocytotic secretory vesicle membranes contributing to Ca^{2+} efflux and the decay of stimulated Ca transients in chromaffin cells and possibly in related neuronal and endocrine secretory cells. It would also be reasonable to assume that recycled vesicles with their captured extracellular Na^+ could continue to remove Ca^{2+} from the cytosol by Na-Ca exchange.

DISCUSSION

The above model of secretory vesicle contributions to Ca^{2+} regulation during secretion is consistent with a number of previous findings, as was previously discussed in detail.[1] First, it is known that a high-capacity Na-Ca exchange mechanism exists in the chromaffin granule membrane.[6-8] Second, we[2] and others[5] have shown that the kinetics of decay of cytosolic Ca^{2+} transients in chromaffin cells is Na^+-dependent, i.e., Na^+ deprivation causes a considerable slowing of the rate of decay of a stimulated Ca^{2+} rise as shown in FIGURE 2. Third, the findings of Ornberg et al.,[9] based on electron probe microanalysis, indicate that there is no detectable Na^+ in mature chromaffin granules in situ in unstimulated cells. However, during exocytotic secretion, the granule membrane, which is now incorporated into the plasma membrane, is exposed to the high Na^+ levels of the extracellular medium and the elevated Ca^{2+} in the cytosol. This should turn on the granule membrane Na-Ca exchanger.

In order for the secretory vesicle exchanger to make a significant contribution to Ca^{2+} efflux from secreting chromaffin cells, the amount of vesicle incorporated into the plasma membrane must be sufficiently abundant; the duration of vesicle membrane exposure to the extracellular medium, i.e., its recycle time must be compatible with the kinetics of decay of cytosolic Ca^{2+} transients and Na^+-dependent Ca^{2+} efflux; and the K_m and V_{max} of the vesicle exchanger must be such as to allow the transport of significant amounts of Ca^{2+}. The small amount of published information that exists for the chromaffin cell plasma membrane and secretory vesicle Na-Ca exchangers[1,6,7,8,14] suggests that they have comparable capacities (per mg membrane protein) for Ca^{2+} transport at cytosolic $[Ca^{2+}]$ under 10 μM and could both contribute to Ca^{2+} efflux. Vesicle recycle half-times have been reported to be on the order of tens of seconds to a few minutes depending on conditions of cell stimulation.[15,16] This would be compatible with the duration of enhanced Ca^{2+} efflux and half-times of decay of evoked cytosolic Ca^{2+} transients[2] during high-level stimulation of chromaffin cells (FIG. 2). Regarding the amount of vesicle membrane that can be inserted into the plasma membrane during stimulated secretion, it can easily exceed that of the entire plasma membrane. Chromaffin secretory vesicles are known to have about a tenfold greater total membrane surface area than the plasma membrane. During maximal stimulation with acetylcholine or a high K^+ depolarizing stimulus, the chromaffin cell may release from 15–25% of its catecholamine content. This would result in more than a doubling of membrane surface area due to vesicle membrane insertion and provide sufficient new Ca^{2+} transporting surface to influence Ca^{2+} efflux and signal decay.

The properties of the chromaffin secretory vesicle Na-Ca exchanger have been reported to be different from the known properties plasma membrane Na-Ca exchangers.[6-8] While plasma membrane exchangers have generally been found to be electrogenic with a 3:1, Na:Ca exchange stoichiometry, chromaffin secretory vesicles have been reported to be electroneutral with a Na:Ca stoichiometry of 2:1.[6-8] Thus, there is the possibility that the secretory vesicle exchanger is a related, but distinct protein from that of the plasma membrane exchanger and may represent a new isoform of exchanger protein on hormone secretory granules and synaptic vesicles. Although this is only an interesting speculation at the present time, the cloning of new Na-Ca exchanger genes from neuronal tissue rich in synaptic vesicles may soon resolve this question.[17]

ACKNOWLEDGMENTS

The authors wish to acknowledge Dr. J. Frank of the UCLA School of Medicine for the fluorescence immunocytochemistry micrographs shown in FIGURE 5 and Dr. K. Philipson for the monoclonal antibody, mAb R3F1, used in those measurements.

REFERENCES

1. JAN, C. R. & A. S. SCHNEIDER. 1992. J. Biol. Chem. **267:** 9695–9700.
2. KAO, L. S. & A. S. SCHNEIDER. 1986. J. Biol. Chem. **261:** 4881–4888.
3. CHERN, Y. J., S. H. CHUEH, Y. J. LIN, C. M. HO & L. S. KAO. 1992. Cell Calcium **13:** 99–106.
4. LIU, P. S. & L. S. KAO. 1990. Cell Calcium **11:** 573–579.
5. GOH, Y. & A. KUROSAWA. 1990. Biochem. Biophys. Res. Commun. **166:** 1346–1351.
6. KRIEGER, BRAUER, H. & M. GRATZL. 1983. J. Neurochem. **41:** 1269–1276.
7. PHILLIPS, J. H. 1981. Biochem. J. **200:** 99–107.
8. SAERMARK, T. 1989. *In* Sodium-Calcium Exchange. T. J. A. Allen, D. Noble & H. Reuter, Eds. 54–65. Oxford University Press. New York.
9. ORNBERG, R. L., G. A. J. KUIJPERS & R. D. LEAPMAM. 1988. J. Biol. Chem. **263:** 1488–1493.
10. HAMPTON, R. Y. & R. W. HOLZ. 1983. J. Cell Biol. **96:** 1082–1088.
11. LADONA, M. G., M. F. BADER & D. AUNIS. 1987. Biochim. Biophys. Acta **927:** 18–25.
12. TANAKA, T., H. YOKOHAMA, M. NEGISHI, H. HAHASHI, S. ITO & O. HAYAISHI. 1987. Biochem. Biophys. Res. Commun. **144:** 907–914.
13. SASAKAWA, N., S. YAMAMOTO, T. NAKAKI & R. KATO. 1988. Biochem. Pharmacol. **37:** 2485–2487.
14. BAILEY, C. A. & J. P. REEVES. 1987. Fed. Proc. **46:** 367.
15. VON GRAFENSTEIN, H., C. S. ROBERTS & P. F. BAKER. 1986. J. Cell Biol. **103:** 2343–2352.
16. NEHER, E. & A. MARTY. 1982. Proc. Natl. Acad. Sci. USA **79:** 6712–6716.
17. LI, Z., S. MATSUOKA, L. V. HRYSHKO, D. A. NICOLL, M. M. BERSOHN, E. P. BURKE, R. P. LIFTON & K. D. PHILIPSON. 1994. J. Biol. Chem. **269:** 17434–17439.

Na^+-Ca^{2+} Exchange in Anoxic/Ischemic Injury of CNS Myelinated Axons

PETER K. STYS AND ISABELLA STEFFENSEN

Loeb Research Institute
Neuroscience, CSB9
Ottawa Civic Hospital
University of Ottawa
1053 Carling Avenue
Ottawa, Ontario, Canada, K1Y 4E9

INTRODUCTION

Myelinated axons of the mammalian central nervous system (CNS) are irreversibly damaged by anoxia/ischemia. Clinical disorders where anoxic/ischemic injury to white matter may result in significant morbidity include stroke, spinal cord injury, ischemic optic neuropathy, and others.[1] As in gray matter, white matter injury is critically dependent on influx of extracellular Ca^{2+}.[2,3] However, the mode of damaging Ca^{2+} entry is different than in neuronal cell bodies and synaptic terminals[2,4] and does not depend on excitotoxin- or voltage-gated Ca^{2+} channels. Recent studies have shown that most of the deleterious Ca^{2+} influx in white matter is mediated by reverse Na^+-Ca^{2+} exchange. The exchanger is recruited to operate in the Ca^{2+} import mode by both Na^+ influx and axonal depolarization. Thus, Na^+-Ca^{2+} exchange blockers, and/or agents that can prevent or delay Na^+ loading and depolarization may prove to be useful therapeutic agents for clinical disorders where white matter is damaged by anoxia/ischemia.

METHODS

We used the *in vitro* rat optic nerve as a representative model of central white matter anoxic/ischemic injury. Properties of this tissue and detailed experimental methods have been described elsewhere.[3,5–7] Briefly, adult Long Evans rats were anesthetized with 20% O_2/80% CO_2 and decapitated. Optic nerves were dissected free and perfused in a modified interface recording chamber with artificial cerebrospinal fluid (CSF) (in mM: NaCl 126, KCl 3.0, $MgSO_4$ 2.0, $NaHCO_3$ 26, NaH_2PO_4 1.25, $CaCl_2$ 2.0, dextrose 10, pH 7.45, 37.0 ± 0.2°C). Under control conditions the tissue was aerated with a 95% O_2/5% CO_2 gas mixture, and *in vitro* anoxia was induced by switching gases to 95% N_2/5% CO_2 for 60 min. Nerves were monitored electrophysiologically by recording the compound action potential (CAP), evoked by applying constant voltage stimulus pulses. The functional integrity of the nerves was assessed quantitatively by computing the area under the CAP. Resting compound membrane potential of the optic nerve bundle was measured directly using the cold grease gap technique.[8] Axoplasmic elemental content was measured with electron probe X-ray microanalysis;[9,10] nerves were taken at set times during *in vitro* anoxia, quench frozen in melting Freon 22 and stored under liquid nitrogen for analysis. Immunohistochemistry with R3F1 antibody (a generous gift from Dr. Ken Philipson) raised against the type I canine cardiac Na^+-Ca^{2+} exchanger was performed using standard techniques.[11,12]

RESULTS AND DISCUSSION

Acute Loss of Excitability during Anoxia

A representative optic nerve CAP under normoxic conditions is shown in FIGURE 1 (panel A, 'control'). The response is characteristically triphasic, with the first peak representing large diameter faster conducting fibers, and peak 3, the smaller, slower myelinated axons. The optic nerve is critically dependent on a continuous supply of O_2 for electrogenesis. The CAP is rapidly abolished after induction of anoxia, with peak 3 decaying first, and the larger diameter fibers being most resistant to the acute effects of anoxia. This likely is due to the less favorable surface-to-volume ratio of small axons, which would result in a more rapid collapse of transmembrane ion gradients, along with a lower resting $[K^+]_i$ [10] (and therefore an already more depolarized resting membrane potential in these fibers). After 60 min of anoxia, and a further 60 min of reoxygenation, CAP magnitude recovers to approximately 20–30% of control, and the waveshape remains severely distorted (FIG. 1). This injury is irreversible, and prolonged reoxygenation produces no further recovery.

The resting membrane potential of fine myelinated CNS axons is extremely difficult to measure directly without causing injury to the fiber. Both the resting value and the potential trajectory during anoxia are important parameters; because the Na^+-Ca^{2+} exchanger is electrogenic, both the rate and direction of Ca^{2+} transport will be affected by membrane potential. Using measurements of axoplasmic Na^+ and K^+ obtained by electron probe X-ray microanalysis,[10] $[K^+]_o$ from ion-selective microelectrodes[13] (also Stys, Waxman and Ransom, unpublished), and estimates of axonal $P_K:P_{Na}$,[8] we calculated the theoretical compound resting membrane potential at rest and at various times during anoxia with the Goldman-Hodgkin-Katz constant field equation. Under normoxic conditions, resting membrane potential is about −83 mV. This value is in close agreement with measured potentials in peripheral fibers, which are reported in the range of −75 to −85 mV.[14-16] The change in potential during anoxia obtained by this method is plotted in FIGURE 1B (bottom graph, open circles). More direct membrane potential measurements were carried out in a grease gap chamber (Lepannen and Stys, unpublished), which yields a set fraction of the true absolute potential.[8,17] These data were normalized to a baseline of −83 mV. The trajectory during anoxia, induced chemically with 2 mM NaCN, is shown superimposed (FIG. 1B, solid line) on calculated theoretical values described above. There is good agreement between these two methods, indicating that these results are likely reasonable estimates of the true membrane potential changes in anoxic optic nerve.

The nerve depolarizes quickly at the onset of anoxia, then more slowly as the 60 min anoxic challenge continues. The initial rapid loss of membrane potential is likely due to a rapid early rise of $[K^+]_o$, followed by a more sustained loss of $[K^+]_i$.[10] The acute loss of the CAP parallels the rapid early depolarization, leading to inactivation of the majority (see below) of Na^+ channels which underlie action potential generation.

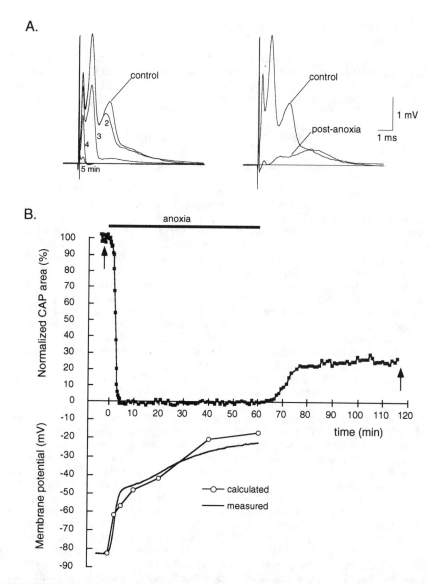

FIGURE 1. (**A**) Representative compound action potential recorded from a rat optic nerve using suction electrodes *in vitro*, exhibits three characteristic peaks (*control*). When anoxia is introduced, the amplitude of the response rapidly declines over several minutes (*left traces*); small fibers (*last peak*) are affected before large fibers (*first peak*). *Traces on the right* illustrate the degree of recovery observed after a 60-min anoxic exposure, followed by 60 min of reoxygenation. Both the size and shape of the potential are severely and irreversibly altered. (**B**) *Top graph* showing area under the compound action potential, normalized to 100%, as a function of anoxia/reoxygenation. There is a rapid loss of excitability at the onset of anoxia, with typically 20–30% recovery of control area after a 60-min exposure. *Bottom graph* illustrates decline of compound resting membrane potential of optic nerve during anoxia. Estimates were obtained by two different methods (see text), one calculated using measurement of internal and external ionic concentrations (*open circles*), the second by direct measurements in a grease gap chamber (*continuous trace*). Both methods yield similar results, reflecting an initial rapid depolarization during the initial 10 min, followed by a more gradual decline.

Ca^{2+}-Mediated Irreversible Injury

The example in FIGURE 1 illustrates the irreversible injury induced by 60 min of anoxia. The mechanisms of this damage were explored in more detail with ion substitution experiments and pharmacological agents, and the results are summarized quantitatively in the bar graph in FIGURE 2. Ca^{2+} is generally thought to play a key role in cell injury in the CNS. Removing Ca^{2+} from the perfusate (with the addition of ethylene glycol-bis-(β-aminoethyl ether)-N,N,N',N'-tetraacetic acid (EGTA)) allowed virtually 100% functional recovery of the CAP after 60 min of anoxia.[3] There are several possible routes of Ca^{2+} entry across the axolemma: voltage- or agonist-gated Ca^{2+} channels, leakage through channels nominally selective for other ions (such as Na^+ channels), reverse Na^+-Ca^{2+} exchange or nonspecific leakage through the axon membrane. Experiments with Ca^{2+} channel blockers including dihydropyridines (nifedipine, nimodipine), polyvalent cations (Mn^{2+}, Co^{2+}, La^{3+}) or ω-conotoxin failed to protect optic nerves from anoxia,[18] indicating that voltage-gated Ca^{2+} channels do not play a significant role in permeating Ca^{2+}. Similarly, ketamine (at concentrations selective for blockade of N-methyl-D-aspartate (NMDA)-gated ionophores) was not protective,[19] suggesting that excitotoxin-gated channels are not involved in white matter injury. Interest-

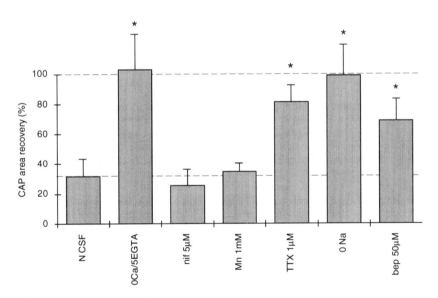

FIGURE 2. Bar graph showing quantitatively the degree of recovery of rat optic nerve compound action potential (CAP) area after 60 min anoxia/60 min reoxygenation. In normal CSF (NCSF) containing 2 mM Ca^{2+}, recovery was approximately 30% of control. Removing Ca^{2+} (+ 5 mM EGTA) allowed 100% recovery of CAP area. Ca^{2+} channel blockers (*e.g.*, nifedipine, Mn^{2+}) were ineffective, but blocking voltage-gated Na^+ channels (TTX 1 μM), or removing Na^+ from the perfusate (0 Na), were both highly protective. These data together suggest that Ca^{2+} plays a critical role in white matter anoxic injury, and that Ca^{2+} influx occurs mainly in a Na^+-dependent manner, likely via reverse Na^+-Ca^{2+} exchange. This idea is supported by the protective effects observed with blockers of Na^+-Ca^{2+} exchange such as bepridil (bep 50 μM) (*$p < 0.0001$ compared to NCSF).

ingly, blocking voltage-gated Na^+ channels with tetrodotoxin (TTX), or bathing the nerves with Li^+- or choline-substituted zero-Na^+ solution were both highly protective.[6]

Taken together, these results indicate not only that Ca^{2+} plays a critical role in mediating anoxic white matter injury, but in addition, the Ca^{2+} entry is dependent on Na^+ influx through Na^+ channels. A plausible unifying hypothesis would be that Ca^{2+} overload occurs via Na^+-Ca^{2+} exchange, driven in the Ca^{2+}-import/Na^+-export mode by a rise in axoplasmic $[Na^+]$ and membrane depolarization. Axoplasmic $[Na^+]$ increases significantly from a baseline of 16 mM to over 80 mM after 20 min of anoxia.[10] Coupled with rapid membrane depolarization (FIG. 1), these conditions would strongly promote Ca^{2+} influx via reverse Na^+-Ca^{2+} exchange.[20,21]

Ca^{2+} Influx Is Mediated by Reverse Na^+-Ca^{2+} Exchange

Several additional lines of evidence implicating exchange-mediated Ca^{2+} overload in anoxic white matter were obtained. Significant protection was seen with pharmacological inhibitors of Na^+-Ca^{2+} exchange such as bepridil (FIG. 2), benzamil and dichlorobenzamil.[6] These compounds are not specific; bepridil for instance has Ca^{2+} channel blocking effects.[22] However, we have shown that Ca^{2+} channels do not play a significant role in white matter injury (see above). Interestingly, the degree of protection afforded by bepridil was consistently below that seen with Na^+ channel blockers[23] or zero-Na^+ perfusate (FIG. 2). Bepridil's inhibition of Na^+-Ca^{2+} is competitive with Na^+, and given the large rise of $[Na^+]_i$ with anoxia, it is possible that for this reason bepridil's blocking efficiency is reduced, allowing Na^+_i-Ca^{2+}_o exchange to proceed. A noncompetitive, selective inhibitor of reverse exchange would likely be more effective.

While zero-Na^+ solution was very protective when applied before the anoxic insult as shown in FIGURE 2, the degree of injury can be significantly modulated depending upon the time at which the Na^+-free solution is applied with respect to anoxia onset. These experiments are illustrated in FIGURE 3, with similar results obtained with choline (which does not permeate through Na^+ channels) or Li^+ (which does permeate through Na^+ channels but substitutes poorly in the Na^+-Ca^{2+} exchange mechanism[24,25]). Perfusing the nerves with Na^+-free solution before (t = -40 or -20 min) anoxia was highly protective, likely because the axoplasm was depleted of Na^+. Since reverse Na^+-Ca^{2+} exchange requires a finite $[Na^+]_i$ ($K_{1/2} \approx 20$–30 mM[26–28]), reducing axoplasmic $[Na^+]$ will inhibit Ca^{2+} influx through this mechanism. Applying zero-Na^+ perfusate under *normoxic* conditions would result in a transient reversal of exchanger operation, causing some Ca^{2+} entry before the axoplasm was depleted of Na^+. This was not injurious, probably because other Ca^{2+} homeostatic mechanisms such as Ca^{2+}-ATPase and mitochondrial buffering were intact under normoxic conditions. In contrast, delaying the introduction of zero-Na^+ solution until after the start of anoxia (t = $+20$ or $+40$ min, FIG. 3) caused significantly more injury than a 60-min period of anoxia with normal $[Na^+]_o$ maintained throughout (dashed line). $[Na^+]_i$ rises significantly in optic nerve axons shortly after the start of anoxia[10] (see above). Introducing zero-Na^+ at this point will expose the fibers to a large *reverse* Na^+ gradient (*i.e.*, $[Na^+]_i = 80$ mM vs $[Na^+]_o = 0$ mM), driving even more Ca^{2+} into the axoplasm through Na^+_i-Ca^{2+}_o exchange than would occur if $[Na^+]_o$ were maintained at the normal level (153 mM) throughout, where the Na^+ gradient only partially collapses (*i.e.*, $[Na^+]_i = 80$

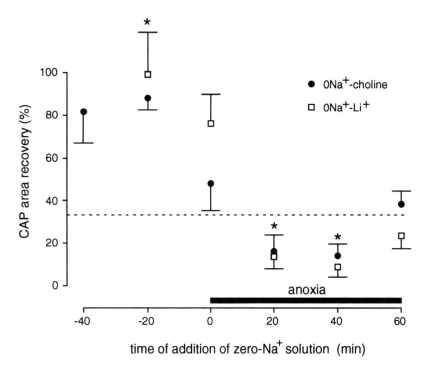

FIGURE 3. Effects of zero-Na^+ solution, applied at various times before or after anoxia, on compound action potential (CAP) recovery. Graph shows CAP area recovery after 60 min of anoxia plotted against the time (in min, with respect to the onset of anoxia) at which zero-Na^+ perfusate (choline- or Li^+-substituted) was applied. Normal CSF was resumed 15 min after the end of anoxia, and post-anoxic readings were taken 1 h after reoxygenation. Recovery was computed as the ratio of CAP area post-anoxia to area pre-anoxia in CSF containing normal $[Na^+]$ (153 mM). Recovery was greatly enhanced when zero-Na^+ CSF was started before anoxia. However, as the introduction of zero-Na^+ CSF was delayed (t = +20 or +40 min), significantly more injury occurred than with normal CSF ($[Na^+]$ = 153 mM) maintained throughout the anoxic period (*dashed line*). Delayed introduction of zero external Na^+, at a time when $[Na^+]_i$ has risen to high levels as a result of anoxia, imposed a strong reverse Na^+ gradient forcing the Na^+-Ca^{2+} exchanger to import even more damaging Ca^{2+}, resulting in greater functional injury (*$p < 0.0001$ compared to control anoxia, *dashed line*). (Modified from Stys et al.[6])

mM vs $[Na^+]_o$ = 153 mM), but is never reversed. This increased Ca^{2+} load is reflected by increased cellular injury and poorer postanoxic CAP recovery as plotted in FIGURE 3.

The data presented above strongly support the idea that Ca^{2+} enters the axoplasm during anoxia, but only recently were Ca^{2+} concentrations directly measured in these small anatomical compartments, confirming the previous predictions. Using electron probe X-ray microanalysis, we have shown that total axoplasmic Ca^{2+} increased about 7-fold during a 60-min anoxic exposure. The increase in total measured Ca^{2+}, expressed in mmol/kg dry weight, is plotted in FIGURE 4A (solid triangles). If Na^+-Ca^{2+} exchange is indeed the dominant Ca^{2+} import mechanism in anoxic CNS axons, we could calculate the expected internal $[Ca^{2+}]$

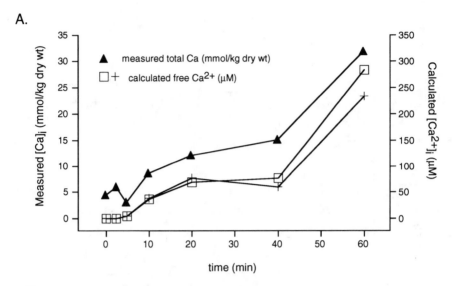
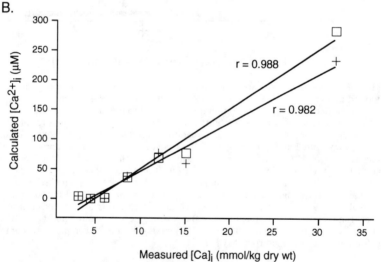

FIGURE 4. (A) Total axoplasmic Ca, as measured by electron probe X-ray microanalysis, rises approximately sevenfold during a 60-min anoxic exposure (*solid triangles*). The concentration of ionized internal Ca^{2+} at Na^+-Ca^{2+} exchanger equilibrium, calculated using estimates of membrane potential and intra- and extracellular ion concentrations, is plotted as a function of anoxia. Two results are shown (*squares* and *crosses*), which yield similar estimates based on two determinations of membrane potential trajectory (see text and FIG. 1B). There is a strong parallel between measured and predicted $[Ca^{2+}]_i$. **(B)** Although total (free + bound) measured [Ca] and calculated free $[Ca^{2+}]$ cannot be directly compared, assuming that free Ca^{2+} changes proportionally to total intracellular Ca^{2+}, we obtain a strong correlation between measured and predicted $[Ca^{2+}]$ (two correlations shown for two predicted $[Ca^{2+}]_i$ shifts; see (A) and text). This result lends further support to the hypothesis that Na^+-Ca^{2+} exchange is the main Ca^{2+} import mechanism in anoxic white matter.

that the exchanger would maintain at equilibrium. Assuming a coupling ratio of 3 Na$^+$: 1 Ca^{2+}:

$$E_{NaCa} = 3E_{Na} - 2E_{Ca}$$

Solving for [Ca^{2+}]$_i$:[29,30]

$$[Ca^{2+}]_i = [Ca^{2+}]_o \exp\left[\frac{FV_m(n-2)}{RT}\right]\left[\frac{[Na^+]_i}{[Na^+]_o}\right]^3$$

where V_m is the resting membrane potential at various times during anoxia. F, R and T have their usual meanings.

We have estimated the change in V_m during anoxia (FIG. 1B), and [Na$^+$]$_i$ is known.[10] Assuming that [Ca^{2+}]$_o$ and [Na$^+$]$_o$ in the extracellular space of the nerve remain relatively stable near the levels of the bathing medium, estimates of ionized axoplasmic [Ca^{2+}] (in μM) that would be maintained if the Na$^+$-Ca^{2+} exchanger were the main determinant, are plotted in FIGURE 4A as a function of anoxic exposure (two plots are shown for these calculations, one for each estimate of V_m trajectory; see above and FIG. 1B). These calculations yield a resting ionized [Ca^{2+}]$_i$ in larger (>2 μm) normoxic optic nerve axons of 115 nM, very close to the value expected in most neuronal cells.[31-33] The result has two important implications. First, the design of axonal ion transport systems with respect to transmembrane ion gradients, resting membrane potential and Ca^{2+} transport is close to optimal, with little wasteful Ca^{2+} cycling across the axolemma, as might occur if the exchanger equilibrium [Ca^{2+}]$_i$ were set far above the actual resting [Ca^{2+}]$_i$. Second, under these conditions the exchanger is poised to readily import Ca^{2+} with even modest membrane depolarization, either physiologically as a result of action potential generation, or under pathological conditions such as persistent anoxia-induced depolarization. Small fibers (<1 μm) yield higher estimates of [Ca^{2+}]$_i$ ($\approx$0.5 μM) and may either possess higher actual baseline Ca^{2+} levels or have somewhat different Ca^{2+} handling mechanisms, so that Na$^+$-Ca^{2+} exchange is not dominant at rest. We favor the latter possibility as the small fibers paradoxically tend to recover more completely following anoxia, in spite of their less advantageous surface-to-volume ratios, yet more modest observed Ca^{2+} accumulation during anoxia.[10]

The calculated data are superimposed on measured total Ca^{2+} illustrating the close concordance between actual measured and theoretical predicted Ca^{2+} levels. A direct comparison between measured and predicted [Ca^{2+}]$_i$ is not possible, however, because electron probe measurements yield total (free + bound) Ca concentrations, whereas calculations based on exchanger thermodynamic equilibrium apply only to ionized species. However, if we assume that ionized axoplasmic Ca^{2+} changes at least in a manner proportional to total Ca^{2+} accumulation as anoxia progresses, then semi-quantitative comparisons are instructive. FIGURE 4B shows the data from panel A (plotted against time) shown as measured total [Ca] plotted against calculated free [Ca^{2+}] (the latter computed for each of the two estimates of membrane potential change shown in FIG. 1B). There is a high degree of correlation ($r > 0.98$) between measured and predicted Ca^{2+}. This tight correspondence provides further support for, 1) the hypothesis that Na$^+$-Ca^{2+} exchange is indeed the main Ca^{2+} influx route, and 2) the assumption that total and free Ca^{2+} vary proportionally during anoxia. If these estimates are correct, it is interesting to note that free [Ca^{2+}]$_i$ rises to extremely high levels (250-300 μM) at the end of the anoxic exposure, which are certainly toxic to the cell. Similarly high Ca^{2+} levels have been observed in hypoxic hippocampal neurons.[34]

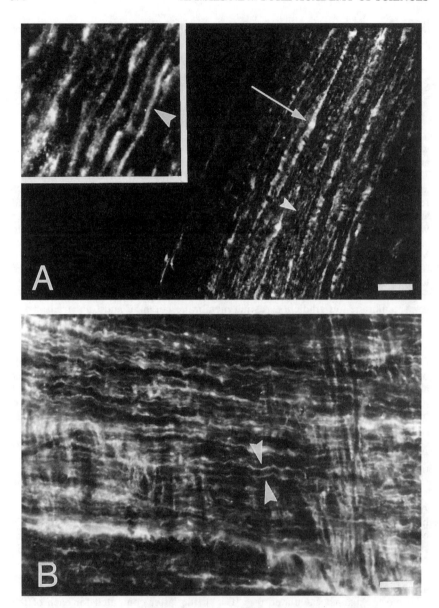

FIGURE 5. Fluorescence images of Na^+-Ca^{2+} exchanger protein immunoreactivity in central white matter tracts. **(A)** A confocal laser image of a longitudinal section of rat optic nerve. Broad linear staining with punctate areas represents glial cell bodies (*arrow*) and their processes. Finer linear profiles demonstrate exchanger distributed continuously along small diameter axons (*arrowhead*). *Inset*: high magnification image showing staining along the internodal region of axons, which may represent axolemma, myelin sheath, or both (*arrowhead*). **(B)** Conventional fluorescence image of a longitudinal section of lateral column from rat spinal cord. Similar fine linear staining indicates axonal localization of Na^+-Ca^{2+} exchange in this white matter region as well (*arrowheads* demarcate outer boundaries of a single axon). *Scale bars*: (A) 63 µm, inset 10 µm, (B) 25 µm.

While the above results provide convincing physiological evidence for the presence of the Na^+-Ca^{2+} exchanger in CNS myelinated axons, its localization has never been directly demonstrated in this tissue. Using a monoclonal antibody raised against the type I exchanger, we recently showed strong immunofluorescence staining in optic nerve (Steffensen and Stys, unpublished). The staining pattern reveals Na^+-Ca^{2+} exchanger immunoreactivity on both glial processes and on finer axonal profiles (FIG. 5A). Similar staining patterns were seen in spinal cord white matter (FIG. 5B), suggesting that this transporter may be ubiquitous in myelinated fibers, and that exchange-mediated Ca^{2+} overload is a general mechanism of white matter injury throughout the CNS. Preliminary results indicate that the Na^+-Ca^{2+} exchanger is distributed continuously along the length of the axon, at both nodal and internodal regions. However, it is not yet clear whether the observed continuous staining reflects exchanger protein along the axolemma, myelin sheath, or both. Studies are ongoing using immunogold techniques to confirm the precise localization at the ultrastructural level.

The "Anoxic Cascade" in White Matter

The sequence of events leading to anoxic/ischemic white matter injury is summarized in FIGURE 6. The initiating event is energy failure as a result of oxygen

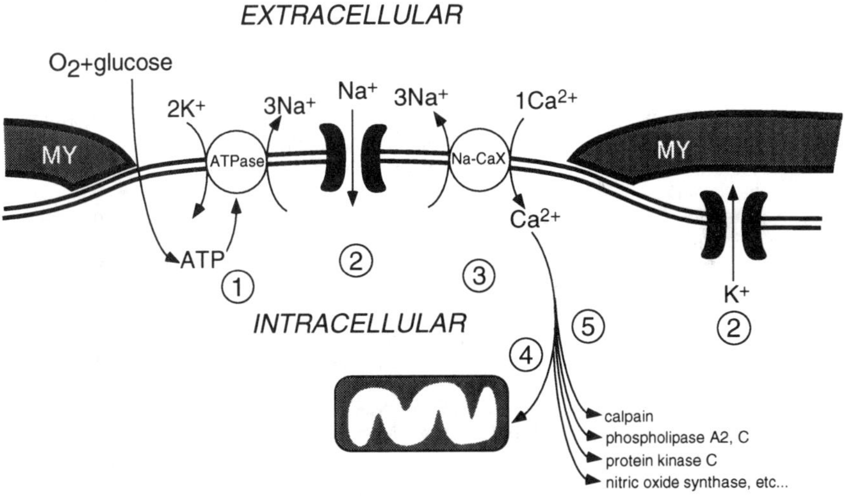

FIGURE 6. Hypothetical sequence of events leading to irreversible injury in CNS myelinated (MY) axons. Anoxia/ischemia produces cellular energy depletion, leading to failure of Na^+-K^+ ATPase (1). This allows unopposed inward leak of Na^+ ions, likely through a non-inactivating subtype of Na^+ channel, and simultaneous axonal depolarization due to efflux of K^+ through K^+ channels and accumulation of this ion in the extracellular space (2). The increase in $[Na^+]_i$ and membrane depolarization both act to drive the Na^+-Ca^{2+} exchanger in the Ca^{2+} import mode, leading to intracellular Ca^{2+} overload (3). Excess Ca^{2+} then leads to mitochondrial injury (4) and damage to the structural integrity of the axon through activation of multiple Ca^{2+}-dependent biochemical systems (5). Although the Na^+-Ca^{2+} exchanger is shown at the Node of Ranvier for convenience, the precise localization of this antiporter in CNS axons is unknown, and is the subject of ongoing studies.

and/or glucose deprivation. This causes failure of Na^+-K^+ ATPase leading to influx of Na^+ ions through voltage gated Na^+ channels, and membrane depolarization due to efflux of K^+ through K^+ channels, with accumulation of this ion in the extracellular space.[13] Interestingly, this inward Na^+ flux continues despite rapid membrane depolarization, and we hypothesize that a subtype of noninactivating Na^+ channel is responsible for this leak.[8,35] The rise of $[Na^+]_i$ coupled with depolarization cause reverse operation of the Na^+-Ca^{2+} exchanger, resulting in a toxic influx of Ca^{2+}, which in turn produces structural damage to the axon through activation of various Ca^{2+}-regulated systems such as calpain, lipases and kinases.[36] Mitochondria are also damaged by excess Ca^{2+} entry, especially at the time of reoxygenation.[10]

This model permits the logical development of therapeutic intervention; for example, interrupting the cascade at the Na^+ entry point with Na^+ channel blockers results in very significant protection, both in anoxic optic nerve *in vitro*[23,37,38] and after *in situ* ischemia.[39] Similarly, blocking the Na^+-Ca^{2+} exchanger is also very protective. Interestingly, after a glutamate challenge to neurons, the exchanger participates in Ca^{2+} homeostasis and appears to function in the Ca^{2+} export mode, so that inhibiting Na^+-Ca^{2+} exchange leads to greater injury.[40,41] This is likely due to a much smaller surface-to-volume ratio in a larger, more spherical neuron, where intracellular ionic perturbations may not be as rapid as in a small diameter axon. A selective blocker of reverse Na^+-Ca^{2+} exchange, perhaps administered with a Na^+ channel blocking agent, could represent a highly effective therapeutic approach aimed at protecting both gray and white matter regions against injury.

REFERENCES

1. STYS, P. K., B. R. RANSOM, J. A. BLACK & S. G. WAXMAN. 1995. Anoxic-ischemic injury in axons. *In* The Axon: Structure, Function and Pathophysiology. S. G. Waxman, J. D. Kocsis & P. K. Stys, Eds. 462–479. Oxford University Press. New York.
2. CHOI, D. W. 1990. Cerebral hypoxia: some new approaches and unanswered questions. J. Neurosci. **10:** 2493–2501.
3. STYS, P. K., B. R. RANSOM, S. G. WAXMAN & P. K. DAVIS. 1990. Role of extracellular calcium in anoxic injury of mammalian central white matter. Proc. Natl. Acad. Sci. USA **87:** 4212–4216.
4. WEISS, J. H., D. M. HARTLEY, J. KOH & D. W. CHOI. 1990. The calcium channel blocker nifedipine attenuates slow excitatory amino acid neurotoxicity. Science **247:** 1474–1477.
5. FOSTER, R. E., B. W. CONNORS & S. G. WAXMAN. 1982. Rat optic nerve: electrophysiological, pharmacological and anatomical studies during development. Dev. Brain Res. **3:** 371–386.
6. STYS, P. K., S. G. WAXMAN & B. R. RANSOM. 1992. Ionic mechanisms of anoxic injury in mammalian CNS white matter: role of Na^+ channels and Na^+-Ca^{2+} exchanger. J. Neurosci. **12:** 430–439.
7. STYS, P. K., B. R. RANSOM & S. G. WAXMAN. 1991. Compound action potential of nerve recorded by suction electrode: a theoretical and experimental analysis. Brain Res. **546:** 18–32.
8. STYS, P. K., H. SONTHEIMER, B. R. RANSOM & S. G. WAXMAN. 1993. Non-inactivating, TTX-sensitive Na^+ conductance in rat optic nerve axons. Proc. Natl. Acad. Sci. USA **90:** 6976–6980.
9. LOPACHIN, R. M., C. M. CASTIGLIA & A. J. SAUBERMANN. 1991. Elemental composition and water content of myelinated axons and glial cells in rat central nervous system. Brain Res. **549:** 253–259.
10. LOPACHIN, JR., R. M. & P. K. STYS. 1995. Elemental composition and water content

of rat optic nerve myelinated axons and glial cells: effects of *in vitro* anoxia and reoxygenation. J. Neurosci. **15:** 6735–6746.
11. MATA, M., D. J. FINK, S. A. ERNST & G. J. SIEGEL. 1991. Immunocytochemical demonstration of Na(+),K(+)-ATPase in internodal axolemma of myelinated fibers of rat sciatic and optic nerves. J. Neurochem. **57:** 184–192.
12. STEFFENSEN, I., M. ANCTIL & C. E. MORRIS. 1993. Neural structures in the receptive field of pleural ganglion mechanosensory neurons of *Aplysia californica*. Cell Tissue Res. **273:** 487–497.
13. RANSOM, B. R., W. WALZ, P. K. DAVIS & W. G. CARLINI. 1992. Anoxia-induced changes in extracellular K^+ and pH in mammalian central white matter. J. Cereb. Blood Flow Metab. **12:** 593–602.
14. BOSTOCK, H. & P. GRAFE. 1985. Activity-dependent excitability changes in normal and demyelinated rat spinal root axons. J. Physiol. (London) **365:** 239–257.
15. MORITA, K., G. DAVID, J. N. BARRET & E. F. BARRETT. 1993. Posttetanic hyperpolarization produced by electrogenic Na(+)-K+ pump in lizard axons impaled near their motor terminals. J. Neurophysiol. **70:** 1874–1884.
16. BARRETT, E. F., K. MORITA & K. A. SCAPPATICCI. 1988. Effects of tetraethylammonium on the depolarizing after-potential and passive properties of lizard myelinated axons. J. Physiol. (London) **402:** 65–78.
17. STAMPFLI, R. 1954. A new method for measuring membrane potentials with external electrodes. Experientia **10:** 508–509.
18. STYS, P. K., B. R. RANSOM & S. G. WAXMAN. 1990. Effects of polyvalent cations and dihydropyridine calcium channel blockers on recovery of CNS white matter from anoxia. Neurosci. Lett. **115:** 293–299.
19. RANSOM, B. R., S. G. WAXMAN & P. K. DAVIS. 1990. Anoxic injury of CNS white matter: protective effect of ketamine. Neurology (Minneapolis) **40:** 1399–1403.
20. MULLINS, L. J., T. TIFFERT, G. VASSORT & J. WHITTEMBURY. 1983. Effects of internal sodium and hydrogen ions and of external calcium ions and membrane potential on calcium entry in squid axons. J. Physiol. (London) **338:** 295–319.
21. REQUENA, J., L. J. MULLINS, J. WHITTEMBURY & F. J. J. BRINLEY. 1986. Dependence of ionized and total Ca in squid axons on Nao-free or high-Ko conditions. J. Gen. Physiol. **87:** 143–159.
22. GALIZZI, J-P., M. BORSOTTO, J. BARHANIN, M. FOSSET & M. LAZDUNSKI. 1986. Characterization and photoaffinity labeling of receptor sites for the Ca^{2+} channel inhibitors *d-cis*-diltiazem, (±)-bepridil, desmethoxyverapamil, and (+)-PN 200-110 in skeletal muscle transverse tubule membranes. J. Biol. Chem. **261:** 1393–1397.
23. STYS, P. K., B. R. RANSOM & S. G. WAXMAN. 1992. Tertiary and quaternary local anesthetics protect CNS white matter from anoxic injury at concentrations that do not block excitability. J. Neurophysiol. **67:** 236–240.
24. BAKER, P. F. 1972. Transport and metabolism of calcium ions in nerve. Prog. Biophys. Mol. Biol. **24:** 177–223.
25. KADOMA, M., J. FROEHLICH, J. REEVES & J. SUTKO. 1982. Kinetics of sodium ion induced calcium ion release in calcium ion loaded cardiac sarcolemmal vesicles: determination of initial velocities by stopped-flow spectrophotometry. Biochemistry **21:** 1914–1918.
26. RAHAMIMOFF, H., D. DAHAN, I. FURMAN, R. SPANIER & M. TESSARI. 1991. Molecular and mechanistic heterogeneity of the Na^+-Ca^{2+} exchanger. Ann. N. Y. Acad. Sci. **639:** 210–221.
27. REQUENA, J., J. WHITTEMBURY & L. J. MULLINS. 1989. Calcium entry in squid axons during voltage clamp pulses. Cell Calcium **10:** 413–423.
28. RASGADO-FLORES, H., E. M. SANTIAGO & M. P. BLAUSTEIN. 1989. Kinetics and stoichiometry of coupled Na efflux and Ca influx (Na/Ca exchange) in barnacle muscle cells. J. Gen. Physiol. **93:** 1219–1241.
29. BLAUSTEIN, M. P. & E. M. SANTIAGO. 1977. Effects of internal and external cations and of ATP on sodium-calcium and calcium-calcium exchange in squid axons. Biophys. J. **20:** 79–111.
30. SHEU, S-S. & H. A. FOZZARD. 1982. Transmembrane Na^+ and Ca^{2+} electrochemical

gradients in cardiac muscle and their relationship to force development. J. Gen. Physiol. **80:** 325–351.
31. BLAUSTEIN, M. P. 1988. Calcium transport and buffering in neurons. Trends Neurosci. **11:** 438–443.
32. MILLER, R. J. 1988. Calcium signalling in neurons. TINS **11:** 415–419.
33. ERECINSKA, M. & I. A. SILVER. 1992. Relationship between ions and energy metabolism: cerebral calcium movements during ischaemia and subsequent recovery. Can. J. Physiol. Pharmacol. **70:** S190–S193.
34. SILVER, I. A. & M. ERECINSKA. 1990. Intracellular and extracellular changes of $[Ca^{2+}]$ in hypoxia and ischemia in rat brain *in vivo*. J. Gen. Physiol. **95:** 837–866.
35. TAYLOR, C. P. 1993. Na^+ currents that fail to inactivate. Trends Neurosci. **16:** 455–460.
36. ORRENIUS, S. & P. NICOTERA. 1995. Mechanisms of calcium-related cell death. Adv. Neurol. In press.
37. STYS, P. K. 1995. Protective effects of antiarrhythmic agents against anoxic injury in CNS white matter. J. Cereb. Blood Flow Metab. **15:** 425–432.
38. FERN, R., B. RANSOM, P. K. STYS & S. G. WAXMAN. 1993. Pharmacological protection of CNS white matter during anoxia: actions of phenytoin, carbamazepine and diazepam. J. Pharmacol. Exp. Ther. **266:** 1549–1555.
39. STYS, P. K. & H. LESIUK. Correlation between electrophysiological effects of mexiletine and ischemic protection of CNS white matter. Neuroscience. In press.
40. MATTSON, M. P., P. B. GUTHRIE & S. B. KATER. 1989. A role for Na^+-dependent Ca^{2+} extrusion in protection against neuronal excitotoxicity. FASEB J. **3:** 2519–2526.
41. ANDREEVA, N., B. KHODOROV, E. STEMASHOOK, E. CRAGOE, JR. & I. VICTOROV. 1991. Inhibition of Na^+/Ca^{2+} exchange enhances delayed neuronal death elicited by glutamate in cerebellar granule cell cultures. Brain Res. **548:** 322–325.

The Sodium-Calcium Exchanger and Glutamate-Induced Calcium Loads in Aged Hippocampal Neurons *In Vitro*[a]

L. R. MILLS

Department of Physiology
University of Toronto
11-430 Playfair Neuroscience Unit
The Toronto Hospital
399 Bathurst Street
Toronto, Ontario, Canada, M5T 2S8

INTRODUCTION

Neuronal Calcium Homeostasis and Brain Aging

Aging is associated with a decline of sensory, motor and cognitive faculties which in their most severe form culminate in profound intellectual deterioration. While the problem of age-related changes in brain function is an active area of biological research, understanding the underlying cellular and molecular basis of aging has proved difficult. One mechanism that has long been regarded as a potential source of age-related alterations in the brain is neuronal calcium regulation.[1-3] The calcium hypothesis of brain aging (see Ref. 3) postulates that changes in the cellular mechanisms that regulate neuronal calcium homeostasis play a critical role in brain aging.

This hypothesis is based upon the recognition that intracellular calcium (Ca_i^{2+}) regulates multiple essential neuronal functions including differential gene expression,[4] excitability,[5] release of neurotransmitters,[6,7] neuronal migration,[8] specific growth cone behaviors and neurite outgrowth,[9,10] but also plays a key role in those cellular changes that lead to degeneration and cell death.[11-13] Given these relationships it is evident that even small alterations in the regulation of Ca_i^{2+} could profoundly alter the nervous system. Such changes could be the basis for the subtle changes in neuronal morphology that occur during development and neuronal plasticity as well as the aberrant morphology observed in aging and a variety of neurodegenerative diseases.[3,14-16]

Implicit in this concept is the idea that brain aging is likely the result of a series of events that, occurring in combination, or in sequence, over a long period, culminate in a disruption in calcium homeostasis. Studies using subcellular preparations have provided direct evidence for age-related changes in multiple components of calcium regulation including calcium channels,[17] the sodium-calcium exchanger[18,19] (but see Ref. 20), calcium uptake by organelles,[21,22] and calcium binding proteins.[23] A variety of indirect evidence also links changes in neuronal calcium regulation and aging.[3,24,25] However, studies assessing neuronal calcium

---

[a] This work was supported by funds from NSERC, the Sandoz Aging Foundation, and the University of Toronto.

regulation in intact cells have been hampered by difficulties in culturing aged and adult neurons from the central nervous system (CNS).

Our laboratory recently developed methods for culturing dissociated hippocampal neurons and hippocampal brain slices from aged rats.[26,27] This article presents data on the role of one of the key mechanisms in calcium regulation, the sodium-calcium exchanger, in the response of aged hippocampal neurons to glutamate.

Calcium homeostasis is a complex process requiring the integration of multiple regulatory processes controlling influx, efflux, sequestration, and release from stores.[28] These mechanisms thus maintain free Ca_i^{2+} concentrations at normal rest levels, despite a 10^{-4}-fold concentration gradient across the membrane, and determine the temporal and spatial characteristics of calcium signals. The sodium-calcium exchanger is one of the major systems that function to extrude calcium from the cell.[28-30] Recent studies (see other chapters in this volume) greatly advanced our understanding of the regulation and function of the sodium-calcium exchanger at both the cellular and molecular level. Although assessing the contribution made by the exchanger to overall calcium regulation is difficult due to the lack of truly selective inhibitors, we have attempted to examine its role in aged neurons in reducing glutamate-induced increases in Ca_i^{2+} by eliminating extracellular sodium.

Glutamate is the major excitatory neurotransmitter in the brain. However, excessive stimulation of neurons by glutamate is lethal, triggering a series of events that culminate in degeneration and eventual cell death.[13] Glutamate neurotoxicity is mediated by increase in Ca^{2+} and can be reduced or prevented by blocking specific classes of glutamate receptors.[31] Glutamate receptors fall into two classes: ionotrophic receptors which have been distinguished by their ability to selectively bind N-methyl-D-aspartic acid (NMDA), kainate, and amino-3-hydroxy-5-methyl-4-isoxazole propionic acid (AMPA), and metabotrophic receptors, with selective binding of trans-1-amino-cyclopentyl-1,3-dicarboxylate (trans-ACPD) and 2-amino-4-phosphonobutyric acid (L-APB) binding. Although NMDA, kainate, AMPA, and metabotrophic receptors have all been implicated in glutamate toxicity, the increase in Ca_i^{2+} that accompanies NMDA receptor stimulation appears to be a key event in its initiation.[31]

RESULTS

Elimination of Extracellular Sodium Slows Neuronal Recovery from a Glutamate-Induced Calcium Challenge

We examined the responses of 14-day cultures of dissociated hippocampal neurons from aged rats to brief glutamate challenges in the presence, and absence, of extracellular sodium (Na^+). Cultures grown in glass bottomed dishes were rinsed in serum-free media, loaded with the calcium indicator dye fluo-3 (4 μg/ml) in HEPES buffered HBSS, and viewed on a Biorad 600 confocal microscope equipped with a Medical systems temperature controller/perfusion unit.[32] Normal saline consisted of a HEPES buffered HBSS (in mM 137 NaCl, 5 KCl, 1.4 $MgCl_2$, 1.8 $CaCl_2$, 5 glucose, 10 $NaHCO_3$, 20 HEPES, 10 μM glycine). In Na^+-free buffer NaCl was replaced by 168 NMDG (N-methyl-D-glucamine) or 144 mM LiCl. Bicarbonate was replaced with 5 mM $KHCO_3$. A neutral density filter of 2 was used on all preparations. Images were taken at 12-second intervals during a 60-second glutamate pulse, at one-minute intervals for the next 9 minutes, and

5-minute intervals thereafter. A glutamate challenge consisted of a brief pulse of glutamate (60 seconds), added to the bath in the presence of normal saline (with Na^+). To insure the glutamate challenge was delivered under identical conditions in both control and Na^+-free experiments, we eliminated extracellular sodium only from the washing buffer (rather than prior to delivery of glutamate). Extracellular Na^+ was replaced with NMDG in most experiments. For repeated glutamate challenges 3 μM glutamate was used since it facilitated a rapid recovery to baseline Ca_i^{2+} levels after the first challenge (see below). Typically cultures received the second identical glutamate challenge 5–15 minutes after the first one. Sister cultures receiving Na^+-containing or Na^+-free buffer during the washout phase were used to evaluate responses to a 100-μM glutamate challenge. In all experiments fluorescence levels (which reflect changes in intracellular calcium (see Ref. 32) were measured in neuronal somata, and are expressed as an average value, plus or minus the standard error of the mean (SEM).

Aged hippocampal neurons respond to a transient 3-μM glutamate challenge with a rapid increase in Ca_i^{2+} concentration followed by a slower decline to rest levels. Ninety percent of neurons (162/180) recovered baseline fluorescence within 5 minutes of a first glutamate challenge. The remaining 10% of neurons (n = 18) that did not recover fully during this period were excluded from further analysis and are not shown in FIGURE 1. All cultures (n = 18) received an identical first challenge but were divided into two groups prior to the second challenge (delivered 15 minutes after the first). In the control group the second challenge was washed out in normal saline; in the other group the second challenge was washed out with Na^+-free buffer. As can be seen in FIGURE 1A the response of control neurons to the first and second glutamate challenges was very similar. Peak fluorescence values were not significantly different and Ca_i^{2+} levels were restored to rest levels within five minutes of receiving the second challenge in 80% of neurons (66/82). In 20% of neurons (16/82), Ca_i^{2+} levels did decline markedly but remained above rest levels for the duration of the experiment (data not shown in FIG. 1A); in these neurons peak Ca_i^{2+} levels were also significantly higher after the second dose (peak fluorescence values ± SEM were 105 ± 14, n = 16) compared to fluorescence values after the first dose 78 ± 9, n = 16).

When Na^+ was eliminated from the washing buffer calcium dynamics during the recovery period after the second challenge differed markedly from the response observed after the first challenge (see FIG. 1B). The extent to which Ca_i^{2+} levels were restored to rest in Na^+-free buffer was variable but in 44/80 or 55% of neurons shown in FIGURE 1B, Ca_i^{2+} concentrations remained significantly above rest for the duration of the experiment. Twenty-eight percent of neurons (22/80) did show full recovery, Ca_i^{2+} levels declining to pretreatment levels (data not shown in FIGURE 1B), but the recovery period was prolonged taking up to 20 minutes in 13% (10/22) of cells, compared to 5 minutes in controls. In 17.5% of neurons (14/80), also not shown in FIGURE 1B, peak calcium levels were significantly higher after the second challenge than the first (158 ± 21, compared to 74 ± 9), and remained elevated for the duration of the experiment. Since recent studies[33] indicate that NMDG and other sodium substitutes were associated with a disruption in intracellular pH regulation and subsequent cytoplasmic acidification we also used LiCl as a sodium substitute. In our experiments substitution with LiCl gave qualitatively similar results to NMDG (data not shown).

These results indicate that when extracellular Na^+ is present in the washing buffer 90% of aged hippocampal neurons in our cultures show full recovery from a glutamate-induced Ca_i^{2+} load. In 80% of neurons Ca_i^{2+} levels were completely restored to rest after a second glutamate challenge. However, in the absence of

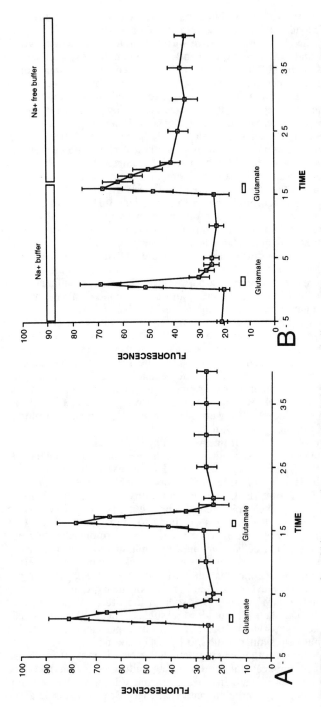

FIGURE 1. Response of aged hippocampal neurons to 3-μM glutamate challenges. Fluorescence levels are expressed as an average value plus or minus the standard error of the mean (SEM). Ninety percent of neurons recovered baseline fluorescence within 5 minutes of a 60-sec glutamate challenge (see text). A second challenge was delivered 15 minutes after the first. The second challenge was washed out in: **(A)** Na^+ buffer or **(B)** Na^+-free buffer.

extracellular Na$^+$ from the washing buffer only 28% of neurons show full recovery after the second challenge.

To examine the effects of larger glutamate challenges sister cultures were used, since preliminary experiments indicated that larger doses of glutamate resulted in sustained elevations for 60 minutes or more of the aged cultures. In control cultures Ca$_i^{2+}$ levels rapidly declined during the initial 5 minutes of washout in Na$^+$ containing buffer although recovery was markedly slower than after the lower dose. By 25 minutes Ca$_i^{2+}$ levels were restored to pre-glutamate levels in 83% of neurons (see FIG. 2). As was observed in experiments using 3 μM glutamate, a subpopulation of neurons did not recover after a single glutamate challenge; in 16% of neurons (not shown in FIG. 2), Ca$_i^{2+}$ levels plateaued at or near peak levels for the duration of the experiment. One percent of neurons showed no rise in Ca$_i^{2+}$ in response to glutamate. In cultures washed in Na$^+$-free buffer peak Ca$_i^{2+}$ levels induced by the glutamate pulse were similar to controls in 87% of neurons (47/55); however, recovery was incomplete and even at one hour Ca$_i^{2+}$ levels remained significantly above baseline levels. In 8/55 neurons (not shown in FIG. 2), glutamate induced a large increase in Ca$_i^{2+}$ but levels did not significantly decline after glutamate was washed out (peak values were 230 + 17 and final values 241 + 18). Similar results were also observed when LiCl was substituted for Na$^+$ rather than NMDG (data not shown).

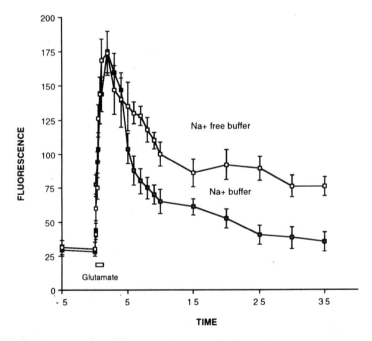

FIGURE 2. Response of aged hippocampal neurons in sister cultures to a 100-μM glutamate challenge (see text). Fluorescence levels are expressed as an average value plus or minus the standard error of the mean (SEM). Glutamate was washed out in normal saline or Na$^+$-free media. In the control cultures Ca$_i^{2+}$ levels were restored to pre-glutamate levels in 83% of neurons. In cultures washed out with Na$^+$-free buffer recovery was incomplete in 87% of neurons; at one hour Ca$_i^{2+}$ levels remained significantly above baseline levels.

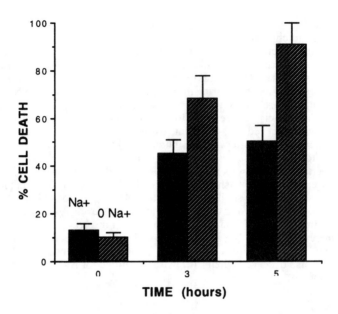

FIGURE 3. Elimination of extracellular Na^+ in the washing buffer potentiates glutamate neurotoxicity in aged hippocampal neurons. Toxicity was evaluated after 200 μM glutamate was washed out using PI in the washing buffer. Percentage toxicity is the average number of PI positive neurons/total number of neurons per dish ± SEM. Eight dishes, 4 control, and 4 washed with sodium-free buffer were used for each time point.

The effects of glutamate on cultured hippocampal neurons range from the selective inhibition of dendritic outgrowth, to, at very high levels, cell death.[11] In the experiments described above propidium iodide (PI) staining revealed some neuronal death within 4 hours of glutamate exposure. These observations prompted us to examine glutamate toxicity in the presence and absence of extracellular sodium. In these experiments a 200-mM glutamate challenge was washed out in normal or Na^+-free buffer for 30 minutes after which the cultures were replaced in the incubator in normal media with serum. (Elimination of Na^+ for a 60-minute period in the absence of glutamate did not increase cell death over the subsequent 8-hour period.) As FIGURE 3 shows, elimination of extracellular Na^+ in the washing buffer did potentiate toxicity at 3 and 5 hours post-glutamate challenge but not at 1 hour. These results suggest that the forward mode of Na^+/Ca^{2+} exchange during the recovery period is critical for cell survival following glutamate exposure. However, it remains possible that cell death is due to other effects of eliminating extracellular sodium (see Ref. 33).

To examine neurons within the context of glia and intact circuitry we cultured hippocampal slices from aged rats using a modification of methods developed by Stoppini *et al.*[34] for neonatal hippocampal slices. Slices from aged rats were cultured on Millipore CM membranes, or in special dishes constructed by us using Millipore CM membrane.[27,35] Slices were transferred to the membranes after slicing and fed by exchanging the media under the membrane. After 7–10 days *in vitro* the cultures, each containing 1–3 slices, were rinsed in serum-free buffer and loaded with 10 μM fluo-3 for 60 minutes at 37°C in the incubator. Since addition of more than 500 μl of medium to the surface of slice cultures for more than 20 minutes causes neuronal cell death most of the fluo-3 solution was added under the membrane. However, 100–200 μl was added on top of the slices. After the incubation period slices were rinsed, and viewed 60–120 minutes later. Successful loading of fluo-3 in neurons in slice cultures as defined by the presence of faintly labelled neurons was variable. Typically neurons can be visualized in the hippocampus using fluo-3 in 95% of slices, but neurons of CA1 can be visualized (see FIG. 4A–D) in only about 30% of slices. When CA1 neurons cannot be visualized the neurons have either died prior to fluo-3 loading during the 14-day culture period or failed for unknown reasons to load fluo-3. These possibilities can be distinguished by using PI to assess the extent of neuronal death induced by NMDA exposure, or by placing slices in a refrigerator at 4°C for 20 minutes. Slices where neurons are alive but did not load with fluo-3 (or died during the fluo-3 loading process) show a defined pattern of PI positive cells throughout CA1 and other hippocampal regions; slices where CA1 neurons have died prior to fluo-3 loading do not show extensive cell death in CA1. These and other experiments revealed that neurons of the dentate gyrus (DG) typically survive the culture procedure better than neurons from areas CA1–4; 95% of slices from aged rats contain numerous DG neurons at 2–3 weeks in culture.

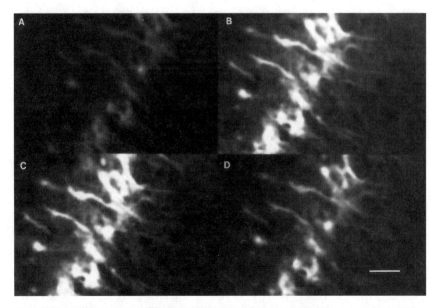

FIGURE 4. NMDA-evoked increases in intracellular calcium in aged CA1 neurons in slice cultures from aged rats. **(A)** Prior to NMDA application fluorescence levels in cell bodies and proximal dendrites in CA1 neurons are low; **(B)** 45 seconds after NMDA challenge was initiated fluorescence levels have risen rapidly as calcium increases; **(C)** peak fluorescence 12 seconds after NMDA washout begins; **(D)** calcium levels have declined markedly from peak values but have not been restored fully to pre-NMDA levels despite a 55-min wash in normal saline. *Scale bar*: 20 microns.

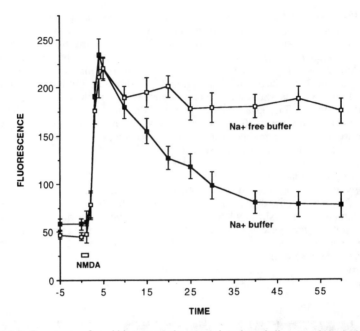

FIGURE 5. Responses of aged hippocampal neurons in cultured slices to a 300-μM NMDA challenge (see text). Fluorescence levels are expressed as an average value plus or minus the standard error of the mean (SEM). Shown are results from CA1 neurons in two slices with similar rates of rise in response to NMDA (see text). In the control slice NMDA causes a rapid increase in Ca_i^{2+} followed by a slow decline during the washout phase with normal saline; n = 29 cells. In the slice perfused with Na-free buffer in the washout phase calcium levels did not return to rest but remained significantly elevated for the duration of the experiment; n = 28 cells.

During the experiments slices were perfused with saline in mM (124 NaCl, 26 NaHCO$_3$, 3 KCl, 1.25 Na$_2$HPO$_4$, 2 CaCl$_2$, 10 D-glucose and glycine) bubbled with 95% O$_2$/5% CO$_2$. For Na$^+$-free buffer Na$^+$ was replaced with equimolar NMDG. In slice cultures NMDA application provided a more reliable increase in intracellular calcium than did glutamate. Experiments on slice cultures used only single doses of NMDA (300 μM) since preliminary experiments showed that neuronal Ca_i^{2+} levels in the slice declined slowly after the drugs were washed out and often plateaued above baseline levels (see FIG. 4A–D). NMDA was added in the bath and the washing buffer was perfused in as it was for the dissociated cultures. Responses to NMDA varied significantly in neurons from slice to slice with respect to initial rates of rise. For this reason we compared responses between slices receiving normal or Na$^+$-free saline in the washing phase by pairing those slices that showed similar rates of rise in response to NMDA (see FIG. 5). Typically, responses were determined in 20–60 neurons within a single area on each slice. FIGURE 5 shows calcium responses in CA1 neurons in a control slice washed with normal saline, and a sister slice washed with Na$^+$-free buffer. In the control slice NMDA causes a rapid increase in Ca_i^{2+}, followed by a slow decline during the washout phase. By 40 minutes post-NMDA, Ca_i^{2+} levels have plateaued just above (see FIG. 5) or at rest values. (Qualitatively similar responses, an initial rise followed by a return to, to close to, rest Ca_i^{2+} levels over a variable time course, were observed in 9/13 other control slices; in 4 control slices neuronal Ca_i^{2+} levels remained elevated after NMDA was washed out.)

NMDA caused a similar rise in neuronal Ca_i^{2+} in a sister slice washed with Na^+-free buffer (see FIG. 5). However, calcium levels declined only modestly and remained significantly elevated for the duration of the experiment. This failure to return to rest calcium levels characterized the majority of slices rinsed with Na^+-free buffer; in 11/15 slices Ca_i^{2+} levels remained at least 50% higher than rest levels for up to 60 minutes. In 3/15 slices neurons showed no recovery, or did not respond to NMDA.

These results demonstrate that CA1 neurons from aged rats can survive in culture and can respond to NMDA. In 71% of slices a transient NMDA exposure caused a transient rise in neuronal Ca_i^{2+} levels which ultimately declined to, or close to, pre-stimuli levels. However, when extracellular Na^+ was eliminated during the washout phase neurons in 80% of the slices did not recover from the NMDA-induced calcium load.

DISCUSSION

Taken together these results suggest that in aging neurons the Na^+-Ca^{2+} exchanger plays a critical role in reducing the Ca_i^{2+} load induced by glutamate. In the absence of extracellular sodium the majority of neurons could not fully recover. During the earliest stages of recovery Ca_i^{2+} presumably remained elevated because the exchanger no longer extrudes calcium, *i.e.*, no longer works in forward mode. Because transport activity depends upon both sodium and calcium gradients and on membrane potential, both of which are altered by glutamate during this period, there may also be a brief contribution by the exchanger working in reverse mode, *i.e.*, bringing calcium in.[39] Recent studies on embryonic cortical neurons suggested that the inability of some neurons to recover from a glutamate challenge followed by a wash in Na^+-free buffer was due to the exchanger working in reverse mode[36] (but see Ref. 33). However, it seems unlikely that the reverse mode exchanger plays a role in the sustained elevation in calcium observed for up to an hour after a transient glutamate exposure.

In the absence of extracellular sodium, neurons did show a significant degree of recovery after both moderate and large glutamate challenges, indicating that other mechanisms of calcium homeostasis can partially compensate for the loss of the exchanger. However, the persistent but stable elevation in Ca_i^{2+} observed long after glutamate has been eliminated suggests that these other mechanisms have an upper limit and cannot fully compensate for the loss of exchanger activity, particularly in the face of larger calcium loads. Alternatively, the absence of extracellular sodium in combination with a large calcium load could cause a sustained change in calcium regulation, or some degree of cell damage, either or both of which could be manifest as a new elevated steady-state Ca_i^{2+} level. The finding that the glutamate toxicity was potentiated in Na^+-free buffer supports the latter possibility although the elevated intracellular calcium levels observed at 40 and 50 minutes post-glutamate challenge may not be directly related to the increased cell death observed 3 and 5 hours after glutamate exposure.

Our results using cultured slices also support our interpretation that the exchanger plays a critical role in the recovery from a calcium load. However, in this case the increased complexity of the slice culture preparation makes it difficult to exclude other indirect effects of reducing extracellular sodium.

These experiments focused exclusively on aged neurons, but similar studies have demonstrated an important role for the sodium-calcium exchanger in restoring calcium levels to rest after a glutamate challenge in embryonic neurons[36,37] (but see Ref. 33). Whether the elimination of extracellular sodium prolongs this recovery phase in young and adult hippocampal neurons to the same degree as it does in aged neurons remains to be determined. Certainly the recovery period following a 100-μM glutamate challenge (with Na$^+$ present in the washing buffer) does appear to be shorter in embryonic hippocampal[33,38] or cortical neurons,[33,36] suggesting that overall calcium regulation may be less effective in aged hippocampal neurons. However, it is also possible that our culture protocols for culturing aged neurons may select for a subpopulation of cells that characteristically demonstrate a prolonged recovery after glutamate challenge.

Our results suggest that the sodium-calcium exchanger plays a critical role in calcium regulation in aged hippocampal neurons. It is as yet unknown whether the activity of the exchanger is in any way compromised in intact aged neurons, although based on results using synaptosome preparations from aged brain,[19] we might predict that the exchanger is impaired. Also unknown are the functional consequences of age-related changes in the exchanger. For example if the sodium-calcium exchanger is less effective on aged neurons we might predict that aged neurons would be more vulnerable to calcium-induced neuronal degeneration and cell death than younger neurons.

Although the causes of 'brain aging' are likely to be multifactorial our central conviction is that changes in neuronal calcium homeostasis are a key event in a final common pathway which leads to age-related dysfunction and cell death. Investigations into the role of the sodium-calcium exchanger are central to our understanding of the relationship between calcium homeostasis and brain aging. Moreover, knowledge of how the sodium-calcium exchanger changes over the lifetime of the nervous system is of relevance to normal development neuroplasticity.

ACKNOWLEDGMENTS

I thank Marina Frantseva and Carmen Tang for technical assistance.

REFERENCES

1. KHACHATURIAN, Z. 1984. Towards theories of brain aging. *In* Handbook of Studies in Old Age. D. W. Kay & G. D. Burrows, Eds. 7–30. Elsevier. New York.
2. GIBSON, G. E. & C. PETERSON. 1987. Calcium and the aging nervous system. *In* Neurobiology of Aging. Vol. 8. Pergamon Press.
3. KHACHATURIAN, Z. S. 1994. The calcium hypothesis revisited. Ann. N. Y. Acad. Sci. **747:** 1–13.
4. BADING, H., D. D. GINTY & M. E. GREENBERG. 1993. Regulation of gene expression in hippocampal neurons by distinct calcium signalling pathways. Science **260:** 181–186.
5. LLINÁS, R. & M. SUGIMORI. 1979. Calcium conductances in Purkinje cell dendrites: their role in development and integration. Prog. Brain Res. **51:** 323–334.
6. KATZ, B. 1969. The Release of Neural Transmitter Substances. Liverpool University Press. Liverpool, England.

7. AUGUSTINE, G. J. & E. NEHER. 1992. Neuronal Ca^{2+} signalling takes the local route. Curr. Opin. Neurobiol. **2**: 302–307.
8. KOMURA, H. & P. RAKIC. 1992. Selective role of N-type calcium channels in neuronal migration. Science **257**: 806–809.
9. KATER, S. B. & L. R. MILLS. 1991. Regulation of growth cone behaviour by calcium. J. Neurosci. **11**(4): 891–899.
10. LANKFORD, K. & P. C. LETOURNEAU. 1987. Evidence that calcium may control neurite outgrowth by regulating the stability of actin filaments. J. Cell Biol. **109**: 1229.
11. SCHANNE, F. A. X., A. B. KANE, E. E. YOUNG & J. L. FARBER. 1979. Calcium dependence of toxic cell death. Science **206**: 700–702.
12. BONDY, S. C. & H. KOMULAINEN. 1988. Intracellular calcium as an index of neurotoxic damage. Toxicology **49**: 35–49.
13. CHOI, D. W. 1988. Glutamate toxicity and diseases of the nervous system. Neuron **1**: 623–634.
14. ROTH, G. S. 1989. Calcium homeostasis and aging: role in altered signal transduction. Ann. N. Y. Acad. Sci. **658**: 68.
15. KATER, S. B., M. P. MATTSON & P. B. GUTHRIE. 1990. Calcium-induced neuronal degeneration: a normal growth cone regulating signal gone awry (?). Ann. N. Y. Acad. Sci. **568**: 252–261.
16. MILLS, L. R. 1991. Neuron-specific and state-specific differences in calcium regulation: their role in the development of neuronal architecture. Ann. N. Y. Acad. Sci. **639**: 312.
17. LANDFIELD, P. W. & T. A. PITLER. Prolonged Ca^+-dependence after hyperpolarizations in hippocampal neurons of aged rat. Science **226**: 1092.
18. MICHAELIS, M. L. 1989. Calcium handling systems and neuronal aging. Ann. N. Y. Acad. Sci. **658**: 89–94.
19. MICHAELIS, M. L. 1994. Ion transport systems and calcium regulation in aging neurons. Ann. N. Y. Acad. Sci. **747**: 407–418.
20. COLVIN, R. A., N. WU, N. DAVIS & C. MURPHY. 1993. Analysis of Na^+/Ca^{2+} exchange activity in human brain: the effect of normal aging. Neurobiol. Aging **14**: 373–381.
21. SHIGENAGA, M. K., T. M. HAGEN & B. N. AMES. 1994. Oxidative damage and mitochondrial decay in aging. Proc. Natl. Acad. Sci. USA **91**: 10771–10778.
22. VITORICA, J. & J. SATRUSTEGUI. 1986. Involvement of mitochondria in the age-dependent decrease in calcium uptake of rat brain synaptosomes. Brain Res. **378**: 36–48.
23. IACOPINO, A. M. & S. CHRISTAKOS. 1990. Specific reduction of calcium-binding protein gene expression in aging and neurodegenerative diseases. Proc. Natl. Acad. Sci. USA **87**: 4078.
24. SMITH, D. O. 1987. Non-uniform changes in nerve-terminal calcium homeostasis during aging. Neurobiol. Aging **8**: 366.
25. ZEE, C. E., VAN DER SCHURRMANN & W. H. GISPEN. 1990. Beneficial effects of nimodipine on peripheral nerve function in aged rats. Neurobiol. Aging **11**: 541.
26. MILLS, L. R., F-H. MANG & C. TANG. 1995. Calcium regulation in cultured hippocampal neurons from aged rats. Neurobiol. Aging. Submitted.
27. MILLS, L. R. & M. FRANTSEVA. 1995. Neuronal and glial development in organotypic cultures of spinal cord and hippocampus from adult and aged rats. J. Neurosci. Res. Methods. Submitted.
28. CARAFOLI, E. 1987. Intracellular calcium homeostasis. Annu. Rev. Biochem. **56**: 395–433.
29. BLAUSTEIN, M. P. 1977. Effects of internal and external cations and of ATP on sodium-calcium and calcium-calcium exchange in squid axons. Biophys. J. **20**: 79–110.
30. BLAUSTEIN, M. P. & R. DIPOLO, Eds. 1991. Sodium-Calcium Exchange: Proceedings of the Second International Conference. Ann. N. Y. Acad. Sci. **639**: 1–667.
31. MICHAELS, R. L. & S. M. ROTHMAN. 1990. Glutamate neurotoxicity *in vitro*: antagonist pharmacology and intracellular calcium concentrations. J. Neurosci. **10**(1): 283–292.
32. MILLS, L. R. 1994. Confocal calcium imaging. *In* Three-Dimensional Confocal Microscopy. J. K. Stevens, L. R. Mills & J. Trogadis, Eds. Academic Press Inc. Orlando, FL.
33. KOCH, A. R. & M. E. BARISH. 1994. Perturbation of intracellular calcium and hydrogen

ion regulation in cultured mouse hippocampal neurons by reduction of the sodium ion concentration gradient. J. Neurosci. **14:** 2585–2593.
34. STOPPINI, L., P. A. BUCHS & D. MULLER. 1991. A simple method for organotypic cultures of nervous tissue. J. Neurosci. Methods **37:** 173–182.
35. PEREZ-VALAZQUEZ, J. L., A. VELUMIAN, P. L. CARLEN, J. F. BECHBERGER, C. C. G. NAUS, M. FRANTSEVA & L. R. MILLS. 1996. The development of astrocytes and neurons in organotypic brain slices lacking connexin 43. NeuroReport.
36. WHITE, R. J. & I. J. REYNOLDS. 1995. Mitochondria and Na^+/Ca^{2+} exchange buffer glutamate induced calcium loads in cultured cortical neurons. J. Neurosci. **15:** 1318–1326.
37. KIEDROWSKI, L., G. BROOKER, E. COSTA & J. T. WROBLEWSKI. 1994. Glutamate impairs neuronal calcium extrusion while reducing sodium gradient. Neuron **12:** 295–300.
38. MILLS, L. R. & R. G. KERR. 1995. Chronic alcohol exposure reduces neuronal survival and potentiates glutamate toxicity in cultured rat fetal hippocampal neurons. Res. Soc. Alchol. Abstr.
39. BLAUSTEIN, M., W. F. GOLDMAN, G. FONTANA, B. KRUEGER & P. YAROWSKY. 1991. Physiological roles of the sodium-calcium exchanger in nerve and muscle. Ann. N. Y. Acad. Sci. **639:** 254.

Release of Catecholamines and Enkephalin Peptides Induced by Reversal of the Na^+-Ca^{2+} Exchanger in Chromaffin Cells[a]

E. P. DUARTE, G. BALTAZAR, P. VERÍSSIMO,
AND A. P. CARVALHO[b]

*Center for Neurosciences of Coimbra
and
Department of Zoology
University of Coimbra
3049 Coimbra Codex, Portugal*

It is well accepted that neurotransmitter release is triggered by the elevation of intracellular free Ca^{2+} concentration ($[Ca^{2+}]_i$) brought about by the arrival of the action potential. Membrane depolarization is required to open up voltage-gated Ca^{2+} channels, but a more direct function of membrane potential in secretion, such as a facilitation of the fusion event, has also been proposed.[1] In neurons, the exocytotic release of aminoacid neurotransmitters stored in small synaptic vesicles requires Ca^{2+} entry through voltage-gated channels,[2,3] whereas the release of neuropeptides, stored in large dense-core vesicles, can be induced in the absence of membrane depolarization by other means of raising $[Ca^{2+}]_i$, such as with Ca^{2+} ionophores.[3]

In this study we compared the effectiveness of Ca^{2+} entering by Na^+-Ca^{2+} exchange with that of Ca^{2+} entering through voltage-gated channels in stimulating the release of noradrenaline, adrenaline and [Leu^5]enkephalin (LEnk) from adrenal chromaffin cells in culture. In these cells, both catecholamines and neuropeptides are stored in chromaffin granules which are homologous to the large dense-core vesicles.[4]

Elevation of the $[Ca^{2+}]_i$ by Reversal of the Na^+-Ca^{2+} Exchanger

The influx of Ca^{2+} by the Na^+-Ca^{2+} exchanger was promoted by reversing the normal inward gradient of Na^+ by first incubating the cells with ouabain to increase intracellular Na^+, followed by using Na^+-free external media (substituted by N-methyl-D-glucamine, NMG^+, or choline, Ch^+). The changes in $[Ca^{2+}]_i$ caused by the reversal of the Na^+ gradient were measured upon the addition of Ca^{2+} to fura-2 loaded cells diluted in Na^+-free, low Ca^{2+} media (5–10 µM free Ca^{2+}, buffered with 0.5 mM EGTA).[5] The initial low level of Ca^{2+} in external media saturated the probe which may have leaked out the cells after the loading, and avoided the Ca^{2+} influx which is known to occur upon Ca^{2+} reintroduction following cell

---

[a] Supported by JNICT.
[b] Corresponding author.

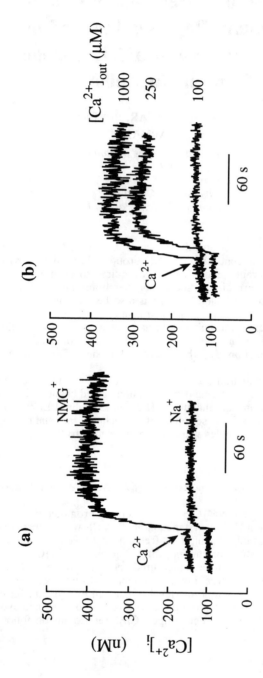

FIGURE 1. Raising intracellular Ca^{2+} concentration ($[Ca^{2+}]_i$) by reversal of the Na^+-Ca^{2+} exchanger. The $[Ca^{2+}]_i$ was determined in cell suspensions with the fluorescent indicator fura-2, as described previously.[5] Chromaffin cells loaded with fura-2 were incubated for 60 min with 50 μM ouabain in Na^+ medium. Aliquots of the cells (50 μl) were then added to 2 ml of low Ca^{2+} (5–10 μM free concentration) NMG^+ medium or Na^+ medium into the cuvette of the spectrofluorimeter. After 2 min of stabilization, $CaCl_2$ was added to achieve 2 mM free Ca^{2+} concentration **(a)**, or the indicated Ca^{2+} concentrations **(b)**. The traces shown are representative of many experiments carried out in several cell preparations.

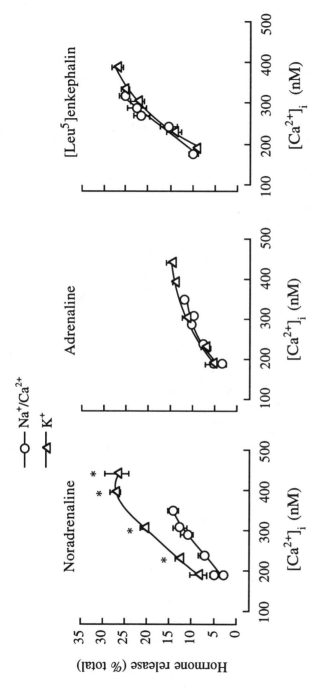

FIGURE 2. Relationship between catecholamine or enkephalin release from chromaffin cells and the average $[Ca^{2+}]_i$ upon reversal of the Na^+-Ca^{2+} exchange or depolarization with 35 mM K^+. Hormone release was measured during 3 min in NMG^+ medium or Ch^+ medium (Na^+-Ca^{2+} exchange), or in Na^+ medium plus 35 mM K^+ (K^+ depolarization) with 100, 250, 500, 1000 or 2000 μM free Ca^{2+}, in cells preincubated with 50 μM ouabain for 60 min, and washed once with medium, either NMG^+, Ch^+, or Na^+, containing the same free Ca^{2+} concentrations. Noradrenaline and adrenaline were analyzed by reversed phase ion-pair HPLC with electrochemical detection.[5] [Leu^5]enkephalin was determined by radioimmunoassay with a commercial antiserum (Peninsula Laboratories). The values of $[Ca^{2+}]_i$ were determined in experiments as those described in FIGURE 1. The data are the means ± SEM of at least three independent experiments performed in triplicate or quadruplicate. * Significantly different from release induced by Na^+-Ca^{2+} exchange ($p < 0.05$) as determined with Student t test.

incubation in Ca^{2+}-free media. Chromaffin cells preincubated with 50 μM ouabain for 60 min undergo an increase in $[Ca^{2+}]_i$ from 130 ± 6 nM to 325 ± 14 nM (n = 15) when placed in NMG^+ or Ch^+ media containing 2 mM Ca^{2+}, but not when placed in Na^+ medium (FIG. 1a), suggesting an influx of Ca^{2+} in exchange for intracellular Na^+ through the Na^+-Ca^{2+} exchanger. The increase in $[Ca^{2+}]_i$ due to the reversal of the Na^+-Ca^{2+} exchanger was of the same order of magnitude as the one observed upon depolarization with 35 mM K^+ (389 ± 31 nM, n = 12).[5]

Effectiveness of Ca^{2+} Entering by the Na^+-Ca^{2+} Exchanger in Inducing the Release of Catecholamines and Enkephalin Peptides from Chromaffin Cells

In order to compare the effectiveness of Ca^{2+} entering by the Na^+-Ca^{2+} exchanger with that of Ca^{2+} entering by channels in stimulating secretion, the $[Ca^{2+}]_i$ was raised to different levels by stimulating the cells, either by reversal of the Na^+-Ca^{2+} exchanger or by K^+ depolarization, in external media with increasing free Ca^{2+} concentrations in the range of 100 to 2000 μM, buffered with 0.5 mM EGTA (FIG. 1b). Catecholamine and LEnk release were measured in the same conditions, and the relationship between the $[Ca^{2+}]_i$ and secretion was evaluated (FIG. 2). We found, for a given value of average $[Ca^{2+}]_i$, a much higher release of noradrenaline when $[Ca^{2+}]_i$ was raised by depolarization than by Na^+-Ca^{2+} exchange. In contrast, similar amounts of adrenaline or of LEnk were released from chromaffin cells when the $[Ca^{2+}]_i$ was raised to the same value by Na^+-Ca^{2+} exchange or K^+ depolarization, indicating that Ca^{2+} entering by the Na^+-Ca^{2+} exchanger is as efficient as Ca^{2+} entering by voltage-gated channels in inducing the release of adrenaline or LEnk. The ratio of noradrenaline to adrenaline release was 1.24 ± 0.23 (n = 23) upon reversal of the Na^+-Ca^{2+} exchange, whereas it was 1.83 ± 0.19 (n = 18) for K^+ depolarization.[5]

Since the measurements of $[Ca^{2+}]_i$ with fura-2 in cell suspensions report only average values of the free $[Ca^{2+}]$ in the cytoplasm, and do not give information about the larger and localized increases in Ca^{2+} in the vicinity of the membrane,[6] the results may suggest that there is a pool of noradrenaline vesicles whose release requires localized increases in intracellular Ca^{2+} in the vicinity of Ca^{2+} channels, whereas the vesicles storing adrenaline and enkephalin peptides can be released outside these domains of high Ca^{2+}.

REFERENCES

1. HOCHNER, B., H. PARNAS & I. PARNAS. 1989. Nature **342:** 433–435.
2. CARVALHO, C. M., C. BANDEIRA-DUARTE, I. L. FERREIRA & A. P. CARVALHO. 1991. Neurochem. Res. **16:** 763–772.
3. VERHAGE, M., H. T. MCMAHON, W. E. J. M. GHIJSEN, F. BOOMSMA, G. SCHOLTEN, V. M. WIEGANT & D. G. NICHOLLS. 1991. Neuron **6:** 517–524.
4. P. DE CAMILLI & R. JAHN. 1990. Annu. Rev. Physiol. **52:** 625–645.
5. DUARTE, E. P., G. BALTAZAR & A. P. CARVALHO. 1994. Eur. J. Neurosci. **6:** 1128–1135.
6. LLINÁS, R., M. SUGIMORI & R. B. SILVER. 1992. Science **256:** 677–679.

Agents That Promote Protein Phosphorylation Increase Catecholamine Secretion and Inhibit the Activity of the Na^+-Ca^{2+} Exchanger in Bovine Chromaffin Cells

L. F. LIN, L-S. KAO,[a] AND E. W. WESTHEAD[b]

[a]*Institute for Biomedical Science*
Academia Sinica
Taiwan
and
Program in Molecular and Cellular Biology
University of Massachusetts
Amherst, Massachusetts 01003

Repeated stimulation of cultured cells from bovine adrenal medulla (chromaffin cells) usually results in steadily decreasing secretory response. We have found that agents that stimulate protein kinases or that inhibit phosphoprotein phosphatases will decrease the rate of desensitization during repeated stimulation by either a nicotinic agonist or by a depolarizing concentration of KCl.[1] An inhibitor of kinases, H7, causes an increase in the desensitization rate. The combination of okadaic acid (an inhibitor of phosphoprotein phosphatase) and 8-Br-cAMP to stimulate protein kinase A (PKA), produces an *increasing* secretory response to repeated stimulation, suggesting that a phosphoprotein is a key protein in the stimulation-secretion pathway and that dephosphorylation of this protein during stimulation lowers response to a subsequent stimulation.

Preliminary data suggested that the rise in cytosolic Ca^{2+} accompanying stimulation is reduced during repeated stimulation. Since the Na^+-Ca^{2+} exchanger is an important element in the maintenance of calcium homeostasis in chromaffin cells,[2,3] we have investigated the role of reversible protein phosphorylation in the activity of the Na^+-Ca^{2+} exchanger of these cells.[4] Cells treated with 1 mM dibutyryl cyclic AMP (dbcAMP), 1 μM phorbol 12, 13-dibutyrate, 1 μM okadaic acid, or 100 nM calyculin A showed lowered Na^+-Ca^{2+} exchange activity and prolonged cytosolic Ca^{2+} transients caused by depolarization. A combination of 10 nM okadaic acid and 1 μM dbcAMP synergistically inhibited Na^+-Ca^{2+} exchange activity. Conversely, 50 μM H7, a protein kinase inhibitor, enhanced Na^+-Ca^{2+} exchange activity. Moreover, we used cyclic AMP-dependent protein kinase and calcium phospholipid-dependent protein kinase catalytic subunits to phosphorylate isolated plasma membrane vesicles and found that the Na^+-Ca^{2+} exchange activity is strongly inhibited by this treatment. These results indicate that reversible protein phosphorylation modulates the activity of the Na^+-Ca^{2+} exchanger and suggest that modulation of the exchanger may play a role in the regulation of secretion.

---

[b] Corresponding author.

The model that we envisage is that on the plasma membrane there are clusters of exocytotic sites, voltage-dependent calcium channels, and sodium-calcium exchangers in close proximity. The secretory response seems to be dependent on a high local calcium concentration near the plasma membrane, rather than on a general cytosolic calcium rise,[5] so activity of the exchanger could be a critical element in modulating the high submembrane calcium transient. The entry of calcium can activate both kinases and protein phosphatases to account for the stimulation-dependent changes in secretory response.

REFERENCES

1. LIN, L. F., K. T. KIM & E. W. WESTHEAD. 1993. Protein phosphorylation at a postreceptor site can block desensitization and induce potentiation of secretion in chromaffin cells. J. Neurochem. **60:** 1491–1497.
2. CHERN, Y. J, S. H. CHUEH, Y. J. LIN, C. M. HO & L-S. KAO. 1992. Presence of Na^+/Ca^{2+} exchange activity and its role in regulation of intracellular calcium concentration in bovine adrenal chromaffin cells. Cell Calcium **31:** 199–106.
3. KAO, L-S. & N-S. CHEUNG. 1990. Mechanism of calcium transport across the plasma membrane of bovine chromaffin cells. J. Neurochem. **54:** 1972–1979.
4. LIN, L. F., L-S. KAO & E. W. WESTHEAD. 1994. Agents that promote protein phosphorylation inhibit the activity of the Na^+/Ca^{2+} exchanger and prolong Ca^{2+} transients in bovine chromaffin cells. J. Neurochem. **64:** 1941–1947.
5. KIM, K. T. & E. W. WESTHEAD. 1989. Cellular responses to Ca^{2+} from extracellular and intracellular sources are different as shown by simultaneous measurements of cytosolic Ca^{2+} and secretion from bovine chromaffin cells. Proc. Natl. Acad. Sci. USA **86:** 9881–9885.

What Mechanisms Are Involved in Ca^{2+} Homeostasis in Hair Cells?

CHRISTIAN CHABBERT,[a] A. SANS,
AND J. LEHOUELLEUR

INSERM U.432
Laboratoire de Neurophysiologie Sensorielle
34095 Montpellier cedex 5, France

INTRODUCTION

In the inner ear, hair cells are responsible in the cochlea for our sensitivity to sounds, in the sacculus and utriculus for our perception of linear accelerations, and in the semicircular canals for our responsiveness to rotatory accelerations. Hair cells are wonderful tools able to detect mechanical stimuli and to transduce them into electrical signals they forward to the brain via afferent chemical synapses. Deflection of their hair bundles open mechanosensitive ionic channels and induce an influx of cations from the endolymph, an extracellular fluid rich in K^+ that bathes the apical surface of the hair cells. This cationic influx depolarizes the cells and effects afferent synaptic transmission by opening voltage-dependent Ca^{2+} channels localized at the presynaptic active zones (for review see Ref. 1).

Ca^{2+} ions are involved in the mechanotransduction process and in the regulation of hair cell activity. In the hair bundle, Ca^{2+} mediates the closure of some transduction channels upon sustained stimulation. This adaptative process allows the cell to continuously reset the point of maximum sensitivity.[2] At the presynaptic active zone, Ca^{2+} ions mediate the release of neurotransmitter and the opening of Ca^{2+}-sensitive K^+ channels involved in frequency tuning.[3,4] At the efferent synapse, the release of acetylcholine increases the cytosolic Ca^{2+} concentration ($[Ca^{2+}]_i$), which in turn induces a hyperpolarization by opening Ca^{2+}-sensitive K^+ channels.[5]

Presence of Mobile Cytosolic Ca^{2+} Buffer

Mechanosensory adaptation, afferent synaptic transmission, frequency tuning, and efferent modulation all require fast and transient variations in $[Ca^{2+}]_i$ in limited cytosolic domains. Such localized $[Ca^{2+}]_i$ variations are the result of focal Ca^{2+} entry across the plasma membrane and strong Ca^{2+} buffering activity in the adjacent cytoplasm. The presence of mobile cytosolic Ca^{2+} buffers, such as Ca^{2+}-binding proteins, has been suggested in frog saccular hair cells to prevent the spread of Ca^{2+} in the cytosol, avoiding cross-talk between these processes.[6]

Requirement of Ca^{2+} Extrusion Mechanisms

The Ca^{2+} entry through the pathways discussed above must be balanced by active extrusion across the plasma membrane. In many cellular systems this

---

[a] Address for correspondence: Laboratory of Sensory Neuroscience, The Rockefeller University, Box 314, 1230 York Ave., New York, NY 10021.

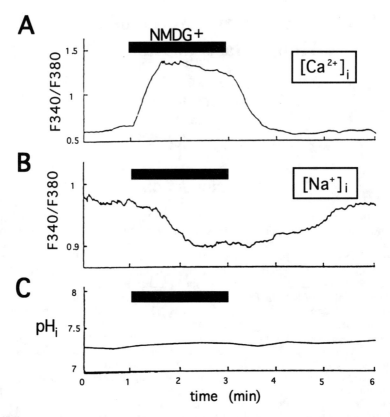

FIGURE 1. Effect of complete removal of external Na^+ ions on $[Ca^{2+}]_i$, $[Na^+]_i$, and pH_i. Replacement of external Na^+ ions by N-methyl-D-glucamine (black bar) increased $[Ca^{2+}]_i$ by 276 ± 89% (n = 46). A representative trace is shown in **(A)**. Similar removal of external Na^+ ions decreased $[Na^+]_i$ by 26 ± 5% (n = 14); a representative trace is shown in **(B)**. The superfusion of Na^+-free solution induced no pH_i variation **(C)**, while control superfusions of acetoxy dicyanobenzol-loaded cells with 25 mM NH_4Cl or Hank's balanced salt solution equilibrated with 5% CO_2 and 95% O_2 induced a pH_i increase and decrease respectively (not shown). Because of the uncertainty in estimating absolute $[Ca^{2+}]_i$, we chose to present fluorescence ratio variations rather than absolute $[Ca^{2+}]_i$ variations. (Adapted from Chabbert et al.[8])

process is mediated by Ca^{2+} pumps or membrane ionic exchangers. While nothing is known about Ca^{2+} pumps in hair cells, the presence of a Na^+-Ca^{2+} exchanger has been detected in cochlear outer hair cells by microfluorimetry.[7] A similar exchanger was more recently detected in type I vestibular hair cells enzymatically isolated from the guinea pig's macula utriculi and crista ampullaris.[8] The effects of the removal of external Na^+ ions have been studied on the cytosolic concentrations of calcium ions ($[Ca^{2+}]_i$), sodium ions ($[Na^+]_i$), and on pH_i in type I hair cells loaded respectively with fura-2, sodium benzofuran isophthalate, and 1,4-diacetoxy-2,3-dicyanobenzol. Complete replacement of external Na^+ ions with

N-methyl-D-glucamine reversibly increased $[Ca^{2+}]_i$ and decreased $[Na^+]_i$ while pH_i remained unchanged (FIG. 1).

Both Ca^{2+} and Na^+ responses were prevented by removing external Ca^{2+} or chelating internal Ca^{2+} with 100 μM BAPTA. The $[Ca^{2+}]_i$ increase evoked by the removal of external Na^+ was reduced by about 55% with the application of 100 μM 5-(N,N-dimethyl) amiloride hydrochloride, an inhibitor of Na^+-coupled transporters. Elevation of the external concentration of Mg^{2+} ions, known to compete with Ca^{2+} ions for the external binding site of Na^+-coupled membrane exchangers, from 0.9 mM to 4 mM prevented both the $[Ca^{2+}]_i$ increase and the $[Na^+]_i$ decrease. In the absence of external K^+, the Na^+-free solution failed to induce a $[Ca^{2+}]_i$ increase, while the replacement of external K^+ restored the $[Ca^{2+}]_i$ response.

DISCUSSION

The $[Ca^{2+}]_i$ increase and $[Na^+]_i$ decrease triggered by external Na^+ removal suggest the presence of coupled Ca^{2+} and Na^+ transport in type I hair cells. Together with the pharmacological inhibition of the responses, these results are consistent with a Na^+-Ca^{2+} exchanger operating in reverse mode. The next step of the study will be the characterization of exchange currents in patch-clamped hair cells.

REFERENCES

1. HUDSPETH, A. J. 1989. Nature **341:** 397–404.
2. EATOCK, R. A., D. P. COREY & A. J. HUDSPETH. 1987. J, Neurosci. **9:** 2821–2836.
3. ROBERTS, W. M., R. A. JACOBS & A. J. HUDSPETH. 1990. J. Neurosci. **10:** 3664–3684.
4. HUDSPETH, A. J. & R. S. LEWIS. 1988. J. Physiol. (London) **400:** 237–274.
5. SHIGEMOTO, T. & H. OHMORI. 1991. J. Physiol. (London) **442:** 669–690.
6. ROBERTS, W. M. 1994. J. Neurosci. **14:** 3246–3262.
7. IKEDA, K., Y. SAITO, A. NISHIYAMA & T. TAKASAKA. 1992. Pflügers Arch. **420:** 493–499.
8. CHABBERT, C., Y. CANITROT, A. SANS & J. LEHOUELLEUR. 1995. Hearing Res. In press.

Immunohistochemical Localization of the Cardiac Sodium-Calcium Exchange Protein in the Inner Ear[a]

P. M. MANCINI[b] AND P. A. SANTI[c,d]

Department of Otolaryngology
[b]*University of Rome "La Sapienza"*
Rome, Italy

[c]*University of Minnesota*
Minneapolis, Minnesota 55455

The mechanisms involved in ion transport and fluid regulation in the inner ear are poorly understood. However, it is well known that the apical surface of the hair cells are bathed in a potassium-rich and sodium-poor fluid called endolymph (FIG. 1A). In addition, this fluid carries a positive standing current that is responsible for the extreme sensitivity of the auditory organ. In the cochlea, endolymph and its electrochemical potential are thought to be produced and maintained by the marginal cells of the stria vascularis. The dark cells appear to be homologous structures in the vestibular system. Several different ion-transporting proteins have been identified[1] on the basolateral membranes of these cells using immunochemical and electrophysiological methods (FIG. 1B), including: sodium/potassium-ATPase, sodium/potassium/chloride cotransporter and chloride channels. Relatively few ion-transporting proteins have been identified on the apical surface of these cells including: a minimum potassium channel and, with the present study, a sodium/calcium exchange protein.

Using antibodies specific for a cardiac sodium/calcium exchanger[2] the protein was immunolocalized in inner ear tissues of the chinchilla. As a positive control, antibodies to the exchanger protein were applied to chinchilla cardiac tissue. The expected pattern of antibody reactivity[3] to the sarcolemma and transverse tubules of the myocytes was observed (FIG. 2A). This confirmed cross-reactivity of the antibodies with chinchilla epitopes and the reliability of our immunohistochemical procedure. In the cochlea, antibodies were reactive to the apical or endolymphatic surface of the marginal cells of the stria vascularis and to the cytoplasm of Hensen's cells of the organ of Corti (FIG. 2B). In the vestibular tissues, the antibodies were reactive to the cytoplasm of the dark cells and transitional epithelial cells of the crista ampularis (FIG. 2C). Immunolocalization of a sodium-calcium exchange protein to the primary transporting epithelial cells (marginal and dark cells) of the inner ear suggest that it has a functional role in sodium-calcium transport in endolymph. However, without functional data its precise role cannot be determined. It is likely that it is one of several calcium-regulating proteins involved in fluid balance between the endolymphatic/perilymphatic compartments. By obtain-

---

[a] This research was supported by a grant from the National Institute on Deafness and Other Communication Disorders (NIDCD).

[d] Corresponding author: Rm. 282, Lions Research Building, Dept. of Otolaryngology, University of Minnesota, Minneapolis, MN 55455. E-mail: Santip@maroon.tc.umn.edu.

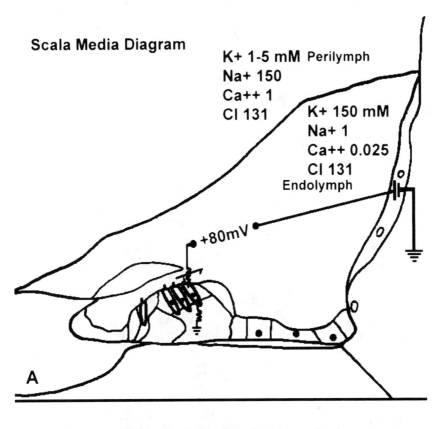

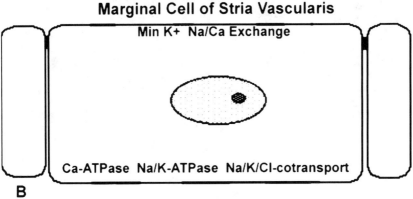

FIGURE 1. (A) Structural and electrochemical features of the scala media of the mammalian cochlea. The epithelium of the stria vascularis is involved in the production of the low sodium/high potassium ion concentration of endolymph and an +80 mV endocochlear potential. This standing current provides the driving force for electromechanical auditory transduction through the hair cells. (B) Location of the ion-transporting proteins associated with the marginal cells of the stria vascularis.

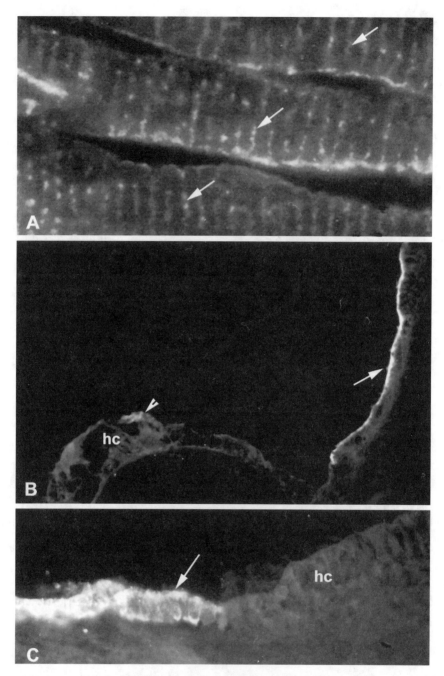

FIGURE 2. (A) By using monoclonal antibody C-2C12[3] and a FITC-conjugated secondary antibody the exchanger protein was immunolocalized on chinchilla cardiac myocyte plasma membrane and transverse tubules (*arrows*). (B) In the cochlea the exchange protein was localized to the apical endolymphatic surface of the marginal cells of the stria vascularis (*arrow*) and to the cytoplasm of Hensen's cells (*arrowhead*) of the organ of Corti. The hair cells (hc) were unreactive. (C) The dark cells and transitional epithelial cells of the crista ampularis in the vestibular system appear to contain the sodium-calcium exchange protein in the cytoplasm of these cells (*arrow*). The hair cells (hc) were unreactive.

ing a more complete knowledge of the molecular composition of the inner ear we will have a better understanding of normal auditory function and the mechanisms involved in certain inner ear diseases, such as Meniere's, that appear to be caused by fluid/ion imbalances.

ACKNOWLEDGMENTS

The authors would like to thank Dr. Kenneth D. Philipson for generously providing antibodies to the cardiac sodium-calcium exchange protein.

REFERENCES

1. WANGEMANN, P., J. LIU & D. C. MARCUS. 1995. Hearing Res. **84:** 19–29.
2. NICOLL, D. A., S. LONGONI & K. D. PHILIPSON. 1990. Science **250:** 562–565.
3. FRANK, J. S., G. MOTTINO, D. REID, R. S. MOLDAY & K. D. PHILIPSON. 1992. J. Cell Biol. **117:** 337–345.

Reversed Mode Na^+-Ca^{2+} Exchange Activated by Ciguatoxin (CTX-1b) Enhances Acetylcholine Release from *Torpedo* Cholinergic Synaptosomes[a]

YVETTE MOROT GAUDRY-TALARMAIN, JORDI MOLGO,
FREDERIC A. MEUNIER, NATHALIE MOULIAN,[b] AND
ANNE-MARIE LEGRAND[c]

Laboratoire de Neurobiologie Cellulaire et Moléculaire
Centre National de la Recherche Scientifique
91198-Gif sur Yvette cedex, France

Ciguatoxin-1b (CTX-1b) is a potent cyclic polyether compound involved in a widespread human seafood poisoning known as ciguatera which develops after consumption of coral reef fish.[1] CTX-1b selectively acts on voltage-sensitive Na^+ channels, in such a way that they are activated at the resting membrane potential, causing tetrodotoxin-sensitive membrane depolarization and repetitive or spontaneous action potentials in excitable cells.[2-4] CTX-1b and brevetoxin-3 bind to site 5 of the neuronal voltage-dependent Na^+ channel protein.[5,6] Numerous physiological effects due to membrane depolarization and increased neurotransmitter release induced by CTX-1b have been reported.[7] However, the mechanisms whereby CTX-1b increases neurotransmitter release are still poorly understood.[7]

In the present study we used pure cholinergic synaptosomes, isolated from the electric organ of *Torpedo marmorata*,[8,9] and a chemiluminescent method for continuous acetylcholine (ACh) detection[9] to investigate the possible involvement of the Na^+-Ca^{2+} exchange system during the action of CTX-1b.

When 10 nM CTX-1b (extracted from poisonous moray eels *Gymnothorax javanicus*, as previously described[1]) was added to a diluted suspension of *Torpedo* synaptosomes incubated in a nominally Ca^{2+}-free medium, no detectable change in ACh release occurred for up to 15 min (FIG. 1A). However, subsequent addition of Ca^{2+} (4 mM) caused, in the continuous presence of CTX-1b, a large ACh release that corresponded to about 20% of the total synaptosomal content (FIG. 1F). CTX-1b-induced ACh release depended on time of exposure in Na^+-containing Ca^{2+}-free medium and on the $[Ca^{2+}]$ used to trigger ACh release.[10]

Tetrodotoxin (TTX) (1 μM) when applied before CTX-1b to synaptosomes kept in a nominally Ca^{2+}-free medium completely prevented CTX-1b-induced ACh release upon Ca^{2+} addition (FIG. 1C). To study further the Na^+ requirement for CTX-1b-induced ACh release $[Na^+]_e$ was replaced, on an equimolar basis, by Li^+.

---

[a] This research was supported, in part, by Grants 92/175 and 94/067 from Direction des Recherches Etudes et Techniques and by an International Cooperation Program of the European Community (CI1 CT 94-0129).

[b] Supported by a fellowship from Laboratoires Servier.

[c] Permanent address: Institut Territorial de Recherches Médicales Louis Malardé, Tahiti, Polynésie Française.

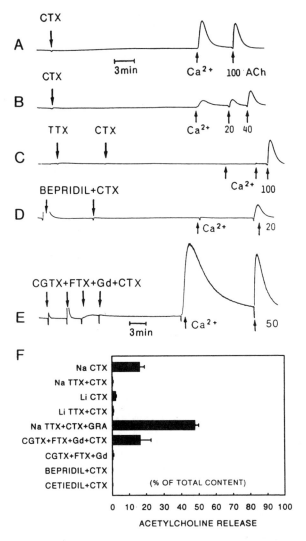

FIGURE 1. (A–E) Effects of 10 nM CTX-1b on ACh release from different synaptosomal preparations incubated in Ca²-free medium for 15 min either in Na^+- (A, C, D, E) or in Li^+-containing medium (B). ACh release was triggered by Ca^{2+} (4 mM) addition to the medium. Notice that pretreatment with TTX (1 μM) (C) and 25 μM bepridil (D) completely prevented the action of CTX-1b on ACh release. Simultaneous blockade of Ca^{2+} channel subtypes by, respectively, 2 μM omega-conotoxin GVIA (CGTX), FTX (1:100,000) and 0.25 mM Gd^{3+} neither prevented nor affected ACh release caused by CTX-1b upon addition of Ca^{2+} (E). *Arrows* indicate addition of toxins, drugs, ACh-standards and Ca^{2+} to the medium. *Numbers* correspond to standard amounts of ACh (pmol) used to calibrate ACh release. *Time calibration* in (A) applied to (A–D). **(F)** The diagram summarizes the results obtained in synaptosomal preparations (n = 4–19) under conditions described in A–E. Results are the mean ± SEM of the amount of ACh released expressed as % of total synaptosomal ACh content. Note that the monovalent ion-ionophore gramicidin-D (0.5 μM) (GRA) could still induce ACh release in the presence of TTX + CTX upon addition of Ca^{2+}.

Under this condition, CTX-1b had no detectable action on ACh release in the Ca^{2+}-free medium (FIG. 1B, F) and triggered much less ACh release upon Ca^{2+} addition (<2% of the total content), indicating that Li^+ can poorly substitute for Na^+. To obviate the possibility that depolarization induced by CTX-1b would activate Ca^{2+} channels and thereby contribute to Ca^{2+}-dependent ACh release caused by the toxin, synaptosomal Ca^{2+} channel subtypes[11] were blocked with omega-conotoxin GVIA (CGTX), FTX, a toxin extracted from the venom of the spider *Agelenopsis aperta*[11] (kindly provided by Dr. R. Llinás) and Gd^{3+}. As shown in FIGURE 1E, F, these agents did not prevent Ca^{2+}-dependent ACh release caused by CTX-1. Thus, it is likely that CTX-1b acts by increasing Na^+ levels which would then enhance Ca^{2+} influx through the reversed operation of the Na^+-Ca^{2+} exchange system. The fact that CTX-1b in Li^+-containing medium caused only a small ACh release is consistent with data showing that Li^+ cannot replace Na^+ in the Na^+-Ca^{2+} exchange process.[12] Finally, several inhibitors of Na^+-Ca^{2+} exchange were tested to verify that Ca^{2+}-dependent ACh release was triggered by the reversed operation of this system. As shown in FIG. 1D, F, bepridil[13] (25 μM) completely prevented Ca^{2+}-dependent ACh release induced by CTX-1b. Similar effects were obtained with 25 μM cetiedil (FIG. 1F) which inhibits activated Na^+-Ca^{2+} exchange in dog erythrocytes (Morot Gaudry-Talarmain, Parker, Calclosure and Orringer, unpublished data). We conclude that CTX-1b activates the reversed operation of the Na^+-Ca^{2+} exchange and allows Ca^{2+} entry into synaptosomes in exchange for Na^+ and, thereby, triggers Ca^{2+}-dependent ACh release.

ACKNOWLEDGMENTS

We thank Drs. M. Israël and S. O'Regan for cogent advice and critical reading of the manuscript.

REFERENCES

1. MURATA, M., A. M. LEGRAND, Y. ISHIBASHI, M. FUKUI & T. YASUMOTO. 1990. J. Am. Chem. Soc. **112:** 4380–4386.
2. BIDARD, J. N., H. P. M. VIJVERBERG, C. FRELIN, E. CHUNGUE, A. M. LEGRAND, R. BAGNIS & M. LAZDUNSKI. 1984. J. Biol. Chem. **259:** 8353–8357.
3. BENOIT, E., A. M. LEGRAND & J. M. DUBOIS. 1986. Toxicon **24:** 357–364.
4. MOLGO, J., J. X. COMELLA & A. M. LEGRAND. 1990. Br. J. Pharmacol. **99:** 695–700.
5. LOMBET, A., BIDARD, J. N. & M. LAZDUNSKI. 1987. FEBS Lett. **219:** 355–359.
6. LEWIS, R. J., M. SELLIN, M. A. POLI, R. S. NORTON, J. K. MACLEOD & M. M. SHEIL. 1991. Toxicon **29:** 1115–1127.
7. MOLGO, J., E. BENOIT, J. X. COMELLA & A. M. LEGRAND. 1992. Ciguatoxin: a tool for research on sodium-dependent mechanisms. *In* Methods in Neuroscience. Neurotoxins. P. M. Conn, Ed. Vol. 8: 149–164. Academic Press. San Diego, CA.
8. MOREL, N., M. ISRAEL, R. MANARANCHE & P. MASTOUR-FRANCHON. 1977. J. Cell Biol. **75:** 43–55.
9. ISRAEL, M. & B. LESBATS. 1981. J. Neurochem. **37:** 1476–1483.
10. MOLGO, J., Y. MOROT GAUDRY-TALARMAIN, A. M. LEGRAND & N. MOULIAN. 1993. Neurosci. Lett. **160:** 65–68.
11. MOULIAN, N. & Y. MOROT GAUDRY-TALARMAIN. 1993. Neuroscience **54:** 1035–1041.
12. TESSARI, M. & H. RAHAMIMOFF. 1991. Biochim. Biophys. Acta **1066:** 208–218.
13. GARCIA, M. L., R. S. SLAUGHTER, V. F. KING & G. J. KACZOROWSKI. 1988. Biochemistry **27:** 2410–2415.

Part IV
Sodium-Calcium Exchange in the Cardiovascular System
Introduction

The heart has long been a popular model for study of *Sodium-Calcium Exchange in the Cardiovascular System*. Much emphasis at the conference was placed on the controversial role of Na^+-Ca^{2+} exchange as a Ca^{2+} influx mechanism in the cardiac tissue. Investigators agree that Na^+-Ca^{2+} exchange is the dominant Ca^{2+} efflux mechanism of the myocardium. Experiments on the role of exchange as a Ca^{2+} trigger mechanism, however, have had conflicting results. Clearly, the exchanger can induce Ca^{2+} release from the sarcoplasmic reticulum under non-physiologic conditions, but the physiologic significance is unclear. Large intracellular Ca^{2+} gradients can exist in myocytes due to the geometry of the diadic cleft region. Modeling of Ca^{2+} fluxes in the cleft region indicate that Ca^{2+} gradients are an integral part of the excitation-contraction coupling process. Possibly, localization of the exchanger in specific domains of the sarcolemma influences the excitation-contraction coupling process. Pharmacological and other interventions have been used to separate the relative roles of sarcoplasmic reticular and sarcolemmal Ca^{2+} fluxes in the contractile process. These experiments indicate that about 20% of the coupling Ca^{2+} is extruded from the cell by the exchanger following each contraction.

Antibodies and DNA probes for the exchanger have been used to examine exchanger levels in different situations. Strikingly, the exchanger is upregulated in the myocardium during the fetal and neonatal periods. Likewise, exchanger transcript levels are elevated during the initial phase of induced hypertrophy. The fate of the exchanger during other myopathies is also being examined.

Calcium in the Cardiac Diadic Cleft

Implications for Sodium-Calcium Exchange

G. A. LANGER AND A. PESKOFF

Departments of Medicine and Physiology
The Cardiovascular Research Laboratory
and
Department of Biomathematics
University of California, Los Angeles School of Medicine
Los Angeles, California 90095

This paper will focus on the work done at the UCLA Cardiovascular Research Laboratory over the past three years directed toward further clarification of subcellular calcium (Ca) movements in the heart, with emphasis on the role played by Na-Ca exchange in these movements.

BACKGROUND

In a cooperative UCLA study with the Institute of Biomembranes at the University of Utrecht, The Netherlands, we defined the constituency and symmetry of the sarcolemmal phospholipids in cultured cardiac cells from neonatal rat ventricle.[1] This remains the only quantitative description of phospholipid asymmetry in heart muscle thus far. It showed that 100% of the anionic phospholipids, phosphatidylserine (PS) and phosphatidylinositol (PI), and 75% of zwitterionic phosphatidylethanolamine (PE) are localized to the inner or cytoplasmic leaflet of the sarcolemma. These three phospholipids are those capable of binding Ca in the membrane. Preliminary evidence indicated, however, that this binding was at a very low affinity (see below). If these sites were, then, in equilibrium with general cytoplasmic Ca concentrations ranging between 0.1 and 1 μM they would bind trivial amounts of Ca and play little, if any, role in control of cellular Ca movements. For no other reason than teleological this seemed unlikely and prompted us to launch a long-term program to define cellular Ca compartmentation with particular attention to sarcolemmal Ca binding.

Calcium Compartmentation

Using newly developed nonperfusion limited ^{45}Ca labeling and washout techniques we kinetically defined three Ca compartments in both adult and cultured neonatal heart cells:[2,3] a fast (t½ exchange <1 sec), a biphasic intermediate (t½ 4 and 19 sec), and a slow (t½ 3–4 min) compartment. Using specific probes the intermediate was localized to the sarcoplasmic reticulum and the slow to the mitochondria. The cellular origin of the fast compartment, accounting for >40% of the cell's exchangeable Ca was not specifically localized but found to be largely lanthanum (La) displaceable. This placed this large, rapidly exchangeable pool at the sarcolemma and/or at sites in rapid equilibrium with the sarcolemma. This prompted us to investigate further the sarcolemmal binding sites.

Sarcolemmal Ca Binding

Using the unique "gas-dissection" technique for instantaneous preparation of highly purified sarcolemmal membrane in high yield from intact, beating cultured cells,[4] we characterized sarcolemmal Ca binding.[5] Two classes of binding sites were determined: (1) K_d = 13 μM, capacity of 7 nmol Ca/mg sarcolemmal protein; (2) K_d = 1.1 mM, capacity 84 nmol/mg. Extraction of the membrane's lipid or treatment with phospholipase C produced almost complete removal of the low-affinity sites (K_d = 1.1 mM). Using the values from the study on phospholipid asymmetry it was calculated that the inner sarcolemmal leaflet phospholipid (PS, PI, PE) could, at saturation, bind >3 mmoles Ca/kg dry wt cell (700 μmols/kg wet wt) or >40% of cellular exchangeable Ca. This is consistent with most of the cell's fast exchangeable Ca ($t^{1/2}$ <1 sec) being localized to phospholipid sites at the inner sarcolemmal leaflet. However, because of the high K_d of the lipids it is required that they are in contact with a subcellular compartment in which Ca is at a much higher concentration than in the general cytosol. This led us to a detailed examination of the region between the junctional SR (JSR) and the inner leaflet of the sarcolemma—the diadic cleft region.

The Diadic Cleft

FIGURE 1 is a schematic of the cleft region. This is the region that contains the "feet" structures or ryanodine receptors,[6] which are the sites from which Ca is released from the JSR. We also localize the Na-Ca exchangers to the diadic cleft on the basis of the study by Frank et al.[7] using fluorescently-labeled monoclonal antibody to the exchanger. The cleft is modeled as a cylindrical space 0.2 μm in radius and, on the basis of the "foot" dimensions, 12 nm in height.[6] If such a junction is placed between two half-sarcomeres in the T tubes, the cleft area to T tube area ratio is 0.47—essentially the same as the ratio measured morphometrically by Page[8] for the rat ventricular cell.

Early modeling of this space[9] produced a surprising result. The average Ca concentration in the cleft was calculated following release from the feet of an amount of Ca sufficient to produce maximal contractile force as reported by Fabiato.[10] 8.8×10^{-20} mol Ca/μm^3 cell water was released at a constant rate over a period of 20 msec. FIGURE 2 shows the result for two conditions—without and with phospholipid anionic sites at the inner sarcolemmal leaflets. Diffusion of Ca within the space was set at 20% the value commonly reported for free diffusion in an aqueous medium ($0.2 \times 5 \times 10^{-6}$ cm^2/sec = 1×10^{-6} cm^2/sec). Without the anionic phospholipid sites, Ca concentration rises to >300 μM within 1–2 msec. With cessation of release the concentration returns to the diastolic level of 100 nM within 1 msec (FIG. 2). Now phospholipid Ca-binding sites are placed at the sarcolemmal inner leaflet at the concentration and K_d(Ca) found experimentally as presented above.[5] The effect is dramatic. Concentration increases more slowly to a lower value of 300 μM over the 20 msec release time—still a marked increase. Upon cessation of release, however, Ca concentration falls very slowly such that it is still 10 μM 60 msec after release stops and is 1 μM over 150 msec after release stops. This demonstrates that though the binding sites have a low affinity (K_d =

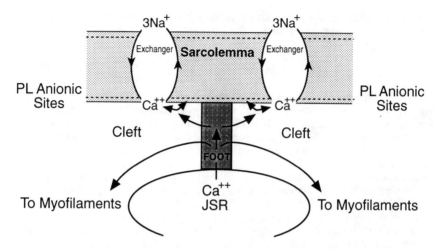

FIGURE 1. Cleft model. Ca is released from the "feet" into the cleft by the process of Ca-induced Ca release. This Ca can be transported out of the cleft via Na-Ca exchange or diffuse out to the myofilaments. The cleft region is modeled as a cylindrical space 0.2 μm in radius and, on the basis of "foot" height, 12 nm in height. The PL anionic sites at the inner sarcolemma are Ca-binding phosphatidylserine, inositol and ethanolamine molecules with $K_d(Ca) = 1.1$ mM at a concentration of 2.3×10^{-19} moles/cleft. The anionic sites, in equilibrium with a fraction of JSR Ca, are the proposed locus of the kinetically-defined Na-Ca exchange-dependent compartment (see text). (From Langer.[25] Reprinted by permission from *Trends in Cardiovascular Medicine*.)

1.1 mM) they produce a marked slowing of diffusion from the region when they are present in high amount. This finding prompted us to consider the effect it would have on Na-Ca exchangers proposed to be localized to the cleft space.

Function of the Na-Ca Exchanger

In order that the Ca concentration in the cell may achieve a steady level, it is necessary that the Na-Ca exchanger remove an amount of Ca equal to that which enters the cell via the "L" channel during each cardiac cycle. At $[Ca]_o = 1$ mM a rat ventricular cell has an influx of approximately 1.5×10^{-16} moles Ca via its "L" channels during a single excitation.[11] *In vivo* minimum heart rate for rats is 300/min[12] giving a channel Ca flux of 4.5×10^{-14} moles/min.

The maximum density and turnover rates for the Na-Ca exchanger are those reported by Hilgemann *et al.*[13] at 400 exchangers per μm² sarcolemma and 5000 cycles/sec. The total membrane surface area of a mature rat cell is approximately 6100 μm², including "T" tubular area.[14] We will first assume homogeneous distribution of the exchangers over the sarcolemmal surface with no special compartmentation of Ca. The exchanger would be exposed to approximately 100 nM $[Ca]_i$ during the diastolic period of the cardiac cycle and, at 1 mM $[Ca]_o$ (generating about 30% max force in the rat[11]), the peak transient $[Ca]_i$ during systole is not more than 1 μM with a mean $[Ca]_i$ of about 0.5 μm during systole.[15] At 300 beats/min approximately 50% of the 200 msec cardiac cycle is occupied by activation and 50% by rest so that the average level of [Ca]$_i$ to which the exchanger would be exposed is certainly not more than 0.5 μM and the average membrane potential is approximately -60 mV during a cycle. At 0.5 μM [Ca]$_i$ and -60 mV (Na$_o$ = 150 mM, Na$_i$ = 0 mM) the inward exchanger current is about 3% of maximum,[16] indicating a turnover of 150 cycles/sec (0.03 × 5000) for each exchanger. The cell has 2.4 × 10^6 exchangers (6100 × 400) and therefore would produce 3.6 × 10^8 cycles/sec (2.4 × 10^6 · 1.5 × 10^2) or 2.16 × 10^{10} cycles/min. Dividing by Avogadro's number (6.02 × 10^{23}) this gives 3.6 × 10^{-14} mols Ca/min transported outward by the exchangers or 1.2 × 10^{-16} mols during each 200 msec cardiac cycle. This is only 80% of the influx value (see above) despite selection of values which would maximize net outward Ca flux in the cell. Clearly homogeneous distribution of the Na-Ca exchangers exposed to the level of [Ca]$_i$ in the general cytosol would leave the rat ventricular cell incapable of sustaining steady state Ca levels even at the lowest ventricular excitation rates.

The major reason that the exchangers do not move enough Ca out of the cell under the conditions above is that they are exposed to a low level (0.5 μM) Ca$_i$ relative to the K$_d$(Ca) of 5–6 μM reported for the exchanger.[17] This problem is obviated if the exchangers are localized to the cleft region (FIG. 1). The slowed diffusion of Ca from this region (FIG. 2) with the attendant prolonged high [Ca]$_i$ will optimize exchanger function. An example illustrates the point: Though the Frank et al.[7] study of exchanger distribution is not quantitative, it appears that the concentration of exchangers in the cleft could be at least twice (800/μm^2) that on the external sarcolemmal surface as measured by Hilgemann et al.[13]. The [Ca]$_i$ in the clefts reaches a maximum of 600 μM at end of the SR release and given the slowed diffusion would saturate the exchanger (K$_d$ = 5–6 μM) for most of the cycle (FIG. 2). During this time the inward exchanger current at the clefts would be 30% maximum.

The cleft area is 885 μm^2 per cell so that each cell would have a total of 7.1 × 10^5 exchangers in the cleft (8.0 × 10^2 · 8.85 × 10^2). Since the turnover rate is 30% maximum (1500 cycles/sec) the total exchanger turnover in the clefts is 1.07 × 10^9 cycles/sec (7.1 × 10^5 · 1.5 × 10^3) or 1.7 × 10^{-15} mols Ca/sec (dividing by Avogadro's number). Since at a heart rate of 300/min there are 5 cycles/sec the exchanger could move 3.4 × 10^{-16} mols of Ca out of the cell with each cycle. Since influx is 1.5 × 10^{-16} mols per cycle (see above) exchanger-mediated efflux

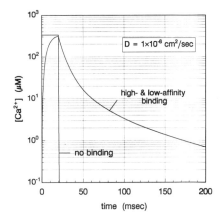

FIGURE 2. Ca diffusion within the cleft. Ca is released from the "feet" in an amount sufficient to produce maximal force[10] over a period of 20 msec. The Ca concentration in the cleft is shown for the case in which no PL anionic sites are present and for the case in which they are present in the quantity experimentally determined.[1] The diffusion rate for Ca within the cleft space itself is set at D = 1 × 10^{-6} cm^2/sec for both conditions. The marked effect of the binding sites is evident. For the case illustrated no Na-Ca exchange is taking place. See FIGURE 3 for operation of the exchanger in the cleft.

at the clefts could easily maintain steady-state cellular Ca levels even at stimulation rates of 600/min.

Inherent in this model is the response of the Na-Ca exchangers to the Ca released into the cleft from the SR even though efflux via the exchangers must match influx via the Ca "L" channels in order that steady-state be maintained. We have therefore calculated the fraction of the SR release which will exit the cell via the exchangers and the fraction exiting the cleft space to the general cytoplasm versus the amount released from the SR. The calculation is made for $K_d(Ca)$ for the exchangers of 5 μM with exchange rate of 30% Vmax (see above) and is illustrated in FIGURE 3. In the rat Ca current increases about 43% as force increases from 5% to 70% maximum.[11] Using the data from Fabiato[10] 5% maximal force requires an SR release of 25 μmols/kg wet wt cells and 70% force requires release of 70 μmols/kg wet wt. The relation depicted in FIGURE 3 shows that at 25 μmols 24% of the release will efflux via the exchanger or 6 μmoles. At 70 μmols 10% of the release will efflux or 7.0 μmoles, an increase of 17%. The percentage of SR release which exits via the exchanger declines as the release increases because the exchanger saturates.

Under most conditions more Ca current produces greater SR release and therefore greater efflux via the exchanger according to our model of the cleft space. Note also that as SR release increases the fraction that leaves the clefts to the cytosol increases from 76 to 90%.

The Na-Ca Exchange-Dependent Ca Compartment

Given the slowed diffusion and large [Ca] gradient predicted by the model within the diadic cleft spaces (FIG. 1), it seemed reasonable that the region would be manifest as a separate, discrete kinetically-defined Ca exchange compartment within the cell. This has proved to be the case.[18-21]

The compartment is reproducibly identifiable in both adult and cultured neonatal ventricular cells. Though the neonatal cells have no "T" system, there is clear evidence of subsarcolemmal SR cisterns with subsarcolemmal clefts in these cells,[22] qualitatively similar to "T" system diadic structures. The discreteness of the compartment is illustrated in FIGURE 4. This shows a nonperfusion limited ^{45}Ca washout of previously ^{45}Ca-labeled adult ventricular cells. Separate washouts are commenced with perfusate that contains zero Na and zero Ca (Na and Ca

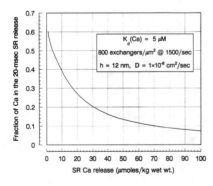

FIGURE 3. Operation of the Na-Ca exchanger. The fraction of Ca transported out of the cleft via Na-Ca exchange versus the amount of Ca released from the "feet." $K_d(Ca)$ for the exchanger = 5 μM operating at a mean -60 mV transsarcolemmal potential over the course of a cardiac cycle. See text for further analysis.

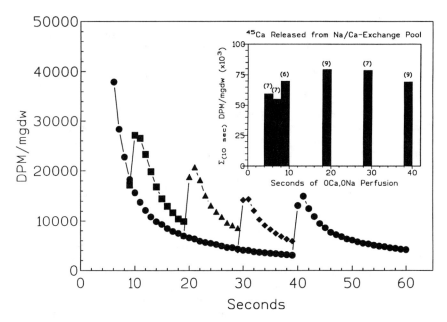

FIGURE 4. The Na-Ca exchange-dependent compartment. Ventricular cells from adult rat were labeled with ^{45}Ca for 30 minutes prior to the ^{45}Ca washouts illustrated. Washout solution contained 0Na, 0Ca for varying periods (9–39 seconds) at which time Na and Ca were returned to the washout solution. Note rapid, large efflux of ^{45}Ca upon activation of Na-Ca exchange. The *inset* shows the mean total ^{45}Ca released over 10 seconds after Na and Ca addition at the times indicated. Note that ^{45}Ca activity released does not decrease with washout time before Na and Ca addition. See text for further discussion. (From Langer & Rich.[18] Reprinted by permission from the *American Journal of Physiology*.)

replaced with choline chloride). This completely inhibits transsarcolemmal Na-Ca exchange. Na and Ca are then returned to reactivate the exchanger at various times during the washout. This results in rapid release of ^{45}Ca into the perfusate at four different (9th, 19th, 29th and 39th seconds) times in the lower graph. The inset shows that the amount released (represented by integration of the transients over a 10-sec period following return of Na and Ca) is independent of the time of reactivation from 5 to 40 seconds. Therefore, ^{45}Ca activity in the compartment remains unchanged, *i.e.*, its Ca remains within the compartment, until Na-Ca exchange is activated. It is then released within a few seconds.[18]

Further work[19–21] defined the compartment's general characteristics: (1) The compartment is absolutely dependent for its exchange on the operation of the Na-Ca exchanger. Its Ca content cannot exit the cell via another route, *e.g.*, the sarcolemmal Ca pump; (2) When the exchanger is "turned-off" (0Na-0Ca perfusion) the compartment accumulates, at $[Ca]_0 = 1$ mM, approximately 500–600 μmols Ca/kg dry wt cells (~100 μmols/kg wet wt); (3) With the exchanger operative the t½ exchange is <600 msec.

If, as seemed most likely to us, the compartment is localized to the diadic cleft region we proposed that a component of JSR Ca and Ca bound to the apposed inner sarcolemmal leaflet (FIG. 1) represent, together, the subcellular locus of the compartment. A recent study[21] gives experimental support to this proposal. We used neonatal cultured cells. Initially, we measured the size of the compartment in intact functional cells, using our nonperfusion limited on-line "scintillation disk" technique.[3,19] The total compartment measured 583 ± 18 (SE) μmols Ca/kg dry wt. cells. Other cells were ^{45}Ca-labeled, under identical conditions, to asymptomatic labeling levels. The cells were then washed with isotope-free solution containing 0Na and 0Ca. This cleared all extracellular ^{45}Ca but retained ^{45}Ca label within the Na-Ca exchange-dependent compartment (see FIG. 4). At this point the cells are placed in a unique "gas-dissection" chamber and, from intact cells, sarcolemmal membrane is instantaneously (within 100 msec) isolated.[1] This permits us to measure the ^{45}Ca bound to the sarcolemma both nonspecifically (with Na and Ca in the wash solution) and specifically to a fraction related to the Na-Ca exchange-dependent compartment (no Na or Ca in the wash solution). The sarcolemmal-bound component of the compartment measured 295 μmols/kg dry wt cells, or 51% of the total compartment.

The compartment was further analyzed using probes that specifically decrease SR Ca content (thapsigargin, caffeine, low-dose ryanodine) which decreased the total cell compartment content and sarcolemmal (SL) binding proportionately, *i.e.,* SL Component/Total Cell Compartment ratio remained unchanged though the total compartment was decreased by as much as 70% by these agents. This indicated that the SL was in equilibrium with another cellular component, probably the SR. If such were the case then an agent which prevented Ca flux from the JSR into the cleft (closed the "feet") should greatly diminish the SL component. Such an agent is high-dose (1 mM) ryanodine.[23] This treatment reduced the SL-bound component of the compartment from 295 to 13 μmoles/kg dry wt and did not change the content of the remaining or putative SR component. The SL/Total Cell ratio fell from 0.51 to 0.05. This gives strong support to the model (FIG. 1) in which the Na-Ca exchange-dependent compartment is localized to the cleft where a component of JSR Ca and a component of SL-bound Ca contribute to the compartment.

It should be noted that we have identified and quantified the compartment under nonsteady-state conditions. We block the exchanger and measure the accumulation of Ca that occurs in a discrete, kinetically-defined compartment. The neonatal cultured cells demonstrate frequent small oscillatory contractions when placed in the 0Na-0Ca solution.[19] These oscillations almost certainly represent Ca leaking from the SR into the clefts which cannot exit the cell via the inhibited exchangers. It then cycles to the myofilaments, back to the SR, out again, etc. The moment Na-Ca is returned and the accumulated Ca in the compartment is released (FIG. 4) the oscillations cease.

CONCLUSIONS

We believe, in intact functional mammalian ventricular cells, that in order for their Na-Ca exchangers to work effectively in the maintenance of steady-state intracellular Ca levels, a large fraction of the exchangers need to be placed in juxtaposition to the subsarcolemmal cleft regions. It is in these subcellular compartments where [Ca] rises to levels which will optimally activate the exchangers with $K_d(Ca) = 5-6$ μM. The elevated [Ca] in the clefts is attributable to the reduced volume into which large amounts of Ca are injected from the JSR but even more so to the delayed diffusion of Ca from the clefts. This delay is due largely to the presence of large amounts of Ca-binding phospholipid (PI, PS and PE) at the inner sarcolemmal leaflet.

In addition to the effect of the diadic cleft space on Ca efflux via the Na-Ca exchanger emphasized in the present paper, it is expected that the space may play a role in early-cycle "reverse" Na-Ca exchange leading to triggering of Ca release,[24] as well as to the role of "L" channel Ca current in this process. These mechanisms are currently under study in our laboratory.

REFERENCES

1. POST, J. A., G. A. LANGER, J. A. F. OP DEN KAMP & A. VERKLEIJ. 1988. Phospholipid asymmetry in cardiac sarcolemma. Analysis of intact cells and "gas dissected" membranes. Biochim. Biophys. Acta **943**: 256–266.
2. LANGER, G. A., T. L. RICH & F. B. ORNER. 1990. Ca exchange under non-perfusion limited conditions in rat ventricular cells: identification of subcellular compartments. Am. J. Physiol. **259**: H592–H602.
3. KUWATA, J. & G. A. LANGER. 1989. Rapid, non-perfusion limited calcium exchange in cultured neonatal myocardial cells. J. Mol. Cell. Cardiol. **21**: 1195–1208.
4. LANGER, G. A., J. S. FRANK & K. D. PHILIPSON. 1978. Preparation of sarcolemmal membrane from myocardial tissue culture monolayer by high velocity gas dissection. Science **200**: 1388–1391.
5. POST, J. A. & G. A. LANGER. 1992. Sarcolemmal calcium binding sites in heart: I. Molecular origin in "gas-dissected" sarcolemma. J. Membr. Biol. **129**: 49–57.
6. RADERMACHER, M., V. RAO, R. GRASSUCCI, J. FRANK, A. P. TIMERMAN, S. FLEISCHER & T. WAGENKNECHT. 1994. Cryo-electron microscopy and three-dimensional reconstruction of the calcium release channel/ryanodine receptor form skeletal muscle. J. Cell Biol. **127**: 411–423.
7. FRANK, J. S., G. MOTTINO, D. REID, R. S. MOLDAY & K. D. PHILIPSON. 1992. Distribution of the Na^+-Ca^{2+} exchange protein in mammalian cardiac myocytes: an immunofluorescence and immunocolloidal gold-labeling study. J. Cell Biol. **117**: 337–345.
8. PAGE, E. 1978. Quantitative ultrastructural analysis in cardiac membrane physiology. Am. J. Physiol. **4**: C147–C158.
9. PESKOFF, A., J. A. POST & G. A. LANGER. 1992. Sarcolemmal calcium binding sites in heart: II. Mathematical model for diffusion of calcium released from the sarcoplasmic reticulum into the diadic region. J. Membr. Biol. **129**: 59–69.
10. FABIATO, A. 1983. Calcium-induced release of calcium from the cardiac sarcoplasmic reticulum. Am. J. Physiol. **295**: C1–C14.
11. WANG, S. Y., L. WINKA & G. A. LANGER. 1993. Role of calcium current and sarcoplasmic reticulum calcium release in control of myocardial contraction in rat and rabbit myocytes. J. Mol. Cell. Cardiol. **25**: 1339–1347.
12. ALTMAN, P. L. & D. S. DITMER, Eds. 1971. Respiration and Circulation Handbook. 340. Fed. of Amer. Soc. for Exp. Biol. Bethesda, MD.
13. HILGEMANN, D. W., D. A. NICOLL & K. D. PHILIPSON. 1991. Charge movement during Na^+ translocation by native and cloned cardiac Na^+/Ca^{2+} exchanger. Nature **352**: 715–718.
14. STEWART, J. M. & E. PAGE. 1978. Improved stereological techniques for studying myocardial cell growth: application to external sarcolemma, T system and intercalated disks of rabbit and rat hearts. J. Ultrastruct. Res. **65**: 119–134.
15. WIER, W. G. 1990. Cytoplasm $[Ca^{2+}]$ in mammalian ventricle: dynamic control by cellular processes. Annu. Rev. Physiol. **52**: 467–485.
16. MATSUOKA, S. & D. W. HILGEMANN. 1992. Steady-state and dynamic properties of cardiac sodium-calcium exchange. Ion and voltage dependencies of the transport cycle. J. Gen. Physiol. **100**: 963–1004.
17. HILGEMANN, D. W., A COLLINS & S. MATSUOKA. 1992. Steady-state and dynamic

properties of cardiac sodium-calcium exchange. Secondary modulation by cytoplasmic calcium and ATP. J. Gen. Physiol. **100:** 933–961.
18. LANGER, G. A. & T. L. RICH. 1992. A discrete Na-Ca exchange-dependent Ca compartment in rat ventricular cells: exchange and localization. Am. J. Physiol **262:** C1149–C1153.
19. POST, J. A., J. H. KUWATA & G. A. LANGER. 1993. A discrete Na^+/Ca^{2+} exchange-dependent, Ca^{2+} compartment in cultured neonatal rat heart cells. Characteristics, localization and possible physiological function. Cell Calcium **14:** 61–71.
20. LANGER, G. A. & T. L. RICH. 1993. Further characterization of the Na-Ca exchange-dependent Ca compartment in rat ventricular cells. Am. J. Physiol. **265:** C556–C561.
21. LANGER, G. A., S. Y. WANG & T. L. RICH. 1995. Localization of the Na/Ca exchange-dependent Ca compartment in cultured neonatal rat heart cells. Am. J. Physiol. **268:** C119–C126.
22. LANGER, G. A., J. S. FRANK & L. M. NUDD. 1979. Correlation of calcium exchange, structure and function in myocardial tissue. Am. J. Physiol. **237:** H239–H246.
23. NAGASAKI, K. & S. FLEISCHER. 1988. Ryanodine sensitivity of the calcium release channel of the sarcoplasmic reticulum. Cell Calcium **9:** 1–7.
24. LEBLANC, N. & J. R. HUME. 1990. Sodium current-induced release of calcium from cardiac sarcoplasmic reticulum. Science **248:** 372–376.
25. LANGER, G. A. 1994. Myocardial calcium compartmentation. Trends Cardiovasc. Med. **4:** 103–109.

Action Potential Duration Modulates Calcium Influx, Na^+-Ca^{2+} Exchange, and Intracellular Calcium Release in Rat Ventricular Myocytes[a]

R. B. CLARK, R. A. BOUCHARD, AND W. R. GILES[b]

Departments of Medical Physiology and Medicine
University of Calgary
3330 Hospital Drive N.W.
Calgary, Alberta, Canada T2N 4N1

INTRODUCTION

Interventions which change the height and/or duration of the cardiac action potential (AP) can significantly modulate contractility.[1–5] Recent work on mammalian cardiac myocytes has shown that changes in stimulation frequency,[6] inhibition of repolarizing potassium currents,[7,8] and stimulation of α_1 and β adrenoceptors[9,10] can produce an increase in AP duration and contraction.

Although these correlations between AP duration and contractility are well established, the mechanism(s) for this positive inotropic effect in mammalian ventricle have been studied in detail only recently (c.f. Ref 7). Increased Ca^{2+} influx through voltage-dependent Ca^{2+} channels,[11–13] altered Ca^{2+} extrusion by the Na^+-Ca^{2+} exchanger,[14–16] or a combination of both[7] could alter the Ca^{2+} supply to the myofilaments. This could also occur as a result of $[Ca^{2+}]_i$ rising indirectly, i.e., by altering sarcoplasmic reticulum (SR) Ca^{2+} loading.[17,18] We[7] have shown that the AP voltage-clamp technique[19,20] can be used in rat ventricular myocytes to identify both Ca^{2+} influx through L-type Ca^{2+} channels (I_{Ca}) and Ca^{2+} efflux by the sarcolemmal Na^+-Ca^{2+} exchanger, which is seen as slow inward tail current (I_{ex}) during repolarization. In these experiments the AP voltage-clamp method has been used to study the roles of the Ca^{2+} current and Na^+-Ca^{2+} exchange in modulation of cardiac contractility by AP duration. Our results show that AP prolongation results in a substantial increase in net Ca^{2+} entry through L-type Ca^{2+} channels. This extra Ca^{2+} influx augments contractility by enhancing SR Ca^{2+} uptake and release; and it is balanced during steady-state stimulation by a corresponding increase in Ca^{2+} extrusion by the sarcolemmal Na^+-Ca^{2+} exchanger.

---

[a] This work was supported by the Medical Research Council of Canada, the Heart and Stroke Foundation of Canada, and the Alberta Heritage Foundation for Medical Research. W.R.G. is an Alberta Heritage Foundation Medical Scientist. R.A.B. was the recipient of a Medical Research Council Postdoctoral Fellowship.

[b] To whom correspondence should be addressed.

METHODS

The methods for isolation of adult rat ventricular myocytes, whole-cell voltage-clamp, measurement of $[Ca^{2+}]_i$, and unloaded cell shortening were described in detail previously.[7] The control solution used to superfuse the cells contained (in mM): NaCl, 140; KCl, 5; Na$^+$ acetate, 2.8; MgSO$_4$, 1; HEPES (4-(2-hydroxyethyl)-1-piperazineethanesulphonic acid), 10; glucose, 10; CsCl, 3; 4-aminopyridine (4-AP), 3; tetrodotoxin (TTX), 0.015; pH was adjusted to 7.4 with NaOH. The concentration of extracellular CaCl$_2$ was varied between nominally zero and 2 mM, as indicated. In zero-Ca^{2+} solution, CaCl$_2$ was replaced with equimolar MgCl$_2$. In some experiments CdCl$_2$ (100 μM) was added to the control solution. 4-AP was dissolved in distilled water and the pH was adjusted to 7.4 using 2N HCl. The Ca^{2+} indicators indo-1 and rhod-2 were obtained from Molecular Probes (Eugene, OR). Membrane potential and currents were recorded at 22°C. The pipette filling solution contained (in mM): Cs$^+$ aspartate, 120; CsCl, 30; HEPES, 5; MgCl$_2$ (6 H$_2$O), 1; Na$_2$ATP, 5; pH was adjusted to 7.1 with CsOH. In some experiments, 100 μM indo-1 (K$^+$ salt) or 300 μM rhod-2 (NH$_3^+$ salt)[21] was added to the pipette solution.

Single cells from rat ventricle were voltage-clamped with either rectangular command signals or one of two different AP waveforms. These action potentials, which were recorded in separate experiments in the absence and presence of 5 mM 4-AP, had durations (at 0 mV) of 5 and 33 ms, and are designated AP-S and AP-L, respectively. For use as voltage-clamp command signals, AP waveforms were delivered from the D/A output of a DT2801A board in a microcomputer, and filtered at 4 kHz prior to delivery to the input of the voltage-clamp amplifier. The experimental apparatus used to measure indo-1 or rhod-2 fluorescence and cell contraction was constructed around a Nikon Diaphot microscope and has been described in our previous papers.[7,22]

RESULTS

A comparison of Ca$^+$ current, I_{Ca}, activated by conventional rectangular depolarizing clamp steps, and by AP-shaped waveforms in an adult rat ventricular myocyte is shown in FIGURE 1. In these experiments Na$^+$ and K$^+$ currents were blocked by using a Cs$^+$-rich micropipette solution (Methods) and by addition of Cs$^+$ (3 mM), 4-AP (3 mM) and TTX (15 μM) to the superfusate. Ca^{2+}-dependent current was identified using the protocol shown at the top. Myocytes were first voltage-clamped in control solution with a fixed rectangular depolarization from −50 to 0 mV (a; FIG. 1, upper panel). The voltage-clamp command signal was then switched abruptly[23] to a short (AP-S) or long (AP-L) AP voltage-clamp waveform which was applied at a fixed frequency until the current reached a steady state (b). The concentration of extracellular Ca^{2+} ($[Ca^{2+}]_o$) was then rapidly (<500 ms) lowered from 2 mM to nominally zero using a multibarrelled local superfusion pipette[23] and currents were again recorded during depolarizations with both the AP (c) and rectangular (d) command waveforms (FIG. 1A). The Ca^{2+}-sensitive difference current was obtained by subtracting the current remaining in the absence of $[Ca^{2+}]_o$ (c, d) from that in control solution (a, b). Very similar results were obtained when cells were superfused with CdCl$_2$ (100 μM) to block I_{Ca}.

To examine the effect of AP duration on I_{Ca}, this voltage clamp protocol was carried out in the same myocyte using both AP-S and AP-L command waveforms

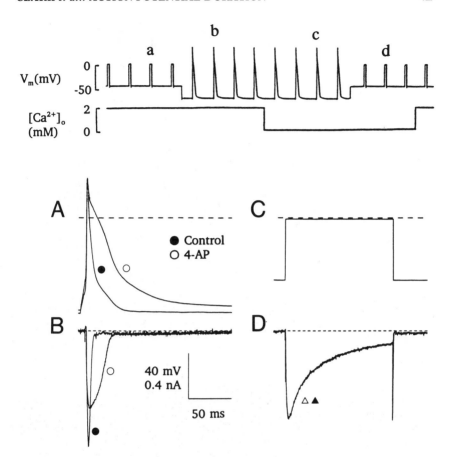

FIGURE 1. Effect of action potential prolongation on whole-cell Ca^{2+} current recorded from an adult rat ventricular myocyte. Action potential (AP) and rectangular voltage-clamp command signals are shown in **(A)** and **(C)**, respectively. The corresponding Ca^{2+}-dependent difference currents in **(B)** and **(D)** were obtained using the stimulation protocol shown at the top (see text). Run-down of currents during the experiment was evaluated and corrected for in all experiments using a scaling factor to normalize currents elicited by a standard depolarizing step from -50 to 0 mV (see text). The result of this procedure is shown in (D), where a correction factor of 1.2 was required to normalize Ca^{2+}-dependent difference currents recorded 3 min (▲) and 10 min (△) after membrane rupture and establishment of whole-cell recording conditions. AP broadening (○) resulted in a decrease of peak I_{Ca}, an increase in its time to peak, and a marked slowing in its return to baseline, when compared with AP-S (●) waveforms. Temperature was $22 \pm 1°C$ in this and all other experiments.

(FIG. 1A). It was necessary to separate changes in I_{Ca} resulting from its time-dependent run-down from alterations due to changes in AP duration. The time-dependent run-down of I_{Ca} was estimated from the scaling factor which was required to normalize Ca^{2+} currents elicited by a fixed depolarizing pulse (-50 to 0 mV; FIG. 1C) applied before and after measuring I_{Ca} with a given AP command

TABLE 1. Effects of Action Potential Prolongation on Ca^{2+}-Dependent Membrane Currents, Indo-1 Transients, and Unloaded Cell Shortening in Rat Ventricular Myocytes[a]

V_{cmd}	I_{Ca} (nA)	Q_{Ca} (pC)	I_{tail} (pA)	$2Q_{tail}$ (pC)	Indo-1 (Peak) ($F_{410/500}$)	CS (Peak) (μm)
AP-S	1.8 ± 0.2	16.2 ± 3	44 ± 3.9	13.2 ± 4	1.18 ± 0.1	3.98 ± 0.36
AP-L	0.87 ± 0.07	33.4 ± 4	64.6 ± 6	28.5 ± 7	1.70 ± 0.2	7.61 ± 0.37
	n = 10	n = 10	n = 10	n = 10	n = 5	n = 5
	p = 0.003	p = 0.001	p = 0.005	p = 0.005	p = 0.01	p = 0.001

[a] Abbreviations: V_{cmd}, AP voltage-clamp command signal; I_{Ca}, peak Ca^{2+}-dependent difference current in nA; Q_{Ca}, integrated charge movement during I_{Ca} in pC; I_{tail}, peak slow inward tail current in pA; Q_{tail}, integrated charge movement during I_{tail} in pC; Indo-1 (Peak), peak systolic indo-1 ratio; CS (Peak), peak unloaded cell shortening in microns (μm); AP-S/AP-L, short and long AP voltage-clamp command signals (Methods); p denotes significance determined by paired t test. All whole-cell current data are given as mean ± SE and are corrected for run-down using the procedure described for FIG. 1. (Adapted from Bouchard et al.[7])

signal (FIG. 1D). This scaling factor was then used to correct Ca^{2+} currents elicited by selected AP voltage-clamp waveforms in a given myocyte. For six different cells, the average scaling factor was 1.24 ± .13 (mean ± SE).

AP prolongation had a number of effects on the Ca^{2+}-dependent difference current (FIG. 1B). There was a decrease in peak inward I_{Ca}, an increase in its time to peak and a marked slowing of its relaxation during repolarization. Integration of Ca^{2+} currents showed that net charge movement during I_{Ca} increased from 16.2 ± 3 pC to 33.4 ± 4 pC (mean ± SEM; n = 10) following the switch from AP-S to AP-L (TABLE 1). Using an estimated single cell volume of 1.68×10^{-11} L and assuming 50% of this total cell volume as the volume of distribution for Ca^{2+} ions,[24] this influx of Ca^{2+} would increase $[Ca^{2+}]_i$ by 6.7 ± .9 and 13.5 ± 1.7 μM per AP-S and AP-L waveform, respectively, in the absence of any intracellular Ca^{2+} buffering.

The marked differences in I_{Ca} recorded with conventional clamp depolarizations and AP waveforms (FIG. 1), and the relatively large Ca^{2+} influxes from each raised questions about the ability of I_{Ca} to trigger/control contraction in rat ventricle. FIGURE 2A shows Ca^{2+}-dependent difference current, $[Ca^{2+}]_i$ transients, and unloaded cell shortening recorded simultaneously during stimulation using an AP-S voltage-clamp command. Peak I_{Ca} was followed after a brief delay (approximately 10 ms) by a rapid rise of $[Ca^{2+}]_i$ and then activation of contraction (cell shortening). Peak systolic $[Ca^{2+}]_i$ occurred approximately 50 ms following the upstroke of the AP. Abrupt removal of $[Ca^{2+}]_o$ (○) using the local superfusion device resulted in complete suppression of both the $[Ca^{2+}]_i$ transient and mechanical activation with 1–3 beats (FIG. 2A, B).

The results in FIGURE 2B demonstrate the effect of AP prolongation on Ca^{2+}-dependent difference current, $[Ca^{2+}]_i$ and cell shortening in the same myocyte. Broadening of the AP resulted in a significant increase in both the magnitude and duration of contraction during steady-state stimulation. Both the time to peak $[Ca^{2+}]_i$, and the time to peak shortening were prolonged, and relaxation was slowed markedly. Peak $[Ca^{2+}]_i$ increased by about 150% and cell shortening increased by approximately 200% (TABLE 1). AP prolongation had no significant effect on the indo-1 fluorescence ratio ($F_{410/500} = 0.8 \pm 0.06$) during diastole, suggesting there was no change in the resting $[Ca^{2+}]_i$.

The Na^+-Ca^{2+} exchanger is essential for maintaining intracellular Ca^{2+} homeostasis in cardiac myocytes.[25–29] FIGURE 3A shows that the Ca^{2+}-dependent membrane currents during AP voltage clamp consist of two distinct phases, an initial large transient component due to the activation of L-type Ca^{2+} channels (a), and a tenfold smaller inward "tail" that activates during late repolarization, and decays during diastole (b). Our previous work,[7] in which rapid replacement of extracellular Na^+ with Li^+ was used to block Na^+-Ca^{2+} exchange activity, demonstrated that these slow tails are due almost entirely to electrogenic activity of the Na^+-Ca^{2+}

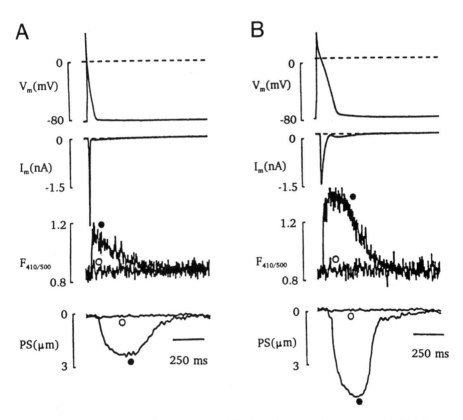

FIGURE 2. Effect of AP prolongation on Ca^{2+}-dependent membrane current, indo-1 fluorescence and unloaded cell shortening. I_{Ca}, $[Ca^{2+}]_i$ and cell shortening were recorded simultaneously. *Top row:* AP voltage-clamp command waveforms used to elicit membrane current and contraction. AP duration at 0 mV was 5 ms (AP-S) and 33 ms (AP-L), in **(A)** and **(B)** respectively. *Second row:* Ca^{2+}-dependent difference current elicited using AP-S and AP-L command signals. *Third and bottom rows:* corresponding indo-1 fluorescence transients and unloaded cell shortening. An abrupt (<500 ms) reduction of $[Ca^{2+}]_o$ from 2 mM (●) to nominally-zero (○) completely suppressed both contraction and $[Ca^{2+}]_i$ transients during stimulation with either AP-S or AP-L voltage-clamp waveforms. (From Bouchard et al.[7] Reprinted by permission from *Circulation Research*.)

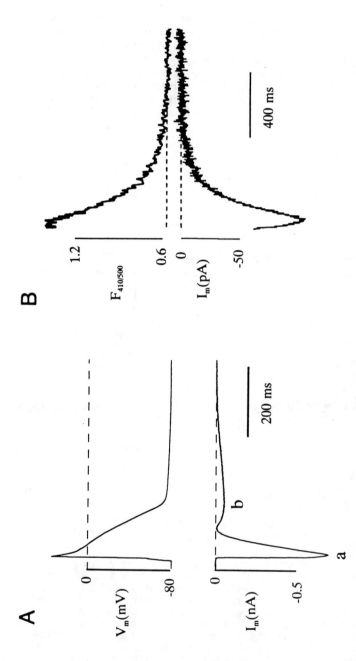

FIGURE 3. Properties of Na^+-Ca^{2+} exchange current during a membrane action potential in a rat ventricular myocyte. **(A)** Comparison of AP voltage-clamp waveform (*top*, AP-L) with the corresponding Ca^{2+}-dependent membrane current (*bottom*). The current waveform consists of two distinct phases, the initial phase (a) resulting from activation of L-type Ca^{2+} channels (I_{Ca}), and a small, late component (b), due to Na^+-Ca^{2+} exchange (I_{ex}). **(B)** Comparison of the time course of decay of the indo-1 transient (*top*) and the Na^+-Ca^{2+} current (*bottom*) in the same myocyte. Note that I_{Ca} is not shown, since it is approximately tenfold larger than I_{ex}. The time courses of decay of the indo-1 transient and I_{ex} were both well described by single exponential functions with time constants of 145 ms and 120 ms, respectively.

exchanger as it extrudes Ca^{2+} from the cell.[22,25-27] If this is the case, then one might expect there also to be a close similarity in the time-course of the decline of the intracellular Ca^{2+} transient and the slow inward tail due to Na^+-Ca^{2+} exchange. This comparison is shown in FIGURE 3B. Both the decline of the intracellular Ca^{2+} transient and the current due to Na^+-Ca^{2+} exchange, I_{ex}, are well described by a single exponential function. In a sample of 5 myocytes, decay of the indo-1 transient had a time constant of 224 ± 6 (mean ± SEM) ms, while in the same cells the decay of I_{ex} had a time constant of 206 ± 8 ms. Although this experimental result does not indicate that the decline of the intracellular Ca^{2+} transient is due to Na^+-Ca^{2+} exchange, it does suggest the pool of intracellular Ca^{2+} which is seen by indo-1 is the same one which strongly modulates Na^+-Ca^{2+} exchange activity.

TABLE 1 provides further evidence that the slow inward tail in fact is due to Na^+-Ca^{2+} exchange. Most of the experimental evidence in mammalian heart indicates that this antiporter process works with a fixed stoichiometry while exchanging on average 3 Na^+ ions for 1 Ca^{2+} ion.[25,27-29] If this is the case, then the quantitative relationship predicted for net charge movements is that the integral of the charge movement during I_{Ca} should be equal to 2× that during the Na^+-Ca^{2+} exchange current, I_{ex}. TABLE 1 shows that this relationship was closely obeyed under our experimental conditions, and also shows that the doubling in net charge movement during I_{Ca} that was produced by changing the AP waveform from AP-S to AP-L was accompanied by an approximately twofold increase in the net charge movement during I_{ex}. This implies that Na^+-Ca^{2+} exchange is the major sarcolemmal transport mechanism controlling Ca^{2+} homeostasis on a beat-to-beat basis in rat ventricle cells.

Previous work on skinned cardiac fibers has shown that SR Ca^{2+} release can be controlled by the rate of change of free Ca^{2+} ($d[Ca^{2+}]_i/dt$) immediately adjacent to SR Ca^{2+} release sites[17] and that Ca^{2+} influx during I_{Ca} can either trigger release of Ca^{2+} from the SR or load it with Ca^{2+} for later release, depending on the kinetics of the simulated I_{Ca} waveform.[17,30] FIGURE 4A shows data obtained from an experiment where the AP voltage-clamp command waveform was switched abruptly from AP-S to AP-L. The myocyte was stimulated at a fixed rate with AP-L for several beats, following which it was switched back to the AP-S signal. These changes in the AP command waveform were made within a single stimulus interval (<5 sec), and the holding potential, AP overshoot and dV/dt of the AP upstroke were identical for both waveforms.

An abrupt switch from an AP-S to an AP-L command resulted in an immediate small increase in peak shortening (FIG. 4A). This was followed by a substantial time-dependent increase in twitch amplitude, which accounted for 90–95% of the total inotropic effect of AP prolongation. Comparison of the I_{Ca} traces marked B1 and B7 (middle panel) with corresponding cell shortening records (bottom panel) shows that the time-dependent increase in cell shortening occurred in the absence of significant changes in I_{Ca}. This result suggests the majority of the time-dependent positive inotropic effect resulting from AP prolongation is due to an enhancement of SR Ca^{2+} loading and release arising from (i) enhanced Ca^{2+} entry and (ii) a delay in the onset of Ca^{2+} extrusion by the Na^+-Ca^{2+} exchanger during slowed repolarization.

The mechanical properties of intact mammalian cardiac muscle and enzymatically dispersed ventricular cells are strongly dependent on the level of SR Ca^{2+} loading. Recent data has shown that both the open probability of purified SR Ca^{2+} release channels and the "gain" of SR Ca^{2+} release in intact myocytes (for review see Ref. 18) may be modulated by the concentration of Ca^{2+} within the SR. In order to exclude any complicating effects on E-C coupling due to overloading of

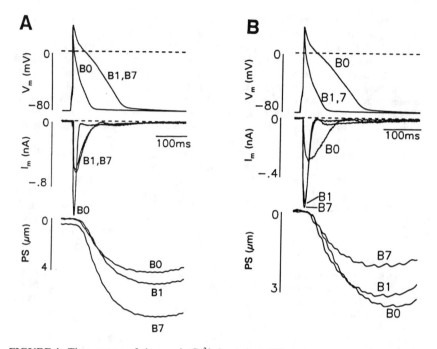

FIGURE 4. Time-course of changes in Ca^{2+}-dependent difference currents and contraction resulting from changes in AP duration. *Top rows:* AP voltage-clamp commands applied at 0.143 Hz to elicit membrane current and contraction. *Middle and lower rows:* corresponding Ca^{2+}-dependent difference currents and cell shortening signals. In both experiments data recorded during the final beat when applying either an AP-S **(A)** or AP-L **(B)** command waveform (B0), and the first (B1) and seventh (B7) beats after switching to an AP-L (A) or AP-S (B) command are superimposed. Note in (A) that the amplitude of the first beat increased slightly immediately following the switch from AP-S to an AP-L despite a twofold reduction in peak I_{Ca}. Continued stimulation using the AP-L waveform resulted in a large time-dependent increase in peak shortening, which occurred in the absence of significant changes in I_{Ca}. The records in (B) shows that an opposite pattern of results obtained when the order of AP voltage-clamp waveforms was reversed, *i.e.*, AP-L waveforms were followed by AP-S waveforms. In this experiment extracellular Ca^{2+} was 1 mM. (From Bouchard et al.[7] Reprinted by permission from *Circulation Research*.)

the SR with Ca^{2+}, additional experiments were performed in which $[Ca^{2+}]_o$ was reduced from 2 to 1 mM, and the sequence of the AP voltage-clamp command waveforms was reversed. FIGURE 4B shows the effect of an abrupt *decrease* in AP duration on Ca^{2+}-dependent difference currents and unloaded cell shortening. This resulted in an immediate *increase* in peak I_{Ca}, a small decrease in cell shortening, and it was followed by a marked time-dependent decrease in twitch amplitude, which occurred without any additional changes in I_{Ca}. Similar results were obtained in 5 different myocytes.

Previous work on vertebrate skeletal[31,32] and cardiac[18] muscle has shown that the rate of Ca^{2+} release from the SR can be approximated by the rate of change of the intracellular Ca^{2+} transient. FIGURE 5 compares I_{Ca} and rhod-2 fluorescence transients during steady-state stimulation with AP-S (○) and AP-L (●) in the same myocyte. The magnitude of the rhod-2 transient elicited by AP-L voltage-clamp commands was increased significantly, but the rate of rise of these transients was markedly decreased. This difference can be seen more clearly in the plots of the first derivative of the rhod-2 transients (denoted dF/dt). AP prolongation resulted in a decrease in the maximum dF/dt, but increased the duration of the dF/dt waveform. The changes in the dF/dt waveform paralleled the changes in the magnitude and duration of Ca^{2+} influx during the AP.

We previously showed[7] that a rapid switch from AP-S to AP-L waveform resulted in a small increase in the amplitude of the first rhod-2 transient following the change in AP waveform, but a marked slowing of the rate of rise which resulted in a large (~50%) decrease in the peak dF/dt. Continued stimulation with AP-L resulted in a substantial increase in the magnitude and the rate-of-rise of the steady-state rhod-2 transient. These changes in the rhod-2 transient occurred in the absence of significant changes in I_{Ca}. These data and the results in FIGURE 5 confirm that most of the positive inotropic effect resulting from AP prolongation

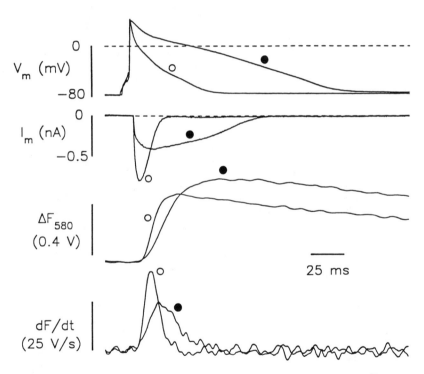

FIGURE 5. Modulation of Ca^{2+} influx, $[Ca^{2+}]_i$ and the first derivative of the Ca^{2+} transient (dF/dt) by AP prolongation. *Top row:* AP-S (O) and AP-L (O) voltage-clamp commands used to elicit membrane current and Ca^{2+} transients. *Second to fourth rows:* corresponding Ca^{2+}-dependent difference currents, rhod-2 fluorescence transients, and the first derivative of rhod-2 fluorescence (dF/dt). Membrane current and Ca^{2+} transients were recorded under steady-state conditions. The average of 5–6 sweeps of each are shown. AP prolongation resulted in a substantial decrease in dF/dt. $[Ca^{2+}]_i$ was 1.5 mM and stimulation frequency was 0.1 Hz throughout.

is due to increased loading of the SR with Ca^{2+}. They also show that the magnitude and time course of Ca^{2+} influx are important parameters in the control of the rate and duration of Ca^{2+} release from the SR.

SUMMARY

The experimental work summarized in this paper and described in more detail in our previous publications[7,8,15,22] demonstrates a very important functional role for Na^+-Ca^{2+} exchange in intracellular Ca^{2+} homeostasis in ventricular myocytes from rat hearts. Ca^{2+} homeostasis in mammalian cardiac myocytes can be considered to be the result of four interactive processes: (i) Ca^{2+} influx through L-type Ca^{2+} channels, (ii) Ca^{2+} release from the SR and its subsequent re-uptake, (iii) intracellular Ca^{2+} buffering, and (iv) Ca^{2+} extrusion across the sarcolemma. Our results demonstrate a number of interesting features of these processes. (1) When the action potential voltage-clamp technique is used to identify the size and time-course of Ca^{2+} fluxes during the action potential, both the peak current and the associated influx of Ca^{2+} are relatively large as was previously demonstrated by Isenberg and his colleagues.[33] (2) Nevertheless, this source of Ca^{2+} is unable, by itself, to produce a significant twitch, which is consistent with previous data from rat ventricle.[34,35] (3) This Ca^{2+} influx, however, does represent the trigger for SR Ca^{2+} release.[36,37] (4) The Na^+-Ca^{2+} exchanger on the SR is able, on average, to extrude all the Ca^{2+} which enters through L-type Ca^{2+} channels, although it provides relatively little Ca^{2+}, i.e., during the course of the normal action potential there is no significant reverse Na^+-Ca^{2+} exchange activity,[22,25,28] at least under our experimental conditions. Our results also suggest that although the L-type Ca^{2+} current cannot by itself trigger *and* control contraction its amplitude, frequency, and time-course can alter the rate and the extent of Ca^{2+} release from the SR.[30] Recently, detailed mathematical formulations and a direct demonstration of some of these phenomena have been published. Stern[39] and Stern and Lakatta[40] predicted more than three years ago that the concentration and the time-course of change in concentration of Ca^{2+} very near the release sites of the SR may be critical determinants of the overall release process. Within the past year Wier and his colleagues[41] and also Lederer et al.[42] have combined electrophysiological measurements with recordings of localized intracellular Ca^{2+} (made using a confocal microscope) and have shown that rapid, and relatively large, but very localized changes in intracellular Ca^{2+} due to Ca^{2+} influx through L-type Ca^{2+} channels are responsible for triggering, and to some extent, controlling the release of Ca^{2+} from the SR. However, it has also been shown that this release depends importantly on the loading or priming state of the SR.[43,44] Perhaps not surprisingly, the massive release of Ca^{2+} from the SR can, itself, alter the pattern of subsequent SR release events[45] (cf. Ref. 46) and the time-course of Ca^{2+} influx through the L-type Ca^{2+} channels.[47]

Thus, although our relatively crude measurements have clearly demonstrated the relationship between L-type Ca^{2+} channel activity and Na^+-Ca^{2+} exchanger function during a normal cardiac action potential in rat ventricle, they fall far short of any delineation of the functional roles of either of these processes in overall Ca^{2+} homeostasis. This additional information can, in principle, be obtained from studies in which cellular microanatomy can be visualized dynamically in conjunction with localized changes in intracellular Ca^{2+} as well as Ca^{2+} of L-type Ca^{2+} channels, SR release, and cell shortening.[48-51]

ACKNOWLEDGMENTS

We thank Mr. S. Somers for help in the manufacture of the fluorescence set-up, and Ms. K. Burrell for excellent secretarial assistance.

REFERENCES

1. Wood, E. H., R. Heppner & S. Weidmann. 1969. Inotropic effects of electric currents. Cir. Res. **24:** 409–445.
2. Morad, M. & Y. Goldman. 1973. Excitation-contraction coupling in heart muscle: membrane control of development of tension. Prog. Biophys. Mol. Biol. **27:** 257–313.
3. Allen, D. G. 1977. On the relationship between action potential duration and tension in cat papillary muscle. Cardiovasc. Res. **11:** 210–218.
4. Wohlfart, B. 1979. Relationship between peak force, action potential duration and stimulus interval in rabbit myocardium. Acta Physiol. Scand. **106:** 395–409.
5. Schouten, V. J. A. 1986. The negative correlation between action potential duration and force of contraction during restitution in rat myocardium. J. Mol. Cell. Cardiol. **18:** 1033–1045.
6. Boyett, M. R. & B. Jewell. 1980. Analysis of the effects of changes in rate and rhythm upon electrical activity in the heart. Prog. Biophys. Mol. Biol. **36:** 1–52.
7. Bouchard, R. A., R. B. Clark & W. R. Giles. 1995. Effects of action potential duration on excitation-contraction coupling in rat ventricular myocytes. Circ. Res. **76:** 790–801.
8. Clark, R. B., R. A. Bouchard, E. Sanchez-Chapula, E. Salinas-Stephanon & W. R. Giles. 1993. Heterogeneity of action potential waveforms and repolarizing potassium currents in rat ventricle. Cardiovasc. Res. **27:** 1795–1799.
9. Fedida, D., A. P. Braun & W. R. Giles. 1993. α_1-Adrenoceptors in myocardium: functional aspects and transmembrane signaling mechanisms. Physiol. Rev. **73:** 469–487.
10. Xiao, R., & E. G. Lakatta. 1993. β_1-Adrenoceptor stimulation and β_2-adrenoceptor stimulation differ in their effects on contraction, cytosolic Ca^{2+}, and Ca^{2+} current in single rat ventricular cells. Circ. Res. **73:** 286–300.
11. Morad, M. & L. Cleeman. 1987. Role of Ca^{2+} channel in development of tension in heart muscle. J. Mol. Cell. Cardiol. **27:** 257–313.
12. Cleeman, L. & M. Morad. 1991. Role of Ca^{2+} channel in cardiac excitation-contraction coupling in the rat: evidence from Ca^{2+} transients and contraction. J. Physiol. **432:** 283–312.
13. Sham, J. S. K., L. Cleemann & M. Morad. 1995. Functional coupling of Ca^{2+} channels and ryanodine receptors in cardiac myocytes. Proc. Natl. Acad. Sci. USA **92:** 121–125.
14. London, B. & J. W. Krueger. 1986. Contraction in voltage-clamped internally perfused single heart cells. J. Gen. Physiol. **88:** 475–505.
15. Bouchard, R. A., R. B. Clark & W. R. Giles. 1993. Regulation of unloaded cell shortening in isolated rat ventricular myocytes by sarcolemmal Na^+/Ca^{2+} exchange. J. Physiol. **469:** 583–599.
16. Bers, D. M., W. J. Lederer & J. Berlin. 1991. Intracellular Ca transients in rat cardiac myocytes: role of Na-Ca exchange in excitation-contraction coupling. Am. J. Physiol. **258:** C944–C954.
17. Fabiato, A. 1985. Simulated calcium current can both cause calcium loading in and trigger calcium release from the sarcoplasmic reticulum of a skinned canine cardiac Purkinje cell. J. Gen. Physiol. **85:** 291–320.
18. Wier, W. G. 1990. Dynamics of control of cytosolic calcium ion concentration. Ann. Rev. Physiol. **52:** 467–485.
19. Doerr, T., R. Denger, A. Doerr & W. Trautwein. 1990. Ionic currents contributing to the action potential in single ventricular myocytes of the guinea-pig studied with action potential clamp. Pflug. Arch. **413:** 599–603.

20. ARREOLA, J., R. T. DIRKSEN, R-C. SHIEH, D. J. WILLIFORD & S-S. SHEU. 1991. Ca^{2+} current and Ca^{2+} transients under action potential clamp in guinea-pig ventricular myocytes. Am. J. Physiol. **261:** C393–C397.
21. MINTA, A., J. P. Y. KAO & R. Y. TSIEN. 1989. Fluorescent indicators for cytosolic calcium based on rhodamine and fluorescein chromophores. J. Biol. Chem. **264:** 8171–8178.
22. BOUCHARD, R. A., R. B. CLARK & W. R. GILES. 1993. Role of Na^+/Ca^{2+} exchange in activation of contraction in rat ventricle. J. Physiol. **472:** 391–413.
23. SHIMONI, Y., R. B. CLARK & W. R. GILES. 1992. Role of an inwardly rectifying K^+ current in the rabbit ventricular action potential. J. Physiol. **448:** 709–727.
24. PAGE, E. 1978. Quantitative ultrastructural analysis in cardiac membrane physiology. Am. J. Physiol. **235:** C147–C158.
25. NOBLE, D., S. J. NOBLE, C. L. BERR, Y. E. EARM, W. K. HO & I. K. SO. 1991. The role of sodium-calcium exchange during the cardiac action potential. Ann. N. Y. Acad. Sci. **639:** 334–354.
26. CAMPBELL, D. L., W. R. GILES, K. ROBINSON & E. F. SHIBATA. 1988. Studies of the sodium-calcium exchanger in bull-frog atrial myocytes. J. Physiol. **403:** 317–340.
27. BRIDGE, J. H. B., J. SMOLLEY & K. SPITZER. 1990. The relationship between charge movements associated with I_{Ca} and I_{NaCa} in cardiac myocytes. Science **248:** 376–378.
28. HILGEMANN, D. 1990. "Best estimates" of physiological Na/Ca exchange function: calcium conservation and the cardiac electrical cycle. *In* Cardiac Electrophysiology: from Cell to Bedside. D. P. Zipes & J. Jalife, Eds. 51–61. W.B. Saunders Co. Philadelphia.
29. EISNER, D. & W. J. LEDERER. 1985. Na-Ca exchange: stoichiometry and electrogenicity. Am. J. Physiol. **248:** C189–C202.
30. FABIATO, A. 1985. Time and calcium dependence of activation and inactivation of calcium induced release of calcium from the sarcoplasmic reticulum of a skinned canine cardiac Purkinje cell. J. Gen. Physiol. **85:** 247–289.
31. MONCK, J. R., I. M. ROBINSON, A. L. ESCOBAR, J. L. VERGARA & J. M. FERNANDEZ. 1994. Pulsed laser imaging of rapid Ca^{2+} gradients in excitable cells. Biophys. J. **67:** 505–514.
32. RIOS, E. & G. PIZARRO. 1991. Voltage sensor of excitation-contraction coupling in skeletal muscle. Physiol. Rev. **71**(3): 849–908.
33. ISENBERG, G. & U. KLOCKNER. 1982. Calcium currents of isolated bovine ventricular myocytes are fast and of large amplitude. Pflueg. Arch. **395:** 30–41.
34. VALDIOMILLOS, M., S. C. O'NEILL, G. L. SMITH & D. A. EISNER. 1989. Calcium-induced calcium release activates contraction in intact cardiac cells. Pflueg. Arch. **413:** 676–678.
35. BERS, D. M., W. J. LEDERER & J. R. BERLIN. 1990. Intracellular Ca transients in rat cardiac myocytes: role of Na-Ca exchange in excitation-contraction coupling. Am. J. Physiol. **258:** C944–C954.
36. NABAUER, M., G. CALLEWAERT, L. CLEEMAN & M. MORAD. 1989. Regulation of calcium release is regulated by calcium, not gating charge, in cardiac myocytes. Science **244:** 800–803.
37. CLEEMANN, L. & M. MORAD. 1991. Role of Ca^{2+} channel in cardiac excitation-contraction coupling in the rat: evidence from Ca^{2+} transients and contractions. J. Physiol. **432:** 283–312.
38. WIER, W. G., T. M. EGAN, J. R. LOPEZ-LOPEZ, & C. W. BALKE. 1994. Local control of excitation-contraction coupling in rat heart cells. J. Physiol. (London) **474:** 463–471.
39. STERN, M. D. 1992. Theory of excitation-contraction coupling in cardiac muscle. Biophys. J. **63:** 497–517.
40. STERN, M. D. & E. G. LAKATTA. 1992. Excitation-contraction coupling in the heart: the state of the question. FASEB J. **6:** 3092–3100.
41. LOPEZ-LOPEZ, J. R., P. S. SHACKLOCK, C. W. BALKE & W. G. WIER. 1995. Local calcium transients triggered by single L-type calcium channel currents in cardiac cells. Science **268:** 1042–1045.
42. CANNELL, M. B., H. CHENG & W. J. LEDERER. 1995. The control of calcium release in heart muscle. Science **268:** 1045–1049.

43. JANCZEWSKI, A. M., H. A. SPURGEON, M. D. STERN & E. G. LAKATTA. 1995. Effects of sarcoplasmic reticulum Ca^{2+} load on the gain function of Ca^{2+} release by Ca^{2+} current in cardiac cells. Am. J. Physiol. **268:** H916–H920.
44. BASSANI, J. W. M., W. YUAN & D. M. BERS. 1995. Fractional SR Ca^{2+} release is regulated by trigger Ca^{2+} and SR Ca^{2+} content in cardiac myocytes. Am. J. Physiol. **268:** C1313–C1329.
45. SIPIDO, K. R., G. CALLEWAERT & E. CARMELIET. 1995. Inhibition and rapid recovery of Ca^{2+} current during Ca^{2+} release from sarcoplasmic reticulum in guinea pig ventricular myocytes. Circ. Res. **76:** 102–109.
46. VALDIVIA, H. H., J. H. KAPLAN, G. C. R. ELLIS-DAVIES & W. J. LEDERER. 1995. Rapid adaptation of cardiac ryanodine receptors: modulation by Mg^{2+} and phosphorylation. Science **267:** 1997–2000.
47. LAMB, G. D. & D. G. STEPHENSON. 1995. Activation of ryanodine receptors by flash photolysis of caged Ca^{2+}. Biophys. J. **68:** 946–948.
48. NIGGLI, E. & P. LIPP. 1995. Subcellular features of calcium signalling in heart muscle: what do we learn? Cardiovasc. Res. **29:** 441–448.
49. SANTANA, L. F., H. CHENG, A. M. GOMEZ, M. B. CANNELL & W. J. LEDERER. 1996. Relation between the sarcolemmal Ca^{2+} current and Ca^{2+} sparks and local control theories for cardiac excitation-contraction coupling. Circ. Res. **78:** 166–171.
50. SHACKLOCK, P. S., W. G. WIER & C. W. BALKE. 1995. Local Ca^{2+} transients (Ca^{2+} sparks) originate at transverse tubules in rat heart cells. J. Physiol. **487:** 601–608.
51. TRAFFORD, A. W., M. E. DIAZ, S. C. O'NEILL & D. A. EISNER. 1995. Comparison of subsarcolemmal and bulk calcium concentration during spontaneous calcium release in rat ventricular myocytes. J. Physiol. **488:** 577–587.

Na-Ca Exchange and Ca Fluxes during Contraction and Relaxation in Mammalian Ventricular Muscle[a]

DONALD M. BERS, JOSÉ W. M. BASSANI,[b]
AND ROSANA A. BASSANI[b]

Department of Physiology
Loyola University Chicago
Stritch School of Medicine
2160 South First Avenue
Maywood, Illinois 60153

INTRODUCTION

The activation of cardiac ventricular muscle contraction requires a large increase in cytoplasmic [Ca] ([Ca]$_i$) allowing Ca to bind to troponin C and thereby activate the myofilaments. During the cardiac action potential Ca influx occurs via the L-type Ca channel, and some amount of Ca influx may also be expected via Na-Ca exchange.[1] In some species sufficient Ca can enter the cell during the action potential to activate substantial contraction in the absence of a functional sarcoplasmic reticulum (SR, *e.g.*, rabbit and guinea pig), while in rat this does not seem to be the case.[2,3] Furthermore, under normal conditions the quantity of Ca entry via Na-Ca exchange is probably small compared to that which enters via Ca current (I_{Ca}), although Ca influx via Na-Ca exchange can be greatly increased by elevation of intracellular [Na] ([Na]$_i$, *e.g.*, Refs. 4, 5).

Nevertheless, when the SR is functional in mammalian ventricular myocytes most of the activating Ca is released from the SR during E-C coupling. While it is clear that L-type I_{Ca} can trigger SR Ca release via Ca-induced Ca-release,[6,7] it is also possible that Ca entry via Na-Ca exchange can trigger SR Ca release.[8–10] The physiological importance of Na-Ca exchange in triggering SR Ca release remains to be clarified.

For ventricular relaxation to occur [Ca]$_i$ must be lowered by transporting Ca out of the cytosol, allowing Ca to dissociate from troponin C. There are four Ca transport systems which compete for cytosolic Ca in cardiac myocytes: 1) the SR Ca-ATPase, 2) the sarcolemmal Na-Ca exchange, 3) the sarcolemmal Ca-ATPase and 4) the mitochondrial Ca uniporter.[11–13] Indeed, in the steady state, the amount of Ca which enters the cytosol (from the extracellular space or the SR) must be restored between beats to prevent net shifts in Ca content. This implies an intrinsic link between Ca fluxes involved in activation of contraction and relaxation.

The theme of this paper is to consider from a quantitative standpoint the relative contributions of the 4 Ca transport systems contributing to relaxation during the normal twitch contraction in various mammalian species. Various meth-

---

[a] This study was supported by a grant from the USPHS, HL30077.

[b] JWMB and RAB are currently at the Dept. and Centro Engenharia Biomédica, Universidade Estadual de Campinas, SP-Brazil.

ods are used to evaluate this competition and as indicated, this also has some implications for E-C coupling.

METHODS

Isometric contraction was measured in ventricular trabeculae (<0.6 mm diameter) dissected from the rabbit right ventricular free wall as described by Bers and Bridge.[14] Muscles were field stimulated to contract at 0.5 Hz and 29°C between interventions. Rapid cooling of the muscle to ~0°C induces a contracture (RCC) which is attributable to the release of Ca from the SR. These RCCs were used both to assess SR Ca content and also to temporarily inhibit Ca transport systems by the low temperature.[14-16] Rapid cooling was accomplished by superfusion at high flow rates (~300 bath volumes/min) with a solenoid switch selecting water jacketed superfusates at the bath inlet. This method produced muscle surface temperature changes which were >90% complete in <500 ms.

Ventricular myocytes were also isolated from rabbit, rat and ferret hearts as previously described.[11-13] Cell shortening and $[Ca]_i$ were measured in individual myocytes using a video edge detector and the fluorescent Ca indicator indo-1, respectively. Indo-1 was loaded by preincubation of cells with the acetoxymethylester form and fluorescence was excited at 365 nm and emission was recorded at 405 and 485 nm. The ratio (R) of fluorescence at these two wavelengths was converted to $[Ca]_i$ based on *in vivo* measurements of minimum and maximum values of R and *in vitro* measurements of K_d.[12]

Whole cell voltage clamp was also used in some of the experiments, essentially as described by Bers *et al*.[17] Pipettes (~1 MΩ resistance) were filled with (in mM): 100 Cs-glutamate, 30 Cs-PIPES, 20 tetraethylammonium chloride, 5–10 Mg-ATP, 0.1 pyruvate, 0.05 indo-1 with pH 7.2 at 23°C.

The normal extracellular solution (NT) contained (in mM): 140 NaCl, 6 KCl, 1 $MgCl_2$, 2 $CaCl_2$, 10 glucose and 5 HEPES at pH 7.40. When rat myocytes were used, the $CaCl_2$ concentration was typically lowered to 1 mM. For experiments at different temperature, the pH of the solutions were adjusted to 7.4 at the experimental temperature.

RESULTS AND DISCUSSION

Rapid Cooling Contractures and Relaxation

We have used several different experimental strategies to evaluate the competition among the four Ca transport systems involved in cardiac relaxation.[11-18] FIGURE 1 shows an approach using RCCs.[14] In this experiment rabbit ventricular muscle was stimulated at 0.5 Hz (29°C) to reach steady state. Then the muscle was rapidly cooled to ~0°C to release the SR Ca and to strongly suppress Ca transport. During the time the muscle was at 0°C the extracellular solution was changed so that upon rapid rewarming in the same solution there was selective inhibition of either the SR Ca reuptake (by inclusion of 10 mM caffeine) or the Na-Ca exchanger (by 0Na, 0Ca + EGTA solution) or both. It can be seen that on the average, inhibition of Na-Ca exchange prolonged the half-time ($t_{1/2}$) of relaxation by about 30%. Inhibition of SR Ca accumulation slowed relaxation by about 70%. When both systems are inhibited (0Na + Caff), relaxation is dramati-

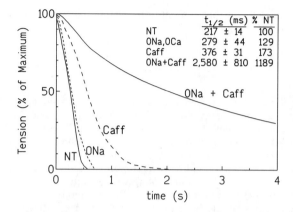

FIGURE 1. Normalized relaxation of isometric force in rabbit ventricular muscle upon rewarming after RCCs. During the RCC at ~1°C (and during rewarming to 30°C) the superfusion solution was changed from a normal Tyrode's (NT) to either 1) prevent Na-Ca exchange by a Na-free, Ca-free solution (0Na), 2) prevent SR Ca sequestration using a NT with 10 mM caffeine (Caff) or 3) inhibiting both transporters simultaneously using a Na-free, Ca-free solution with 10 mM caffeine (0Na + Caff). Relaxation half-times and percents of the NT value for pooled results are shown in the *inset*. (From Bers & Bridge.[14] Reprinted by permission from *Circulation Research*.)

cally slowed (*i.e.*, by more than 1000%) and is often incomplete even after 20 sec. This result indicates that the SR Ca-pump and to a lesser extent sarcolemmal Na-Ca exchange can produce cardiac relaxation. It also shows that other systems (such as the sarcolemmal Ca-ATPase or mitochondrial Ca uptake) are too slow to account for cardiac relaxation. There was also no voltage dependence of relaxation by the SR, whereas relaxation attributed primarily to Na-Ca exchange was slowed by depolarization, as expected for this electrogenic exchanger.

RCCs were also used in a different way to assess this competition in guinea pig ventricular myocytes by Bers *et al.*[16] and in rabbit ventricular myocytes and muscle by Hryshko *et al.*[18] In those experiments paired RCCs were performed where Ca extrusion via Na-Ca exchange was prevented during the rewarming relaxation of the first RCC. This resulted in no change in the amplitude of RCC or the associated Ca_i transients. This indicated that when the Na-Ca exchange was inhibited, virtually all of the Ca released by the SR was taken back up by the SR. On the other hand, if the Na-Ca exchange was allowed to function, the second RCC was 23 ± 3% smaller than the first in rabbit and 36% in guinea pig ventricular myocytes, and comparable results of 27% were found in multicellular rrabbit ventricular muscle.[18] These results indicate that the Ca responsible for 23–36% of relaxation in these cells is extruded by Na-Ca exchange (see also TABLE 1).

Ca Removal during Caffeine-Induced Contractures and Twitches in Rabbit and Rat Myocytes

Rapid application of 10 mM caffeine appears to release all of the SR Ca leading to a large contraction and Ca_i transient. Sustained exposure to caffeine during the

TABLE 1. Ca Transport during Ventricular Myocyte and Muscle Relaxation

Species	Ref.	Temp. (°C)	Contribution to Relaxation (%)		
			Na-Ca X	SR	Slow
Rabbit [Ca]$_i$	29	35	27	70	3
	29	25	23	74	3
	12	22	28	70	2
Rat v. [Ca]$_i$	12	22	7	92	1
Ferret [Ca]$_i$	13	22	29	64	4
Rabbit relax	29	25	28	70	2
	29	35	21	76	3
Ferret relax	29	25	30	63	7
	29	35	28	67	5
Cat relax	29	25	47	51	2
	29	35	32	66	2
Paired RCCs					
Rabbit relax	18	29	23	76	1
Rabbit muscle relax	18	29	27	73	
Guinea pig v. [Ca]$_i$	16	30	36	64	

time that [Ca]$_i$ is declining also prevents Ca accumulation into the SR (but unlike RCCs does not prevent Ca transport by other mechanisms). By studying the [Ca]$_i$ decline and relaxation during such caffeine-induced contractures (CafC) information can be obtained about the other Ca removal systems (Na-Ca exchange, sarcolemmal Ca-ATPase and mitochondrial uniporter).

In a series of experiments summarized in FIGURE 2, we measured relaxation

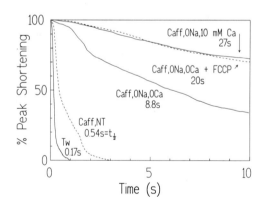

FIGURE 2. Normalized cell relaxation relengthening in single rabbit ventricular myocytes. The conditions were either 1) having all Ca transporters functional during a steady state twitch (Tw), 2) preventing SR Ca uptake during a caffeine-induced contracture (CafC) in NT (Caff, NT), 3) additionally inhibiting Na-Ca exchange in 0Na, 0Ca solution (Caff, 0Na, 0Ca) and a) either inhibiting mitochondrial Ca uptake with 1 μM FCCP + 1 μM oligomycin so only sarcolemmal Ca-ATPase was functional (Caff, 0Na, 0Ca + FCCP) or b) additionally inhibiting the sarcolemmal Ca-pump by elevating [Ca]$_o$ to 10 mM after predepletion of [Na]$_i$, so that only mitochondrial Ca uptake was functional (Caff, 0Na, 10Ca). The traces and mean t$_{1/2}$ values are based on experiments described by Bassani et al.[11]

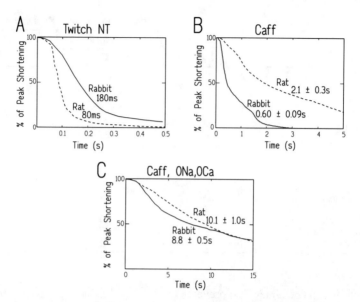

FIGURE 3. Normalized mechanical relaxation of **(A)** steady-state twitches and caffeine contractures in control **(B)** and 0Na, 0Ca solution **(C)** recorded in rat and rabbit ventricular myocytes. Mean $t_{1/2}$ values for each type of contraction are also indicated (after Bassani et al.[12]).

of twitch vs CafC in rabbit ventricular myocytes where one or more of the other Ca transport systems were inhibited.[11] That is, Na-Ca exchange was inhibited by 0Na, 0Ca (with EGTA), mitochondrial Ca uptake was inhibited by exposure to the protonophore uncoupler, FCCP (1 μM, which is expected to dissipate the mitochondrial membrane potential) and Ca extrusion via the sarcolemmal Ca-ATPase was limited by elevating $[Ca]_o$ after predepletion of $[Na]_i$ (by superfusion for 5 min in 0Na, 0Ca solution) to prevent Ca entry via Na-Ca exchange when $[Ca]_o$ was increased during the CafC in the absence of $[Na]_o$.

Again, in FIGURE 2 it can be seen that inhibition of both SR Ca uptake and Na-Ca exchange slowed relaxation ($t_{1/2}$ = 8.8 ± 1.0 sec). When either the mitochondrial system or the sarcolemmal Ca-ATPase were also inhibited relaxation was slowed another 2–3 fold. Based on simplistic assumptions and the $t_{1/2}$ values for relaxation, we concluded that with respect to the Na-Ca exchange that the SR Ca uptake was 2–3-fold faster and the sarcolemmal Ca-ATPase and mitochondrial Ca transport were 37 to 50 times slower respectively.[11] Even at this superficial level this might suggest that 2/3 to 3/4 of the Ca during relaxation in rabbit goes to the SR, with most of the rest extruded via Na-Ca exchange.

In further studies[12–13] we wanted to be somewhat more quantitative about this issue and also explore possible species differences in this aspect of Ca regulation by multiple systems. FIGURE 3C shows the normalized relaxation from CafC where the Na-Ca exchange is also inhibited by 0Na, 0Ca solution, immediately before and during the CafC. The relaxation was not appreciably different in rat vs rabbit cells ($t_{1/2}$ = 9–10 sec). This indicated that the combined function of the slow systems (sarcolemmal Ca-ATPase and mitochondria uptake) were comparable in these species.

When the Na-Ca exchanger was allowed to function during CafC in control solution (FIG. 3B), the rabbit myocytes relaxed much faster than rat (almost 4-fold), possibly reflecting a stronger Na-Ca exchange in rabbit myocytes. When all four Ca removal systems were allowed to participate in [Ca]$_i$ decline during a twitch (FIG. 3A), the situation reversed, such that relaxation in the rat myocytes was about twice as fast as in rabbit. This may indicate a faster SR Ca accumulation in the rat, which more than compensates for the slower Na-Ca exchange. This is also consistent with comparative SR Ca transport measurements in a more isolated system, where SR Ca transport rate was ~2 times faster in rat ventricular myocytes than rabbit.[19] Similar results were observed for Ca$_i$ transients in these cells.[12]

A limitation with the type of experiments in FIGURE 3 is that twitches are being compared with CafC. Thus we developed experimental protocols where either the SR Ca-ATPase or the sarcolemmal Na-Ca exchange could be selectively blocked during a normal twitch activated by an action potential.[12] FIGURE 4 shows Ca$_i$ transient during twitches in rabbit and rat myocytes where the SR Ca-ATPase was completely blocked by thapsigargin (but with normal SR Ca load). After steady state stimulation cells were incubated for 5 min in 0Na, 0Ca solution, with 2.5 μM thapsigargin included during the last 2 min just before switching back to NT and stimulation of a twitch. This specific protocol allowed the SR Ca load to remain at the normal level, but also produced complete inhibition of SR Ca uptake (as tested by attempts to reload the SR after Ca depletion).[20] Of course only one test twitch can be given and after much longer rest times in thapsigargin solution, the SR Ca content gradually declines, even in 0Na, 0Ca.[21] There are two key features of note in FIGURE 4. First, the Ca$_i$ transients are larger after the SR Ca-pump is blocked. This is consistent with the idea that rapid Ca transport by the SR Ca-pump normally limits the peak of the Ca$_i$ transient, and this seems to be true for both rabbit and rat myocytes. Second, the time constant (τ) of [Ca]$_i$ decline is prolonged

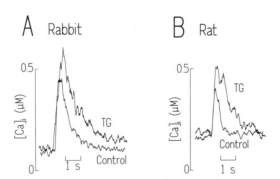

FIGURE 4. [Ca]$_i$ transients recorded during electrically-stimulated twitches in rabbit **(A)** and rat **(B)** ventricular myocytes before (control) and after treatment with 2.5 μM thapsigargin (TG). Stimulation was applied 10 s after switching to control solution after 5–7 min pre-perfusion with 0Na, 0Ca solution. TG treatment (2 min exposure) was performed during this pre-perfusion period and SR Ca load was maintained despite complete block of the SR Ca-pump. (From Bassani et al.[12] Reprinted by permission from the *Journal of Physiology*).

in the presence of thapsigargin. In rabbit the $[Ca]_i$ decline is only slowed by a factor of 2 (τ increases from 0.496 ± 0.034 s to 0.978 ± 0.120 s), where in rat thapsigargin slows $[Ca]_i$ decline by a factor of 9 (τ increases from 0.181 ± 0.008 s to 1.66 ± 0.30 s). This is certainly consistent with the foregoing results suggesting that SR Ca uptake is more critical in rat.

Similar experiments were done with selective inhibition of the Na-Ca exchange. In this case cells were first depleted of $[Na]_i$ by superfusion for 5–7 min in 0Na, 0Ca solution to allow the Na-pump to extrude Na and prevent Ca influx via Na-Ca exchange when $[Ca]_o$ is subsequently elevated (even to 100 mM) in the continued absence of $[Na]_o$. Twitches in this case were activated in Na-free solution with Li in place of Na so that the measured action potentials were essentially normal and Na-Ca exchange in either direction was completely prevented. For rabbit cells this produced qualitatively similar results to those for thapsigargin in FIGURE 4, although somewhat smaller in magnitude. That is, the amplitude of the Ca_i transient was increased by 48% and the τ of $[Ca]_i$ decline was prolonged by 45% (from 406 ± 28 ms to 588 ± 54 ms). For rat the effects of Na-Ca exchange inhibition were very modest. There was no increase in the amplitude of the Ca_i transient and there was only a 20% slowing of $[Ca]_i$ decline (consistent with voltage clamp results in rat ventricular myocyte[17]). These results indicate that the Na-Ca exchanger is considerably more important in rabbit than rat myocytes as inferred above.

When the two procedures above were combined to produce simultaneous block of both the Na-Ca exchange and the SR Ca-ATPase, the rate of $[Ca]_i$ decline in both rabbit and rat myocytes was the same ($\tau \sim 12$ sec). This is the same τ measured for the $[Ca]_i$ decline during CafC in 0Na, 0Ca in these same cells. However, the amplitude of the Ca_i transients during twitches with the SR Ca-pump and Na-Ca exchanger blocked is only about half of that observed during CafC with the Na-Ca exchanger blocked. This is consistent with the measurements that about half of the SR Ca content is released during a normal twitch.[20,22]

Strong Sarcolemmal Ca-ATPase in Ferret Ventricular Myocytes

In rabbit ventricular myocytes we evaluated the contribution of mitochondria and the sarcolemmal Ca-ATPase to the slow decline of $[Ca]_i$ observed during CafC in 0Na, 0Ca.[11] Using high $[Ca]_o$ (after $[Na]_i$ depletion) to limit Ca extrusion via the sarcolemmal Ca-pump and FCCP to inhibit mitochondrial uptake, we demonstrated that about half of the slow flux went through each of these systems (with the sarcolemmal Ca-pump being slightly faster). Furthermore, when all four Ca transport systems were inhibited during a CafC in 0Na, 10 mM $[Ca]_o$ and FCCP, relaxation and $[Ca]_i$ decline in rabbit ventricular myocytes was virtually abolished. This can be seen in the top trace in FIGURE 5A. This type of experiment made it convincing that these were the only four Ca transport systems which are important to consider in relaxation and that our inhibitory approaches seem to work (at least in rabbit ventricle).

When we extended these approaches to ferret ventricular myocytes we were surprised to find that relaxation of CafC in 0Na, 0Ca was remarkably fast ($t_{1/2} \sim$ 1.8 s vs 9–10 sec in rabbit and rat[12,13]). In addition, the protocol which virtually abolished relaxation in rabbit ventricular myocytes (CafC in 0Na, 10 mM $[Ca]_o$ and FCCP) only increased the $t_{1/2}$ of relaxation in ferret myocyte to ~4 sec[13] (and see top Ca_i transient curve in FIG. 5B). We first ruled out the possibility that this

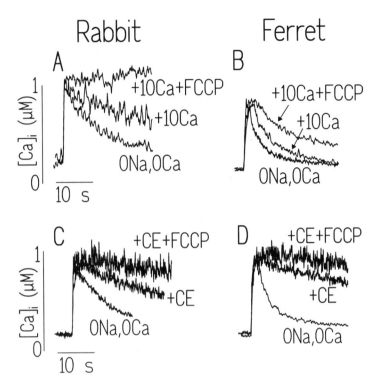

FIGURE 5. [Ca]$_i$ transients in rabbit (**A** and **C**) and ferret ventricular myocytes (**B** and **D**) obtained during SR Ca release by 10 mM caffeine in Na-free solution. The control conditions in each panel were a CafC in 0Na, 0Ca. In (A) and (B) CafC were repeated during inhibition of the sarcolemmal Ca-ATPase with 10 mM Ca (+10Ca) and after additional inhibition of mitochondrial Ca uptake by FCCP (+10Ca + FCCP). In (C) and (D) the CafC were repeated after carboxyeosin loading to inhibit the sarcolemmal Ca-pump (+CE) and after additional inhibition of the mitochondrial Ca uptake by FCCP (+CE + FCCP). (From Bassani et al.[25] Reprinted by permission from *Pflügers Archiv*.)

rapid relaxation in ferret was due to incomplete inhibition of SR Ca uptake, Na-Ca exchange, mitochondrial Ca uptake or any other saturable internal pool.[13] The likeliest explanation then seemed that the thermodynamic approach to limit the sarcolemmal Ca-ATPase by increasing [Ca]$_o$ (even to 100 mM), which appeared to work well in the rabbit and rat, was not able to suppress this system in ferret.

Gatto and co-workers[23,24] recently demonstrated that the plasma membrane Ca pump is strongly inhibited by several fluorescein analogues, among them eosin (tetrabromofluorescein) and carboxyeosin (at μM levels). We used carboxyeosin as an alternative means to inhibit the sarcolemmal Ca-ATPase, loading it into myocytes by incubation with a cell permeant esterified form of carboxyeosin, which can be trapped in cells in the same way as the familiar Ca indicator acetoxymethylesters. FIGURE 5C shows that loading rabbit ventricular myocytes with carboxyeosin produced similar effects to elevating [Ca]$_o$ to

10 mM during CafC ± FCCP (e.g., compare FIG. 5A vs 5C).[25] In ferret myocytes carboxyeosin produced the same effects as either high $[Ca]_o$ or carboxyeosin did in rabbit myocytes (±FCCP, compare FIGS. 5C and 5D). Thus, the sarcolemmal Ca-pump seems likely to be responsible for the much faster relaxation observed in ferret cells after block of SR Ca accumulation and Na-Ca exchange transport. Since the thermodynamic [Ca] gradient was less effective in limiting Ca extrusion in the ferret myocytes, the sarcolemmal Ca-pump in these cells may also have different fundamental characteristics from that in rabbit ventricular myocytes. Indeed, the more powerful sarcolemmal Ca-ATPase in ferret can be appreciated by the larger difference between the 0Na, 0Ca and +CE curves in FIGURE 5C vs 5D.

Quantitative Analysis of Ca Fluxes during Relaxation

Experiments like those described above and in FIGURES 4 and 5 have allowed us to go a step further in quantitative analysis of Ca flux by and competition among the various systems.[12-13] First the free $[Ca]_i$ can be converted to total cytoplasmic [Ca] ($[Ca]_t$), using the passive myoplasmic buffering characteristics measured by Hove-Madsen and Bers[26] and assuming that this buffering is in rapid equilibrium. Then differentiation of $[Ca]_t$ with respect to time ($d[Ca]_t/dt$) provides the rate of Ca transport from the myoplasm during relaxation. This transport rate must be the sum of the individual transport rates given by

$$d[Ca]_t/dt = J_{SR} + J_{Na\text{-}CaX} + J_{Slow} - L \quad (1)$$

where the J terms refer to flux through the SR Ca-ATPase, Na-Ca exchange and the combined slow transporters respectively and L is a constant Ca leak into the cytoplasm (assumed to be small compared to other fluxes during $[Ca]_i$ decline). For simplicity J_{SR}, $J_{Na\text{-}CaX}$ and J_{Slow} can be empirically described as simple [Ca] dependent fluxes of the form

$$J_x = \frac{V_{max}}{1 + (K_m/[Ca]_i)^n} \quad (2)$$

We first fit J_{slow} by using the decline of $[Ca]_i$ during a twitch in 0Na, 0Ca + TG (or a caffeine-induced contracture in 0Na, 0Ca) where J_{SR} and $J_{Na\text{-}CaX}$ are zero. Then the determined V_{max}, K_m and n for J_{Slow} are held constant to determine the set of parameters which best describe either J_{SR} or $J_{Na\text{-}CaX}$ (using a twitch $[Ca]_i$ transient in either 0Na or TG respectively) leaving out the appropriate term in Eq. 1. Finally we can simulate the action of all systems working simultaneously during a normal twitch by using the free $[Ca]_i$ during that twitch to calculate the instantaneous individual fluxes through each system. Moreover, integration of the fluxes allows the calculation of how the individual systems contribute to the total Ca flux during twitch relaxation.

FIGURE 6 shows the results of such calculations based on measured Ca_i transients. It can be seen that during a normal twitch the fractions of Ca transported by the SR, Na-Ca exchange and slow systems are 70, 28 and 2% respectively in rabbit and 92, 7 and 1% in rat myocytes. This 28% estimate of Ca flux by Na-Ca exchange in rabbit agrees with previous estimates described above.[14,18] The 7% value for rat agrees with similar experimental results in that species by Negretti

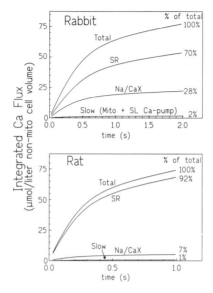

FIGURE 6. Integrated Ca flux during a normal twitch in rabbit and rat myocytes. Free $[Ca]_i$ during relaxation was based on the mean values for resting and peak $[Ca]_i$ and the τ of $[Ca]_i$ decline. This free $[Ca]_i$ was then converted to $[Ca]_t$ using passive myoplasmic Ca buffering measured by Hove-Madsen and Bers,[26] including $[\text{indo-1}]_i = 50\ \mu M$ and $K_{in} = 250$ nM. After differentiation $(d[Ca]_t/dt)$ the $[Ca]_i$ dependence of each transport system in Eq. 1 was sequentially fit to Eq. 2 as described in the text. Values obtained for J_{SR}, $J_{Na\text{-}CaX}$ and J_{Slow} respectively were: V_{max} (in $\mu mol \cdot 1$ nonmitochondrial volume$^{-1} \cdot s^{-1}$) = 81, 46 and 3.9 for rabbit and 207, 27 and 4 for rat; K_m (in nM) = 264, 316, and 362 in rabbit and 184, 257 and 268 in rat; n = 3.7, 3.7, 3.2 in rabbit and 3.9, 3.4 and 3.5 in rat. Then the Ca flux through each system was calculated for the $[Ca]_i$ transient decline during a normal twitch. To convert values in units of $\mu mol \cdot$ liter nonmitochondrial volume^{-1} to units of $\mu mol \cdot$ kg wet wt^{-1}, the V_{max} values (and buffer concentrations) should be divided by 2.5.[31] (After Bassani et al.[12])

et al.,[27] but is smaller than previous indirect estimates based on time constants of relaxation or $[Ca]_i$ decline.[17,28]

Temperature Effects on Relaxation and $[Ca]_i$ Decline

All of the above myocyte experiments were done at room temperature. Since all of the membrane transport systems are expected to change at body temperature, it is fair to ask how this relative competition among Ca transport systems is altered at different temperatures (e.g., approaching physiological). Puglisi et al.[29] recently performed such experiments and analysis for rabbit, ferret and cat ventricular myocytes at 25 and 35°C. While all the Ca transport systems are faster at 35°C than at 25°C (with apparent Q_{10} in our experiments generally in the range of 1.7–2.5), the relative contribution of the Na-Ca exchange or the SR Ca-ATPase to $[Ca]_i$ decline or relaxation was essentially not changed. That is, for rabbit ventricular myocytes, analysis like that shown in FIGURE 6 at 25 and 35°C were

70:27:3% and 74:23:3% for SR:Na-Ca exchange:lumped slow systems respectively. Moreover, there was not a very great difference between rabbit, ferret and cat (except that the sarcolemmal Ca-ATPase was notably stronger in the ferret as shown above). Some of the values derived from our different studies on this issue are listed in TABLE 1.

Steady-State Cellular Ca Balance

The focus above has been on measuring Ca removal from the myoplasm during relaxation. In the steady state, the total amount of Ca extruded from the cell during each cardiac cycle must be the same as the amount which enters over a cardiac cycle, otherwise there would be a progressive net change in cellular Ca and this would not correspond to steady state. Thus, if 28% of the cytosolic Ca removed during relaxation leaves the cell via Na-Ca exchange, then this same amount of Ca must enter the cell. That is, 28% of the amount of Ca responsible for activating contraction must enter the cell during a steady-state twitch.

Delbridge et al.[30] have carried out voltage clamp experiments in rabbit ventricular myocytes to assess this issue. The integral of the Ca current was used to assess the amount of Ca influx at a steady-state twitch activated by depolarization from -40 to 0 mV (under conditions where Ca influx via Na-Ca exchange was limited by low $[Na]_i$). They also integrated the Na-Ca exchange current activated during a CafC to assess the SR Ca content (accounting for the 7% of SR Ca release which expected to be transported by the mitochondria and sarcolemmal Ca-ATPase under these conditions[12]). Assuming that 43% of the SR Ca content was released during a steady state twitch,[20] the Ca current contributed $23 \pm 2\%$ of the activating Ca. This is in very good agreement with the estimates described above concerning the amount of Ca extruded via Na-Ca exchange during a steady-state twitch. Considering the shorter action potential in rat ventricular myocytes, it is quite possible that less Ca enters in this species, consistent with the smaller Ca extrusion seen in FIGURE 6.

An interesting side note is that the less direct conclusions drawn from contraction/relaxation and RCC studies agree surprisingly well with more direct detailed analysis of $[Ca]_i$ transients even in quantitative terms (see TABLE 1). This may make some initial experiments possible with simpler technology. It also indicates that Ca removal from the myoplasm is a very important rate limiting step in the process of myocardial relaxation.

In conclusion, we have evaluated the dynamic interaction of the four key Ca transport systems which are responsible for relaxation in cardiac muscle. In mammalian ventricular myocytes it seems clear that SR Ca uptake is the dominant Ca removal process. However, this dominance differs in different species. For example, in rabbit the SR Ca-pump transports only 2.5 times the amount transported by the Na-Ca exchange, whereas in rat ventricular myocytes this value is closer to 14. Furthermore, this difference is probably due to differences in both the SR Ca-ATPase and Na-Ca exchange.

SUMMARY

There are four cellular Ca transport systems which compete to remove Ca from the myoplasm in mammalian ventricular myocytes. These are 1) the SR Ca-

ATPase, 2) the sarcolemmal Na-Ca exchange, 3) the sarcolemmal Ca-ATPase and 4) the mitochondrial Ca uniporter. Using multiple experimental approaches we have evaluated the dynamic interaction of these systems during the normal cardiac contraction-relaxation cycle. The SR Ca-ATPase and Na-Ca exchange are clearly the most important, quantitatively; however, the relative roles vary in a species-dependent manner. In particular, the SR is much more strongly dominant in rat ventricular myocytes, where ~92% of Ca removal is via SR Ca-ATPase and only 7% via Na-Ca exchange during a twitch. In other species (rabbit, ferret, cat, and guinea pig) the balance is more in the range of 70% SR Ca-ATPase and 25–30% Na-Ca exchange. Ferret ventricular myocytes also exhibit an unusually strong sarcolemmal Ca-ATPase. During the steady state the same amount of Ca must leave the cell as enters over a cardiac cycle. This implies that 25–30% of the Ca required to activate contraction must enter the cell, and experiments demonstrate that this amount of Ca may be supplied by the L-type Ca current.

ACKNOWLEDGMENTS

We are pleased to acknowledge Drs. Larry V. Hryshko, Leif Hove-Madsen, José L. Puglisi, John H. B. Bridge, and Kenneth W. Spitzer for contributions to the body of work described here. Ms. Melanie Robinson and Beth Tumilty also provided valuable technical assistance.

REFERENCES

1. BERS, D. M. 1991. Excitation-Contraction Coupling and Cardiac Contractile Force. (Single author monograph.) 1–258. Kluwer Academic Press. Dordrecht, Netherlands.
2. SUTKO, J. L. & J. T. WILLERSON. 1980. Ryanodine alteration of the contractile state of rat ventricular myocardium. Comparison with dog, cat and rabbit ventricular tissues. Circ. Res. **46:** 332–343.
3. BERS, D. M. 1985. Ca influx and SR Ca release in cardiac muscle activation during postrest recovery. Am. J. Physiol. **248:** H366–H381.
4. BERS, D. M. 1987. Mechanisms contributing to the cardiac inotropic effect of Na-pump inhibition and reduction of extracellular Na. J. Gen. Physiol. **90:** 479–504.
5. BERS, D. M., D. M. CHRISTENSEN & T. X. NGUYEN. 1988. Can Ca entry via Na-Ca exchange directly activate cardiac muscle contraction? J. Mol. Cell. Cardiol. **20:** 405–414.
6. BEUCKELMANN, D. J. & W. G. WIER. 1988. Mechanism of release of calcium from sarcoplasmic reticulum of guinea pig cardiac cells. J. Physiol. **405:** 233–255.
7. FABIATO, A. 1985. Time and calcium dependence of activation and inactivation of calcium-induced release of calcium from the sarcoplasmic reticulum of a skinned canine cardiac Purkinje cell. J. Gen. Physiol. **85:** 247–290.
8. LEBLANC, N. & J. R. HUME. 1990. Sodium current-induced release of calcium from cardiac sarcoplasmic reticulum. Science **248:** 372–376.
9. LEVI, A. J., K. W. SPITZER, O. KOHMOTO & J. H. B. BRIDGE. 1994. Depolarization-induced Ca entry via Na-Ca exchange triggers SR release in guinea pig cardiac myocytes. Am. J. Physiol. **266:** H1422–H1433.
10. KOHMOTO, O., A. J. LEVI & J. H. B. BRIDGE. 1994. Relation between reverse sodium-calcium exchange and sarcoplasmic reticulum calcium release in guinea pig ventricular cells. Circ. Res. **74:** 550–554.
11. BASSANI, R. A., J. W. M. BASSANI & D. M. BERS. 1992. Mitochondrial and sarcolemmal Ca transport can reduce $[Ca]_i$ during caffeine contractures in rabbit cardiac myocytes. J. Physiol. **453:** 591–608.

12. BASSANI, J. W. M., R. A. BASSANI & D. M. BERS. 1994. Relaxation in rabbit and rat cardiac cells: species-dependent differences in cellular mechanisms. J. Physiol. **476:** 279–293.
13. BASSANI, R. A., J. W. M. BASSANI & D. M. BERS. 1994. Relaxation in ferret ventricular myocytes: unusual interplay among calcium transport systems. J. Physiol. **476:** 295–308.
14. BERS, D. M. & J. H. B. BRIDGE. 1989. Relaxation of rabbit ventricular muscle by Na-Ca exchange and sarcoplasmic reticulum Ca-pump: ryanodine and voltage sensitivity. Circ. Res. **65:** 334–342.
15. BRIDGE, J. H. B. 1986. Relationships between the sarcoplasmic reticulum and transarcolemmal Ca transport revealed by rapidly cooling rabbit ventricular muscle. J. Gen. Physiol. **88:** 437–473.
16. BERS, D. M., J. H. B. BRIDGE & K. W. SPITZER. 1989. Intracellular Ca transients during rapid cooling contractures in guinea-pig ventricular myocytes. J. Physiol. **417:** 537–553.
17. BERS, D. M., W. J. LEDERER & J. R. BERLIN. 1990. Intracellular Ca transients in rat cardiac myocytes: role of Na/Ca exchange in excitation-contraction coupling. Am. J. Physiol. **258:** C944–C954.
18. HRYSHKO, L. V., V. M. STIFFEL & D. M. BERS. 1989. Rapid cooling contractures as an index of SR Ca content in rabbit ventricular myocyte. Am. J. Physiol. **257:** H1369–H1377.
19. HOVE-MADSEN, L. & D. M. BERS. 1993. SR Ca uptake and thapsigargin sensitivity in permeabilized rabbit and rat ventricular myocytes. Cir. Res. **73:** 820–828.
20. BASSANI, J. W. M., R. A. BASSANI & D. M. BERS. 1993. Twitch-dependent SR Ca accumulation and release in rabbit ventricular myocytes. Am. J. Physiol. **265:** C533–C540.
21. BASSANI, R. A. & D. M. BERS. 1995. Rate of diastolic Ca release from the sarcoplasmic reticulum of intact rabbit and rat ventricular myocytes. Biophys. J. **68:** 2015–2022.
22. BASSANI, J. W. M., W. YUAN & D. M. BERS. 1995. Fractional SR Ca release is altered by trigger Ca and SR Ca content in cardiac myocytes. Am. J. Physiol. **268:** 1313–1319.
23. GATTO, C. & M. A. MILANICK. 1993. Inhibition of the red blood cell calcium pump by eosin and other fluorescein analogues. Am. J. Physiol. **264:** C1577–C1586.
24. GATTO, C., C. C. HALE & M. A. MILANICK. 1995. Eosin, a potent inhibitor of the plasma membrane Ca pump, does not inhibit the cardiac Na-Ca exchanger. Biochemistry **34:** 965–972.
25. BASSANI, R. A., J. W. M. BASSANI & D. M. BERS. 1995. Relaxation in ferret ventricular myocytes: role of the sarcolemmal Ca ATPase. Pflüg. Arch. **430:** 573–579.
26. HOVE-MADSEN, L. & D. M. BERS. 1993. Passive Ca buffering and SR Ca uptake in permeabilized rabbit ventricular myocytes. Am. J. Physiol. **264:** C677–C686.
27. NEGRETTI, N., S. C. O'NEILL & D. A. EISNER. 1993. The relative contributions of different intracellular and sarcolemmal systems to relaxation in rat ventricular myocytes. Cardiovasc. Res. **27:** 1826–1830.
28. CRESPO, L. M., C. J. GRANTHAM & M. B. CANNELL. 1990. Kinetics, stoichiometry and role of the Na-Ca exchange mechanism in isolated cardiac myocytes. Nature **345:** 618–621.
29. PUGLISI, J. L., R. A. BASSANI, J. W. M. BASSANI, J. N. AMIN & D. M. BERS. 1996. Temperature and the relative contributions of Ca transport systems in cardiac myocyte relaxation. Am. J. Physiol. In press.
30. DELBRIDGE, L. M., J. W. M. BASSANI & D. M. BERS. 1996. Steady-state twitch Ca fluxes and cytosolic Ca buffering in rabbit ventricular myocytes. Am. J. Physiol. **39:** C192–C199.
31. FABIATO, A. 1983. Calcium-induced release of calcium from the cardiac sarcoplasmic reticulum. Am. J. Physiol. **245:** C1–C14.

The Roles of the Sodium and Calcium Current in Triggering Calcium Release from the Sarcoplasmic Reticulum

M. B. CANNELL, C. J. GRANTHAM, M. J. MAIN,
AND A. M. EVANS

*Department of Pharmacology and Clinical Pharmacology
St. George's Hospital Medical School
Cranmer Terrace
London SW17 ORE, England*

INTRODUCTION

Cardiac muscle contraction is a consequence of a 10–20-fold increase in intracellular calcium concentration ($[Ca^{2+}]_i$) that results from a calcium influx across the surface membrane as well as calcium release from intracellular stores (the sarcoplasmic reticulum or SR). While the SR provides most of the calcium, the influx across the sarcolemma is still extremely imporant since it not only contributes to the increase in $[Ca^{2+}]_i$ but also "triggers" the release of calcium via the "calcium-induced calcium release mechanism" (CICR). Two mechanisms have been identified which produce this calcium influx. The first is via calcium channels, predominantly the "L-type" or dihydropyridine-sensitive calcium channel, whose activity in the T-tubular system has been demonstrated recently.[1] The second mechanism is the voltage dependent Na-Ca exchanger whose role in producing a trigger calcium influx is currently the cause of some debate. While Levi *et al.*[2] have argued strongly that the Na-Ca exchange may be more important than the calcium current, it is worth noting that several investigators have been unable to demonstrate Na-Ca exchange-induced calcium release from the SR. This is in marked contrast to the SR calcium release produced by the calcium current which is routinely observed by all investigators. This difference immediately suggests that the role of the exchanger in triggering SR calcium release is less important than the calcium current. Nevertheless, since the first report of calcium current independent SR calcium release,[3] it has been clear that the Na-Ca exchanger *can* trigger SR release, so the pertinent question is not whether the exchanger can trigger SR calcium release but rather what is its relative role under *physiological* conditions?

Na-Ca exchange-induced release has been observed under abnormal conditions such as "calcium-overload" produced by sodium-pump inhibition;[3] depolarization to +100 mV,[4] and at less positive potentials in myocytes perfused internally with high (20 mM) sodium concentrations.[5] Under more physiological conditions, Levi *et al.*[2,6] demonstrated that, when the measured calcium current was blocked, SR calcium release could be evoked by reverse mode Na-Ca exchange. Furthermore, these authors proposed that the Na-Ca exchange may be more important in triggering CICR than the calcium current. Additional support for this proposal has come from observations of sodium-current (I_{Na}) triggered CICR, which was explained by I_{Na} increasing the concentration of sodium at the cytoplasmic surface of the exchanger which, in turn, would accelerate calcium influx via the exchanger to a point where CICR was activated.[7–9]

Recently, cardiac E-C coupling has been observed at the subcellular level, and it is now known that the cardiac calcium transient is the summation of discrete elementary SR calcium release events (called calcium sparks).[1,10] This observation has given new insight into E-C coupling and it now seems that E-C coupling is under local control, with the trigger for SR calcium release occurring in a micro domain, similar to the fuzzy space discussed by Lederer et al.[11] Thus for the Na-Ca exchanger to make a significant contribution to E-C coupling it must be able to produce an increase in $[Ca^{2+}]_i$ in the fuzzy space comparable to that produced by the sarcolemmal calcium channel current. In this paper we will first reexamine the ability of the exchanger to produce calcium release under normal conditions and then consider, from an analytical standpoint, the possible role of the exchanger in E-C coupling.

RESULTS AND DISCUSSION

In a recent series of experiments, Lipp and Niggli[8] showed that, after a train of conditioning pulses (designed to load the SR with calcium), it was possible to evoke a sizable calcium transient by depolarizing from −80 to −40 mV (which should minimize calcium current activation) in the presence of the calcium channel inhibitor verapamil. However, when we repeated their experiment, we were unable to demonstrate such a sodium-current induced calcium transient (FIG. 1). It should be noted that this result is not explained by an inability of the SR to release calcium, since a normal calcium transient could be produced by activating the calcium current. Assuming that the results of Lipp and Niggli[8] are not explainable by the spurious activation of unblocked calcium channels by voltage escape during the sodium current, how can these very different results be reconciled? To answer this question we should consider the factors that will determine the ability of the Na-Ca exchange to produce the calcium influx needed to trigger SR calcium release.

The reversal potential of the Na-Ca exchange is determined both by the sodium (E_{Na}) and calcium (E_{Ca}) electrochemical gradients. In addition, the driving force (but not necessarily the kinetics) for calcium entry on the exchanger is determined by the difference between membrane potential (E_m) and the exchanger equilibrium potential (E_{Na-Ca}):

$$\text{Driving force} = E_m - E_{Na-Ca} = E_m - 3E_{Na} + 2E_{Ca}$$

To evaluate this equation we need to know the levels of internal sodium and calcium in the region that E-C coupling takes place. Unfortunately, this region (the diadic space) is too small to allow direct measurement of these quantities. Nevertheless, an estimate of the trigger calcium level has been obtained from analysis of the properties of calcium sparks. It has been estimated that E-C coupling results from a 2.2×10^4 increase in calcium spark rate,[10] and if the probability of activating an SR calcium release channel is proportional to the square of the calcium level,[10,12] then the local $[Ca^{2+}]_i$ that activates SR calcium release must be 15 times the resting level of $[Ca^{2+}]_i$. Assuming a resting level of $[Ca^{2+}]_i$ between 60–100 nM, it is possible to estimate that the local $[Ca^{2+}]_i$ must be ~9–15 μM in the region where the calcium trigger for E-C coupling is sensed (which we will assume is the diadic space). Such high levels of calcium can only be achieved by

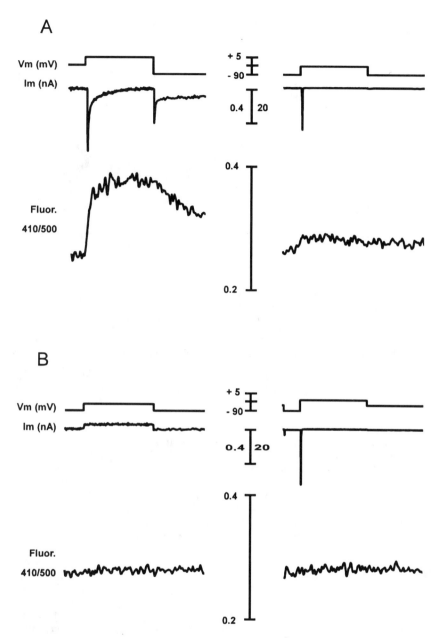

FIGURE 1. Verapamil blocks both I_{Ca} and I_{Na}-induced $[Ca^{2+}]_i$ transients. **(A)** Control transients activated by steps from −40 to +5 mV (*left*) to activate the I_{Ca} and −90 to −50 mV (*right*) to activate I_{Na}. Note the small transient evoked by the "selective" activation of I_{Na}. However, 10 μM verapmil blocks both transients, despite the lack of effect on I_{Na} **(B)**. The $[Ca^{2+}]_i$ transient was reported by ratio of indo-1 fluorescence at 410 & 500 nm. The indo-1 (25 μM) was introduced into the cell by diffusion from the patch pipette used to voltage clamp the cell. The pipette filling solution contained (in mM)): CsOH, 100; aspartate, 100; CsCl, 30; NaCl, 5; HEPES, 10; Mg-ATP, 5; (pH 7.2 with CsOH); indo-1 0.025 (K-salt).

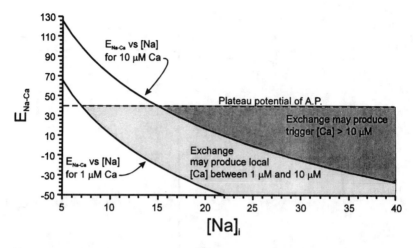

FIGURE 2. Relationship between the reversal for Na-Ca exchange and internal sodium to achieve trigger calcium levels of 1 and 10 μM $[Ca^{2+}]_i$. The plateau potential of the action potential is indicated by the *dotted line*. To meet or exceed a trigger level of 10 μM, the exchanger would have to operate in the *dark shaded area* of the diagram, showing that internal sodium would have to be greater than 17 mM to achieve such a trigger calcium level. The *light grey area* shows the region for trigger calcium levels between 1 and 10 μM $[Ca^{2+}]_i$ which would require sodium levels between 7 and 17 mM. Note that the lower trigger level would imply that the peak of the normal $[Ca^{2+}]_i$ transient would be sufficient to trigger further SR release, which is not believed to occur under normal conditions.[1,21]

the exchanger at high levels of internal sodium or at positive membrane potentials (under resting conditions there is a ~50 mV driving force for calcium *extrusion* by the exchanger.[13] Thus, for the exchanger to act as a major trigger for E-C coupling, it would have to both reverse its direction of transport and raise $[Ca^{2+}]_i$ in the diadic space by a factor of 15 during the period over which E-C coupling takes place (~10 ms). However, we will not initially consider the kinetics of the exchanger to explore the ability of the exchanger to achieve such a trigger level, since the exchanger could never raise $[Ca^{2+}]_i$ beyond the level predicted by the thermodynamics of the exchange reaction. Put another way, the thermodynamics of the transport reaction places absolute limits on the levels of $[Ca^{2+}]_i$ that can be achieved by the exchanger.

FIGURE 2 shows the relationship between E_{Na-Ca} and the internal sodium level for the exchanger to achieve a local trigger calcium level of 10 μM. It is notable that, at 10 mM internal sodium, the membrane potential would have to be about +70 mV for the exchanger to produce a trigger calcium of 10 μM—in good agreement with the results of Sham *et al.*[4] At 20 mM internal sodium, the trigger calcium would be achieved for membrane potentials greater than about +20 mV, and Nuss and Houser[5] observed triggered contractions between +40 and +60 mV (note that the sodium pump would have tended to reduce the level of internal sodium in the experiments of Nuss and Houser[5] so the agreement between the thermodynamic analysis and experimental results may be even closer). However, at −40 mV, the test potential used by Lipp and Niggli[8] and ourselves, the internal sodium concentration would have to be about 38 mM to result in such a trigger calcium level. This is much greater than the internal sodium level used in the patch pipette (5 mM), so our failure to observe a Na-Ca exchange evoked contraction seems quite consistent with thermodynamic requirements of the exchange.

As pointed out by Lederer et al.,[11] the sodium current would have to flow into a small fraction of the cell to increase sodium levels sufficiently to produce a large trigger calcium influx via Na-Ca exchange. In fact, it would be necessary for the sodium current to increase the local sodium level by about 33 mM for the result of Lipp and Niggli[8] to be explained with these absolute thermodynamic constraints on the behavior of the exchanger.

It has been estimated that the local volume in which E-C coupling is triggered is about 0.2% of cell volume.[10] If we were to assume that all the sodium current flowed *only* into this region of the cell, a 30-nA sodium current in a 30-pl cell for 5 ms would produce an increase of 25 mM in local sodium. Although such a calculation seems to imply that the increase in local sodium might be sufficient to provide the required trigger calcium level, it is important to appreciate that this type of calculation will grossly overestimate the true levels of sodium that will occur during the sodium current. This problem arises because the sodium that enters via the sodium current must diffuse from the site of entry. If sodium did not diffuse from the mouth of the sodium channel, then that region would have to be (effectively) electrically insulated and therefore unable to respond to the change in membrane potential. A connected point is that the region would be electrically silent and not therefore contribute to the recorded current. The problem in evaluating the likely changes in local sodium concentration can be overcome by solving the reaction/diffusion equations for the sodium entering via a point source (the sodium channel).

To further examine the possible local increase in internal sodium concentration associated with the activation of sodium channels, the diadic cleft was modeled as a thin disk-like, 10 nm thick (the space between the SR and T-tubule membranes) and 100 nm in diameter. As shown in FIGURE 3, activation of a sodium channel with a 4-pA single channel current would result in an appreciable increase in local sodium concentration. The increase is strongly dependent on the diffusion coefficient for sodium and the values on the curves correspond to diffusion coefficients (D) of 50%, 21% and 7% of that observed in free solution. Although the model considers only one sodium channel per diad, reported sodium current densities suggest that there will be only ~5 sodium channels per μm^2 so it seems unlikely that the sodium channels would be close enough to each other to alter the calculation results markedly. Placing the channels very close together (*i.e.*, within 20 nm) will start to reduce the single channel current as local sodium accumulation will reduce the driving force for sodium entry (this effect gets larger as the diffusion coefficient decreases). The calculated increase in local sodium is *only* significant for the time that the sodium channel is open. For example, 10 nm from the channel, the sodium concentration takes only 8 μs to fall to 6.5 mM after channel closure (with the lowest diffusion coefficient modelled). Even if the Na-Ca exchanger were able to respond *kinetically* to such a short-term increase in sodium level, it is clearly extremely difficult to explain the production of a 10 μM trigger at 40 mV via the Na-Ca exchange mechanism unless the Na-Ca exchanger is packed around the sodium channel. As noted above, larger reductions in the diffusion coefficient will reduce the single channel current as sodium will accumulate around the channel and reduce the driving force for sodium entry. Thus, the possible magnitude of the influx via sodium channels will be limited by the diffusion of sodium from the diadic space rather than by the single channel current and the number of channels in the diad.

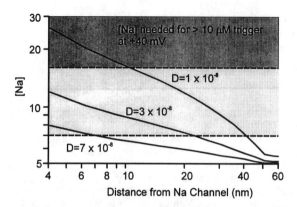

FIGURE 3. Calculations of the sodium levels that occur in the diad during the activation of a 4-pA sodium channel situated in the center of the diadic space (modelled as a circular region 100 nm in diameter and 10 nm high) at steady state. Diffusion coefficients of 7×10^{-6}, 3×10^{-6} and 1×10^{-6} cm^2 s^{-1} are shown. Note that both axes have logarithmic scales. It is clear that sodium levels above 17 mM (needed to achieve a trigger calcium level of 10 μM [Ca^{2+}]$_i$ at +40 mV) only exist adjacent to the sodium channel itself. Sodium levels needed to exceed a trigger of 1 μM [Ca^{2+}]$_i$ at +40 mV exist across most of the diad, but such levels exist only while the sodium channel is open. Unless the diffusion of sodium is much more restricted than modelled here (which would barely increase the local [Na] but would increase the persistence of the elevated [Na]), it is clear that such short lived increases in sodium are hardly sufficient to provide the proposed trigger [Ca^{2+}]$_i$ levels.

The above analyses are predicated on the assumption that the trigger calcium level is 10 μM, but there is evidence that under some conditions the trigger calcium level can be less than this. For example, under conditions of calcium overload (where the resting level of calcium is increased) propagating waves of SR calcium release are observed.[1,14–16] Since the calcium transient reaches a peak of about 1 μM, the observation of propagating SR release suggests that the required trigger calcium level could be about an order of magnitude lower during calcium overload. To achieve such a trigger calcium level at -40 mV with the exchanger, internal sodium would have to be about 18 mM, a level much closer to that which can be achieved by sodium channel activation (cf. FIG. 3). Thus the results of Lipp and Niggli[8] can be explained if SR calcium release in their experiments was more sensitive to the local calcium level than expected from the results of Cannell *et al.*[10] In addition, the inevitable voltage escape during the sodium current would make the trigger calcium level more attainable.

As noted above, our analysis is based on the assumption that the exchanger is kinetically capable of approaching thermodynamic equilibrium over the physiological time scale during which E-C coupling takes place (10 ms). With a *maximum* turnover number of $\sim$20,000 s^{-1} (see Hilgemann in this volume), each exchanger could introduce no more than 200 calcium ions into the diadic space in 10 ms. This can be compared to the flux associated with a calcium channel which, for a single channel current of 0.3 pA,[17] would introduce about 10,000 calcium ions over the same period. Thus if the exchanger were operating at its *maximum* rate, it is clear that the exchanger is a far slower mechanism for producing a trigger calcium influx. However, this upper limit estimate ignores the problem of relative exchanger density, which can be circumvented by direct comparison of exchanger and calcium current densities.

Kimura, Noma and Irisawa[18] reported that (at 35°C) the Na-Ca exchange current was ~2.5 pA/pF at 30 mM internal sodium and a test potential of +10 mV (their Fig. 1). For comparison, the maximum calcium current we have observed is 25 pA/pF. This result suggests that the Ca^{2+} influx via I_{Ca} may be about five times larger than that due to the exchanger operating conditions designed to stimulate calcium influx. In connection with this point, Cannell[19] reported that, with 8 mM internal Na, the rate of calcium influx via the exchanger at +10 mV was 0.04 μM s^{-1}. The current equivalent for this rate of exchange would be 18 pA (assuming a cell volume of 30 pl with buffering power of 150 μM bound/μM free). For such a cell, the cell membrane area would be about 120 pF, which suggests an exchanger flux equivalent to a 0.3 pA/pF calcium current. It is notable that this estimate is in good agreement with the measurements of Kimura, Noma and Irisawa[18] (assuming a cubic dependence of exchanger rate on internal sodium). Thus the exchanger does not appear capable of producing a calcium influx as large as that due to calcium channels even when intracellular sodium is raised. However, exchanger fluxes can be markedly increased by "deregulation" which can be produced by proteolysis. When the exchanger is deregulated, the maximum currents are about 40 pA/pF which suggests that the exchanger could then become a significant source of trigger calcium. Under such conditions, we would expect calcium influx via the exchanger to be able to trigger a large calcium release from the SR.

From all of the above considerations, we suggest that the exchanger is unlikely to be as potent a trigger mechanism as the calcium current under normal physiological conditions. Thus reverse-mode Na-Ca exchange would have a minor role in triggering CICR. Our failure to detect sodium current-induced calcium release should not be taken as evidence that the exchanger *cannot* trigger calcium release from the SR, but rather that such release is not a robust phenomenon, relying heavily on other factors such as the SR calcium content, temperature and local sodium accumulation. In contrast, calcium release is routinely observed during calcium current activation—even under depotentiated conditions.[20] In addition, when low concentrations of cadmium or organic calcium channel blockers are applied, which barely affect the sodium current or Na-Ca exchange, SR calcium release is greatly reduced.[10] Nevertheless, the Na-Ca exchanger may become more important when the cell calcium content (load) increases, since any increase in cytosolic Ca will inhibit I_{Ca} and promote Na-Ca exchange (as a result of an increase in internal sodium and the catalytic effect of internal calcium). Since an increase in the sensitivity of CICR may also occur in these conditions, reverse-mode Na-Ca exchange may then be able to act as an effective trigger for CICR, as first observed by Berlin et al.[3]

REFERENCES

1. CHENG, H., W. J. LEDERER & M. B. CANNELL. 1993. Calcium sparks: elementary events underlying excitation-contraction coupling in heart muscle. Science **262:** 740–744.
2. LEVI, A. J., P. BROOKSBY & J. C. HANCOX. 1993. One hump or two? The triggering of calcium release from the sarcoplasmic reticulum and the voltage dependence of contraction in mammalian cardiac muscle. Cardiovasc. Res. **27:** 1743–1757.
3. BERLIN, J. R., M. B. CANNELL & W. J. LEDERER. 1987. Regulation of twitch tension

in sheep cardiac Purkinje fibers during calcium overload. Am. J. Physiol. **253:** H1540–H1547.
4. SHAM, J. S. K., L. CLEEMANN & M. MORAD. 1992. Gating of the cardiac Ca^{2+} release channel: the role of the Na^+ current and the Na^+-Ca^{2+} exchange. Science **255:** 850–853.
5. NUSS, H. B. & S. R. HOUSER. 1992. Sodium-calcium exchange-mediated contractions in feline ventricular myocytes. Am. J. Physiol. **263:** H1161–H1169.
6. LEVI, A. J., P. BROOKSBY & J. C. HANCOX. 1993b. A role for depolarisation induced calcium entry on the Na-Ca exchange in triggering intracellular calcium release and contraction in rat ventricular myocytes. Cardiovasc. Res. **27:** 1677–1690.
7. LEBLANC, N. & J. R. HUME. 1990. Sodium current-induced release of calcium from cardiac sarcoplasmic reticulum. Science **248:** 372–376.
8. LIPP, P. & E. NIGGLI. 1994. Sodium current-induced calcium signals in isolated guinea-pig ventricular myocytes. J. Physiol. **474:** 439–446.
9. LEVESQUE, P. C., N. LEBLANC & J. R. HUME. 1994. Release of calcium from guinea-pig cardiac sarcoplasmic reticulum induced by sodium-calcium exchange. Cardiovasc. Res. **28:** 370–378.
10. CANNELL, M. B., H. CHENG & W. J. LEDERER. 1994. Spatial non-uniformity's in $[Ca^{2+}]_i$ during excitation-contraction coupling in cardiac myocytes. Biophys. J. **67:** 1942–1956.
11. LEDERER, W. J., E. NIGGLI & R. W. HADLEY. 1990. Sodium-calcium exchange in excitable cells: fuzzy space. Science **248:** 283.
12. GYORKE, S. & M. FILL. 1993. Ryanodine receptor adaptation: control mechanism of Ca^{2+}-induced Ca^{2+} release. Science **260:** 807–809.
13. CRESPO, L. N., C. J. GRANTHAM & M. B. CANNELL. 1990. Kinetics, stoichiometry and role of the Na-Ca exchange in isolated cardiac myocytes. Nature **345:** 618–621.
14. BERLIN, J. R., M. B. CANNELL & W. J. LEDERER. 1989. I_{TI} in single rat cardiac ventricular cells: relationship to fluctuations in intracellular calcium. Circ. Res. **65:** 115–126.
15. TAKAMATSU, T. & W. G. WIER. 1990. Calcium waves in mammalian heart: quantification of origin, magnitude, waveform and velocity. FASEB J. **4:** 1519–1525.
16. LIPP, P. & E. NIGGLI. 1993. Microscopic spiral waves reveal positive feedback in subcellular calcium signalling. Biophys. J. **65:** 2272–2276.
17. ROSE, W. C., C. W. BALKE, W. G. WIER & E. MARBAN. 1992. Macroscopic and unitary properties of physiological ion flux through L-type Ca^{2+} channels in guinea-pig heart cells. J. Physiol. **456:** 267–284.
18. KIMURA, J., A. NOMA & H. IRISAWA. 1986. Na-Ca exchange current in mammalian heart cells. Nature **319:** 596–599.
19. CANNELL, M. B. 1991. Contribution of sodium-calcium exchange to calcium regulation in cardiac muscle. Ann. N. Y. Acad. Sci. **639:** 428–443.
20. HAN, S., A. SCHIEFFER & G. ISENBERG. 1994. Ca^{2+} load of guinea-pig ventricular myocytes determines efficacy of brief Ca^{2+} currents as trigger for Ca^{2+} release. J. Physiol. **48:** 411–421.
21. O'NEILL, S. C., J. G. MILL & D. A. EISNER. 1990. Local activation of contraction in isolated rat ventricular myocytes. Am. J. Physiol. **258:** C1165–C1168.

Evidence That Reverse Na-Ca Exchange Can Trigger SR Calcium Release[a]

SHELDON LITWIN,[b] OSAMI KOHMOTO,[c]
ALLEN J. LEVI,[d] KENNETH W. SPITZER,[b] AND
JOHN H. B. BRIDGE[b,e]

[b]*Nora Eccles Harrison Cardiovascular Research and
Training Institute
University of Utah
Salt Lake City, Utah 84112*

[c]*University of Tokyo
Tokyo, Japan*

[d]*University of Bristol
Bristol, United Kingdom*

INTRODUCTION

It is now widely accepted that in heart muscle contraction occurs when a small rise of Ca in the vicinity of the SR causes a much larger release of Ca from the interior of the SR,[4] which then binds to troponin and activates contraction. This process is referred to as Ca-induced Ca release (CICR). It appears that the extent of release of Ca from the SR is graded with the extent of rise of the Ca that induces release.[5] Some information on the molecular basis of this process in now available. Large Ca release channels containing ryanodine receptors are found in the SR membrane.[14] These channels are gated by Ca. Thus when Ca rises in the vicinity of an SR release channel a site on the channel is occupied and this causes an increase in the probability of opening of the release channel. Why the channel does not respond in a regenerative fashion to the Ca that passed through it is unclear although some plausible explanations are now available.[18]

The origin of the Ca that triggers SR Ca release is currently under investigation. It seems to be widely accepted that the L-type Ca current is capable of triggering SR Ca release. This was first inferred by London and Krueger.[12] Since then a number of people have adduced evidence that the Ca current provides sufficient Ca to induce release from the SR. Most recently Lopez-Lopez *et al.*[13] demonstrated that in cells treated with verapamil the probability of detecting local Ca transients (sparks) with a confocal microscope exhibited a bell shaped dependence on voltage. This is consistent with the idea that these local Ca transients are tightly linked to the flux of Ca through Ca channels. These authors conclude that local Ca transients

---

[a] Supported in part by National Institutes of Health Grant HL42357, the Nora Eccles Treadwell Foundation, Sankyo Life Science Foundation, and the Department of Physiology, School of Medical Sciences, University of Bristol, UK.

[e] Address for correspondence: John H. B. Bridge, PhD, University of Utah, CVRTI, Bldg. 500, Salt Lake City, Utah 84112.

are not linked to Na-Ca exchange. However, a number of investigators have obtained evidence that under a variety of circumstances an additional pathway for Ca influx, *i.e.*, the Na-Ca exchange is capable of triggering SR Ca release. First Berlin *et al.*[1] suggested that under circumstances where the SR was overloaded in sheep Purkinje fibers SR Ca release could be triggered by the Na-Ca exchange. A number of authors have adduced evidence that when intracellular Na is elevated the expected relationship between triggered events and voltage shifts from a bell shape to a more sigmoid shape.[9,15,19] The results seem to suggest that as the potential becomes more positive Na-Ca exchange begins to contribute significantly as a trigger for SR Ca release. In addition some authors have obtained evidence that activation of inward Na current can produce sufficient Na entry to enhance triggering by reverse Na-Ca exchange.[7,8,11]

We recently were able to demonstrate that under a variety of circumstances the Na-Ca exchange is able to trigger SR Ca release and hence contractions.[6,9] If both the Na-Ca exchange and the Ca current are capable of triggering SR Ca release simultaneously, it is important to be able to selectively inhibit one of the putative triggers without affecting the other. For example to demonstrate triggering by Na- Ca exchange it is important to be able to selectively remove the Ca current. The drug nifedipine can be used to accomplish this. However, if the drug is applied slowly the SR content can be depleted. We therefore developed methods to apply nifedipine rapidly so that the Ca current can be rapidly and completely blocked. Any remaining triggered contractions are inferred to be triggered by the Na- Ca exchange. Several results that we have obtained suggest that even when intracellular Na is 10 mM rapid inhibition of the Ca current leaves contractions that are triggered by the Na-Ca exchange.

Na-Ca Exchange Can Trigger SR Ca Release

In most of the experiments described below we investigated isolated ventricular myocytes under voltage clamp either at room temperature or 30°C. It is possible to investigate the contribution of either Na-Ca exchange or Ca current to contraction by voltage clamping cells with single microelectrodes and holding them at -40 mV to inactivate Na current. These cells are then depolarized to potentials that lie between -30 and $+60$ mV. In most of the experiments to be described cell shortening was used as a surrogate for SR Ca release. Cell shortening can be conveniently measured with a video-based motion detector.[17] A number of our results depend crucially on being able to rapidly change extracellular solutions.[16] In particular we attempted to completely block Ca current before any depletion of the SR Ca can take place. This is essential if one is to investigate the contribution of the Na-Ca exchange to triggering. Without a functioning Ca current the SR can deplete significantly within half a dozen beats. It therefore becomes extremely difficult to distinguish the effect of Ca channel blockers on the release trigger and SR Ca depletion.

Relationships between Shortening, Voltage and I_{Ca}

The L-type Ca current exhibits a well-known bell shaped dependence on voltage. If we assume that this Ca current is the sole trigger for SR Ca release then the voltage dependence of triggered events should also be bell shaped. This has already been observed.[2] Since the extent of SR Ca release is believed to be graded

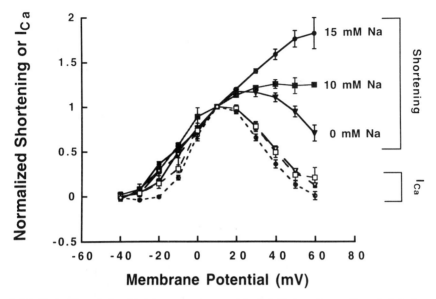

FIGURE 1. The relationship between voltage and I_{Ca} is bell shaped regardless of dialyzing Na concentration. However the relationship between shortening and voltage depends upon the concentration of dialyzing Na. When dialyzing Na is nominally zero mM the relationship between voltage and the extent of shortening approaches a bell shape.

with the size of the trigger, it is reasonable to expect the magnitude of triggered contractions to be similar at two voltages (*e.g.*, −20 and +40 mV) where these two currents are of similar magnitude. If on the other hand the release process possesses some intrinsic voltage dependence that is different from the voltage dependence of the Ca current then a simple bell shaped relationship between triggered contractions and voltage might not exist. It is also possible that the shape of the relationship between the magnitude of the Ca current and the magnitude of SR Ca release (the gain of the process) changes with voltage[20] or some other process besides the Ca currrent is involved in triggering SR Ca release. These processes might also cause the relationship between triggered events and voltage to depart from a simple bell shape. In addition if the Na-Ca exchange is involved in triggering SR Ca release, we expect the magnitude and the shape of the relationship between voltage and triggered contractions to show a clear dependence on intracellular Na.

We first investigated the relationship between the extent of shortening, I_{Ca} and voltage in rabbit ventricular myocytes under voltage clamp. Cells were held at −40 mV and clamped to various potentials between −40 and +60 mV. Between each test pulse the cell was subject to conditioning pulses from −40 to 10 mV every 3 seconds to establish steady state SR Ca loading. This procedure was repeated on cells dialyzed with nominally zero, 10 of 20 mM pipette Na. It is clear that while the relationship between Ca current and voltage remains bell shaped regardless of the concentration of dialyzing Na, the relationship between the extent of shortening and voltage depends upon the concentration of dialyzing Na (FIG. 1).

With zero Na in the pipette cellular contractions diminished with voltage at

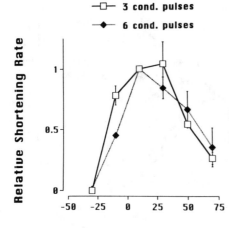

FIGURE 2. Shortening voltage relationships obtained in the absence of a Na gradient. Relationships were measured after 3 or 6 conditioning pulses which filled the SR to differing extents. Despite this the relationship between shortening and voltage was bell shaped.

potentials that are positive to 10 mV. This bell-shaped relationship between voltage and contractions suggests that most of the SR Ca release is triggered by I_{Ca}. With 10 mM Na in the pipette the relationship between contractions and voltage did not assume a simple bell shape. At potentials greater than 10 mV contractions reached a plateau. With 15 mM Na in the pipette the shortening rate continued to increase at potentials positive to 10 mV. It is therefore clear that some process that depends upon the value of dialyzing Na influences the shape of the shortening voltage relationship.

As intracellular Na is increased we expect intracellular Ca to increase concomitantly. This is because under these circumstances the efflux of Ca by Na-Ca exchange is reduced and influx is increased. Some of this net gain of cellular Ca is acquired by the SR. It might be argued that if the SR Ca load is increased then the shape of the shortening voltage relationship might be altered in such a way that at positive potentials the gain of the release process is increased. There is some precedent for the idea that the gain of the release process does depend upon voltage.[20] We investigated this possibility by voltage clamping cells with pipettes deficient in Na. We constructed relationships between triggered contractions current and voltage by voltage clamping cells from −30 mV to +60 mV in the absence of a transmembrane Na gradient. Between measurements at each voltage the extracellular Na was restored and steady state contractions were established to ensure a consistent loading of the sarcoplasmic reticulum. If after removing the Na gradient the cell is stimulated a few times the SR Ca load can be increased and under these circumstances the relationship between current, shortening and voltage can again be established. In these experiments either 3 or 6 conditioning pulses were applied in the absence of a Na gradient to vary SR Ca load. Regardless of the SR Ca content we found that the relationship between voltage and shortening was bell shaped in the absence of a Na gradient. We conclude from this experiment that the voltage dependence of triggered contractions was independent of SR Ca content under these circumstances (FIG. 2).

We conclude that the departure from a bell-shaped relationship that is observed with varying amounts of dialyzing Na is most likely to be due to the presence of Na-Ca exchange. It is quite possible that at positive potentials as the value of the Ca current declines it is replaced by reverse Na-Ca exchange which triggers SR

Ca release. This process should increase with increasing dialyzing Na and would account for the fact that in the presence of a Na gradient the magnitude of contractions either declines more slowly or continues to increase when the Ca current is clearly declining. We will now describe a body of direct evidence that suggests that the Na-Ca exchange is indeed capable of triggering SR Ca release under a variety of conditions.

Effect of Exchanger Inhibitory Peptide on SR Ca Release

We investigated the possibility that the Na-Ca exchange was capable of triggering SR Ca release under circumstances where intracellular Na was assumed to be approximately normal (10 mM) and when it was elevated to 20 mM or reduced to 0 mM.[6] We voltage clamped Guinea pig ventricular cells with microelectrodes that contained 20 mM Na and held them at a potential of −120 mV, which is negative to the reversal potential of the Na-Ca exchange even when intracellular Na is 20 mM. Just before an experimental trial we held the cell at −40 mV and

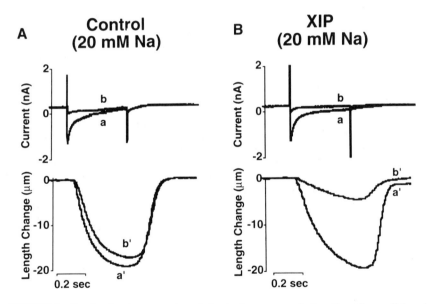

FIGURE 3. Graphs showing currents and shortening measured when the cell was dialyzed with 20 mM sodium. **(A)** *Top*, L-type calcium currents. a indicates current measured after the cell was abruptly superfused with 20 μM nifedipine. *Bottom*, cell shortening. a' indicates shortening measured at the same time as current a; b', shortening measured at the same time as current b. **(B)** *Top*, L-type calcium currents measured after the cell was dialyzed with 20 μM exchange inhibitory peptide (XIP) for 15 minutes. a indicates current measured after depolarization from −40 mV; b, current measured after the cell was abruptly superfused with nifedipine. *Bottom*, cell shortening. a' indicates shortening measured at the same time as current a; b', shortening measured at the same time as current b. In the presence of XIP, shortening becomes dependent on calcium current. (From Chin *et al.*[3] Reprinted by permission from *Circulation Research*.)

then depolarized the cell to +10 mV for 300 msec. This activated a large Ca current and an accompanying large contraction (FIG. 3A). We then switched the superfusing solution to one containing 20 μM nifedipine and again depolarized the cell to +10 mV for 300 msec. This produced a dramatic reduction in the magnitude of the Ca current but had little effect on the magnitude of the contractions, which we assumed were produced by Ca that was discharged from the sarcoplasmic reticulum.

We conclude that either a very small residual Ca current can under these circumstances trigger large contractions or alternatively some other process, most likely the Na-Ca exchange could trigger Sr Ca release. We tested the possibility that the Na-Ca exchange was involved in triggering with exchanger inhibitory peptide (XIP).[10] This substance is also known to inhibit Na-Ca exchange current.[3] We dialyzed the cells with a pipette solution containing 20 μM XIP and repeated the foregoing experiment, *i.e.*, we activated contractions in the presence of XIP and after rapidly applying nifedipine to block I_{Ca}. Under these circumstances most of the contraction was inhibited (FIG. 3B). Thus when the cells were dialyzed with 20 mM Na, blockade of the Ca current failed to inhibit contraction. However, in the presence of substances that block both Na-Ca exchange and calcium current it is possible to block most of the contraction. The simplest interpretation of this result is that there are in fact two triggers for SR Ca release. One is I_{Ca} and the other is Na-Ca exchange. If the cell is dialyzed with 20 mM Na the presence of either trigger is sufficient to activate contraction.

We conducted similar experiments with cells that were dialyzed with 10 mM Na. We found that under these circumstances approximately 30% of the contraction could be activated when the Ca current was largely blocked (FIG. 4). This fraction of the contraction appears to be inhibited by XIP and as such is probably triggered by reverse Na-Ca exchange.

Finally we dialyzed cells with pipette solutions containing no Na. Under these circumstances we expected that only Ca current would be available to trigger contractions. We therefore activated contractions with voltage clamp pulses from −40 to +10 mV before and after rapidly inhibiting Ca current with nifedipine. It is clear from FIG. 5 that when Ca current is the only available trigger most of the contraction is inhibited by nifedipine. Most of the XIP-inhibitable contractions are therefore dependent upon intracellular Na as expected.

It might be argued that with elevated intracellular Na there is sufficient Ca entry by reverse Na-Ca exchange to directly activate the contractile elements. We investigated this possibility with two drugs that prevent the SR from sequestering Ca. When heart cells are treated with a combination of ryanodine and thapsigargin the SR is prevented from storing Ca. In the presence of these two drugs we established that depolarizing the cell from −40 to +10 mV activated a small contraction. However, in the presence of 20 μM nifedipine which completely abolished I_{Ca} also completely eliminated contraction. It should be pointed out that under these circumstances we expect reverse Na-Ca exchange to be functioning. We therefore conclude that when the SR is not functioning reverse Na-Ca exchange cannot directly activate the contractile elements and this suggests that it does indeed act as a trigger. This trigger is particularly potent when intracellular Na is elevated.

Additional Evidence That the Na-Ca Exchange Can Trigger SR Ca Release

In a series of experiments designed to demonstrate that Na-Ca exchange can indeed trigger contraction, Kohmoto *et al.*[9] activated contraction under circum-

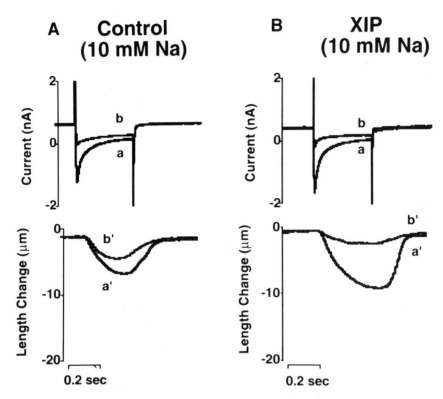

FIGURE 4. Graphs showing currents and shortening measured when the cell was dialyzed with 10 mM sodium. **(A)** *Top*, currents measured in the absence, a, of 20 μM nifedipine and, b, in its presence. *Bottom*, cell shortening, a', measured at the same time as current a; b', shortening measured at the same time as current b after nifedipine treatment. **(B)** *Top*, L-type calcium currents measured after the cell was dialyzed with 20 μM nifedipine. *Bottom*, a' indicates shortening measured at the same time as current measured in the presence of 20 μM nifedipine. *Bottom*, a' indicates shortening measured at the same time as current a; b', shortening measured at the same time as current b after nifedipine treatment. (From Chin et al.[3] Reprinted by permission from *Circulation Research*.)

stances where reverse exchange is likely to occur vigorously. Guinea pig ventricular cells were voltage clamped and held at −40 mV with single microelectrodes that contained 10 mM Na. First the cell was subject to a train of conditioning pulses of 400 msec duration at 0.33 Hz. This activated L type Ca currents and accompanying contractions. Next the rapid solution switcher was used to switch to Na-free solution for 100 msec. The switch started 50 msec before and lasted to 50 msec after the depolarizing clamp pulse. In four cells a switch to Na-free solution increased the amplitude of the phasic contraction by 30 ± 3.1% (FIG. 6).

This experiment suggests that when reverse Na-Ca exchange is enhanced a larger Sr Ca release and hence contraction is triggered. It might be argued that the brief exposure to a solution deficient in Na caused reverse Na-Ca exchange to occur to an extent that resulted in an increase in SR Ca content. Were this to

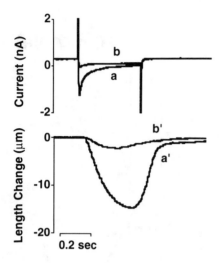

FIGURE 5. Currents and shortening measured when the cell was dialyzed with sodium-deficient solution. *Top*, a and b indicate calcium current measured before and after nifedipine treatment. *Bottom*, a' and b' indicate shortening corresponding to currents a and b. Note that when the cell is not dialyzed with sodium, contraction becomes strikingly dependent on calcium current. (From Chin et al.[3] Reprinted by permission from *Circulation Research*.)

occur enhanced contractions due to depolarization in the absence of Na might be incorrectly assigned to triggering by Na-Ca exchange when they were in fact due to an enlarged SR Ca pool. We therefore tested this possibility by removing and then returning Ca for 50 msec before depolarizing the cell. The ensuing depolarization produced no increase in contraction, from which we conclude that the exchange did not produce its effect by loading SR Ca stores (FIG. 7.)

Several additional results seem to suggest that the Na-Ca exchange can indeed function as a trigger for SR Ca release. By applying the Ca channel antagonist nifedipine between depolarizing pulses we were able to block Ca current (FIG. 8). Despite this when we depolarized the cell from -40 to $+10$ mV we elicited a considerable contraction.

This contraction was activated despite the absence of any measurable Ca current. The most likely explanation for this was that the contraction was triggered by reverse Na-Ca exchange. In a similar series of experiments we activated contractions and action potentials. When we rapidly applied nifedipine and then activated an action potential we significantly reduced the duration of the action potential. The accompanying contraction, although of reduced magnitude, was not abolished. Shortening the action potential is certainly consistent with removal of the Ca current. From this we infer that an alternative process, presumably the Na-Ca exchange, produced triggered contractions. These contractions are not abolished by 100 μM nickel, which indicates that they were not activated by t type Ca currents, and they were completely abolished by nickel, which is consistent with idea that they were triggered by reverse Na-Ca exchange.

SUMMARY

Several results suggest that the Na-Ca exchange can function as a trigger promoting SR Ca release and ensuing contractions. First, if the Ca current was the sole trigger for contraction we would expect the relationship between triggered

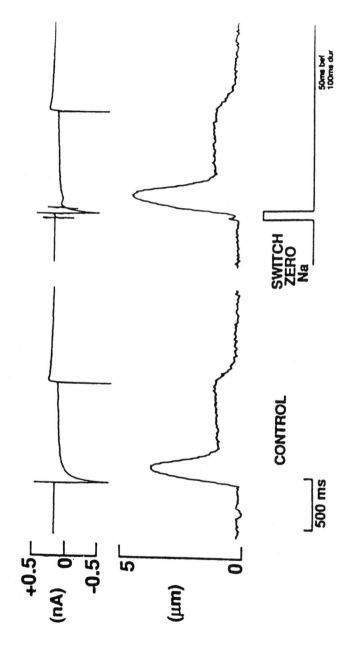

FIGURE 6. Effect on phasic twitch of a switch to Na-free solution just before depolarization. The rapid switch to Na-free solution was made 50 ms before start of the test pulse. The switch to Na-free solution lasted for a total of 100 ms so that Na-free solution was removed 50 ms after stepping to +10 mV. (From Levi et al.[9] Reprinted by permission from the *American Journal of Physiology*.)

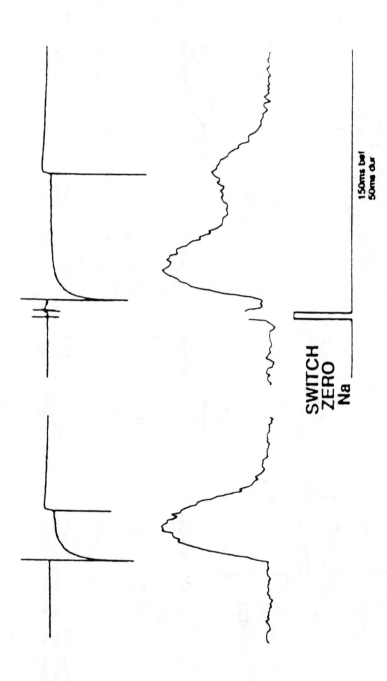

FIGURE 7. Effect on phasic twitch of a switch to Na-free solution just before depolarization. The switch to Na-free solution was made for 50 ms and then normal Na was applied for 100 ms before the test depolarization. (From Levi et al.[9] Reprinted by permission from the *American Journal of Physiology*.)

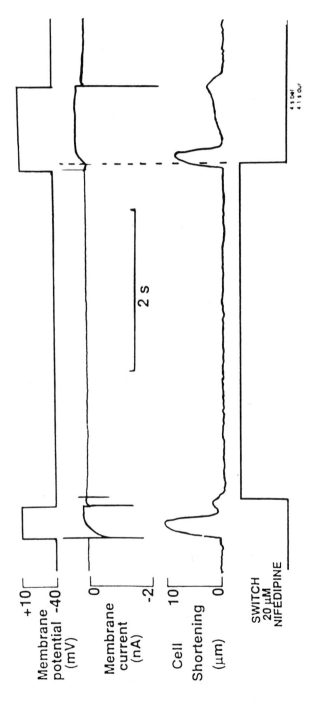

FIGURE 8. Effect of blocking I_{Ca} in the interval between depolarizing pulses. A rapid switch to a solution containing 20 μM nifedipine was made between successive voltage-clamp pulses. Before the 1-second test pulse to +10 mV, a train of 6 conditioning pulses from −40 to +10 mV was applied to establish a steady state level of SR Ca load. The final pulse in this train is shown at the beginning of the record. Each conditioning pulse elicited I_{Ca} and a phasic twitch. Nifedipine was applied 4 seconds before the start of the test pulse and its application was continued until 100 ms after the start of the test pulse. Even though I_{Ca} was abolished completely by nifedipine, there was still a substantial phasic twitch elicited by the test pulse. (From Levi et al.[9] Reprinted by permission from the *American Journal of Physiology*.)

contractions and voltage to be similar to the relationship between Ca current and contraction. When Na is present in the pipette this is not observed. Between −40 and +10 mV the relationships between contractions and voltage and current and voltage are similar. At potentials positive to 10 mV the Ca current declines as expected but contractions either decline much more slowly or continue to increase depending upon the concentration of intracellular Na. In addition, we have observed that contractions can be activated when Ca current is largely or completely blocked. Since these contractions are sensitive to the presence of ryanodine and thapsigargin they appear to be triggered by Na-Ca exchange. Also, contractions that are activated in the presence of nifedipine are sensitive to the Na-Ca exchange inhibitor XIP. Finally, rapid removal of extracellular Na apparently stimulates enough reverse exchange to enhance triggering of SR Ca release without affecting the SR content.

It is clear that the shape of the shortening voltage relationship depends upon the concentration of dialyzing Na. This is likely to occur for two reasons. Either the shape of the shortening voltage relationship depends upon the extent to which Na-Ca exchange contributes a trigger for SR Ca release or alternatively the shape of the shortening voltage relationship depends upon SR Ca content. The latter is known to depend upon the Na concentration. In addition it is now established that the gain of SR Ca release is influenced by SR content. However, we studied triggered contractions in the absence of a Na gradient when the only available trigger is the Ca current. We measured triggered contractions over a range of voltages between −30 and +60 mV. Between each measurement we reestablished the Na gradient and activated a series of conditioning pulses to standardize the SR Ca content. Just before a test pulse we removed extracellular Na and activated either 3 or 6 pulses to produce two different SR Ca loads (in the absence of a Na gradient entering Ca cannot be extruded and therefore changes the SR Ca content). Regardless of the number of prepulses in the absence of a Na gradient the shortening voltage relationship was similar and bell shaped. From this we conclude that the shape of the relationship between shortening and voltage does not depend upon SR Ca content. Therefore, we conclude that the asymmetry in the shortening voltage relationship that depends upon intracellular Na is due to a contribution of reverse Na-Ca exchange.

It is too early to say what the physiological significance (if any) of triggering by reverse exchange actually is. However, it does seem likely that it might provide a powerful inotropic mechanism. For example intracellular Na might be expected to change with heart rate and to be elevated at higher heart rates. Presumably this increased intracellular Na would tend to favor triggering by reverse exchange and would therefore enhance contractility at a time when it would be most required.

REFERENCES

1. BERLIN, J. R., M. B. CANNELL & W. J. LEDERER. 1987. Regulation of twitch tension in sheep cardiac Purkinje fibers during calcium overload. Am. J. Physiol. Heart Circ. Physiol. **253:** H1540–H1547.
2. BEUCKELMANN, D. J. & W. G. WIER. 1988. Mechanism of release of calcium from sarcoplasmic reticulum of guinea-pig cardiac cells. J. Physiol. **405:** 233–255.
3. CHIN, T. K., K. W. SPITZER, K. D. PHILIPSON & J. H. B. BRIDGE. 1993. The effect of exchanger inhibitory peptide (XIP) on sodium-calcium exchange current in guinea pig ventricular cells. Circ. Res. **72:** 497–503.
4. FABIATO, A. 1983. Calcium-induced release of calcium from the cardiac sarcoplasmic reticulum. Am. J. Physiol. **245:** C1–C14.

5. FABIATO, A. 1985. Time and calcium dependence of activation and inactivation of calcium-induced release of calcium from the sarcoplasmic reticulum of a skinned canine cardiac Purkinje cell. J. Gen. Physiol. **85:** 247–289.
6. KOHMOTO, O., A. J. LEVI & J. H. B. BRIDGE. 1994. Relation between reverse sodium-calcium exchange and sarcoplasmic reticulum calcium release in guinea pig ventricular cells. Circ. Res. **74:** 550–554.
7. LEBLANC, N. & J. R. HUME. 1990. Sodium current-induced release of calcium from cardiac sarcoplasmic reticulum. Science **248:** 372–376.
8. LEVESQUE, P. C., N. LEBLANC & J. R. HUME. 1991. Role of reverse-mode Na^+-Ca^{2+} exchange in excitation-contraction coupling in the heart. Ann. N. Y. Acad. Sci. **639:** 386–397.
9. LEVI, A. J., K. W. SPITZER, O. KOHMOTO & J. H. B. BRIDGE. 1994. Depolarization-induced Ca entry via Na-Ca exchange triggers SR release in guinea pig cardiac myocytes. Am. J. Physiol. Heart Circ. Physiol. **266:** H1422–H1433.
10. LI, Z., D. A. NICOLL, A. COLLINS, D. W. HILGEMANN, A. G. FILOTEO, J. T. PENNISTON, J. N. WEISS, J. M. TOMICH & K. D. PHILIPSON. 1991. Identification of a peptide inhibitor of the cardiac sarcolemmal Na^+-Ca^{2+} exchanger. J. Biol. Chem. **266:** 1014–1020.
11. LIPP, P. & E. NIGGLI. 1994. Sodium current-induced calcium signals in isolated guinea-pig ventricular myocytes. J. Physiol. (London) **474:** 439–446.
12. LONDON, B. & J. W. KRUEGER. 1986. Contraction in voltage-clamped internally perfused single heart cells. J. Gen. Physiol. **88:** 475–505.
13. LÓPEZ-LÓPEZ, J. R., P. S. SHACKLOCK, C. W. BALKE & W. G. WIER. 1995. Local calcium transients triggered by single L-type calcium channel currents in cardiac cells. Science **268:** 1042–1045.
14. MEISSNER, G. 1994. Ryanodine receptor/Ca^{2+} release channels and their regulation by endogenous effectors. Annu. Rev. Physiol. **56:** 485–508.
15. NUSS, H. B. & S. R. HOUSER. 1992. Sodium-calcium exchange-mediated contractions in feline ventricular myocytes. Am. J. Physiol. Heart Circ. Physiol. **263:** H1161–H1169.
16. SPITZER, K. W. & J. H. B. BRIDGE. 1989. A simple device for rapidly exchanging solution surrounding a single cardiac cell. Am. J. Physiol. Cell Physiol. **256:** C441–C447.
17. STEADMAN, B. W., K. B. MOORE, K. W. SPITZER & J. H. B. BRIDGE. 1988. A video system for measuring motion in contracting heart cells. IEEE Trans. Biomed. Eng. **35:** 264–272.
18. STERN, M. D. 1992. Theory of excitation-contraction coupling in cardiac muscle. Biophys. J. **63:** 497–517.
19. VORNANEN, M., N. SHEPHERD & G. ISENBERG. 1994. Tension-voltage relations of single myocytes reflect Ca release triggered by Na/Ca exchange at 35°C but not 23°C. Am. J. Physiol. Cell Physiol. **267:** C623–C632.
20. WIER, W. G., T. M. EGAN, J. R. LÓPEZ-LÓPEZ & C. W. BALKE. 1994. Local control of excitation-contraction coupling in rat heart cells. J. Physiol. (London) **474:** 463–471.

Regulation of Sodium-Calcium Exchange in Intact Myocytes by ATP and Calcium

R. A. HAWORTH AND A. B. GOKNUR

Department of Anesthesiology
University of Wisconsin
B6/387 Clinical Science Center
600 Highland Avenue
Madison, Wisconsin 53792

INTRODUCTION

Two significant properties of the Na-Ca exchanger revealed by studies on squid axon are that the exchanger is controlled by ATP[1,2] and by intracellular Ca (Ca_i).[3,4] The exchanger in heart is likewise controlled by ATP[5,6] and by Ca_i.[6-9] Given the role which Ca_i plays in excitation-contraction coupling, and the role which Na-Ca exchange could play in this coupling, not only in Ca efflux[10-13] but also in Ca influx,[14-17] understanding the regulation of exchange activity by Ca_i is clearly of great importance. Also, the failure of hearts to recover function following a period of ischemia is correlated with an increased uptake of Ca from the perfusate, most likely via Na-Ca exchange.[18] Since heart tissue undergoing ischemia suffers a loss of ATP followed by limited regeneration of ATP on reperfusion,[19] it is important from the standpoint of pathophysiology to understand the role of ATP and intracellular Ca in the regulation of the exchanger in the heart.

Much information about Na-Ca exchange in heart has been gained from kinetic studies of Na-Ca and Ca-Ca exchange in sarcolemmal vesicles.[20-24] In these vesicles, however, control of Na-Ca exchange by both Ca^{25} and ATP[26] is limited. Control of the exchanger in heart by ATP has been shown in intact cells[5] and in giant excised patches,[6,27] and control by Ca has been shown in dialyzed cells,[7-9] in giant excised patches,[6,28] and in intact cells.[29,30] The ease of controlling the environment on both sides of the membrane with giant excised patches has allowed the regulation by ATP and Ca_i to be investigated in considerable detail.[6,27,28] These studies have found evidence for a strong intracellular Na-dependent inactivation of the exchanger which ATP or intracellular Ca are able to relieve. The broad features of this inactivation have also now been observed in dialyzed whole myocytes.[31]

A concern with such measurements, however, is that the intrusion necessary to achieve such control comes at the price of potentially altering the properties of labile control mechanisms. There is evidence that the control mechanisms of the Na-Ca exchanger are easily altered. Single perfused cells show an extremely high affinity for the action of Ca at the intracellular regulatory site,[9,32] whereas excised patches show an affinity for Ca at this site which is very much lower.[6] For this reason it is valuable to examine the ATP concentration dependence of the exchanger, and the regulation of the exchanger by Ca_i, in cells which are intact. Some of the results of such investigations in our laboratory are reviewed here.

METHODS

Cell Isolation

Heart cells were isolated from female retired breeder rats according to our original method,[33] as recently modified.[34] The modification used was condition 5 in Table 2 of that paper: the perfusion buffers contained 25 mM N-2-hydroxyethylpiperazine-N'-ethanesulfonic acid (HEPES), adjusted to pH 7.4 with NaOH, in place of bicarbonate, plus basal Eagle medium amino acids. Ca (1 mM) was restored to the recirculating perfusate 15 min after enzyme addition. This method gave a high yield of cells with a high percentage (74.3 ± 6.0) of rod-shaped cells in the presence of 1 mM Ca.[34]

Experimental Medium

Cells were suspended in a medium containing (mM) NaCl 118, KCl 4.8, HEPES 25, KH_2PO_4 1.2, $MgSO_4$ 1.2, $CaCl_2$ 1.0, pyruvic acid 5, glucose 11, and (μM) insulin 1, adjusted to pH 7.4 with NaOH. Suspensions were maintained aerobic by equilibration with air in a shaking incubator at 37°C.

Isotope Uptake by Na-Loaded Cells

Cells with tritiated water (1 μCi/ml) were loaded with Na in experimental medium without Ca and plus 0.25 mM EGTA by incubation for 30 min at 37°C with ouabain (1 mM). To measure ^{45}Ca uptake, either Ca containing ^{45}Ca (0.1 μCi/ml) was added to the cell suspension, or cells were added to different media containing Ca and ^{45}Ca, at time zero. Aliquots were removed at time intervals for centrifugation through bromododecane. To measure ^{22}Na efflux, ^{22}Na was added with the ouabain, and cells were diluted after the 30 min incubation, at time zero on the figures, into the unlabelled media described in the text. Aliquots were removed at time intervals and centrifuged, as above. To measure rates of ^{22}Na uptake, ^{22}Na was added to the concentrated cells after the 30 min incubation. Aliquots were removed at time intervals, diluted, and centrifuged immediately.

ATP Measurements

Cell samples taken at the indicated times were extracted with an equal volume of ice cold 16% perchloric acid. The acid extracts were neutralized with potassium bicarbonate and frozen. Thawed samples were centrifuged, and HPLC analysis was performed on an aliquot of the supernatant using a Whatman Partisil 10SAX anion exchange column.

Cell Morphology

Aliquots of cell suspension removed at the times shown were fixed by mixing with an equal volume of cold 2% glutaraldehyde, a slide was made, and the percentage of cells with a rod-shaped configuration as revealed by optical microscopy was counted. Cells were considered to be rod-shaped if they retained their relaxed sarcomere length, in the region of 1.8 μm.[35]

RESULTS AND DISCUSSION

Activation by ATP

When isolated cells in suspension containing 1 mM Ca are exposed to ^{45}Ca, a rapid labelling of the cells is observed, which quickly levels off (FIG. 1). Addition of rotenone (4 μM, an inhibitor of mitochondrial respiration) plus p-trifluoromethoxyphenylhydrazone (FCCP, 2 μM, an uncoupler of oxidative phosphorylation, which also induces a high mitochondrial ATPase activity) results in cell contracture and depletion of ATP down from 17.08 ± 2.26 to 0.63 ± 0.11 nmol/mg within 8 minutes. This ATP depletion, however, is not accompanied by any uptake of ^{45}Ca, until the Ca ionophore A23187 is added (FIG. 1). This shows that in spite of a large inwardly directed Ca gradient no massive influx of Ca occurs in cells on ATP depletion. Since the Na-Ca exchanger in squid axon was known to be regulated by ATP, and Na-Ca exchange was known to be active in heart sarcolemma, this observation (FIG. 1) led us to investigate the extent to which exchange activity in the heart might also be dependent on ATP.[5]

When cells were loaded with Na in the absence of Ca by incubation with ouabain for 30 minutes, such cells showed a large Ca uptake when 1 mM Ca was added back, which was strongly inhibited by treatment of cells with rotenone plus FCCP (FIG. 2). The uptake also did not occur in a low Na medium (FIG. 2). This Na-dependent Ca uptake thus appeared to be inhibited by ATP depletion. However, rotenone and FCCP will also tend to limit cellular Ca uptake by preventing mitochondrial Ca uptake, regardless of their action to deplete cytosolic ATP.

The site of action of the inhibition of Ca uptake was further investigated by monitoring the level of intracellular free Ca using the Ca sensitive indicator quin2. Na-loaded cells loaded with quin2 showed a fast rate of rise of intracellular Ca on addition of 1 mM Ca to cells in suspension, while similar cells which had also been ATP depleted with rotenone plus FCCP showed a very low rate of rise (FIG. 3). This shows that the inhibition of cellular Ca uptake by this treatment occurred primarily at the sarcolemma and not because of mitochondrial inhibition. This conclusion has been further supported by the observation that treatments such as oligomycin plus FCCP, which eliminate mitochondrial Ca uptake but allow substantial levels of cytosolic ATP levels to be maintained, do not affect the initial rate of Ca uptake by Na-Ca exchange.[36]

When Na-loaded cells were diluted into a medium containing no Na, the rate of Ca uptake by Na-Ca exchange was increased (FIG. 4), as expected from a competition between extracellular Na and Ca for occupancy of the exchanger transport sites. Under these conditions there was still a measurable rate of Ca entry after ATP depletion (FIG. 4). This suggests that the exchanger is not completely inactivated by ATP depletion.

The observation of inhibition of exchanger-mediated Ca uptake by ATP depletion of myocytes caused us to investigate how much ATP was needed to activate the exchanger in intact cells, and how ATP depletion might affect exchanger-mediated Ca uptake under conditions like ischemia.

We found previously[35] that, when cells are incubated under conditions which simulate ischemia (anoxic incubation of pelleted cells without glucose at 37°C), they one by one undergo contracture, and the time course of the decline in %

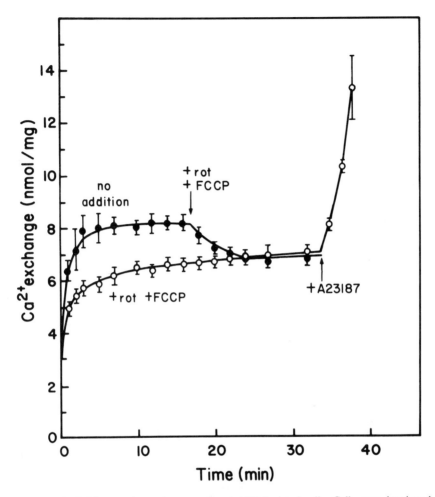

FIGURE 1. Calcium exchange by normal and ATP-depleted cells. Cells were incubated (2.4 mg/ml) in experimental medium containing 1 mM $CaCl_2$. ^{45}Ca (0.1 μCi/ml) was added at t = 0. Half of the cells were treated with rotenone (4 μM) plus FCCP (2 μM) for 8 min before addition of ^{45}Ca, and then 33 min later A23187 (1.5 μg/ml) was added. Aliquots (0.5 ml) were removed 30 s before ^{45}Ca addition for ATP analysis. Cell ^{45}Ca content was measured in 0.5 ml aliquots removed at times shown. Rotenone (4 μM) plus FCCP (2 μM) was also added to normal cells 14 minutes after ^{45}Ca. (From Haworth et al.[5] Reprinted by permission from *Circulation Research*.)

rod-shaped cells (cells with a sarcomere length >1.8 μm) parallels the decline in ATP measured in the resuspended cell pellet (FIG. 5). In this experiment oligomycin was also added, in order to prevent any ATP resynthesis by oxidative phosphorylation when the cells were resuspended. From this observation we concluded that the ATP level in individual cells remained at a high plateau level during ischemia until some point (probably at the point of glycogen depletion) at which it declined

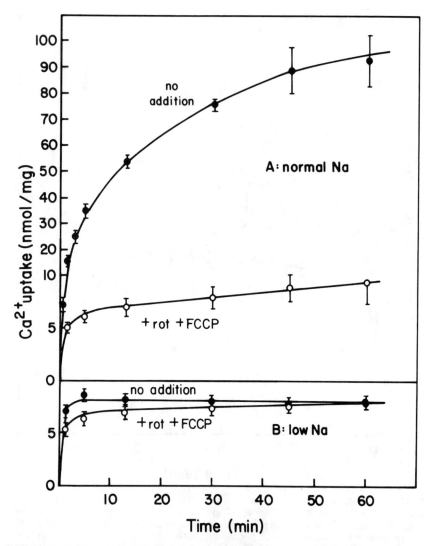

FIGURE 2. Sodium and ATP dependence of massive calcium uptake by ouabain-treated cells. Cells were washed and resuspended in an experimental medium without calcium. Ouabain (1 mM) was added 30 min before calcium and rotenone (4 μM) plus FCCP (2 μM) 6 min before calcium. Aliquots were removed 15 s before addition and analyzed for ATP: normal Na, normal ATP = 13.7 ± 1.9 nmol/mg; normal Na plus rotenone/FCCP = 0.52 ± 0.04 nmol/mg; low Na, normal ATP = 13.8 ± 2.7 nmol/mg; low Na plus rotenone/ FCCP = 1.7 ± 0.5 nmol/mg. (From Haworth et al.[5] Reprinted by permission from *Circulation Research*.)

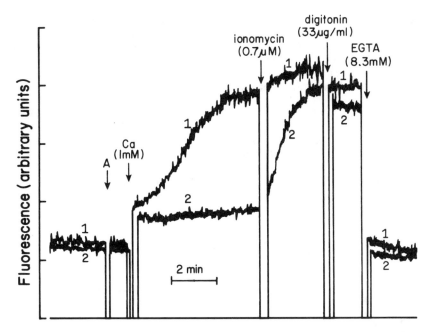

FIGURE 3. Rate of rise of [Ca]$_i$ in Na-loaded cells, measured by quin2. Cells were loaded with quin2 then exposed to 1 mM ouabain ± oligomycin (36.4 μM) and iodoacetate (0.91 mM) for 30 min. ATP depletion was essentially complete after 15 min under these conditions.[5] CaCl$_2$ (1 mM) was added at the time shown. *Trace 1*, cells plus dye, with ATP; *trace 2*, cells plus dye, ATP depleted; *traces 3 and 4*, cells without dye, with and without ATP. (From Haworth *et al.*[5] Reprinted by permission from *Circulation Research*.)

catastrophically to near zero, when contracture occurred. This conclusion was subsequently supported by Bowers *et al.*[37] by luciferase measurements of ATP decline in single myocytes exposed to metabolic inhibitors. This behavior of individual cells complicates the attempt to fractionally deplete cells of ATP in order to determine the ATP dependence of exchanger activation. Indeed, when Ca uptake was measured in suspensions of cells loaded with Na by EDTA treatment[38] after various times of incubation under these conditions, a linear dependence of Ca uptake rate was observed on both % rod-shaped cells and ATP levels measured on the cell suspensions (FIG. 6). In light of the heterogeneity of ATP levels inferred for this condition, this suggests that the measured rate of Ca uptake is the result of some cells with ATP which take up a lot of Ca and some cells without ATP which take up very little.

In an attempt to create a state of fractional ATP content in all cells we examined the condition where cells are allowed to all undergo contracture by anoxic incubation for 1 hour in tubes as above, but in the absence of mitochondrial inhibitors. These cells all go through a state of ATP depletion to near zero levels, and on resuspension and reoxygenation are all able to quickly resynthesize ATP from the fraction of their nucleotide pool which has not been degraded to adenosine and inosine. When Ca uptake was measured under these conditions, a high uptake rate was found in reoxygenated cells which had only one quarter to one third of their original ATP content (FIG. 7, open triangles). Inclusion of rotenone during the incubation prevented ATP resynthesis on cell resuspension, and these cells showed little Ca uptake (FIG. 7, filled triangles). This implies that the Km for ATP is low compared with the concentration of ATP in normal cells: the best fit curve in FIGURE 7 gives a Km of about 2 nmol/mg, just 10% of normal cellular ATP content. We find an intracellular water content of about 2 μl/mg protein. The Km would thus be about 1 mM, which contrasts with the value of 7.5 to 10 mM found in giant excised patches for the dependence of exchange current stimulation on MgATP concentration.[27] The result also suggests that ATP depletion after ischemia plus reperfusion is unlikely to limit the rate of Ca uptake by Na-Ca exchange in the whole heart if at least one quarter of the ATP is restored.

A final noteworthy feature of these experiments was that the method of EDTA treatment we developed for these experiments to quickly equilibrate cells with Na also had the effect of equilibrating intracellular and extracellular pH. The

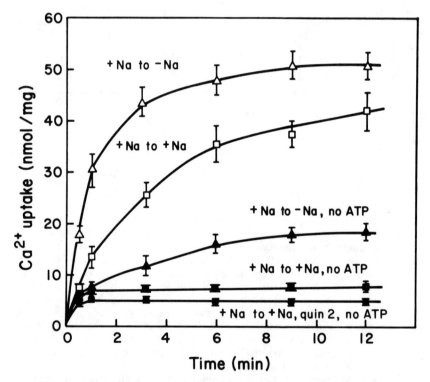

FIGURE 4. Stimulation of ^{45}Ca uptake by dilution into a medium without sodium. Cells were exposed to ouabain 30 min before dilution and oligomycin/iodoacetate (see FIG. 3) was added 15 min before the dilution. At t = 0, they were diluted into 4 volumes of medium with or without sodium, containing 1 mM ^{45}Ca-labelled CaCl$_2$. Some cells were also loaded with quin2 prior to treatment with ouabain and oligomycin/iodoacetate. The medium without sodium had sodium replaced with choline and pH was adjusted to 7.4 with tetramethylammonium hydroxide. The dilution buffers also contained ouabain (1 mM). (From Haworth *et al.*[5] Reprinted by permission from *Circulation Research.*)

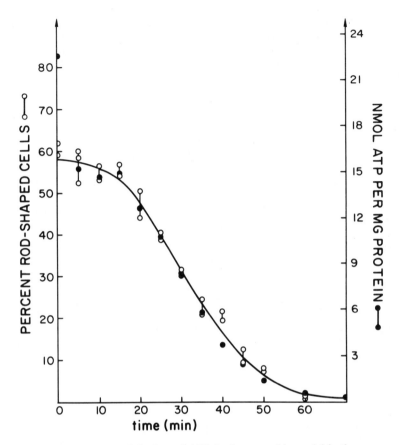

FIGURE 5. Contracture and the loss of ATP in the anaerobic model in the presence of oligomycin. Note that the time zero was taken before the addition of oligomycin (6 nmol/mg protein), but the clock was started when the tubes were set to incubate at 37°C. Protein concentration was 2.67 mg/ml. (From Haworth et al.[35] Reprinted by permission from *Circulation Research*.)

strong inhibition of exchanger-mediated Ca uptake by ATP depletion still seen under these conditions shows that this inhibition is not primarily mediated by intracellular acidosis. This is significant because ATP depletion does result in intracellular acidosis, and intracellular acidosis does inhibit Na-Ca exchange.[39,40]

Activation by Ca

The activation of the Na-Ca exchanger by intracellular Ca was first observed by isotopic studies in squid axon,[3,4] but was also seen in dialyzed heart cells, as a dependence of exchanger-mediated outward current on intracellular Ca.[7] In the course of seeking to understand the control of Na fluxes across the sarcolemma of Na-loaded cells we came to conclude that such fluxes in heart cells were mainly mediated by the Na-Ca exchanger, under the control of activation by intracellular Ca. Moreover, in normal intact cells at rest the exchanger appeared to be almost inactive, and then became activated during stimulated beating, either directly or indirectly by Ca which entered the cell through Ca channels.

When isolated Na-loaded cells labelled with ^{22}Na were diluted into unlabelled medium, the presence of a low amount of Ca (0.2 mM free) in the dilution medium induced a rapid rate of efflux of ^{22}Na (FIG. 8A). The additional presence of verapamil blocked activation of ^{22}Na efflux (FIG. 8A). Similarly, when ^{22}Na with or without 0.2 mM Ca was added to concentrated cells, the rate of ^{22}Na uptake was strongly activated when Ca was present (FIG. 8B). This suggested that Ca may be entering cells by Ca channels to activate Na-Na exchange activity.

This possibility was tested further by examining the reversibility of the Ca-induced activation. Labelled Na-loaded cells were activated by adding 0.2 mM Ca for 2 minutes and were then diluted into unlabelled medium containing either 0.2 mM Ca, 0.2 mM Ca plus verapamil, or excess EGTA. Cells diluted into medium with Ca showed a very rapid rate of ^{22}Na efflux which this time was unaffected by verapamil (FIG. 9). This suggests that verapamil does not inhibit the Ca-induced ^{22}Na flux directly, but rather that the activating action of Ca is at an intracellular site, which Ca reaches by entry through Ca channels. Cells diluted into EGTA showed a rapid initial rate of ^{22}Na efflux, but the efflux rate then became inhibited (FIG. 9). This suggests that removal of extracellular Ca does affect the activated rate of ^{22}Na efflux, but not immediately. The fast initial rate of ^{22}Na efflux into the EGTA medium shows that the ^{22}Na flux is indeed Na-Na exchange, and not the sum of Na-Ca and Ca-Na exchange, since the latter could not occur in the absence of extracellular Ca. The full reversibility of the Ca-induced activation was then shown by incubating activated cells with excess EGTA for 10 minutes

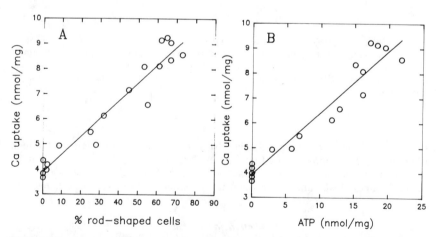

FIGURE 6. Graphs showing calcium uptake after 1 min by Na-Ca exchange in cells that were incubated with oligomycin and had a heterogeneous distribution of ATP. See text for experimental design. **(A)** Dependence on percentage of rod-shaped cells. **(B)** Dependence on ATP. Data are from three experiments. (From Haworth and Goknur.[38] Reprinted by permission from *Circulation Research*.)

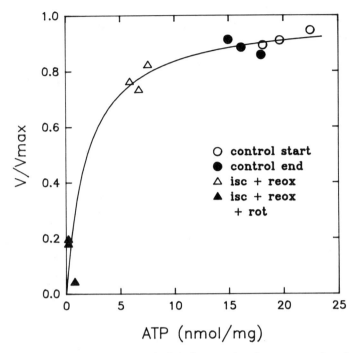

FIGURE 7. Restoration of calcium uptake in cell suspensions by reoxygenation after simulated ischemia. isc, ischemia; reox, reoxygenation; rot, rotenone. The incubation conditions corresponding to each symbol are described in the text. ATP dependence of ^{45}Ca uptake. Data are from three experiments. (From Haworth & Goknur.[38] Reprinted by permission from *Circulation Research*.)

before dilution into EGTA medium: the rate of ^{22}Na efflux had fully returned to its original low level (FIG. 9). This again is consistent with the action of Ca at an intracellular site, since reversal of activation will presumably require time for the intracellular Ca to leave.

Further experiments showed that the stimulated rate of ^{22}Na efflux shown by cells activated by Ca required extracellular Na, and was not only inhibited but also reversed by dichlorobenzamil, the Na-Ca exchange inhibitor. From these results we concluded that intracellular Ca was activating Na-Na exchange through the Na-Ca exchanger.[29] A similar activation by cytosolic Ca of Na-Na exchange through the exchanger has been observed in squid axon.[4]

The above results caused us to investigate the status of exchanger activation in normal cells, not Na loaded, suspended in medium with physiological levels of Ca.

When normal cells in medium containing 1 mM Ca were equilibrated with ^{22}Na and diluted into unlabelled medium, the observed rate of ^{22}Na efflux was strongly stimulated when the dilution medium was subjected to electric field stimulation (FIG. 10A). When ^{22}Na was added to unlabelled cells, the rate of ^{22}Na uptake was found to be similarly increased by electric stimulation (FIG. 10B), while total cellular Na was unaffected (FIG. 10C). The kinetics of these Na fluxes are remarkably similar to those observed in Na-loaded cells (FIG. 8). Further experiments showed that the effect of electrical stimulation was reversible within a few seconds of ceasing stimulation, required entry of extracellular Ca through Ca channels, and was unaffected by ouabain. A similar activation of Na fluxes could be achieved through depolarization with high extracellular KCl. We concluded that these Ca-dependent Na fluxes were caused by activation of the Na-Ca exchanger by Ca entering the cells during excitation, and that in cells at rest the exchanger was essentially unactivated.[30]

Loss of Activation by Ca in ATP-Depleted Cells

When cells loaded with Na and ^{22}Na as in FIGURE 8 were depleted of ATP by treatment with rotenone (3 μM) plus FCCP (0.2 μM) for 8 minutes and were then diluted into unlabelled medium, the rate of ^{22}Na efflux was the same low rate whether or not the cells were diluted into a medium containing 0.2 mM Ca (FIG. 11A). This shows that ATP depletion prevents the activating effect of extracellular Ca. Moreover, the rate of ^{22}Na efflux into medium without Ca was also strongly inhibited compared with the rate for cells with normal ATP (FIG. 11A). This implies that the basal Ca independent rate of Na-Na exchange was also inhibited

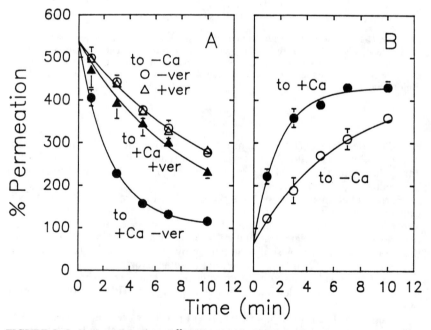

FIGURE 8. Induction of increased ^{22}Na flux in Na-loaded cells by calcium. **(A)** Efflux. The preventive action of verapamil (ver, 10 μM) is also shown. **(B)** Influx. Experimental procedure is described in the text. (From Haworth *et al.*[29] Reprinted by permission from *Circulation Research*.)

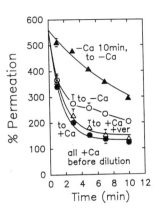

FIGURE 9. Reversal of calcium-induced ^{22}Na efflux in sodium loaded cells by EGTA. Verapamil (10 μM) fails to reverse, whereas EGTA incubation causes reversal. Line connecting data for efflux into EGTA medium (*open circles*) is a cubic spline interpolation. Experimental procedure is described in the text. (From Haworth *et al.*[29] Reprinted by permission from *Circulation Research*.)

by ATP depletion. ATP depletion could well prevent access of Ca to the intracellular site of activation, by inhibition of Ca channels. The effect of ATP depletion on cells which were already activated was therefore also examined. When Na-loaded cells were activated by adding 0.2 mM excess Ca 1 minute before addition of rotenone plus FCCP, the rate of ^{22}Na efflux on dilution into unlabelled medium was again strongly inhibited by the ATP depletion treatment (FIG. 11B). In this case the cells already contained sufficient Ca to activate cells with normal ATP (FIG. 8), but it was clearly insufficient to activate the ATP-depleted cells. In other experiments, we also found that addition of A23187, a Ca ionophore which mediates Ca entry into ATP-depleted cells (FIG. 1), still fails to activate ^{22}Na efflux from ATP-depleted cells, while such treatment is capable of inducing activation in cells with normal ATP whose Ca uptake has been inhibited with verapamil as in

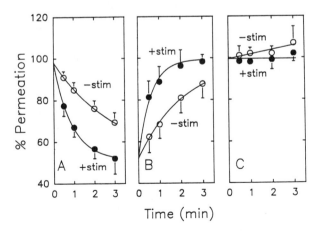

FIGURE 10. Effect of electrical stimulation (stim) on rates of ^{22}Na efflux, uptake, and total cellular sodium. **(A)** ^{22}Na efflux. **(B)** Total cellular sodium. Experimental procedure is described in the text. (From Haworth & Goknur.[30] Reprinted by permission from *Circulation Research*.)

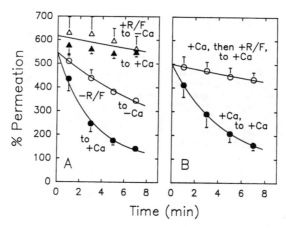

FIGURE 11. Prevention and reversal of Ca-induced ^{22}Na efflux in Na-loaded cells by ATP depletion. **(A)** Prevention; **(B)** reversal. Experimental procedure is described in the text.

FIGURE 1 (data not shown). These results suggest that in ATP-depleted cells not only is exchanger activation by ATP lost but also exchanger activation by Ca. Measurements of intracellular pH with DMO (see Ref. 38 for method) showed a pH of 7.4, the same as extracellular. The loss of Ca activation therefore cannot be accounted for by acidosis. ATP was found to increase the affinity of the exchanger for activating Ca by up to a factor of 10 in giant membrane patches.[27] Another likely contributor to loss of Ca sensitivity here is an expected rise in intracellular Mg concentration. Ischemic rat hearts show a 3-fold increase in intracellular Mg concentration which correlates with ATP depletion,[41] and Mg competes effectively with Ca for binding to the Ca binding domain of the cardiac Na-Ca exchanger.[42] A 3-fold increase in free Mg concentration would be expected to increase the Km for Ca binding almost 3-fold. Given that conditions of ischemia where ATP depletion normally occurs also result in considerable acidosis, and acidosis is a further powerful inhibitor of Ca-induced activation,[39,40] it is clear that the Na-Ca exchanger is regulated to switch off under these conditions. This may be a protective mechanism to limit Ca influx. However, it is also clear from the results reviewed here that the exchanger quickly reactivates under conditions of reoxygenation, and in the presence of Na loading can quickly cause a Ca overload even though cellular ATP levels are substantially reduced from normal.

SUMMARY

Regulation of Na-Ca exchange activity by ATP and by intracellular Ca (Ca_i) has been studied in suspensions of intact Na-loaded adult rat cardiac myocytes using ^{45}Ca uptake and exchange of ^{22}Na.

ATP depletion of Na-loaded myocytes results in a strong inhibition of the Na-Ca exchanger, manifested as a strong inhibition of intracellular Na-dependent Ca uptake. Ca uptake by Na-loaded cells in the course of ATP depletion can be very heterogeneous because of the heterogeneity amongst cells of the extent of ATP depletion. This can result in a false measure of the dependence of exchanger activity on cell ATP content. Under conditions intended to maximize the uniformity of cell ATP content amongst cells we found a half maximal rate of Ca uptake with a cell ATP content of 1.96 nmol/mg, about 10% of the normal cell ATP level. The results suggest that ATP depletion after ischemia plus reperfusion is unlikely to limit the rate of Ca uptake by Na-Ca exchange in the whole heart if at least one quarter of the ATP is restored.

Ca addition to myocytes loaded with Na in the absence of Ca results in a strong activation of the Na-Ca exchanger at an intracellular site, manifested as a large activation of Na-Na exchange activity. A similar activation of the exchanger is observed in cells with a normal level of intracellular Na, suspended in a medium containing physiological levels of Ca, when the cells are stimulated to beat by application of an electric field. This suggests that regulation of the exchanger by Ca_i is important physiologically, in the regulation of excitation-contraction coupling.

Cells depleted of ATP show not only a strongly inhibited rate of Na-Ca exchange and Na-Na exchange, but also a strongly reduced degree of activation by Ca_i, even in ATP-depleted cells with no acidosis. This could result from the combined effect of ATP loss and an elevated intracellular Mg concentration on Ca binding affinity at the regulatory site.

REFERENCES

1. BLAUSTEIN, M. P. & E. M. SANTIAGO. 1977. Effects of internal and external cations and of ATP on sodium-calcium and calcium-calcium exchange in squid axons. Biophys. J. **20:** 79–111.
2. DIPOLO, R. 1977. Characterization of the ATP-dependent calcium efflux in dialyzed squid giant axons. J. Gen. Physiol. **69:** 795–813.
3. ALLEN, T. J. A. & P. F. BAKER. 1985. Intracellular Ca indicator quin-2 inhibits Ca^{2+} inflow via Na^+/Ca^{2+} exchange in squid axon. Nature **315:** 755–756.
4. DIPOLO, R. & L. BEAUGE. 1987. Characterization of the reverse Na/Ca exchange in squid axons and its modulation by Ca_i and ATP. J. Gen. Physiol. **90:** 505–525.
5. HAWORTH, R. A., A. B. GOKNUR, D. R. HUNTER, J. O. HEGGE & H. A. BERKOFF. 1987. Inhibition of calcium influx in isolated adult rat heart cells by ATP depletion. Circ. Res. **60:** 586–594.
6. HILGEMANN, D. W. 1990. Regulation and deregulation of cardiac Na^+-Ca^{2+} exchange in giant excised sarcolemmal membrane patches. Nature **344:** 242–245.
7. KIMURA, J., A. NOMA & H. IRISAWA. 1986. Na/Ca exchange current in mammalian heart cells. Nature **319:** 596–597.
8. KIMURA, J., S. MIYAMAE & A. NOMA. 1987. Identification of sodium-calcium exchange current in single ventricular cells of guinea pig. J. Physiol. **384:** 199–222.
9. NODA, M., M. NAKAO, R. N. SHEPHERD & D. C. GADSBY. 1988. Dependence on $[Ca]_i$, $[Ca]_o$, and $[Na]_i$ and block by 3′4′-dichlorobenzamil, of outward Na/Ca exchange current in guinea pig ventricular myocytes (abstract). J. Mol. Cell. Cardiol. **20**(Suppl. 4): S40.
10. REUTER, H. & N. SEITZ. 1968. The dependence of calcium efflux from cardiac muscle on temperature and external ion composition. J. Physiol. **195:** 451–470.
11. SCHOUTEN, V. J. A. & H. E. D. J. TER KEURS. 1985. The slow repolarization phase of the action potential in rat heart. J. Physiol. **360:** 13–25.
12. EGAN, T. M., D. NOBLE, S. J. NOBLE, T. POWELL, A. J. SPINDLER & V. W. TWIST. 1989. Sodium-calcium exchange during the action potential in guinea-pig ventricular cells. J. Physiol. **411:** 639–661.
13. BRIDGE, J. H. B., J. R. SMOLLEY & K. W. SPITZER. 1990. The relationship between charge movements associated with I_{Ca} and I_{Na-Ca} in cardiac myocytes. Science **248:** 376–378.

14. LEBLANC, N. & J. R. HUME. 1990. Sodium current-induced release of calcium from cardiac sarcoplasmic reticulum. Science **248:** 372–376.
15. LEVI, A. J., P. BROOKSBY & J. C. HANCOX. 1993. One hump or two? The triggering of calcium release from the sarcoplasmic reticulum and the voltage dependence of contraction in mammalian cardiac muscle. Cardiovasc. Res. **27:** 1743–1757.
16. LIPP, P. & E. NIGGLI. 1994. Sodium current-induced calcium signals in isolated guinea-pig ventricular myocytes. J. Physiol. **474**(3): 439–446.
17. LEVI, A. J., K. W. SPITZER, O. KOHMOTO & J. H. B. BRIDGE. 1994. Depolarization-induced Ca entry via Na-Ca exchange triggers SR release in guinea pig cardiac myocytes. Am. J. Physiol. **266:** H1422–H1433.
18. TANI, M. 1990. Mechanisms of Ca^{2+} overload in reperfused ischemic myocardium. Annu. Rev. Physiol. **52:** 543–559.
19. REIMER, K. A., M. L. HILL & R. B. JENNINGS. 1981. Prolonged depletion of ATP and of the adenine nucleotide pool due to delayed resynthesis of adenine nucleotides following reversible myocardial ischemic injury in dog. J. Mol. Cell. Cardiol. **13:** 229–239.
20. SLAUGHTER, R. S., J. L. SUTKO & J. P. REEVES. 1983. Equilibrium calcium-calcium exchange in cardiac sarcolemmal vesicles. J. Biol. Chem. **258:** 3183–3190.
21. REEVES, J. P. & J. L. SUTKO. 1979. Sodium-calcium ion exchange in cardiac membrane vesicles. Proc. Natl. Acad. Sci. USA **76:** 590–594.
22. REEVES, J. P. & J. L. SUTKO. 1983. Competitive interactions of sodium and calcium with the sodium-calcium exchange system of cardiac sarcolemmal vesicles. J. Biol. Chem. **258:** 3178–3182.
23. PHILIPSON, K. D. & A. Y. NISHIMOTO. 1980. Na/Ca exchange is affected by membrane potential in cardiac sarcolemma vesicles. J. Biol. Chem. **255:** 6880–6882.
24. PHILIPSON, K. D. 1984. Interaction of charged amphiphiles with Na/Ca exchange in cardiac sarcolemma vesicles. J. Biol. Chem. **259:** 13999–14002.
25. REEVES, J. P. & P. PORONNIK. 1987. Modulation of Na^+-Ca^{2+} exchange in sarcolemmal vesicles by intravesicular Ca^{2+}. Am. J. Physiol. **252:** C17–C23.
26. CARONI, P. & E. CARAFOLI. 1983. The regulation of the Na^+-Ca^{2+} exchanger of heart sarcolemma. Eur. J. Biochem. **132:** 451–460.
27. COLLINS, A., A. V. SOMLYO & D. W. HILGEMANN. 1992. The giant cardiac membrane patch method: stimulation of outward Na^+-Ca^{2+} exchange current by MgATP. J. Physiol. **454:** 27–57.
28. HILGEMANN, D. W., A. COLLINS & S. MATSUOKA. 1992. Steady-state and dynamic properties of cardiac sodium-calcium exchange. J. Gen. Physiol. **100:** 933–961.
29. HAWORTH, R. A., A. B. GOKNUR & D. R. HUNTER. 1991. Control of the Na/Ca exchanger in isolated heart cells. I. Induction of Na/Na exchange through the Na/Ca exchanger of isolated heart cells by Ca. Circ. Res. **69:** 1506–1513.
30. HAWORTH, R. A. & A. B. GOKNUR. 1991. Control of the Na/Ca exchanger in isolated heart cells. II. Beat-dependent activation in normal cells by Ca_i. Circ. Res. **69:** 1514–1524.
31. MATSUOKA, S. & D. W. HILGEMANN. 1994. Inactivation of outward Na/Ca exchange current in guinea pig ventricular myocytes. J. Physiol. **476:** 443–458.
32. MIURA, Y. & J. KIMURA. 1989. Sodium-calcium exchange current. J. Gen. Physiol. **93:** 1129–1145.
33. HAWORTH, R. A., D. R. HUNTER & H. A. BERKOFF. 1980. The isolation of Ca^{2+}-resistant myocytes from the adult rat. J. Mol. Cell. Cardiol. **12:** 715–723.
34. HAWORTH, R. A., A. B. GOKNUR, T. F. WARNER & H. A. BERKOFF. 1989. Some determinants of quality and yield in the isolation of adult heart cells from rat. Cell Calcium **10:** 57–62.
35. HAWORTH, R. A., D. R. HUNTER & H. A. BERKOFF. 1981. Contracture in intact isolated adult rat heart cells: Role of Ca^{2+}, ATP and compartmentation. Circ. Res. **49:** 1119–1128.
36. HAWORTH, R. A., A. B. GOKNUR & H. A. BERKOFF. 1989. Inhibition of Na/Ca exchange by general anesthetics. Circ. Res. **65:** 1021–1028.
37. BOWERS, K. C., A. P. ALLSHIRE & P. H. COBBOLD. 1992. Bioluminescent measurement

in single cardiomyocytes of sudden cytosolic ATP depletion coincident with rigor. J. Mol. Cell. Cardiol. **24:** 213–218.
38. HAWORTH, R. A. & A. B. GOKNUR. 1992. ATP dependence of calcium uptake by the Na/Ca exchanger of adult heart cells. Circ. Res. **71:** 210–217.
39. HILGEMANN, D. W., S. MATSUOKA, G. A. NAGEL & A. COLLINS. 1992. Steady-state and dynamic properties of cardiac sodium-calcium exchange. J. Gen. Physiol. **100:** 905–932.
40. DOERING, A. E. & W. J. LEDERER. 1993. The mechanism by which cytoplasmic protons inhibit the sodium-calcium exchanger in guinea-pig heart cells. J. Physiol. **466:** 481–499.
41. MURPHY, E., C. STEENBERGEN, L. A. LEVY, B. RAJU & R. E. LONDON. 1989. Cytosolic free magnesium levels in ischemic rat heart. J. Biol. Chem. **264:** 5622–5627.
42. LEVITSKY, D. O., D. A. NICOLL & K. D. PHILIPSON. 1994. Identification of the high affinity Ca-binding domain of the cardiac Na/Ca exchanger. J. Biol. Chem. **269:** 22847–22852.

Functional Roles of Sodium-Calcium Exchange in Normal and Abnormal Cardiac Rhythm

DENIS NOBLE,[a] JEAN-YVES LeGUENNEC,[b] AND
RAIMOND WINSLOW[c]

[a]*University Laboratory of Physiology*
Parks Road
Oxford OX1 3PT, United Kingdom

[b]*Laboratoire de Physiologie des Cellules Cardiaques et*
Vasculaires
EP21 CNRS
Parc de Grandmont
Tours, France

[c]*Department of Biomedical Engineering*
Johns Hopkins University
Baltimore, Maryland

INTRODUCTION

This paper briefly reviews recent work on the role of Na-Ca exchange in normal cardiac activity, and its role in the initiation of arrhythmias produced by ectopic beats generated in regions of sodium and calcium overload.

Normal Rhythm

Sodium-Calcium Exchange during Atrial-Type Action Potentials

It is well established that in the case of action potentials with the triangular waveform typical of atrial cells (and of some ventricular cells such as those of the rat), we can clearly distinguish two phases of net calcium movement. The first phase corresponds to the initial spike of the action potential and the rapid repolarization to around −30 mV. During this time, calcium flows in through the calcium channels while the inward mode of the Na-Ca exchange is relatively inactive, waiting for intracellular calcium to rise to activate it. During the second phase, *i.e.*, the low late plateau, the situation is reversed. The calcium current is deactivated by the rapid repolarization, while the rise in intracellular calcium activates calcium efflux via the exchange. The late low plateau is therefore almost fully maintained by the net inward charge carried by the exchange.[1–3] These results have been fully reconstructed using models of multicellular atrial tissue[4] and isolated atrial cells.[5]

Na-Ca Exchange during Interrupted Ventricular Action Potentials

This analysis, however, cannot apply to action potentials, like those in ventricular cells of guinea pig, human and many other species, in which the initial depolarization is followed by a long plateau at positive voltages. Since the inward mode of the exchange is weak at positive potentials,[6] it would be expected to carry very much less current. In fact, the net exchange activity could be outward, corresponding to a net entry of calcium at least at the beginning of the high plateau. The exchange would then be helping the calcium current to take calcium into the cell and according to some recent experimental evidence it could be sufficient on its own in this mode to trigger SR calcium release.[7,8]

Can we nevertheless show that the sodium-calcium exchanger does contribute significant current to this form of action potential? This question was first answered by Egan et al.[9] by artificially interrupting the action potential at various times to impose a voltage clamp to a voltage negative to −30 mV. Just as in atrial cells, this produces a slow inward flow of current that depends on external sodium and intracellular calcium in the way expected for Na-Ca exchange. Further confirmation of this interpretation is obtained by plotting the envelope of the tail current amplitudes, which reflects the time course of the activating parameter, *i.e.*, internal calcium.

Reconstruction Using Computations

These experiments established that, were the ventricular cell to repolarize as quickly as an atrial cell, it would activate comparable Na-Ca exchange currents to those observed in atrial cells. This information can then be used to attempt to determine the actual time course of the exchange current during the plateau itself.[9]

However, the problem with this kind of calculation is that, because the exchanger stoichiometry involves 3 Na ions, its activity is extremely sensitive to the level of intracellular sodium. When $[Na]_i$ is very low (*i.e.*, the energy of the sodium gradient is large) the exchange current becomes net inward very early during the calcium transient. When $[Na]_i$ is high the energy available to the exchanger from the sodium gradient is greatly reduced and the current is then largely net outward almost throughout the duration of the plateau. It becomes inward only during the rapid phase of repolarization. This is a highly dynamic and finely balanced situation, which has great functional significance for the nature of calcium balance in the heart.[10] But it makes extrapolation from the experimental results on tail currents very difficult. It is therefore necessary to look for more direct interventions during the action potential plateau itself to determine how strongly sodium-calcium exchange contributes to the action potential characteristics.

Direct Perturbation of Na-Ca Exchange during the Plateau

This can be done by perturbing external sodium very rapidly at various times during the action potential plateau.[11] Rapid switching of solution from 140 to 70 mM sodium (replaced by lithium) was achieved using a solenoid operated movement of a double-barreled pipette from which the two solutions flowed onto the cell.[12] By refining the experimental technique the half time of the solution switch was reduced to only 7 msec. which is very small compared to the duration of the plateau. It is therefore possible to perturb the activity of the Na-Ca exchanger very rapidly indeed at various times during the plateau.

Using this method it was possible to show that there is a large contribution of Na-Ca exchange to determining the duration of the ventricular action potential. Switching to low (70 mM) sodium early in the plateau shortens the action potential by around 20%. As the switch is applied later during the plateau the effect weakens. This shows that the exchange current is exerting an effect throughout the action potential plateau. Computations using a ventricular cell model reported at the previous International Meeting on Sodium-Calcium Exchange[13] reproduce these experimental results very well (FIG. 1).

Abnormal Rhythm

Ectopic Beating in a Single Cell Model

These results from experimental and theoretical work on normal cardiac beats provide some confidence that the models correctly describe the intensity and dynamics of sodium-calcium exchange. It is therefore worthwhile exploring how far the models succeed in predicting the role of sodium-calcium exchange in abnormal rhythm. At the previous meeting, it was shown that the atrial cell model correctly predicts the sequence of changes following sodium pump inhibition.[13]

The immediate effect on removing the outward pump current is a small prolongation of the action potential. As internal sodium increases the sodium-calcium exchanger is deprived of the energy gradient it normally uses to keep calcium out of the cell. The action potential then greatly shortens as the net exchanger activity moves from being predominantly inward to outward, a process that also acts to produce a rise in intracellular calcium. Initially, virtually all this increased calcium is stored by the sarcoplasmic reticulum leading to larger calcium release (and therefore larger contractions) with very little rise in resting calcium. Above a certain level (around 15 mM) of internal sodium, however, the SR can no longer hold all the calcium entering the cell. At this stage, calcium acts to trigger spontaneous release of SR calcium in an oscillatory manner and resting calcium levels rise significantly. This is the state of calcium-overload and it is well known to be arrhythmogenic. The mechanism of arrhythmia in such conditions is also well known. Each rise in internal calcium activates inward membrane current, either by activating the inward mode of the sodium-calcium exchanger or by activating nonspecific cation channels. In ventricular cells, the evidence is that the major component of such oscillatory (transient inward) current is generated by sodium-calcium exchange since the current-voltage relation for the transient inward current is very similar to that of the inward mode of the sodium-calcium exchanger.[14] If this inward current is sufficiently large it may trigger ectopic beats.

FIGURE 2 shows that the atrial cell model correctly describes such ectopic beating. In this computation, internal sodium was allowed to rise to 16.24 mM, well within the overload range. The first action potential generated by the model was initiated by a stimulus. Notice that the rise in intracellular calcium *follows* the calcium current and the action potential upstroke. By contrast, all the subsequent beats, which occur at a frequency of around 2 Hz, are spontaneous. The rise in intracellular calcium *precedes* each action potential and there is a substantial inward Na-Ca exchange current that succeeds in triggering each of these ectopic beats.

Initiation and Propagation of Ectopic Beats in Large Network Models

The atrial cell model has been incorporated into models of very large 2 and 3 dimensional network models, incorporating up to 4 million model cells to represent

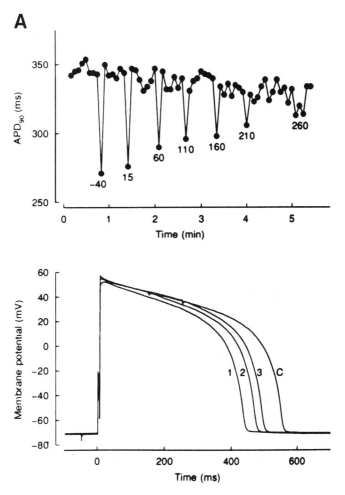

FIGURE 1. (A) Shortening of action potential in guinea-pig ventricular cell following sudden reduction of $[Na]_o$ from 140 to 70 mM at various times during the plateau. The *top panel* shows the measured action potential durations in control recordings and during sudden and brief reductions in $[Na]_o$. The figures below the trace show the time of the solution switch taking the action potential upstroke as zero. The *bottom panel* shows superimposed action potential recordings. **(B)** Reconstruction of this experiment using a model of the guinea pig ventricular cell. The *top panel* reconstructs the action potential changes, the *middle panel* shows the computed Na-Ca exchange current, while the *bottom panel* shows the computed intracellular calcium.[11]

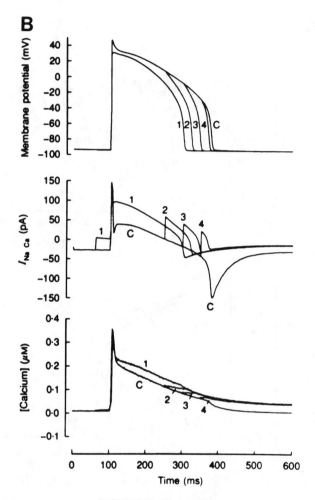

FIGURE 1. (*Continued*)

a sheet or block of cardiac tissue sufficiently large to begin to answer a very important functional question.[15,16] This is how large an area of sodium and calcium overloaded cells is sufficient to initiate a propagated ectopic beat and so be the cause of arrhythmia in the whole heart?

The network models are constructed by linking each model cell to its nearest neighbours (4 in a 2-dimensional sheet, 6 in the case of a 3-dimensional block) via resistances representing gap junctions. First, the junctional conductance necessary to achieve the correct conduction velocity (60 cm s^{-1}) in a network of normal cells was determined. This was found to be 1 uS, which corresponds to 20,000 gap junction channels of 50 pS unit conductance. A region of sodium- and calcium-overloaded cells was then introduced into the center of the net. The number of such cells was varied, as was the junctional conductance, and the number of

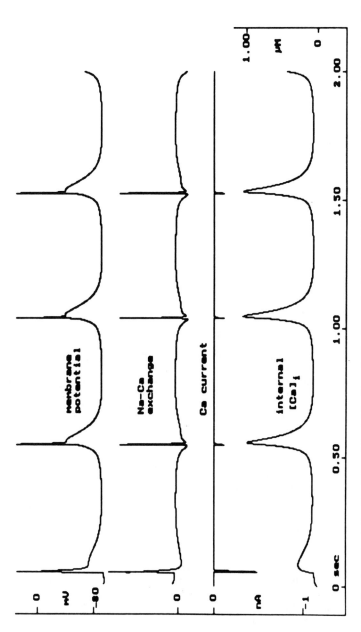

FIGURE 2. Regular ectopic beating induced in atrial cell model[5] by increasing $[Na]_i$ to 16.24 mM. This was achieved by 90% block of the Na-K pump for a period of 300 seconds. The first action potential in the train shown here was initiated by a current pulse of 1.3 nA for 2 msec. This triggers the normal sequence of events: inward calcium current precedes both the rise of $[Ca]_i$ and the inward phase of i_{NaCa}. Subsequent beats are spontaneous and arise from internal calcium oscillations. The rise in $[Ca]_i$ then precedes the action potential and the much-reduced calcium current. Note: for the sake of clarity, the baseline for plotting Na-Ca exchange current was shifted upwards compared to that for the calcium current.

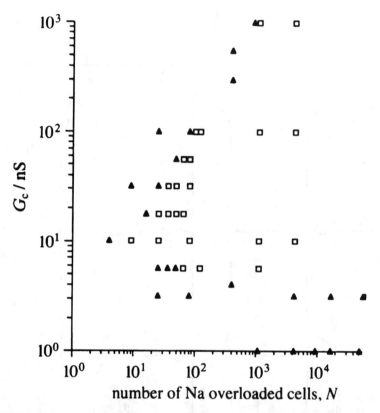

FIGURE 3. Results of varying the number of Na-overloaded cells (N, *abscissa*) situated at the center and cell-to-cell coupling conductance (G_c, *ordinate*) in a 512 × 512 atrial cell model network. *Filled triangles* mark G_c-N pairs for which spontaneous, propagating beats were not obtained. *Open squares* mark G_c-N pairs producing spontaneous propagating beats.[16]

overloaded cells required to initiate sustained propagated ectopic beating was determined for various values of junctional conductance. Clearly, if the 'damaged' area is too small then, even though the calcium oscillations within this region would be large enough to initiate action potentials in isolated cells (as in FIG. 2), a propagated beat will not occur because hyperpolarizing current from the surrounding normally polarized resting cells will prevent excitation from occurring.

FIGURE 3 shows the results obtained for a 25-dimensional network of 512 × 512 model cells, a total number of 262,144 cells. Open squares indicate combinations that produced propagated ectopic beats. Filled triangles indicate combinations for which the overload area failed to trigger a propagated beat. What is striking is that for all coupling conductances except for unrealistically low values (less than 5 nS, corresponding to fewer than 100 channels) a surprisingly small area of overloaded cells can initiate propagated beats. Even at the 'normal' level of 1000 nS, as few as 1000 cells can initiate a propagated disturbance. At lower coupling conductances (and it is likely that coupling conductances are reduced in ischemic tissue, for example) even smaller areas are sufficient. Thus reducing the coupling conductance by a factor of 10 also reduces the number of overloaded cells required to around 100.

Similar studies were done with a 3-dimensional network of 128 × 128 × 32 atrial cells, i.e., a total of just over 0.5 million cells. In this case, a spherical volume of cells of diameter 20 (around 4,000 cells) was sufficient to initiate a propagated beat.[16]

These figures correspond to extremely small areas of tissue and they would readily explain why some fatal cardiac arrhythmias can be triggered with no prior indications in the electrocardiogram: the so-called 'silent' arrhythmias. Volumes of tissue corresponding to fewer than 5,000 cells (a sphere only about 1 mm in radius) would hardly be likely to produce significant deformation of the global ECG.

They also raise an interesting and fundamental question concerning the initiation of rhythm in the heart. Normal rhythm initiated by the SA node is generated by a very much larger number of pacemaker cells (over 100,000). Moreover, successful initiation of an excitation wave that propagates beyond the node itself and invades the atrium depends on fine anatomical detail (interdigitations between sinus and atrial tissue) to provide the required safety factor for propagation.[17]

CONCLUSION

The calcium overload ectopic beat mechanism is therefore a comparatively robust pacemaker. Why is this? One possible explanation may lie in the molecular nature of the sodium-calcium exchange protein and its manifestation in the overall current-voltage relation generated by this transport mechanism. The inward mode of the exchange displays a roughly exponential dependence on membrane potential, with the inward current greatly increasing as the membrane is hyperpolarized.[6] The equations chosen for representing the exchanger in the models described in this paper also show this property. This is precisely the property required for this mechanism, in combination with intracellular calcium oscillations that do not require membrane potential changes to initiate them, to strongly resist inhibition by hyperpolarizing current from surrounding normally polarized resting tissue. In fact, hyperpolarization of calcium-overloaded cells undergoing calcium oscillations will only succeed in *increasing* the depolarizing current they generate.

Paradoxically, and unfortunately, a transport mechanism that is ideally suited to its role in the rapid achievement of calcium balance on a beat-to-beat basis during normal cardiac rhythm is also ideally suited to the generation of arrhythmia in the heart.

REFERENCES

1. MITCHELL, M. R., T. POWELL, D. A. TERRAR & V. W. TWIST. 1984. Calcium-activated inward current and contraction in rat and guinea-pig ventricular myocytes. J. Physiol. **391:** 545–560.
2. SCHOUTEN, V. J. A. & H. E. D. J. TER KEURS. 1985. The slow repolarization phase of the action potential in rat heart. J. Physiol. **360:** 13–25.
3. EARM, Y. E., W. K. HO & I. S. SO. 1990. Inward current generated by sodium-calcium exchange during the action potential in single atrial cells of the rabbit. Proc. R. Soc. Lond. Ser. B **240:** 61–81.
4. HILGEMANN, D. W. & D. NOBLE. 1987. Excitation-contraction coupling and extracellu-

lar calcium transients in rabbit atrium: reconstruction of basic cellular mechanisms. Proc. R. Soc. Lond. Ser. B **230:** 163–205.
5. EARM, Y. E. & D. NOBLE. 1990. A model of the single atrial cell: relation between calcium current and calcium release. Proc. R. Soc. Lond. Ser. B **240:** 83–96.
6. KIMURA, J., S. MIYAMAE & A. NOMA. 1987. Identification of sodium-calcium exchange current in single ventricular cells of guinea-pig. J. Physiol. **384:** 199–222.
7. LEVI, A. J., P. BROOKSBY & J. C. HANCOX. 1993. A role for depolarization-induced calcium entry on the Na-Ca exchange in triggering intracellular calcium release in rat ventricular myocytes. Cardiovasc. Res. **27:** 1677–1690.
8. LEVI, A. J., K. W. SPITZER, J. H. B. BRIDGE & O. KOHMOTO. 1994. Depolarization-induced Ca entry via Na-Ca exchange triggers SR release in guinea-pig cardiac myocytes. Am. J. Physiol. **266.** In press.
9. EGAN, T., D. NOBLE, S. J. NOBLE, T. POWELL, A. J. SPINDLER & V. W. TWIST. 1989. Sodium-calcium exchange during the action potential in guinea-pig ventricular cells. J. Physiol. **411:** 639–661.
10. NOBLE, D. & G. C. L. BETT. 1993. Reconstructing the heart: a challenge for integrative physiology. Cardiovasc. Res. **27:** 1701–1712.
11. LEGUENNEC, J-Y. & D. NOBLE. 1994. Effects of rapid changes of external Na^+ concentration at different moments during the action potential in guinea-pig myocytes. J. Physiol. **478:** 493–504.
12. SPITZER, K. & J. BRIDGE. 1989. A simple device for rapidly exchanging solution surrounding a single cardiac cell. Am. J. Physiol. **256:** C441–C447.
13. NOBLE, D., S. J. NOBLE, G. C. L. BETT, Y. E. EARM, W. K. HO & I. S. SO. 1991. The role of sodium-calcium exchange during the cardiac action potential. Ann. N. Y. Acad. Sci. **639:** 334–353.
14. FEDIDA, D., D. NOBLE, A. C. RANKIN & A. J. SPINDLER. 1987. The transient inward current, i_{TI}, and related contraction in guinea-pig ventricular myocytes. J. Physiol. **392:** 523–542.
15. WINSLOW, R. L., A. VARGHESE, D. NOBLE, C. ADLAKHA & A. HOYTHYA. 1993. Generation and propagation of ectopic beats induced by spatially localized Na-K pump inhibition in atrial network models. Proc. R. Soc. Lond. Ser. B **254:** 55–61.
16. WINSLOW, R. L., D. CAI, A. VARGHESE & Y-C. LAI. 1995. Generation and propagation of normal and abnormal pacemaker activity in network models of cardiac sinus node and atrium. Chaos, Solitons and Fractals **5:** 491–512.
17. WINSLOW, R. L. & H. JONGSMA. 1995. Role of tissue geometry and spatial localization of gap junctions in generation of the pacemaker potential. J. Physiol. **487:** 126P.

The Exchanger and Cardiac Hypertrophy[a]

DONALD R. MENICK,[b] KIMBERLY V. BARNES,
USHA F. THACKER,[c] MYRA M. DAWSON,
DIANE E. McDERMOTT, JOHN D. ROZICH,
ROBERT L. KENT, AND GEORGE COOPER IV

Cardiology Division, Department of Medicine
and the
Gazes Cardiac Research Institute
Medical University of South Carolina
and the
Ralph H. Johnson Department of
Veterans Affairs Medical Center
Charleston, South Carolina

INTRODUCTION

The Na-Ca exchanger plays a major role in Ca^{2+} efflux and therefore in the control and regulation of intracellular calcium in the cardiac muscle cell, or cardiocyte. We recently observed that hemodynamic load induces a rapid 2–4-fold increase in Na-Ca exchanger message levels. This finding is very remarkable and suggests that the expression of the exchanger, which is so critical to calcium homeostasis in the heart, is acutely sensitive to changes in cardiac load.

Rapid Upregulation of the Na-Ca Exchanger in Response to Pressure Overload

While congestive heart failure (CHF) is a multifactorial, progressive disease involving many different pathophysiologies, it begins with the initially compensatory hypertrophic response of the heart to increased load.[1-3] To understand the reasons that initially compensatory hypertrophy so frequently deteriorates into CHF, sense must be made at the molecular level of a complex cascade of hemodynamic, neurogenic, autocrine, and paracrine events that both regulate cardiocyte mass and, eventually, affect cardiocyte contractile function. In this context, cardiac hypertrophy is the initial response in a cascade leading towards congestive heart failure. While the mechanisms of the cardiocyte response to either normal

---

[a] This research was supported in part by National Institutes of Health Grant HL48788.
[b] Corresponding author: Donald R. Menick, Ph.D., Cardiology Division, Medical University of South Carolina, 171 Ashley Avenue, Charleston, South Carolina, 29425-2221.
[c] Current address: Department of Biochemistry, UMDNJ, 185 South Orange Avenue, Newark, NJ 07103.

or abnormal hemodynamic load are poorly understood, they must include many signaling activation pathways leading from the cell surface to the genome, including the rapid induction of immediate early genes such as c-*fos* and c-*myc*,[4,5] as well as the later up- and downregulation of a multitude of genes for myofibrillar, calcium transport and binding, and other cytosolic and membrane proteins.[6] However, the immediate early gene response is rather ubiquitous, and the role of the gene products in the hypertrophy of terminally differentiated cardiocytes remains obscure. Thus the linkage to downstream events of the multitude of early transcriptional and/or posttranscriptional events which may be relevant to the cardiocyte response to load are largely unknown. Therefore, there is a need to distinguish from these ubiquitous responses physiologically relevant initial markers of cardiocyte load.

In order to identify physiologically significant markers of cardiac load, we have developed an *in vivo* model of acute right ventricle (RV) pressure overload that allows the left ventricle (LV) to serve as an internal control from the same heart.[7] General anesthesia is administered through use of parenteral drugs without significant perturbations in hemodynamic parameters such as systemic blood pressure and heart rate. A single arterial catheter is placed in the animal after anesthesia is established. This is followed by the transvenous passage of a specially designed balloon-tipped catheter with one pressure port at the catheter tip and one more proximal port which enables direct recording of both pulmonary artery (PA) pressure and RV pressure (FIG. 1). By changing the degree of inflation and position of the catheter within the PA, selected pressures can be generated in the RV with careful attention directed to avoiding alterations in systemic hemodynamics. Importantly, this model allows comparison between the pressure overloaded RV and the normally loaded LV as a same animal genetic control. After maintaining the RV pressure overload for variable times, total RNA was extracted from free walls of the RV and LV and Northern blots were performed. As expected, the immediate early genes, c-*fos*, c-*myc* and c-*zif*, are induced within an hour of pressure overload (not shown).

Importantly, there is also a rapid increase in the expression of the Na-Ca exchanger.[8] By one hour the level of exchanger mRNA was increased 2–4-fold in the pressure overloaded RV when compared with the level in normally loaded LV from the same heart (FIG. 2). The increase in exchanger message level was sustained for at least 4 hours. To determine whether there was a corresponding increase in the amount of exchanger protein, microsomal fractions were prepared from LV and RV free wall after 48 hours of chronic RV pressure overload. The chronic pressure overload of the feline RV was surgically created by placing a constricting band around the pulmonary artery as previously described.[9] Equivalent amounts of microsomal protein from pressure overload RV and same animal normal LV were Western blotted with a polyclonal antibody specific for the Na-Ca exchanger (gift of G. Lindenmayer, MUSC, Charleston, SC). Chronic maintenance of RV pressure overload for 48 hours induced a significant increase in the level of exchanger in the membrane (FIG. 3).

Upregulation of the Exchanger in Cultured Cardiocytes

Based on the upregulation of the Na-Ca exchanger in the whole animal model, Northern blot analysis was carried out on the more readily manipulated feline adult cardiocytes and neonatal rat cardiocytes. Phenylephrine (PE) was used as an established model of stimulating α-agonist receptor pathways to induce neonatal

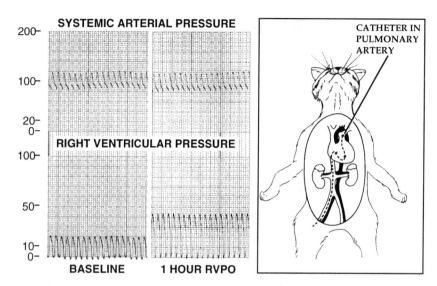

FIGURE 1. Schematic illustration of the *in vivo* model of pulmonary artery balloon-induced right ventricular pressure overload (RVPO) showing systemic arterial and RV pressure recordings from a 1-hour right RVPO cat. Pressure overload was created by passing a catheter through the femoral vein into the pulmonary artery with fluoroscopic guidance and inflating the balloon-tipped catheter in the pulmonary artery. The catheter had two pressure ports, one at the distal end and the second 1.5 cm proximal to the first. RV pressures were measured at the start and continuously while the balloon was inflated for either 1 or 4 hours. Balloon inflation did not change the continuously measured systemic arterial pressure. (From Rozich *et al.*[7] Reprinted by permission from the *Journal of Molecular and Cellular Cardiology*.)

cardiocyte hypertrophy. The Na-Ca exchanger was upregulated 3–5-fold in neonatal cardiocytes with PE stimulation within one hour and remained upregulated for several hours (FIG. 4). Because mechanical loading of isolated papillary muscles enhanced Na^+ influx[10] and Ca^{2+} influx was enhanced in pressure overloaded hearts,[11] we examined whether exchanger message was induced in response to Na^+ or Ca^{2+} influx in isolated cardiocytes. Na^+ influx can be initiated in cardiocytes by treatment with veratridine, which also causes a secondary increase in myoplasmic calcium.[12]

Much like hemodynamic load, veratridine treatment of adult cardiocytes accelerated the rates for total protein synthesis and specifically for myosin synthesis.[8] Veratridine treatment of both neonatal and adult isolated cardiocytes also induced the expression of the immediate early genes, c-*zif* and c-*fos* (not shown), and results in the same 2–4-fold rapid increase in exchanger message as observed in the hemodynamically loaded RV (FIG. 5).[8] Importantly, insulin treatment of adult cardiocytes, which does increase the rates of general protein synthesis, does not induce an increase in the exchanger message level (FIG. 5). Therefore, our initial data appear to support the idea that induction of exchanger message may be fairly specific (*e.g.*, response to load and/or Na^+ or Ca^{2+} influx) and not a general shock response.

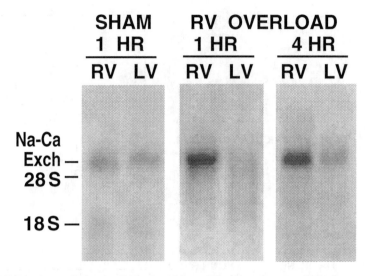

FIGURE 2. Autoradiographs of Northern blots of Na-Ca exchanger mRNA from 1- and 4-hours RVPO and SHAM cats. There is an increase in RV versus LV Na-Ca exchanger mRNA at each time point in RVPO but not in SHAM. Each lane was loaded with 15 µg of total RNA from paired RV and LV tissue from the same cat, confirmed as equal by subsequent measurement of 28S rRNA; film exposure was 3 days.

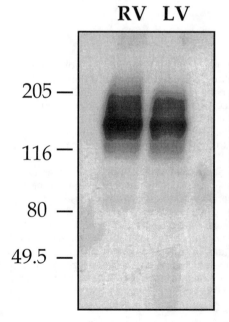

FIGURE 3. Western blot for Na-Ca exchanger in pressure overloaded ventricle. 20 µg of microsomal protein extracted from 48-hour pressure overloaded RV free wall and same heart control LV free wall were separated on 7.5% sodium dodecyl sulfate (SDS) polyacrylamide gel. Proteins were blotted electrophoretically onto PVDF membrane. Blot was reacted with anti-Na-Ca exchanger (gift of George Lindenmayer, MUSC, Charleston, SC). Bound antibodies were detected using horseradish peroxidase-conjugated anti-rabbit immunoglobulin G (IgG) antibodies and exposed to film after treatment with detection reagents from ECL (enhanced chemiluminescense, Amersham).

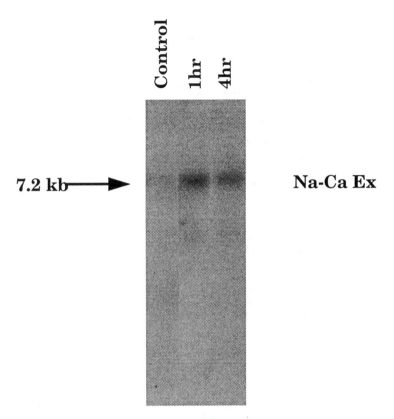

FIGURE 4. Phenylephrine stimulation of neonatal cardiocytes upregulates the Na-Ca exchanger message. Northern blot analysis of Na-Ca exchanger message levels in control and in 1- and 4-hour phenylephrine (0.1 mM)-treated rat neonatal cardiocytes.

Cloning of the Na-Ca Exchanger cDNA

In order to further characterize expression and the mechanism of hemodynamic, PE or veratridine induction of message level, we isolated the feline Na-Ca exchanger cDNA. The feline Na-Ca exchanger clone contains 252 nucleotides of the 5'-untranslated region (UTR), the entire open reading frame, and 259 bases of the 3'-UTR. The Na-Ca exchanger is very highly conserved across species. It has a 92% and 89% nucleotide identity with the canine[13] and bovine[14] clones. The amino acid identity between these three species is >98%, and five of the 15 differences lie in the 32 amino terminal signal sequence.

The 252 base 5'-UTR has a second out-of-frame initiation codon with an open reading frame coding for 17 aa before reaching a stop codon. The upstream initiation codon (TTGATGC) has reasonable conformity to the consensus Kozak sequence (PuCCAUGG).[15] Once again this feature is conserved with respect to the bovine sequence,[14] which has two possible upstream initiation sites, with one found in the context of a Kozak consensus sequence. The 3'-UTR of the cat pNCX7 clone is only 259 nucleotides long. It does not contain a polyadenylation signal or a poly A tail. The mature transcript is ~7 kb long; therefore our 3.4-kb clone only contains a small portion of the 3'-UTR.

The Na-Ca Exchanger Message Half-Life

The short segment of the 3'-UTR is extremely interesting. There is an AU-rich region containing a single Shaw-Kamen sequence AUUUA, which has been demonstrated to be important to message stability in lymphokines, cytokines, and immediate early genes. This region is highly conserved in feline, canine, bovine, and human clones. The feline and canine sequences have a 100% identity for our entire 259 base 3'-UTR (FIG. 6) and an 88% identity with the human and bovine sequences. The rapid induction, the very long 3'-UTR, the out-of-frame upstream initiation codon, and the AUUUA element point to a message that has low abundance and is highly regulated and possibly has a rapid rate of turnover.

In order to determine whether the increase in Na-Ca exchanger message levels is due to transcriptional or posttranscriptional regulation, message half-life studies were carried out. Neonatal cardiocytes pretreated with veratridine for 4 hours and control (no veratridine) cells were treated with actinomycin D to halt transcription and RNA isolated at selected time points. Northern analysis revealed that Na-Ca exchanger message half-life is 6.0 hours in veratridine-stimulated cells and 5.4 hours in the control cells (FIG. 7). This difference in half-life is much less than the 2–4-fold changes seen in message levels with veratridine stimulation indicating

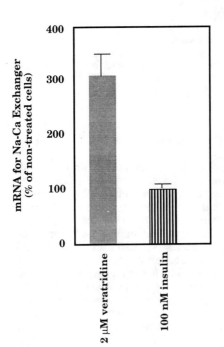

FIGURE 5. Veratridine tripled exchanger mRNA levels in cultured cardiocytes, while insulin had no effect. Adult cardiocytes were treated with 2 μM veratridine or 100 nM insulin for 4 hours and their RNA extracted, blotted, and hybridized with [^{32}P] cDNA probe for the exchanger.

```
Cat:  3191  AGGAACAATCAGATATAGTAAATTTATATATATACGTATATATATACATAAAATATTA  3250
            ||||||||||||||||||||||||||||||||||||||||||||||||||||||||||
Dog:  2941  AGGAACAATCAGATATAGTAAATTTATATATATACGTATATATATACATAAAATATTA  3000

Cat:  3251  TGTATAATGAACAGAGAGGAAACTGACATTTGTCATGTTCACTTACCTGCTGATGGAATCCA  3310
            ||||||||||||||||||||||||||||||||||||||||||||||||||||||||||||
Dog:  3001  TGTATAATGAACAGAGAGGAAACTGACATTTGTCATGTTCACTTACCTGCTGATGGAATCCA  3060

Cat:  3311  GCTTCAAGAGCATACTCTGTACTAGGGCTGAAGTGAGAAACCATCACCTCCCATTCCCAG  3370
            ||||||||||||||||||||||||||||||||||||||||||||||||||||||||||
Dog:  3061  GCTTCAAGAGCATACTCTGTACTAGGGCTGAAGTGAGAAACCATCACCTCCCATTCCCAG  3120

Cat:  3371  GGGCGTCATCACATTGAACAAGGCATGGAGGCAGGGCATCTTTGCAGCTCAGCCTAGAA  3430
            ||||||||||||||||||||||||||||||||||||||||||||||||||||||||||
Dog:  3121  GGGCGTCATCACATTGAACAAGGCATGAGGAGGCAGGGCATCTTTGCAGCTCAGCCTAGAA  3180

Cat:  3431  GGACTGTGTTCTGGAATTC  3449
            |||||||||||||||||||
Dog:  3181  GGACTGTGTTCTGGAATTC  3199
```

FIGURE 6. 3'-UTR of the feline and canine[16] cDNA clones for the cardiac Na-Ca exchanger. Following the termination codon the sequences are identical for at least 258 nucleotides.

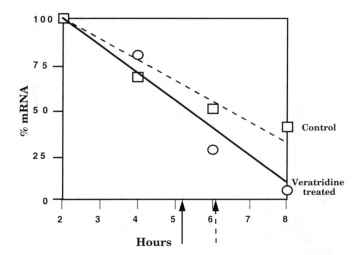

FIGURE 7. Message half-life of the Na-Ca exchanger. Neonatal cardiocytes control (□) and veratridine stimulated (○) for 4 hours were treated with actinomycin D (5 μg/ml) at time 0. The cells were scraped at different time points, RNA was extracted and Northern blots were carried out. The level of Na-Ca exchanger mRNA remaining was plotted as percent of time 0 transcript level versus time. The message half-life for veratridine-pretreated cells was ~5 hours, while the untreated cells was ~6 hours.

that the majority of regulation is at the transcriptional level and not at the posttranscriptional level.

Mapping the 5'-UTR of Exchanger Isoforms

Rapid amplification of cDNA 5'-ends (5'-RACE) was utilized to identify and clone two distinct Na-Ca exchanger 5'-UTR isoforms from the feline heart. The first 5'-UTR cardiac isoform 1 extends 122 bases 5' of the translational start site. The second 5'-UTR isolated by 5'-RACE has 428 bases 5' of the translational start site. The sequencing of these clones revealed that both have identical sequences from −34 base 3' through the open reading frame to the +250 primer site. The 5' most 122 bases of the cardiac isoform 2 is identical to the cardiac isoform 1 sequence (FIG. 8). Extensive screening of our cardiac cDNA libraries yielded clones identical to our 5'-RACE clones. In addition, a third 5'-UTR isoform was identified with a sequence identical to the first 117 bp (H4) 5' of −34 position of our second 5'-RACE clone (cardiac isoform 2) but had a unique 5'-end of an additional 106 bases (FIG. 9; H3). Northern blots were probed with each of the unique sequences (H1, H2, H3 and H4) found in our three putative cardiac isoforms to confirm their expression in the heart and screen their tissue distribution. The H1, H3 and H4 probes hybridized at a relatively higher level than the H2 probe. Because H1 is common to isoform 1 and 2 and H4 is common to isoform 2 and 3 but H3 is unique to isoform 2, it appears that isoform 1 and 3 are expressed at higher levels in the heart than isoform 2. The feline cardiac isoform 1 (exon H1) is expressed

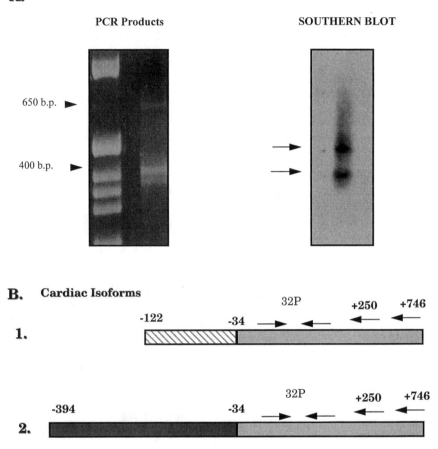

FIGURE 8. Polymerase chain reaction (PCR) products from round one of 5'-RACE in the heart. After thirty-five rounds of amplification using a cDNA specific primer 250 bases downstream of the translational start site, 25 µl of the PCR reaction were loaded on a 1.5% agarose gel. After electrophoresis, two bands were seen approximately 400 and 650 base pairs in size (**A**). A Southern blot was performed using a radioactive PCR probe generated from the first 150 bases of the exchanger open reading frame. These bands were later subcloned, sequenced and found to contain distinctly different sequences representing two different isoforms of the exchanger (**B**).

in both heart and brain. It is interesting to note that the feline cardiac isoform 1 has a 68% homology with the rat Br2 isoform which is also expressed in the heart and brain.[16] The cardiac isoform 1 also has an 84% homology with the 5'-end (−268 through −218) of the bovine p17 clone.[14] The feline cardiac isoform 3 (exon H3-H4) is also expressed in both heart and brain but isoform 2 (exon H1-H2-H4) has been detected only in the heart. None of the cardiac isoforms has been detected in the kidney.

Only one 78-nucleotide species (K1) was detected with 5'-RACE of the feline kidney. It has no homology with the rat kc1 isoform.[16] 5'-RACE was used to analyze the Na-Ca exchanger transcripts from brain. Four unique species were identified in brain. The feline Br1 isoform extends 123 nucleotides 5' of −34 splice site and has a 60% homology with the rat Br1 isoform. The Br2 isoform was identical to the kidney isoform but extends 5' for another 203 nucleotides. Br3 extends 102 nucleotides and Br4 extends 66 nucleotides 5' of the −34 bp splice site. The Br3 and Br4 5'-UTR isoforms have no homology with any other known sequence.

A. Cardiac 5' UTR Isoforms

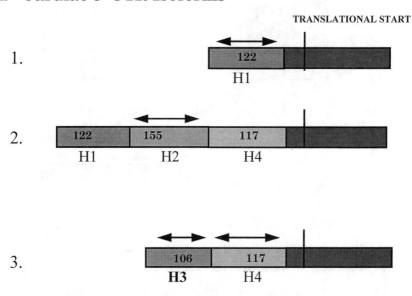

B. Northern blots

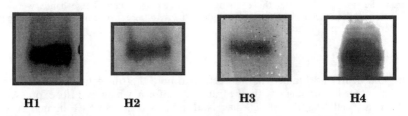

FIGURE 9. Diagram of three cardiac isoforms of the 5'-UTR of the exchanger. **(A)** The sizes of cardiac isoforms 1, 2, and 3 are approximately 122, 394, and 223 bp, respectively. **(B)** All three have been confirmed to be expressed in the heart through Northern blot analysis using radioactive PCR probes generated from each of the unique regions.

A.

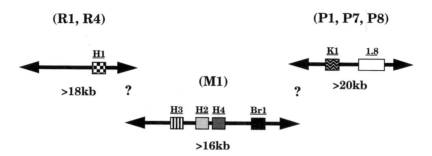

B.

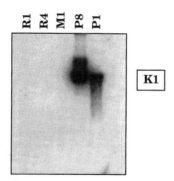

FIGURE 10. Schematic representation of the feline Na-Ca exchanger genomic clones. (A) The exons are represented by the *boxes*. (B) Southern blot of genomic clones probed with a sequence specific for the feline K1 isoform.

Cloning of the 5'-End of the Feline Na-Ca Exchanger Gene

Feline genomic clones containing the 5'-end of the Na-Ca exchanger gene were isolated to characterize transcriptional elements which regulate cardiac-specific and inducible expression during myocardial cell hypertrophy and examine the mechanism of alternative splicing that gives rise to the 5'-UTR exchanger isoforms. Polymerase chain reaction (PCR)-generated probes corresponding to each of the putative heart exons (H1, H2, H3 and H4) and one corresponding to the 5'-end of the open reading frame (+201 to +601) were used to screen our feline genomic library. This yielded several positive clones corresponding to three distinct genomic regions (FIG. 10). Two positive clones (R1, R4) ranging in size from 16–19 kb contain the H1 exon. A second clone (M1; ~16 kb) contains the H2, H3 and

H4 exons. Three other positive genomic clones (P1, P7, P8) contain the 1.8-kb exon sequence from the -32 position through the $+1826$ position of the open reading frame. At this time, we are mapping our genomic clones and are not sure whether the R, M and P clones are contiguous. There is greater than 2 kb of intron 3' of the 1.8-kb exon, and the P1, P7 and P8 clones do not contain any of the other downstream coding region exons. PCR analysis maps H3 approximately 100 bases 5' of H4. Southern blots indicate that K1 (which is identical to Br2) maps to the P1, P7 and P8 clones and Br1 maps to the M1 clone (FIG. 10).

We are currently mapping these genomic clones in order to characterize the Na-Ca exchanger gene promoter(s). In addition to our observation of the upregulation of the exchanger in the early stages of hypertrophy,[8] others have reported increases in exchanger activity late in cardiac hypertrophy[17] and increases in exchanger mRNA and protein levels in end-stage heart failure.[18] Therefore, the study of the molecular mechanisms which govern the regulation of a gene whose product is critical to cardiac calcium homeostasis should lend physiologically significant insight into the mechanism of early hypertrophy growth which leads to failure.

REFERENCES

1. COOPER, G., R. J. TOMANECK, J. C. EHRHARDT & M. L. MARCUS. 1981. Chronic progressive pressure overload of the cat right ventricle. Circ. Res. **48:** 488–497.
2. COOPER, G. 1987. Cardiocyte adaptation to chronically altered load. Annu. Rev. Physiol. **49:** 501–518.
3. CHIEN, K. R., Z. HONG, K. U. KNOWLTON, W. MILLER-HANCE, M. VAN-BILSEN, T. X. O'BRIEN & S. M. EVANS. 1993. Transcriptional regulation during cardiac growth and development. Annu. Rev. Physiol. **55:** 77–95.
4. IZUMO, S., B. NADAL-GINARD & V. MAHDAVI. 1988. Proto-oncogene induction and reprogramming of cardiac gene expression produced by pressure overload. Proc. Natl. Acad. Sci. USA **85:** 339–343.
5. MULVAGH, S. L., L. H. MICHAEL, M. B. PERRYMAN, R. ROBERTS & M. D. SCHNIEDER. 1987. A hemodynamic load *in vivo* induces cardiac expression of c-*myc* cellular oncogene. Biochem. Biophys. Res. Commun. **147:** 627–636.
6. SWYNGHDAUW, B. 1986. Developmental and functional adaptation of contractile proteins in cardiac and skeletal muscles. Physiol. Rev. **66:** 710–714.
7. ROZICH, J. D., M. A. BARNES, P. G. SCHMID, M. R. ZILE, P. J. MCDERMOTT & G. COOPER IV. 1995. Load effects on gene expression during cardiac hypertrophy. J. Mol. Cell. Cardiol. **27:** 485–499.
8. KENT, R. L., J. D. ROZICH, P. L. MCCOLLAM, D. E. MCDERMOTT, U. F. THACKER, D. R. MENICK, P. J. MCDERMOTT & G. COOPER IV. 1993. Rapid expression of the Na^+-Ca^{2+} exchanger in response to cardiac pressure overload. Am. J. Physiol. **265** (Heart Circ. Physiol. 34): H1024–H1029.
9. COOPER, G., IV, R. L. KENT, C. E. UBOH, E. W. THOMPSON & T. A. MARINO. 1985. Hemodynamic versus adrenergic control of cat right ventricular hypertrophy. J. Clin. Invest. **75:** 1403–1414.
10. KENT, R. L., J. K. HOOBER & G. COOPER IV. 1989. Load responsiveness of protein synthesis in adult mammalian myocardium: role of cardiac deformation linked to sodium influx. Circ. Res. **64:** 74–85.
11. HANEDA, T., P. A. WATSON & H. E. MORGAN. 1989. Elevated aortic pressure, calcium uptake, and protein synthesis in rat heart. J. Mol. Cell. Cardiol. **21**(Suppl. 1): 131–138.
12. FOSSET, M., J. DE BARRY, M-C. LENOIR & M. LAZDUNSKI. 1977. Analysis of molecular aspects of Na^+ and Ca^{2+} uptakes by embryonic cardiac cells in culture. J. Biol. Chem. **252:** 6112–6117.
13. NICOLL, D. A., S. LONGONI & K. D. PHILIPSON. 1990. Molecular cloning and functional expression of the cardiac sarcolemmal Na^+-Ca^{2+} exchanger. Science **250:** 562–565.

14. ACETO, J. F., M. CONDRESCU, C. KROUPIS, H. NELSON, N. NELSON, D. NICOLL, K. D. PHILIPSON & J. P. REEVES. 1992. Cloning and expression of the bovine cardiac sodium-calcium exchanger. Arch. Biochem. Biophys. **298:** 553–560.
15. KOZAK, M. 1989. The scanning model for translation: an update. J. Cell Biol. **108:** 229–241.
16. LEE, S-L., A. S. L. YU & J. LYTTON. 1994. Tissue-specific expression of Na^+-Ca^{2+} exchanger isoforms. J. Biol. Chem. **269:** 14849–14852.
17. NAKANISHI, H., N. MAKINO, T. HATA, H. MATSUI, K. YANO & T. YANAGA. 1989. Sarcolemmal Ca^{2+} transport activities in cardiac hypertrophy caused by pressure overload. Am. J. Physiol. **257:** H349–H356.
18. STUDER, R., H. REINECKE, J. BILGER, T. ESCHENHAGEN, M. BÖHM, G. HASENFUSS, H. JUST, J. HOLTZ & H. DREXLER. 1994. Gene expression of the cardiac sodium-calcium exchanger in end-stage human heart failure. Circ. Res. **75:** 443–453.

Na-Ca Exchange in Circulating Blood Cells

J. P. GARDNER[a,b,c] AND M. BALASUBRAMANYAM[a,b]

[a]Hypertension Research Program, MSB-F464
Departments of [b]Pediatrics and [c]Physiology
University of Medicine and Dentistry–
New Jersey Medical School
185 South Orange Avenue
Newark, New Jersey 07103

INTRODUCTION

Despite significant advances in characterizing the physiology and molecular biology of the Na-Ca exchanger in different tissues, relatively little progress has been made in defining the presence and/or role of this transport system in circulating blood cells. There is even debate whether the Na-Ca exchanger is present in T lymphocytes. In this report we summarize the current status of Na-Ca exchange in erythrocytes, platelets and white blood cells, including neutrophils, monocytes and lymphocytes. We have included results of a Na-dependent Ca uptake assay to examine the presence of Na-Ca exchange activity in leukemic or transformed cell lines. The second portion of this paper discusses our evidence that peripheral lymphocytes exhibit Na-Ca exchange activity that, under specific conditions, appears enhanced following intracellular Ca store depletion.

METHODS

Cell Lines and Isolation of Peripheral Lymphocytes

K-562 erythroleukemic cells, THP-1 monocytes, Jurkat lymphoblasts (EC-6) and YAC-1 T lymphoma cells were from American Type Tissue Collection. Epstein-Bar virus- (EBV) transformed B cell lines were a generous gift from Dr. Winfried Siffert, Universitätsklinikum Essen, Germany, and from NIGMS human genetic mutant cell repository (repository no. GM10850). Molt-4 lymphocytes and HL-60 cells were gifts from Dr. George Studzinsky, UMDNJ-New Jersey Medical School. Peripheral blood lymphocytes were obtained from ACD-treated blood from healthy volunteers by Ficoll-Hypaque density gradient centrifugation as described;[1] for some experiments, CD4$^+$ T cells were isolated using magnetic activated cell sorting (MACS), in which beads were conjugated to CD4$^+$ mAb and T cells were positively selected using a *miniMACS* separator and type MS column (Miltenyi Biotec Inc.).

Na-Dependent Ca Uptake Assay

Cells retaining the Na-Ca exchanger can mediate Ca influx by "reverse mode" Na-Ca exchange activity (*i.e.*, intracellular Na (Na$_i$)-dependent Ca influx). We determined the presence of Na-Ca exchange activity in cultured cells and peripheral lymphocytes by assying ^{45}Ca influx in Na-containing and Na-free media. Ouabain pretreatment was used to inhibit the Na-K-ATPase, increase Na_i and decrease the inwardly directed Na gradient that inhibits Ca entry by "forward mode" Na-Ca exchange. Thus, a greater accumulation of Ca in Na-free conditions is evidence of Na-Ca exchange. This assay has been used successfully to identify Na-Ca exchange activity in several different cell types, including Chinese hamster ovary (CHO) cells transfected with the bovine cardiac Na-Ca exchanger,[2] and is a reliable protocol for detecting the presence of this transport system.

Cells were preincubated with or without ouabain (0.1 mM for peripheral lymphocytes and 0.4 mM for cultured cells) in Na-medium containing (in mM): 140 NaCl, 5 KCl, 1 $MgCl_2$, 1 $CaCl_2$, 20 HEPES, 10 glucose and 0.1% BSA (pH 7.4) for 30 min. Ca uptake was initiated by diluting aliquots of cells (5 μl, 1 × 10^8/ml) 20-fold into Na- or Na-free medium (isosmotic N-methyl-D-glucamine (NMDG) substitution for NaCl) containing 10 μCi/ml ^{45}CaCl$_2$ (final extracellular [Na] = 6.7 mM in Na-free medium). For cells receiving ouabain pretreatment, ouabain was also included in the assay buffer. Ca uptake was stopped by addition of 5 ml ice-cold HEPES-medium (40 mM HEPES, pH 7.0, 100 mM $MgCl_2$ and 10 mM $LaCl_3$); extracellular radioactivity was removed by rapid filtration of cells on 0.45 μ filters as described.[1] Experiments with peripheral lymphocytes showed a linear increase in Ca uptake over an initial 30–40 s (see FIG. 2), the assays shown in FIGURE 1 were performed at a single (*i.e.*, 30 s) time point. Each value represents the mean of triplicate measurements from 3–5 experiments. Background counts (^{45}Ca bound to filters after filtration in the absence of cells) were subtracted from all uptake determinations.

Fluorescence Measurements

Peripheral lymphocytes were incubated for 30 min at 37°C with 2 μM fura-2 AM in Na-medium, with or without 0.1 mM ouabain as indiated. To remove external fura-2, cells were centrifuged and resuspended in nominally Ca-free Na-medium (medium with $CaCl_2$ omitted), and injected into cuvettes containing 3 ml nominally Ca-free Na-medium or Ca-free, Na-free (isosmotic LiCl substitution for NaCl) medium. Fluorescence was monitored in stirred cells at 37°C in a fluorolog II spectrofluorometer after addition of $CaCl_2$ to bring extracellular Ca to 1 mM. Excitation and emission wavelengths were 340/380 nm and 505 nm, respectively. Calibration of intracellular Ca (Ca_i) concentration and determination of autofluorescence (subtracted from all fluorescence measurements prior to calculations) was as described.[1]

Mn was utilized as a Ca surrogate to study thapsigargin (TG)-mediated Ca entry in fura-2 loaded cells. In this assay, Mn entering the cell binds to fura-2 and quenches its fluorescence. Quenching was monitored at 360 nm. Mn uptake was performed in cells exposed to Na- or Na-free medium containing 30 nM TG, 0.5 mM $MnCl_2$ and 0.1–3 μM $LaCl_3$. Results were normalized using 360 nm values obtained prior to addition of $MnCl_2$.

RESULTS AND DISCUSSION

Erythrocytes

The Na-Ca exchanger has been described in dog[3] and ferret[4] erythrocytes, and a potential role for this transporter in cell volume regulation has been proposed.[3]

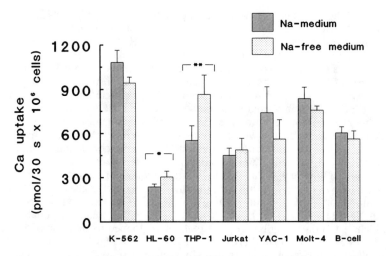

FIGURE 1. Na-dependent Ca uptake in ouabain-treated cells. Cells were treated for 30 min at 37°C in Na-medium containing 0.4 mM ouabain and Ca uptake determined after 30 s incubation in $^{45}CaCl_2$-containing Na-medium or Na-free medium NMDG substituted for NaCl), as described in Methods. Bars represent mean ± SE for 3–5 experiments; each individual experiment was performed in duplicate. The data for B cells were obtained using the NIGMS/GM10850 cell line; similar results were obtained with 3 other cell lines provided by Dr. Siffert. Student's t test was performed using Sigma Plot; statistical significance is indicated by *asterisks* (*$p < 0.03$. **$p < 0.01$).

Red cells are devoid of nuclei, intracellular organelles and, in these animals, the Na-K pump, and have served as a useful cell model for studying the kinetics and substrate specificity of Na-Ca exchange.[4,5] One report examining a murine erythroleukemic cell line suggested the Na-Ca exchanger was involved in increased Ca uptake and committment to erythroid differentiation.[6] However, no Na-Ca exchange activity has been observed in human red cells,[3] and we have been unable to detect Na-Ca exchange activity in undifferentiated, erythroid blast cells of the human K-562 cell line (FIG. 1).

Platelets

In contrast to red cells, human platelets exhibit Na-Ca exchange,[7] and this feature has made them the subject of several physiological and pathophysiological ion transport studies.[7-12] An observation originally established in platelets and later extended to other cells expressing Na-Ca exchange activity was Na-loading (produced by ouabain preincubation) could lead to increased Ca content of intracellular stores.[8] This finding suggested an important interaction of the Na-Ca exchanger with intracellular Ca stores and Ca_i regulation. A second, critical finding by Kimura *et al.*[12] was that the platelet exchanger is K-dependent and cotransports K with Ca. These qualities characterize the transporter as a retinal rod type exchanger. Whether the platelet exchanger is an isoform of the rod Na-Ca + K exchanger, or a different molecular entity, remains to be determined.

Neutrophils

Human neutrophils comprise approximately 50% of all white cells and are separated from other leukocytes by dextran sedimentation followed by Ficoll-Hypaque density gradient centrifugation. Simchowitz and co-workers proposed the Na-Ca exchanger mediated the respiratory burst, chemotactic and phagocytic responses of activated neutrophils.[13,14] These responses were attributed to increased Ca_i resulting from enhanced reverse mode Na-Ca exchange activity. Na-depletion and amiloride analogs like benzamil inhibited ^{45}Ca influx, and ouabain enhanced ^{45}Ca uptake,[13] suggesting that the exchanger is present in these cells. Furthermore, we have demonstrated enhanced Na-dependent Ca uptake in promyelocytic leukemic HL-60 cells (FIG. 1). These cells can differentiate to either granulocyte-like or monocyte/macrophage-like cells, and the demonstration of significant increases in ^{45}Ca influx in Na-free conditions supports the observation that Na-Ca exchange is present in neutrophils. Of greater concern is the exclusive use of amiloride analogs to link enhanced Na-Ca exchange activity to neutrophil function. These analogs inhibit Na-coupled transporters, Na-K-ATPase activity, voltage-gated Na, K and Ca channels, numerous receptor systems, protein kinases and protein, RNA and DNA synthesis.[15] Confirmation by independent means of the physiological responses due to Na-Ca exchange-mediated increases in Ca_i would strengthen the role(s) proposed for this transport system in mediating neutrophil responses following activation.

Monocytes and Lymphocytes

These cells are often studied in clinical protocols examining transport phenomena,[1] and are separated from other peripheral blood cells by centrifugation over Ficoll-Hypaque. Although the Na-Ca exchange and its physiological function have not been examined *per se* in circulating monocytes, Donnadieu and Trautmann[16] demonstrated that reversal of the Na gradient (by substituting Li for Na) increased Ca_i and/or delayed the return of Ca_i to basal levels in stimulated murine macrophages. The role proposed for the Na-Ca exchanger in these cells was the facilitation of Ca_i extrusion following an agonist-induced Ca_i transient. We have observed significant Na-Ca exchange activity in the human THP-1 monocytic cell line compared to HL-60 and other cultured or circulating cells (FIG. 1). In accord with data obtained from other cells expressing a Na-Ca exchanger, intracellular Ca stores in THP-1 monocytes were increased after 1 or 2 hr incubation with ouabain. Unlike the platelet exchanger, Na-dependent Ca uptake was insensitive to extracellular [K] (data not shown).

Lymphocytes comprise 25–35% of all leukocytes; T cells are the predominant cell type and B cells and monocytes exist as 10–25% of the cell population obtained after density gradient centrifugation. Some groups have used transformed or leukemic lymphocyte cell lines for Na-Ca exchange studies, as these cultured cells serve as readily accessible and pure populations of T or B cells. Simply put, the status of Na-Ca exchange in cultured T and B cell lines is unresolved. On one side of the issue, Na-Ca exchange activity in lymphocytes has been detected in Jurkat T cells[17] and murine YAC-1 T lymphoma cells,[18] and reported as an anecdotal finding in adenosine triphosphate (ATP)-stimulated EBV-transformed B cells.[19] The work with T cells utilized amiloride analogs and bepridil to identify Na-Ca exchange activity, and the criticisms listed previously for these mechanism-based inhibitors would also apply to lymphocyte studies. However, at least one laboratory has observed that Na substitution results in elevated Ca_i in Jurkat T and B cells and is enhanced in ouabain-treated T cells (personal communication, D. Schulze, Univ. of Maryland). In contrast, our investigations have shown no evidence of Na-Ca exchange in these cell lines, Molt-4 T lymphoblasts or EBV-transformed B cells (FIG. 1). We have also employed a fura-2 based assay to examine Na-dependent

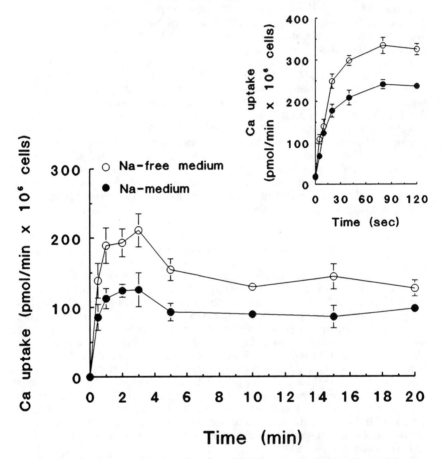

FIGURE 2. Effect of extracellular Na on Ca uptake in ouabain-treated peripheral lymphocytes. Lymphocytes were assayed for Ca uptake as described in legend to FIGURE 1 and Methods. Symbols represent mean ± SE for 5 experiments; each individual experiment was performed in quadruplicate and averaged. Where error bars are not shown, the SE was smaller than the symbol. Repeated measures analysis of variance (ANOVA) was performed using the general linear model procedure of SAS (Version 6.0, SAS Inst.) and indicated significant differences ($p = 0.0125$) in Ca uptake values obtained in Na- vs Na-free medium. *Inset*: Ca influx in lymphocytes from a single individual. For this experiment, only the background counts were subtracted from data points. Linear regressions for the first 4 data points (*i.e.*, 0, 5, 10 and 20 s) were: Na-medium, $y = 8.0(x) + 26.7$, $r = 0.98$; and for Na-free medium, $y = 10.9(x) + 33.7$, $r = 0.99$. Symbols represent mean ± SE for quadruplicate measurements. (From Balasubramanyam *et al.*[1] Reprinted by permission from *The Journal of Clinical Investigation*.)

Ca extrusion, and were unable to detect Na-Ca exchange activity in these cells. These negative data are in agreement with microfluorimetry experiments, in which Na substituted solutions failed to modulate Ca_i in Jurkat T cells and a human T cell subclone.[16,20] If one were to assume the exchanger is present in cultured lymphocytes, the discrepancies obtained by different investigators suggest detection of Na-Ca exchanger is dependent on the method used to assay Na-Ca exchange, and/or exchange activity may be modified *in vitro* or by cell-dependent transformation processes. Moreover, cell lines like Jurkat T cells may not always display uniform properties when maintained in different laboratories.

Na-Ca Exchange in Peripheral Lymphocytes

Na-Ca exchange has been detected in membrane vesicle preparations from lymphocytes from rabbits[20] and mature, peripheral lymphocytes from hu-

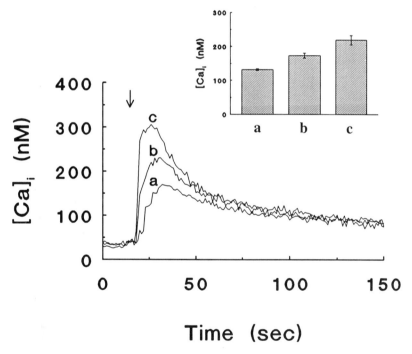

FIGURE 3. Intracellular Ca store size following Na pump inhibition with ouabain in peripheral lymphocytes. Lymphocytes were pretreated with 0.1 mM ouabain for a) 0 hrs, b) 1 hr, or c) 2 hrs in 1 mM Ca-containing Na-medium. Cells were loaded with fura 2-AM during the last 30 minutes of ouabain treatment. Cells were resuspended in Na-medium containing 0.3 mM EGTA and 0.1 mM ouabain and injected into cuvettes containing similar buffer; at 20 s (*arrow*), 5 μM ionomycin and 100 nM TG were added. *Inset*: Peak ionomycin/TG-induced $[Ca]_i$ of cells preincubated with ouabain for a) 0, b) 1 or c) 2 hrs and treated as in main figure. Initial $[Ca]_i$ in ouabain-treated cells injected into Ca-free Na-medium were: 28.8 ± 4.8, 34.0 ± 6.5 and 33.5 ± 5.6 nM for 0, 1, and 2 hrs ouabain treatment, respectively. Results are the mean ± SE of 3 experiments.

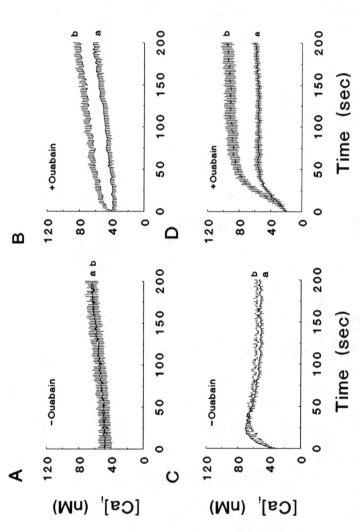

FIGURE 4. Effect of Na gradient and EGTA pretreatment on Ca_i profile. Lymphocytes were incubated for 30 min with fura-2 AM at 37°C in Na-medium containing 1 mM $CaCl_2$ (**A, C**) or 0.3 mM EGTA (**B, D**), with or without 0.1 mM ouabain as indicated. For experiments, cells were resuspended in nominally Ca-free Na-medium and injected into cuvettes containing 3 mls of nominally Ca-free a) Na-medium (line a), or b) Na-free (LiCl substituted for NaCl) medium (line b). Just prior to fluorescence monitoring, $CaCl_2$ was added (final extracellular [Ca] = 1 mM) to the cuvette. Results are from 5 (A and B), 8 (C) and 10 (D) experiments; in this and FIGURE 5C, vertical bars denote SE. Additional experimental details are given in Ref. 1. (From Balasubramanyam et al.[1] Reprinted by permission from *The Journal of Clinical Investigation*.)

mans.[1,22,23] On the basis of experiments in which amiloride analogs or bepridil inhibited lymphocyte growth, Wacholtz et al.[22] proposed that ligand activation of the Na-Ca exchanger could produce increases in Ca_i sufficient to drive lymphocyte proliferation. Other lines of evidence have indirectly implied the existence of Na-Ca exchange in lymphocytes. For example, RNA for Na-Ca exchanger isoforms has been detected in spleen, thymus and lymph nodes from rats.[24] An earlier investigation by our group showed extracellular Na enhanced Ca extrusion, suggesting the presence of Na-Ca exchange activity in peripheral lymphocytes. We therefore examined Na-dependent Ca uptake and Ca_i regulation in ouabain-treated lymphocytes under conditions designed to promote Ca entry. Since lymphocyte activation is accompanied by Ca mobilization from intracellular stores and Ca entry from the extracellular space, we also explored the concept that intracellular Ca store depletion results in Ca entry via the Na-Ca exchanger. The following studies were performed with human peripheral lymphocytes and $CD4^+$ isolated T cells.

Peripheral lymphocytes exhibit a high resting $[Na]_i$ (~20 mM, Refs. 1, 25, although see Ref. 16); half-hour ouabain treatment results in an approximate doubling of Na_i.[1] We first measured ^{45}Ca influx in ouabain-treated cells to determine the presence and relative activity of the Na-Ca exchanger in this population of predominantly mature T cells (FIG. 2). Ca uptake increased rapidly during the first 30 s, followed by a plateau phase (2–3 min) and subsequent decline to levels 60–80% of maximum Ca entry. Ca influx was increased by 30% in cells suspended in Na-free medium compared to Na-medium. In addition, early (5–20 s) measurements suggest an approximate 35% increase in the rate of Ca entry occurs in Na-free conditions (inset, FIG. 2). The decline in Ca uptake was slightly greater in cells suspended in Na-free medium and could be explained by loss of Na_i during the assay. However, additional experiments with 5(6)-carboxyeosin, a Ca pump inhibitor,[26] suggest Ca extrusion across the plasma membrane effects this decrease (data not shown). Despite the relatively high basal $[Na]_i$, enhanced Ca uptake was not observed if ouabain was omitted from the assay.[1] Thus, enhanced Ca influx mediated by a Na-Ca exchange mechanism is observed in peripheral lymphocytes under conditions of increased Na_i and decreased extracellular Na.

Similar to other cells expressing the Na-Ca exchange, ouabain treatment alters lymphocyte $[Ca]_i$ and intracellular Ca store size. Fura-2-loaded cells treated with 0.1 mM ouabain showed only slight increases in basal Ca_i. However, as shown in FIGURE 3, intracellular stored Ca was dramatically increased by 31.3% (1 hr) and 65.9% (2 hrs) following ouabain treatment. These data suggest the amount of Ca stored in lymphocytes is significantly increased in the presence of ouabain. As these results were obtained in cells incubated in 140 mM Na-medium throughout the ouabain treatment period, it appears inhibition of forward mode Na-Ca exchange activity modulates intracellular store Ca homeostasis. An extension of this finding is that enhancement of agonist-stimulated Ca mobilization in peripheral lymphocytes would be expected when Na_i is increased.

We further examined the effect of the Na gradient and intracellular Ca stores on Ca influx. Ouabain pretreatment and exposure to low Na-medium resulted in elevated Ca_i (FIG. 4, top panel). For cells loaded in Ca-free medium, a procedure that reduces the Ca content of intracellular stores by approximately 70%,[1] the Ca_i response was significantly greater than cells with intact intracellular Ca stores (FIG. 4, bottom panel). These results suggest that the Na-Ca exchanger could modulate Ca influx following cell activation and Ca store depletion. This phenomenon was further studied by treating lymphocytes with TG, a sarco(endo)plasmic Ca-ATPase (SERCA) inhibitor. Because depletion of intracellular stores by EGTA

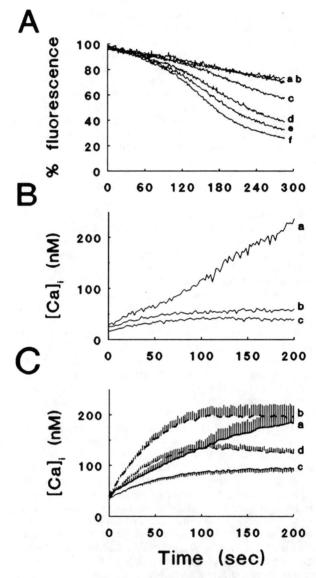

FIGURE 5. Na-Ca exchange dependent increase in Ca_i in the absence of SDCI-mediated Ca uptake. Inhibition of TG-evoked Mn uptake by $LaCl_3$. Lymphocytes were exposed to 30 nM TG and/or $LaCl_3$ in Na-medium and monitored at the isosbestic point of fura-2, 360 nm. Traces: a) vehicle (6 μl DMSO), b) TG plus 3 μM La, c) TG plus 1 μM La, d) TG plus 0.3 μM La, e) TG plus 0.1 μM La, f) TG alone. Initial 360 nm fluorescence values prior to addition of Mn were $1.41 \pm 0.04 \times 10^6$ cps (n = 6). Results are representative of 3 experiments. **(B)** Inhibition of TG-evoked Ca entry by $LaCl_3$. Lymphocytes were resuspended in nominally Ca-free Na-medium and injected into cuvettes containing 30 nM TG; and a) 1 mM Ca, Na-medium, b) 1 mM Ca, Na-medium plus 3 μM La, or c) 0.3 mM EGTA,

treatment and exposure of cells to TG activates Ca entry via store-dependent Ca channels,[27] we utilized La as a potent blocker to determine the extent of Ca uptake by Na-Ca exchange pathways in TG-treated lymphocytes. TG-stimulated Ca entry was monitored by assaying Mn uptake. In the presence of 1 mM Ca, low doses of LaCl$_3$ (1–3 μM) inhibit store-dependent Ca influx (SDCI) pathways without affecting Na-Ca exchange activity. (Note: [La] at 10 μM or higher is required to inhibit the Na-Ca exchanger, and furthermore, Mn is not transported by the Na-Ca exchanger). LaCl$_3$ was thus used to distinguish TG-stimulated Ca (and Mn) entry by channel and Na-Ca exchange pathways. In FIGURE 5A, TG-stimulated Mn uptake and its inhibition by La are shown. Mn entry was sensitive to La-inhibition; at 3 μM La, Mn quenching of fura-2 fluorescence was not different from control (TG-free) Mn uptake. Moreover, La inhibited Ca entry in TG-stimulated, ouabain-free lymphocytes (FIG. 5B). TG-induced increases in Ca$_i$ in cells in La-containing medium (trace b) were similar to cells suspended in Ca-free medium (trace c): both responses were significantly lower than the TG-induced increase in Ca$_i$ in cells suspended in Ca-containing, La-free medium (trace a). Thus, 3 μM La inhibits TG-stimulated Ca uptake in the presence of 1 mM [Ca]. Finally, the effect of La on TG-stimulated increases in Ca$_i$ in ouabain-treated cells is shown in FIGURE 5C. La reduced, but did not abolish, the increase in Ca$_i$ in both Na- (cf. traces a and c) and Na-free medium (cf. traces b and d); inhibition by La was more pronounced in Na- than in Na-free medium. TG increased the rate of Ca entry and [Ca]$_i$ by 1–1.5-fold in Na-free medium, even in the absence of Ca entry via Ca channel pathways (cf. traces c and d). Similar results were obtained when La was replaced with 50 μM SKF 96365, an SDCI inhibitor. The results indicate that in peripheral lymphocytes, TG enhances Ca entry through a La-insensitive pathway that is stimulated by elevated Na$_i$ and reduced extracellular Na (*i.e.*, the Na-Ca exchanger).

In most cases, CD4$^+$ lymphocytes are the predominant subset of T cells obtained after Ficoll-Hypaque density gradient centrifugation. We performed preliminary studies examining the presence of Na-Ca exchange in MACS-separated CD4$^+$ cells. CD4$^+$ cells exhibit Na-dependent Ca uptake (FIG. 6A), albeit at a much lower rate than unfractionated lymphocytes. The effects of SERCA inhibition on Ca influx were also examined. Unlike the data obtained with peripheral lymphocytes, CD4$^+$ cells exposed to TG failed to show an elevated Ca$_i$ profile in Na-free conditions compared to cells in Na-medium (traces a and b). However, if SDCI is blocked with 3 μM La, TG-treatment of cells results in an approximate doubling of Ca$_i$ compared to cells in Na-medium (FIG. 6B, cf. traces d and c). These data suggest that although the Ca$_i$ profile attributed to the lymphocyte Na-Ca exchanger may include some B cell and monocyte responses, it is unlikely that small amounts

---

Na-medium. [Ca]$_i$ in response to TG (measured as basal [Ca]$_i$ at 0 s subtracted from peak [Ca]$_i$ achieved at ~100 sec) were 42.5 ± 7.4 nM and 35.5 ± 5.3 nM for cells in Ca-containing, Na-medium plus LaCl$_3$ (*trace b*) and EGTA-Na-medium (*trace c*), respectively. Results are representative of 6 experiments. **(C)** Effect of ouabain pretreatment and extracellular Na on TG-induced Ca$_i$ profile in the presence of LaCl$_3$. Lymphocytes were pretreated with ouabain in Na-medium, resuspended in nominally Ca-free Na-medium and injected into cuvettes containing ouabain, 30 nM TG and a) Na-medium (*solid line*), b) Na-free medium (*dashed line*), c) Na-medium plus 3 μM La (*solid line*), or d) Na-free medium plus 3 μM La (*dashed line*). Immediately prior to the experiment, CaCl$_2$ (final extracellular [Ca] = 1 mM) was added. Results are from 6 experiments. (From Balasubramanyam *et al.*[1] Reprinted by permission from *The Journal of Clinical Investigation*.)

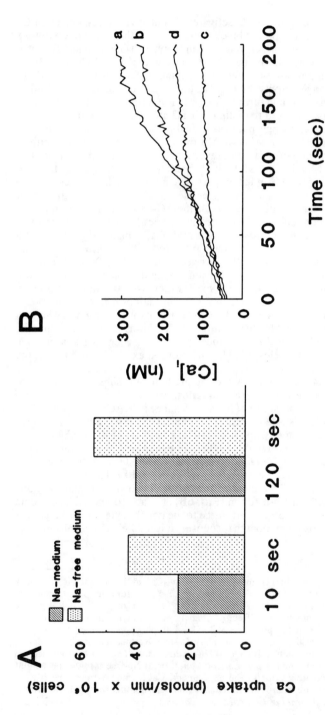

FIGURE 6. CD4[+] lymphocytes exhibit Na-Ca exchange-mediated Ca uptake and TG-induced increases in Ca$_i$. **(A)** Na-Ca exchange-mediated Ca uptake. CD4[+] lymphocytes were obtained using CD4[+] mAb and the MACS isolation system; cells were treated for 30 min at 37°C in Na-medium containing 0.1 mM ouabain and ^{45}Ca influx assayed at indicated times in the presence of Na- or Na-free medium. Fluorescence-activated cell sorter (FACS) analysis with CD14 and CD45 mAbs indicated ~93% of MACS-isolated cells were lymphocytes, and ~0.9% and 3.6% of cells were granulocytes and monocytes, respectively. Each value represents the mean of duplicate measurements. **(B)** Effect of ouabain pretreatment and extracellular Na on TG induced Ca$_i$ profile in the presence of LaCl$_3$. Cells were pretreated with ouabain in Na-medium, resuspended in nominally Ca-free Na-medium and injected into cuvettes containing ouabain, 30 nM TG and a) Na-medium, b) Na-free medium, c) Na-medium plus 3 μM La, or d) Na-free medium plus 3 μM La. Immediately prior to the experiment, CaCl$_2$ (final extracellular [Ca] = 1 mM) was added.

of monocytes are responsible for Na-dependent Ca influx. Although these results need to be confirmed, they are consistent with the idea that $CD4^+$ lymphocytes exhibit Na-Ca exchange and may be the predominant cell type responsible for Na-dependent Ca uptake in peripheral lymphocyte preparations.

SUMMARY

The experiments with peripheral lymphocytes raise two provocative questions: is SDCI composed of Ca influx via both a Ca channel and Na-Ca exchanger?, and what is the role of Na-Ca exchange in lymphocytes? In regard to the first issue, the potential for this dual Ca influx pathway exists, inasmuch as both Ca store depletion (by exposure of cells to EGTA) and TG-treatment initiated Ca influx that was enhanced following reversal of the Na gradient. These data could be interpreted to suggest a role for Ca influx via the exchanger during lymphocyte activation. However, our ability to demonstrate Na-Ca exchange activity was facilitated by the removal of Ca sequestering or extrusion mechanisms, including SERCA Ca pumps and forward mode Na-Ca exchange. Thus, it seems likely that under physiological conditions the primary function of the exchanger is to mediate Ca efflux. In this regard, it might play a role in lymphocyte activation by limiting net Ca entry during the sustained phase of Ca mobilization. Since sustained Ca entry is critical for Ca-dependent processes including interleukin-2 production, exchange activity would be an important modulator of this process. Changes in membrane potential, intracellular [Na] and cytosolic pH could therefore regulate Ca_i through its effects on Na-Ca exchange activity. Future challenges include defining the role of the Na-Ca exchange in Ca_i homeostsis and characterizing its function in lymphocyte populations.

REFERENCES

1. BALASUBRAMANYAM, M., C. ROHOWSKY-KOCHAN, J. P. REEVES & J. P. GARDNER. 1994. Na^+/Ca^{2+} exchange mediated calcium entry in human lymphocytes. J. Clin. Invest. **94:** 2002–2008.
2. PIJUAN, V., Y. ZHUANG, L. SMITH, C. KROUPIS, M. CONDRESCU, J. F. ACETO, J. P. REEVES & J. B. SMITH. 1993. Stable expression of the cardiac sodium-calcium exchanger in CHO cells. Am. J. Physiol. **264:** C1066–C1074.
3. PARKER, J. C. 1989. Sodium-calcium and sodium-proton exchangers in red blood cells. Methods Enzymol. **173:** 292–300.
4. MILANICK, M. A. 1989. Na/Ca exchange in ferret red blood cells. Am. J. Physiol. **256:** C390–C398.
5. MILANICK, M. A. & M. D. S. FRAME. 1991. Kinetic models of Na-Ca exchange ion ferret red blood cells; interaction of intracellular Na, extracellular Ca, Cd and Mn. Ann. N. Y. Acad. Sci. **639:** 604–615.
6. SMITH, R. L., I. G. MACARA, R. LEVENSON, D. HOUSMAN & L. CANTLEY. 1982. Evidence that a Na^+/Ca^{2+} antiport system regulates murine erythroleukemia cell differentiation. J. Biol. Chem. **257:** 773–780.
7. BRASS, L. F. 1984. The effect of Na^+ on Ca^{2+} homeostasis in unstimulated platelets. J. Biol. Chem. **259:** 12571–12575.
8. ROEVENS, P. & D. DE CHAFFOY DE COURCELLES. 1990. Ouabain increases the calcium concentration in intracellular stores involved in stimulus-response coupling in human platelets. Circ. Res. **67:** 1494–1502.
9. HAYNES, D. H., P. A. VALANT & P. N. ADJEI. 1991. Calcium extrusion by the sodium-calcium exchanger of the human platelet. Ann. N. Y. Acad. Sci. **639:** 592–603.

10. COOPER, R. S., N. SHAMSI & S. KATZ. 1989. Intracellular calcium and sodium in hypertensive patients. Hypertension **9:** 224–229.
11. MAZZANTI, L., R. A. RABINI, F. EMANUELA, P. FUMELLI, E. BERTOLI & R. DE PIRRO. 1990. Altered cellular Ca^{2+} and Na^+ transport in diabetes mellitus. Diabetes **39:** 850–854.
12. KIMURA, M., A. AVIV & J. P. REEVES. 1993. K^+-dependent Na^+-Ca^{2+} exchange in human platelets. J. Biol. Chem. **268:** 6874–6877.
13. SIMCHOWITZ, L. & E. J. CRAGOE, JR. 1988. Na^+-Ca^{2+} exchange in human neutrophils. Am. J. Physiol. **254:** C150–C164.
14. DALE, W. E. & L. SIMCHOWITZ. 1991. The role of Na^+-Ca^{2+} exchange in human neutrophil function. Ann. N. Y. Acad. Sci. **639:** 616–630.
15. KACZOROWSKI, G., R. S. SLAUGHTER, V. F. KING & M. L. GARCIA. 1989. Inhibition of sodium-calcium exchange: identification and development of probes of transport activity. Biochim. Biophys. Acta **988:** 287–302.
16. DONNADIEU, E. & A. TRAUTMANN. 1993. Is there a Na^+/Ca^{2+} exchanger in macrophages and in lymphocytes? Pflugers Arch. **414:** 448–455.
17. WACHOLTZ, M. C., E. J. CRAGOE, JR. & P. E. LIPSKY. 1993. Delineation of the role of a Na^+Ca^{2+} exchanger in regulating intracellular Ca^{2+} in T cells. Cell. Immunol. **147:** 95–109.
18. KRAUT, R. P., A. H. GREENBERG, E. J. CRAGOE, JR. & R. BOSE. 1993. Pyrazine compounds and measurement of cytosolic Ca^{2+}. Anal. Biochem. **214:** 413–419.
19. LOHN, M. & F. MARKWARDT. 1995. Purinoceptor-mediated Ca^{2+}-signaling in human B-lymphocytes. Biophys. J. **68:** A229 (abstract).
20. DONNADIEU, E., G. BISMUTH & A. TRAUTMANN. 1992. Calcium fluxes in T lymphocytes. J. Biol. Chem. **267:** 25864–25872.
21. UEDA, T. 1983. Na^+-Ca^{2+} exchange activity in rabbit lymphocyte plasma membranes. Biochim. Biophys. Acta **734:** 342–346.
22. WACHOLTZ, M. C., E. J. CRAGOE, JR. & P. E. LIPSKY. 1992. A Na^+-dependent Ca^{2+} exchanger generates the sustained increase in intracellular Ca^{2+} required for T cell activation. J. Immunol. **149:** 1912–1920.
23. BALASUBRAMANYAM, M., M. KIMURA, A. AVIV & J. P. GARDNER. 1993. Kinetics of Ca^{2+} transport across the lymphocyte plasma membrane. Am. J. Physiol. **265** (Cell Physiol. 34): C321–C327.
24. LEE, S-L., A. S. L. YU & J. LYTTON. 1994. Tissue-specific expression of Na^+-Ca^{2+} exchanger isoforms. J. Biol. Chem. **269:** 14849–14852.
25. TEPPEL, M., S. KUHNAPFEL, G. THEILMEIER, C. TEUPE, R. SCHLOTMANN & W. ZIDEK. 1994. Filling state of intracellular Ca^{2+} pools triggers trans plasma membrane Na^+ and Ca^{2+} influx by a tyrosine kinase-dependent pathway. J. Biol. Chem. **269:** 26239–26242.
26. GATTO, C. & M. A. MILANICK. 1993. Inhibition of the red blood cell calcium pump by eosin and other fluorescein analogues. Am. J. Physiol. **264:** C1577–C1586.
27. ZWEIFACH, A. & R. S. LEWIS. 1993. Mitogen-regulated Ca^{2+} current of T lymphocytes is activated by depletion of intracellular Ca^{2+} stores. Proc. Natl. Acad. Sci. USA **90:** 6295–6299.

Effects of External Mg^{2+} on the Na-Ca Exchange Current in Guinea Pig Cardiac Myocytes

JUNKO KIMURA

Department of Pharmacology
Fukushima Medical College
Fukushima 960-12, Japan

INTRODUCTION

Na-Ca exchange plays an important physiological role as a regulator of intracellular Ca^{2+} concentration ($[Ca^{2+}]_i$) in cardiac myocytes.[1,2] Since the stoichiometry is 3:1, the Na-Ca exchange generates a membrane current, which can operate as an inward or an outward current depending on the membrane potential, internal and external $[Ca^{2+}]$ and $[Na^+]$.[3,4] Among the external ions physiologically present, the effect of $[Na^+]_o$ and $[Ca^{2+}]_o$ on the Na-Ca exchange current has so far been investigated.[5,6] Mg^{2+}_o has been known to inhibit the Na-Ca exchanger in cardiac sarcolemmal vesicles.[7,8] However, there have been only a few quantitative reports on the blocking effect of Mg^{2+} in intact cardiac cells. In the present study, the effect of Mg^{2+}_o on the Na-Ca exchange current was investigated quantitatively in guinea pig single ventricular cells.

METHODS

Single ventricular cells of the guinea pig heart were obtained as described previously.[5] The whole-cell clamp was performed using a patch clamp amplifier (Act ME, Tokyo). The external solution for measuring the outward exchange current contained (mM): NaCl 150, $MgCl_2$ 1, 5, 10 or 30, $CaCl_2$ 1, 2, 5 or 10, ouabain 0.02, verapamil 0.002 and ryanodine 0.005, HEPES 5 (pH 7.2). The pipette solution contained (mM): NaCl 20, CsOH 90, aspartic acid 90, MgATP 5, K_2CrP 5, BAPTA 20, $CaCl_2$ 10, HEPES 20 (pH 7.2, pCa 6.85). The external solution for recording the inward exchange current contained (mM): LiCl or NaCl 140, $MgCl_2$ 1 or 10, HEPES 5 (pH 7.2 with CsOH). Ouabain, verapamil and ryanodine were also added as described above. The pipette solution for the inward exchange current contained (mM): CsCl 30, CsOH 90, aspartic acid 50, $MgCl_2$ 3, MgATP 5, K_2CrP 5, BAPTA 20, $CaCl_2$ 16, HEPES 20 (pH 7.2, pCa 6.24).

The current-voltage (I-V) relation was obtained by ramp pulses with the speed of 170 mV/250 ms. The kinetic values were obtained by fitting the curves for the Michaelis-Menten equation by the Marquardt method using a program "Kotaro" (Sankaido, Tokyo). Initial values for the iteration of the Marquardt method was obtained by Hanes-Woolf plot ([S]/i versus [S]). The numerical data are expressed as mean ± SD.

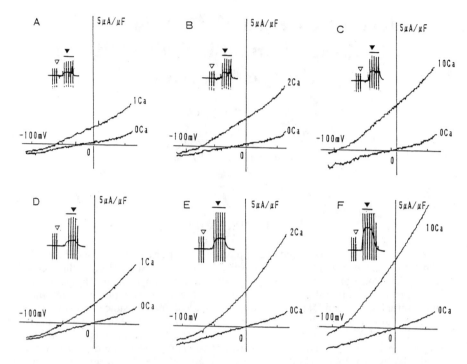

FIGURE 1. I-V curves of the control (0Ca) and the outward Na-Ca exchange current at 10 mM $[Mg^{2+}]_o$ **(A–C)** and at 1 mM $[Mg^{2+}]_o$ **(D–F)**. The exchange current was induced by $[Ca^{2+}]_o$ indicated on the *right* of each curve. The *inset* shows the chart record of the current traces in response to ramp pulses. *Open arrowheads* indicate the control plotted and *filled arrowheads* the current in the presence of $[Ca^{2+}]_o$. The *bar* indicates the period of $[Ca^{2+}]_o$ application. The interval between the regular ramp pulses was 3 s. The exchange current is smaller at higher $[Mg^{2+}]_o$ at each $[Ca^{2+}]_o$.

RESULTS

Effects of $[Mg]_o$ on the Outward Na-Ca Exchange Current

To examine the effect of $[Mg^{2+}]_o$ on the Na-Ca exchange, an outward exchange current was measured in the presence of four different $[Mg^{2+}]_o$, 1, 5, 10 or 30 mM. FIGURE 1 shows representative I-V curves at 1 and 10 mM $[Mg^{2+}]_o$. The outward exchange current was induced by briefly elevating $[Ca^{2+}]_o$ from nominally free to 1, 2, 5 or 10 mM at each $[Mg^{2+}]_o$. The ordinate of FIGURE 1 indicates the current density which was obtained by dividing the current magnitude by the capacitance of the cell. This procedure enables one to compare the current magnitude in different cells. The current density was smaller at 10 mM $[Mg^{2+}]_o$ than that at 1 mM at each $[Ca^{2+}]_o$, indicating that external Mg^{2+} has an inhibitory effect on the Na-Ca exchange current.

The concentration-response curves between $[Ca^{2+}]_o$ and the current density at the four different $[Mg^{2+}]_o$ are plotted in FIGURE 2A. The current density became

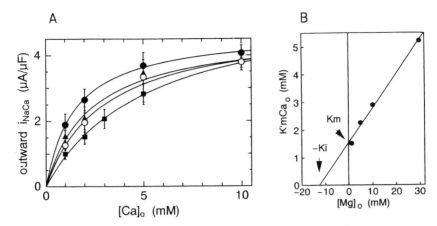

FIGURE 2. (A) Concentration-response curves between $[Ca^{2+}]_o$ and the outward Na-Ca exchange current at four different $[Mg^{2+}]_o$: 1 mM $[Mg^{2+}]_o$ (●, n = 5, 5, 7, 5 at 1, 2, 5, 10 mM $[Ca^{2+}]_o$, respectively), 5 mM $[Mg^{2+}]_o$ (▲, n = 6, 6, 5, 6), 10 mM $[Mg^{2+}]_o$ (○, n = 6, 6, 3, 6), and 30 mM $[Mg^{2+}]_o$ (■, n = 6, 6, 4, 5 at 1, 2, 3 and 5 mM $[Ca^{2+}]_o$, respectively). The exchange current was measured as the difference between the current in the presence of $[Ca^{2+}]_o$ and the control at +50 mV. **(B)** $K'mCa_o$ versus $[Mg^{2+}]_o$ yields a true $KmCa_o$ of 1.55 mM and Ki of 12.5 mM $[Mg^{2+}]_o$.

progressively lower as $[Mg^{2+}]_o$ became higher. To identify the mode of inhibition by $[Mg^{2+}]_o$, the apparent I_{max} values (I'_{max}) and the apparent Km values for $[Ca^{2+}]_o$ ($K'mCa_o$) at each $[Mg^{2+}]_o$ were obtained by fitting the curves by the Michaelis-Menten equation $i = I'_{max} * [Ca^{2+}]_o/(K'mCa_o + [Ca^{2+}]_o)$ using a computer program "Kotaro," which incorporates the Marquardt method for the nonlinear curve fitting. TABLE 1 summarizes the values of $K'mCa_o$ and I'_{max} at the four different $[Mg^{2+}]_o$. It can be seen that as $[Mg^{2+}]_o$ increases, $K'mCa_o$ increases significantly while the I'_{max} values do not change appreciably. This result indicates that external Mg^{2+} is a competitive inhibitor of the Na-Ca exchanger against external Ca^{2+}. The Ki value, which is $[Mg^{2+}]_o$ that doubles the slope of the $1/i$ versus $1/[Ca^{2+}]_o$, can be obtained by plotting the $K'mCa_o$ values against $[Mg^{2+}]_o$ from the following equation for the competitive inhibition.

$$K'm = (Km/Ki) * [I] + Km \qquad \text{(Segel,}^9 \text{ p. 109)}$$

TABLE 1. Kinetic Values of the Outward Na-Ca Exchange Current[a]

$[Mg^{2+}]_o$ (mM)	$K'mCa_o$ (mM)	I'_{max} ($\mu A/\mu F$)	n
1	1.52 ± 0.19	4.74 ± 0.17	22
5	2.27 ± 0.26	4.72 ± 0.19	23
10	2.91 ± 0.50	4.95 ± 0.33	21
30	5.23 ± 1.42	5.74 ± 0.96	21

[a] Apparent Km and I_{max} values obtained at various $[Mg^{2+}]_o$. Values are mean ± SD. n indicates the number of data.

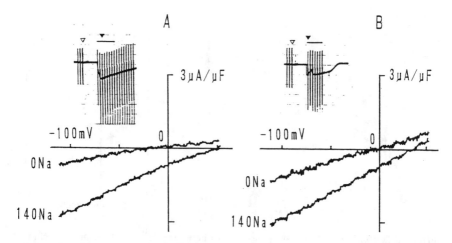

FIGURE 3. Representative I-V curves of the control at 0 mM $[Na^+]_o$ and the inward Na-Ca exchange current induced by 140 mM $[Na^+]_o$ at 1 mM **(A)** and at 10 mM **(B)** $[Mg^{2+}]_o$. The *inset* is a chart recording of the currents. The *bar* above indicates the period 140 mM $[Na^+]_o$ was superfused.

where [I] is a concentration of an inhibitor, in this case $[Mg^{2+}]_o$. As shown in FIGURE 2B, the $K'mCa_o$ values obtained above by the Marquardt method could be fitted by a line which intersected with the abscissa at a Ki value of 12.5 mM. The intersection on the ordinate indicates the Km value without external Mg^{2+}, which is 1.55 mM. Using these values, the $[Mg^{2+}]_o$ required for 50% inhibition, $[I]_{0.5}$, can be calculated by the following equation.

$$[I]_{0.5} = (1 + [S]/Km) * Ki \qquad \text{(Segel,}^9 \text{ p. 106)}$$

where [S] indicates a concentration of a substrate, in this case $[Ca^{2+}]_o$. Inserting the values of Km = 1.55 mM and Ki = 12.5 mM, $[I]_{0.5}$ is approximately 21 mM at 1 mM $[Ca^{2+}]_o$.

Effect of External Mg^{2+} on the Inward Na-Ca Exchange Current

Whether $[Mg^{2+}]_o$ also inhibits the inward Na-Ca exchange current was investigated. The inward exchange current was induced by briefly changing external cation from 140 mM $[Li^+]_o$ to 140 mM $[Na^{2+}]_o$ in the presence of 570 nM $[Ca^{2+}]_i$ (pCa 6.24) in the pipette solution (FIG. 3). The current densities measured at −100 mV were 1.93 ± 0.27 $\mu A/\mu F$ (n = 16) at 1 mM $[Mg]_o$ and 1.89 ± 0.23 $\mu A/\mu F$ (n = 9) at 10 mM $[Mg^{2+}]_o$. These values were not significantly different between 1 and 10 mM $[Mg^{2+}]_o$. This result suggests that $[Mg^{2+}]_o$ up to 10 mM does not affect the inward exchange current as potently as it does the outward exchange current.

DISCUSSION

The present study indicates that external Mg^{2+} competitively inhibits cardiac outward $Na_i = Ca_o$ exchange current, since the $K'mCa_o$ value progressively increased as $[Mg^{2+}]_o$ increased, while the I'_{max} value did not change significantly. The Ki value of $[Mg^{2+}]_o$ is 12.5 Mm. This conclusion agrees with that of previous workers but with lower Ki values. Wakabayashi and Goshima[10,11] found a competitive inhibition of Mg^{2+} with the Ki value of 1–5 mM on the Na^+-dependent Ca^{2+} uptake in cultured myocardial cells from mouse fetus heart, neuroblastoma cells, and cultured chick heart cells. Ledvora & Hegyvary[8] also reported that 5 mM $MgCl_2$ inhibited 50% Na-Ca exchange in dog heart sarcolemma vesicles. Even lower Ki values of $[Mg^{2+}]_o$ were reported in smooth muscle cells and renal epithelial cells by Smith's group.[12,13] According to them, Mg^{2+}_o competitively blocks Ca^{2+} influx with a Ki value of 93 ± 7 μM in cultured smooth muscle cells from rat aorta[12] and 90 ± 10 μM in renal epithelial cells.[13] However, these low Ki values shifted to a higher value of 5 mM, when the main cation of the external solution was changed from N-methyl D-glucamine to K^+. Thus it seems that the potency of the Mg^{2+} block depends on the species of the external cation.

Comparing the effect of various cations, Philipson's group has shown that Mg^{2+} was the least potent inhibitor among various divalent and trivalent cations in cardiac sarcolemmal vesicles.[14] Ni^{2+} is a reversible inhibitor of the Na-Ca exchange current.[3] Ehara et al.[15] concluded that Ni^{2+} was a mixed type inhibitor of the outward exchange current. Thus different cations may have different modes of inhibition.

Fractional inhibition (i) of Na-Ca exchange current by $[Mg^{2+}]_o$ can be calculated from the following equation.

$$i * 100 (\%) = [I]/([I] + Ki * (1 + [S]/Km)) \quad \text{(Segel,}^9 \text{ p. 105)}$$

Inserting the Ki value of 12.5 mM and $KmCa_o$ of 1.5 mM, if $[Ca^{2+}]_o$ is 2 mM, the inhibition of the exchange is approximately 6.5% at 2 mM $[Mg^{2+}]_o$ and 15% at 5 mM $[Mg^{2+}]_o$. The inhibition was significant for the Ca^{2+}-influx mode but less for the efflux mode, since the inward exchange current was not much affected by $[Mg^{2+}]_o$ up to 10 mM at 140 mM $[Na^+]_o$. This difference may be beneficial for the function of the exchanger to facilitate Ca^{2+} extrusion and prevent Ca influx, although the percent inhibition by $[Mg^{2+}]_o$ is small. This property may also explain an arrhythmogenic tendency under hypomagnesemia. Thus $[Mg^{2+}]_o$ is, although weak, a modulatory factor of the Na-Ca exchanger.

ACKNOWLEDGMENTS

I would like to thank Dr. Akira Takai from Nagoya University for a suggestion on the statistical analysis, Dr. Ikuro Sato from Miyagi Prefectural Cancer Center for providing me with the computer program "Kotaro," which he has developed, and Professor Hironori Nakanishi and Dr. Isao Matuoka for reading the manuscript.

REFERENCES

1. ALLEN, T. J. A., D. NOBLE & H. REUTER, Eds. 1989. Sodium-Calcium Exchange. Oxford University Press, Oxford.
2. BLAUSTEIN, M. P., R. DIPOLO & J. P. REEVES, Eds. 1991. Sodium-Calcium Exchange. Annals of the New York Academy of Sciences, Vol. 639.

3. KIMURA, J., S. MIYAMAE & A. NOMA. 1987. Identification of sodium-calcium exchange current in single ventricular cells of guinea-pig. J. Physiol. **84:** 199–222.
4. BEUCKELMANN, D. J. & W. G. WIER. 1989. Sodium-calcium exchange in guinea-pig cardiac cells: exchange current and changes in intracellular Ca. J. Physiol. **414:** 499–520.
5. MIURA, Y., & J. KIMURA. 1989. Sodium-calcium exchange current. Dependence on internal Ca and Na and competitive binding of external Na and Ca. J. Gen. Physiol. **93:** 1129–1145.
6. MATSUOKA, S. & D. W. HILGEMANN. 1992. Steady-state and dynamic properties of cardiac sodium-calcium exchange. J. Gen. Physiol. **100:** 963–1001.
7. TROSPER, T. L. & K. PHILIPSON. 1983. Effects of divalent and trivalent cations on Na^+-Ca^{2+} exchange in cardiac sarcolemmal vesicles. Biochim. Biophys. Acta **731:** 63–68.
8. LEDVORA, R. F. & C. HEGYVARY. 1983. Dependence of Na^+-Ca^{2+} exchange on monovalent cations. Biochim. Biophys. Acta **729:** 123–136.
9. SEGEL, I. H. 1975. Enzyme Kinetics. John Wiley & Sons, Inc. New York.
10. WAKABAYASHI, S. & K. GOSHIMA. 1981. Kinetic studies on sodium-dependent calcium uptake by myocardial cells and neuroblastoma cells in culture. Biochim. Biophys. Acta **642:** 158–172.
11. WAKABAYASHI, S. & K. GOSHIMA. 1981. Comparison of kinetic characteristics of Na^+-Ca^{2+} exchange in sarcolemma vesicles and cultured cells from chick heart. Biochim. Biophys. Acta **645:** 311–317.
12. SMITH, J. B., E. J. CRAGOE & L. SMITH. 1987. Na^+/Ca^{2+} antiport in cultured arterial smooth muscle cells. J. Biol. Chem. **262:** 11988–11994.
13. LYU, R. M., L. SMITH & J. B. SMITH. 1991. Sodium-calcium exchange in renal epithelial cells: dependence on cell sodium and competitive inhibition by magnesium. J. Membr. Biol. **124:** 73–83.
14. BERS, D. M., E. D. PHILIPSON & A. Y. NISHIMOTO. 1980. Sodium-calcium exchange and sidedness of isolated cardiac sarcolemmal vesicles. Biochim. Biophys. Acta **601:** 358–371.
15. EHARA, T., S. MATSUOKA & A. NOMA. 1989. Measurement of reversal potential of Na^+-Ca^{2+} exchange current in single guinea-pig ventricular cells. J. Physiol. **410:** 227–249.

Ca^{2+} Influx via Na-Ca Exchange and I_{Ca} Can Both Trigger Transient Contractions in Cat Ventricular Myocytes

A-M. VITES[a] AND J. A. WASSERSTROM

Department of Medicine
Northwestern University Medical School
The Feinberg Cardiovascular Research Institute
Chicago, Illinois 60611

INTRODUCTION

The possibility has been raised that fast sodium current may activate contractions in heart by reversing the Na-Ca exchanger.[1,2] The traditional role of the exchanger is thought to be one of preserving resting calcium. To examine the possibility that calcium influx via reverse mode Na-Ca exchange can effectively trigger calcium release from the SR to produce a transient contraction in mammalian heart, we used isolated cat ventricular myocytes under voltage clamp conditions. Experiments were performed at 34°C because the Q_{10} of the exchanger is approximately 4[3] and the exchanger was directly reversed by holding the membrane potential at very positive voltages. The contractions produced by the reversed Na-Ca exchange (I_{Na-Ca}) were compared to those produced by L-type calcium currents (I_{Ca}).

MATERIALS AND METHODS

Suction pipettes ($\approx$2 MΩ) were filled with (mM): 120 K-aspartate, 25 KCl, 0.5 MgCl$_2$, 4 diK-ATP, 0.06 EGTA, 6 NaCl, 20 HEPES; pH = 7.2 with KOH. Cells were superfused at 34°C with Normal Tyrode's (mM): 140 NaCl, 5.4 KCl, 0.5 MgCl$_2$, 10 HEPES, 0.4 NaH$_2$PO$_4$, 11 glucose, 1.8 CaCl$_2$; pH = 7.4 with NaOH. Cell shortening was taken as an indication of contraction and was shown as downward deflections. The membrane current and voltage were measured using the discontinuous single electrode "switch" voltage clamp mode. The myocyte image was aligned lengthwise with the raster lines of the video edge detector on the monitor to measure changes in cell length (ΔL).

Replenishing the Sarcoplasmic Reticulum (SR) with Calcium

Myocytes were conditioned with prepulses to replenish the SR stores with calcium prior to a test voltage step (V_{test}) to trigger a contraction. This protocol

---

[a] Corresponding author: Ana-María Vites, Ph.D., Reingold ECG, Morton 2-694, Northwestern University Medical School, 303 East Chicago Ave., Chicago, Illinois 60611.

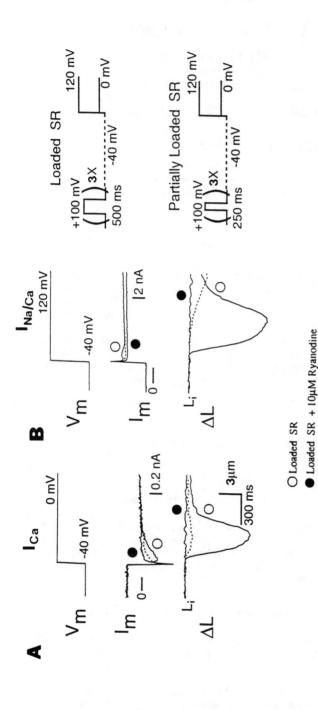

FIGURE 1. Contractions in a cardiac myocyte triggered by **(A)** I_{Ca} and **(B)** I_{Na-Ca}. A step V_{test} to 0 mV from $V_H = -40$ mV activated I_{Ca} and a V_{test} to +120 mV activated reverse Na-Ca exchange (I_{Na-Ca}). Membrane currents (I_m) and cell contractions (downward deflections representing myocyte shortening, ΔL) were recorded in the same cell under three different conditions: (1) the cell was preconditioned with three 500-ms pulses to +100 mV; (2) the conditioning pulses were 250 ms in duration and (3) the cell was conditioned with three prepulses of 500 ms in duration but this time after exposure to 10 μM ryanodine. When the myocyte was conditioned with 500-ms pulses the peak ΔL produced by I_{Ca} and I_{Na-Ca} were 12.6 μm and 13.5 μm, respectively (○). Note that the time to peak was prolonged when contractions were activated at very positive potentials, which is likely due to calcium persistently entering the cell via the reversed exchanger. When the myocyte was conditioned with 250-ms pulses the contractions were a lot smaller (*dashed lines*). Ryanodine prevented both contractile responses even though the cell was conditioned with 500-ms pulses (●). Note the transient inward current after the peak I_{Ca} (scale is 0.2 nA) and the transient outward current in I_m in response to the step to +120 mV (scale is 2 nA). Activation of these currents seemed dependent on the degree of calcium stored in the SR, suggesting that they are activated by released calcium.

insured that SR calcium stores were returned to the same state before a contraction was tested. The lack of after-contractions confirmed that these cells were not calcium overloaded.

RESULTS

Shortening was triggered by test voltage steps (V_{test}) ranging from -30 mV to $+140$ mV, from a holding potential (V_H) of -40 mV after the SR was replenished with calcium. There was a close correlation between the degree of shortening and the degree of SR loading. Myocyte shortenings, under two different loading conditions, are shown in FIGURE 1. Under "loaded" conditions (○), a V_{test} from -40 mV to 0 mV activated I_{Ca} and a transient contraction (A). Under the same conditions a V_{test} from -40 mV to $+120$ mV (B) produced a transient contraction equivalent to that produced by I_{Ca}, which relaxed while the membrane voltage was held at $+120$ mV. When the duration of the conditioning pulses was reduced so that the SR was "partially loaded," the contraction produced by I_{Ca} was very

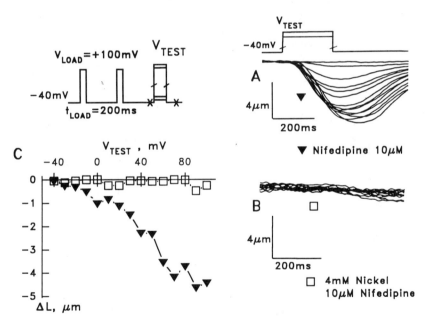

FIGURE 2. Nifedipine-insensitive contractions occur at positive potentials and are prevented by Ni⁺. The transient contractions remaining in the presence of 10 μM nifedipine start activating at $V_{test} > +50$ mV (see panel B). Myocytes were conditioned with two steps to $+100$ mV to replenish the SR with calcium and a test step was applied (-30 mV to $+100$ mV) from a holding potential of -40 mV. **(A)** Myocyte contractions in the presence of nifedipine in response to the various V_{test}. **(B)** The contractions after subsequent exposure to 4 mM Ni⁺ in response to the various V_{test} in the same myocyte. **(C)** The voltage dependence of contractions in the presence of nifedipine (10 μM) and after subsequent addition of 4 mM Ni⁺. Similar results were obtained in all 5 cells tested.

small and transient, and that produced by I_{Na-Ca} was a steady increase in the degree of shortening as if calcium were gradually accumulating in the cytoplasm. Note that the slow activation of shortening may be accentuated by SR calcium uptake. In this cell 10 μM ryanodine (●) prevented the contractions produced by I_{Ca} and I_{Na-Ca}. Note that the rate of the slowly developing shortening, at $V_{test} = +120$ mV, was reduced even further to the point that it was negligible at this time scale. Similar results were obtained in all eight cells where several degrees of SR loading were tested.

In order to distinguish I_{Na-Ca} from I_{Ca} triggering, 10 μM nifedipine was used to block I_{Ca}. Contractions produced by very positive V_{test} ($>+60$ mV) were not prevented by nifedipine as shown in the example in FIGURE 2. Contractions triggered by various V_{test} in the presence of nifedipine are shown in (A) and the voltage dependence of contraction is shown in panel C (▼). Addition of Ni^+ to the bath prevented all the nifedipine-insensitive contractions (panels B and C (□)). This is consistent with the inhibitory action of Ni^+ on the exchanger.

CONCLUSIONS

Transient contractions can be triggered at potentials which activate I_{Ca} as well as at more positive potentials where the reversed Na-Ca exchange is the most likely candidate to produce calcium influx. Activation of contraction by I_{Ca} or reverse I_{Na-Ca} appeared similar in that their magnitude was greatly affected by the degree of SR calcium replenishment and by their sensitivity to ryanodine. The rates of shortening appeared to be similar and they both relaxed while the membrane was being held at positive potentials. These experiments indicate that the reverse mode of the exchanger appears to have the capacity to trigger rapid CICR from the SR as is the case with I_{Ca}. Our results may be an indication that the Na-Ca exchange proteins have preferential access to the release sites as do the calcium channels, but there are controversial reports on this matter. While there is no convincing evidence that the exchanger plays a role in cardiac EC-coupling, the fact that the initial step in an action potential is fast Na^+ influx, one would expect the exchanger to reverse in response to the sudden increase in sodium. For example, a change from 6 mM to 24 mM in subsarcolemmal $[Na^+]$ could produce a shift in the E_{Na-Ca} from -8 mV to -117 mV. The initial depolarization during *phase 0* of an action potential, which is positive to E_{Na-Ca}, may also contribute to the reversal of the Na-Ca exchanger. This possibility would open a whole new mode of achieving EC-coupling in cardiac myocytes under physiological conditions.

REFERENCES

1. LEBLANC, N. & J. R. HUME. 1990. Sodium current-induced release of calcium from cardiac sarcoplasmic reticulum. Science **248:** 372-376.
2. LEVI, A. J., P. BROOKSBY & J. C. HANCOX. 1993. One hump or two? The triggering of calcium release from the sarcoplasmic reticulum and the voltage dependence of contraction in mammalian cardiac muscle. Cardiovasc. Res. **27:** 1743-1757.
3. KIMURA, J., S. MIYAMAE & A. NOMA. 1987. Identification of sodium-calcium exchange current in single ventricular cells of guinea-pig. J. Physiol. **384:** 199-222.

Demonstration of an Inward Na^+-Ca^{2+} Exchange Current in Adult Human Atrial Myocytes

GUI-RONG LI AND STANLEY NATTEL

Research Center
Montreal Heart Institute
Montreal, Quebec, Canada H1T 1C8

In the heart, Na^+-Ca^{2+} exchange has been considered to play a major role in extruding Ca^{2+} and maintaining a low intracellular Ca^{2+} level.[1] Since the exchange ratio is $3Na^+:1Ca^{2+}$, Na^+-Ca^{2+} exchange is electrogenic.[2,3] Ca^{2+} released from the sarcoplasmic reticulum (SR) store is believed to contribute importantly to the electrogenic Na^+-Ca^{2+} exchange current (I_{Na-Ca}).[4] It has been demonstrated that I_{Na-Ca} can contribute to action potential duration (APD) in cardiac muscles from a variety of animal species;[5,6] however, whether I_{Na-Ca} contributes to APD in human cardiac muscles has not been carefully evaluated.[7] The goal of the present study is to determine whether an electrogenic I_{Na-Ca} is present in adult human atrial myocytes, and if so, whether it contributes to APD.

METHODS

Atrial cells were enzymatically isolated from specimens of human right atrial appendage obtained from patients undergoing coronary bypass surgery, and whole-cell patch clamp techniques were used to record ionic currents and action potential (AP). Cs^+ (for current recording) or K^+ (for AP recording) were the predominant cations (130 mM) in the pipette solution, which also included 0.05 mM EGTA. The cells were superfused with Tyrode solution (containing 2 mM Ca^{2+}), and temperature was maintained at $36 \pm 0.5°C$.

RESULTS

A large inward tail current was detected upon repolarizing to voltages between -120 (257 ± 37 pA) and -20 (125 ± 21 pA, n = 6) mV, after a brief (4 ms) voltage step to $+30$ mV from a holding potential of -70 mV. The voltage dependent activation of the inward tail current was studied in the presence of 10 μM verapamil to block I_{Ca} with 4 ms steps from -70 mV to between -50 and $+40$ mV, and showed a peak current at $+10$ and $+20$ mV (149 ± 26 and 154 ± 28 pA, respectively, n = 6). This inward current was significantly suppressed by reduction of external $[Na^+]$ by replacement with choline (FIG. 1) or Li^+, or addition of the I_{Na-Ca} blocker Ni^{2+} (50 μM) (FIG. 2). The addition of 2 μM ryanodine significantly decreased the inward tail current in 5 cells, indicating a role for Ca^{2+} release from the SR. Furthermore, exposure to ryanodine or reduction of external $[Na^+]$ by substitution with Li^+ shortened APD.

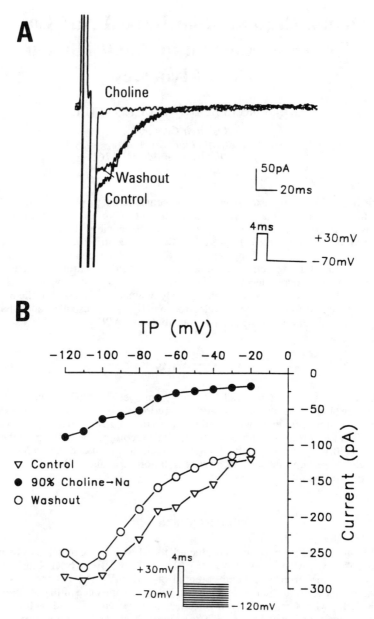

FIGURE 1. Effect of reduction of external [Na$^+$] on the inward tail current in a representative cell. The external [Na$^+$] was reduced to 12.6 mM by substitution with choline. **(A)** The inward tail current recorded in the absence and presence of external Na$^+$ replacement with choline. **(B)** I-V relation of the tail current. Similar results were obtained in a total of 3 cells.

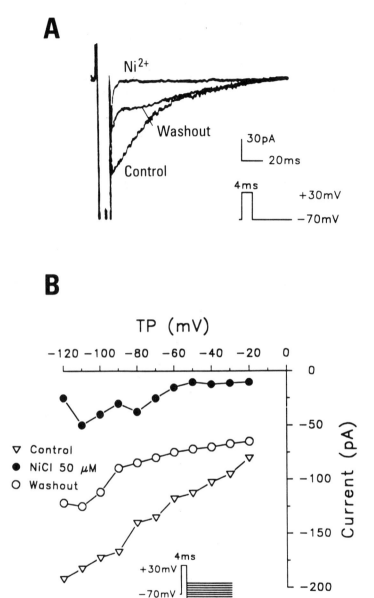

FIGURE 2. The inward tail current was significantly decreased by Ni^+. **(A)** The current was recorded in a human atrial cell in the absence and presence of 50 μM NiCl. **(B)** I-V relation of the inward tail currents. Similar results were obtained in 4 other cells.

Results indicate that (1) brief depolarization steps (4 ms) from −70 mV activated a large inward tail current upon repolarization in human atrial myocytes; (2) the tail current amplitude increased as repolarization voltage became more negative, and voltage-dependent activation peaked at +10 and +20 mV; (3) the tail current is dependent on external [Na$^+$] and intracellular Ca^{2+} release from the SR, and is reversibly inhibited by a low concentration of Ni^{2+}. Therefore, the inward tail appears to be generated by the Na$^+$-Ca^{2+} exchanger, which contributes to APD.

In conclusion, $I_{Na\text{-}Ca}$ is present in human atrial cells, is elicited by depolarization and appears to contribute importantly to APD.

REFERENCES

1. PHILIPSON, K. D. 1985. Ann. Rev. Physiol. **47:** 561–571.
2. PITTS, B. J. R. 1979. J. Biol. Chem. **254:** 6232–6235.
3. REEVES, J. P. & C. C. HALE. 1984. J. Biol. Chem. **259:** 7733–7739.
4. HILGEMANN, D. W. & D. NOBLE. 1987. Prog. R. Soc. Lond. B **230:** 163–205.
5. EGAN, T. M., D. NOBLE, S. J. NOBLE, T. POWELL & A. J. SPINDER. 1989. J. Physiol. (London) **411:** 639–661.
6. EARM, Y. E., W. K. HO & I. S. SO. 1990. Prog. R. Soc. Lond. B **240:** 61–81.
7. CORABOEUF, E. & J. NARGEOT. 1993. Cardiovasc. Res. 1713–1725.

Intracellular pH Is Insensitive to Changes in Intracellular Calcium Concentration in Isolated Rat Ventricular Myocytes[a]

K. W. DILLY, A. E. DOERING,[c] W. A. ADAMS,
C. AUSTIN,[b] AND D. A. EISNER

Department of Veterinary Preclinical Sciences
[b]*Department of Physiology*
University of Liverpool
P.O. Box 147
Liverpool L69 3BX, England

[c]*CVRL*
UCLA School of Medicine
Los Angeles, California 90024-1760

Studies using multicellular preparations of cardiac muscle have found a close relationship between the control of intracellular calcium ($[Ca^{2+}]_i$) and pH (pH_i)[1] such that an increase in $[Ca^{2+}]_i$ causes an intracellular acidification.[2,3] Two hypotheses have been proposed for this acidosis, i) displacement of H^+ from intracellular buffers by Ca^{2+} and ii) accumulation of lactate influencing pH_i. We have investigated this further in single cells by measuring pH_i using the fluorescent indicator SNARF and increasing $[Ca^{2+}]_i$ by a) electrical stimulation, b) increased extracellular calcium concentration ($[Ca^{2+}]_o$) and c) increased $[Ca^{2+}]_o$ plus 50 μM ouabain, an inhibitor of the membrane Na/K pump, which causes an increase in $[Na^+]_i$ resulting in an increased $[Ca^{2+}]_i$ via the action of Na-Ca exchange.[4] Transient increases in $[Ca^{2+}]_i$ produced by electrical stimulation did not produce any detectable change of pH_i. Ouabain caused an acidification, presumably by increased $[Na^+]_i$ inhibiting Na-H exchange. The additional application of 10 mM Ca^{2+} in the presence of ouabain caused no further intracellular acidification. Measurements of $[Ca^{2+}]_i$ using indo-1 showed that ouabain and high $[Ca^{2+}]_o$ caused an increase similar in magnitude to peak systolic $[Ca^{2+}]_i$ which was considerably greater than that produced by ouabain alone. We have also measured $[Ca^{2+}]_i$ while changing pH_i by the extracellular application of the weak base NH_4Cl. On increasing pH_i from 7.15 to 7.20 no change in $[Ca^{2+}]_i$ was observed. Preliminary experiments in which $[Ca]_i$ and pH_i were measured simultaneously using the above indicators support these findings (FIG. 1). The fact that elevating $[Ca^{2+}]_i$ decreases pH_i in multicellular but not single cell preparations could be explained if the acidification is due to extracellular accumulation of lactate influencing pH_i in multicellular preparations, whereas in single cells any lactate produced may diffuse away freely. Previous studies in which the application of cyanide produces an acidification in whole heart but not in isolated cells support this hypothesis.[5]

---

[a] Supported by the British Heart Foundation and the Wellcome Trust.

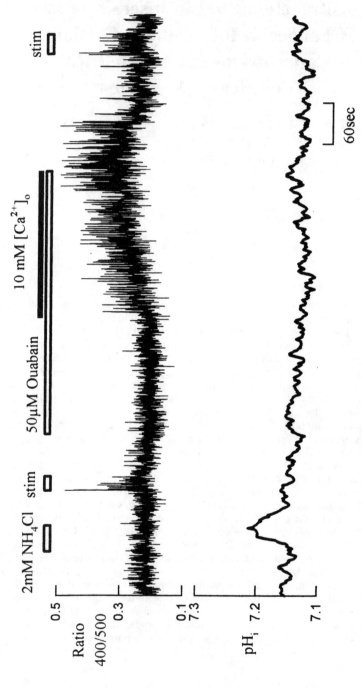

FIGURE 1. The effects of electrical stimulation, NH_4Cl (2 mM), ouabain (50 μM) and elevating $[Ca^{2+}]_o$ on $[Ca^{2+}]_i$ and pH_i measured simultaneously.

REFERENCES

1. BERS, D. M. & D. ELLIS. 1982. Pflügers Arch. **393:** 171–178.
2. ALLEN, D. G., D. A. EISNER, P. G. MORRIS, J. S. PIROLO & G. L. SMITH. 1986. J. Physiol. **376:** 121–141.
3. VAUGHAN-JONES, R. D., W. J. LEDERER & D. A. EISNER. 1983. Nature **301:** 522–524.
4. REUTER, H. & N. SEITZ. 1968. J. Physiol. **195:** 451–470.
5. EISNER, D. A., G. L. SMITH & S. C. O'NEILL. 1993. Basic Res. Cardiol. **88:** 421–429.

Immunolocalization of the Na^+-Ca^{2+} Exchanger in Cardiac Myocytes

J. S. FRANK, F. CHEN,[a] A. GARFINKEL, E. MOORE,[b] AND K. D. PHILIPSON

Department of Pediatrics
UCLA School of Medicine
3754 Macdonald Research Building
Los Angeles, California 90024-7045
and
[b]*University of British Columbia*
Canada

The Na^+-Ca^{2+} exchanger is a major pathway for transmembrane flux of Ca^{2+} in cardiac cells and thus plays a key role in regulating [Ca^{2+}] in cardiac myocytes.[1] Immunolabeling in adult cardiac myocytes showed localization of the Na^+-Ca^{2+} exchanger to the peripheral sarcolemma and especially in the T tubules.[2] We used immunofluorescent labeling in isolated rat and guinea pig myocytes to collect confocal optical sections through the thickness of the cells. With a combination of volume rendering and gradient based shading for surface enhancement, we were able to produce 3D images of the T tubules and peripheral sarcolemma. The exchanger was localized along the T tubules with discrete regions of higher density. To examine the distribution of Na^+-Ca^{2+} exchanger in relationship to the ryanodine receptors in the T tubules, we obtained 3D images of rat cells labeled with antibodies to these molecules at 0.25 μM intervals in the digital imaging microscope. The results showed ~40% overlap of the exchanger with the ryanodine receptor. In electron microscope studies with immunogold, labeling was sparse but more gold was present in the T tubules and in the intercalated discs than on the peripheral sarcolemma.

In addition, co-localization of Na^+-Ca^2 exchanger with ankyrin was also examined in adult and neonatal myocardial cells. Ankyrin is a peripheral membrane protein which links integral membrane proteins to other elements of the cytoskeleton. Previous studies indicated that ankyrin is associated with Ca^{2+} transport proteins such as the DHP receptor and the Na^+-Ca^{2+} exchanger. The results showed that in adult cells, co-localization of these two proteins occurred. Neonatal ventricular myocytes less than 10 days of age have not yet developed T tubules. Ankyrin labeling of neonatal cells at 5 days of age was mainly present at the Z disc, while the Na^+-Ca^{2+} exchanger was only present on the peripheral sarcolemma. These results show that ankyrin is present at the Z disc before T tubules form in the immature cells. Ankyrin is thus available to immobilize the Na^+-Ca^{2+} exchanger during development of T tubules. Labeling of the adult rabbit peripheral sarcolemma, although less than neonatal sarcolemma, was greater than that in adult rat or guinea pig. This indicates a possible species variation of Na^+-Ca^{2+} exchanger density distribution. The age-related change in distribution of the Na^+-

---

[a] Corresponding author.

Ca^{2+} exchanger may reflect different interactions with the cytoskeleton at various development stages.[3,4]

REFERENCES

1. PHILIPSON, K. D. 1990. The cardiac Na^+-Ca^{2+} exchanger. *In* Calcium and the Heart. G. A. Langer, Ed. 85–108. Raven Press. New York.
2. FRANK, J. S., G. MOTTINO, D. REID, R. S. MOLDAY & K. D. PHILIPSON. 1992. Distribution of the Na^+-Ca^{2+} exchange protein in mammalian cardiac myocytes: an immunofluorescence and immunocolloidal gold-labeling study. J. Cell Biol. **117:** 337–345.
3. ARTMAN, M. 1992. Sarcolemmal Na^+-Ca^{2+} exchange activity and exchanger immunoreactivity in developing rabbits. Am. J. Physiol. **263** (Heart Circ. Physiol. **32**): H1506–H1513.
4. CHEN, F., G. MOTTINO, T. S. KLITZNER, K. D. PHILIPSON & J. S. FRANK. 1995. Distribution of the Na^+-Ca^{2+} exchange protein in developing rabbit myocytes. Am. J. Physiol. **268** (Cell **37**):

Sodium-Calcium Exchange Expression in Ischemic Rabbit Hearts

MALCOLM M. BERSOHN[a]

Cardiology Section
Department of Medicine
West Los Angeles VA Medical Center
and
University of California
Los Angeles, California 90073

INTRODUCTION

I previously investigated the effects of global myocardial ischemia in rabbit hearts on Na^+-Ca^{2+} exchange activity measured in highly purified sarcolemmal vesicles. The Vmax of Na^+-Ca^{2+} exchange measured under initial velocity conditions was unaffected by 20 minutes of ischemia but was significantly decreased to 80% of control after 30 minutes and to 50% of control after 60 minutes of ischemia.[1,2] The Ca^{2+} affinity was unchanged in ischemia.

Recovery of normal sarcolemmal activity following ischemia might depend on resynthesis of Na^+-Ca^{2+} exchanger protein and would require the presence of adequate mRNA. Therefore, I have investigated the effects of global ischemia on the levels of Na^+-Ca^{2+} exchanger (NCX1) mRNA in rabbit hearts.

METHODS

The left ventricles from pentobarbital anesthetized male New Zealand rabbits were divided into 6 pieces approximately 0.4 g each, one for control and 5 for different ischemic periods. Ischemic tissue was incubated at 37°C in glucose-free Tyrode solution under a nitrogen atmosphere for 10 to 60 minutes. Following the ischemic period, tissue was immediately frozen in liquid nitrogen for later preparation of total RNA by CsCl sedimentation. The yield of RNA was not changed in ischemic tissue.

Total RNA was electrophoresed (20 µg/lane) and blotted onto charged nylon membranes. The blots were hybridized successively with random-primed ^{32}P labeled probes made from cDNA coding for rabbit heart NCX1, for human glyceraldehyde phosphate dehydrogenase (GAPDH), and for human heat-shock protein 70 (HSP70). Autoradiograms were quantitated by densitometry.

RESULTS AND DISCUSSION

On examining the raw data in TABLE 1, there appears to be some reduction in the levels of all 3 mRNAs examined as a function of the ischemic duration.

---

[a] Address for correspondence: Cardiology Section (111E), West Los Angeles VA Medical Center, 11301 Wilshire Blvd., Los Angeles, CA 90073.

TABLE 1. mRNA Quantitation by Densitometry of Autoradiograms[a]

	Ischemia Duration					
	0 min	10 min	20 min	30 min	45 min	60 min
mRNA						
NCX1	37 ± 10	23 ± 9	21 ± 10	18 ± 10	15 ± 9	15 ± 8
HSP70	86 ± 19	70 ± 14	73 ± 10	69 ± 8	63 ± 10	71 ± 9
GAPDH	82 ± 16	50 ± 12	49 ± 11	40 ± 9	39 ± 13	42 ± 11

[a] Raw data, mean ± SE of 6 experiments, arbitrary units.

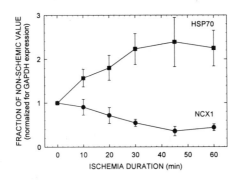

FIGURE 1. Na^+-Ca^{2+} exchanger and HSP70 mRNA in ischemia. Data are the mean ± SE of the densitometer output from scanned autoradiograms normalized for the GAPDH levels in each experiment before normalizing for the control value in each experiment.

Thus there may be some nonspecific degradation of all mRNAs during ischemia. However, normalizing for the GAPDH level should eliminate such nonspecific effects. The data in FIGURE 1 were normalized for the GAPDH level in each sample and additionally normalized for the control value for each mRNA species to facilitate comparison of different mRNAs. There is a striking difference in the effects of ischemia on NCX1 mRNA, which still decreases dramatically after normalizing for GAPDH, and mRNA for the stress protein HSP70, which increases relative to GAPDH levels. Thus the decrease in NCX1 message is not entirely a nonspecific finding of mRNA degradation, because it persists after data have been normalized for GAPDH, and HSP70 increased rather than decreased. Therefore, the recovery of normal sarcolemmal Na^+-Ca^{2+} exchange activity after even a brief ischemic period may be delayed by altered transcription that could limit synthesis of new exchanger protein.

REFERENCES

1. BERSOHN, M. M., K. D. PHILIPSON & J. Y. FUKUSHIMA. 1982. Am. J. Physiol. **242:** C288–C295.
2. BERSOHN, M. M. & K. D. PHILIPSON. 1982. Circulation **66**(Suppl II): II-290. (Abstract)

Ontogeny and Hormonal Regulation of Cardiac Na^+-Ca^{2+} Exchanger Expression in Rabbits

SCOTT R. BOERTH, WILLIAM A. COETZEE, AND
MICHAEL ARTMAN[a]

New York University School of Medicine
New York, New York 10016

As a consequence of deficiency of the sarcoplasmic reticulum (SR) at birth, the immature mammalian heart is relatively more dependent upon transsarcolemmal Ca^{2+} fluxes for regulating contraction and relaxation. It has been proposed that the cardiac sarcolemmal Na^+-Ca^{2+} exchanger (NCX) assumes a dominant role in regulating intracellular Ca^{2+} in the late fetal and early newborn rabbit heart. Previous studies demonstrate that at the time of birth, NCX steady-state mRNA and protein levels are approximately 2.5 times greater than adult values.[1,2] NCX expression decreases rapidly after birth, reaching adult levels by about three weeks of age.

Recent experiments were performed to characterize the magnitude of NCX current in single myocytes isolated from rabbits 1–21 days old.[3] NCX current density was highest in newborns (1–4 days old) and with increasing age, NCX current density decreased at both positive and negative membrane potentials. Most of the decrease occurred within the first few days after birth and NCX current density reached adult levels by approximately three weeks of age.[3] These results parallel the postnatal changes observed for NCX mRNA and protein.[1,2]

Factors influencing the normal postnatal downregulation of NCX expression remain largely uncharacterized. In rabbits, thyroid hormone levels rise dramatically at the time of birth, peak by 2–3 weeks of age and then decline slightly to adult levels. This normal thyroid hormone surge has been implicated by others in the postnatal upregulation of the Sarco(Endo)plasmic Reticulum Ca^{2+}-ATPase (SERCA). The time course of the thyroid hormone changes and upregulation of SERCA expression parallel the temporal sequence of the postnatal downregulation of NCX expression. Therefore, we speculated that changes in thyroid hormone status might be responsible for reciprocal regulation of NCX and SERCA2a expression during the first 3 weeks after birth.

Initial experiments were performed to characterize the effects of hypothyroidism (induced by 4 weeks of PTU treatment) and hyperthyroidism (l-thyroxine administration for 1 week) on NCX and SERCA2a expression in adult rabbits. Standard Northern analysis was performed using poly(A^+) RNA isolated from

---

[a] Corresponding author: Pediatric Cardiology, FPO Suite 9-V, NYU Medical Center, 530 First Ave., New York, NY 10016.

ventricles of rabbits of various ages and following various treatments. Blots were hybridized with a guinea pig cardiac NCX cDNA (from K. D. Philipson) and a rabbit SERCA2a cDNA (from D. H. MacLennan). Steady-state mRNA levels were quantitated by Northern slot blot analysis as described previously from our laboratory.[2]

FIGURE 1 illustrates that *hypothyroidism* increased expression of NCX and decreased SERCA2a expression in ventricular myocardium of adult rabbits. Conversely, *hyperthyroidism* decreased NCX expression and increased SERCA2a expression. These results support the concept that thyroid hormone reciprocally regulates NCX and SERCA2a expression.

Subsequently experiments were performed to determine if hypothyroidism from birth would prevent the normal postnatal downregulation of NCX expression. Immature rabbits were made hypothyroid by administration of PTU from gestational day 25 through 21 days after birth. FIGURE 2 illustrates that steady-state NCX mRNA levels at 21 days of age in the hypothyroid PTU-treated animals were essentially unchanged from levels measured at birth in untreated euthyroid controls. Therefore, blocking the postnatal surge in thyroid hormone prevented the normal postnatal downregulation of cardiac NCX expression.

In conclusion, high levels of expression and activity of cardiac NCX at birth suggest that sarcolemmal Na^+-Ca^{2+} exchange may play a relatively greater role in regulating myocardial intracellular calcium concentrations during late fetal and early newborn maturation. The alterations in NCX steady-state mRNA levels induced by changes in thyroid hormone status indicate that thyroid hormone may be important in the regulation of cardiac NCX expression. Results from hypothyroid immature rabbits suggest that the thyroid hormone surge

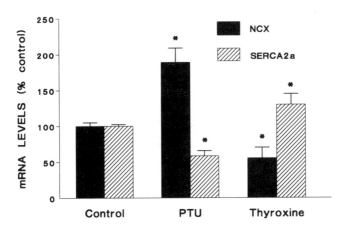

FIGURE 1. Effects of changes in thyroid hormone status on NCX and SERCA2a expression in ventricular myocardium from adult rabbits. Steady-state mRNA levels were normalized to the euthyroid untreated control animals (n = 4; serum T_4 = 2.32 ± 0.20 μg/dl). PTU indicates hypothyroid group (4 weeks of PTU administration; n = 4; serum T_4 = 0.73 ± 0.09 μg/dl). Thyroxine indicates hyperthyroid group (200 μg/kg l-thyroxine intramuscularly daily for 1 week; n = 3; serum T_4 > 24 μg/dl). * different from control ($p < 0.05$; ANOVA).

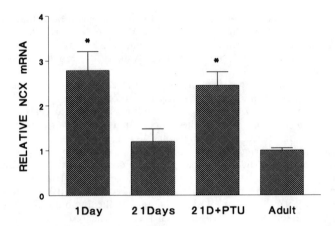

FIGURE 2. Effect of hypothyroidism from birth on NCX expression in ventricular myocardium from immature rabbits. Steady-state mRNA levels were normalized to normal adult values (n = 3 in each group). 1 Day and 21 Days indicate postnatal age. 21 D + PTU indicates 21 day old animals hypothyroid from birth due to administration of PTU from gestational day 25 through 21 days of age. Serum T_4 concentration was significantly lower in the 21-day-old PTU-treated animals (2.23 ± 0.38 µg/dl) compared with normal euthyroid 21-day-old rabbits (9.40 ± 1.09 µg/dl; $p < 0.05$; t-test). * different from adult ($p < 0.05$; ANOVA).

after birth is important for the normal postnatal downregulation of NCX expression.

REFERENCES

1. ARTMAN, M. 1992. Sarcolemmal Na^+-Ca^{2+} exchange activity and exchanger immunoreactivity in developing rabbit hearts. Am. J. Physiol. **263:** H1506–H1513.
2. BOERTH, S. R., D. B. ZIMMER & M. ARTMAN. 1994. Steady-state mRNA levels of the sarcolemmal Na^+-Ca^{2+} exchanger peak near birth in developing rabbit and rat hearts. Circ. Res. **74:** 354–359.
3. ARTMAN, M., H. ICHIKAWA, M. AVKIRAN & W. A. COETZEE. 1995. Na^+-Ca^{2+} exchange current density in cardiac myocytes from rabbits and guinea pigs during postnatal development. Am. J. Physiol. **268:** H1714–H1722.

Functional Relevance of an Enhanced Expression of the Na^+-Ca^{2+} Exchanger in the Failing Human Heart

M. FLESCH, F. PÜTZ, R. H. G. SCHWINGER, AND M. BÖHM

[a]Clinic III for Internal Medicine
University of Cologne
Joseph Stelzmann-Str. 9
50924 Cologne, Germany

INTRODUCTION

In animal models of cardiac hypertrophy and failure, an increased activity of the Na^+-Ca^{2+} exchanger has been observed.[1,2] Only recently, it was reported that Na^+-Ca^{2+} exchanger mRNA and protein levels are increased in failing compared to nonfailing myocardium.[3] The functional relevance of these findings has remained unexplained so far. It has been suggested that the increased expression of the Na^+-Ca^{2+} exchanger in the failing heart is of functional relevance because enhanced extrusion of Ca^{2+} via this mechanism serves as a compensatory adaptation for the impaired function of the sarcoplasmic reticulum in the failing heart.[3] Alternatively, the increased expression of the Na^+-Ca^+ exchanger could be important because it allows an increased influx of Ca^{2+} into the myocyte via the exchanger. In order to study the functional relevance of changes in exchanger activity, Na^+-Ca^{2+} exchanger mRNA and protein levels were determined in failing and in nonfailing human left ventricular myocardium and related to the inotropic effect of the Na^+ channel activator BDF 9148. BDF 9148 is known to increase the intracellular Na^+ concentration by prolonging the open state of sarcolemmal Na^+ channels. It has been suggested that this leads to an activation of Na^+-Ca^{2+} exchanger activity.[4] For further characterization of the failing myocardium, SR Ca^{2+} mRNA levels were determined.

METHODS

Experiments were performed on left ventricular myocardium from patients with end-stage heart failure (NYHA IV) due to idiopathic dilated cardiomyopathy (n = 8) or ischemic cardiomyopathy (n = 6) and from 8 nonfailing donor-hearts (NF). For Northern blot analysis 10 µg total RNA per lane were separated by gel electrophoresis. Nylon membranes were successively hybridized with a 500-bp cDNA fragment (EcoRI-EcoRI) encoding for rat heart Na^+-Ca^{2+} exchanger

---

[a] Corresponding author.

(K. D. Philipson, Los Angeles)[5] and a 2-kb SR Ca^{2+}ATPase cDNA fragment (BamHI-BamHI, D. H. MacLennan, Toronto). Na^+-Ca^{2+} exchanger protein levels were analyzed by Western blot analysis using an antiserum against the canine Na^+-Ca^{2+} exchanger (K. D. Philipson, Los Angeles).[6] Myocardial force of contraction was determined on isolated, electrically driven (1 Hz, 5 ms,

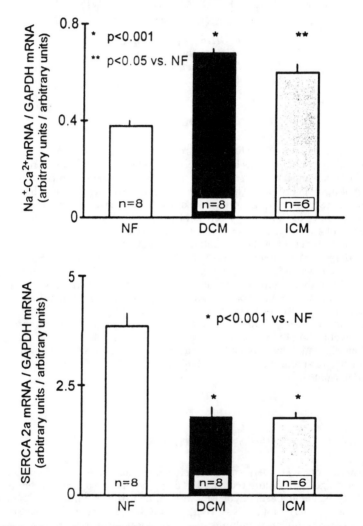

FIGURE 1. Mean values (± SEM) for the mRNA levels of the sarcolemmal Na^+-Ca^{2+} exchanger (Na^+-Ca^{2+}) and the sarcoplasmic reticulum Ca^{2+}-ATPase (SERCA 2a) in left ventricular myocardium from nonfailing (NF) and failing human hearts due to idiopathic dilated cardiomyopathy (DCM) and idiopathic cardiomyopathy (ICM). For standardization, values obtained by densitometric analysis of Northern blot hybridization signals were related to densitometric values of GAPDH mRNA hybridization signals.

TABLE 1. Inotropic Effects of BDF 9148 and Isoprenaline on Failing and Nonfailing Human Left Ventricular Myocardium[a]

		NF (n = 8)	NYHA IV (n = 9)
BDF 9148	Maximal increase in FOC (mN)	6.2 ± 0.97	6.22 ± 0.49
	EC_{50} values (µmol/l)	1.65 (1.3–3.0)	0.175 (0.155–0.22)*
Isoprenaline	Maximal increase in FOC (mN)	8.1 ± 1.3	2.2 ± 0.3*
	EC_{50} values (µmol/l)	0.016 (0.01–0.02)	0.06 (0.04–0.08)*

[a] NF = nonfailing human left ventricular papillary muscle strips; NYHA IV = end-stage failing human left ventricular papillary muscle strips; EC_{50} values = concentration producing half maximal effects (range); *$p < 0.05$ vs NF.

intensity 10–20% greater than threshold, 37°C) left ventricular papillary muscle strips.

RESULTS

Northern blot analysis of the Na^+-Ca^{2+} exchanger mRNA steady state levels in failing and nonfailing human left ventricular myocardium revealed a single hybridization signal at the expected position corresponding to a mRNA size of 7.1 kb. Mean values were significantly decreased by 79% ($p < 0.001$) in dilated cardiomyopathy and by 58% ($p < 0.05$) in ischemic cardiomyopathy compared to nonfailing myocardium (FIG. 1, upper panel). Immunochemical detection of the Na^+-Ca^{2+} exchanger protein showed two immunoreactive bands in all specimens at a position corresponding to a molecular weight of 120 kDa and 70 kDa. For quantification of Na^+-Ca^{2+} exchanger protein levels, densitometric values of both bands were added. Mean values were increased by 35.5% ($p < 0.001$) in dilated cardiomyopathy and by 20.4% ($p < 0.05$) in ischemic cardiomyopathy compared to nonfailing myocardium, respectively. Mean values for relative SR Ca^{2+} ATPase mRNA levels were decreased by 54% in dilated cardiomyopathy ($p < 0.001$) and in ischemic cardiomyopathy ($p < 0.001$), when compared to nonfailing myocardium (FIG. 1, lower panel). Functional experiments revealed that BDF 9148 (0.03–30 µmol/l) increased the force of contraction in failing and nonfailing papillary muscle strips in a concentration-dependent manner, the maximal positive inotropic response to BDF 9148 being similar in both groups. The potency of BDF 9148 to increase force of contraction was greater ($p < 0.01$) in failing than in nonfailing myocardium as judged by the EC_{50} values (0.175 (0.155–0.22) µmol/l, NYHA IV; 1,65 (1.3–3.0) µmol/l, NF) (TABLE 1). In contrast, the maximal positive inotropic effect of isoprenaline was significantly decreased in failing myocardium compared to nonfailing myocardium, as was the potency of isoprenaline ($p < 0.05$) (TABLE 1).

CONCLUSIONS

The observation that in failing human myocardium mRNA as well as protein levels of the sarcolemmal Na^+-Ca^{2+}-exchanger are increased whereas SR Ca^{2+} mRNA levels are significantly reduced suggests that the increased expression of the Na^+-Ca^{2+} exchanger in the failing myocardium could be useful for accessory diastolic Ca^{2+} extrusion compensating the reduced sequestration of Ca^{2+} into the sarcoplasmic reticulum.[7] However, the increase in exchanger molecules could also lead to an increased Ca^{2+} influx via the exchanger when the intracellular Na^+ concentration is increased. Since increased Ca^{2+} influx via the exchanger in response to an increase in the intracellular Na^+ concentration has been suggested to contribute to the modulation of cardiac contractility,[8–10] this would explain the enhanced potency of the Na^+ channel activator BDF 9148 in failing compared to nonfailing myocardium. Thus, the increased Na^+-Ca^+ exchanger mRNA and protein levels in the failing human myocardium and consequently increased exchanger activity seem to be of functional relevance for the modulation of cardiac force of contraction.

REFERENCES

1. DAVID-DUFILHO, M., M-G. PERNOLLET, H. L. SANG, P. BENLIAN, M. DE MENDONCA, M-L. GRICHOIS, M. CIRILLO, P. MEYER & M-A. DEVYNCK. 1986. Active Na^+ and Ca^{2+} transport, Na^+-Ca^{2+} exchange, and intracellular Na^+ and Ca^{2+} content in young spontaneously hypertensive rats. J. Cardiovasc. Pharmacol. **8**(Suppl. 8): S130–S135.
2. NAKANISHI, H., N. MAKINO T. HATA, H. MATSUI, K. YANO & T. YANAGA. 1989. Sarcolemmal Ca^{2+} transport activities in cardiac hypertrophy caused by pressure overload. Am. J. Physiol. **257:** H349–H356.
3. STUDER, R., H. REINECKE, J. BILGER, T. ESCHENHAGEN, M. BÖHM, G. HASENFUß, H. JUST, J. HOLTZ & H. DREXLER. 1994. Gene expression of the cardiac Na^+-Ca^{2+} exchanger in end-stage human heart failure. Circ. Res. **75:** 443–453.
4. SCHWINGER, R. H. G., M. BÖHM, C. MITTMANN, K. LA ROSÉE & E. ERDMANN. 1991. Evidence for a sustained effectiveness of sodium-channel activator in failing human myocardium. J. Mol. Cell. Cardiol: **23:** 461–471.
5. NICOLL, D. A., S. LONGONI & K. D. PHILLIPSON. 1990. Molecular cloning and functional expression of the cardiac sarcolemmal Na^+-Ca^{2+} exchanger. Science **250:** 562–565.
6. PHILIPSON, K. D., S. LONGONI & R. WARD. 1988. Purification of the cardiac Na^+-Ca^{2+} exchange protein. Biochim. Biophys. Acta. **945:** 298–306.
7. BEUCKELMANN, D. J., M. NÄBAUER & E. ERDMANN. 1992. Intracellular calcium handling in isolated ventricular myocytes from patients with terminal heart failure. Circulation **85:** 1046–1055.
8. BERS, D. M. 1987. Mechanisms contributing to the cardiac intropic effect of Na-pump inhibition and reduction of extracellular Na. J. Gen. Physiol. **90:** 479–504.
9. LEVI, A. J., P. BROOKSBY & J. C. HANCOX 1993. A role for depolarisation induced calcium entry on the Na-Ca exchange in triggering intracellular calcium release and contraction in rat ventricular myocytes. Cardiovasc. Res. **27:** 1677–1690.
10. HARRISON, S. M. & M. R. BOYETT. 1995. The role of the Na^+-Ca^{2+} exchanger in the rate dependent increase in contraction in guinea pig ventricular myocytes. J. Physiol. **482**(3): 555–566.

Role of the Cardiac Sarcolemmal Na^+-Ca^{2+} Exchanger in End-Stage Human Heart Failure

HANS REINECKE, ROLAND STUDER,
ROLAND VETTER,[a] HANJÖRG JUST, JÜRGEN HOLTZ,[b]
AND HELMUT DREXLER

Cardiology and Angiology
Internal Medicine III
University of Freiburg
Breisacher Strasse 33
D-79106 Freiburg, Germany

[a]*Max Delbrück Center*
Berlin-Buch, Germany

[b]*Institute for Pathophysiology*
University of Halle
Halle, Germany

INTRODUCTION

Diastolic and systolic dysfunction appear to be related to altered Ca^{2+} handling of the cardiac myocyte (reviewed in Ref. 1). Disturbed Ca^{2+} handling might also affect influx and efflux of other ions, and thereby could increase the arrhythmogenic potential of the failing heart. In this context, the cardiac sarcolemmal Na^+-Ca^{2+} exchanger represents an important mechanism of Ca^{2+} transport across the sarcolemma. The Na^+-Ca^{2+} exchanger uses the electrochemical Na^+ gradient to extrude Ca^{2+} in a ratio of 3 Na^+ to 1 Ca^{2+}, thereby generating an inward current. In the face of a depressed expression and function of the SR Ca^{2+}-ATPase in the failing human heart,[2-5] the Na^+-Ca^{2+} exchanger might play a compensatory role in providing increased Ca^{2+} efflux thereby limiting intracellular diastolic Ca^{2+} overload. Therefore, the present study was designed to investigate the gene expression (mRNA and protein) of the Na^+-Ca^{2+} exchanger in conjunction with measurements of the Na^+-Ca^{2+} exchange activity in patients with end-stage heart failure.

METHODS

Patients I (Northern blot analysis). Left ventricular samples from patients undergoing heart transplantation: coronary artery disease (CAD, n = 11), dilated cardiomyopathy (DCM, n = 13), mean age 51 ± 8 years. Nonfailing controls from organ donors: NF (n = 7), mean age 45 ± 12 years. *Patients II* (Western blot analysis and $^{45}Ca^{2+}$ uptake). CAD (n = 6), DCM (n = 5), mean age 51 ± 13; nonfailing controls (NF, n = 6), mean age 41 ± 21 years. All patients with heart failure were grouped in NYHA functional class IV. Northern and slot blot analysis were performed by using a guinea pig Na^+-Ca^{2+} exchanger cDNA probe and, as

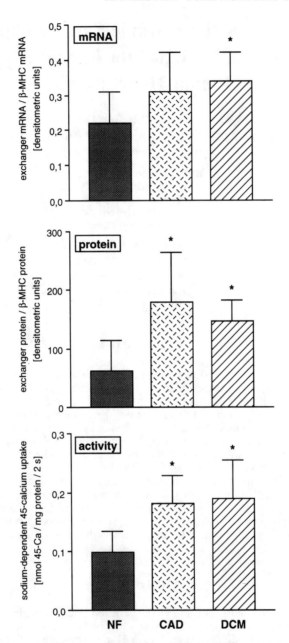

FIGURE 1. Bar graphs showing mRNA level, protein level and activity of the sodium-calcium exchanger in patients with coronary artery disease (CAD) and dilated cardiomyopathy (DCM) compared to nonfailing controls (NF). * $=p < 0.05$, ANOVA followed by Student-Newman-Keuls test.

a standard for a myocyte-specific mRNA, human β-myosin heavy chain (β-MHC) cDNA probe. Western blot analysis was performed by using a Na^+-Ca^{2+} exchanger antiserum and, as a standard for a myocyte-specific protein, a β-MHC monoclonal antibody. The functional activity of the Na^+-Ca^{2+} exchanger was determined by measuring the Na^+-dependent $^{45}Ca^{2+}$ uptake into membrane vesicles prepared from left ventricular samples.

RESULTS

FIGURE 1 (top) shows that the mRNA level of the Na^+-Ca^{2+} exchanger normalized to β-MHC mRNA was significantly increased in DCM. In the case of CAD the increase of Na^+-Ca^{2+} exchanger mRNA per β-MHC mRNA did not quite reach statistical significance. The protein level of the Na^+-Ca^{2+}-exchanger protein per β-MHC protein (FIG. 1, middle) and the functional activity (FIG. 1, bottom) was significantly increased both in CAD and DCM as compared to nonfailing controls ($\approx 160\%$ and $\approx 85\%$, respectively, versus nonfailing control).

CONCLUSIONS

The present findings demonstrate that the increased expression of the Na^+-Ca^{2+} exchanger protein is translated into enhanced activity in human heart failure, suggesting that the enhanced activity is due to an increase in the number of functional exchanger molecules rather than to enhanced exchange rate by preexisting exchanger molecules. We hypothesize that the increased exchanger expression and function in failing hearts may, in part, compensate for depressed SR Ca^{2+}-ATPase function. While increased Na^+-Ca^{2+} exchange in the failing heart is likely to limit diastolic intracellular Ca^{2+} overload and to improve relaxation, it may contribute to delayed afterdepolarizations and, therefore, may be associated with an increased risk of arrhythmias. Thus, the altered expression and functional activity of the Na^+-Ca^{2+} exchanger may represent one example for a functional alteration at the cellular level of cardiomyocytes which might be involved in the well-known increased incidence of arrhythmias in patients with severe chronic heart failure.

REFERENCES

1. ARAI, M., H. MATSUI & M. PERIASAMY. 1994. Sarcoplasmic reticulum gene expression in cardiac hypertrophy and failure. Circ. Res. **74:** 555–564.
2. MERCADIER, J.-J., A.-M. LOMPRÉ, P. DUC, K. R. BOHELER, J. B. FRAYSSE, C. WISNEWSKY, P. D. ALLEN, M. KOMAJDA & K. SCHWARTZ. 1990. Altered sarcoplasmic reticulum Ca^{2+}-ATPase gene expression in human ventricle during end-stage heart failure. J. Clin. Invest. **85:** 305–309.
3. STUDER, R., H. REINECKE, J. BILGER, T. ESCHENHAGEN, M. BÖHM, G. HASENFUSS, H. JUST, J. HOLTZ & H. DREXLER. 1994. Gene expression of the cardiac sodium-calcium exchanger in end-stage human heart failure. Circ. Res. **75:** 443–453.
4. LIMAS, C. J., M. OLIVARI, I. F. GOLDENBERG, T. B. LEVINE, D. G. BENDITT & A. SIMON. 1987. Calcium uptake by cardiac sarcoplasmic reticulum in human dilated cardiomyopathy. Cardiovasc. Res. **21:** 601–605.
5. GWATHMEY, J. K., L. COPELAS, R. MACKINNON, F. J. SCHOEN, M. D. FELDMAN, W. GROSSMAN & J. P. MORGAN. 1987. Abnormal intracellular calcium handling in myocardium from patients with end-stage heart failure. Circ. Res. **61:** 70–76.

Lanthanum Provides Cardioprotection by Modulating Na^+-Ca^{2+} Exchange[a]

NILANJANA MAULIK, ARPAD TOSAKI,
RICHARD M. ENGELMAN, GORACHAND CHATTERJEE,
AND DIPAK K. DAS

Cardiovascular Division
Department of Surgery
University of Connecticut School of Medicine
Farmington, Connecticut 06030

INTRODUCTION

Intracellular Ca^{2+} overloading is a well-recognized feature in myocardial ischemic and reperfusion injury.[1] The exact mechanism for Ca^{2+} influx into the cell remains controversial, but it is generally accepted that both Ca^{2+} slow channels and Na^+-Ca^{2+} exchangers play a role in the pathophysiology of myocardial ischemic and reperfusion injury.[2] It is known that acidosis induced by anaerobic metabolism leads to enhanced exchange of intracellular H^+ for Na^+ which in turn becomes instrumental for the intracellular Ca^{2+} accumulation. This hypothesis is well supported by the observations that inhibitors of Na^+-H^+ exchangers such as amiloride can inhibit the accumulation of Ca^{2+}.[2]

Approximately 41% of the total exchangeable Ca^2 is localized in a kinetic compartment defined by rapid compartment, and this Ca^{2+} pool is rapidly displaced by lanthanum (La^{3+}), a known inhibitor of Ca^{2+}-binding proteins.[3] La^{3+} is also a potent inhibitor of Ca^{2+} entry into mitochondria via the uniport system.[4] La^{3+} has also been found to nonspecifically block the Ca^{2+} channels.

In this study, we examined the effects of La^{3+} on myocardial rhythm disturbances associated with ischemia and reperfusion. In an attempt to examine its mechanism of actions, we also examined the myocardial contents of Na^+, K^+, Ca^{2+} and Mg^{2+}.

METHODS

Isolated Perfused Rat Heart Preparation and Measurements of Arrhythmias

Male Sprague-Dawley rats weighing approximately 350 gm were anesthetized with intraperitoneal pentobarbital (60 mg/kg) followed by heparin (500 IU/kg; i.v) administration. After thoracotomy, hearts were excised, aorta cannulated and the hearts purfused with Krebs-Henseleit bicarbonate buffer using Langendorff technique for 5 min at a perfusion pressure of 100 cm H_2O.[5] Perfusion was then switched to working mode as described previously.[5] An epicardial ECG was recorded by a polygraph throughout the experiment by placing two silver elec-

---

[a] This study was supported by National Institutes of Health Grants HL 34360 and HL 22559 and a grant from the American Heart Association.

trodes attached directly to the heart. The ECGs were analyzed to determine the incidence of ventricular fibrillation (VF) and ventricular tachycardia (VT) as described elsewhere.[5]

Measurements of Myocardial Ion Contents

Electrolyte content of the heart was measured using a Perkin-Elmer atomic absorption spectrophotometer as described previously.[5] In short, at the end of reperfusion the hearts were rapidly cooled to 0–5°C by submersion in and perfused for 5 min with ice-cold ion-free solution containing 100 mM trishydroxymethylaminomethane and 220 mM sucrose to washout ions from extracellular space and to arrest the activities of membrane enzymes responsible for ion transport. Hearts were dried for 48 hr at 100°C and ashed at 550°C for 20 min. The ash was dissolved in 5 ml 3M HNO_3 and diluted 10-fold with deionized H_2O. The wavelengths used for the measurements of Na^+, K^+, Ca^{2+} and Mg^{2+} were 330.3, 404.4, 422.7 and 286.0 nm, respectively.

RESULTS

Effects of La^{3+} on Arrhythmias

Thirty minutes of transient normothermic global ischemia followed by reperfusion was chosen in order to assess the antiarrhythmic effects of La^{3+}. FIGURE 1 shows that at 0.05 and 0.1 mg/l concentrations of La^{3+}, a significant reduction in the increase of sustained and total (sustained + nonsustained) VF was observed. Thus, La^{3+} at 0.05 and 0.1 mg/l reduced the incidence of sustained and total VF from nontreated control values of 93% and 100% to 33% ($p < 0.05$) and 42% ($p < 0.05$), 50% ($p < 0.05$) and 50% ($p < 0.05$), respectively (FIG. 1A). The incidence of VT followed the same pattern (FIG. 1B). Increasing concentrations of La^{3+} (1 and 5 mg/l) failed to reduce the incidence of VF and VT indicating that at higher concentrations the antiarrhythmic effect of La^{3+} is lost.

Effects of La^{3+} on Na^+, K^+, Ca^{2+} and Mg^{2+}

As described in Methods, we examined the effects of La^{3+} on the ionic composition of heart during ischemia and reperfusion. Since the results of FIGURE 1 demonstrate that at 0.05 mg/l La^{3+} exerts a potent antiarrhythmic effect against reperfusion-induced arrhythmias, we selected this concentration to study the changes in myocardial ion concentrations. At 0.05 mg/l, La^{3+} did not change the baseline concentrations of Na^+, K^+, Ca^{2+} or Mg^{2+} (TABLE 1). After 30 min of normothermic global ischemia followed by 30 min of reperfusion, La^{3+} significantly reduced myocardial Na^+ and Ca^{2+} gains and prevented K^+ and Mg^{2+} loss.

DISCUSSION

The results of this study demonstrate that La^{3+} (0.05 mg/l) reduced the incidence of ventricular fibrillation associated with the reperfusion of ischemic myocar-

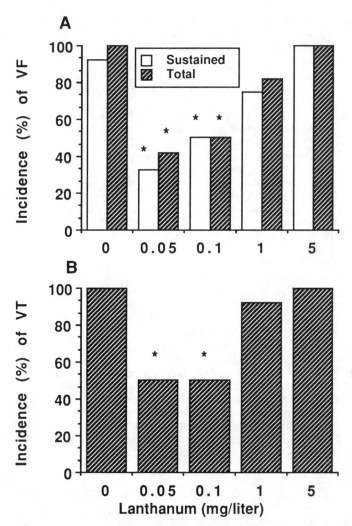

FIGURE 1. The incidence of reperfusion-induced ventricular fibrillation (VF) is shown **(A)** as sustained (*open columns*) and total (*hatched columns*) VF after 30 min of ischemia followed by reperfusion in untreated and La^{3+}-treated hearts. The incidence of ventricular tachycardia is shown in **(B)**. Comparisons were made to the untreated control group. Results are expressed as means ± SEM of n = 12 rats per group. * $p < 0.05$.

TABLE 1. Effects of La^{3+} (0.05 mg/l) on Myocardial Ion Contents (mmol/kg dry wt) in the Pre- and Post-Ischemic Myocardium[a]

Ions	Before Ischemia		After Ischemia	After Reperfusion
	Before La^{3+}	After La^{3+}		
Control				
Na^+	30 ± 4	—	94 ± 6	93 ± 5
K^+	300 ± 8	—	200 ± 9	207 ± 7
Ca^{2+}	1.3 ± 0.2	—	5.1 ± 0.5	4.7 ± 0.4
Mg^{2+}	18 ± 1.2	—	7.2 ± 0.6	8.5 ± 0.5
La^{3+}				
Na^+	28 ± 3	31 ± 4	56.5 ± 2*	59 ± 1.9*
K^+	294 ± 6	301 ± 9	252 ± 6*	250 ± 6*
Ca^{2+}	1.4 ± 0.2	1.2 ± 0.2	2.64 ± 0.2*	2.43 ± 0.3*
Mg^{2+}	19 ± 1.5	18 ± 1.6	12.1 ± 0.4*	11.9 ± 0.6*

[a] Results are expressed as mean ± SEM of n = 12 per group. Comparisons were made with the untreated group at fixed time points. * $p < 0.05$.

dium suggesting that it is beneficial to heart in reducing ischemic/reperfusion injury. Our results further indicate that La^{3+} reduced the accumulation of Na^+ and Ca^{2+} in heart in concert with the enhancement of K^+ and Mg^{2+} loss. It is well recognized that the transport of Ca^{2+} in heart involves a carrier-mediated transmembrane exchange of Na^+ and Ca^{2+}. La^{3+} was previously shown to block the Ca^{2+} entry and Ca^{2+} efflux via the plasma membrane Ca^{2+}-ATPase.[6] Since La^{3+} can displace Ca^{2+} from the extracellular pools in the heart, it was used in an attempt to modulate the injury originating from the calcium paradox.[7] The inhibition of electrically stimulated ^3H-norepinephrine release by La^{3+} was found to be associated with the activation of stabilizing outward K^+ current.[8] A recent study showed the ability of La^{3+} to decrease total cell calcium.[9] Our results, thus, support the previous findings and further demonstrate that La^{3+} can reduce the incidence reperfusion arrhythmias simultaneously modulating the Na^+-Ca^{2+} exchanger.

REFERENCES

1. LIU, X., R. M. ENGELMAN, Z. WEI, D. BAGCHI, J. A. ROUSOU, D. NATH & D. K. DAS. 1993. Attenuation of myocardial reperfusion injury by reducing intracellular calcium overloading with dihydropyridines. Biochem. Pharmacol. **45:** 1333–1341.
2. OTANI, H. 1993. Role of calcium in the pathogenesis of myocardial reperfusion injury. In Pathophysiology of Reperfusion Injury. D. K. Das, Ed. 181–220. CRC Press. Boca Raton, FL.
3. LANGER, G. A., T. L. RICH & F. B. ORNER. 1990. Calcium exchange under nonperfusion limited conditions in rat ventricular cells: identification of subcellular compartments. Am. J. Physiol. **259:** H592–H602.
4. MELA, L. 1969. Inhibition and activation of calcium transport in mitochondria. Effect of lanthanides and local anesthetic drugs. Biochemistry **8:** 2481–2486.
5. TOSAKI, A., G. A. CORDIS, P. SZERDAHELYI, R. M. ENGELMAN & D. K. DAS. 1994. Effects of preconditioning on reperfusion arrhythmias, myocardial functions, formation of free radicals, and ion shifts in isolated ischemic/reperfused rat hearts. J. Cardiovasc. Pharmacol. **23:** 365–373.
6. PANDOL, S. J., M. S. SCHOEFFIELD, C. J. FIMMEL & S. MUALLEM. 1987. The agonist-sensitive calcium pool in the pancreatic aciner cell. J. Biol. Chem. **262:** 16963–16968.

7. KAYGISIZ, Z., M. Z. AKIN, A. BASARAN, H. V. GUNES & B. TIMURALP. 1991. The effect of lanthanum on the isolated perfused rat heart at different extracellular calcium concentrations. Acta Physiol. Hungarica **78:** 191–200.
8. PRZYWARA, D. A., S. V. BHAVE, A. BHAVE, P. S. CHOWDHURY, T. D. WAKADE & A. R. WAKADE. 1992. Activation of K^+ channels by lanthanum contributes to the block of transmitter release in chick and rat sympathetic neurons. J. Membr. Biol. **125:** 155–162.

Sodium-Calcium Balance in Coronary Angiography

Experimental Experience with Isotonic Iodixanol

PER JYNGE,[a,d] GEIR FALCK,[a] HANS K. PEDERSEN,[b]
JAN O. G. KARLSSON,[c] AND HELGE REFSUM[c]

[a]*Department of Pharmacology and Toxicology*
University of Trondheim, Norway

[b]*Department of Physiology*
University of Tromsø, Norway

[c]*Nycomed Imaging AS*
Oslo, Norway

Coronary injection of X-ray contrast media (CM) may be associated with cardiac side effects like arrhythmia, contractile failure and coronary spasm. Many factors contribute. One particularly important factor is that CM induces rapid bidirectional shifts of extracellular ions, particularly of Na and Ca, which may upset Na-Ca exchange and cellular Ca control.

We perfused isolated, paced and global hearts from rat, guinea pig and rabbit, and measured left ventricular developed pressure (LVDP) during transient washout (5–7 sec) of ions and equilibration with CM.[1] Mean LVDP values during perfusion with modified Krebs-Henseleit buffers (80% reduction in Ca, Na, K, Mg) or with nonionic media like normosmolal mannitol and slightly hyperosmolal iohexol were recorded in TABLE 1.

TABLE 1. Left Ventricular Developed Pressure (LVDP) in Rat, Guinea Pig, and Rabbit Perfused Hearts

	LVDP (%)		
Medium	Rat	Guinea Pig	Rabbit
Low Ca (0.25 mM)	36	57	68
Low Na (30 mM)	184	148	160
Low K (0.9 mM)	62	118	114
Low Mg (0.12 mM)	106	103	103
Mannitol	179	137	133
Iohexol (150 mgI/ml)	148	127	128

Main conclusions from the study[1] were that:

- Ionic shifts affect myocardial Na-Ca exchange and cellular Ca control within a time scale employed in coronary angiography.

[d] Correspondence to: Per Jynge, Department of Pharmacology and Toxicology, University of Trondheim, Medisinsk-Teknisk Senter, 7005 Trondheim, Norway.

- A nonionic CM like iohexol induces a positive inotropic effect in parallel with mannitol which has been regarded as an inert sugar.

Modern CM like iohexol are monomeric (single triiodinated benzene ring compound) and lowosmolal (<600–1000 mOsm/kg H_2O) when used in clinical formulations (300–350 mgI/ml). Iodixanol, a new nonionic dimer, produces hyposmolal CM without any additive. Supplemented with NaCl, iodixanol represents the first generation of normosmolal CM. Effects on LVDP of adding low Ca to 3 formulations of iodixanol and NaCl were investigated in isolated rat hearts.[2] FIGURE 1 presents main findings during coronary bolus perfusion (2–5 sec).

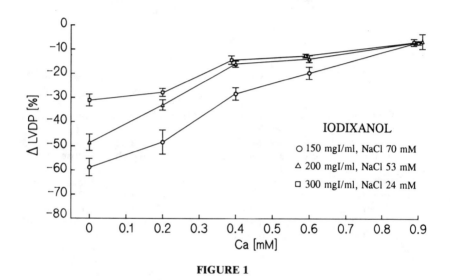

FIGURE 1

Main conclusions from the study[2] were that:

- Iodixanol-based CM show a Na-dependent depression of LVDP that is reversed by adding low Ca.
- Iodixanol-based CM may benefit from a small electrolyte supplement based on a balance between Na and Ca.

Other studies have shown that Visipaque (Nycomed Imaging AS), containing iodixanol 320 mgI/ml, NaCl 19 mM and $CaCl_2$ 0.3 mM, prevented negative inotropy in rat hearts[3] and reduced the incidence of ventricular arrhythmias in *in vivo* dog hearts.[4]

REFERENCES

1. JYNGE, P., H. BLANKSON, J. SCHJØTT, T. HOLTEN & A. N. ØKSENDAL. 1995. Invest. Radiol. **30:** 173–180.
2. JYNGE, P., T. HOLTEN & A. N. ØKSENDAL. 1993. Invest. Radiol. **28:** 20–25.
3. BLANKSON, H., T. HOLTEN, A. N. ØKSENDAL & P. JYNGE. Acta Radiol. In press.
4. PEDERSEN, H. K., E. A. JACOBSEN & H. REFSUM. 1994. Acad. Radiol. **1:** 136–144.

Properties of the Na^+-Ca^{2+} Antiport of Heart Mitochondria[a]

DENNIS W. JUNG, KEMAL BAYSAL, AND
GERALD P. BRIERLEY[b]

Department of Medical Biochemistry
The Ohio State University
Hamilton Hall
1645 Neil Avenue
Columbus, Ohio 43210-1218

There is considerable evidence that the concentration of free Ca^{2+} ($[Ca^{2+}]$) in the matrix of the mitochondrion contributes to the regulation of metabolism and function in myocardial cells.[1,2] Matrix $[Ca^{2+}]$ is maintained by a balance between influx on the Ca uniport and efflux via nNa^+-Ca^{2+} antiport.[3,4] A better understand-

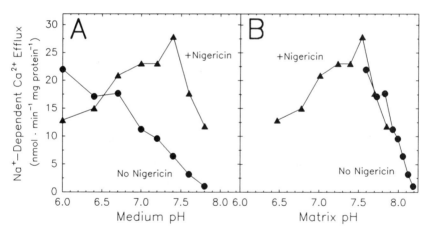

FIGURE 1. Rate of Na^+-Ca^{2+} antiport in the presence and absence of nigericin as a function of medium pH (**A**) and matrix pH (**B**) for isolated heart mitochondria respiring with succinate. The medium contained KCl (0.1 M), succinate (3 mM), oligomycin (2 μg/ml), rotenone (1 μg/ml) and either 15 mM Pipes (pH < 7.0) or HEPES (pH > 7, 25°C). Following mitochondrial Ca^{2+} loading (20 μM $CaCl_2$), ruthenium red (0.5 μM) was added to block the Ca^{2+} uniport (followed by nigericin (1 μM), if added), and Ca^{2+} efflux was initiated by addition of NaCl (20 mM). Ca^{2+} efflux rates were determined using antipyrylazo III (100 μM), and matrix pH was measured using cSNARF-1 as detailed in Reference 7. Note the steep sensitivity of the antiport reaction to matrix pH in the physiological range of pH 7.4 to 8.

[a] Supported by U.S. Public Health Services Grant HL09364-29.
[b] Corresponding author.

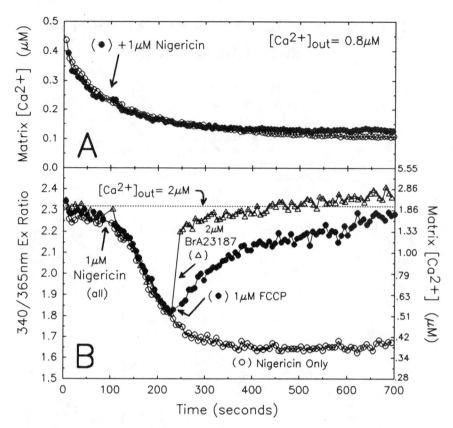

FIGURE 2. In the presence of nigericin, heart mitochondria maintain a significant [Ca^{2+}] gradient **(A)** which is sensitive to collapse by uncoupler or BrA23187 **(B)**. Mitochondria were equilibrated with fura-2 and added to a medium containing KCl (0.1 M), HEPES (15 mM, pH 7.35, 25°C), rotenone (1 μg/ml), oligomycin (2 μg/ml), cyclosporin A (1 μM), NaCl (20 mM), and EGTA/CaCl$_2$ to buffer the medium [Ca^{2+}] at 0.8 μM (A) or 2 μM (B). Ruthenium red (1 μM) was added 10 s after the mitochondria and fluorescence measurements were then started. Matrix [Ca^{2+}] was calculated as described.[10] Note that the [Ca^{2+}] scale in (B) is nonlinear.

ing of these transporters is essential if we are to have a complete picture of how matrix [Ca^{2+}] is controlled. Based largely on a single study showing that A23187 did not alter the steady state established by the Na$^+$-Ca^{2+} antiport,[5] there has been a consensus that this antiporter promotes an electroneutral exchange of 2 Na$^+$ for 1 Ca^{2+}.[3] In support of this concept, a partially purified, reconstituted 110-kDa protein with Na$^+$-Ca^{2+} antiport activity promoted only electroneutral antiport.[6] This report calls attention to recent studies from our laboratory on the stoichiometry of the nNa$^+$-Ca^{2+} antiport.

We found that nigericin strongly stimulated Na$^+$-dependent Ca^{2+} efflux from heart mitochondria suspended at neutral pH in a KCl medium (FIG. 1).[7] This effect was found to be due to the ability of nigericin to equilibrate matrix pH (pH$_m$) with medium pH by K^+/H^+ exchange. The optimal pH_m for Na^+/Ca^{2+} antiport was near pH_m 7.5 regardless of the external pH. These experiments indicate that the antiport is asymmetric and strongly dependent on pH_m, and suggest that pH_m may contribute to the regulation of matrix $[Ca^{2+}]$. The high rates of Na^+-induced Ca^{2+} efflux seen in the presence of nigericin (FIG. 1) occurred when ΔpH and ΔpNa were vanishingly small and $\Delta\Psi$ was elevated. This suggests that the antiporter may respond to $\Delta\Psi$ and not promote electroneutral Na^+-Ca^{2+} exchange.

In a collaborative study with Drs. Karlene and Thomas Gunter we used a null-point treatment to measure the $[Ca^{2+}]$ gradients maintained under conditions in which virtually all Ca^{2+} movement could be attributed to the Na^+-Ca^{2+} antiport.[8] At equilibrium the nNa^+-Ca^{2+} antiport will maintain a $[Ca^{2+}]_{out}:[Ca^{2+}]_m$ gradient equal to $10^{((2-n)\Delta\Psi/59 + n\Delta pNa)}$.[5,8,9] It was found that the Ca^{2+} gradients maintained at equilibrium under many different conditions of ΔpH were 15- to 100-fold greater than those predicted for a passive electroneutral (n = 2) antiport but were compatible with an exchange of $3Na^+$ per Ca^{2+}.[8]

These observations led us to reexamine the protocols of Brand[5] using fluorescent probes to monitor $[H^+]$ and $[Ca^{2+}]$ gradients.[10] Respiring heart mitochondria, suspended in KCl and treated with ruthenium red to block the Ca^{2+} uniport, extrude Ca^{2+} via the Na^+-Ca^{2+} antiport (Na^+-dependent and diltiazem-sensitive) and establish a large $[Ca^{2+}]$ gradient (FIG. 2).[10] If this gradient is maintained by electroneutral antiport, addition of nigericin (which virtually eliminates ΔpH and ΔpNa, but increases $\Delta\Psi$) should abolish this gradient. However, nigericin had little or no effect on the gradient (FIG. 2A). The gradient is abolished when $\Delta\Psi$ is dissipated by uncoupler, or when the exogenous electroneutral exchanger BrA23187 is added (FIG. 2B).

These results indicate the nNa^+-Ca^{2+} antiport is a passive mechanism mediating "downhill" electrophoretic exchange, most likely 3 Na^+ per Ca^{2+}, although the possibility remains that the stoichiometry may be variable. An electrophoretic antiport may be advantageous physiologically in that it would permit effective Ca^{2+} cycling between the matrix and cytosol when ΔpH (and ΔpNa) are low.[3]

REFERENCES

1. McCormack, J. G., A. P. Halestrap & R. M. Denton. 1990. Physiol. Rev. **70:** 391–425.
2. Hansford, R. G. 1991. J. Bioenerg. Biomembr. **23:** 823–855.
3. Crompton, M. 1985. Curr. Top. Membr. Transp. **25:** 231–276.
4. Gunter, T. & D. R. Pfeiffer. 1990. Am. J. Physiol. **258:** C755–C786.
5. Brand, M. D. 1985. Biochem. J. **229:** 161–166.
6. Li, W., Z. Shariet-Madar, M. Powers, X. Sun, R. D. Lane & K. D. Garlid. 1992. J. Biol. Chem. **267:** 17983–17989.
7. Baysal, K., G. P. Brierley, S. Novgorodov & D. W. Jung. 1991. Arch. Biochem. Biophys. **291:** 383–389.
8. Baysal, K., D. W. Jung, K. K. Gunter, T. E. Gunter & G. P. Brierley. 1994. Am. J. Physiol. **266:** C800–C808.
9. Gunter, K. K., M. J. Zuscik & T. E. Gunter. 1991. J. Biol. Chem. **266:** 21640–21648.
10. Jung, D. W., K. Baysal & G. P. Brierley. 1995. J. Biol. Chem. **270:** 672–678.

Na-Ca Exchange Studies in Sarcolemmal Skeletal Muscle[a]

H. GONZALEZ-SERRATOS,[b] D. W. HILGEMANN,[c]
M. ROZYCKA,[b] A. GAUTHIER,[b] AND
H. RASGADO-FLORES[d]

[b]*Department of Physiology*
School of Medicine
University of Maryland
660 West Redwood Street
Baltimore, Maryland 21201

[c]*Department of Physiology*
University of Texas Southwestern Medical Center
5323 Harry Hines Boulevard
Dallas, Texas 75235

[d]*Department of Physiology and Biophysics*
FUHS/The Chicago Medical School
3333 Green Bay Road
North Chicago, Illinois 60064

INTRODUCTION

In skeletal muscle, the main buffering system to maintain cytosolic free Ca^{2+} concentration ($[Ca^{2+}]_i$) at rest (*i.e.*, $<\times 10^{-7}$ M) is the sarcoplasmic reticulum (SR). During excitation, the amount of Ca^{2+} released by the SR depends on the amount of Ca^{2+} stored in the SR which in turn is the consequence of the availability of Ca^{2+} in the cytosol. Due to the fact that the SR has a limited capacity to store Ca^{2+} and since there is a constant influx of "leak" Ca^{2+} at rest,[1,2] a continuous extrusion of Ca^{2+} at the same rate as the "leak" influx must exist to maintain $[Ca^{2+}]_i$ constant during steady state conditions. Without such a Ca^{2+} efflux mechanism a continuous Ca^{2+} accumulation will produce contractures or muscle deterioration[3] once the SR capacity to store Ca^{2+} is reached. It has been shown that intra- and extracellular Na^+ concentrations ($[Na^+]$) affect Ca^{2+} fluxes,[4] which have been interpreted as evidence of a Na-Ca exchanger. Furthermore, transverse tubular vesicles isolated from frog skeletal muscle were found to exhibit substantial Na-Ca exchange activity.[5,6] We have shown that post-fatigue caffeine contractures are larger than pre-fatigue ones and that various $[Na^+]_o$ lead to different $[Ca^{2+}]_i$ levels. We attributed this to the exchanger activity.[7,8] In the above experiments, the activity of the exchanger was obscured since the SR Ca^{2+} pump and the calmodulin (CaM) regulated Ca^{2+} sarcolemmal or tubular pump remained functional. The present experiments were undertaken to further characterize the Na-Ca exchanger in the sarcolemma of skeletal muscle cells by assessing the effects

---

[a] Supported by NIH grants: R01-NS17048 to H.G-S., R01-HL51323 to D. W. H. and R01NS28563 to H. R-F. H. Rasgado-Flores is an established investigator of the AHA.

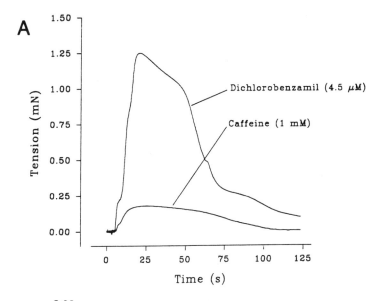
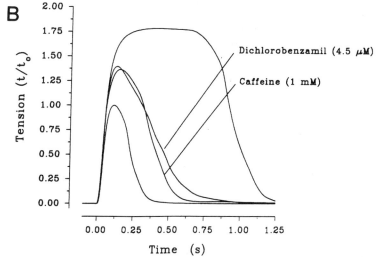

FIGURE 1. Effect of caffeine and or dichlorobenzamil on tension development at 3°C. **(A)** Contractures elicited with caffeine and caffeine plus DCB. Superfusion of the fiber with a Ringer solution containing 1 mM caffeine alone induced the smaller transient contracture (*bottom trace*). This contracture is small in relation to maximal tetanic tension or twitch tension. Superfusion of the fiber with Ringer solution containing 1 mm caffeine plus 4.5 μM DCB induced a contracture similar in magnitude to peak twitch tension (*top trace*). This second contracture was elicited several minutes after the small contracture produced with caffeine alone had ended. They are shown superimposed here for purpose of comparison. **(B)** *Bottom* and *top traces* without labels correspond to a twitch and a 70-Hz, 0.7-s duration tetanus elicited in normal Ringer. The twitch was obtained first and then three minutes later the tetanus. *Middle traces* correspond, as the labels indicate, to twitches elicited in the presence of 1 mM caffeine added 5 minutes earlier to the Ringer solution and in the presence of 4.5 μM DCB added 10 minutes after the removal of the caffeine containing Ringer solution. pH 7.2.

of inhibiting the SR and/or the CaM-dependent Ca pump on tension development and by directly measuring the Na-Ca exchanger currents.

Mechanical Measurements

The experiments were performed on single muscle fibers isolated from the tibialis anterior of the frog *Rana pipiens*.[7] To assess the contribution of the Na-Ca exchanger to extrude cytosolic Ca^{2+}, the SR Ca pump was partially inhibited by cooling the cells to 3°C. The addition of 1 mM caffeine at this temperature (FIG. 1A) induced a small contracture (not seen at 20°C). Under the same conditions (3°C plus caffeine), adding 4.5 μM dichlorobenzamil amiloride (DCB) induced an approximately 7-fold larger contracture of similar duration. Twitches produced in either caffeine or DCB alone were potentiated (FIG. 1B). In experiments done at 22°C, the Na-Ca exchanger activity was assessed by inhibiting the SR Ca pump with 20 μM cyclopiazonic acid (CPA). CPA in the presence of extracellular Na^+ and Ca^{2+} induced small and prolonged contractures (half the magnitude of twitch tension) reflecting a partial inhibition of the SR pump. Withdrawal of extracellular Na^+ and Ca^{2+} did not induce contractures until CPA was added. Under these conditions, CPA induced powerful and prolonged contractures, not accompanied by changes in membrane potential. In these experiments the exchanger activity was obscured by the CaM-regulated Ca^{2+} pump which is driven by the H^+ gradient.[9] We tested its contribution by increasing pH from 7.2 to 9.5. At high (but not at low) pH, 1 mM caffeine-induced small contractures and twitches were larger and substantially prolonged. Upon addition of DCB, the DCB-induced contractures became large and continuous. Under these conditions, twitches were potentiated and prolonged in duration from ~0.1 to between 0.8 to 1 s.

Electrophysiology

To further test for the presence of a sarcolemmal Na-Ca exchanger, we assessed for the presence of exchanger currents. The electrophysiological experiments were performed on giant inside-out excised sarcolemmal patches formed as described previously[10,11] from detached blebs obtained from collagenase-treated mouse psoas muscle. The solutions used were similar to the ones used previously for heart membranes.[10] As shown in FIGURE 2A, application of 90 mM $[Na^+]_i$ induced a small outward current which decayed over ~20 s to a low steady state level. After treating the cytoplasmic surface with α-chymotrypsin, the current induced by Na_i^+ became stationary at approximately the magnitude of the previous peak current. Na^+ dependence of the Na_i-induced outward current after α-chymotrypsin treatment is shown in FIGURE 2B. Results are fitted to a Hill equation with a slope (n) similar to that found for cardiac Na-Ca exchange and a half-maximal concentration (k) × 2 higher than that found for cardiac exchange current. The outward current was inhibited by Ni_i and by cytoplasmic DCB as shown in FIGURE 2C and D, respectively.

CONCLUSIONS

The above experiments further support the evidence of a Na-Ca exchanger operating in the Ca^{2+} efflux mode in skeletal muscle. The activity of this exchanger

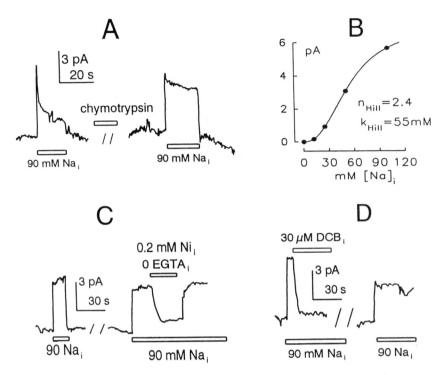

FIGURE 2. Sodium-calcium exchange current in inside-out excised patches from psoas muscle. Pipette inner diameters were 10–20 μm. Solutions were the same as those used to isolate sodium-calcium exchange currents in cardiac membrane. The extracellular [Ca^{2+}] concentration was 4 mM; cytoplasmic free Ca^{2+} was 1.4 μM (10 mM EGTA, 6 mM added calcium, pH 7.0). On the cytoplasmic side, Na was replaced by cesium to achieve the indicated cytoplasmic sodium concentrations (Na_i). Chloride was replaced with 2-(N-morpholino)ethanosulphonic acid (MES) in some experiments to test for the possible contamination of results by chloride currents; results for putative exchange current were not evidently changed. **(A)** Probable outward exchange current. Application of 90 mM Na_i induced a small outward current which decayed over approximately 20 s to a low steady state level. After treating the cytoplasmic surface with 1 mg/ml α-chymotrypsin for 1 min, the current induced by Na_i became stationary (*e.g.*, did not decrease over time) at approximately the magnitude of the previous peak current. **(B)** Na dependence of the Na_i-induced outward current after α-chymotrypsin treatment. Results are fitted with a Hill equation. The slope (n_{Hill}) of the Na_i dependence is similar to that of cardiac sodium-calcium exchange current, while the half-maximum Na_i concentration (k_{Hill}) is more than 2-times higher than that found for cardiac exchange current. **(C)** Inhibition of the putative outward exchanger current by cytoplasmic application of nickel (Ni_i). As indicated, a cytoplasmic solution was superfused which contained 0.2 mM Ni_i and no EGTA. The current was inhibited by about 85%. **(D)** Inhibition of the putative outward exchange current by cytoplasmic application of 30 μM DCB (DCB_i). DCB_i completely inhibited the putative outward current in seconds, and the effect was largely reversible.

produces extrusion of the Ca^{2+} that constantly leaks into these cells.[8] When the Na-Ca exchanger activity is set by various $[Na^+]_o$ and/or $[Na^+]_i$, different $[Ca^{2+}]_i$ levels are reached which eventually are reflected in tension development. The Na-Ca exchanger thus contributes to maintain the SR Ca^{2+} concentration level and thereby $[Ca^{2+}]_i$ in a steady state.

REFERENCES

1. BIANCHI, C. P. & A. M. SHANES. 1959. J. Gen. Physiol. **42**: 803–815.
2. CURTIS, B. A. 1966. J. Gen. Physiol. **50**: 225–267.
3. WROGEMAN, K. & S. D. J. PENA. 1976. Lancet. **27**: 672–673.
4. COSMOS, E. E. & E. J. HARRIS. 1961. J. Gen. Physiol. **44**: 1121–1130.
5. DONOSO, P. & C. HIDALGO. 1989. Biochim. Biophys. Acta **978**: 8–16.
6. HIDALGO, C., F. CIFUENTES & P. DONOSO. 1991. Ann. N. Y. Acad. Sci. **639**: 483–497.
7. GARCIA, M. C., H. GONZALEZ-SERRATOS, J. P. MORGAN, C. PERREAULT & M. ROZYCKA. 1991. J. Musc. Res. Cell Motil. **12**: 412–424.
8. CASTILLO, E., H. GONZALEZ-SERRATOS, H. RASGADO-FLORES & M. ROZYCKA. 1991. Ann. N. Y. Acad. Sci. **639**: 554–557.
9. SHWIENING, C. J., H. J. KENNEDY & R. C. THOMAS. 1993. Proc. R. Soc. London B. **253**: 285–289.
10. HILGEMANN, D. W. 1989. Pflügers Arch. ges Physiol. **415**: 247–249.
11. COLLINS, A., A. V. SOMLYO & D. HILGEMANN. 1992. J. Physiol. **454**: 27–57.

Lorin J. Mullins,
Professor of Biophysics

Lorin J. Mullins, Professor of Biophysics

A Life Dedicated to the Study of the Interaction of Ions with Excitable Membranes

JAIME REQUENA[a]

"Simón Bolívar Chair"
The Physiological Laboratory
Cambridge University
Cambridge CB2 3EG, England

INTRODUCTION

Lorin J. Mullins (FIG. 1) was an extraordinary individual and a world-class biophysicist. During his long and fruitful scientific career he combined theoretical and practical work of the highest quality. He was a model of both intellectual mastery and experimental dexterity. I would like to honor him by writing about his science, particularly of those years during which I was fortunate to enjoy his friendship and undertake scientific research alongside him.

I first saw him during the 1967 Symposium at the Venezuelan Institute of Scientific Research (IVIC) in Caracas, an event that consolidated the launching in the international scientific arena of its Biophysics Centre.[1] However, that was not his first visit to Caracas. Indeed, he had been there countless times before, either as an invited lecturer to scientific meetings or as an advisor to the Directors of IVIC. Lorin first visited Venezuela around 1955, in his words, "during the good old PJ (Pérez Jimenez) times" (referring to the military dictator who ruled Venezuela during the 1950s). At that time, a black Cadillac limousine used to whisk him right from the door of the Super Constellation airplane to the office of the founder of IVNIC, on top of the hills of Pipe, near San Antonio de los Altos, in the outskirts of Caracas. He used to joke about our style of government since, with the removal of the dictator and the advent of the Venezuelan democracy in 1958, he was forced to go through customs, like everybody else, and carry his suitcase to a friend's car. For all this help that he gave to our scientific establishment, the government of Venezuela gave Lorin the coveted Order of Andrés Bello (First Class) in 1982, and in 1988 he was granted the Order of Francisco de Miranda (also First Class).

Wartime

Lorin Mullins was born on September 23, 1917 in San Francisco, California. He was originally trained as a physical chemist (1937) and received his PhD (1940) from the University of California at Berkeley. His first paper was published in *Science*[2] one year before he was awarded his doctorate. He studied ion transport in *Nitella* under S. C. Brooks, who was primarily interested in how ions and

---

[a] Address for correspondence: Apartado 80383, Prados del Este, Caracas 1080 A, Venezuela.

FIGURE 1. Lorin J. Mullins, Professor of Biophysics (1917–1993).

molecules penetrated animal cells. Brooks saw an advantage in employing these giant unicellular algae to pioneer in permeability studies the use of radioisotopes that were just beginning to be produced in the cyclotron at Berkeley. Mullins learned from Brooks the importance of innovation in experimental work and the need to tackle problems with a multidisciplinary approach. These were to become the hallmarks of his science in later life.

His first job (1940–1943) was as an instructor in medical physiology at the University of Rochester under Wallace Fenn, who was dedicated to the study of ion permeability in muscle and in red blood cells. Together, they were the first to prove that erythrocytes were truly permeable to both K^+ and Na^+.[3,4] From Fenn, Mullins learned the need for a rigorous analysis of the nature of the problem in order to benefit from carefully planned experiments.

World War II cut short his stay in Rochester. During his tour of duty, he had postings in the US to train others to pilot the huge B-17 Flying Fortresses although he spent some time flying combat missions over Europe. After the war, Mullins left the Army with the rank of Major. The residents of Woods Hole even today remember him walking around town or sitting in the evenings in Captain Kidd, wearing his army-issued green flying jacket. Fortunately for Mullins, a few years back, when the garment gave up, Luz Whittembury was with us in Woods Hole, ready to rescue it. She was the wife of his dear friend Jose Whittembury then at the Universidad Cayetano Heredia in Lima, Peru (now at Case Western Reserve University) and while we worked away doing experiments, she was writing her excellent creole cookbook on Mullins's ancient mechanical typewriter. After surgery, worthy of an entry in the *Guinness Book of World Records*, the jacket was brought back into service again. He rarely went places without that garment, especially after retirement, when it was his faithful companion around Baltimore airfields.

Ion Selectivity

After the war, Mullins was a Guggenheim Fellow in Europe and a Merck Fellow at Johns Hopkins University in Baltimore, where he worked with D. W. Bronk and F. Brink on the permeation of radioactive ions into frog nerve. From 1947 he worked for a couple of years with G. B. von Hevesy in Niels Bohr's laboratory in Copenhagen and in the Zoological Station in Naples. In 1950, he joined Purdue University.

For some time a new branch of science was in the process of creation and the appearance in 1952 of the Hodgkin and Huxley theory on nerve excitation[5–9] was the turning point that marked the coming of age of Biophysics. With this theory the problem of understanding the nervous impulse moved from the exclusive realm of physics; that is, from the domain of electrical measurements to one in which chemistry played a prime role.

The big question, then, was the nature of ion selectivity processes because, during excitation, the membrane changes transitorily from a state favoring the permeation of potassium ions (K^+) over sodium ions (Na^+) to the reverse state. Other major questions in those years were how the membrane potential controls the kinetics of the selectivity processes, and how the dissipative process which leads to Na^+ entry and equivalent K^+ loss is reversed. These three problems were at the core of Lorin Mullins' scientific curiosity over the past 40 years.

Ten years after the Hodgkin and Huxley papers appeared, the accepted explanation of why the excitable membrane at rest is mainly selective to K^+ was that this ion is smaller than the Na^+. This view was derived from the Eisenman theory[10] of selectivity based upon ion binding to fixed charge in glass surfaces, a model that quickly caught the eye of the scientific establishment and became firmly rooted. Then, as now, research that questioned the entrenched belief received little encouragement. In that sense, Mullins's experimental attempt to put forward an alternative explanation for the question of selectivity resembled a crusade, as we shall see.

For Mullins it seemed essential to think about the nature of ions in solution, and to develop a physically reasonable view of how ions might escape from this environment into another such as the excitable membrane, that was sufficiently polarizable to serve as a substitute for its hydration shell. Mullins tackled the problem of calculating the rather large binding energy of water molecules to Na^+ and K^+, and was able to show that a little over half the hydration energy was in the layers of water beyond the first shell. He visualized an alkali cation with one complete water shell as an entity that could penetrate a pore,[11] while a totally dehydrated ion could penetrate only if the "lining" of the pore replaced the hydration shell that the ion possessed in the water phase, so that the change in the energy of solvation was reasonably small. Moreover, the rate of penetration of each ion would be proportional to how well it fitted the pore: that is, big ions would have a low permeability because they were larger then the pore and small ions would have a low permeability because of difficulties in establishing "adequate bonding" within the pore. If Na^+ and K^+ penetrated either as the dehydrated ion, or with one water shell, then a sodium ion would not fit a potassium ion-sized pore as well as would a potassium ion itself because the crystal radius of sodium ions is smaller than that of potassium ions. According to purely geometric and electrostatic concepts, ions like Rb^+ and Cs^+ ought to be more permeable than K^+ because their apparent hydrated radii are less than that of K^+. FIGURE 2 is an attempt by Mullins to describe the effect of apparent ionic radii on pore permeation.[12]

Mullins decided to put this theoretical exercise to the test. By showing that both Cs^+ and Na^+ influx were smaller than that of K^+ in frog skeletal muscle fibers,[13] he proved that something was not right in the reigning selectivity hypothesis. To Mullins, either a relatively small polypeptide that had a polarizable core, and space for the ion, or a protein that formed a channel across the membrane, were the two main possibilities to explain the process of ion permeation through a hydrophobic excitable membrane. It has to be stressed that these views were developed by Mullins decades before helical polypeptide channels were known. Indeed, structures such as the gramicidin channel with its interior lined with hydrophilic groups, and its exterior surface covered with hydrophobic residues, started to emerge only in the late 1970s.[14]

These studies led Mullins to postulate that during excitation the mean pore size in the membrane could be changed from a large size that was suited to the passage of K^+ to a smaller one that would allow only the movement of Na^+, both ions moving following their electrochemical gradient.[15] Mullins campaigned hard for the single channel hypothesis with a variable sized pore to fit both Na^+ and K^+ transit during excitation.[16]

At that time, no inhibitors of Na^+ or K^+ excitation currents were available. It was the discovery of drugs such as tetrodotoxin (TTX) which blocked Na^+ current,[17] that definitively tilted the balance in favor of the separate ion channel hypothesis during excitation: each channel being of the optimal size to fit its ionic species. This is the one occasion I can recall that Mullins was caught off target. However, his stand on this issue in no way invalidates his idea

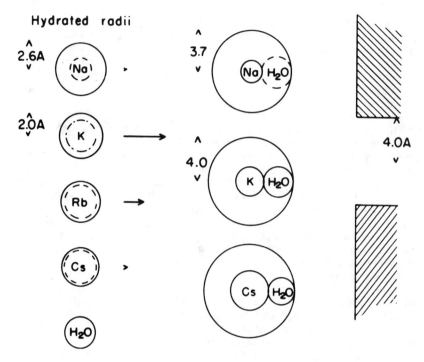

FIGURE 2. Calculated "hydrated" radii of alkali metal cations (*left*), compared to a calculated size of those ions with a single shell of hydration. (From Mullins.[12])

behind the partial ionic hydration hypothesis to explain ion movement through voltage-gated pores.

Excitation in Plants

While Hodgkin and Huxley had clearly shown the crucial role of external Na (Na_o) in bringing about electrical excitability in the nerve, Mullins found it difficult to understand the nature of the similar process in excitable algal cells. Indeed, excitability in *Chara* did not apparently depend on a specific ion in the external medium without which excitation was lost. At Purdue, with the help of an energetic graduate student, Cornelius Gaffey, he attempted to identify an ion that would produce excitability in *Chara*. They presumed that some anion flowing from the cell, specifically Cl^-, was available just outside the excitable membrane during depolarization and that this was somehow responsible for promoting further cell depolarization.

In the early 1950s, experimental verification of this hypothesis was intrinsically difficult because the long-lived chloride isotope available had only a very low specific activity, while the short-lived isotope was impractical to work with. Fortunately, Mullins had friends in Oak Ridge National Laboratory who could supply some ^{36}Cl that had gone into the pile in 1946 and had, therefore, a large gain in specific activity. With a thirty-day loading period and a fairly refined counting technique, it proved possible to show that there was a massive outpouring of Cl^- during a single action potential in *Chara*. Parallel experiments with ^{42}K showed that there was an equivalent efflux of this ion. From these results he concluded that electrical excitation in the plant cell involved first the efflux of chloride that produced the depolarization, followed by the efflux of K^+ that repolarized the cell.

In his words, "these simple and satisfying findings were not received by the scientific establishment with any great enthusiasm."[18] He was distressed to learn that the paper describing this work was rejected by the *Journal of General Physiology* because one of its editors who had developed a rival theory was chosen as the referee. He therefore sent the manuscript to the *Journal of Physiology* and Bernard Katz, who acted as its reviewer, considered it a clear demonstration of how electrical excitation in plants takes place. Richard Keynes, at the time also an editor, said it was undoubtedly the first botanical paper ever published by the *Journal of Physiology*.[19] Years later, a similar situation arose over a paper Mullins and I wrote together, but this time it was the *Journal of Physiology* that rejected the manuscript for a reason similar to that before. It was finally accepted by Paul Cranefield, a lifelong friend of Lorin's and colleague on the editorial board of the *Journal of General Physiology*.[20] It seems, thus, that the score in the everlasting match of British *vs* US science is even, as far as the Mullins publication record is concerned!

Mullins would continue to work on the problem of excitability in plants with D. S. Mailman in *Nitella*[21] but in the 1960s his main scientific focus shifted back to the problem of excitability in animal cells.

Anesthesia

The other side of excitation is its blockade. Mullins' interest in this topic was generated by a paper by Brink and Posternak,[22] who showed that while it was possible to stop conduction in a ganglion, axons passing through it were not affected by homologous series of alcohols up to n-octanol.

Earlier, Ferguson[23] had suggested that organic chemical molecules influenced biological functions if they were present as a constant fraction of saturation of the compound in aqueous solution; this he related to a constant thermodynamic activity of the compound acting in the biological phase being affected. This suggestion was a step forward from the much simpler Meyer-Overton partition coefficient hypothesis.[24]

In a homologous series of compounds, the thermodynamic activity required to produce a given anesthetic effect increases with the number of methylene groups, up to a point. Beyond that point, the addition of methylene groups leads to a loss of effect. This has been thought simply to result from the increasing difficulty of dissolving compounds in the aqueous phase. While a correlation can be made between water solubility and the speed with which compounds produce their effects, this argument, kinetic in nature, ignores the fact that given enough time, the system ought to come to equilibrium. Nevertheless, compounds beyond a specific number of carbon atoms do not show effects even though they might be highly lipid-soluble. An example of this curious phenomenon, christened the Cut-Off Effect,[25] can be seen in FIGURE 3.

While still at Purdue, Mullins studied this problem and after a theoretical breakthrough on his part, and the development by F. McBee, a chemist colleague at

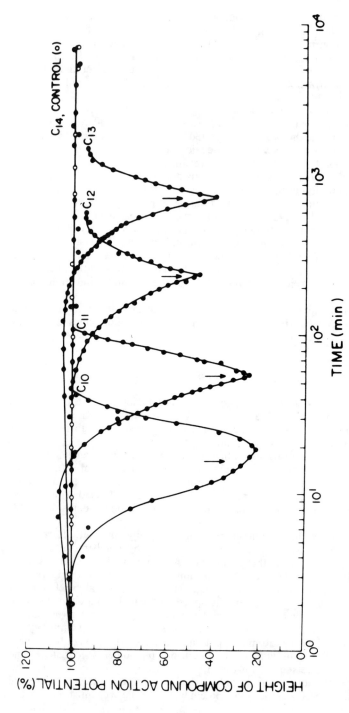

FIGURE 3. An example of a cut-off effect in bundles of a few axons from sciatic nerves of the toad *Bufo marinus*. Time course (logarithmic scale) of the height of the compound action potential (on a percent basis) is plotted as a function of various members of the normal secondary (substituted at carbon 5) homologous series of aliphatic alcohols. Each code represents the homologue containing the indicated number of carbon atoms. The control curve is for Ringer's solution (○), while the other curves are for Ringer's solution saturated with the test molecule (●). *Arrows* mark the time when the Ringer's solution containing the alcohol was switched to Ringer's alone in order to test the reversibility of the phenomenon. (From Requena *et al.*[25] Reprinted by permission from the *Journal of Membrane Biology*.)

Purdue, of a method for preparing perfluorinated hydrocarbons, Mullins postulated 1) that anesthetic molecules would work if they were not too large to fit into the interstices of the phospholipid structure, and if the molecule had a reasonable attractive force between itself and the hydrophobic domains that it encountered in the bilayer, and 2) that anesthetic action could be forecast by using the parameters of the Hildebrand theory of solutions. Further refinements were needed to overcome certain difficulties, such as that of compounds having similar molecular volumes and very close internal attractive forces (and therefore almost the same boiling point) but which exhibited different thresholds for anesthesia. To solve this paradox, Mullins looked into the mechanism by which the anesthetic molecules could be introduced into the bilayer structure such that steric considerations were brought into play as the molecular size of the perturbing agent was increased. The resulting theory overcame the shortcomings of both the Meyer-Overton and the Ferguson hypothesis. It constitutes his famous Critical Volume Hypothesis of Anesthesia,[26] published in 1954 but still valid today! At the time of its publication, he was surprised that his suggestions were received with so much enthusiasm. Basically, his hypothesis had a predictive value for drug design: halothane, the widely used general anesthetic of today, was designed with the Mullins hypothesis in mind.

Although dedicated to developing other lines of research in the domain of ion transport, as fruitful as his work in anesthesia, he kept a vigilant eye on the field. Years later he was very impressed by the work of my PhD supervisor, Denis Haydon and his group at Cambridge University, who had measured the capacitance (and therefore, the thickness) of the axolemma in squid axons under the effect of n-alkanes.[27] They showed that when the membrane is thickened by such additions, the Na^+ currents are reduced. But more to the point, this finding in a biological preparation mimicked those recorded in artificial bilayer systems. The most likely explanation was that the thickening of the membrane was caused by hydrocarbons entering the core of the phospholipid structures. Yet other studies of Haydon showed just as clearly that although alcohols did not thicken the bilayer they did inhibit Na^+ currents.[28] Here, a thickening in the hydrophobic region was unlikely since the polar groups of long-chain alcohols lie at the membrane/water interface. Mullins made a point of stressing that these findings clearly suggested that the inhibition of Na^+ channels may come about by different mechanisms. In other words, there is no unique way to induce anesthesia in nerve tissue. This militates against a unified theory of general anesthesia.

I cannot forget the occasion when this view was originated. It was on a day when Lorin was visiting my laboratory in IVIC, where Denis and I were doing our very first membrane thickness experiments in Venezuelan axons under voltage clamp. After a few fizzing axons, we decided to rethink the problem and headed for my mother's house, convinced that a good glass of scotch and soda on the rocks on the verandah would help immensely to reveal our mistake! On our way to Terepaima my Range Rover caught fire and I almost lost my visitors! Not that I had to do much to save Denis's precious cargo, for in nanoseconds his briefcase was flying out of the window onto the pavement, to be followed by Denis and Lorin clutching his bundle of notes and books! A passing army truck produced a fire extinguisher that saved the car and me. Lorin then did a wonderful first aid repair connecting a power line to the starter motor, and in no time we arrived home and embarked on solving the mystery of CO_2 generation in frying axons. We soon concluded that it had a lot to do with inexperience! Next day, while we planned a new set of experiments, the Range Rover was recabled, and that was the start of a most rewarding friendship between Denis Haydon and Lorin Mullins, my two distinguished friends, tutors and colleagues.

Ion Transport in Nerve

While at Purdue, Mullins had as a PhD student, Ray Sjodin, who worked on anesthesia by looking at oscillations in membrane potential induced by xylene and benzene.[29] A solid friendship developed and when Mullins was appointed Professor of Biophysics and then Chairman of the Department of Biophysics at the University of Maryland at Baltimore in 1959, he asked Sjodin to join him as his first assistant professor.

As a supervisor, he was somewhat atypical in that he applied a "sink or swim" approach in his doctoral guidance. He was constantly demanding from his students (and collaborators) precision in the experimental data, rigor in its analysis and excellence in the research. All that did not mean that he did not care for his students or the younger generation; Mullins was always scouting out talent and pointing students to career opportunities. A good example is that of a young Belgian graduate student who arrived to do his PhD in 1965. His work in Baltimore and in later life made Mullins very proud of the archetype of his research student and friend. Paul DeWeer did his PhD on the effect of ADP and orthophosphate on Na^+ efflux from squid axon[30] and after a short period as assistant professor with him, moved on but retained Mullins' love for ions, membranes and their interactions. Another research student of Mullins was Richard Moore with whom he shared his passion for sailing and flying. Antonio Frumento from Argentina also worked with Mullins and was the first of many Latin American scientists that he trained and with whom he collaborated.

From 1959 up to retirement, his attention focused on how the ionic movements caused by excitation were reversed. At the Marine Biological Laboratory in Woods Hole, Mullins tried to solve the third question related to the problem of excitation using as a model the squid giant axon, available there during the unpopular spring season. This was customarily followed by a time of writing up every summer holiday in MBL's unique library.

In the late 1950s, measurement of cation fluxes in squid axons initially involved immersing the axon in radioactive seawater for a short time and then rinsing and extruding axoplasm, and counting this for influx measurements. If immersion in the radioisotope solution was prolonged, then the efflux of the loaded ion could be measured. A refinement in efflux measurement was to inject the isotope with a microinjector and thus avoid extracellular isotopic contamination. Using this technique, Hodgkin and Keynes[31] were able to make efflux measurements of ^{45}Ca.

A decisive improvement in fluxes measurements in squid axon was internal dialysis. This technique, developed by Mullins and his lifelong collaborator Jack Brinley, made it possible to obtain quantitative measurements of ion transport because dialysis allows the experimenter to control at will the internal environment from which the ion was moving in giant cells.[32] He met Brinley through the review process of Brinley's NIH grant proposal, which he had found to be extremely interesting. Brinley, then a young researcher at Johns Hopkins University, wanted to study cation fluxes in lobster nerve, and Mullins was doing the same, but in squid giant axons. So it was natural that, being in the same town with the same goals, they teamed up together to develop this most ingenious technique. A schematic representation of a dialysis setup is shown in FIGURE 4. Intracellular dialysis

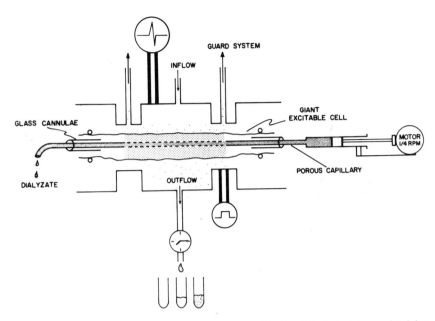

FIGURE 4. Schematic diagram of the apparatus used in the dialysis technique of Brinley and Mullins. A giant squid axon is tied at both ends to glass cannulae. Through one of these a dialysis capillary is inserted longitudinally such that a porous region is placed in the middle of the axon length. The intracellular-like media pumped by a motor-driven syringe passes through the dialysis capillary and can be collected for biochemical analysis or influx measurement. In the center region of the axon, the bathing solution can be collected (efflux configuration) as the outflow from a coupled syringe pump that is continuously exchanging a pool of external solution in contact with the region of the axon under dialysis. A system of liquid guard coupled pumps helps to delimit this pool. External electrodes permit to stimulate and monitor electrical nerve activity.

quickly became a routine procedure of considerable significance in looking at a variety of transport processes.[33] For the sodium pump,[34] it has provided the strongest support for the hypothesis that no substrate other than ATP or deoxy ATP can energize it (FIGURE 5). It has been used, also, to measure K^+ transport[35] and also to study the movement of H^+ [36] and Cl^- [37] ions.

After the highly successful measurements of Na^+ and K^+ fluxes in squid axons by dialysis, it was natural that these sorts of measurements, both efflux and influx, would be extended to divalent cations such as Ca^{2+}. This was accomplished by Reinaldo DiPolo, in the early 1970s.[38] However, in terms of influx assessment, from the very first measurements, Mullins worried that not all the Ca^{2+} entering the axon could be collected by the internal dialysis capillary, as will be discussed below.

On their part, Mullins and Brinley following DiPolo showed that the magnitude of radiotracer ^{45}Ca fluxes in dialyzed squid axons depends on several ionic factors and on the membrane potential, with these variables themselves interacting.[39] For example, Ca^{2+} efflux is increased by hyperpolarization, but only if extracellular Na_o is present. This sort of finding suggested to Mullins that the coupling ratio

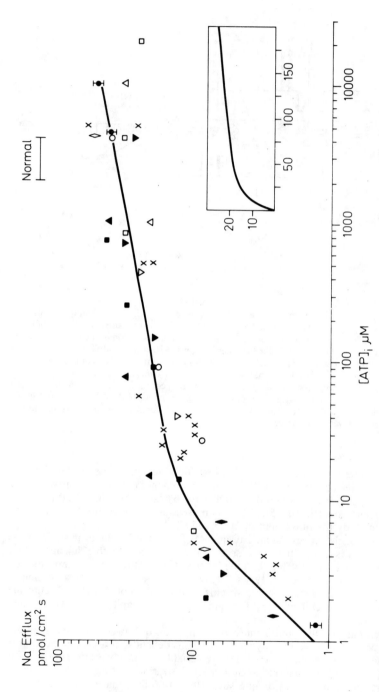

FIGURE 5. Relationship between [ATP] and Na$^+$ efflux in dialysed squid axons shown on log-log plot. *Horizonal bar* indicates range of [ATP] in axoplasm extruded from freshly dissected axons taken from living squids. *Inset* shows data for range 0 to 200 μM ATP, plotted on linear scale. (From Brinley & Mullins.[34] Reprinted by permission from the *Journal of General Physiology*.)

of the Na-Ca exchange might possibly be variable and that when the energy in the Na gradient is not in great demand it becomes electroneutral, in a 2 to 1 exchange mode, but capable of moving into a electrogenic mode of coupling with values of 3, 4 or 5, when so required. During the Second International Conference on Na-Ca Exchange, sponsored by the New York Academy of Sciences, Mullins campaigned hard for this idea devoting his keynote speech to this issue.[40] I believe this was the last paper he wrote containing no experimental data.

In 1974 the group formed by Lorin Mullins and Jack Brinley, with Teresa Tiffert and Tony Scarpa from the US, decided to team up with Reinaldo Dipolo and me in IVIC. The objective was to launch a massive attack on the problem of regulation of intracellular ionized Ca ($[Ca^{2+}]_i$) in nerve cells. It was decided that the work should be done in Woods Hole, using techniques that exploited the advantages given by the big size of the northern squid while, in Caracas, experiments should be done that took advantages of facilities present in the Biological Station at Mochima, namely, axons that were smaller but plentiful.

The first study of this program found a new application for the dialysis setup where the dialysis capillary was used as a means to confine an optical Ca indicator; a small drop of the photoluminescent protein aequorin was introduced into it and placed under its porous region. Measurements showed that free Ca was ca 30 nM and given the position of the Ca probe this was considered the $[Ca^{2+}]_i$ in the centre of the axon. Parallel studies using arsenazo III as Ca indicator injected into squid axons, and after a sizable correction for Mg^{2+} interference observed in the spectrophotometric absorbance measurement, gave for the $[Ca^{2+}]_i$ 60 nM. The optical signal being integrated over the whole of the axoplasm represented the average axoplasmic concentration. This study[41] showed, among other things, that unlike monovalent cations in axoplasm, the free Ca is very low and that it is at best a small fraction of the known Ca content of the cell. Indeed, the Ca content of fresh axons was about 50 μmol/kg axoplasm,[42] and so it was clear that only one thousandth of the axoplasmic Ca is free in the axon.

Further, in 1977 Brinley, Tiffert, Scarpa and Mullins[43] reported that in a squid axon, poisoning the mitochondria in a variety of ways leads to only very small changes in ionized Ca thus showing, for the first time, that in $vivo$ mitochondria are not a significant storage site for Ca at physiological levels of $[Ca]_i$. Buffering by mitochondria occurs only at concentrations of Ca well exceeding the physiological level (ca 1000 nM). Metabolic inhibitors that block electron transport or collapse the proton gradient in mitochondria have little effect on Ca buffering at resting physiological levels of Ca. It became clear to Mullins, then, that both radiotracer flux and Ca concentration measurements need to take into account the realities of intracellular Ca buffering, a point which he continuously recalled to researchers in the field.

This program lasted for a couple of years, and eventually Lorin, Pepe Whittembury and I, together with the occasional help of Teresa Tiffert, David Eisner, Guy Vassort, Tony Scarpa or Jack Brinley, worked in partnership for the next fourteen years, concentrating our resources on the detection with optical probes of levels of $[Ca^{2+}]_i$. Concurrently, Reinaldo DiPolo and Louis Beaugé, a former associate of Mullins in Maryland, carried on their research on Ca transport in squid giant axons, looking at fluxes measured with the dialysis technique of radioactive isotopes.

His experiments, and reflections, on the Na-Ca exchange mechanism led Mullins in 1977 to publish a very rare type of paper in the *Journal of General Physiology*, his theoretical model for the exchanger.[44] This is still a source of inspiration for all researchers working in the field.

The advances in the field of Ca transport brought about by Mullins and his collaborators were very significant indeed. Since I was involved it would not be proper to enlarge more on these, so I will limit myself to several episodes that reveal Mullins' character, his talent and intuition, or the fun of being involved with him.

The first set of ingenious results on which I wish to comment were those obtained in axons in which the luminous response representing the [Ca] signal was confined to a region of the axoplasm that was, thus, isolated or "dissected out."[45] The injection of phenol-red together with aequorin into axons results in a high probability that photons emitted by aequorin reacting with Ca^{2+} in the core of the axoplasm will be absorbed by the dye before they escape from the axon; photons produced by the aequorin reaction at the periphery of the axoplasm are much less likely to be absorbed. This technique thus favored observation of changes in Ca^{2+} occurring at the periphery of the axon. With this paper Mullins set the pace for studies of ion movement at the subcellular level, a program of research widely followed today, thanks to the advent of personal computing and imaging processing. FIGURE 6 attempts to show the locus of Ca^{2+} as measured with aequorin under various conditions employed to isolate an intracellular region.

One of the issues on Ca transport that troubled Mullins most was the existence of a "late" Ca channel in squid axon,[46,47] described by the late Peter Baker and collaborators at Cambridge. This experimental finding did not fit his scheme as one of the mechanisms promoting Ca entry into the cell. His uneasiness had something to do with the fact that the experimental observation was mainly made on "sensitized" axons; according to the Plymouth terminology this referred to axons isolated from refrigerated mantles from animals killed at sea. His intuition and the small print in those Baker papers commenting on experiments in axons from "live" squid (most probably at the instigation of A. L. Hodgkin), prompted the campaign to search for the channel in fresh squid axons either from Mochima or Woods Hole.

Taking due notice that the level of internal sodium ion concentration ($[Na^+]_i$) in a "fresh" axon was of the order of 10 to 20 mM and that this level could rise in "sensitized" axons up to 80–100 mM, Mullins thought it was obviously desirable that an explanation would be found if the phenomenon was studied as a function of the $[Na^+]_i$. With the aid of a fairly refined optical detection setup, it was possible to show that at low $[Na^+]_i$ the "channel" could not be detected. However, in axons that had artificially raised the level of $[Na^+]_i$, it was possible to record a Ca signal associated with the so-called "late" Ca channel and revealed during a sudden extracellular-induced K_o^+ depolarization. The typical "late" Ca channel signal was simply Ca^{2+} entering through the reversal operation of the Na-Ca exchange mechanism.[48,49] Indeed, the magnitude and kinetics of the normalized Ca signals due to K_o-induced depolarization depended upon the $[Na^+]_i$ in a fashion indistinguishable from that observed for similarly normalized Ca signal but from experiments where Na_o was replaced by an inert cation (FIG. 7). This last condition being considered as the demonstration of Na-Ca exchange. Once this set of results was published, the life of this strange channel ended, removing one source of distraction in the analysis of the factors contributing to the entry of Ca^{2+} in squid axon during depolarization.

Those experiments paved the way for the last set of results obtained with Mullins that I would like to recall. They concern the observation of Ca movement through the exchanger in response to controlled changes in membrane voltage. In the early 1980s, this was a very active research frontier, with several groups attempting to show alterations in Ca influx or efflux as a consequence of controlled

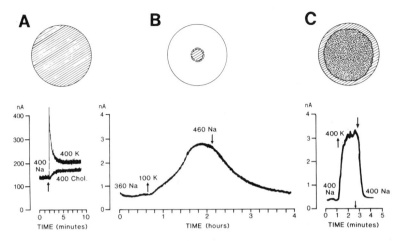

FIGURE 6. Locus of Ca measured with aequorin. Three separate methods of using aequorin are compared. **(A)** Aequorin is injected in a squid axon and its molecules are uniformly distributed in the axoplasm (*dashed region*). Shown below is the Ca signal (measured as photocurrent) in response to a 450-mM K_o^+ depolarization in 50 mM Ca seawater. Note that the response is fast, the time scale is in minutes and large, and the current is in hundreds of nA. **(B)** Aequorin is confined to a porous dialysis capillary located in the center of a squid axon and it is sensing the deep region of the axoplasm. Shown below is the Ca signal (measured as photocurrent) in response to a 100-mM K_o^+ seawater depolarization. Note now that the time scale is in hours and the current is in nA. **(C)** Both aequorin and phenol-red are injected into a squid axon and the Ca signal now originates from the subaxolemic region. Shown below is the Ca signal (measured as photocurrent) in response to a 450-mM K_o^+ depolarization in 3 mM Ca seawater. Note that the response is fast, the time scale is in minutes but small, and the current is nA, with no initial spike as in (A). This is most probably due to the low Ca_o employed, 3 mM. Notwithstanding, the Na-Ca exchange mechanism ought to be saturated at that level of Ca_o.

electrically induced changes in membrane potential. We got into the race with some disadvantages, the worst being the departure of Jack Brinley who had gone to work for NIH. He was the "gadgeterian" in our team. Pepe Whittembury took over the role but it is well known that he prefers to mend equipment rather than to buy it. Thus our voltage clamp apparatus was supposedly inferior to that of our competitors. Despite that, Mullins's daily dose of encouragement at 11 pm sharp at the Captain Kidd in Woods Hole, NIH faith that kept us going for a couple of seasons, and the gold-coated injector cum voltage-clamp-electrode developed by Pepe Whittembury[50] made the difference! Indeed, we were able to show changes in $[Ca^{2+}]_i$, as measured with optical indicators, in axons under voltage clamp control.[51,52] FIGURE 8 shows a good example of the data collected on Ca movement during voltage clamp pulses.

In the field of Ca research, Mullins was, as usual, very controversial in his analysis of experimental results. He was a scientist who did not like to neglect details and who wished to have a coherent picture that embraced all available facts. For example, in dialysed axons, $[Ca^{2+}]_i$ needs to be higher than 0.5 μM to promote the entry of Ca^{2+} under Na_o-free conditions[53] or with depolarization. On the other hand, under similar test conditions there is a large Ca^{2+} entry dependent upon Na_i[35] in intact axons with a normal $[Ca^{2+}]_i$, which could be four times lower

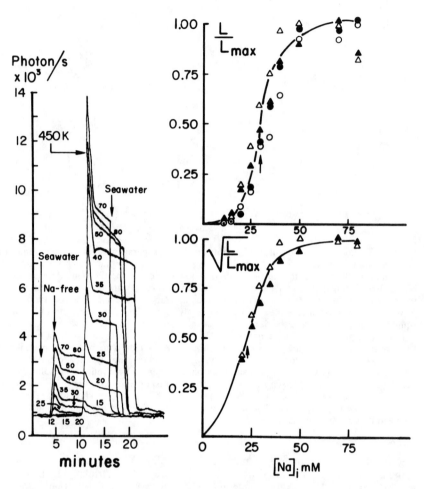

FIGURE 7. The dependence on Na_i of the magnitude of the Ca^{2+} signal produced by the removal of Na_0, followed immediately by a K depolarizaiton. The *left panel* shows the time course of records of the aequorin signal superimposed for various Na_i levels, all in the same axon and for 10 mM Ca_o. Records were synchronized at the onset of depolarization. The *right panel* shows, as a function of Na_i, either the normalized (as a fraction of the maximum) magnitude of the increment in light (*top part*) or the normalized square root (as a fraction of the maximum) (*bottom part*) of the increment registered for the light signal. In the latter case, points were omitted for the low light emission range since this must be expected to be linear with [Ca]. The experimental points were: the peak of the rapid phasic response (▲) and the tonic or plateau (△) reached during a K depolarization and the peak (●) and the plateau (○) observed during the Tris (Na-free) epidose. The axon was injected with aequorin only. Axon diameter, 0.600 mm. Membrane potential, −54 mV. (From Requena et al.[48] Reprinted by permission from the *Journal of General Physiology*.)

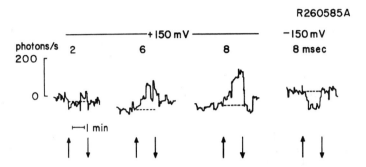

FIGURE 8. Time course of aequorin luminescence (in photons/s) from an axon tested with trains of voltage clamp pulses. The record shows the effect of trains (*shadowed areas during dashed lines*) of 2, 6 and 8 msec duration at constant amplitude (150 mV applied for 90 s at 10/s). At the *extreme right* is the response to a single train of hyperpolarizing pulses (8 msec in duration and −150 mV in amplitude). External seawater contained 3 mM Ca_o and 150 mM of each of K, Na and Tris. 1 mM of 3, 4-diaminopyridine was also added to the external solution. The axon, injected with TEA, had an initial Vm of −55 mV and was prestimulated to raise the $[Na^+]_i$ by 45 mM. The calibration bar is placed above a luminescence level of 1100 photons/s. Axon diameter 600 μm. (From Requena et al.[52] Reprinted by permission from *Cell Calcium*.)

than the minimum value required in dialysed axons. For Mullins, the reason for this difference in net Ca entry and the isotopically measured Ca influx was either that the internal dialysis system did not capture all the entering Ca^{2+}, or that the use of Ca buffers and artificial internal conditions somehow inactivated a good deal of the carrier.[54]

Mullins last piece of experimental work was reported during the Second International Conference on Na-Ca Exchange.[55] Squid giant axons were injected simultaneously with the Ca^{2+} indicators Fura-2 and aequorin. Fura-2 was calibrated *in situ* by measuring limiting fluorescence with a time-sharing multiple wavelength spectrofluorometer. The average $[Ca^{2+}]_i$ obtained with Fura-2 for resting axons was 184 nM. Mullins concluded that the sensitivity of Fura-2 to $[Ca^{2+}]_i$ is at least as great as that of aequorin, thus permitting its use in the characterization of Ca homeostasis mechanisms such as Na-Ca exchange. It was found, however, that for conventional voltage clamp experiments requiring an internal current electrode, Fura-2 is not a convenient Ca^{2+} probe, since electrode reactions in the axoplasm denature the dye. A comparison of aequorin luminescence with Fura-2 fluorescence demonstrated, as he had predicted in his 1979 review,[56] that around physiological levels of $[Ca^{2+}]_i$, light output by aequorin is linear with $[Ca^{2+}]$ is shown in FIGURE 9. This set of results validated, retroactively, quite a number of experimental results based upon aequorin measurements that, for years, were subjected to the criticism of the unknown stoichiometry of the aequorin reaction with Ca leading to light production.

Sodium and Calcium Ion Movement in the Heart

Lorin Mullins' extensive investigations of the Na-Ca exchange mechanism in squid axon convinced him that the change in the Na electrochemical gradient caused by electrical activity could promote the entry of Ca^{2+} by the reversed operation of the counter transport mechanism.[57] Mullins foresaw that this finding, derived from the domain of axonology, was very relevant to cardiac physiology, and using his gifted imagination embarked on a crusade to sell his theory to the researchers entrenched in that bunker.

This theory was set out in a most successful book, *Ion Transport in Heart*, published in 1981 and dedicated to another lifelong friend, Silvio Weidmann on the occasion of his 60th birthday.[58] In essence, Mullins argued that the presence of Ca^{2+} currents during excitation implies that the Ca^{2+} lost during the resting interval of the duty cycle must also be detectable. And this needs to be so, because Ca movement is affected by the operation of the electrogenic exchange mechanism according to the direction of the coupled Na electrochemical gradient. Thus, there must exist a potential value, which he called the "reversal potential," where no electrical current is generated by the exchange mechanism operation. He postu-

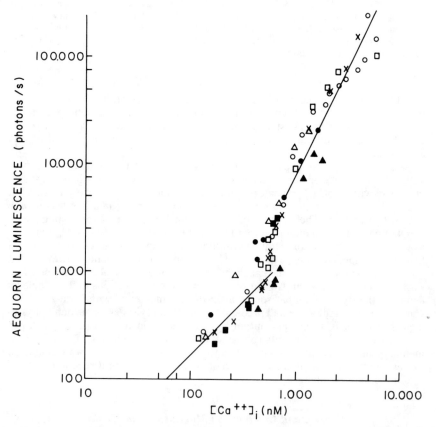

FIGURE 9. *In vivo* calibration of aequorin luminescence (log scale in photons/s) as a function of $[Ca^{2+}]_i$ (in nM) as computed from Fura-2 fluorescence ratio (340 nm/380 nm). (From Requena *et al.*[55] Reprinted by permission from the New York Academy of Sciences.)

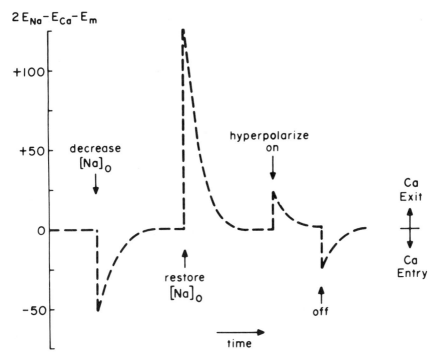

FIGURE 10. Thermodynamic parameter ($2E_{Na} - E_{Ca} - E_m$) plotted as a function of time for two conditions: (a) $[Na]_o$ is decreased such that the initial change in E_{Na} returns to its initial value. Restoration of $[Na]_o$ leads to a Ca^{2+} exit until $[Na^+]_i$ returns to its original value. (b) Hyperpolarizing the membrane to 25 mV leads to a change in E_{Ca} such that $[Ca^{2+}]_i$ declines. Depolarization has an opposite effect. (From Mullins.[58] Reprinted by permission from Raven Press.)

lated that in cardiac cells this potential is halfway between the resting potential and the plateau of the action potential. This being so, during electrical activity the exchanger must generate currents that are alternately inward and outward as schematically represented in FIGURE 10.

His theory was at first ignored or rejected by cardiac physiologists only to be endorsed later by a large group of colleagues who were convinced by its beauty, simplicity and relevance. And although in some of its fine details it might have been oversimplified and overstated, time and further experiments have shown how right Mullins was to bring into the world of cardiac voltage clampers the hypothesis of a Na-Ca exchanger mechanism assisting Ca transport in the heart. Today, his little book is a cornerstone of the cardiac physiology paradigm, essential reading that has provoked more research papers than he could ever have dreamed of.

EPILOGUE

In 1988 Lorin Mullins retired from the Chair of Biophysics of the University of Maryland and his love for experiments was replaced by his passion for flying.

Nevertheless, he kept up an active scientific life, participating in symposia, attending conferences and doing the odd experiment. At the end of 1992, he came to Venezuela for the last time, for a Na-Ca Exchange Training Workshop organized in his honor by Reinaldo DiPolo and Luis Beaugé for the Centro Internacional de Cooperación Científica Simón Bolívar of the Instituto Internacional de Estudios Avanzados (IDEA) in Caracas. Ironically, a military coup d'état erupted during the workshop and lectures had to be shared with visions of aircraft engaged in combat over the city. Some people think that our democracy started to pass away that day. Biophysics, his science, appeared, as well, to be disappearing with the coming of age of "molecular physiology." I felt then that this was also the beginning of Lorin's final days. He died shortly afterwards, in his home in Chestertown, MD on April 14, 1993, accompanied to the last minute by Rowena, his wife and steadfast companion of fifty years. He leaves one daughter and one son: Carla and Andrew Mullins.

REFERENCES

1. Proceedings of a symposium on Cell Membrane Biophysics. 1968. J. Gen. Physiol. **51** (#5.2).
2. MULLINS, L. J. & S. C. BROOKS. 1939. Radioactive ion exchange in living protoplasm Science **90**: 256.
3. MULLINS, L. J., W. O. FENN, T. R. NOONAN & L. HAEGE. 1941. The permeability of erythrocytes in radioactive potassium. Am. J. Physiol. **135**: 93.
4. FENN, W. O., T. R. NOONAN, L. J. MULLINS & L. HAEGE. 1941. The exchange of radioactive potassium with body potassium. Am. J. Physiol. **135**: 149–163.
5. HODGKIN, A. L., A. F. HUXLEY & B. KATZ. 1952. Measurements of current-voltage relations in the membrane of the giant axon of *Loligo*. J. Physiol. (London) **116**: 424–448.
6. HODGKIN, A. L. & A. F. HUXLEY. 1952a. Currents carried by sodium and potassium ions through the membrane of the giant axon of *Loligo*. J. Physiol. (London) **116**: 449–472.
7. HODGKIN, A. L. & A. F. HUXLEY. 1952b. The components of membrane conductance in the giant axon of *Loligo*. J. Physiol. (London) **116**: 473–496.
8. HODGKIN, A. L. & A. F. HUXLEY, 1952c. The dual effect of membrane potential on sodium conductance in the giant axon of *Loligo*. J. Physiol. (London) **116**: 497–506.
9. HODGKIN, A. L. & A. F. HUXLEY, 1952d. A quantitative description of membrane current and its application to conduction and excitation in nerve. J. Physiol. (London) **117**: 500–544.
10. EISENMAN, G. 1962. Cation selective glass electrodes and their mode of operation. Biophys. J. **2**: 259–323.
11. MULLINS, L. J. 1956. The Structure of Nerve Cell Membranes. Pub. No. 1. American Institute of Bioological Sciences.
12. MULLINS, L. J. 1961. The macromolecular properties of excitable membranes. Ann. N.Y. Acad. Sci. **94**: 390.
13. MULLINS, L. J. 1959. The penetration of cations in muscle. J. Gen Physiol. **42**: 817–829.
14. HLADKY, S. B. & D. A. HAYDON. 1972. Ion transfer across lipid membranes in the presence of gramicidin. Biochim. Biophys. Acta **274**: 294–312.
15. MULLINS, L. J. 1959. An analysis of conductance changes in squid axoms. J. Gen. Physiol. **42**: 1013–1035.
16. MULLINS, L. J. 1968. A single channel or a dual channel mechanism for nerve excitation. J. Gen. Physiol. **52**: 550–552.
17. NARAHASHI, T., J. W. MOORE & W. R. SCOTT. 1964. Tetrodotoxin blockage of sodium conductance increase in lobster giant axons. J. Gen. Physiol. **47**: 965–974.
18. MULLINS, L. J. 1989. Propagation of electrical impulses. *In* Membrane Transport:

People and Ideas. D. C. Tosteson Ed. American Physiological Society. Bethesda, MD.
19. GAFFEY, C. T. & L. J. MULLINS. 1958. Ion fluxes during the action potential of chara. J. Physiol. (London) **144:** 505–524.
20. MULLINS, L. J. & J. REQUENA. 1979. Calcium measurement in the periphery of an axon. J. Gen. Physiol. **74:** 403–413.
21. MAILMAN, D. S. & L. J. MULLINS. 1966. The electrical measurement of chloride fluxes in *Nitella*. Austra. L. Biolog. Sci. **19:** 385–397.
22. BRINK, F. J. & J. M. POSTERNAK. 1948. Thermodynamics analysis of the relative effectiveness of narcotics. J. Cell Comp. Physiol. **32:** 387–396.
23. FERGUSON, J. 1939. The use of chemical potentials as indices of toxicity. Proc. R. Soc. Lond. Ser. B **127:** 387–404.
24. MEYER, H. H. & H. HEMMI. 1935. Beitrage zur Theorie des Narkose III. Biochem. Z. **277:** 39–71.
25. REQUENA, J., M. E. VELAZ, J. R. GUERRERO & J. D. MEDINA. 1985. Isomers of long-chain alkane derivatives and nervous impulse blockage. J. Membr. Biol. **84:** 229–238.
26. MULLINS, L. J. 1954. Some physical mechanisms in narcosis. Chem. Rev. **54:** 289–323.
27. HAYDON, D. A., J. REQUENA & B. W. URBAN. 1980. Some effects of aliphatic hydrocarbons on electrical capacity and ionic currents of the squid axon membrane. J. Physiol. (London) **309:** 229–245.
28. HAYDON, D. A. & B. W. URBAN. 1983. The action of alcohols and other non-ionic surface-active substance on the Na current of the squid axon. J. Physiol. (London) **341:** 411–428.
29. SJODIN, R. A. & L. J. MULLINS. 1958. Oscillatory behavior of the squid axon membrane potential. J. Gen. Physiol. **42:** 39–47.
30. DEWEER, P. 1970. Effects of intracellular adenosine-5'-diphosphate and orthophosphate on the sensitivity of sodium efflux from squid axon to external sodium and potassium. J. Gen. Physiol. **56:** 583–620.
31. HODGKIN, A. L. & R. D. KEYNES. 1957. Movements of labelled calcium in squid axons. J. Physiol. 138, 253–281.
32. BRINLEY, F. J. & L. J. MULLINS. 1967. Sodium extrusion by internally dialysed squid axons. J. Gen. Physiol. **50:** 2303–2331.
33. MULLINS, L. J., F. J. BRINLEY, JR., S. G. SPANGLER & R. F. ABERCROMBIE. 1977. Magnesium efflux in dialysed squid axon. J. Gen. Physiol. **69:** 389–400.
34. BRINLEY, F. J. & L. J. MULLINS. 1968. Sodium fluxes in internally dialysed squid axons. J. Gen. Physiol. **52:** 181–211.
35. MULLINS, L. J. & F. J. BRINLEY. 1967. Potassium fluxes in dialysed squid axons. J. Gen. Physiol. **53:** 704–740.
36. BORON, W. F. & P. DEWEER. 1976. Intracellular Ph transients in squid giant axons caused by CO_2, NH_3 and metabolic inhibitors. J. Gen. Physiol. **67:** 91–112.
37. RUSSELL, J. M. 1984. Chloride in the squid giant axon. Curr. Top. Membr. Transp. **22:** 177–193.
38. DIPOLO, R. 1973. Calcium efflux from internally dialysed squid giant axons. J. Gen. Physiol. **62:** 575–589.
39. MULLINS, L. J. & F. J. BRINLEY, JR. 1975. The sensitivity of calcium efflux from squid axons to changes in membrane potential. J. Gen. Physiol. **65:** 135–152.
40. MULLINS, L. J. 1991. Is stoichiometry constant in Na-Ca exchange? Ann. N. Y. Acad. Sci. **639:** 96–98.
41. DIPOLO, R., J. REQUENA, F. J. BRINLEY, JR., L. J. MULLINS, A. SCARPA & T. TIFFERT. 1976. Ionized calcium concentrations in squid axons. J. Gen. Physiol. **67:** 433–467.
42. REQUENA, J., L. J. MULLINS & F. J. BRINLEY, JR. 1979. Calcium content and net fluxes in squid giant axons. J. Gen. Physiol. **73:** 327–342.
43. BRINLEY, F. J., JR., T. TIFFERT, A. SCARPA & L. J. MULLINS. 1977. Intracellular calcium buffering capacity in isolated squid axons. J. Gen. Physiol. **70:** 355–384.
44. MULLINS, L. J. 1977. A mechanism for Na/Ca transport. J. Gen. Physiol. **70:** 681–695.
45. MULLINS, L. J. & J. REQUENA. 1979. Ca measurement in the periphery of an axon. J. Gen. Physiol. **74:** 393–413.

46. BAKER, P. F., H. MEVES & E. G. RIDGWAY. 1973a. Effects of manganese and other agents on the calcium uptake that follow depolarization of squid axons. J. Physiol. (London), **231:** 511–526.
47. BAKER, P. F., H. MEVES & E. G. RIDGWAY. 1973b. Calcium entry in response to maintained depolarization of squid axons. J. Physiol. (London) **231:** 527–548.
48. REQUENA, J., L. J. MULLINS, J. WHITTEMBURY & F. J. BRINLEY, JR. 1986. Dependence of ionized and total Ca in squid axons on Nao-free or high Ko conditions. J. Gen. Physiol. **87:** 143–159.
49. MULLINS, L. J. & J. REQUENA. 1981 "The "late" Ca channel in squid axons. J. Gen. Physiol. **78:** 683–700.
50. VANWAGONER, D. & J. WHITTEMBURY. 1985. An improved current electrode/injection capillary for large cells voltage clamp. J. Physiol. (London) **365:** 8p.
51. MULLINS, L. J., J. REQUENA & J. WHITTEMBURY. 1985. Ca entry in squid axons during voltage clamp pulses in mainly Na/Ca exchange. Proc. Nat. Acad. Sci. USA **82:** 1847–1851.
52. REQUENA, J., J. WHITTEMBURY & L. J. MULLINS. 1989. Ca entry in squid axons during voltage clamp pulses. Cell Calcium **10:** 413–423.
53. DIPOLO, R. 1979. Calcium influx in internally dialysed squid giant axons. J. Gen. Physiol. **73:** 91–113.
54. MULLINS, L. J. & J. REQUENA. 1989. Comparing sodium-calcium exchange as studied with isotopes with measurements of $[Ca^{2+}]_i$. *In* Na-Ca Exchange. J. Allen, D. Nobel & H. Reuter, Eds. 246–260. Oxford University Press. Oxford.
55. REQUENA, J., J. WHITTEMBURY, A. SCARPA, F. J. BRINLEY, JR. & L. J. MULLINS. 1991. Intracellular ionized calcium changes in squid giant axons monitored by Fura-2 and aequorin. Ann. N. Y. Acad. Sci. **639:** 112–125.
56. REQUENA, J. & L. J. MULLINS. 1979. Ca movement in nerve fibers. Quart. Rev. Biophys. **12:** 371–460.
57. MULLINS, L. J. 1979. The generation of electron currents in cardiac fibers by Na/Ca exchange. Am. J. Physiol. **236**(C): 103–110.
58. MULLINS, L. J. 1981. Ion Transport in Heart. Raven Press. New York.

Subject Index

Acidic residues, mutagenesis of, 92
Acidosis
   intracellular pH drop in, 233
   loss of Ca activation and, 476
Action potential, duration and contractility of, 417
Action potential duration (APD), electrogenic $I_{Na\text{-}Ca}$ and, 525
Activation, Ca in ATP-depleted cells and, 474
5′-Adenylimidodiphosphate (AMP-PNP), mechanism of Mg-ATP and, 152
Adrenal chromaffin cell, secretory vesicle contributions in, 356
Alkali metal ions, Ca^{2+} uptake and, 304
Alpha repeats
   intramolecular homology and, 20
   mutagenesis studies and, 88
Alpha-chymotrypsin, Na-Na exchange activated by, 236
Alpha-thrombin, phosphorylation and, 254
Alternative splicing, 46
   isoforms generated by, 115
   kidney $Na^+\text{-}Ca^{2+}$ exchanger and, 61
Alternative splicing of mRNA, frog heart and, 41
Aluminum, ATP effect and, 154
Aminophospholipid translocase ('flippase'), giant membrane patch technique and, 136
Anesthesia, Lorin Mullins and, 567
Angiotensin II, phosphorylation and, 254
Ankyrin
   actin cytoskeleton and, 75–76
   T tubules and, 532
Anoxia, CNS myelinated axons injured by, 366
Antibodies
   site-specific antipeptide, 29
   specificity and, 323
Antibody binding, ^{3}H-monoclonal, 259
Antibody recognition, 46
Antisense oligodeoxynucleotides (ODNs)
   cardiac myocytes and, 119
   DCT cells and, 116
   exchanger mRNA and, 93
Arrhythmias, effects of La^{3+} on, 547
Aspartate residues, mutations at, 90
Astrocytes
   localization of $Na^+\text{-}Ca^{2+}$ exchanger in, 324
   $Na^+\text{-}Ca^{2+}$ exchanger in, 36
Asymmetry of bidirectional Ca movements, 230

ATP
   aminophospholipid translocase and, 137
   regulation of exchanger by, 42
ATP depletion, exchanger-mediated Ca uptake affected by, 466
ATP stimulation, Na^+ gradient-dependent Ca^{2+} uptake and, 282
ATP/ADP ratio, cellular, 205
ATP-binding concensus site, frog heart and, 42
ATPase, protein kinases and, 258
Atrial myocytes, adult human, electrogenic $I_{Na\text{-}Ca}$ in, 525
Atrial-type action potentials, cardiac rhythm and, 480
Axonic cascade, white matter and, 375
Axons, myelinated, 366

Barnacle muscle cells, voltage dependence in, 236
Bell-shaped relationship, Na gradient and, 454
Bell-shaped temperature curve, 227
Bepridil, anoxic injury and, 370
Beta-agonists, mammalian heart and, 39
Binding sites, affinity of, 354
Bovine retinal rod Na-Ca+K exchanger, 336
Brain aging, neuronal calcium homeostasis and, 379
Brain isoforms, NCX feline, 123

C-Myc, dexamethasone and, 266
Caffeine, contractures induced by, 432
Calcemic hormones, 293
Calcium
   axonal injury and, 369
   caged, 98
   effects of La^{3+} on, 547
   intra-axonal, 371
   intracellular, pH and, 529
Calcium activation, secondary, 75
Calcium binding site, $^{45}Ca^{2+}$ overlay technique and, 25
Calcium channels, ciguatoxin-1b and, 404
Calcium compartmentation, 408
Calcium concentration jump, 290
Calcium current
   calcium release and, 443
   contraction and, 440
   I_{Ca}, 417
Calcium efflux, transfected CHO cells and, 79

Calcium homeostasis
  exchanger in kidney and, 68
  neuronal, 379
Calcium imaging, 295
^{45}Ca overlay, Ca^{2+} binding site and, 25
Calcium overload, end-stage heart failure and, 543
Calcium pump, XIP alteration and, 286
Calcium reabsorption, kidney Na^+-Ca^{2+} exchanger and, 59
Calcium regulation
  giant excised patch technique and, 22
  multiple functional states of exchanger and, 163
Calcium sequestration within ROS disks, 342
Calcium translocation site, protons at, 186
Calcium transport, bell-shaped temperature curve and, 228
Calcium-induced calcium release (CICR), reverse mode of exchanger and, 524
Capacitance measurements, exchanger electrogenicity and, 148
Carboxyeosin, sarcolemmal ATPase and, 437
Cardiac glycosides, additional secondary action of, 359
Cardiac hypertrophy, hemodynamic load and, 489
Cardiac isoforms, 498
  feline NCX, 122
Cardiac myocytes
  contractions in, 522
  effects of NCX1 antisense oligonucleotides on, 119
  external Mg^{2+} and, 515
Cardiac sarcolemmal NCX, end-stage human heart failure and, 543
Cardiac sarcolemmal vesicles, XIP in, 173
Cardiac sodium-calcium exchange expression, regulation of, 537
Cardiovascular system, Na^+-Ca^{2+} exchange in, 407
Catecholamines, reversal of NCX and, 391
Cations, external monovalent, 279
Cellular and cardiac physiology, relevance of findings *in vitro* to, 232
Cellular Ca balance, steady-state, 440
Cellular function of Na^+-Ca^{2+} exchanger, 12
Cetiedil, 405
Charge movements
  direct measurement of, 145
  giant membrane patch technique and, 136
Chinese hamster ovary cells, calcium homeostasis in, 73
Chromaffin cells
  functions of, 356
  protein phosphorylation and, 395
  reversal of NCX in, 391
Chromaffin granules, 356

Chymotrypsin, current-voltage relationships and, 144
Chymotrypsin digestion, proton inhibition and, 186
Ciguatoxin-1b, 404
Consecutive translocation, 352
Contraction
  cardiac myocytes and, 522
  myocardial force of, 540
Contractures, rapid cooling, 431
Coronary angiography, sodium-calcium balance in, 551
Coupling ratio, rat brain exchanger and, 313–314
Current transient, 352
Current-voltage relationships
  mutagenesis studies and, 91
  sodium translocation and, 143
Cyclic AMP, NCX activity and, 263
Cyclopiazonic acid (CPA)
  Ca^{2+} and, 321
  contractures induced by, 558
Cytoplasmic calcium ($[Ca]_i$) decline, 439
Cytoplasmic pH, 185
Cytoskeleton
  actin, 75
  ATP-dependent processes and, 153
Cytosolic Ca^{2+} oscillations, 356
Cytosolic factor, MgATP stimulation and, 214
Cytosolic free Ca^{2+}, changes in, 337

Density of exchange sites, vertebrate photoreceptors and, 346
Detailed balance, principle of, 272
Dexamethasone, NCX mRNA and, 265
Diadic cleft
  calcium in, 409
  sodium increase and, 447
Dialysis, prolonged, 213
Dialyzing Na, bell-shaped relationship and, 454
Dichlorobenzamil amiloride (DCB), twitch produced in, 558
Digital imaging, 320
1,25-Dihydroxy-vitamin D_3, kidney Na^+-Ca^{2+} exchanger and, 70
DiOC, cells labeled with, 326
Distal convoluted tubule (DCT) cells, 115
DM-nitrophen, 290
Drosophila Na^+-Ca^{2+} exchanger, structure of, 52

EC coupling
  calcium trigger for, 444
  cardiac, 524

SUBJECT INDEX

Ectopic beats, initiation and propagation of, 482
Electrogenic Na^+-Ca^{2+} exchange current (I_{Na-Ca}), action potential duration and, 525
Electrogenicity
  retinal photoreceptors and, 354
  transporter, 148
Electroneutral exchange, voltage sensitivity and, 236–237
Endolymph, inner ear and, 400
Endoplasmic reticulum, transfected CHO cells and, 82
Energy buffer system, 205
Epithelial cell, downregulation of NCX activity by PMA in, 262
Erythrocytes, dog and ferret, 503
Exchange current noise, 151
Exchanger inhibitory peptide (XIP)
  autoregulatory function of, 172
  cysteine replacements and, 286
  inactivation process and, 28
  modifications of, 284
  SR Ca release and, 455
Exchanger inhibitory peptide domain, possible mechanism for, 180
Exchanger superfamily, 21
Excised patches, retinal photoreceptors and, 349
Excitation in plants, Lorin Mullins and, 566
Excitation-contraction, heart, 8
Excitation-contraction coupling, 417
Exocytosis, Ca^{2+} efflux and, 360
Exons/introns, sequence analysis of, 49
Expression, functional, 47
Extracellular pH, effect of changes in, 195
Extracellular sodium, glutamate-induced calcium challenge and, 380

FCCP, mitochondrial membrane potential and, 434
5′ splicing site, 111
5′-RACE (rapid amplification of DNA 5′-ends), 63
Flash-photolysis, 98
Flippase, *see* Aminophospholipid translocase
Fluorescence immunocytochemistry, chromaffin cells and, 359
FMRFa (Phe-Met-Arg-Phe-NH_2), 288
Forskolin, NCX mRNA and, 263
Forward exchange, 350
Frog heart, peculiarities of, 39
Functional activity, end-stage heart failure and, 543
Functional expression, 47

Fura-2, correlation between ^{45}Ca uptake and, 74
Future, Na^+-Ca^{2+} exchanger and, 13

Gating, transporter, 139
Gene expression, end-stage heart failure and, 543
Gene structure, human NCX1, 103
Genomic clone
  5′-end exons and, 65
  rabbit, 49
Giant membrane patch, 136
Glia
  external monovalent cations and, 279
  Na^+-Ca^{2+} exchanger identified in, 12
Glutamate residues, mutations at, 90
Glutamate-induced calcium loads, aged hippocampal neurons and, 379
Glycine residues, mutations at, 90
Growth factors
  glucocorticoids and, 264
  phosphorylation and, 249
Guinea pig myocytes, voltage sensitivity in, 237
Guinea pig single ventricular cells, Mg^{2+} and, 515

Hair cells, frog saccular, 397
Heart
  excitation-contraction in, local control of, 8
  guinea pig, single ventricular cells of, 515
  human, failing, 539
  ion transport in, Lorin Mullins and, 577
  ischemic rabbit, 534
  novel alternatively spliced isoform of NCX in, 129
  transgenic mouse, 126
Heart failure, human, end-stage, 543
Heart mitochondria, Na^+-Ca^{2+} antiport of, 553
Heart rate, intracellular Na and, 462
Heat-shock protein, ischemia duration and, 535
HeLa cells, clone RBE-1 transfected, 30
Hippocampal neurons, aged, 379
Homology of transporters, structural, 53
Human NCX1 gene, exon composition of, 104
Human NCX2 gene, exon composition of, 105

Immunofluorescence, 323
  clone human Na^+-Ca^{2+} exchanger and, 47
Inactivation, Na^+-dependent, 27
  transfected CHO cells and, 75

Influx, sodium and calcium, 297
Inner ear, cardiac NCX protein in, 400
1,4,5-Inositol triphosphate (InsP$_3$), transfected CHO cells and, 77
In situ hybridization, kidney Na$^+$-Ca^{2+} exchanger and, 60
Insulin-producing cells, 132
Intact cells
  how much ATP needed to activate exchanger in, 466
  Mg^{2+} and, 515
Intracellular calcium
  action potential duration and, 417
  direction of exchange and, 115
Intracellular sodium, trigger and, 456
Ion selectivity, Lorin Mullins and, 564
Ion binding sites, electroneutral exchange and, 237
Ischemia
  ATP depletion and, 466
  intracellular pH drop in, 233
Isoforms, 49
  novel alternatively spliced, 129
  specific Na$^+$-Ca^{2+} exchanger, 29
I-V relation, whole-cell exchange current and, 166

Kidney
  NCX1 in, 48
  rat Na$^+$-Ca^{2+} exchanger gene in, 58–59
  sodium-calcium exchanger and, 9
Kidney isoforms, NCX feline, 123
Kinetic model, development of, 278
Kinetic parameters, detailed balance and, 273
Kinetic studies, transient state, 94
Kinetics
  NCX current, Ca^{2+} concentration jump and, 290
  outward Na$^+$-Ca^{2+} exchange current and, 517
  rat brain synaptosomes and, 300
  technical advances in, 135

Lanthanum, cardioprotection provided by, 546
Linearly independent cycles, 273
Local control, excitation-contraction in heart and, 8
Lymphocytes
  peripheral, 507
  sodium-calcium exchanger and, 12

Magnesium
  effects of La^{3+} on, 547
  external, 515

Magnesium ATP (MgATP), phosphoarginine difference from, 200
Magnesium ATP stimulation, nerve cytosolic factor required for, 210
Magnesium concentration, loss of Ca sensitivity and, 476
Mechanical stimuli, hair cells and, 397
Mechanism
  exchange, 142
  technical advances in, 135
Membrane potential
  axonal, 367
  Ca^{2+} fluxes and, 311
Mitochondria
  caffeine and, 436
  transfected CHO cells and, 82
Mitotic metaphase cells, 296
Mobile cytosolic Ca^{2+} buffers, hair cells and, 397
Modulation, growth factors and, 249
Modulation reactions, exchanger, 137
Molecular actions, predicting therapeutic efficacy from, 1
Molecular biology, progress in, 19
Molecular investigations of the Na$^+$-Ca^{2+} exchanger, 13
Monocytes, transport phenomena and, 505
Monovalent cations, absence of, 167
Mouse cardiac NCX, cloning of, 126
mRNA, ischemic duration and, 534
Mutagenesis studies, 86
Myocardial rhythm disturbances, effects of La^{3+} on, 546
12-Myristate 13-acetate (PMA), phosphorylation and, 254

N-terminal segment, protein in HeLa cells and, 32
NACA3, 132
NACA7, 132
NACA8, 129
Nematode *C. elegans*, cDNA clone in, 107
Nerve, ion transport in, Lorin Mullins and, 570
Nerve terminals, physiological roles of Na$^+$-Ca^{2+} exchanger in, 314
Neural system, Na$^+$-Ca^{2+} exchange in, 299
Neuronal Na$^+$-Ca^{2+} exchanger, local control feature in, 8
Neurons, distribution of Na$^+$-Ca^{2+} exchanger in, 328
Neuropeptides, reversal of NCX and, 391
Neutrophils, human, 505
Newborn maturation, myocardial intracellular calcium concentrations during, 537
Nickel, contractions and, 523
Nifedipine, contractions and, 523

Nigericin, Na^+-dependent Ca^{2+} efflux stimulated by, 554
Noise analysis, methods of, 151
Northern blot, kidney Na^+-Ca^{2+} exchanger and, 60

Oligodeoxynucleotides, antisense, 93
Optic nerve, anoxia and, 367
Osteoblastic cells, calcemic hormones in, 293
Ouabain, rat aortic myocytes and, 321
Outward Na^+-Ca^{2+} exchange current, effects of $[Mg]_o$ on, 516

P-type ATP-binding site, 43
Pancreatic B cells, 288
Parathyroid hormone, kidney Na^+-Ca^{2+} exchanger and, 70
Partial reactions, deprotonation and, 223
Patch clamp, retinal photoreceptors and, 347
Peptides, inhibition and, 174
pH
 cardiac Na^+-Ca^{2+} exchange and, 182
 intracellular, calcium and, 529
pH regulation, role of, 3
pH titration curves, 220
Phenylephrine, cardiocyte hypertrophy and, 490–491
Phorbol ester
 Ca^{2+} fluxes and, 310
 downregulation of NCX mRNA by, 261
Phosphatidic acid, exchange current and, 155
Phosphatidylcholine, ATP effect and, 155
Phosphatidylinositols, exchange current and, 155
Phosphatidylserine, exchange current and, 155
Phosphoarginine, modulation of Na^+-Ca^{2+} exchange fluxes and, 199
Phospholipid environment, XIP and, 177
Phospholipid messengers, regulation and, 154
Phosphorylation
 effects of ATP on exchange currents and, 75
 exchange process not regulated by, 353
 growth factors and, 249
 MgATP stimulation and, 214
 protein, catecholamine secretion and, 395
 regulation by, 37
Phylogeny of transporter sequences, 53
Physiological conditions, triggering SR release under, 443
Plasma membrane, antibodies raised against, 325
Plateau, action potential, 481
Platelet-derived growth factor, phosphorylation and, 250
Platelets, human, 504
Potassium, effects of La^{3+} on, 547

Potassium ions, interactions between Ca^{2+} and Na^+ ions and, 338
Primary neurons, effects of NCX1 antisense oligonucleotides on, 119
Protein kinase, plasma membrane calcium ATPase and, 258
Protein kinase C, renal epithelial cells and, 260
Proton inhibition, outward Na^+-Ca^{2+} exchange current and, 195
Protonated-deprotonated forms, 225
Putative NCX gene, 107

Q$_{10}$, 439

Rate-limiting pathways, 219
Rate-limiting sodium and calcium transport, 221
Rate-limiting step, Ca^{2+} transport as, 350
Reconstitution, XIP inhibition and, 175
Regulation
 rat brain synaptosomes and, 300
 technical advances in, 135
 transfected CHO cells and, 75
Relaxation, species differences in, 432–433
Retinal photoreceptors, Na^+-Ca^{2+},K^+ exchange sites in, 346
Reverse Na^+-Ca^{2+} exchange
 Ca^{2+} and K^+ accumulation during, 349–350
 SR calcium release and, 451
RHE-1, RBE-2 and, 31
Rhythm, cardiac, abnormal, 482
RNA splicing, 49
RNase protection, 65

Sarcolemmal Ca binding, 409
Sarcolemmal Ca-ATPase, caffeine and, 436
Sarcolemmal phospholipids, 408
Sarcolemmal skeletal muscle, Na^+-Ca^{2+} exchange studies in, 556
Sarcoplasmic reticulum
 action potential contractility and, 417
 Ca^{2+} stored in, 321
 replenishing, with calcium, 521
 triggering calcium release from, 443
Sarcoplasmic reticulum Ca^{2+} concentration level, 560
Sarcoplasmic reticulum Ca content, 440
Sarcoplasmic reticulum Ca^{2+} release, reverse Na^+-Ca^{2+} exchange and, 451
Sarcoplasmic reticulum Ca^{2+}-ATPase (SERCA 2a), failing human heart and, 540

Secretory vesicle Na-Ca exchanger, 356
Serine residues, mutations at, 88
Shortening, I_{Ca} and voltage and, 453
6-Kb cDNA, 114
Skeletal muscle, sarcolemmal, 556
Smooth muscle cells, Na^+-Ca^{2+} exchanger and, 12
Sodium, effects of La^{3+} on, 547
Sodium channel, axonal injury and, 370
Sodium channel activator, failing human heart and, 539
Sodium gradient-dependent Ca^{2+} uptake, ATP stimulation of, 282
Sodium sea water Ca^{2+}-free (NaSW0Ca) solution, 237
Sodium translocation site, cytoplasmic pH and, 185
Sodium-calcium antiport, properties of, 553
Sodium-calcium balance, coronary angiography and, 551
Sodium-calcium exchange current
 outward, 516
 separation of, 100
Sodium-calcium exchange protein, XIP in, 171
Sodium-calcium exchange-dependent Ca compartment, 412
Sodium-calcium exchanger
 cardiac sarcolemma, 217
 cloning of, 110
 function of, 410
 other regulators of, 3
Sodium-calcium exchanger gene, feline, 121
Sodium-calcium exchanger isoform expression, 124
Sodium-calcium exchanger message half-life, 494
Sodium-calcium exchanger upregulation, cultured myocytes and, 490
Sodium-calcium gene, feline, 499
Sodium-dependent ^{45}Ca efflux, chromaffin cells and, 356
Sodium-dependent inactivation (I_1)
 multiple functional states of exchanger and, 160
 physiological pH and, 192
 stimulation and inhibition of, 140
Sodium-independent calcium dependent inactivation (I_2), multiple functional states of exchanger and, 164
Sodium-induced outward current, cytoplasmic dichlorobenzamil amiloride (DCB) and, 558
Sodium-potassium pump, mutagenesis of residues and, 91
Sodium-sodium exchange, α-chymotrypsin and, 236

Splicing products, 113
Squid axons
 phosphoarginine in, 199
 voltage sensitivity in, 237
Squid optic nerve, MgATP stimulation and, 208
Stable isotopes, ion microscopy imaging of, 295
Store-dependent Ca influx (SDCI)
 $LaCl_3$ and, 511
 transfected CHO cells and, 77
Sulfosuccinimidyl acetate (SNA), peptide 1 modified with, 285
Synaptosomes, rat brain, 300

T tubules, ankyrin present at, 532
Temperature, relaxation and, 439
Temperature-dependent curves, 226
Tension development, various $[Na^+]_o$ and/or $[Na^+]_i$ and, 560
Thapsigargin
 SR Ca load and, 435
 transfected CHO cells and, 77
Therapeutic efficacy, molecular actions and, 1
Therapeutic intervention, possibilities for, 4
Thermodynamic constraints, rate coefficients and, 277
Threonine residues, mutations at, 88
Thrombin, NCX mRNA and, 267
Thyroid hormone, rabbit, 536
Tissue-specific expression, 49
Transcription, kidney, brain, and heart, and, 68
Transient current, 292
Transport cycle, exchanger, 138
Trigger, reverse Na^+-Ca^{2+} exchange and, 456
Triggering calcium release, calcium current in, 443
Trypsin, I_1- and I_2-inactivations and, 165
Turnover, 292
Turnover number, retinal photoreceptors and, 346
Twitch, dichlorobenzamil amiloride (DCB) and, 558

V anadate, phosphoarginine and, 206
Vascular smooth muscle cells
 localization of Na^+-Ca^{2+} exchanger in, 324
 phosphorylation in, 249
 sodium-calcium exchanger mRNA and activity in, 268

SUBJECT INDEX

Ventricle, action potential duration and contractility in, 417
Ventricular action potentials, cardiac rhythm and, 480
Ventricular cells, Mg^{2+} and, 515
Veratridine, cardiomyocytes and, 491
Voltage clamp
 antisense oligodeoxynucleotides and, 95
 electroneutral modes of exchange and, 237
Voltage dependence, barnacle muscle cells and, 236

Voltage sensitivity, rat brain exchanger and, 313–314
Voltage-dependent steps, location of, 2

X-ray contrast media, coronary injection of, 551
Xenopus oocytes, mouse cardiac NCX in, 126

Z disk, ankyrin present at, 532

Index of Contributors

Adams, W. A., 529–531
Allen, C. J., 286–287
Artman, M., 536–538
Austin, C., 529–531

Baazov, D., 217–235
Balasubramanyam, M., 502–514
Baltazar, G., 391–394
Barnes, K. V., 121–125, 489–501
Bassani, J. W. M., 430–442
Bassani, R. A., 430–442
Baysal, K., 553–555
Beaugé, L., 199–207, 208–216, 279–281, 282–283
Berberián, G., 208–216, 282–283
Bers, D. M., 430–442
Bersohn, M. M., 534–535
Bindels, R. J. M., 58–72
Bland, K. S., 119–120
Blaustein, M. P., 300–317, 318–335
Boerth, S. R., 536–538
Böhm, M., 539–542
Bollen, A., 132–133
Borin, M. L., 318–335
Bouchard, R. A., 417–429
Bridge, J. H. B., 451–463
Brierley, G. P., 553–555

Cannell, M. B., 7–18, 443–450
Carafoli, E., 37–45, 103–109, 110–114
Carvalho, A. P., 391–394
Chabbert, C., 397–399
Chandra, S., 295–298
Chatterjee, G., 546–550
Chen, F., 532–533
Cheng, H., 7–18
Chernaya, G., 73–85
Chumakov, I., 103–109
Clark, R. B., 417–429
Coetzee, W. A., 536–538
Condrescu, M., 73–85
Cook, O., 29–36
Cooper, G., IV, 489–501

Das, D. K., 546–550
Dawson, M. M., 121–125, 489–501
Delgado, D., 208–216
Denison, H. A., 284–285
Desantiago, J., 236–248
Dilly, K. W., 529–531
DiPolo, R., 199–207, 208–216
Doering, A. E., 182–198, 529–531
Drexler, H., 543–545

Duarte, E. P., 391–394
duBell, W., 7–18, 46–57

Eisner, D. A., 182–198, 529–531
Engelman, R. M., 546–550
Espinosa-Tanguma, R., 236–248
Evans, A. M., 443–450

Falck, G., 551–552
Flesch, M., 539–542
Fontana, G., 300–317
Frank, J. S., 86–92, 532–533
Friedman, P. A., 115–118
Furman, I., 29–36

Gabellini, N., 110–114
Gardner, J. P., 502–514
Garfinkel, A., 532–533
Gatto, C., 284–285, 286–287
Gauthier, A., 556–560
Gesek, F. A., 115–118
Giles, W. R., 417–429
Goknur, A. B., 464–479
Gonzalez-Serratos, H., 556–560
Gourlet, P., 288–289
Grantham, C. J., 443–450
Guerini, D., 37–45

Hale, C. C., 171–181, 284–285
Hartung, K., 290–292
Haworth, R. A., 464–479
He, S., 7–18, 46–57
Herchuelz, A., 132–133, 288–289
Hilgemann, D. W., xiii–xiv, 136–158, 159–170, 556–560
Holgado, A., 279–281
Holtz, J., 543–545
Hryshko, L. V., 20–28, 86–92

Islam, S., 119–120
Iwamoto, T., 249–257
Iwata, T., 37–45, 110–114

Jan, C-R., 356–365
Juhaszova, M., 318–335
Jung, D. W., 553–555
Just, H., 543–545
Jynge, P., 551–552

Kao, L-S., 395–396
Kappl, M., 290–292
Karlsson, J. O. G., 551–552

Kasir, J., 29–36
Kent, R. L., 489–501
Khananshvili, D., 217–235
Kieval, R., 7–18
Kilav, R., 58–72
Kim, I., 126–128
Kimura, J., 515–520
Kirby, M. S., 46–57
Kofuji, P., 7–18, 46–57
Kohmoto, O., 451–463
Kraev, A., 37–45, 103–109
Krieger, N. S., 293–294

Langer, G. A., 408–416
Lattanzi, D., 129–131
Lebrun, P., 288–289
Lederer, W. J., 7–18, 46–57, 182–198
Lee, C. O., 126–128
Lee, H-W., 258–271
Lee, S-L., 58–72
Lee, W-S., 58–72
Legrand, A-M., 404–406
Leguennec, J-Y., 480–488
Lehouelleur, J., 397–399
Levi, A. J., 451–463
Levitsky, D. O., 20–28
Li, G-R., 525–528
Lin, L. F., 395–396
Lindenmayer, G. E., 318–335
Lipp, P., 93–102
Litwin, S., 451–463
Low, W., 29–36
Luo, S., 7–18, 46–57
Lytton, J., 58–72

Main, M. J., 443–450
Mancini, P. M., 400–403
Matsuoka, S., 20–28, 86–92, 159–170
Maulik, N., 546–550
McDermott, D. E., 489–501
Menick, D. R., 121–125, 489–501
Meunier, F. A., 404–406
Michaelis, M. L., 119–120
Milanick, M. A., 284–285, 286–287
Mills, L. R., 379–390
Molgo, J., 404–406
Moore, E., 532–533
Morot Gaudry-Talarmain, Y., 404–406
Morrison, G. H., 295–298
Moulian, N., 404–406

Nattel, S., 525–528
Navangione, A., 346–355
Naveh-Many, T., 58–72
Neubauer, C. F., 7–18
Nicoll, D. A., 20–28, 86–92

Niggli, E., 93–102
Noble, D., 1–6, 480–488

Pedersen, H. K., 551–552
Peskoff, A., 408–416
Philipson, K. D., xiii–xiv, 20–28, 86–92, 159–170, 532–533
Pütz, F., 539–542

Rahamimoff, H., 29–36
Rasgado-Flores, H., 236–248, 556–560
Reeves, J. P., 73–85
Refsum, H., 551–552
Reilly, R. F., 129–131
Reinecke, H., 543–545
Requena, J., 562–582
Rispoli, G., 346–355
Rogers, T. B., 7–18
Rogowski, R. S., 300–317
Rojas, H., 208–216
Rozich, J. D., 489–501
Rozycka, M., 556–560
Ruknudin, A., 7–18, 46–57

Sans, A., 397–399
Santi, P. A., 400–403
Santiago, E. M., 318–335
Schneider, A. S., 356–365
Schnetkamp, P. P. M., 336–345
Schulze, D. H., 7–18, 46–57
Schwaller, B., 93–102
Schwinger, R. H. G., 539–542
Sellers, P. H., 272–278
Shigekawa, M., 249–257
Shimizu, H., 318–335
Silver, J., 58–72
Smith, J. B., 258–271
Smith, L., 258–271
Spitzer, K. W., 451–463
Steffensen, I., 366–378
Studer, R., 543–545
Stys, P. K., 366–378
Svoboda, M., 132–133
Szerencsei, R. T., 336–345

Takahashi, K., 119–120
Thacker, U. F., 489–501
Tie, J., 236–248
Tosaki, A., 546–550
Tucker, J. E., 336–345

Valdivia, C., 46–57
Van Baal, J., 58–72
Vandermeers, A., 288–289
Van Eylen, F., 132–133, 288–289
Vassort, G., xiii–xiv

INDEX OF CONTRIBUTORS

Vatashski, R., 29–36
Vellani, V., 346–355
Veríssimo, P., 391–394
Vetter, R., 543–545
Vites, A-M., 521–524

Wagg, J., 272–278
Wakabayashi, S., 249–257
Wasserstrom, J. A., 521–524
Weil-Maslansky, E., 217–235

Weiss, J. N., 20–28
Westhead, E. W., 395–396
White, K. E., 115–118
Winslow, R., 480–488
Wisel, S., 46–57

Xu, W-Y., 284–285, 286–287

Yip, R. K., 318–335